Solutions Manual and
Study Guide for

Organic Chemistry

Joseph M. Hornback

Balasingam Murugaverl

Brooks/Cole Publishing Company
ITP® An International Thomson Publishing Company

Pacific Grove • Albany • Belmont • Bonn • Boston • Cincinnati • Detroit • Johannesburg • London
Madrid • Melbourne • Mexico City • New York • Paris • Singapore • Tokyo • Toronto • Washington

Senior Developmental Editor: *Keith Dodson*
Editorial Assistant: *Nancy Conti*
Marketing Team: *Heather Woods, Deanne Brown*
Production Editor: *Mary Vezilich*
Cover Design: Roy R. Neuhaus

For more information, contact:

BROOKS/COLE PUBLISHING COMPANY
511 Forest Lodge Road
Pacific Grove, CA 93950
USA

International Thomson Publishing Europe
Berkshire House 168-173
High Holborn
London WC1V 7AA
England

Thomas Nelson Australia
102 Dodds Street
South Melbourne, 3205
Victoria, Australia

Nelson Canada
1120 Birchmount Road
Scarborough, Ontario
Canada M1K 5G4

International Thomson Editores
Seneca 53
Col. Polanco
11560 México, D. F., México

International Thomson Publishing Japan
Hirakawacho Kyowa Building, 3F
2-2-1 Hirakawacho
Chiyoda-ku, Tokyo 102
Japan

International Thomson Publishing Asia
60 Albert Street
#15-01 Albert Complex
Singapore 189969

International Thomson Publishing GmbH
Königswinterer Strasse 418
53227 Bonn
Germany

Printed in Canada

5 4 3 2

ISBN 0-534-35259-6

Table of Contents

Chapter 1
A SIMPLE MODEL FOR CHEMICAL BONDS

1.1 Lewis structures depict the electrons in the outer shell (valence shell) as dots around the symbol of the element. The number of valence shell electrons of an atom is the same as its group number in the periodic table.

a) $:\underset{\cdot\cdot}{\ddot{\mathrm{Br}}}\cdot$ b) $\cdot\mathrm{Ca}\cdot$ c) $\cdot\underset{\cdot}{\dot{\mathrm{Ge}}}\cdot$

1.2 Calcium chloride and sodium sulfide are binary metal-nonmetal compounds. Electrons are exchanged between the metal and the nonmetal atoms to form charged atoms (ions) that satisfy the octet rule. These ions with opposite charges form ionic compounds.

a) $\cdot\mathrm{Ca}\cdot \quad + \quad 2\ :\underset{\cdot\cdot}{\ddot{\mathrm{Cl}}}\cdot \longrightarrow \mathrm{Ca}^{2\oplus} \quad + \quad 2\ :\underset{\cdot\cdot}{\ddot{\mathrm{Cl}}}:^{\ominus}$

b) $2\ \mathrm{Na}\cdot \quad + \quad \cdot\underset{\cdot\cdot}{\ddot{\mathrm{S}}}\cdot \longrightarrow 2\ \mathrm{Na}^{\oplus} \quad + \quad :\underset{\cdot\cdot}{\ddot{\mathrm{S}}}:^{2\ominus}$

1.3 Use Table 1.1 in the text to answer this problem. Carbon has four valence shell electrons and forms four bonds to satisfy the octet rule. Elements chlorine and bromine each have seven valence shell electrons and will form one bond. The element hydrogen forms one bond.`

a)
```
       ..
      :Cl:
  ..   ..  ..
 :Cl: C :Cl:
  ..   ..  ..
      :Cl:
       ..
```

b) $\mathrm{H}:\underset{\cdot\cdot}{\ddot{\mathrm{Br}}}:$

1.4 This problem is similar to 1.3. Hydrogen prefers to form one bond and sulfur prefers to form two bonds.

$$\mathrm{H}:\underset{\cdot\cdot}{\ddot{\mathrm{S}}}:\mathrm{H}$$

1.5 The stability of a species can be determined by examining whether each atom satisfies the octet rule. In this problem the carbon atom does not satisfy the octet rule, so the species is unstable.

1.6 a) octet rule satisfied, stable b) ten electrons around N, unstable.

1.7 Follow the guidelines shown below for writing Lewis structures of a molecule.

(1) Start with the neutral atoms in the molecule. If the molecule is charged, add (if negative) or subtract (if positive) one additional electron for each unit of charge.
For example, in CH_2O, carbon has four valence electrons, each hydrogen has one, and oxygen has six. The compound is not charged so we do not change the number of electrons.

$$\cdot\dot{\underset{\cdot}{C}}\cdot \quad + \quad 2\ H\cdot \quad + \quad \cdot\ddot{\underset{\cdot\cdot}{O}}\cdot$$ (total of 12 valence electrons)

(2) Write a skeletal arrangement of the atoms by bonding the atoms other than hydrogen together. The central atom is generally a nonmetal other than hydrogen , oxygen or halogen.

$$\begin{array}{c} :\dot{O}: \\ \vdots \\ \cdot\underset{\cdot}{C}\cdot \end{array}$$

Here 10 of the total 12 valence electrons have been used.

(3) Distribute the remaining electrons as bonds to hydrogens. Keep track of electrons and check to make sure that the octet rule is satisfied. If there are not enough electrons to satisfy the octet rule, form additional bonds between atoms. Remember additional electrons cannot be added because this would change the charge on the species. (See worked examples in Figures 1.4, 1.5, 1.6, and 1.9 in the text for more detail.)

:O:
H:C:H

In this case there are not enough electrons to satisfy the octet rule for the carbon and the oxygen. Since the carbon and the oxygen need one more electron each to satisfy the octet rule, it is necessary to form one additional bond between the two atoms

:O:
::
H:C:H

Check to see that the octet rule is satisfied.

a) H:C:C:C:H (each C also bonded to H above and H below)   b) H:C:::C:H   c) H:C:H with :Ö: double-bonded above C   d) H:N:O:H with H above N

1.8

$$H{:}\ddot{O}{:}H \;+\; H{:}\ddot{Cl}{:} \longrightarrow H{:}\ddot{O}^{\oplus}{:}H\ (\text{with } H \text{ below O}) \;+\; {:}\ddot{Cl}{:}^{\ominus}$$

1.9 Formal charge = [valence electrons in the atom] - [# of unshared electrons] - $\frac{1}{2}$ [# of shared electrons]

where the number of valence electrons is the same as the group number of the atom. Unshared electrons are those electrons in lone pairs. Shared electrons are those electrons involved in bonding.

Total charge (TC) of the molecule is the sum of all the formal charges of atoms in that molecule.

For example, in (a);

FC of carbon = 4 - 0 - (8/2) = 0

FC of oxygen = 6 - 2 - (6/2) = +1

a) Total Charge = TC = +1

b) TC = -1

c) TC = -1

e) TC = 0

f) TC = 0

1.10 The octet rule is an important criterion for estimating the stability of a compound represented by a particular Lewis structure. Formal charges can be used to further refine the estimate of stability. Other things being equal, the structure with fewer formal charges is more stable.
In H-Cl-O, the formal charge on chlorine is +1, on oxygen is -1, and on hydrogen is 0, whereas the formal charges on each of the atoms in H-O-Cl is zero. Therefore H-O-Cl is more stable due to less formal charges.

1.11

The actual structure is a resonance hybrid of these two structures. The actual bonds between the carbon and both of the oxygens are identical.

1.12 Use electronegativity values in Table 1.2 to answer this problem. A bond dipole is due to unequal electronegativities of the two atoms involved in a bond. A bond is polarized so that the negative end of the dipole is on the more electronegative element. An arrow pointing from the positive end of the dipole to the negative end is used to show the direction of polarization.

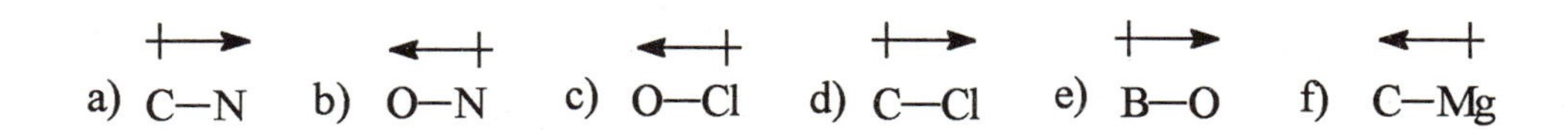

1.13 Use Figure 1.11 in the text to predict the geometries. Remember that multiple bonds count as one pair of electrons for the purpose of VSEPR theory.

a) linear b) trigonal planar c) tetrahedral

1.14 a) tetrahedral at C
bent at O (tetrahedral)
b) trigonal planar at C
bent at N (trigonal planar)
c) trigonal planar at C, bent at O (trigonal planar)

1.16 First determine the geometry of the molecule by VSEPR theory. Then find the individual bond dipoles of the molecule. The overall dipole moment is the vector sum of the individual bond dipoles.

a) $H_2C{=}\ddot{O}{:}$ b) $H_3C{-}F$ c) $H{-}CH_2{-}C{\equiv}N$

1.17 Ionic bonds are formed when metals combine with non-metals. Covalent bonds are formed between two non-metal elements.

a) $K^{\oplus}$ $:\ddot{Cl}:^{\ominus}$

(ionic)

b) $:\ddot{Cl}:\ddot{N}:\ddot{Cl}:$ / $:\ddot{Cl}:$

(covalent)

1.18

Like N, phosphorus has 5 electrons and prefers to form three bonds. PH_3 should be pyramidal.

$H:\ddot{P}:H$ / H

1.19

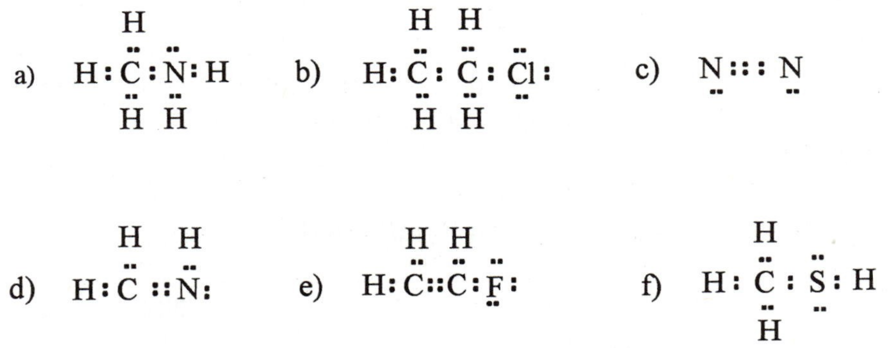

a) H:C:N:H (with H above C; H below C and N; lone pair on N)

b) H:C:C:Cl: (with H above and below each C; lone pairs on Cl)

c) N:::N (lone pair on each N)

d) H:C::N: (with H above C and N; lone pair on N)

e) H:C::C:F: (with H above each C; lone pairs on F)

f) H:C:S:H (with H above and below C; lone pairs on S)

1.20

a) H—N—N≡N: (formal charges: -1, +1, 0)

b) H—N=N=N: (formal charges: 0, +1, -1)

c) H—C—N≡N: with H on C (formal charges: -1, +1, 0)

d) H—C(H)(H)—C(=O)—O: (formal charges: C 0, carbonyl O 0, O -1)

e) H—O=C—H with H on C (formal charges: O +1, C 0)

f) H—B—H with H above and below B (formal charge: B -1)

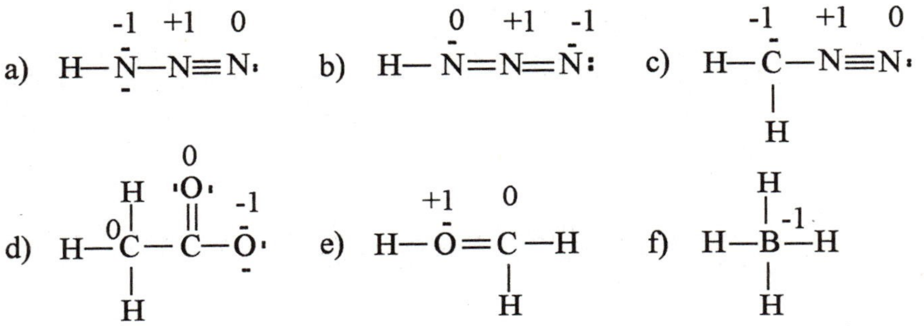

1.21 The two structures are resonance structures because the positions of all atoms are identical. Only the position of a lone pair and a bonding pair of electrons are changed.

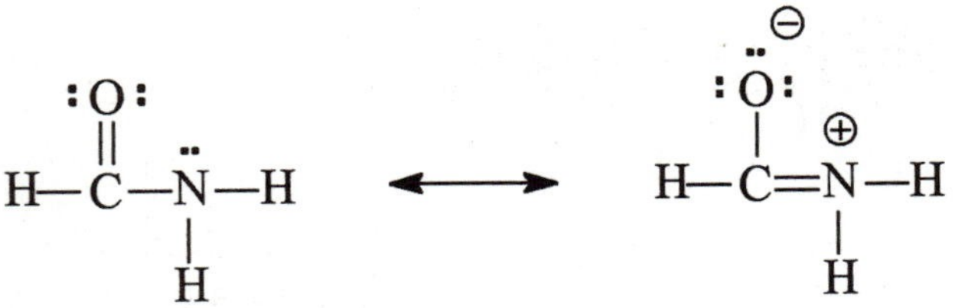

H—C(=O:)—N(H)—H ⟷ H—C(—:O:⊖)=N⊕(H)—H

Since the resonance structure on the left has formal charges of zero on all the atoms, we would expect it to be more stable.

1.22 Although the atoms in this structure have formal charges, the structure satisfies the octet rule. Therefore carbon monoxide is stable.

$$\overset{-1}{:C}\equiv\overset{+1}{O:}$$

1.23 Follow the example in Figure 1.15 to predict the dipole moment. The overall dipole moment of a molecule can be predicted by estimating the net direction of bond dipoles in the molecule.

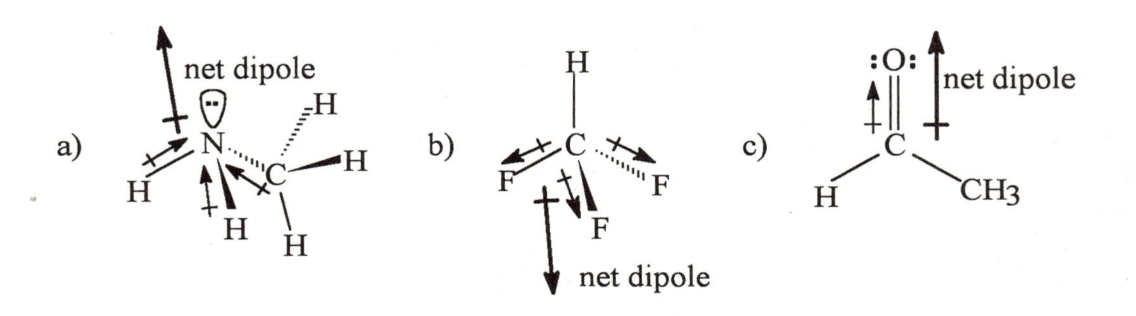

1.24

a) H–C(H)=Ö⊕–C(H)(H)–H

trigonal planar (C); trigonal planar (O); tetrahedral (C)

b) H–C(H)=C=C(H)–H

trigonal planar; linear; trigonal planar

1.25 a) All atoms in a Lewis structure should satisfy the octet rule. In the resonance structure on the left the sulfur atom has only one covalent bond and needs six more electrons to satisfy the octet rule.

$H_3C–C(=\ddot{O}:)–\ddot{S}:^{\ominus}$ ⟷ $H_3C–C(–\ddot{O}:^{\ominus})=\ddot{S}:$

1.26

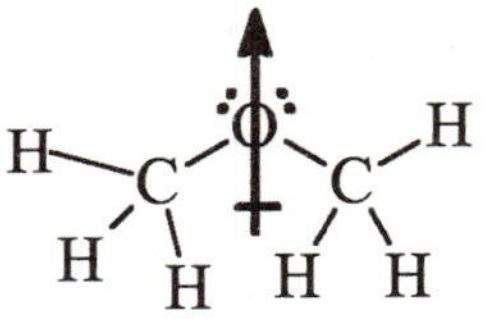

tetrahedral geometry at oxygen

1.27

1.28 The bonds will be polarized towards the nitrogen atom. The **N** atom in the ammonia molecule has a zero formal charge. In the ammonium cation the nitrogen atom carries a formal charge of plus one. This makes the **N** of the ammonium cation more electronegative and its hydrogens have a larger partial positive charge than the respective atoms in ammonia.

1.29 Ionic compounds normally exhibit extremely high melting points compared to covalent compounds. Since $CaCl_2$ is ionic, one would expect it to have a much higher melting point than 68°C. Therefore it would be wise not to trust your lab partner and look it up yourself.

1.30 The formal charge on the aluminum atom is -1 [FC_{Al} = 3 - 0 - (8/2)]; those on the Cl atoms are zero [FC_{Cl} = 7 - 6 - (2/2)]. The shape of this anion, with 4 electron pairs (bonds) around Al, is tetrahedral.

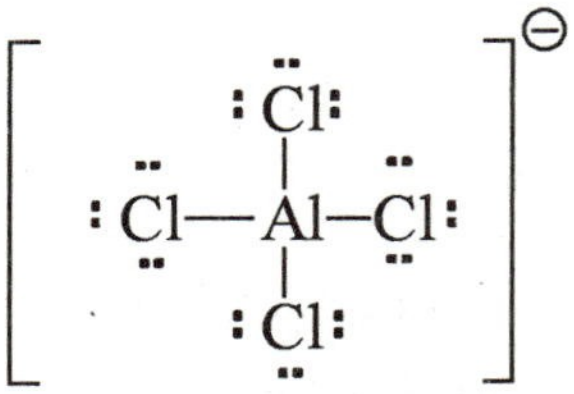

1.31 In the cyanate anion the negative charge is distributed between the nitrogen and the oxygen atoms. The shape of this anion is linear.

$$:\ddot{O}=C=\ddot{N}:^{\ominus} \longleftrightarrow {}^{\ominus}:\ddot{\underset{..}{O}}-C\equiv N:$$

1.32

a) 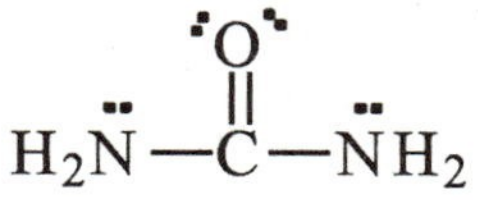 $H_2\ddot{N}-C(=\ddot{O})-\ddot{N}H_2$

urea

b) 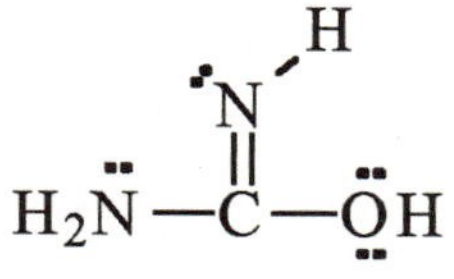 $H_2\ddot{N}-C(=\ddot{N}H)-\ddot{O}H$

an isomer of urea

1.33 Deviations from the octet rule are observed in ions and molecules with atoms from periods other than the second in the periodic table. Atoms of period 2 elements can accommodate only 8 valence electrons; hence the octet rule. Atoms in periods beyond 2 can hold more than 8 electrons in their valence shell. Phosphorus is a period 3 element can accommodate more than 8 electrons in its valence shell, which allows it to form more than 3 bonds whereas nitrogen (a period 2 element) has a capacity for only 8 electrons in its outer shell, limiting it to 3 bonds.

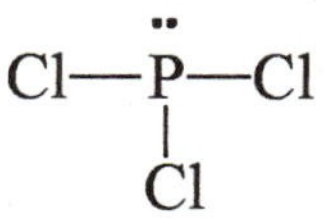

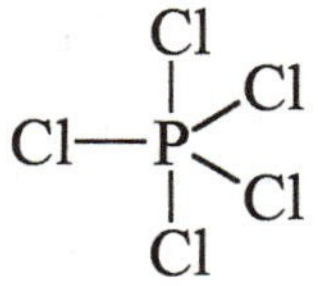

$$Cl-\ddot{P}(Cl)-Cl \qquad PCl_5$$

1.34 The absence of the two unpaired electrons in the unstable species NH_3 results in a formal charge of +2 on the nitrogen.

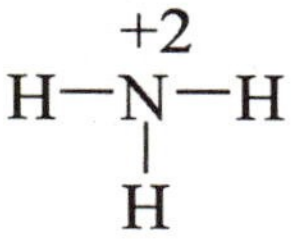

$$H-\overset{+2}{N}(H)-H$$

1.35 a) and b)

$$\underset{-1}{:\ddot{O}:}-\overset{+1}{\ddot{O}}=\underset{0}{\ddot{O}} \longleftrightarrow \underset{0}{\ddot{O}}=\overset{+1}{\ddot{O}}-\underset{-1}{:\ddot{O}:}$$

c) There are three regions of electron density around the central oxygen. According to the VSEPR model, three regions of electron density around a central atom are arranged in a trigonal planar geometry. The resulting molecular geometry of ozone is bent.

d) The actual structure of ozone is a hybrid of the two resonance structures. Therefore the bonds of ozone are identical.

1.36

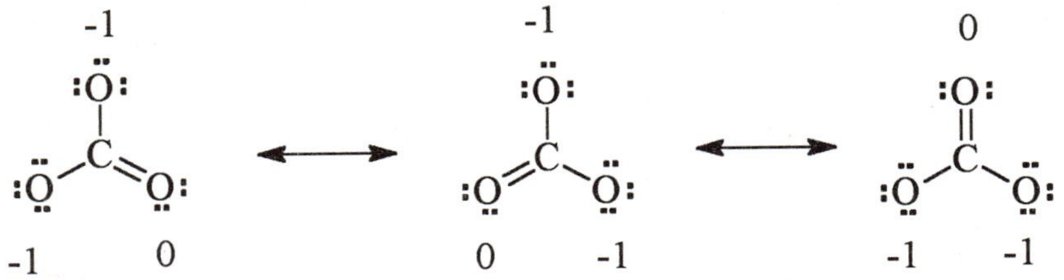

The shape of the carbonate anion is trigonal planar.
The actual structure of this ion is a hybrid of the three resonance structures. The actual distribution of electrons in each of the C-O bonds is an average of that shown by the three Lewis structures. Therefore all the oxygen atoms in this molecular ion are identical.

1.37 a) The charge on this species is +1.

b) Since there are only three regions of electron density around the central carbon atom, the geometry is trigonal planar.

c) Because there are only 6 electrons around the carbon, the octet rule is not satisfied and the cation is not stable.

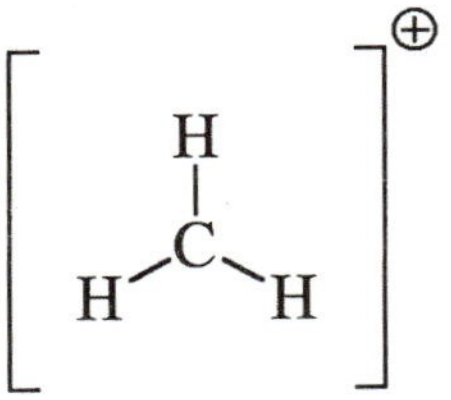

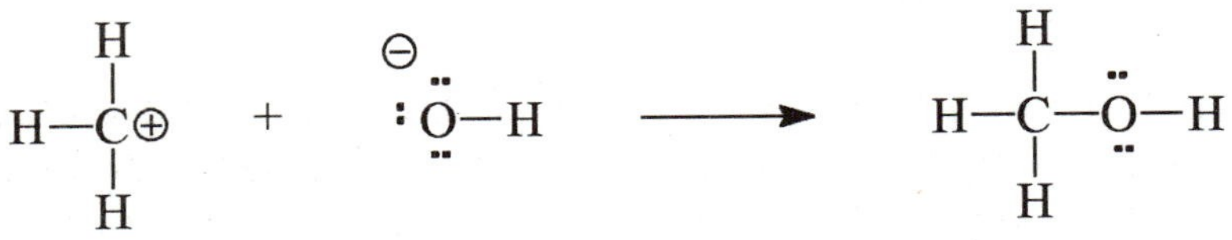

1.38 A bond dipole depends on the absolute value of the difference in electronegativities between the two atoms. In the case of FCl and ICl, the absolute electronegativity differences are very similar. Therefore the dipole moments of these molecules are also similar. However, the direction of the bond dipole is different.

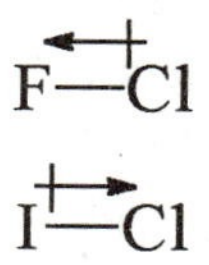

F—Cl

I—Cl

1.39 The shape of PCl_5 is trigonal bipyramid. Although there are bond dipoles at each P-Cl bond, there is no net dipole because the individual bond dipoles cancel.

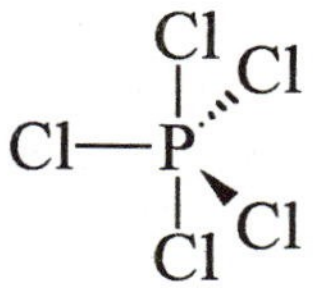

1.40 The actual structure of benzene is a resonance hybrid of two resonance structures. Each of the C-C bonds in benzene is between a double bond and a single bond.

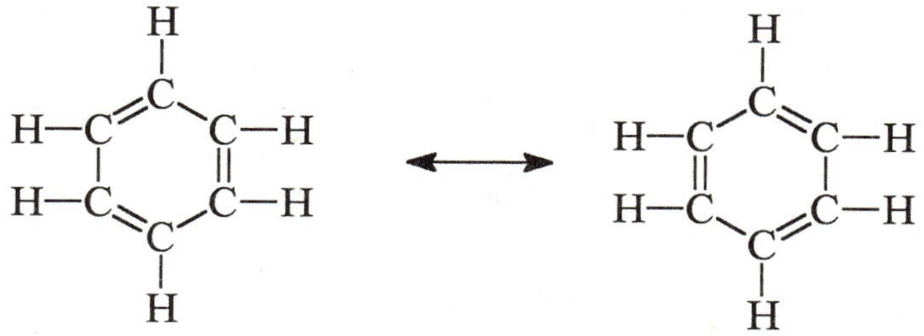

After completing this chapter, you should be able to:

1. Write the best Lewis structure for any molecule or ion. This includes determining how many electrons are available, whether multiple bonds are necessary, and satisfying the octet rule if possible. For complex molecules, however, the connectivity must be known.
Problems 1.2, 1.3, 1.4, 1.7, 1.17, 1.18, 1.19, 1.26, 1.30, 1.31.

2. Calculate the formal charge on any atom in a Lewis structure. (In fact, you should be starting to recognize the formal charges on some atoms in some situations without doing a calculation.)
   Problems 1.9, 1.20, 1.30.

3. Estimate the stability of a Lewis structure by whether it satisfies the octet rule and by the number and the distribution of the formal charges in the structure.
   Problems 1.5, 1.6, 1.10, 1.21, 1.22, 1.37.

4. Recognize some simple cases in which resonance is necessary to describe the actual structure of a molecule. However, a better understanding of resonance will have to wait until Chapter 3.
   Problems 1.11, 1.25, 1.27, 1.31, 1.35, 1.36, 1.40.

5. Arrange the atoms that are of most interest to organic chemistry in order of their electronegativities and assign the direction of the dipole of any bond involving these atoms.
   Problems 1.12, 1.38.

6. Determine the shape of a molecule from its Lewis structure by using VSEPR theory.
   Problems 1.13, 1.14, 1.23, 1.24, 1.26, 1.30.

7. Determine whether a compound is polar or not and assign the direction of its dipole moment.
   Problems 1.15, 1.16, 1.26, 1.39.

Chapter 2
ORGANIC COMPOUNDS--- A FIRST LOOK

2.1 Molecules that satisfy the octet rule are likely to be stable. Another factor that needs to be considered when estimating the stability is the formal charges in a given structure. Any structure that does not satisfy the octet rule is not very stable. The presence of formal charges is less destabilizing. For example, the structure in part (a), satisfies the octet rule at all atoms. However, one of the carbon atoms has a formal charge. Therefore this molecule is less stable.

a) less stable   b) stable   c) less stable
d) stable   e) stable   f) stable   g) stable

2.2

```
                                        H                    H
                                        |                    |
                                     H--C--H              H--C--H
   H  H  H  H  H               H  H     |  H             H   |   H
   |  |  |  |  |               |  |     |  |             |   |   |
H--C--C--C--C--C--H         H--C--C-----C--C--H       H--C---C---C--H
   |  |  |  |  |               |  |     |  |             |   |   |
   H  H  H  H  H               H  H     H  H             H   |   H
                                                          H--C--H
                                                             |
                                                             H
```

2.4

```
                                                            H                   H
                                         H                  |                   |
                                         |               H--C--H             H--C--H
   H  H  H  H                   H  H     O  H            H  |  H             H  |
   |  |  |  |                   |  |     |  |            |  |  |             |  |
H--C--C--C--C--O--H          H--C--C-----C--C--H      H--C--C--C--O--H    H--C--C--O--H
   |  |  |  |                   |  |     |  |            |  |  |             |  |
   H  H  H  H                   H  H     H  H            H  H  H             H  |
                                                                             H--C--H
                                                                                |
                                                                                H

                                                               H
                                                               |
                                                            H--C--H
   H  H  H     H                H  H     H  H               H  |     H
   |  |  |     |                |  |     |  |               |  |     |
H--C--C--C--O--C--H          H--C--C--O--C--C--H         H--C--C--O--C--H
   |  |  |     |                |  |     |  |               |  |     |
   H  H  H     H                H  H     H  H               H  H     H
```

2.5 a) same b) same c) isomers d) same

2.6 For a hydrocarbon with the formula C_nH_x, the degree of unsaturation (DU) is equal to $\frac{(2n+2) - x}{2}$.

a) DU = 0

b) DU = 2

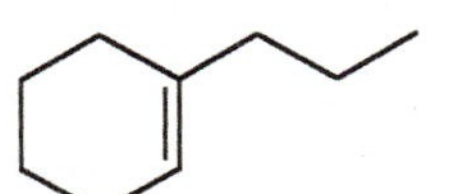

c) DU = 4

2.7 a) same b) same c) isomers d) same
e) same f) isomers

2.9 When calculating DU for formulas that have atoms other than C and H, use the following set of rules:

1. Add the number of halogens in the formula to number of hydrogens in the formula.
2. Ignore the number of oxygens in the formula.
3. Add the number of nitrogens in the formula to the maximum number of hydrogens.

For example in C_6H_8ClN, the DU = $\frac{[2(6) + 2 + 1 - (8 + 1)]}{2} = 3$.

a) DU = 1

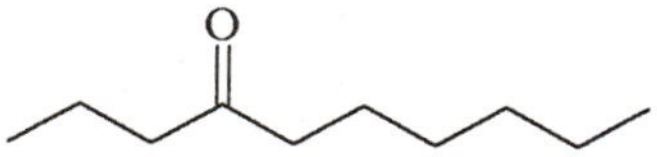

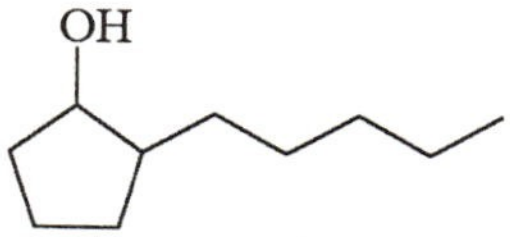

b) DU = 3

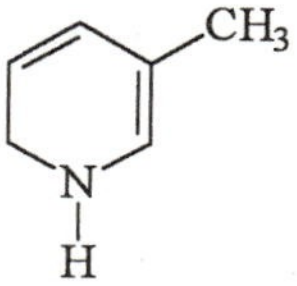

c) DU = 0

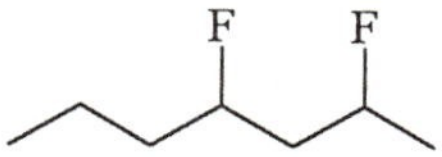

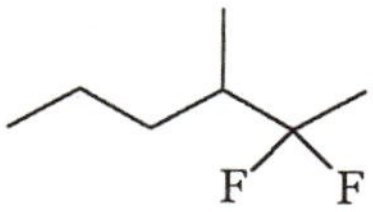

d) DU = 3

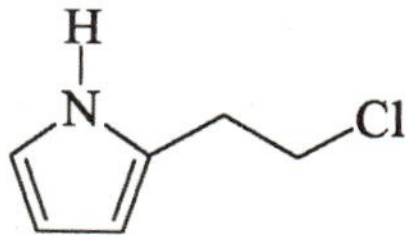

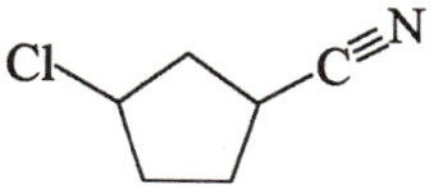

e)DU = 3

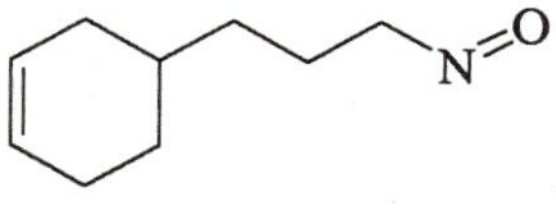

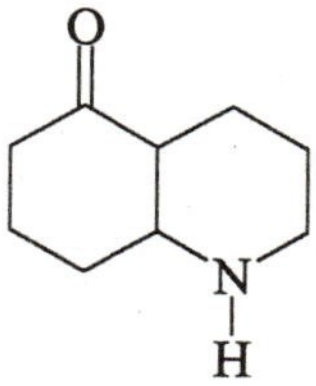

2.10 Intermolecular forces of attraction result from charge-charge interactions between two molecules. These charge-charge interactions can be classified into four basic kinds.

1. **ion-ion interaction**: attractive force between two oppositely charged ions. This is the strongest interaction of all.
2. **ion-dipole interaction**: attractive forces between an ion and a polar molecule. This interaction is much weaker than an ion-ion interaction.
3. **hydrogen bonding**: a special type of dipole-dipole interaction between a hydrogen bonded to an electronegative atom in a molecule and an electronegative atom in another polar molecule. This is weaker than ion-dipole, but stronger than dipole-dipole.

4. **van der Waals forces**:

i) dipole-dipole: interaction between two polar molecules. This is weaker than hydrogen bonding.

ii) dipole-induced dipole: interaction between a polar molecule and a non-polar molecule. This is weaker than dipole-dipole.

iii) London forces: interaction between two non-polar molecules. This is the weakest interaction of all.

a) London b) van der Waals c) ion-ion
d) van der Waals and hydrogen bonding

2.11

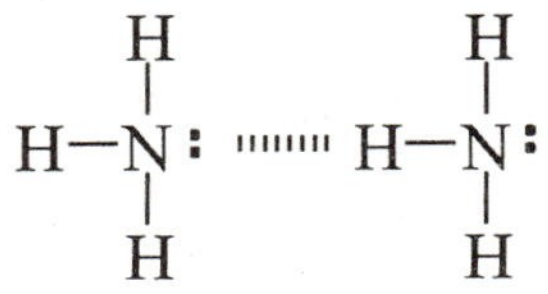

```
    H            H
    |            |
H—N:  ''''''  H—N:
    |            |
    H            H
```

2.12 The stronger the intermolecular attractive forces are, the higher the melting point is. KBr is an ionic compound and will have strong ion-ion interactions. CH_3Br is a covalent molecule and has only weak van der Waals forces of attraction. Therefore KBr will exhibit a higher melting point.

2.13 Boiling points also depend on the magnitude of the intermolecular attractive forces. The molecule $CH_3CH_2CH_2OH$ will have the higher boiling point because it has hydrogen bonding and van der Waals forces of interactions. The ether has only van der Waals forces of attraction.

2.14 The two compounds have a polar, hydrophilic part and a non-polar, hydrophobic part. The effects of two parts are in competition. Water is a polar compound and will dissolve polar compounds. As the non-polar part of the molecule gets larger, its solubility in water will decrease. The compound $CH_3CH_2CH_2CH_2CO_2H$ will have a higher solubility because it has a smaller non-polar part.

2.15 a) ether b) alcohol c) carboxylic acid

d) amide e) ester f) arene and aldehyde

2.16

a) OH

b)

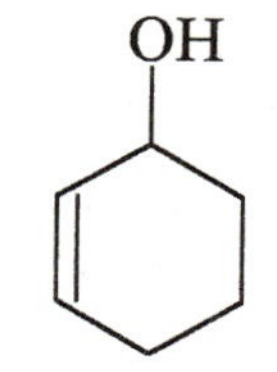

c) O Cl

d)

e) CN O H

f) O

2.17

a) b) c)

d) e) f)

2.18 a) alcohol b) alkene and alcohol c) ketone and alkyl chloride
d) arene e) nitrile and aldehyde f) alkene and ketone

2.19 a) arene and carboxylic acid b) alcohol and alkyne
c) alkene and aldehyde d) carboxylic acid e) ester
f) nitrile and ketone

2.20 a) same b) isomers c) same d) same e) isomers
f) isomers

2.21

a) DU = 0 O OH OH

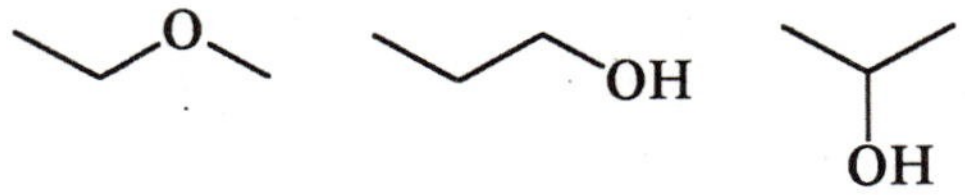

b) DU = 0 Cl Cl Cl Cl

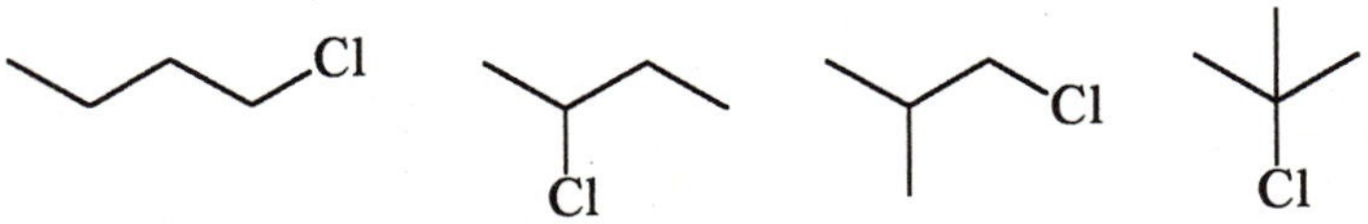

c) DU = 1

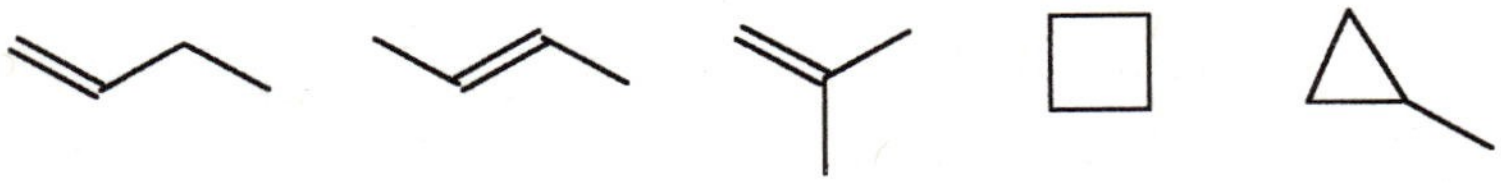

d) DU = 0

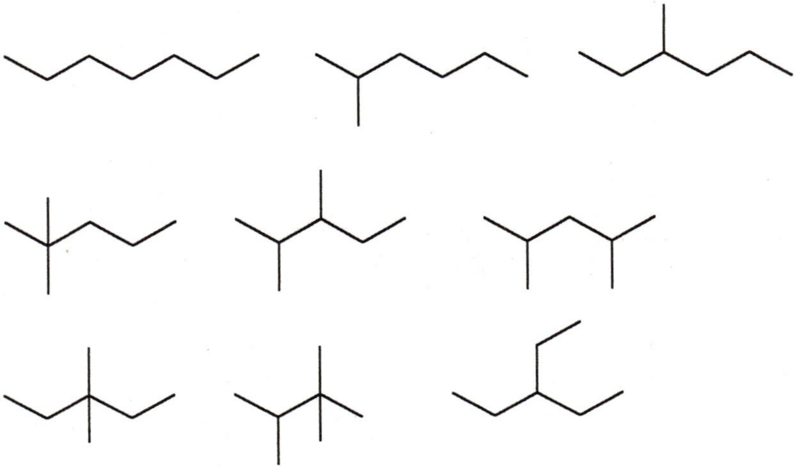

2.22 The DU for $C_4H_8O = 1$

a) 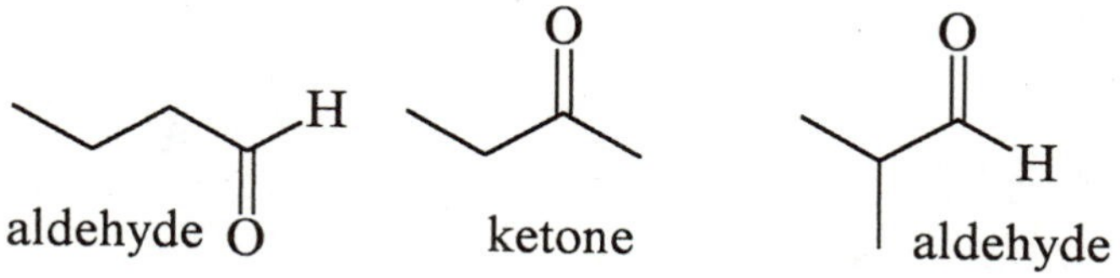 aldehyde ketone aldehyde

b)

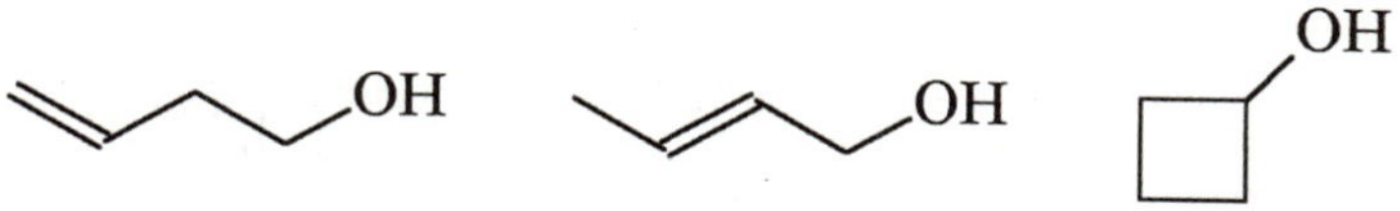

c)

2.23 The DU for $C_5H_{10} = 1$

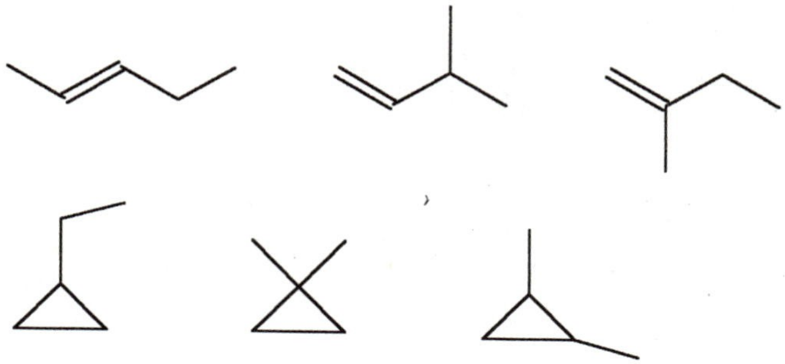

2.24

a) DU = 2 (structure: pentanenitrile, CN) b) DU = 2 (structure: cycloheptanone, O)

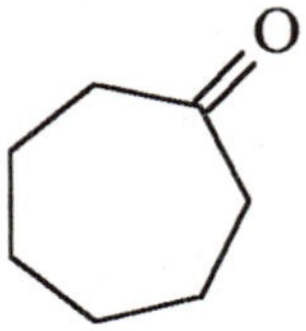

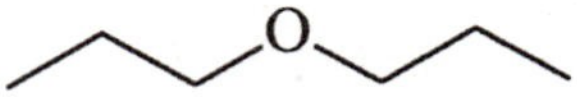

c) DU = 0 (structure: dipropyl ether, O) d) DU = 4 (structure: toluene)

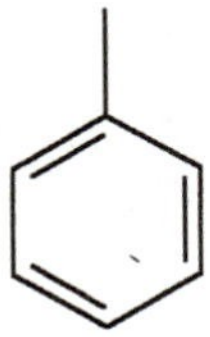

2.25 The functional group nitrile (CN) has a triple bond between the C and the N atoms, which would require a DU of 2. The DU for $C_8H_{17}N$ = 1. Therefore it is not possible to have a nitrile functional group.

2.26

a) 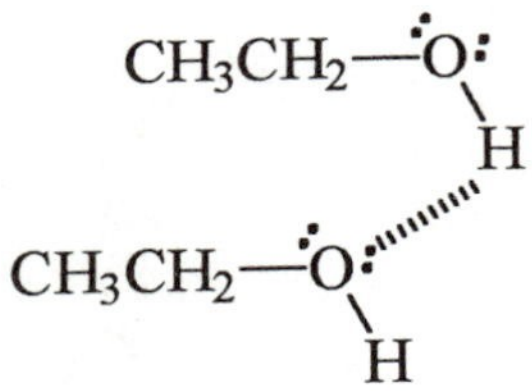 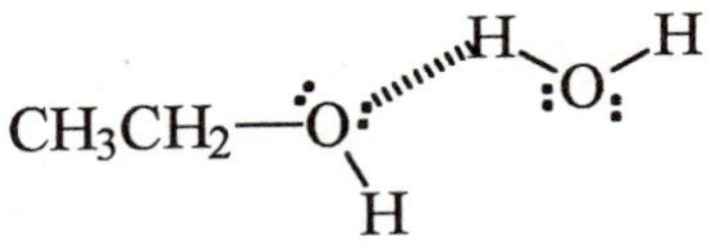 CH_3CH_2—O—H ····· O(CH_2CH_3)—H

CH_3CH_2—O(H) ····· H—O—H

CH_3CH_2—O—H ····· O(H)—H

H—O—H ····· O(H)—H

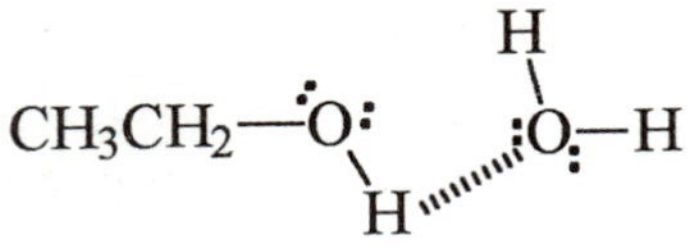

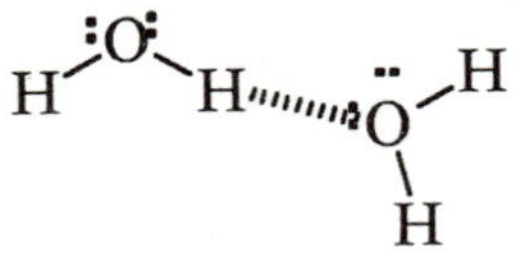

b) 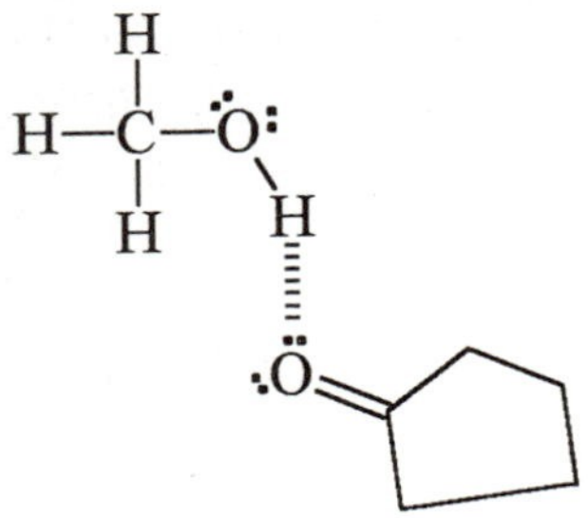 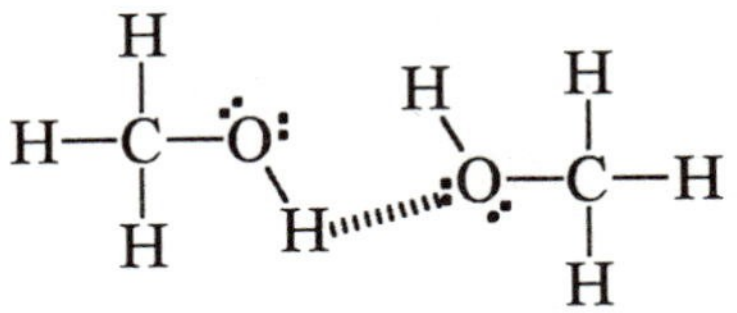 H_3C—O—H ····· O=(cyclopentanone)

H_3C—O—H ····· O(H)—CH_3

2.27 The shape of a compound has a dramatic effect on its physical properties. The effect of shape of a molecule on its boiling point is quite different from the effect on its melting point. Solids are more ordered. More symmetrical molecules can pack into the crystal lattice better, have larger attractive forces, and therefore exhibit higher melting points. Liquids on the other hand are less ordered than solids. Rod-shaped molecules have more surface area than spherical molecules of similar molecular mass, and therefore have slightly higher boiling points due to increased London forces. Therefore the rod-shaped isomer will exhibit the lower melting point ($-90^{\circ}C$), and the higher boiling point ($117^{\circ}C$) of these two isomeric alcohols.

29 **2.28**

a) Both compounds are non-polar and are rod shaped. The compound with 10 carbons has a higher molecular mass (also surface area) than the one with 8 carbons, so it has a higher boiling point.

b) These two compounds are isomers, and have similar shapes. The alcohol $CH_3CH_2CH_2OH$ has a higher boiling point because it has both hydrogen bonding and van der Waals forces of interactions. The other molecule has only van der Waals forces of attraction.

c) Both compounds have similar shapes. The ether CH_3OCH_3 is higher boiling because the dipole-dipole forces of this polar molecule will hold the molecules together more strongly than the London forces in the non-polar compound $CH_3CH_2CH_3$.

2.29 The isomer with the higher melting point has a more symmetrical shape, which allows it to pack into the crystal lattice better.

33 **2.30** Hexane is a non-polar solvent and will best dissolve non-polar compounds. As the non-polar character of a compound increases, its solubility in hexane will increase. Therefore the salt with the non-polar hydrocarbon chain will be more soluble.

2.31 Like dissolves like. Both benzene and hexane are non-polar compounds, and are miscible. Water is a polar compound. Therefore, benzene and water are immiscible.

2.32 These two isomers have different functional groups. One is a carboxylic acid and the other is an ester. The carboxylic acid is miscible with water because it is more polar and can hydrogen bond with water.

2.33

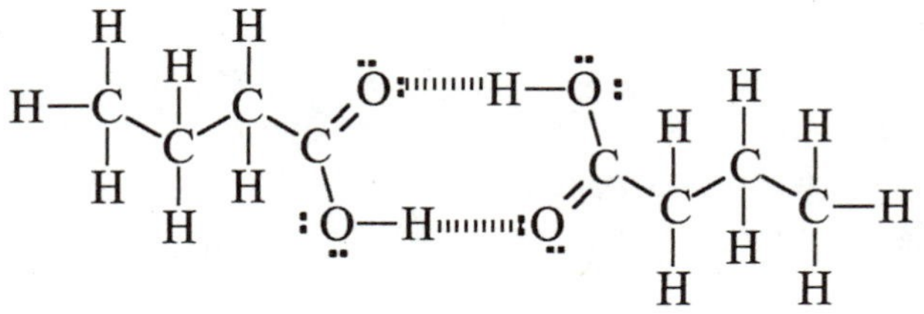

```
      H     H
      |  H  |        ..        ..
  H—C    |    C      O:''''''H—O:      H
      |    C    |  C//           \    H  |   H
      H    |   H   \              C    |  C    |
           H      :O—H''''''':O//   C   |  C—H
                   ..        ..     |  H   |
                                    H      H
```

2.34 Use Table 2.1 in the text for calculating total bond energies of compounds. The compound with the aldehyde functional group has 4 C-H bonds, 1 C-C bond, and 1 C=O bond, giving it a total bond energy of 646 kcal/mol (4x98 + 81 +173). The other isomer has 3 C-H bonds, 1 C=C bond, 1 C-O bond, and 1 O-H bond, giving a total bond energy of 627 kcal/mol (3x98 + 145 + 79 + 109). Therefore the aldehyde is more stable.

2.35

CH_2=CH_2	+ H-H	$\longrightarrow$ CH_3-CH_3
537kcal/mol	104kcal/mol	669kcal/mol

The total bond energy of the product, CH_3-CH_3 (669 kcal/mol) is larger than the reactants; CH_2=CH_2 (537 kcal/mol), and H_2 (104 kcal/mol). This makes the product more stable than the reactants, and therefore, the reaction is energetically favorable.

2.36

Isomers of C_5H_{12} Isomers of $C_5H_{11}Cl$

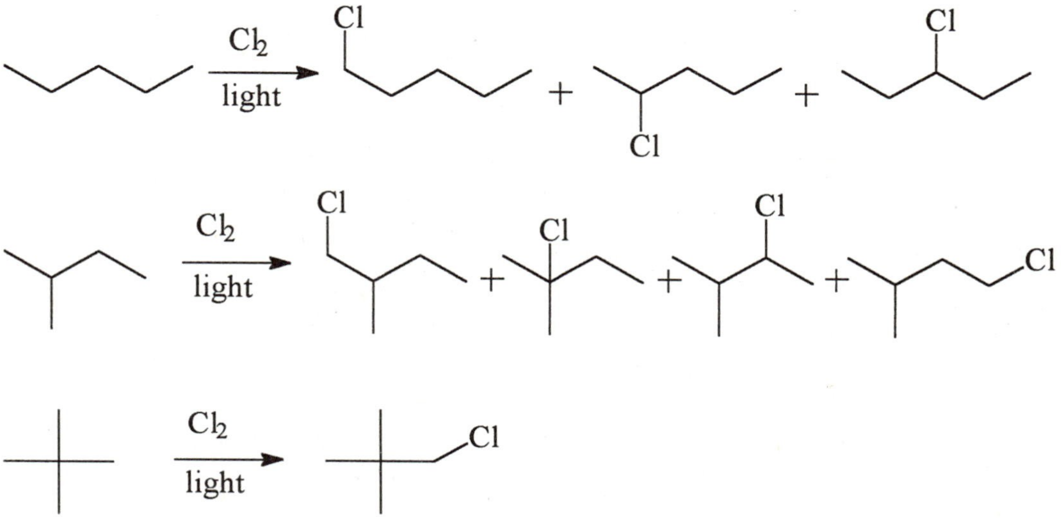

Cl_2, light

Cl Cl Cl

Cl_2, light

Cl Cl Cl Cl

Cl_2, light

Cl

Only the straight chain isomer gives three isomers of $C_5H_{11}Cl$.

2.37 The DU of C_6H_{12} = [(2x6+2) - 12]/2 = 1

All of the carbons of the isomer that gives only one $C_6H_{11}Cl$ must be identical.

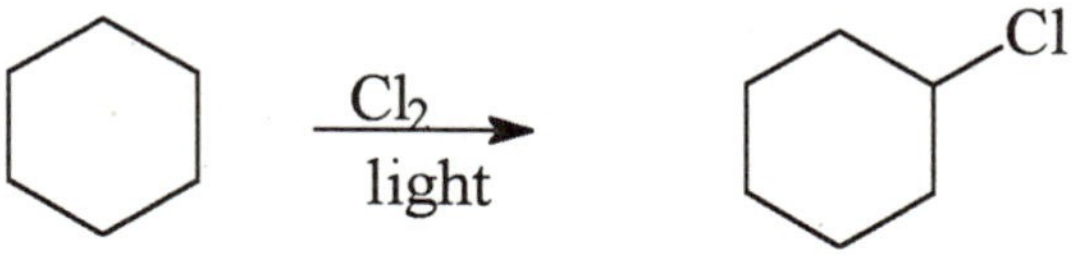

Cl_2, light

Cl

C_6H_{12} $C_6H_{11}Cl$

All of the other possible isomers of C_6H_{12} give more than one $C_6H_{11}Cl$.

2.38

From Chapter 1, $\mu = (e)x(d)$. Because F is more electronegative than Cl, e is large for CH_3F than for CH_3Cl. But the C-Cl bond (1.78 Å) is longer than the C-F bond (1.38 Å) so *d* is larger for CH_3Cl than for CH_3F

After completing this chapter, you should be able to:

1. Quickly recognize the common ways in which atoms are bonded in organic compounds. You should also recognize unusual bonding situations and be able to estimate the stability of molecules with such bonds.
   Problem 2.1.

2. Know the trends in bond strengths and bond lengths for the common bonds.

3. Recognize when compounds are structural isomers and be able to draw structural isomers for any formula.
   Problems 2.2, 2.3, 2.4, 2.5, 2.7, 2.20, 2.21, 2.22, 2.23, 2.36, 2.37.

4. Calculate the degree of unsaturation for a formula and use it to help draw structures for that formula.
   Problems 2.6, 2.8, 2.9, 2.24, 2.25.

5. Draw structures using any of the methods we have seen. You should also be able to examine a shorthand representation for a molecule and recognize all of its features.
   Problems 2.16, 2.17.

6. Examine the structure of a compound and determine the various types of intermolecular forces that are operating. On this basis, you should be able to crudely estimate the physical properties of the compound.
   Problems 2.10, 2.11, 2.12, 2.13, 2.14, 2.26, 2.27, 2.28, 2.29, 2.30, 2.31, 2.32, 2.33.

7. Recognize and name all of the important functional groups.
   Problems 2.15, 2.18, 2.19.

Chapter 3
ORBITALS AND BONDING

3.1

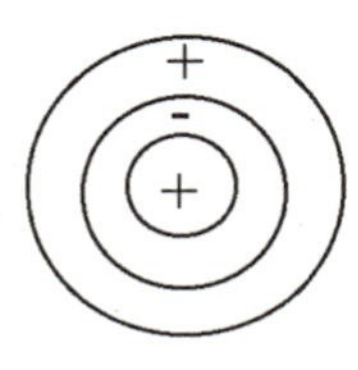

3s

Like the 2s orbital, the 3s orbital is spherical in shape but larger. The 3s orbital has two spherical nodes whereas the 2s orbital has only one node.

3.2 An orbital energy diagram for an atom shows the relative energies of orbitals and how electrons are distributed in these orbitals. Remember to put the electrons in the lowest energy orbital available while observing the Pauli exclusion principle and Hund's rule.

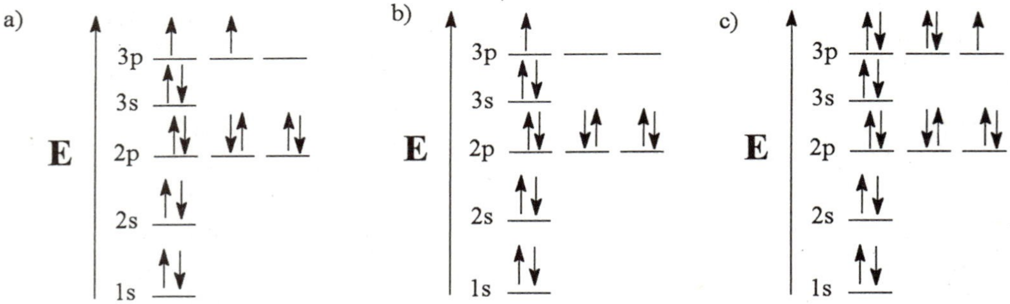

a) E: 3p, 3s, 2p, 2s, 1s

b) E: 3p, 3s, 2p, 2s, 1s

c) E: 3p, 3s, 2p, 2s, 1s

3.3 Any electron arrangement that is in violation of any one of the basic rules mentioned above in problem 3.2 is higher in energy than the **ground state** electron configuration. The higher energy electron configuration of atom is called an **excited state**.

a) In this electron configuration one of the three electrons in the singly filled 2p orbitals has its spin opposite to that of the other two electrons.

This electron configuration is in violation of Hund's rule and therefore is an excited state.

b) The lower energy 2s orbital is only half filled. Therefore this is an excited state.

3.4

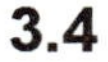

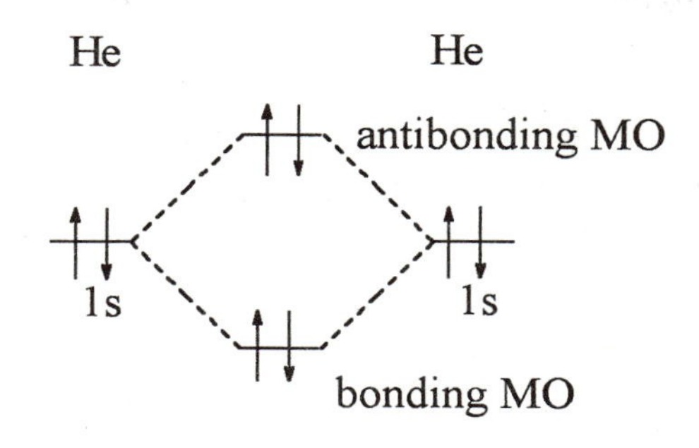

3.5

Osp3 $\sigma_{Csp3\ +\ H1s}$

H—C—Ö—C—H (each C bearing two H)

$\sigma_{Csp3\ +\ Osp3}$

The unshared electrons on the oxygen are in nonbonding sp^3 hybridized atomic orbitals.

3.6 1) $\sigma_{Csp2\ +\ H1s}$ 2) $\sigma_{Csp3\ +\ H1s}$ 3) $\sigma_{Csp3\ +\ Csp2}$
4) $\sigma_{Csp2\ +\ Nsp2}$, $\pi_{C2p\ +\ N2p}$ 5) $\sigma_{Csp3\ +\ Nsp2}$ 6) $\sigma_{Csp3\ +\ H1s}$

3.8 1) $\sigma_{Csp2\ +\ H1s}$ 2) $\sigma_{Csp3\ +\ H1s}$ 3) $\sigma_{Csp3\ +\ Csp2}$ 4) $\sigma_{Csp2\ +\ Csp2}$, $\pi_{C2p\ +\ C2p}$
5) $\sigma_{Csp2\ +\ Csp}$ 6) $\sigma_{Csp\ +\ Csp}$, 2 $\pi_{C2p\ +\ C2p}$ 7) $\sigma_{Csp\ +\ H1s}$

3.9 a) 1) $\sigma_{Csp3\ +\ H1s}$ 2) $\sigma_{Csp3\ +\ Csp2}$ 3) $\sigma_{Csp2\ +\ H1s}$ 4) $\sigma_{Csp2\ +\ Osp2}$, $\pi_{C2p\ +\ O2p}$
b) 1) $\sigma_{Csp3\ +\ Csp2}$ 2) $\sigma_{Csp2\ +\ Csp2}$, $\pi_{C2p\ +\ C2p}$ 3) $\sigma_{Csp2\ +\ Csp}$
4) $\sigma_{Csp\ +\ Nsp}$, 2 $\pi_{C2p\ +\ N2p}$ 5) $\sigma_{Csp2\ +\ H1s}$
c) 1) $\sigma_{Csp2\ +\ Osp2}$, $\pi_{C2p\ +\ O2p}$ 2) $\sigma_{Csp\ +\ H1s}$ 3) $\sigma_{Csp\ +\ Csp}$, 2 $\pi_{C2p\ +\ C2p}$
4) $\sigma_{Csp2\ +\ Csp}$ 5) $\sigma_{Csp3\ +\ Csp2}$ 6) $\sigma_{Csp3\ +\ H1s}$
d) 1) $\sigma_{Csp3\ +\ H1s}$ 2) $\sigma_{Csp3\ +\ Csp3}$ 3) $\sigma_{Csp3\ +\ Nsp3}$ 4) $\sigma_{Nsp3\ +\ H1s}$

3.10

b) The π bonds are not conjugated because they are separated by CH_2 groups on each end.

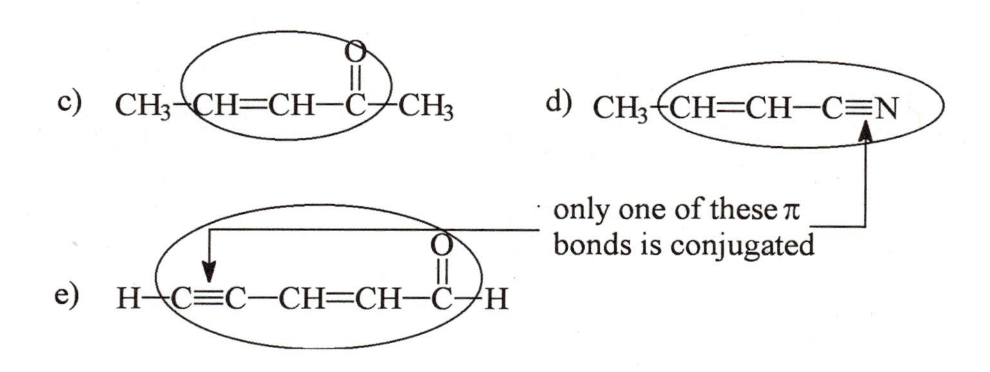

f) The π bonds are not conjugated because they are separated by a CH_2 group.

3.11 An atom with an unshared pair of electrons that is next to a π bond will assume a hybridization that puts the unshared electrons in a p orbital so that the electrons can be conjugated with the p orbitals of the π bond.

b) 1) sp^3 2) sp^2 3) sp^2 4) sp^3
c) 1) sp^2 2) sp^2 3) sp^2 4) sp^3
d) 1) sp^2 2) sp^2

3.12 In each resonance structure the nuclei of atoms are in identical positions; only the electrons are moved.

a) These are not resonance structures because the H atom has moved from one oxygen in the left structure to the other in the right structure.
b) These are not resonance structures because a H atom has moved from the oxygen in the left structure to the carbon in the right structure.
c) These are resonance structures.
d) These are not resonance structures because a H atom has moved from the right terminal carbon in the structure on the left to the second carbon from the left in the structure on the right..

3.13 Each resonance structure must have the same number of electrons and the same total charge.

b) Two electrons are missing from the structure on the right.
c) The structure on the right has 5 bonds to C and 2 more electrons.
d) In the structure on the right, the center N has only 6 electrons.
e) The C bonded to O has 5 bonds (10 electrons) in the right structure.

3.14 The relative stability of resonance structures can be estimated by their adherence to the octet rule, the number of formal charges, and the location of the charges.

b) The right structure is less important because it has formal charges and the octet rule is not satisfied at the positively charged carbon atom.
c) The first two structures are equally important; the last structure is less important because it has more formal charges and the octet rule is not satisfied at the N.

3.15

c) The first three structures below are of identical stability and contribute equally to the resonance hybrid. The last structure can be neglected. The ion has a large amount of resonance stabilization.

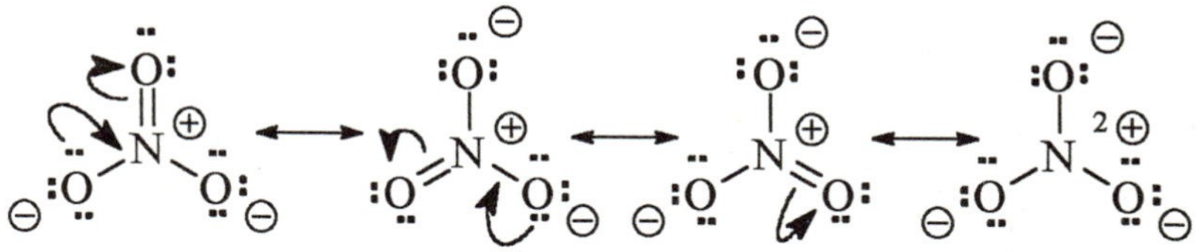

3.16

a)

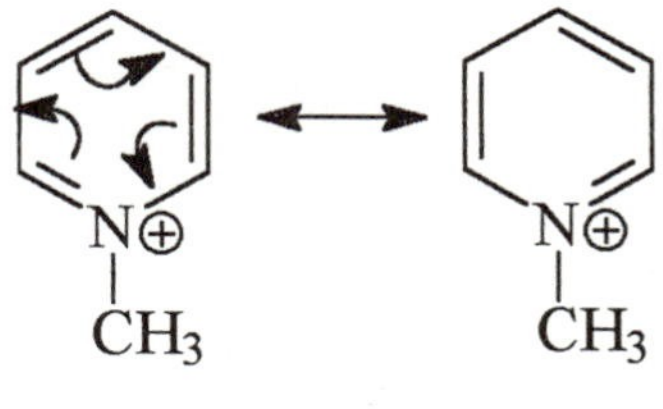

b) CH_3-N ⟷ CH_3-N

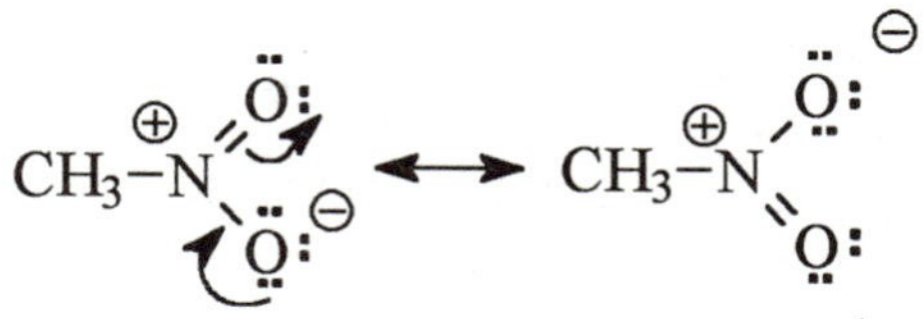

c)

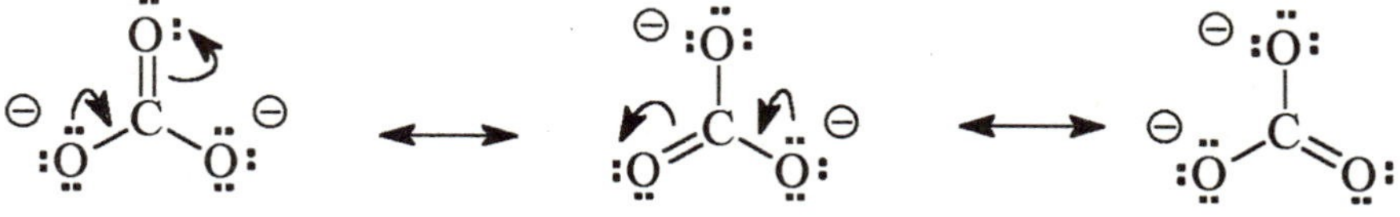

d)

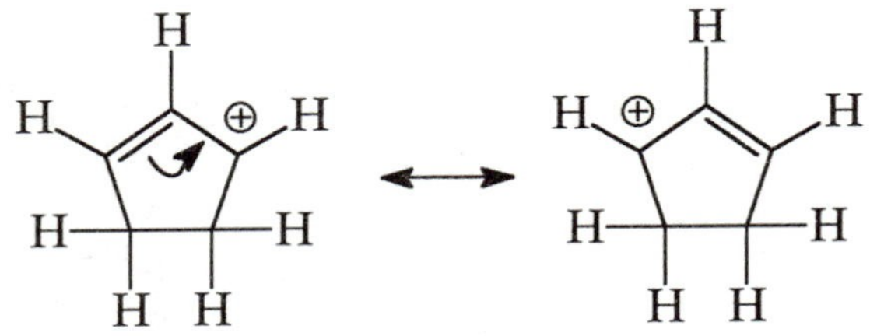

e)

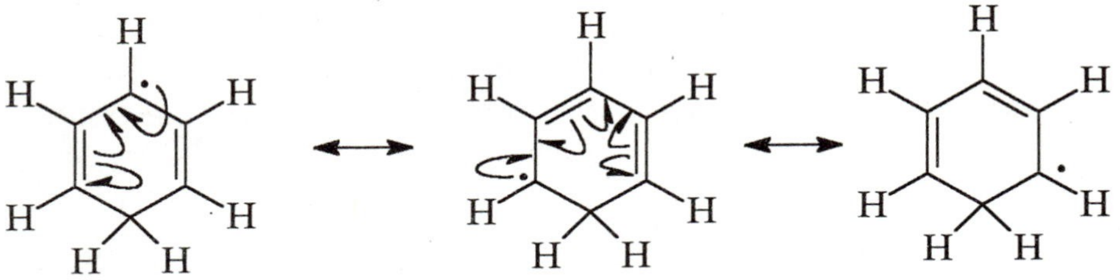

3.17

a)

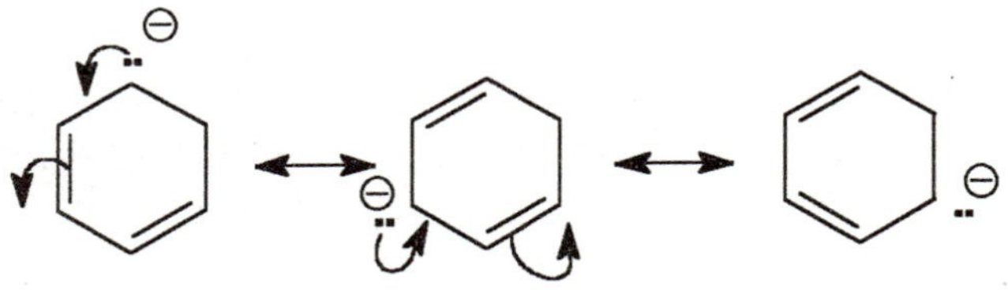

Each of these structures make similar contributions to the resonance hybrid because they are of about the same stability. Because there are three important resonance structures, this anion has a large resonance stabilization.

e)

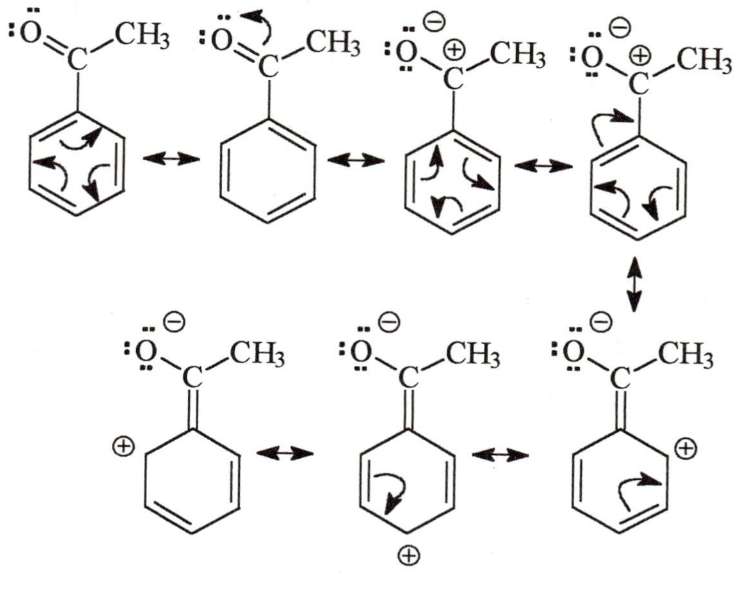

The first two resonance structures have no formal charges and are equal in stability. These two structures contribute more to the resonance hybrid. The other five resonance structures have formal charges and are of similar stability. These five resonance structures contribute less to the resonance hybrid than the first two. Because there are seven resonance structures this compound has a large resonance stabilization.

f)

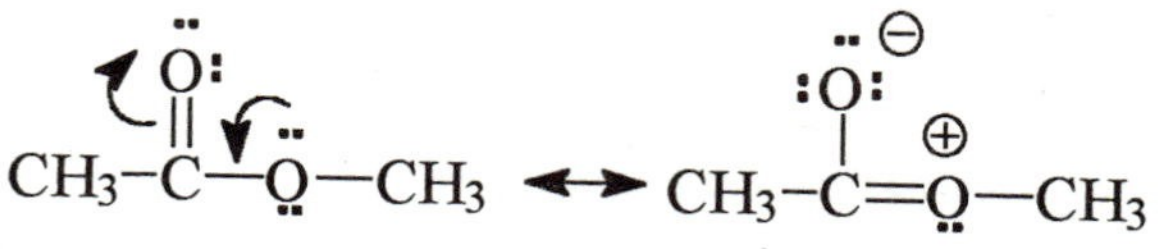

$CH_3-C-O-CH_3 \leftrightarrow CH_3-C=O-CH_3$

The structure on the left contributes more to the resonance hybrid because it does not have any formal charges. This ester has only a small resonance stabilization.

3.18

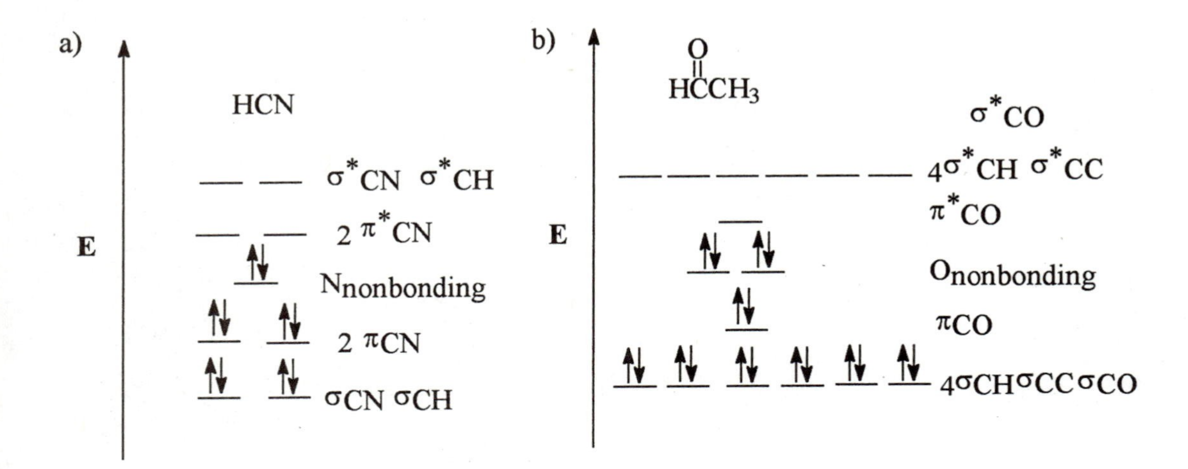

a)
E
HCN
σ*CN σ*CH
2 π*CN
Nnonbonding
2 πCN
σCN σCH
b)
E
O
HCCH3
σ*CO
4σ*CH σ*CC
π*CO
Ononbonding
πCO
4σCH σCC σCO

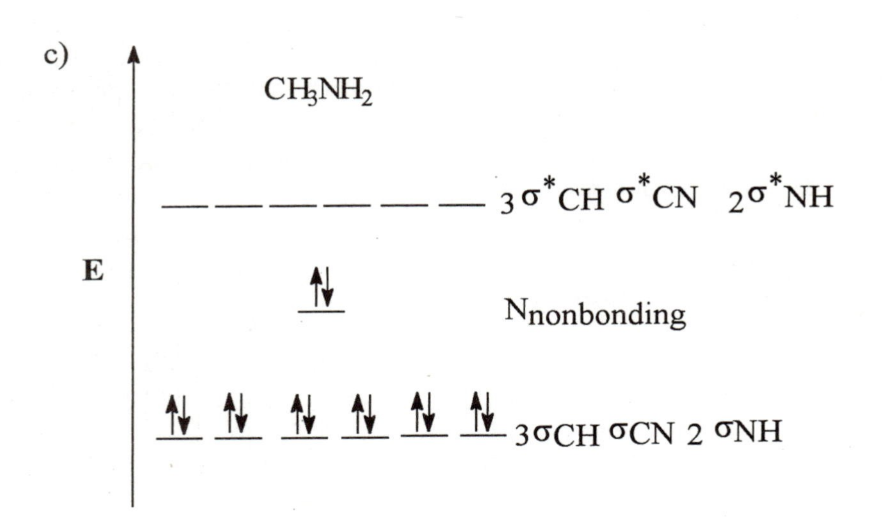

c)
E
CH3NH2
3 σ*CH σ*CN 2σ*NH
Nnonbonding
3σCH σCN 2 σNH

3.19

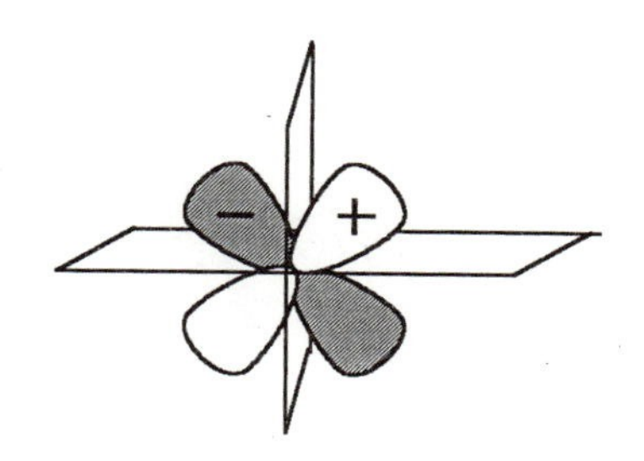

3.20

a) $CH_3{-}NH_2$
sp3 sp3

b) $CH_2{=}CHCH_2C{\equiv}N$
sp sp
sp2 sp2 sp3

c) sp3
sp2 sp3
sp2 sp3

d) O sp2
sp2
sp3 sp3
sp3 sp3
sp3

e) sp
sp3 sp3
sp
OH
sp3

f) H sp2 NH
sp2 sp2
O sp3 sp3
sp2

3.21

a) 1. $\sigma_{Csp2+H1s}$, 2. $\sigma_{Csp2+Csp2}$ and π_{Cp+Cp},
3. $\sigma_{Csp2+Csp2}$, 4. $\sigma_{Csp2+Csp3}$
5. $\sigma_{Csp3+H1s}$, 6. $\sigma_{Csp2+Osp2}$ and π_{Cp+Op}

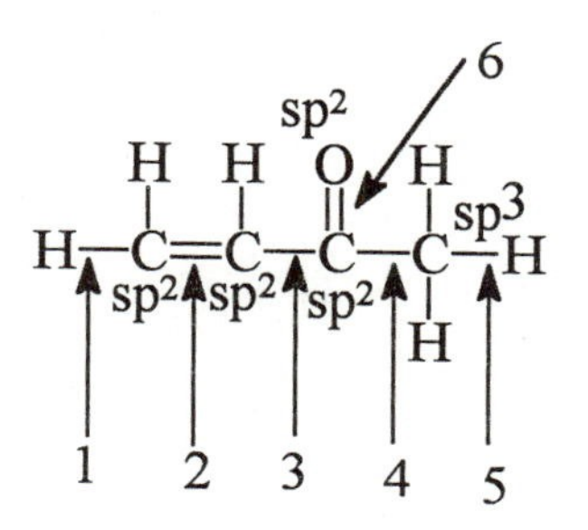

b) 1. $\sigma_{Csp3+H1s}$, 2. $\sigma_{Csp3+Csp3}$,
3. $\sigma_{Csp3+Osp3}$, 4. $\sigma_{Osp3+H1s}$.

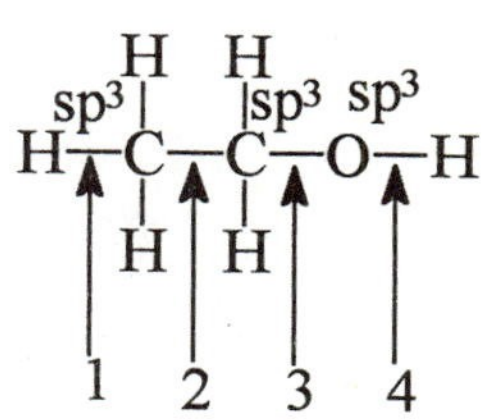

c) 1. $\sigma_{Csp3+Csp}$, 2. $\sigma_{Csp+Nsp}$ and two π_{Cp+Np}.

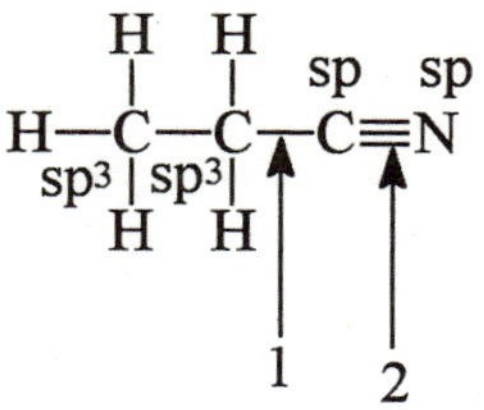

H H
sp sp
H—C—C—C≡N
sp^3 | sp^3 |
H H
1 2

d) 1. $\sigma_{Csp3+Csp}$, 2. $\sigma_{Csp+Csp}$ and two π_{Cp+Cp},
3. $\sigma_{Csp+Csp2}$, 4. $\sigma_{Csp2+H1s}$,
5. $\sigma_{Csp2+Osp2}$ and π_{Cp+Op}.

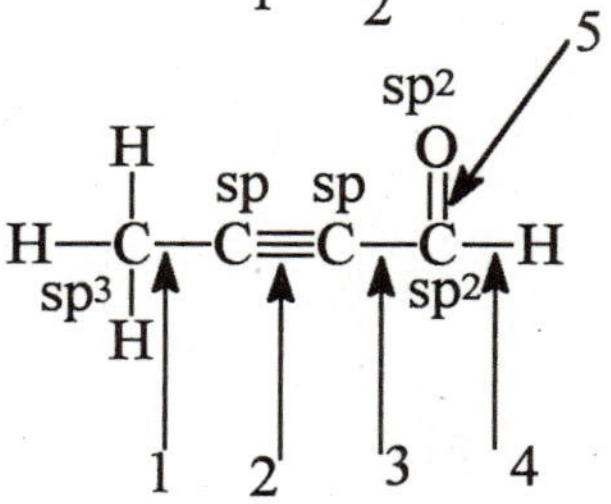

sp^2 5
H O
sp sp ||
H—C—C≡C—C—H
sp^3 | sp^2
H
1 2 3 4

3.22

a) 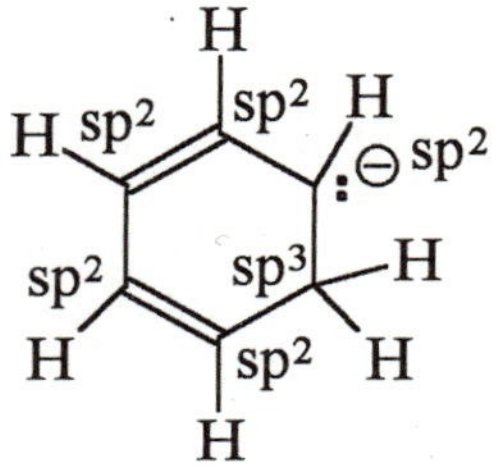 H sp^2 | sp^2 H | ⊖ sp^2 | sp^2 | sp^3 H | H | sp^2 H | H | H

b) N | N | all sp^2

c) sp^3 sp^3 | H_3C CH_3 | N sp^2 | sp^2

d) O | all sp^2

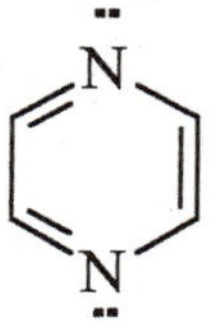

e) N: | N | H | all sp^2

f) $\dot{C}H_2$ | all sp^2

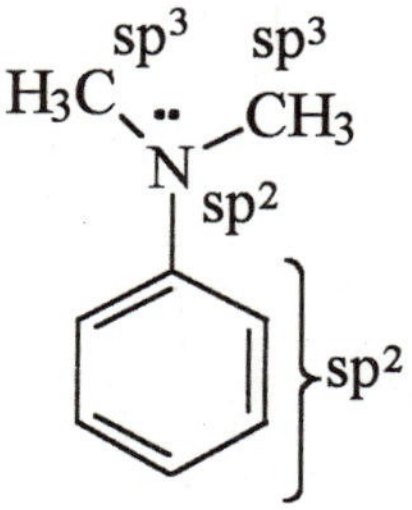

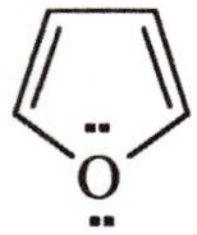

g) sp^2 | O: | CH_3C—OH | sp^3 sp^2 sp^2

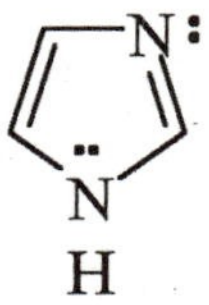

h) sp^2 | O: | CH_3C—$NHCH_3$ | sp^3 | sp^3 sp^2 sp^2

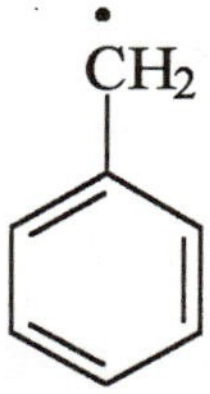

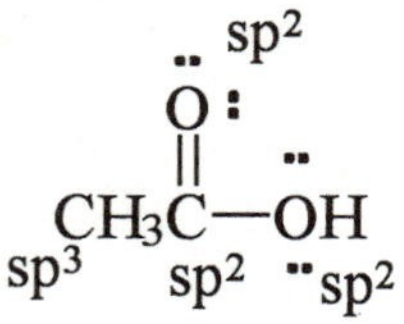

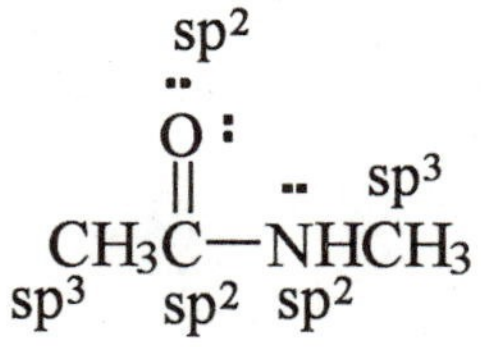

3.23

a) 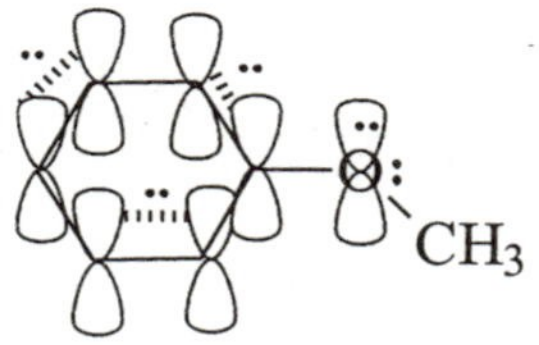 b) 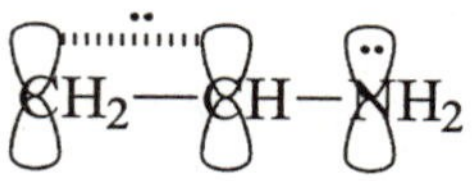 $CH_2—CH—NH_2$ c) 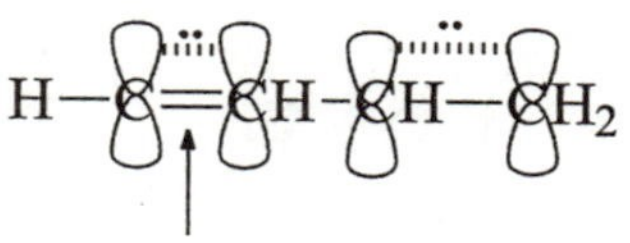 $H—C{=}CH—CH—CH_2$

The p orbitals of the second π bond are perpendicular to these orbitals and are not part of the conjugated system.

3.24

a)

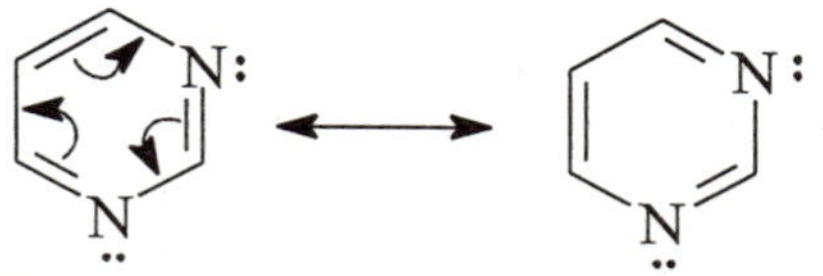

The unshared electron pairs on the nitrogens are not part of the conjugated system.

b)

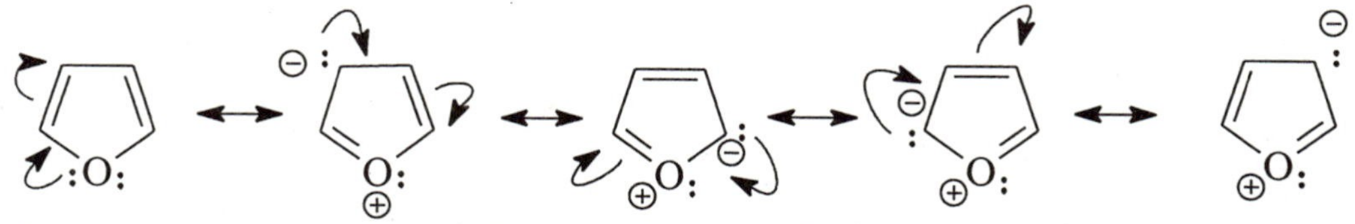

One unshared electron pair on the oxygen is part of the conjugated system and is involved in resonance.

c) 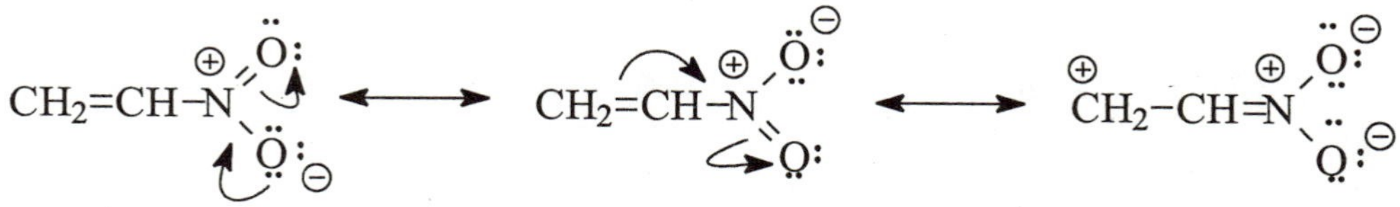 $CH_2{=}CH{-}N^{\oplus}(=O)O^{\ominus}$ ⟷ $CH_2{=}CH{-}N^{\oplus}(O^{\ominus})=O$ ⟷ $^{\oplus}CH_2{-}CH{=}N^{\oplus}(O^{\ominus})O^{\ominus}$

d) $CH_3-C(=\overset{\oplus}{\ddot{O}}-H)(-\ddot{O}-H)$ ⟷ $CH_3-C(-\ddot{O}-H)(=\ddot{O}^{\oplus}-H)$

e) $H-C{\equiv}C-\ddot{C}H_2^{\ominus}$ ⟷ $H-\ddot{C}^{\ominus}{=}C{=}CH_2$

The unshared electron pair on the carbon is conjugated with one of the pi bonds of the triple bond. The p orbitals of the other pi bond are perpendicular to these orbitals and are not involved in resonance.

3.25 The actual structure of a compound most resembles the most stable resonance structure. A more stable resonance structure contributes more to the resonance hybrid and is said to be more important. The relative stability of resonance structures can be estimated by their adherence to the octet rule, and the number and location of formal charges. If a structure is much less stable than other structures, then it is not an important contributor to the resonance hybrid.

a) The structure in the middle has three formal charges and the N has only six electrons. The N atom is in violation of the octet rule in the structure on the right.
b) The second structure is not a resonance structure for the first structure, because a hydrogen atom has moved.
c) The N atom in the second structure has only six electrons.
d) The second structure violates the octet rule at the carbon.

3.26

a)

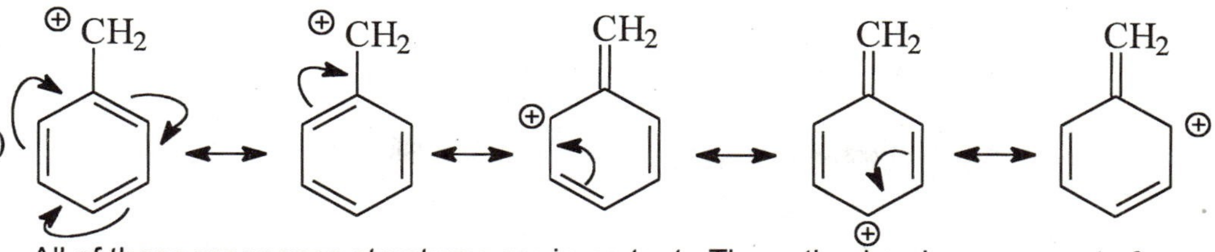

All of these resonance structures are important. The cation has large amount of resonance stabilization.

b) 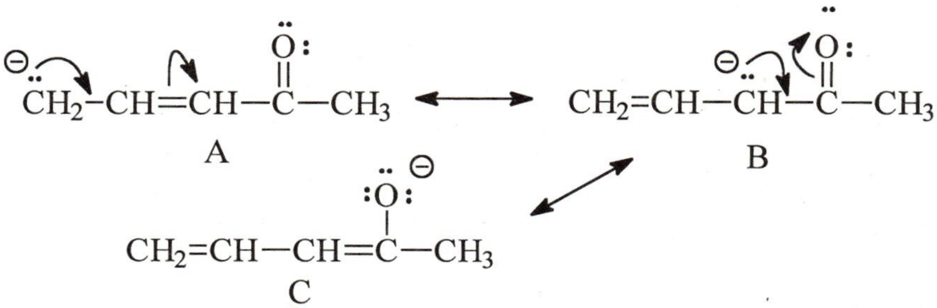 $CH_2-CH=CH-C(=O)-CH_3$ (A) $\longleftrightarrow$ $CH_2=CH-CH-C(=O)-CH_3$ (B) $\longleftrightarrow$ $CH_2=CH-CH=C(O^-)-CH_3$ (C)

Resonance structure C contributes more to the resonance hybrid because the negative charge is on the oxygen atom. Structures A and B are similar in stability. This anion has large amount of resonance stabilization.

c)  $CH_3-C(=O)-NH_2$ $\longleftrightarrow$ $CH_3-C(O^-)=NH_2^+$

The second structure contributes less to the resonance hybrid because it has formal charges. The first structure provides a more accurate picture of the compound. This compound has a small amount of resonance stabilization.

d) $CH_3{-}\overset{\overset{\large :\ddot{O}:}{||}}{C}{-}\ddot{N}H^{\ominus}$ ⟷ $CH_3{-}\overset{\overset{\ominus\,:\ddot{O}:}{|}}{C}{=}\ddot{N}H$

The second structure contributes slightly more to the resonance hybrid because the negative charge is on the more electronegative atom. This anion has a large resonance stabilization.

e) $\overset{\ominus}{\ddot{C}}H_2{-}CH{=}CH{-}\overset{\oplus}{N}(={\ddot{O}}:)(-\ddot{O}:^{\ominus})$ ⟷ $CH_2{=}CH{-}\overset{\ominus}{\ddot{C}}H{-}\overset{\oplus}{N}(={\ddot{O}}:)(-\ddot{O}:^{\ominus})$

A ⟷ B

$CH_2{=}CH{-}CH{=}\overset{\oplus}{N}(-\ddot{O}:^{\ominus})(-\ddot{O}:^{\ominus})$

C

Resonance structure C is somewhat more stable than A and B because the negative charges are on the oxygen atoms. This compound has considerable amount of resonance stabilization.

f) $CH_3{-}\overset{\overset{:\ddot{O}}{||}}{C}{-}\overset{\ominus}{\ddot{C}}H{-}\overset{\overset{\ddot{O}:}{||}}{C}{-}CH_3$ ⟷ $CH_3{-}\overset{\overset{:\ddot{O}}{||}}{C}{-}CH{=}\overset{\overset{\ominus\,:\ddot{O}:}{|}}{C}{-}CH_3$

A ⟷ B

$CH_3{-}\overset{\overset{\ominus\,:\ddot{O}:}{|}}{C}{=}CH{-}\overset{\overset{\ddot{O}:}{||}}{C}{-}CH_3$

C

The resonance structures B and C are equal in stability, and contribute more to the resonance hybrid than A. This anion has large resonance stabilization.

3.27 One of the lone pairs of electrons on the oxygen atom of this carbocation is conjugated with the empty p orbital of the positively charged carbon. The resonance structure on the right is very important because the octet rule is satisfied at both C and O.

$H_2\overset{\oplus}{C}{-}\ddot{O}{-}CH_3$ ⟷ $H_2C{=}\overset{\oplus}{\ddot{O}}{-}CH_3$

3.28

a) The anion on the left is more stable because the negative charge can be delocalized on to the oxygen atom. The other anion is not conjugated and therefore has no resonance stabilization.

$$CH_3-\overset{\overset{:\ddot{O}}{\|}}{C}-\overset{\ominus}{\ddot{C}}H-CH_3 \longleftrightarrow CH_3-\overset{\overset{\ominus\,:\ddot{O}:}{|}}{C}=CH-CH_3$$

b) The anion on the right is stabilized by resonance whereas the anion on the left is not because it is not conjugated.

$$\overset{\ominus}{\ddot{C}}H_2-C\equiv N: \longleftrightarrow CH_2=C=\overset{\ominus}{\ddot{N}}:$$

3.29 The p orbital of the positively charged carbon of the carbocation on the left is conjugated with the p orbitals on the carbons of the benzene ring. Therefore this carbocation is resonance stabilized. The carbocation on the right has no resonance stabilization because of the sp^3 hybridized carbon that separates the p orbital of the positive carbon from the p orbital of the carbon of the benzene ring.

$\overset{\oplus}{C}H-CH_3$ ⟷ $CH-CH_3$ ⟷ $CH-CH_3$ ⟷ $CH-CH_3$

3.30

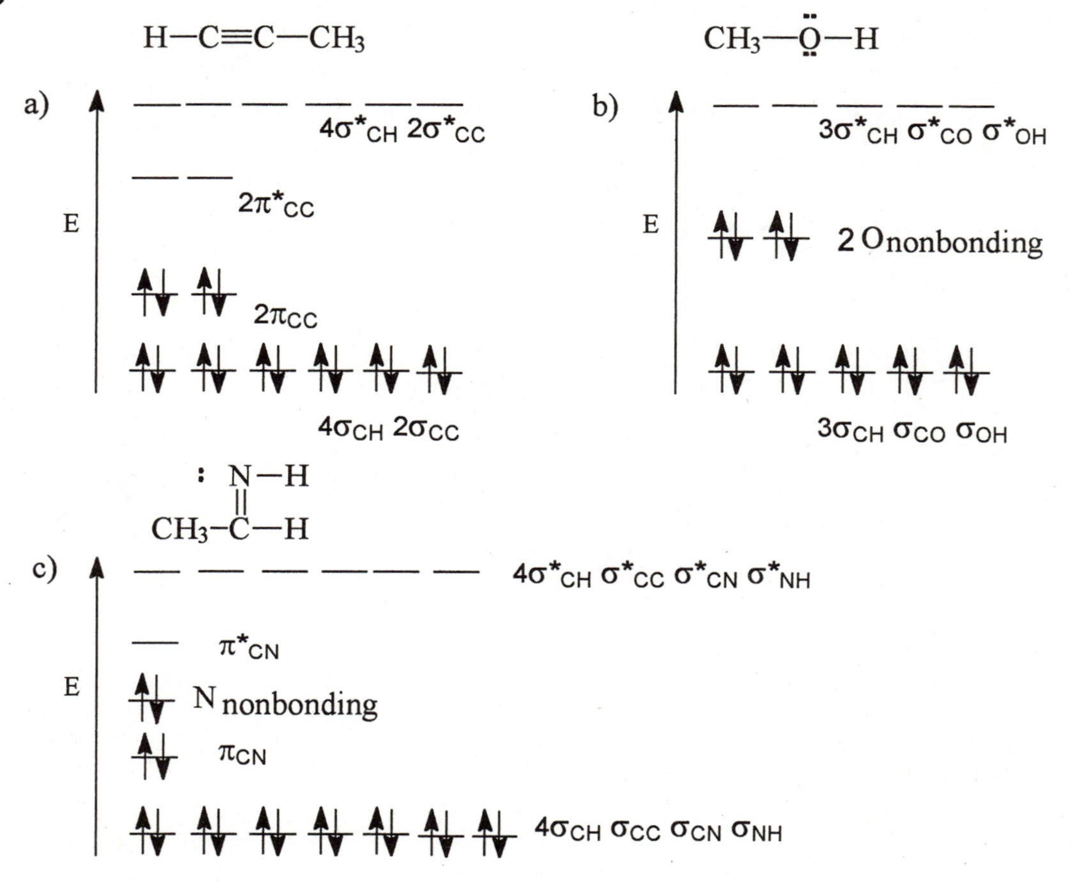

3.31 The lowest energy excited state of $CH_2{=}CH_2$ has one electron promoted from the π MO to the π* MO (don't worry about the spin for now).

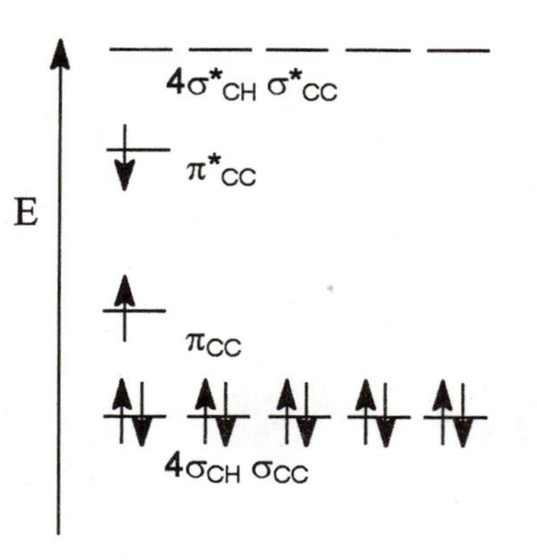

3.32 The additional electron will go into the empty sigma antibonding MO of the H_2 molecule causing some destabilization. However, the species will still have a bond between the two hydrogens. Because the bond is destabilized due to the electron in the antibonding MO, It should be weaker and longer than the bond of H_2 (approximately one-half of a bond).

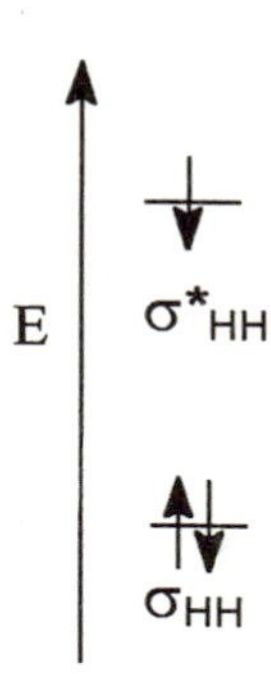

3.33 In the excited state of H_2, the unstable σ^* and the stable σ molecular orbitals are half-filled. Therefore there is no bond between the hydrogen atoms. The effect of one electron in bonding MO is approximately canceled by the effect of one electron in the antibonding MO.

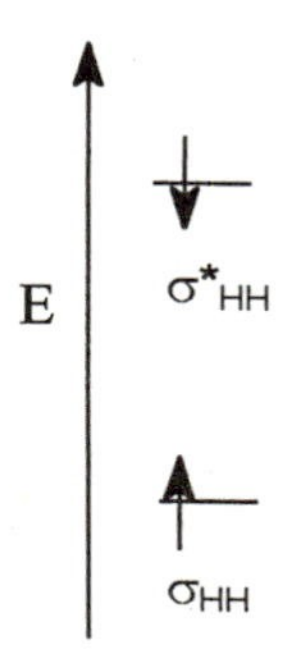

3.34 The two p orbitals on the central carbon (sp hybridized) are perpendicular to each other. One of these perpendicular p orbitals is parallel to the p orbital on one of the terminal carbons and the other is parallel to the p orbital on the other carbon. Thus, the p orbitals of one π bond are perpendicular to the p orbitals of the other π bond. The molecule is not conjugated

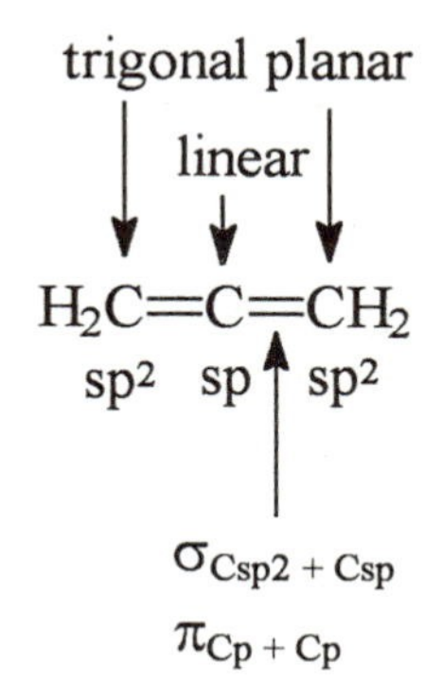

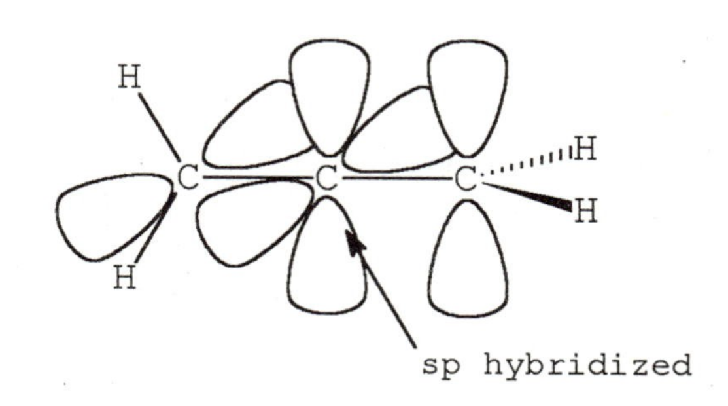

3.35

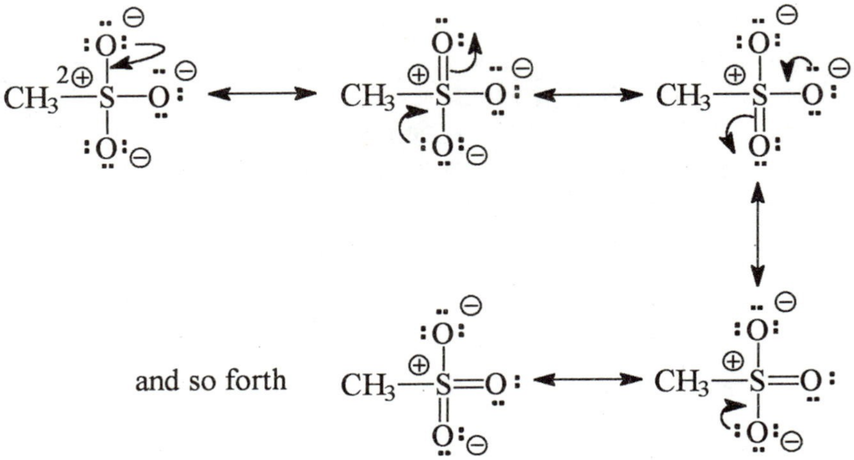

CH_3 CH_3 CH_3

and so forth CH_3 CH_3

3.36

shortest bond

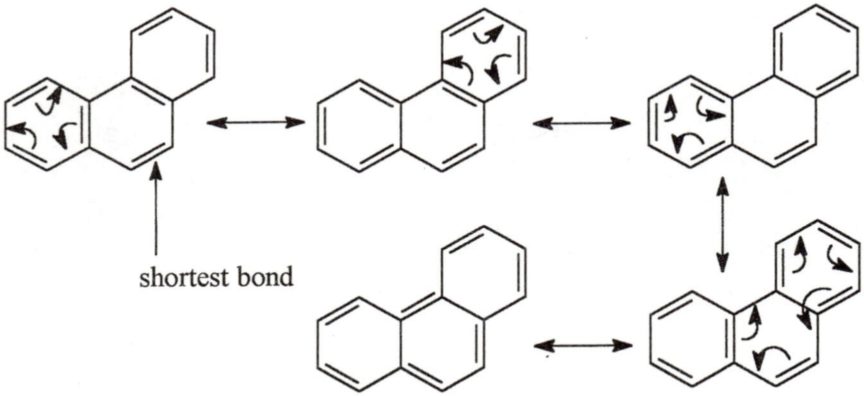

The five resonance structures are quite similar and are expected to make nearly equal contributions to the resonance hybrid. However, the number of times that a particular C-C bond appears as a double bond in these structures varies. The shortest bond is a double bond in four of the resonance structures and a single bond in only one. All of the other bonds are single bonds in at least two of the structures.

3.37

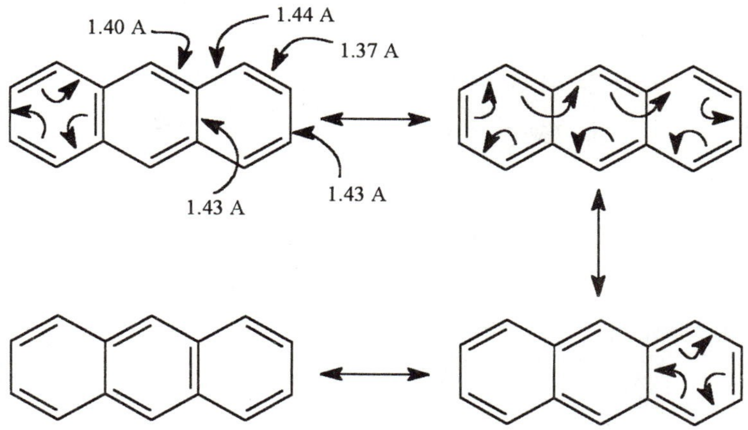

1.40 A 1.44 A 1.37 A 1.43 A 1.43 A

The longest bonds (1.44Å and 1.43Å) are single in 3 structures and double in one structure. The bond of intermediate bond length (1.40Å) is single in 2 structures and double in 2 structures. The shortest bond (1.37Å) is single in one structure and double in 3 structures.

3.38 The odd electron in this radical is in a p orbital, so the species is conjugated. It has two important resonance structures. The odd electron is located on a different carbon in the two resonance structures, providing two different sites for the coupling with the chlorine radical.

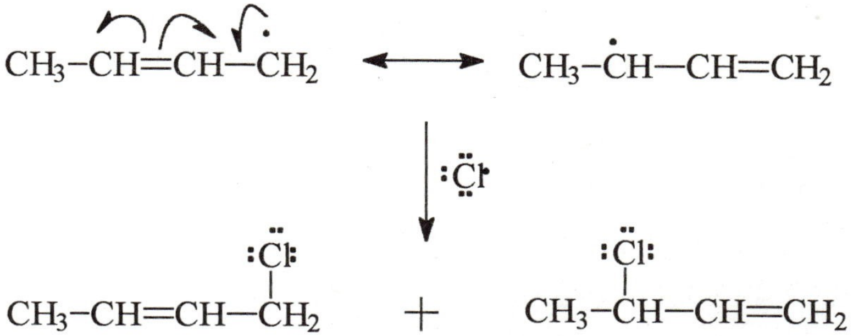

$CH_3{-}CH{=}CH{-}\dot{C}H_2 \longleftrightarrow CH_3{-}\dot{C}H{-}CH{=}CH_2$

$\downarrow$:Cl·

$CH_3{-}CH{=}CH{-}CH_2Cl$ + $CH_3{-}CHCl{-}CH{=}CH_2$

3.39

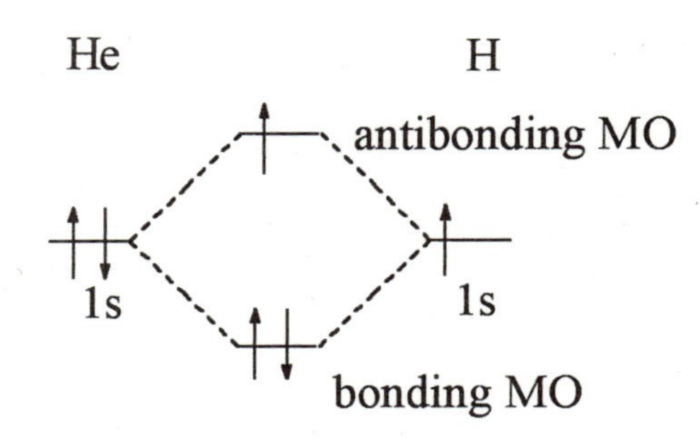

There are a total of three electrons in the two MOs of the HeH molecule, two electrons are in the stable bonding MO and one is in the unstable antibonding MO. The molecule is predicted to be more stable than the separate atoms on the basis of this simple picture.

After completing this chapter, you should be able to:

1. Assign the ground-state electron configuration for simple atoms.
   Problems 3.2, 3.3.

2. Identify any bond as sigma or pi.
   Problem 3.34.

3. Draw pictures for various sigma and pi bonding and antibonding MOs.

4. Identify the hybridization of all atoms of a molecule.
   Problems 3.6, 3.8, 3.9, 3.11, 3.20, 3.22.

5. Identify the type of molecular orbital occupied by each electron pair in a molecule and designate the atomic orbitals that overlap to form that MO.
   Problems 3.5, 3.7, 3.8, 3.9, 3.21.

6. Draw the important resonance structures for any molecule. Assign the relative importance of these structures and estimate the resonance stabilization energy for the molecule.
   Problems 3.12, 3.13, 3.14, 3.15, 3.16, 3.17, 3.24, 3.25, 3.26, 3.27, 3.28, 3.29, 3.35, 3.36, 3.37, 3.38.

7. Show a MO energy level diagram for all the orbitals for any molecule.
   Problems 3.18, 3.30, 3.31, 3.32, 3.33, 3.39.

Chapter 4
Proton Transfer --- A Simple Reaction

4.1 According to the Bronsted-Lowry definition, an acid is a proton donor and a base is a proton acceptor. Compounds that have both a hydrogen and an unshared pair of electrons can potentially react as either an acid or a base, depending on the reaction conditions. For example, water, has both unshared pairs of electrons and hydrogens. Therefore it can act like an acid or a base.

b) both c) acidd) base e) both f) both
g) both

4.2 Conjugate acids are the protonated products of the reactant bases. For example, addition of a proton to the base hydroxide anion, produces its conjugate acid, H_2O.

a) 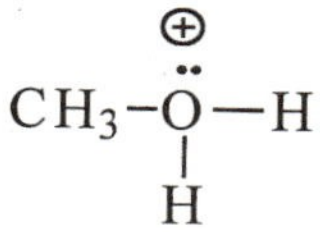 $CH_3-\overset{\oplus}{\ddot{O}}-H$ (with H bonded below O) b) 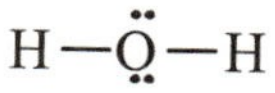 $H-\ddot{\underset{..}{O}}-H$ c) 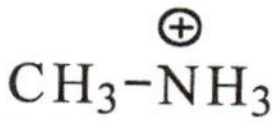 $CH_3-\overset{\oplus}{N}H_3$

4.3 Conjugate bases are the deprotonated products of the corresponding reactant acids.

a) 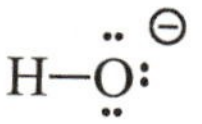 $H-\ddot{\underset{..}{O}}:^{\ominus}$ b) 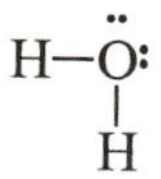 $H-\ddot{O}:$ (with H bonded below O) c) 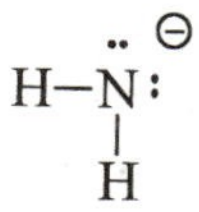 $H-\ddot{N}:^{\ominus}$ (with H bonded below N) d) 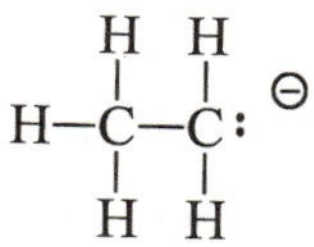 $H-C(H)(H)-C(H)(H):^{\ominus}$

4.4 The reaction of an acid with a base is in equilibrium with the conjugate acid and the conjugate base. Remember, in an acid-base reaction the conjugate acid is the protonated form of the corresponding base and the conjugate base is the deprotonated form of the corresponding acid.

	base		acid		conjugate acid		conjugate base
a)	$:\ddot{N}H_2^{\ominus}$	+	$H-\ddot{O}-H$	⇌	$:NH_3$	+	$^{\ominus}:\ddot{O}-H$
b)	$CH_3\ddot{O}:^{\ominus}$	+	$H-\ddot{O}^{\oplus}(H)-H$	⇌	$CH_3\ddot{O}-H$	+	$H-\ddot{O}-H$

4.5 According to the Lewis acid-base definition, an acid is an electron pair acceptor and a base is an electron pair donor.

a) Lewis acid　b) Lewis base　c) Lewis acid
d) Lewis base　e) both

4.7 The acidity constant, K_a, is a measure of the acidic strength of a compound. The **pK_a** is by definition **$-\log K_a$**. As the strength of the acid increases, the **K_a** increases and the **pK_a** decreases. For example acetylene in (b) has a larger pK_a than water. Therefore it is a weaker acid than water.

a) stronger　b) weaker　c) stronger　d) weaker

4.8 The weaker the acid, the stronger its conjugate base. If the K_a of a compound's conjugate acid is larger than that of water (or the pK_a of the compound's conjugate acid is smaller than that of water), then the compound is a stronger base than hydroxide ion.

a) stronger　b) stronger　c) weaker　d) weaker

4.9 The equilibrium favors the formation of the more stable compounds. In the case of acid-base reactions, the weaker acid and the weaker base are favored. For example, in (a) the acid acetylene has a lower pK_a than the acid ethane. This makes acetylene a stronger acid than ethane. Therefore the equilibrium favors the products.

b) favors reactants　c) favors products

4.10 The equilibrium favors the weaker acid and the weaker base.
a) favors products b) favors products

4.11

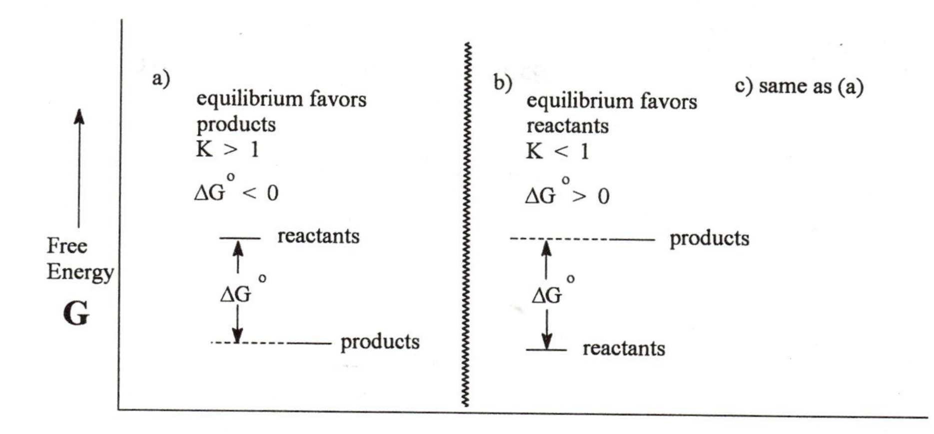

4.12 b)

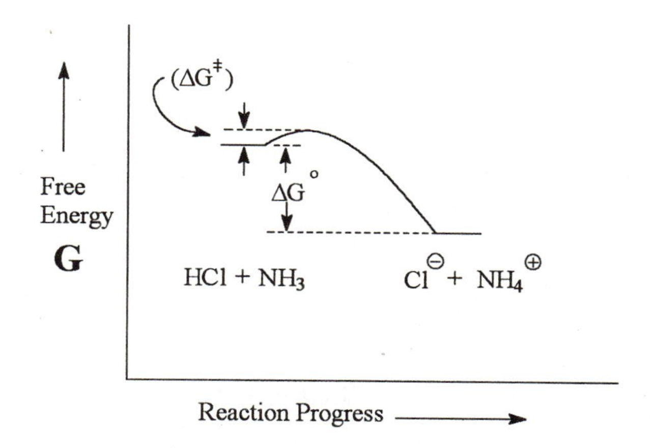

4.13 The acidity of a hydrogen is affected by the electronegativity and the size of the atom to which it is bonded. Electronegativity is the dominant factor when comparing atoms in the same row of the periodic table. The acidity increases from left to right in a row of the periodic table because the stability of the conjugate base increases as the electronegativity of the atom increases. When comparing atoms of the same column, the controlling factor is the size. The larger atoms have a weaker bond to hydrogen, making the removal of the proton easier. Therefore the acidity increases from top to bottom in a column of a periodic table.

a) HCl b) $PH_4^{\oplus}$ c) H_2S

4.14 Remember, a stronger base is a produced from a weaker acid. For example, H_2O, the conjugate acid of OH^-, is a weaker acid than H_2S, the conjugate acid of HS^-. Therefore OH^- is a stronger base than HS^-.

a) HO^- b) CH_3NH^-

4.15 Hydrogens attached to the more electronegative elements are more acidic.

b) CH_3CH_2O–(H) c) CH_3S–(H)

4.16 The acidity of a compound depends on the stability of its conjugate base. In carboxylic acids the inductive effect (effect of a nearby dipole) of an electron withdrawing group will destabilize the acid, but stabilize the conjugate base, resulting in a stronger acid.

a) CHF_2CO_2H (More electron withdrawing groups causes a larger stabilization of the conjugate base, resulting in a stronger acid.)
b) CHF_2CO_2H (F is more electronegative than Br.)
c) $CH_3OCH_2CO_2H$ (The OCH_3 group is an electron withdrawing group.)

4.17 The hybridization of the atom bonded to the hydrogen has a large effect on the acidity of the hydrogen. The higher the s character of a hybridized orbital higher its stability. Conjugate bases with unshared electrons in sp hybridized orbitals are more stable than those with electrons in the sp^2 or sp^3 hybridized orbitals. Therefore, the hydrogen bonded to the sp orbital is more acidic.

$CH_3CH_2C{\equiv}C$–(H)

4.18 Delocalization and stabilization of the unshared electron pair of the conjugate base makes a stronger acid. The nitro group is more effective in increasing the acidity when it is at the para position than when it is at the meta position. In the resonance structures of the conjugate base of the meta isomer of nitrophenol, the electron pair is never on the carbon atom to which the nitro group is attached, so the electrons cannot be delocalized onto the oxygen of the nitro group. In the para isomer, however, the electrons are placed on the carbon bonded to the nitro group, providing additional resonance stabilization.

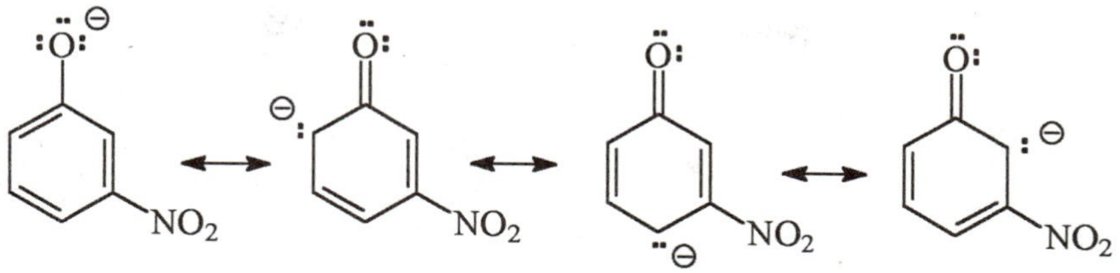

4.19

b) The compound on the right is a stronger acid because the conjugate base of this compound has two electron withdrawing groups conjugated to the carbon with the unshared pair of electrons, providing additional inductive and resonance stabilization. The conjugate base of the other compound has only one electron withdrawing group conjugated to carbon with the unshared pair of electrons.

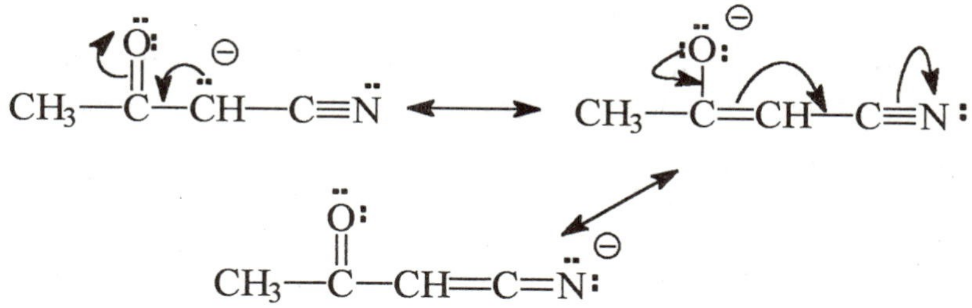

c) The compound on the left is more acidic because of two factors. One is the inductive stabilization of the conjugate base due to the electron-withdrawing group. The other is the additional resonance stabilization of the conjugate base. The unshared electrons of the conjugate base can be delocalized on to the oxygen atom of the electron-withdrawing group. The conjugate base of phenol does not have the additional inductive or resonance stabilization.

d) This is similar to problem 4.18. The para isomer (left) is more acidic because the electron-withdrawing group is more effective in stabilizing the conjugate base when substituted on the para position than it is when substituted on the meta position.

e) The compound on the right is more acidic due to inductive and resonance stabilization of the conjugate base. The conjugate base of CH_3CH_3 has no such stabilization.

4.20

a) The compound on the right is a weaker base because the nitro group exerts both an electron-withdrawing inductive effect as well as a resonance effect that makes the pair of electrons on the nitrogen atom of the amino group less available to react as a base.

b) The anion on the left is the weaker base because it has a larger resonance stabilization (two additional resonance structures) than does the anion on the right (one additional resonance structure).

4.21 Remember, the equilibrium favors the formation of the weaker acid and the weaker base. The larger the pK_a, the weaker the acid.

a) Ammonia (pK_a = 38) is a weaker acid than a ketone (pK_a = 20), so the equilibrium favors the products.

b) Dimethylsulfoxide (pK_a = 38) is a weaker acid than an alcohol (pK_a = 16), so the equilibrium favors the products.

c) HCl (pK_a = -7) is a much stronger acid than phenol (pK_a = 10), so the equilibrium favors the reactants.

4.22 The Lewis base is referred to as a nucleophile. A nucleophile is an electron rich species that seeks an electron poor site. The Lewis acid is referred to as an electrophile. It seeks an electron rich site.

a) $CH_3CH_2{-}Br$ (electrophile) + $:\overset{\ominus}{C}{\equiv}N:$ (nucleophile) $\longrightarrow$ $CH_3CH_2{-}CN + Br^{\ominus}$

b) $CH_3{-}I$ (electrophile) + $:\overset{\ominus}{\ddot{O}}{-}H$ (nucleophile) $\longrightarrow$ $CH_3{-}OH + I^{\ominus}$

c) $CH_3{-}\overset{\ominus}{\ddot{O}}:$ (nucleophile) + $CH_3CH_2CH_2{-}I$ (electrophile) $\longrightarrow$ $CH_3CH_2CH_2{-}OCH_3 + I^{\ominus}$

4.23 The conjugate acid is the protonated form of the base.

a) $C_6H_5NH_3^{\oplus}$ (benzene ring bearing $NH_3^{\oplus}$) b) cyclopentane bearing $:\ddot{O}H$ c) $CH_3CH_2CH_2\overset{\oplus}{\ddot{O}}H_2$

d) $CH_3C{\equiv}CH$ e) (isopropyl)–$\ddot{N}H$–(isopropyl)

4.24 The stability of the conjugate base depends on the stabilization of the unshared electron pair produced by the removal of a proton from the corresponding acid. Remember, the stability of the base depends on such factors as the electronegativity and the hybridization of the atom where the electrons are located, resonance delocalization of the basic electrons, and the inductive effect of nearby groups.

a) $CH_3CH_2\ddot{\underset{..}{O}}:^{\ominus}$ b) $^{\ominus}:\ddot{\underset{..}{O}}-\overset{O}{\overset{||}{C}}-CH_2CH_2OH$ c) $H_2NCH_2CH_2\ddot{\underset{..}{O}}:^{\ominus}$

4.25 According to Lewis acid-base definition, an acid is an electron pair acceptor. Any compound that has an unfilled valence orbital is a potential Lewis acid.

a) The boron atom in BCl_3 has an empty p orbital that can accept electrons. Therefore it is a Lewis acid.
b) Methane is not a Lewis acid since there are no empty orbitals in the valence shell.
c) The positively charged carbon atom of this species is sp^2 hybridized, and has an empty p orbital to accept electrons. Therefore it is a Lewis acid.

4.26 According to Lewis definition, a base is an electron pair donor. Any compound that has an unshared electron pair can behave as a base.

a) This is a Lewis base because it has two unshared pairs of electrons.
b) This is not a Lewis base because it does not have any unshared pairs of electrons.
c) This compound has an unshared pair of electrons. Therefore it is a Lewis base.
d) This cation does not have any unshared pairs of electrons. It is not a Lewis base.
e) This compound is a Lewis base.

4.27 pK_a is = $-\log K_a$
a) 3.76 b) 50

4.28 $K_a = 10^{-pKa}$ (antilog of $-pK_a$)
a) 1×10^{-25} b) 4.9×10^{-10}

4.29
a) In both cases the conjugate bases are stabilized by an inductive effect and by resonance. However, the compound on the right is the stronger acid because the hydrogen is removed from the more electronegative oxygen atom.
b) The hydrogen on sulfur is more acidic than the one on oxygen, so the compound on the left is a stronger acid.
c) The compound on the right is the stronger acid because the conjugate base produced by the removal of a proton from the CH_3 group is stabilized by resonance. There is no such stabilization in butane.
d) The structure on the right is the stronger acid because the CF_3 group is a strong electron withdrawing group. The methyl group is weakly electron donating and destabilizes the conjugate base.
e) The compound on the right is more acidic because the nitrogen in this molecule is sp^2 hybridized. The unshared pair of electrons of the conjugate base will more stable in a sp^2 hybridized orbital

4.30 See the reasoning in problem 4.20. The stronger the base, the higher the tendency to donate electrons. In general, factors that stabilize the basic electrons will decrease the basic strength.
a) The anion on the left is a stronger base because the basic electron pair is more available for covalent bonding. In the other anion the electron pair is stabilized by the nitro group due to additional resonance delocalization and an inductive effect.

b) Acid strength increases down a column of the periodic table. Therefore base strength decreases down a column and the nitrogen compound is a stronger base than the phosphorus compound. Therefore, the compound on the right is a stronger base.

c) The inductive stabilization of the anion by chlorine is more than that by bromine due to the higher electronegativity of Cl. Therefore the anion on the left is a stronger base.

d) Oxygen and nitrogen are in the same row of the periodic table. Because oxygen is more electronegative than nitrogen, the electrons are less available for bonding. Therefore the anion on the right is a stronger base.

4.31 The acidic hydrogen is the one bonded to the oxygen for all these compounds. Therefore the acidic strength of these compounds now depends on the factors that stabilize their conjugate bases.

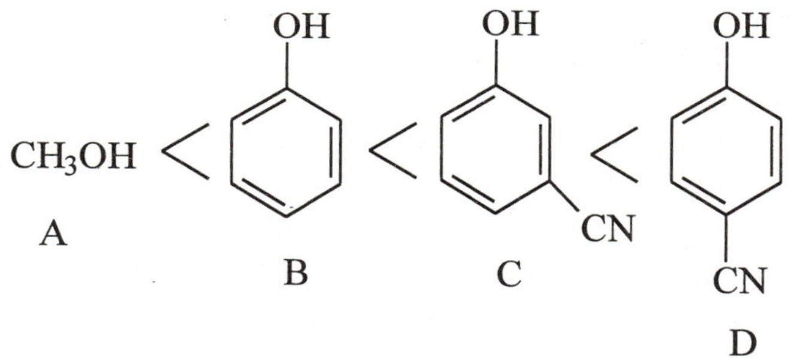

OH OH OH

CH_3OH < < <

CN

A B C CN

D

Compound B is a stronger acid than A because of resonance stabilization of the conjugate base. Compound C is a stronger acid than B because of the inductive withdrawing effect of the CN group. Compound D is a stronger acid than C because of the additional resonance stabilization of the conjugate base.

4.32 The basic electron pair in all these compounds is the one on the nitrogen atom. Stabilization of the basic electrons due to resonance or to an inductive effect decreases the basic strength.

Compound A is less basic than B because the carbon-oxygen double bond is one of the best groups at stabilizing (by resonance) a pair of electrons on an adjacent atom. Compound B is less basic than C because of additional resonance stabilization by the NO_2 group. Compound C is less basic than D because of resonance stabilization of the basic electrons.

$$\underset{\text{A}}{CH_3\overset{O}{\overset{\|}{C}}\ddot{N}H_2} < \underset{\text{B}}{p\text{-}O_2NC_6H_4\ddot{N}H_2} < \underset{\text{C}}{C_6H_5\ddot{N}H_2} < \underset{\text{D}}{C_6H_{11}\ddot{N}H_2}$$

4.33

$$\underset{\text{weakest acid}}{CH_3\overset{O}{\overset{\|}{C}}OH} < HO\overset{O}{\overset{\|}{C}}CH_2CH_2\overset{O}{\overset{\|}{C}}OH < \underset{\text{strongest acid}}{HO\overset{O}{\overset{\|}{C}}-\overset{O}{\overset{\|}{C}}OH}$$

4.34

a) The products are favored because the conjugate acid is weaker than the reactant acid.
b) The equilibrium favors the products because ammonia is a weaker acid than ethyne.
c) The reactants are favored because the conjugate acid, an alcohol, is a stronger acid than the reactant acid, ethyne.

4.35

a) CH_3CH_3 is a weaker acid than CH_3CN, because the conjugate base of CH_3CN is stabilized by resonance, so the reactants are favored.

$$CH_3\,CH_3 + \overset{\ominus}{\ddot{C}}H_2CN \rightleftharpoons \overset{\ominus}{\ddot{C}}H_2CH_3 + CH_3CN$$

b) The conjugate acid is stronger due to the inductive effect of the CF_3 group, so the reactants are favored.

$$CF_3\overset{:\ddot{O}}{\overset{\|}{C}}\ddot{\underset{..}{O}}:^{\ominus} + CH_3\overset{:\ddot{O}}{\overset{\|}{C}}\ddot{\underset{..}{O}}H \rightleftharpoons CF_3\overset{:\ddot{O}}{\overset{\|}{C}}\ddot{\underset{..}{O}}H + CH_3\overset{:\ddot{O}}{\overset{\|}{C}}\ddot{\underset{..}{O}}:^{\ominus}$$

c) The reactant acid is stronger because N is more electronegative than C, so the products are favored.

$$CH_3\overset{O}{\overset{\|}{C}}NH_2 + CH_3\overset{O}{\overset{\|}{C}}CH_2^{\ominus} \rightleftharpoons CH_3\overset{O}{\overset{\|}{C}}NH^{\ominus} + CH_3\overset{O}{\overset{\|}{C}}CH_3$$

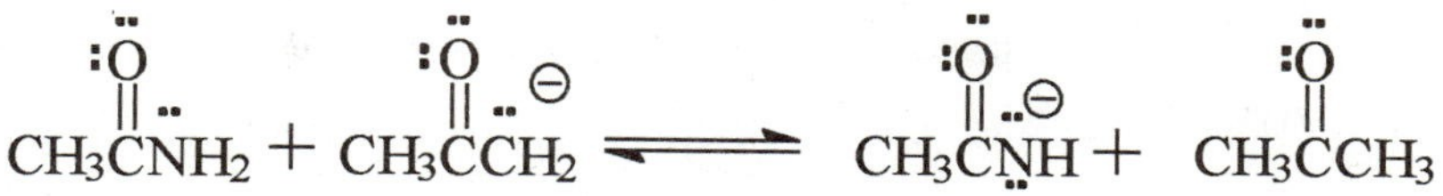

d) The products are favored because the reactant acid is stronger due to the resonance stabilization of its conjugate base..

$$CH_3CH_2CH_2\overset{O}{\overset{\|}{C}}OH + {}^{\ominus}OH \rightleftharpoons CH_3CH_2CH_2\overset{O}{\overset{\|}{C}}O^{\ominus} + H_2O$$

4.36

a) A sulfonic acid is a stronger acid than a carboxylic acid.

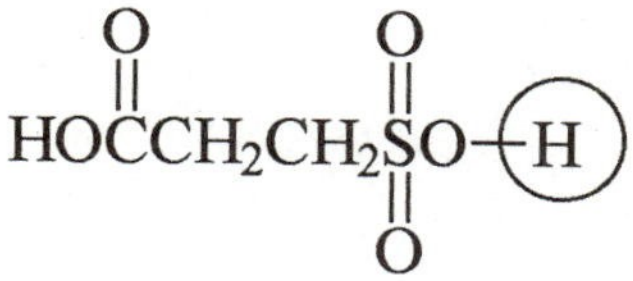

$$HO\overset{O}{\overset{\|}{C}}CH_2CH_2\underset{O}{\underset{\|}{\overset{O}{\overset{\|}{S}}}}O\text{–}(H)$$

b) The circled hydrogen is more acidic because the conjugate base is resonance stabilized.

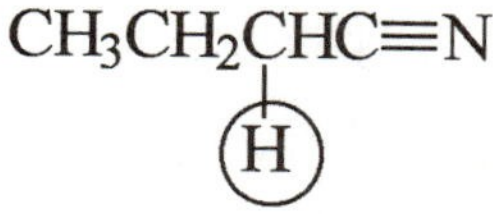

$$CH_3CH_2\underset{(H)}{\underset{|}{C}}HC{\equiv}N$$

c) A carboxylic acid is a stronger acid than a phenol.

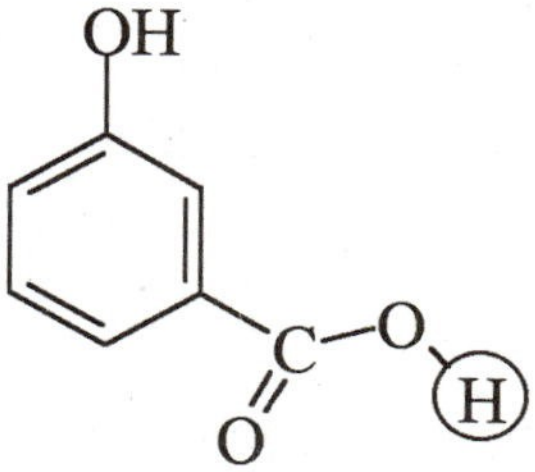

OH
C—O—(H)
O

d) A phenol is a stronger acid than an alcohol

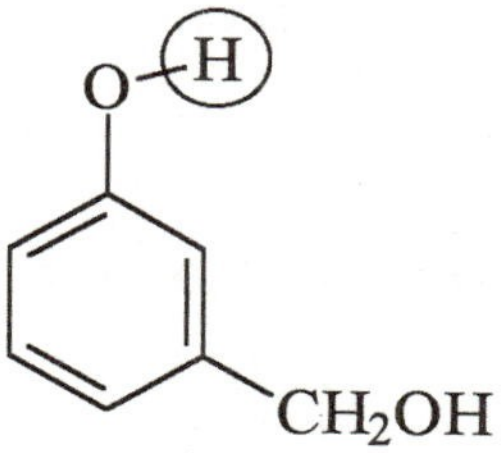

O—(H)
CH₂OH

e) The circled hydrogen is more acidic because the conjugate base is stabilized by resonance involving both carbonyl groups.

$$CH_3\overset{O}{\overset{\|}{C}}\underset{\text{(H)}}{\underset{|}{C}}H\overset{O}{\overset{\|}{C}}OCH_2CH_3$$

4.37

a) The equilibrium favors the products.

$$\underset{pK_a \approx 9}{CH_3\overset{O}{\overset{\|}{C}}CH_2\overset{O}{\overset{\|}{C}}CH_2CH_3} + :\ddot{\underset{..}{O}}^{\ominus}CH_2CH_3 \rightleftharpoons CH_3\overset{O}{\overset{\|}{C}}\overset{\ominus}{\ddot{C}}H\overset{O}{\overset{\|}{C}}CH_2CH_3 + \underset{pK_a \approx 16}{H\ddot{\underset{..}{O}}CH_2CH_3}$$

b) The equilibrium favors the products.

$$\underset{pK_a \approx 10}{CH_3CH_2NO_2} + CH_3\ddot{\underset{..}{O}}:^{\ominus} \rightleftharpoons CH_3\overset{\ominus}{\ddot{C}}HNO_2 + \underset{pK_a \approx 16}{CH_3\ddot{\underset{..}{O}}H}$$

c) The equilibrium favors the products.

$$\underset{pK_a \approx 25}{CH_3\overset{O}{\overset{\|}{C}}OCH_3} + {}^{\ominus}:\ddot{N}(\overset{CH_3}{\overset{|}{C}}HCH_3)_2 \rightleftharpoons {}^{\ominus}\ddot{C}H_2\overset{O}{\overset{\|}{C}}OCH_3 + \underset{pK_a \approx 38}{H\ddot{N}(\overset{CH_3}{\overset{|}{C}}HCH_3)_2}$$

d) The conjugate acid is a stronger acid because the F atom is closer to the reaction site, so the equilibrium favors the reactants. .

$$FCH_2CH_2\overset{O}{\overset{\|}{C}}OH + CH_3\underset{F}{\underset{|}{C}}H\overset{O}{\overset{\|}{C}}\ddot{\underset{..}{O}}:^{\ominus} \rightleftharpoons FCH_2CH_2\overset{O}{\overset{\|}{C}}\ddot{\underset{..}{O}}:^{\ominus} + CH_3\underset{F}{\underset{|}{C}}H\overset{O}{\overset{\|}{C}}\ddot{\underset{..}{O}}H$$

e) The equilibrium favors the products.

$$CH_3CH(OH)COOH + NaOH \rightleftharpoons CH_3CH(OH)CO_2^{-}\,Na^{+} + H_2O$$

$pK_a \approx 4$ $\qquad$ $pK_a = 15.7$

f) The equilibrium favors the products.

$$(CH_3)_3C{-}O^{-} + H_2O \rightleftharpoons (CH_3)_3C{-}OH + {}^{-}OH$$

$pK_a = 15.7$ $\qquad$ $pK_a = 19$

4.38 $AlCl_3$ is the Lewis acid because Al has an empty p orbital, which can accept an electron pair. $CH_3CH_2OCH_2CH_3$ is the Lewis base because it has lone pairs of electrons to donate.

4.39 In both compounds the basic electrons are on a nitrogen atom. The unshared electron pair of the structure on the left is in a sp^2 hybridized orbital, while in the structure on the right they are in a sp^3 hybridized orbital. The sp^2 orbital is lower in energy than the sp^3 orbital, so the structure on the left is a weaker base.

4.40 The unshared electron pair on the nitrogen of this compound is in a p orbital which is conjugated with the carbonyl p orbitals. These electrons are stabilized by resonance. If the nitrogen is protonated, this resonance stabilization is lost. The electron pairs on the oxygen atom are in atomic orbitals that are not involved in resonance.

4.41

$CH_3—C(=\ddot{O}:)—\ddot{C}H_2^{\ominus}$ (A) $\longleftrightarrow$ $CH_3—C(—\ddot{O}:^{\ominus})=CH_2$ (B) $\rightleftharpoons$ $CH_3—C(=\ddot{O}:)—CH_3$ (C) + $CH_3—C(—\ddot{O}H)=CH_2$ (D)

The anion has two resonance structures, A and B. Protonation of the anion on the carbon produces the conjugate acid C and protonation on the oxygen produces the conjugate acid D.

4.42

a) Isomer B is conjugated and has resonance stabilization.
b) The conjugate base of A is resonance stabilized. Two of the three contributing resonance structures are as follows. Protonation at one of the carbons bearing a partial negative charge produces A and protonation at the other produces B.

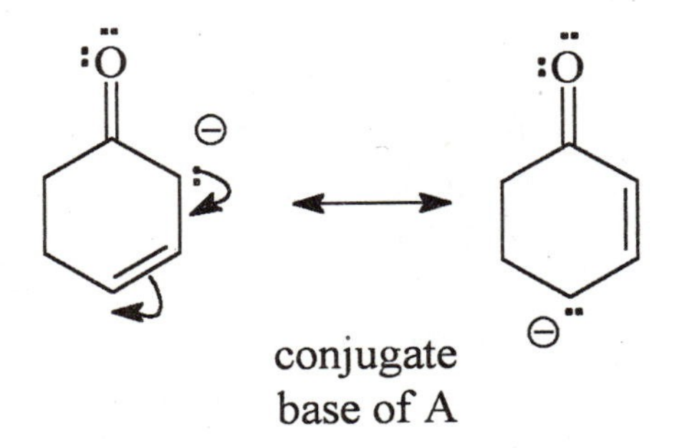

4.43

a) The methoxy group, CH_3O, has an electron-withdrawing inductive effect which destabilizes the acid and stabilizes the conjugate base. Therefore the inductive effect of the methoxy group should increase the acidity of compound D.

b) The methoxy group has an electron-donating resonance effect which destabilizes the conjugate base, weakening the acidity of D.

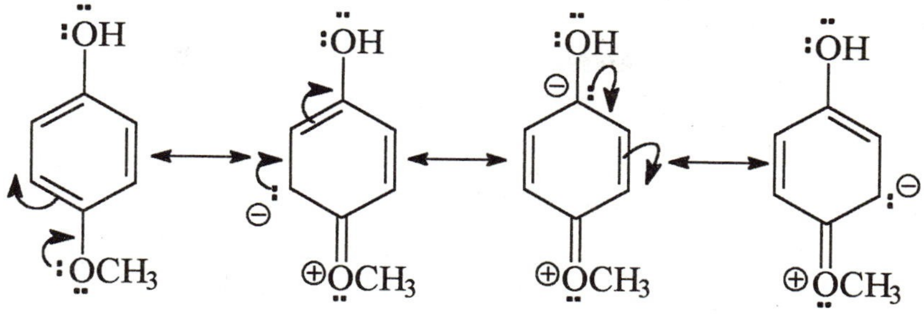

c) Yes, the resonance effect is slightly stronger than the inductive effect so D is slightly weaker acid than C.

4.44 The compounds at the beginning of Table 4.2 are very strong acids and the compounds at the end of this table are extremely weak acids. The equilibrium constants of the very strong acids are very large while those of extremely weak acids are very small. Very large and very small pK_a values cannot be measured accurately because the concentrations of either the reactants or the products are extremely small. Thus, the equilibrium constants of these compounds are determined by some indirect method, and are only approximate values. Therefore the pK_a values of these compounds cannot be determined very precisely, so they are listed without any figures right of the decimal place.

4.45 E = electrophilic site N = nucleophilic site

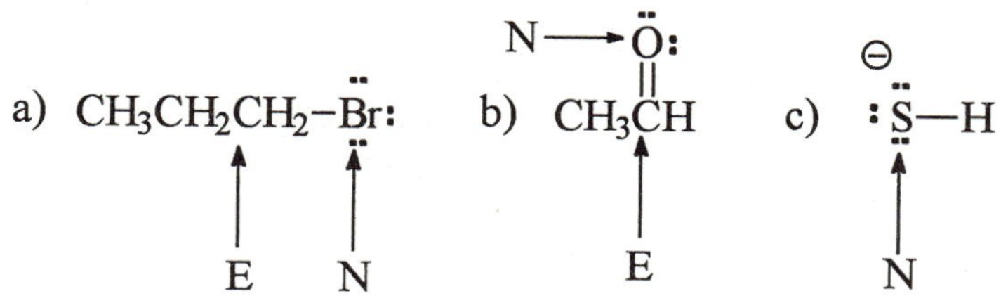

a) $CH_3CH_2CH_2-Br$: (E at the carbon bonded to Br; N at Br)

b) $CH_3CH=O$ (N at O; E at the carbonyl carbon)

c) $^{-}$:S–H (N at S)

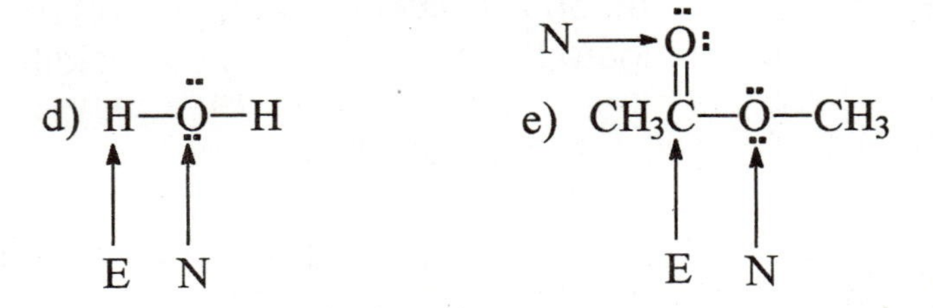

4.46

a) $$CH_3CH_2I + CH_3CH_2C(=O)O^{-} \longrightarrow CH_3CH_2C(=O)OCH_2CH_3 + I^{-}$$

electrophile nucleophile

b) $$CH_3NH_2 + CH_3Br \longrightarrow CH_3N^{+}H_2CH_3 + Br^{-}$$

nucleophile electrophile

c) $$C_6H_5O^{-} + CH_3CH_2CH_2Cl \longrightarrow C_6H_5OCH_2CH_2CH_3 + Cl^{-}$$

nucleophile electrophile

After completing this chapter you should be able to:

1. Write an acid-base reaction for any acid and base.
   Problems 4.1, 4.2, 4.3, 4.4, 4.35, 4.37.

2. Recognize acid or base strengths from K_a or pK_a values and use these to predict the position of an acid-base equilibrium.
   Problems 4.6, 4.7, 4.8, 4.9, 4.10, 4.21, 4.34, 4.35.

3. Predict and explain the effect of the structure of the compound, such as the atom bonded to the hydrogen, the presence of an electron-donating or -withdrawing group, hydrogen bonding, the hybridization of the atom attached to the hydrogen, or resonance on the strength of an acid or base.
   Problems 4.13, 4.14, 4.16, 4.18, 4.29, 4.30, 4.37, 4.38, 4.43.

4. Using the same reasoning, arrange a series of compounds in order of increasing or decreasing acid or base strength.
   Problems 4.19, 4.20, 4.31, 4.32, 4.33.

5. Identify the most acidic proton in a compound.
   Problems 4.15, 4.17, 4.24, 4.36.

6. Recognize substitution reactions.
   Problems 4.22, 4.46.

7. Recognize some nucleophiles and electrophiles.
   Problem 4.22, 4.45, 4.46.

Chapter 5
FUNCTIONAL GROUPS AND NOMENCLATURE

5.1 Follow steps 1 through 5 outlined in Section 5.4 in your text for the naming these alkanes.

Find the longest, continuous carbon chain in the compound. The number of carbons in this backbone determines the root. If there is more than one chain of equal length, choose the one with the more branches. For example, in structure (e) there are two different 6 carbon chains, one with two branches and the other with three branches. Choose the one with the three branches as the root chain.

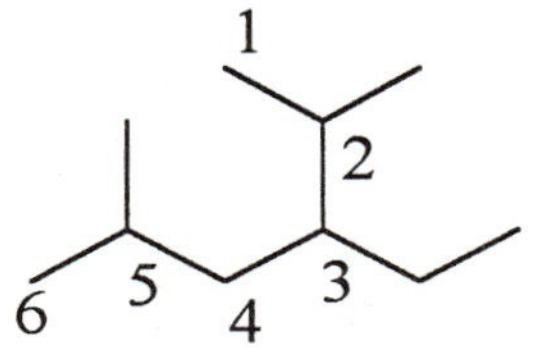

1
2
6 5 4 3

Number the root carbon chain starting from the end that will give the lower number to the first branch. If the first branch occurs at equal distances from either end, then choose the end that will give the lower number to the second branch. In the above example, the first branch occurs at equal distance from both ends, so the correct numbering is as shown. Then name the groups attached to the root .

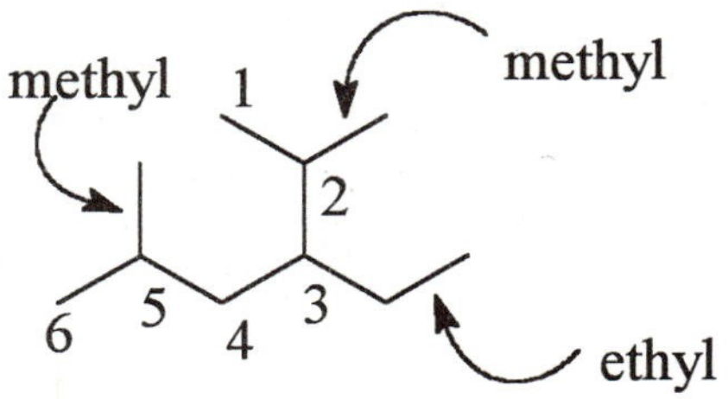

methyl 1 methyl
2
6 5 4 3 ethyl

Assemble the name as a single word in the following order: *number-group root suffix.*
The correct IUPAC name for the structure in (e) is 3-ethyl-2,5-dimethylhexane.

a) 2-methylpentane

b) 2-methylpentane

c) 2,4-dimethylhexane

d) 5-ethyl-3-methyl-5-propylnonane

e) 3-ethyl-2,5-dimethylhexane

f) 3-ethyl-2,6-dimethylheptane

5.2 When drawing the structure from a name, start from the tail end of the name and work backwards to the head of the name. For example in 2,2,5-trimethylheptane, the suffix ***ane*** indicates that the compound is an alkane and the root ***hept*** indicates that the longest continuous carbon chain is seven carbons long. The numbers indicate the positions of the three methyl groups attached to the carbon backbone. First draw the carbon chain indicated by the root and number the chain starting at either of the ends. Then attach the appropriate groups at the correct positions.

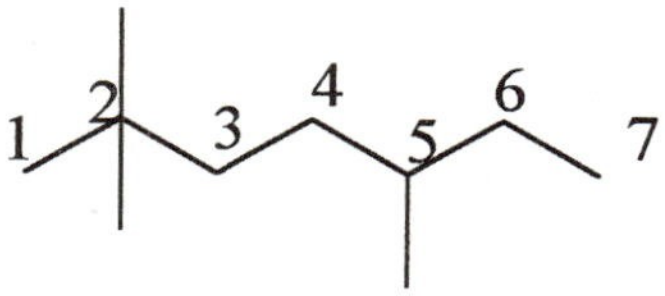

1 2 3 4 5 6 7

b)

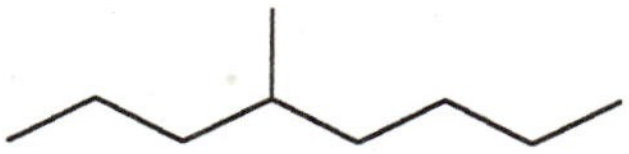

c)

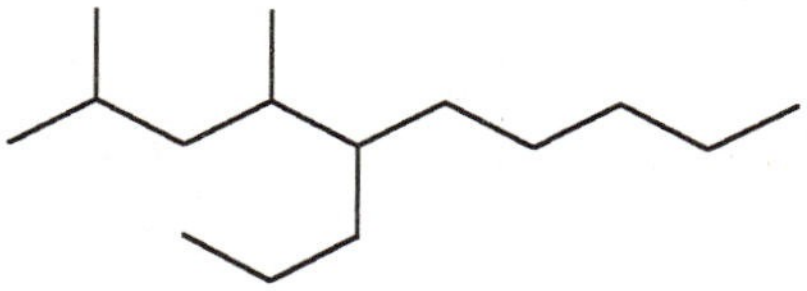

5.3 To answer this problem, first draw the structure suggested by the name. Then name the structure according IUPAC rules as described in solution **5.1**. For example the structure for the name given in (b) is as follows:

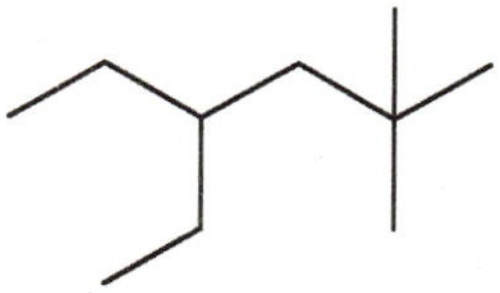

Examination of the structure shows that the root carbon chain has been numbered from the wrong end. The proper numbering is as follows:

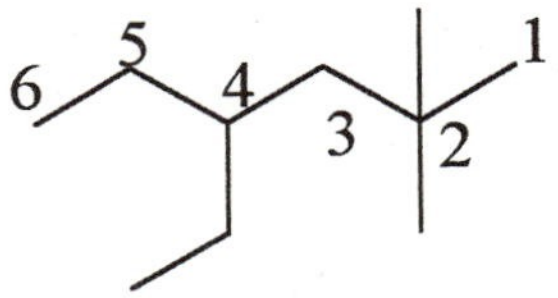

The correct name for the structure in (b) is 4-ethyl-2,2-dimethylhexane.

c) The name must indicate the positions of both methyl groups, so the number "2" must be repeated in the name. The correct name is 2,2-dimethylpentane.

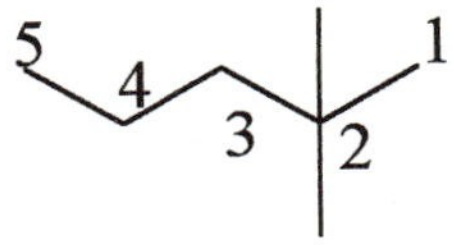

5.4 When naming complex groups, remember that the carbon that is attached to the root chain always gets the number 1. Therefore the longest chain in the complex group must include this carbon and the numbering must also begin at this carbon. Once the longest carbon chain is identified and numbered, the naming is straightforward. Do not forget to use the suffix ***yl*** and to put the name in parentheses.

b) In this complex group the longest chain has 5 carbons and there is a methyl group on carbon 2, so the name is (2-methylpentyl).

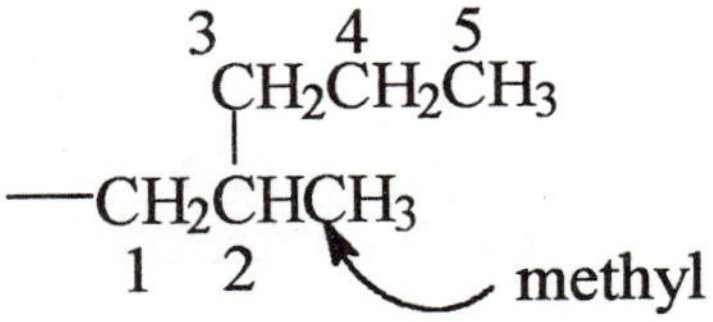

$$\begin{array}{l} \quad\quad\quad ^{3}CH_2{}^{4}CH_2{}^{5}CH_3 \\ \quad\quad\quad | \\ -\,^{1}CH_2{}^{2}CHCH_3 \leftarrow \text{methyl} \end{array}$$

c) (1-methylpropyl) d) (2,2-dimethylpropyl)

5.5 a) 4-(1-methylethyl)heptane

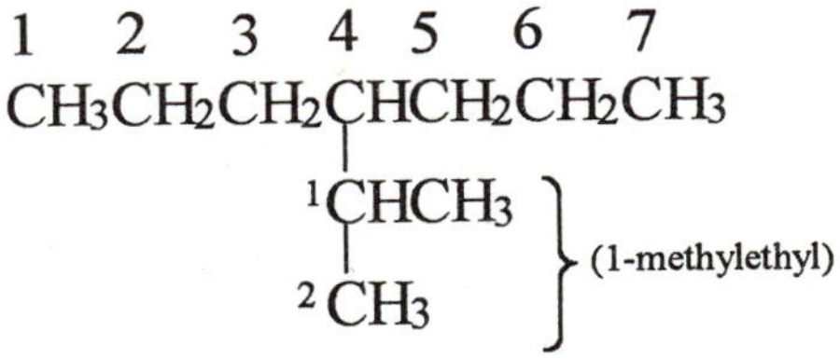

$$\begin{array}{l} ^{1}CH_3{}^{2}CH_2{}^{3}CH_2{}^{4}CH{}^{5}CH_2{}^{6}CH_2{}^{7}CH_3 \\ \quad\quad\quad\quad | \\ \quad\quad\quad ^{1}CHCH_3 \\ \quad\quad\quad\quad | \\ \quad\quad\quad ^{2}CH_3 \end{array} \Bigg\} \text{(1-methylethyl)}$$

b) 5-(1,2-dimethylpropyl)decane

5.6 Refer to solution 5.2.

b) The name 3-ethyl-7methyl-5-(1-methylpropyl)undecane indicates that the root is undecane (11) and the substituents are 3-ethyl, 7-methyl and 5-(1-methylpropyl).

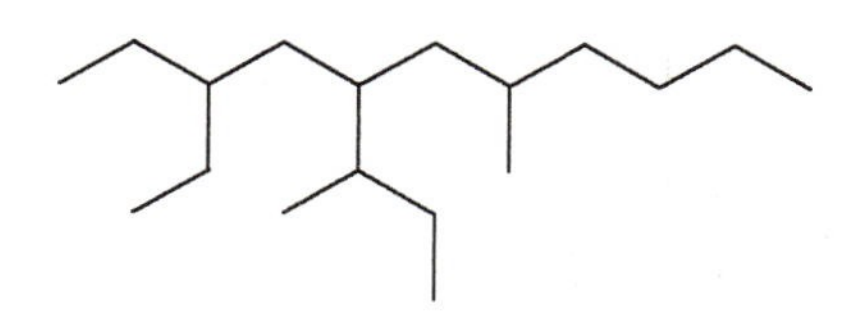

5.7 Carbon atoms in a compound can be designated according to how many other carbons are directly bonded to them. A primary carbon (p or 1°) is bonded to one other carbon, a secondary carbon (s or 2°) is bonded to two other carbons, tertiary carbon (t or 3°) is bonded to three carbons, and a quaternary carbon (q or 4°) is bonded to four carbons.

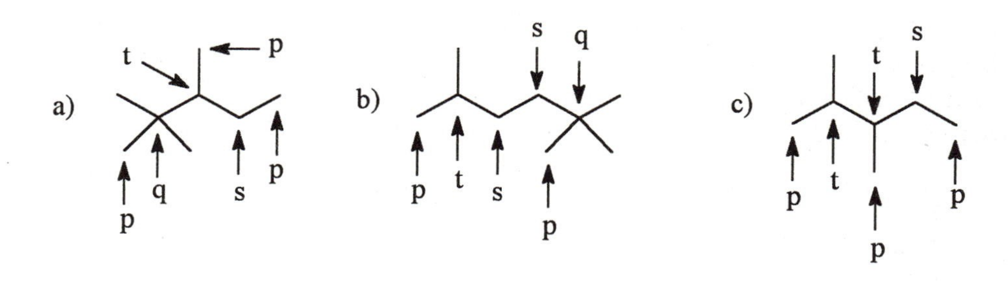

5.8 4-isopropylheptane

5.9

5.10 The procedure for naming cycloalkanes is very similar to that used for alkanes. Here the root is a cyclic alkane and the prefix ***cyclo*** is attached to the root to indicate the ring.

a) The ring has 5 carbons, so the root is cyclopentane. To keep the numbers as low as possible, begin at one of the methyl groups and proceed by the shortest possible path to the other methyl group. The correct name for this compound is 1,2-dimethylcyclopentane.

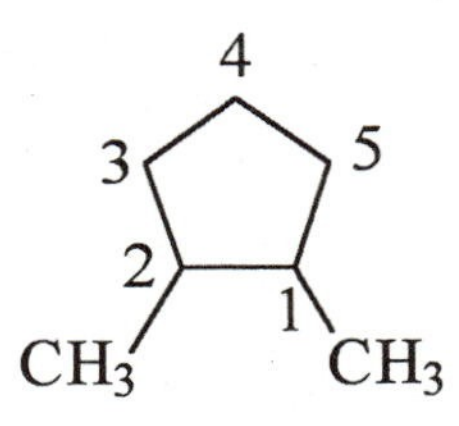

b) (1-methylpropyl)cyclohexane. Here no number is needed to locate the group because all positions of the ring are identical.

c) 5-cyclopentyl-2-methylheptane. There are 7 carbons in the longest alkyl chain while the ring has only 5 carbons, so the ring is named as a group.

d) 1-ethyl-3,5-dimethylcyclooctane

5.11

a) 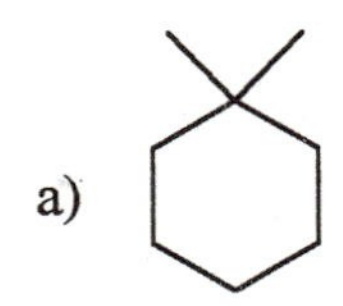 b) 

5.12 The naming procedure for alkenes is similar to that used for alkanes except for the following :

1) The longest continuous chain must include the double bonded carbons.
2) The suffix for alkenes is ***ene***.
3) The root is numbered from the end that gives the lowest possible number to the first carbon of the double bond. The number of the first carbon is used in designating the position of the double bond.

b) The longest carbon chain containing the double bond has 6 carbons, so the root name is hexene. The root is numbered from the end that will give the lowest possible number to the double bond. The correct name of this compound is 2,4-dimethyl-2-hexene.

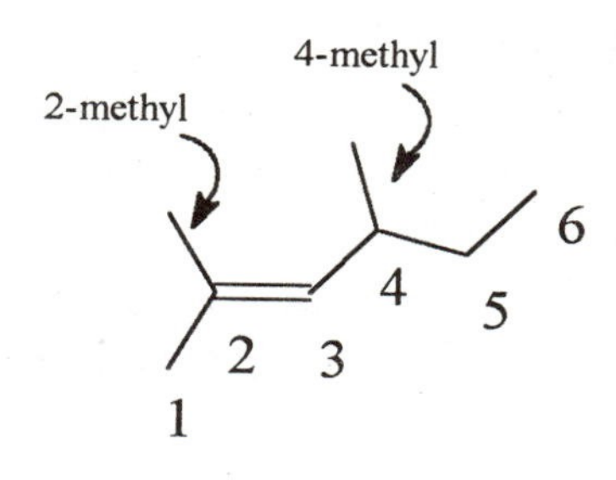

c) Here the root is a 7 carbon ring with three double bonds, so the root name is cycloheptatriene. The name of the compound is 2-methyl-1,3,5-cycloheptatriene.

d) 3-ethyl-1,2-dimethylcyclopentene

5.13

a) b) c)

5.14 The procedure for the naming of alkynes similar to that used for alkenes, except the suffix is ***yne***.

a) 3-isopropyl-1-heptyne
b) 2-methylpent-1-en-3-yne
c) 3-(2-methylpropyl)-1,4-hexadiyne

5.15

a) $HC{\equiv}CCH_2CH_2CH_3$ b)

5.16 Alkyl halides are named as alkanes with the halogen as a substituent.

a) 5-bromo-2,4,4-trimethylheptane
b) 1-chloro-3-ethyl-1-methylcyclopentane

5.17

Br

5.18 Alcohols are given the name of the hydrocarbon from which they are derived, with the suffix ***ol***. The longest chain must include the carbon bearing the hydroxy group and it is numbered to give the lowest possible number to this carbon (similar to alkenes and alkynes).

a) 2-butanol b) 3-methyl-3-hexanol c) 3-cyclopentyl-1-propanol

d) 3-bromo-3-methylcyclohexanol

5.19

a) 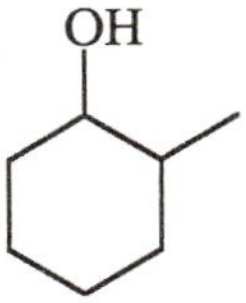 OH

b) 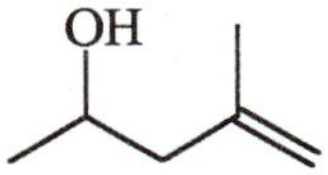 OH

5.20 For simple ethers, common names are often used. In these, each of the alkyl group is named, followed by the word ether. Complex ethers need to be named using the IUPAC system. In this systematic naming, the smaller alkyl group and the oxygen are designated as an *alkoxy* substituent on the larger group.

a) ethyl methyl ether

$CH_3OCH_2CH_3$

methyl ethyl

b) 1-chloro-3-methoxycyclopentane. Methoxy is the smaller alkyl group and the root is a 5 carbon cycloalkane with a chlorine bonded to one of the carbons.

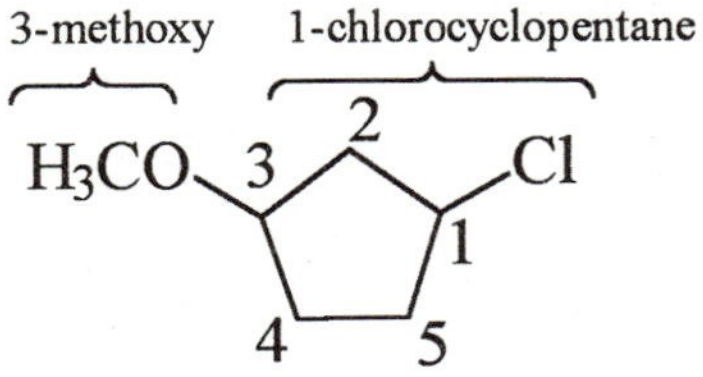

3-methoxy 1-chlorocyclopentane

H_3CO 3 2 Cl 1 4 5

5.21 In the common names of amines, the suffix ***amine*** is appended to the name of the alkyl group. The systematic nomenclature for amines is very similar to that used in naming alcohols except for the suffix ***amine***.

a) This is a simple amine, so we will use the common naming system. Here the a propyl group is attached to the nitrogen, so the name of the compound is propylamine.

b) The ethyl group attached to the nitrogen is given the prefix *N*-. The name of this compound is *N*-ethylcyclopentylamine.

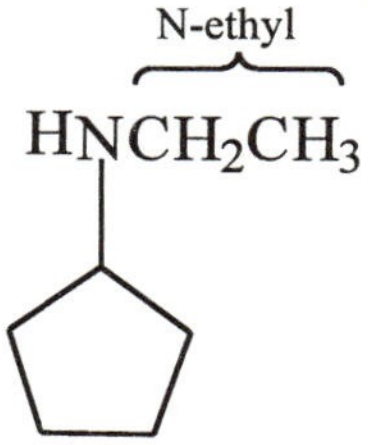

N-ethyl

$HNCH_2CH_3$

5.22 a) *N*,5-dimethyl-2-hexanamine

b) This compound has both an alcohol and an amine functional group. Only one of these can be designated with the suffix, and the other must be named as a group. The hydroxy group has higher priority than the amino group and is used to determine both the suffix and the numbering. The name of this compound is 5-amino-2-hexanol.

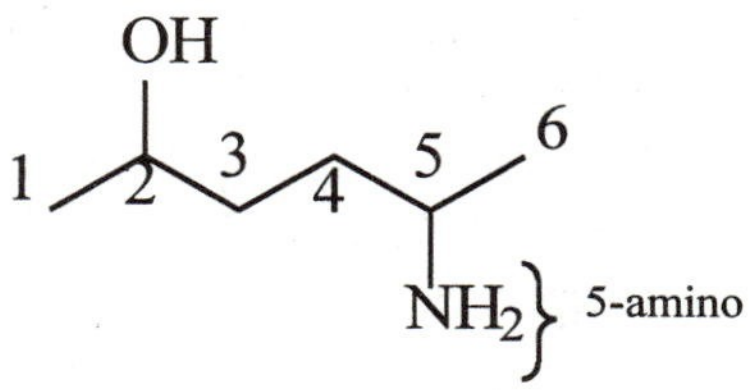

OH

1 2 3 4 5 6

NH_2 5-amino

5.23

a) 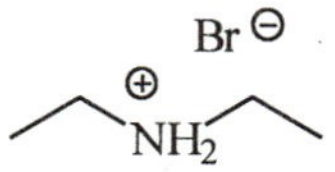 $Br^{\ominus}$ $\overset{\oplus}{N}H_2$

b) 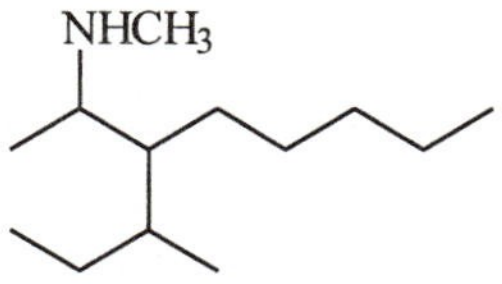 $NHCH_3$

5.24 a) 3,4-dimethylheptane
b) 4-ethyl-2-hexene
c) 3-chloro-1-pentyne
d) ethylcyclopentane
e) 3-methyl-2-pentanol
f) ethyl propyl ether
g) diethylamine

5.25

a) 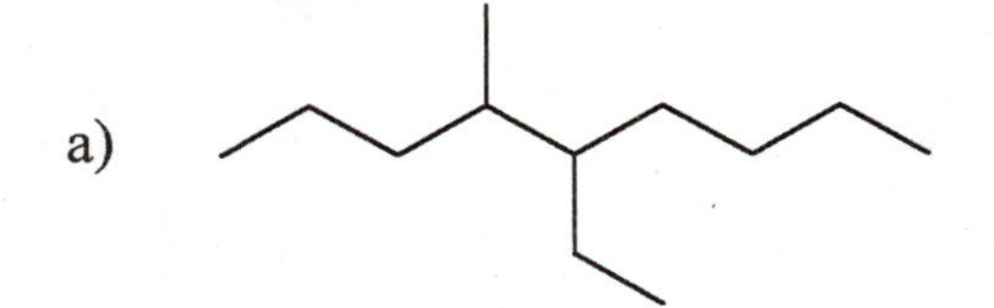

b)

c) HO, $-CH_3$

d)

e) H_3C, CH, CH_2CH_3

f) CH_3, CH_3C-OH, CH_3

5.26

hexane

2-methylpentane

3-methylpentane

2,2-dimethylbutane

2,3-dimethylbutane

5.27

a) 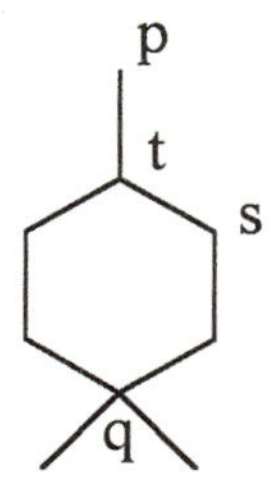
p
t
s
q

b) 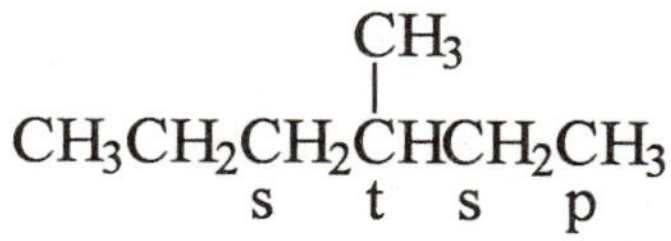
CH3
CH3CH2CH2CHCH2CH3
s t s p

5.28

a) CH_3CH_2OH
a primary alcohol

b) 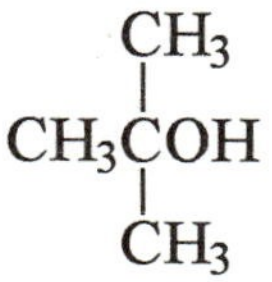
CH3
CH3COH
CH3
a tertiary alcohol

c) 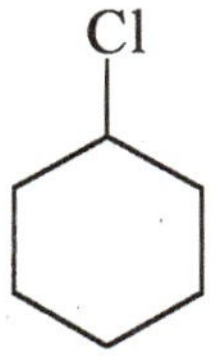
Cl
a secondary alkyl chloride

d) 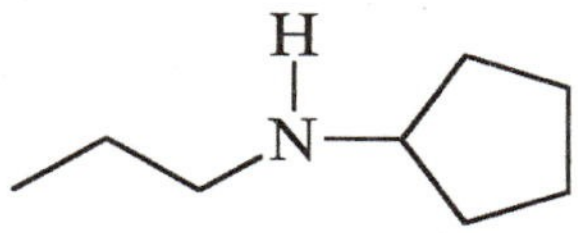
H
N
a secondary amine

5.29 a) The longest chain has 6 carbons. The correct name is 3-methyl-3-hexene.

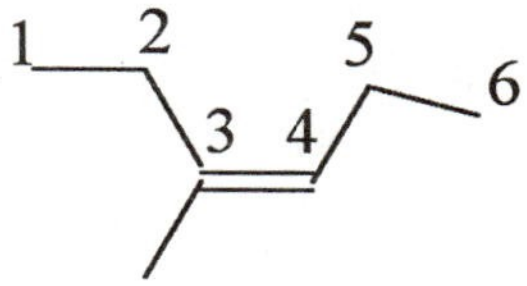
1
2
3
4
5
6

b) The smallest possible number should be given to the first branch. The correct name is 2,7,8-trimethydecane.

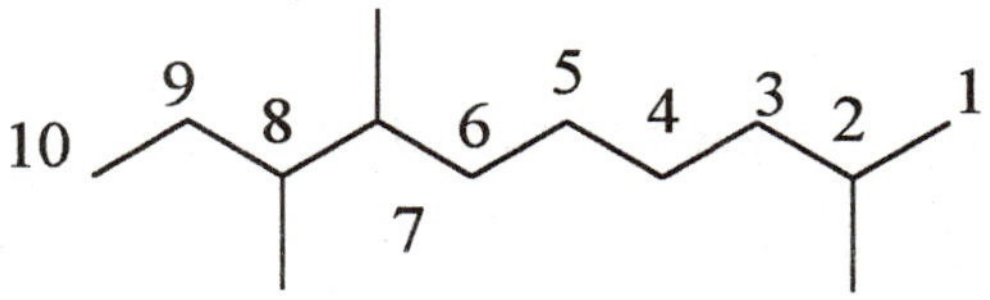
10
9
8
7
6
5
4
3
2
1

c) The ring has 6 carbons, so the root is cyclohexane. The correct name is propylcyclohexane.

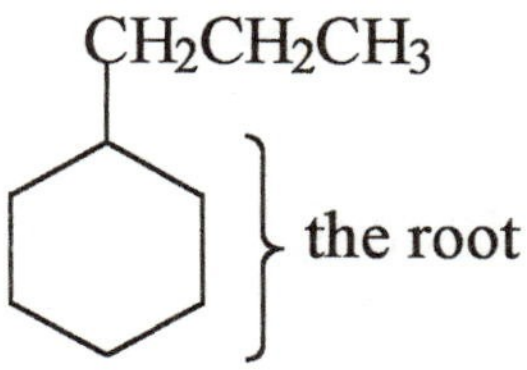
CH2CH2CH3
the root

d) The branches on the root should get the smallest numbers possible. The correct name is 1,2,4-trimethylcyclopentane.

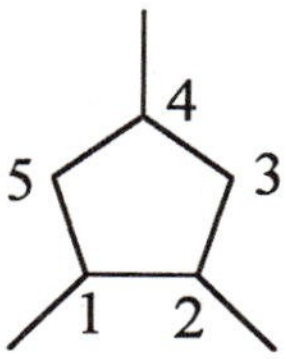

e) The double bond has more priority than the chlorine, so the correct name is 3-chloropentene.

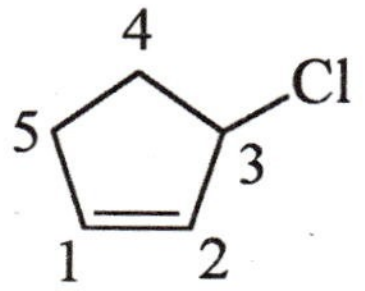

f) The alcohol has a higher priority than the triple bond, so the correct name is 4-pentyn-2-ol.

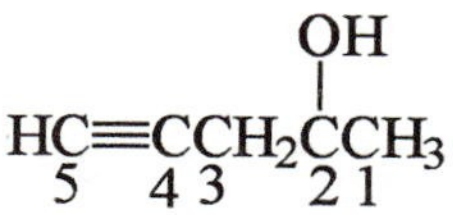

$$\underset{5}{HC}\equiv\underset{4}{C}\underset{3}{CH_2}\underset{2}{\overset{OH}{\overset{|}{C}}}\underset{1}{CH_3}$$

g) The carbon of the alkyl substituent attached to the ring should get the number 1 and the group name is placed in parentheses. The correct name is 2-(3-methylbutyl)cyclopentanol.

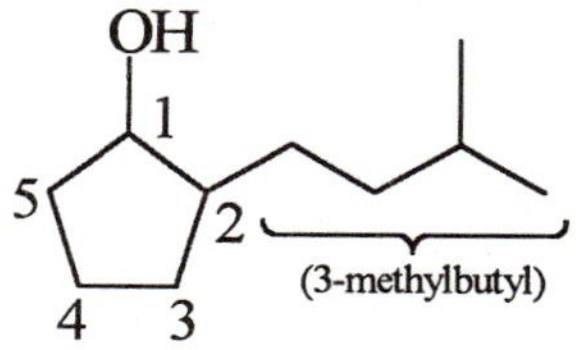

(3-methylbutyl)

h) The amine has priority over the double bond and the methyl group attached to the nitrogen should get a prefix N-. The correct name is N-methyl-3-butenamine.

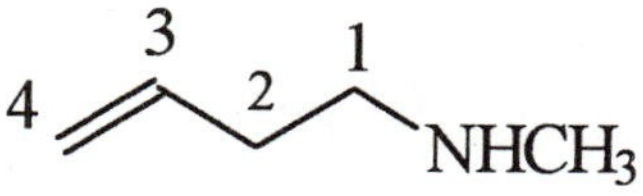

$NHCH_3$

5.30 a) 2,3,6-trimethyloctane b) 1,1,2,3-tetramethylcyclohexane
c) 2,3,4,4-tetramethylcyclohexene
d) 3-bromocyclohexene e) 3-cyclopenten-1-ol
f) pent-1-en-4-yne g) 5-methyl-5-hexen-3-ol
h) 4-(1-methylbutyl)-3-cyclohexen-1-ol
i) 5,5-dichloro-2-hexyne j) 3-methoxycyclohexanol
k) 3-isopropylcycloheptylamine
l) butoxycyclopentane

5.31

a) 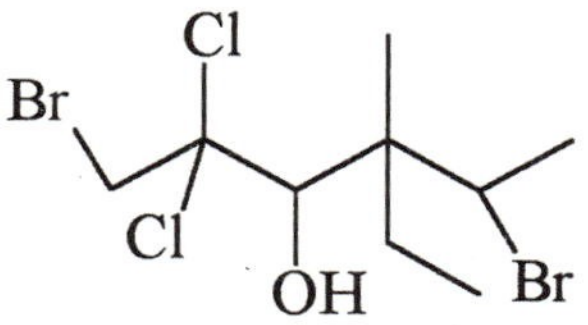 Br, Cl, Cl, OH, Br

b) 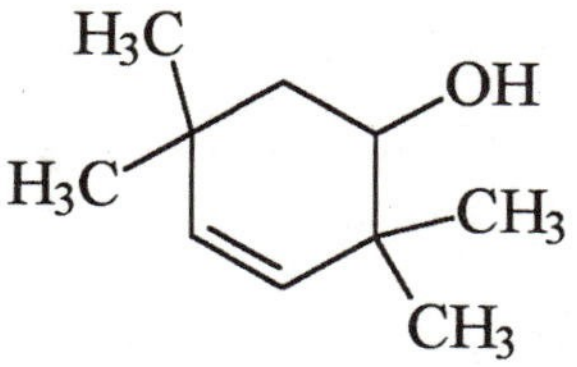 H_3C, H_3C, OH, CH_3, CH_3

c) 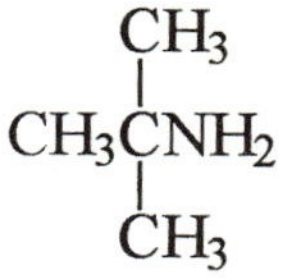 CH_3 / CH_3CNH_2 / CH_3

d) 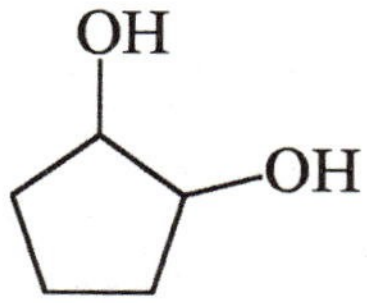 OH, OH

e)  O

f) 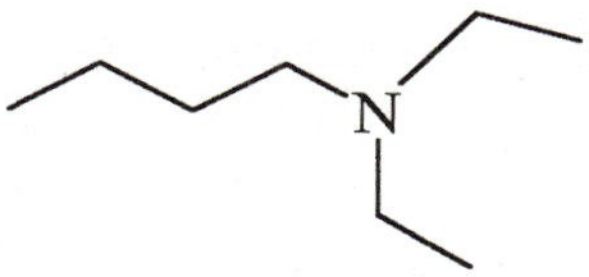 N

g) 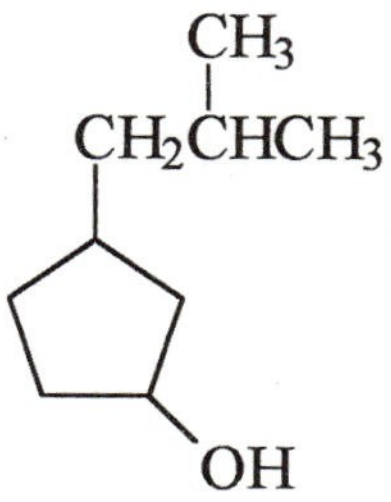 CH_3 / CH_2CHCH_3 / OH

h)

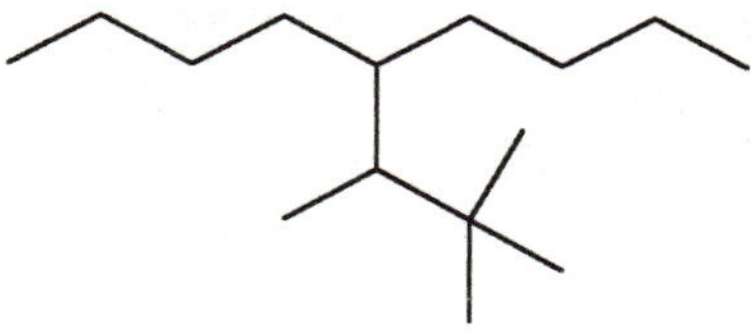

5.32 a) The melting point of a compound depends on how well the individual molecules can pack into the crystal lattice. The cyclic shape of cyclopentane allows it to pack better than the straight chain of pentane. Therefore, cyclopentane has a higher melting point (-94°C) than pentane (-130°C).
b) Intermolecular attractive forces are much larger in 1-pentanol than in pentane because of polarity and hydrogen bonding of the hydroxy group. Therefore 1-pentanol has a higher melting point.

5.33 The boiling point of liquids depends largely on the polarity, surface area, and molecular mass of the molecule.

a) Nonane and octane are nonpolar hydrocarbons of similar molecular geometry. However, nonane has the higher boiling point because it has a larger molecular mass than octane.
b) The polarity of an alkene is not much different from that of an alkane. Therefore, the physical properties of an alkene are similar to those of the corresponding alkane. The boiling points of nonane and 3-nonene are approximately the same.
c) 1-Nonanol has the higher boiling point because it is capable of hydrogen bonding.
d) Propylamine is higher boiling because it can hydrogen bond. Trimethylamine does not have any hydrogens bonded to the nitrogen and is incapable of hydrogen bonding.
e) Cyclopentanol is higher boiling because it can form hydrogen bonds, whereas diethyl ether cannot.
f) The physical properties of alkynes are very similar to those of alkenes and alkanes with the same number of carbons. Therefore 1-butene and 1-butyne will exhibit very similar boiling points.
g) 1-Pentanol is higher boiling because it can hydrogen bond.
h) Cyclopentanol and cyclopentylamine are both polar, both can form hydrogen bonds, and they have similar molecular masses. However, because oxygen is more electronegative than nitrogen, cyclopentanol forms stronger hydrogen bonds. Therefore, cyclopentanol has a higher boiling point (141°C) than cyclopentylamine (108°C).

5.34 When two immiscible liquids are mixed together, the denser liquid forms the lower layer. Chloroform is denser than water because of its more massive chlorine atoms, so it forms the lower layer.

5.35 Remember, like dissolves like. Water is very polar and can hydrogen bond, so the solubility of a compound in water depends on the polarity of the compound.

a) 1-Butanol has a higher solubility in water because it has a H bonded to the O and is can hydrogen bond with water. 1-Chlorobutane is less polar.
b) 1-Butanol has a higher solubility in water because it has a smaller nonpolar hydrophobic hydrocarbon chain than 1-hexanol.
c) Diethylamine is more soluble in water than pentane because it is polar and can hydrogen bond with water. Pentane is nonpolar and therefore hydrophobic.

5.36 Amines often have unpleasant, fishy odors. Therefore it is possible that the fishy smelling unknown compound has an amine functional group. Amines are bases, so they react with strong acids to form water soluble salts. Testing the solubility of the unknown compound in dilute aqueous hydrochloric acid is one way of confirming the presence of the amine functional group.

After completing this chapter, you should be able to:

1. Provide the systematic (IUPAC) name for an alkane.
   Problems 5.1, 5.3, 5.26.

2. Draw the structure of an alkane whose name is provided.
   Problems 5.2, 5.6.

3. Name a complex group.
   Problems 5.4, 5.5.

4. Name a cycloalkane, an alkene, an alkyne, an alkyl halide, an alcohol, an ether, or an amine.
   Problems 5.10, 5.12, 5.14, 5.16, 5.18, 5.20, 5.21, 5.22, 5.24, 5.29, 5.30.

5. Draw the structure of a compound containing one of these functional groups when the name is provided.
   Problems 5.11, 5.13, 5.15, 5.17, 5.19, 5.23, 5.25, 5.31.

6. Predict the approximate physical properties of a compound containing one of the above functional groups.
   Problems 5.32, 5.33, 5.34, 5.35.

Chapter 6
STEREOCHEMISTRY

The use of models is an invaluable aid in understanding the material presented in this chapter. You are strongly encouraged to use models to solve the problems in this chapter.

6.1 Geometrical isomers are stereoisomers that differ in the placement of groups about the double bond. In order for an alkene to exhibit geometrical isomerism, the two groups on each end of the double bond must be different.

a) $CH_3CH_2(H)C{=}C(CH_3)(H)$ and $CH_3CH_2(H)C{=}C(H)(CH_3)$

b) none $CH_3CH_2(H)C{=}CH_2$ — These substituents (H, H) are identical

c) none These groups are identical (CH_3, CH_3) $(CH_3)_2C{=}C(H)(CH_3)$

d) $Cl(CH_3)C{=}C(H)(CH_2CH_3)$ and $Cl(CH_3)C{=}C(CH_2CH_3)(H)$

6.2 The stability of geometrical isomers depends on the steric strain energy associated with the molecule. Steric strain destabilizes a molecule by forcing it to deviate from its optimum bonding geometry. The geometrical isomer with the larger groups on opposite sides (*trans*) of the double bond is more stable.

a) $CH_3CH_2(H)C{=}C(CH_3)(H)$ [CH_3CH_2 and CH_3 cis] and $CH_3CH_2(H)C{=}C(H)(CH_3)$ [CH_3CH_2 and CH_3 trans]

b) $CH_3((CH_3)_3C)C{=}C(H)(CH_3)$ [$(CH_3)_3C$ and CH_3 cis] and $CH_3((CH_3)_3C)C{=}C(CH_3)(H)$ [$(CH_3)_3C$ and CH_3 trans]

In both cases the right isomer is more stable because the larger groups are *trans*.

6.3 The **Cahn-Ingold-Prelog sequence rules** are used to assign priorities to groups. These rules use the atomic numbers of the atoms attached to the carbons of the double bond to determine the priority of groups.

Rule 1: Of the two atoms attached to one carbon of the double bond, the one with the higher atomic number has the higher priority.

Rule 2: If the two atoms attached to the carbon are the same, compare the atoms attached to them, in order of decreasing priority. The decision is made at the first point of difference.

Rule 3: Multiple-bonded atoms are treated as though they are equivalent to the same number of single bonds to the same atoms.

a) Carbon has higher priority than hydrogen. $—CH_2CH_3$

b) The carbon bonded to O,O,and O has higher priority than carbon bonded to O,O, and N. $—C(=O)OH$

c) These two groups differ only by the bonding arrangement at the third carbon. The third carbon bonded to C, H, and H has higher priority than that bonded to three hydrogens. $—CH_2CH_2CH_2CH_3$

d) The cyano group is treated as though the carbon is bonded to three nitrogens, so it has higher priority. $—C{\equiv}N$

e) The carbon on the aromatic ring is treated as though it is bonded to three other carbons, whereas the carbon of the isopropyl group is bonded to two carbons and a hydrogen.

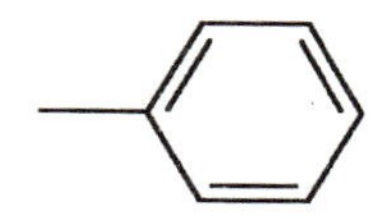

6.4 The first step in the process of assigning the ***E*** or ***Z*** designation to alkenes is to assign priorities to the groups attached to the carbons of the double bond. If the high priority groups are on the same side of the double bond, the alkene gets the designation ***E***. If these groups are opposite sides of the double bond, then the configuration is designated as ***Z***.

a) The atoms attached to the left carbon of the double bond are both carbons. The upper C is bonded to three Hs. The lower C is attached to two Hs and a C. Therefore, the lower group has the higher priority.

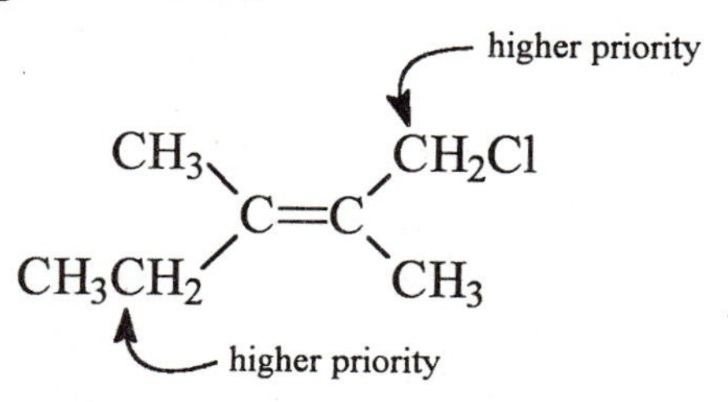

The atoms attached to the right carbon of the double bond are also both carbons. However, the upper group has higher priority because the carbon is bonded to a chlorine. The higher priority groups are on opposite sides of the double bond, so this is the *E* isomer.

b) Remember, double and triple bonds that are part of a group are treated as though they are two or three single bond to the same atom. The higher priority groups are on opposite sides of the double bond, so the compound is the *E* isomer.

d) The compound is the *Z* isomer.

6.5 Conformations are the various shapes that a molecule can assume by rotations about single bonds. The relative stabilities of these conformations can be estimated by examining the torsional and steric strains associated with each of the conformations. Newman projections offer a convenient way for examining and assessing the relative stabilities of these conformations. For propane, all the eclipsed conformations are of the same energy while all the staggered conformations are of the same energy.

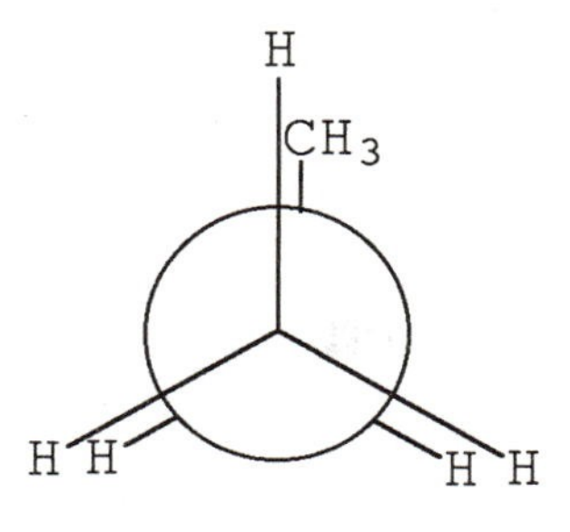

eclipsed conformation

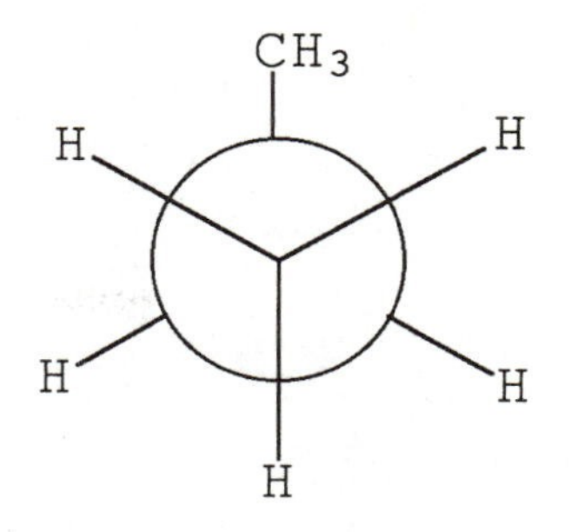

staggered conformation

The eclipsed conformation has a higher energy due to torsional strain (electron-electron repulsion of the eclipsed bonds) and steric strain (due to eclipsing CH_3 and H. These strains are minimized in the staggered conformation, so it is lower in energy.

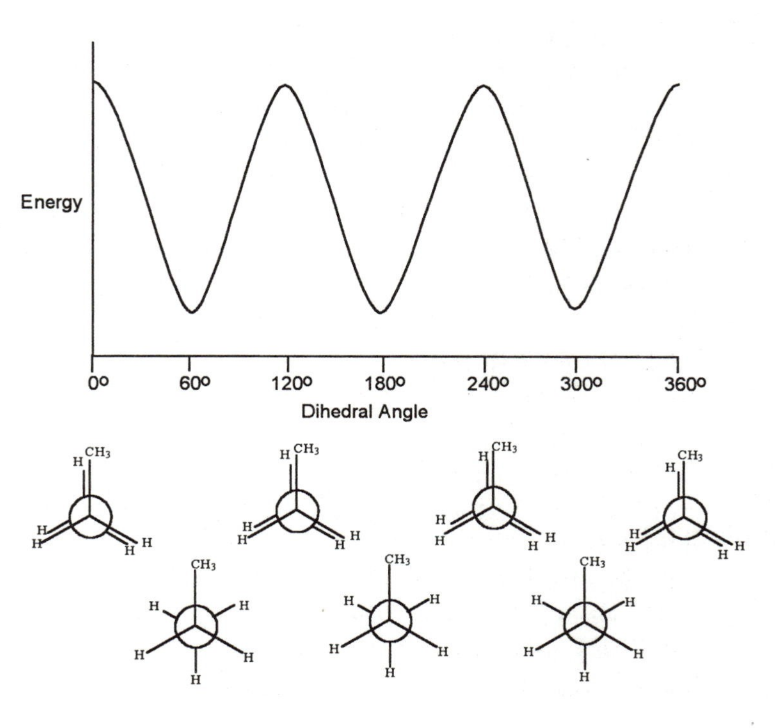

6.7

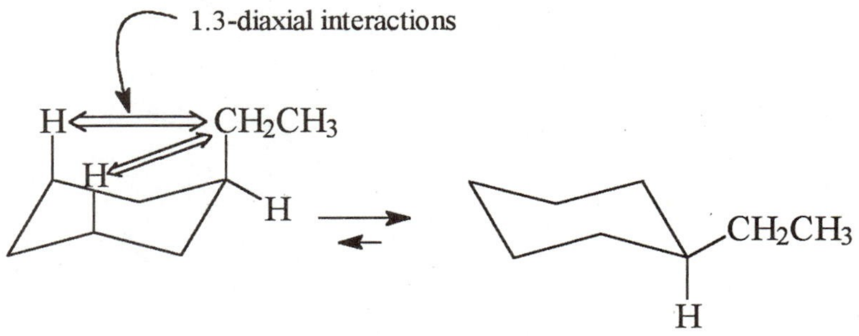

1.3-diaxial interactions

H ⟺ CH_2CH_3

H

H ⇌ CH_2CH_3

H

The conformation on the left has steric crowding between the axial ethyl group and the two axial hydrogens. This steric strain, known as 1,3-diaxial interactions, destabilizes this conformer. Actually this steric interaction is similar to the gauche interaction in butane. In general, substituents larger than hydrogen prefer to be equatorial on a cyclohexane ring to avoid 1,3-diaxial interactions. The conformer on the right is produced by the ring flip, and does not have any 1,3-diaxial interactions because the ethyl group is equatorial. Therefore, the conformation on the right is more stable.

6.9 To answer this question we need to know the axial strain energies caused by the 1,3-diaxial interactions between an axial substituent and the two axial hydrogens on the same side of the cyclohexane ring. These axial strain energies are listed in Table 6.2 in the text. The lower the axial strain energy caused by a substituent, the larger the concentration of the conformation with the substituent in the axial position will be at equilibrium.

a) The axial destabilization energy of the CN group is smaller than that of the CH_3 group. Therefore, cyanocyclohexane will have more of the conformation with the cyano group axial present at equilibrium.
b) Ethylcyclohexane will have more of the conformation with the ethyl group axial present at equilibrium because the phenyl group has a larger axial destabilization energy than the ethyl group.
c) Chlorocyclohexane will have more of the conformation with the axial Cl present at equilibrium because the ethyl group has a larger axial destabilization energy.

6.10 Stereoisomers are compounds that have the same chemical formula and connectivity but different arrangement of the bonds in space. In cycloalkanes, they are similar to geometrical isomers in that they cannot interconvert without breaking a bond.

a) CH_3 / CH_3 and CH_3 / CH_3

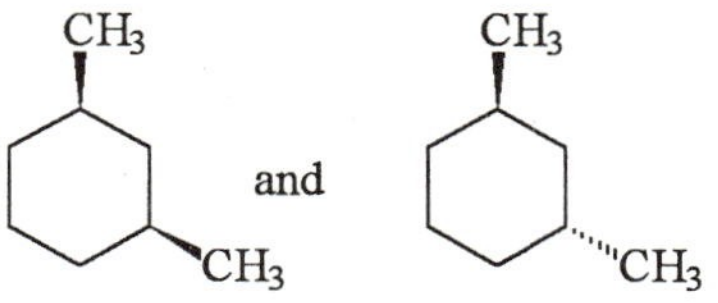

b) CH_2CH_3 / CH_2CH_3 and CH_2CH_3 / CH_2CH_3

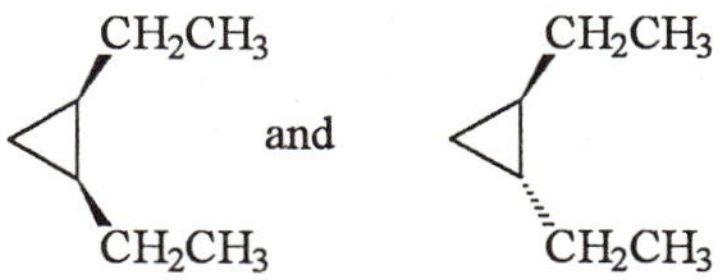

c) Cl / CH_3 and Cl / CH_3

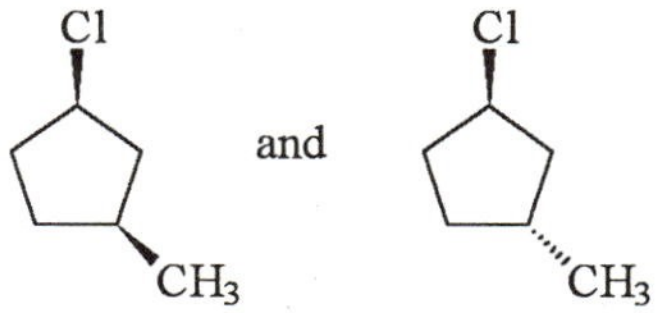

6.12 The flat representation of a cyclohexane ring shown below shows the relation of axial and equatorial bonds. The bonds on one side of the ring (above or below) are *cis* and they alternate between axial and equatorial as one proceeds around the ring. Similarly, bonds on opposite sides of the ring are *trans* .

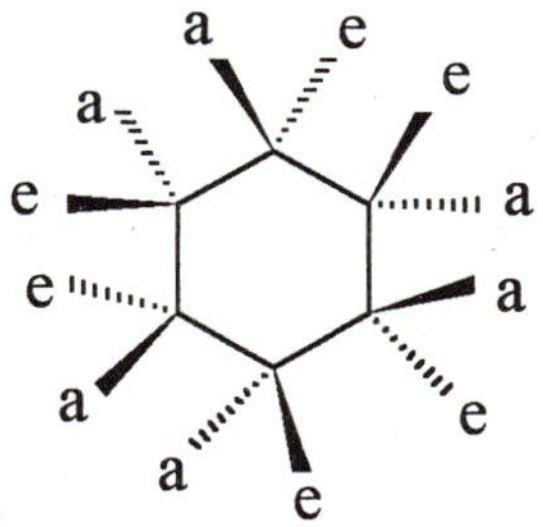

a e e a a e e a a e e a

a) In *cis*-1,4-dimethylcyclohexane one methyl group is axial and one is equatorial. The ring flip of one conformation converts the axial methyl to equatorial and the equatorial methyl to axial.

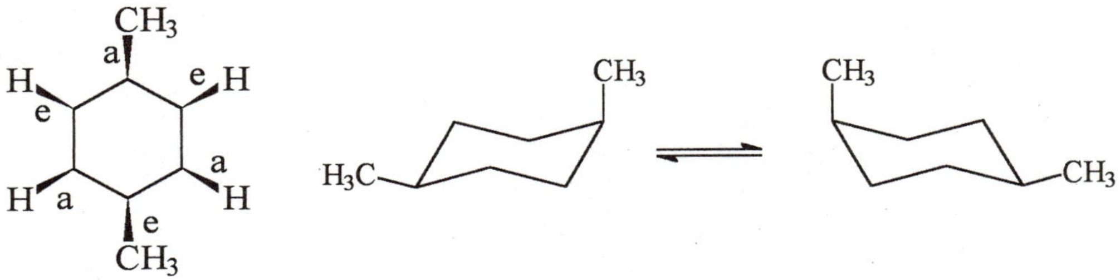

CH_3 a, H e, e H, a H, H a, e CH_3 — H_3C / CH_3 ⇌ CH_3 / CH_3

b) In one conformation of the trans isomer both methyl groups are axial and in the other conformation, produced by the ring flip, both methyl groups are equatorial.

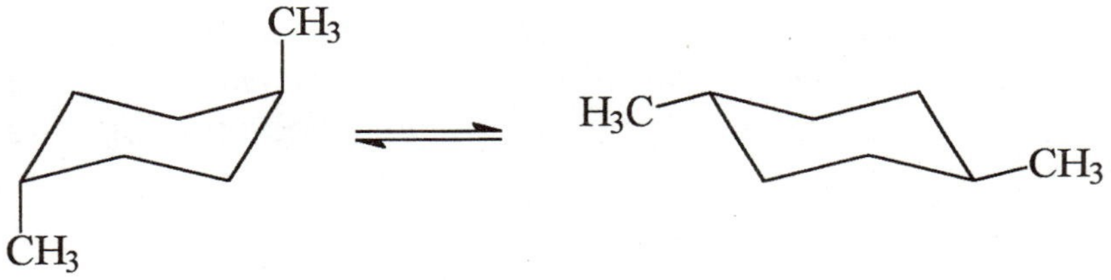

CH_3 / CH_3 ⇌ H_3C / CH_3

c) The *trans*-isomer is more stable because it has a conformation with both methyl groups equatorial. Steric interactions are at a minimum when bulky substituents are in equatorial positions.

d) The more stable conformation of the more stable isomer is the conformation of the *trans*-isomer shown in (b) with both methyls equatorial.

6.13

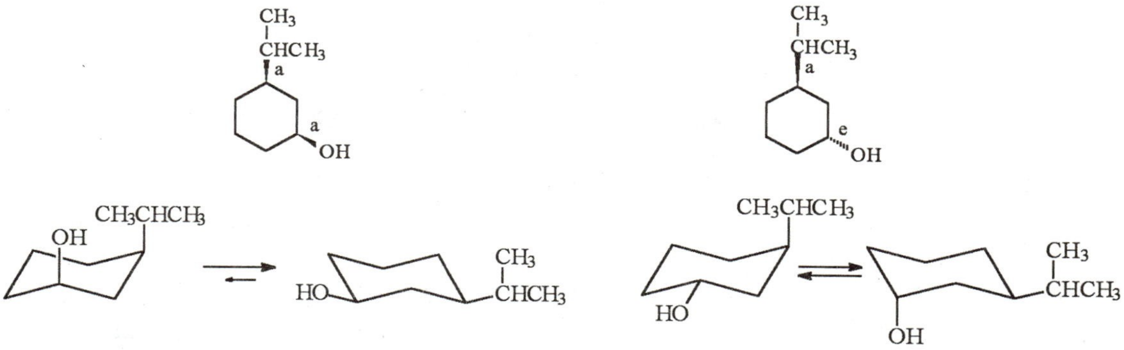

cis-3-isopropylcyclohexanol   trans-3-isopropylcyclohexanol

a) The conformation with both groups equatorial is the more stable conformation of the *cis*-stereoisomer. The conformation with the isopropyl group equatorial and the hydroxy group axial is the more stable conformation of the *trans*-stereoisomer.

b) The axial strain energy for various substituents listed in Table 6.2 are actually the energy differences between the two conformations of the cyclohexane with the corresponding substituent. The *cis*-isomer exists almost entirely in the conformation with both substituents equatorial. The lower energy conformation for the *trans*-isomer [0.9 kcal/mol (3.8 kJ/mol)] is the one with the isopropyl group equatorial because the axial destabilization energy for isopropyl is larger than that for the OH group. Therefore, the *cis*-stereoisomer is more stable than the *trans*-stereoisomer by 0.9 kcal/mol (3.8 kJ/mol).

6.14 In the representation of a cyclohexane chair conformation, the relative geometry of bonds can be easily identified using the following approach:

All bonds projecting closer to the top of the page at each carbon are on the top side of the ring and are cis to each other. Similarly, the bonds projecting closer to the bottom of the page at each carbon are on the bottom side of the ring and are cis to each other but trans to the bonds on the top of the ring.

b)

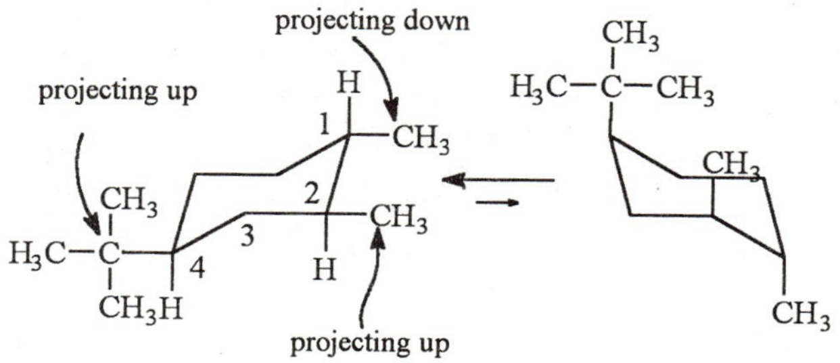

The methyl on C-2 is projecting towards the top of the page while the methyl on C-1 is projecting towards the bottom of the page, so the methyls are *trans.* The *t*-Bu on C-4 and the methyl on C-2 are projecting towards the top of the page, so they are *cis* to each other. All the substituents in this conformation are equatorial, therefore the conformation shown on the left is more stable. In the conformations of all the other possible stereoisomers of this compound, at lest one of the substituents will be axial. Therefore, the stereoisomer shown is most stable.

c)

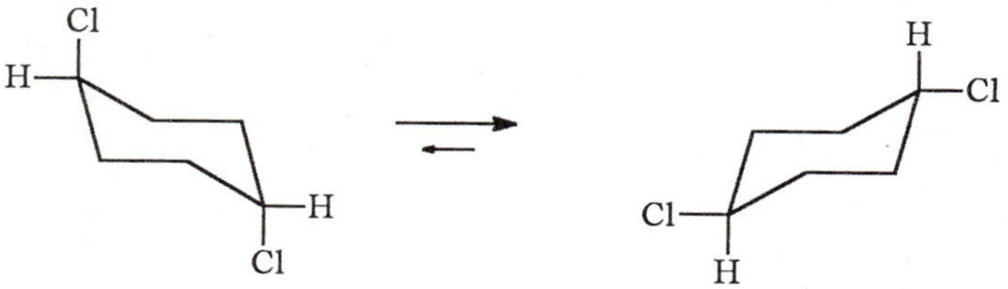

The chlorines are on opposite sides of the ring, so they are *trans.* The chlorines are also axial in the conformation on the left. The conformation produced by the ring flip (right) is more stable because both chlorines are equatorial. In the other stereoisomer of this compound the chlorines are in a *cis* geometry, and at least one of the chlorines will be axial in each conformation. Therefore, the stereoisomer shown is more stable.

d)

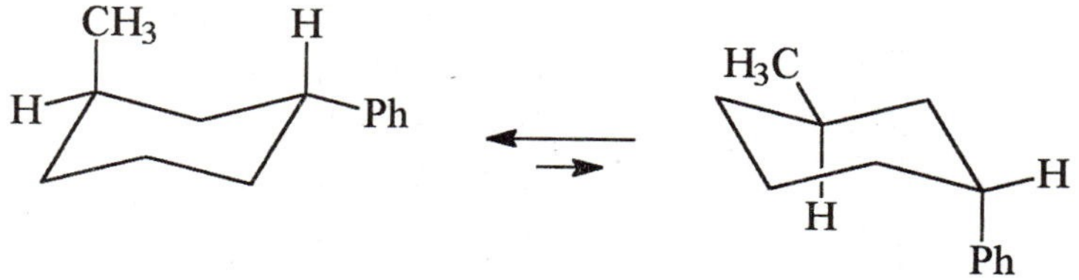

The phenyl and methyl groups are *trans* to each other. The conformation on the left is more stable because the axial destabilization energy of the methyl group is smaller than that for the phenyl group. In the *cis* isomer of this compound, the more stable conformation will have both phenyl and methyl groups equatorial. Therefore the *cis*-stereoisomer is more stable.

6.15 One of the ways of determining whether an object is chiral is to examine its symmetry. Although there are several different symmetries an object may posses, the only kind we will be concerned with is the plane of symmetry or mirror plane. Any object that has a plane of symmetry is *achiral.*

a) achiral b) chiral c) chiral d) achiral e) chiral f) chiral

6.16 Chiral centers are carbon atoms in a compound that are attached to four different groups. Such carbons are also referred to as asymmetric carbons or chiral carbons. Carbons that are doubly bonded or triply bonded are not chiral. The presence of chiral centers in a molecule does not necessarily mean that the molecule is chiral. The final way to determine if a molecule is chiral is to examine whether the molecule has a plane of symmetry. Any molecule that has a plane of symmetry is not chiral.

a) This compound has no chiral centers. Therefore, it is achiral.
b) This molecule is not chiral because it does not have any chiral centers. Remember, a doubly bonded carbon is not chiral.
c) This molecule has one chiral center, so it is chiral.

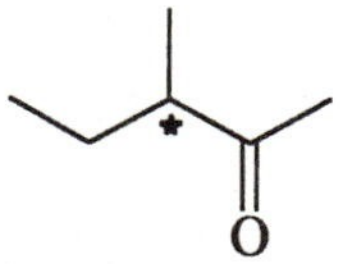

O

d) This molecule has one chiral center, so it s chiral.

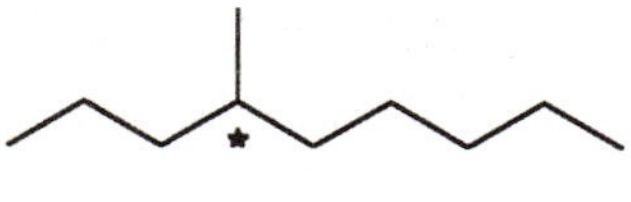

e) The carbon bonded to the chlorine is bonded to a hydrogen and bonded to two other carbons that are part of the ring. Proceeding around the ring in one direction, the second carbon encountered is attached to two methyl groups, while in the other direction it is the fourth carbon. Therefore these groups are not identical. This molecule is chiral.

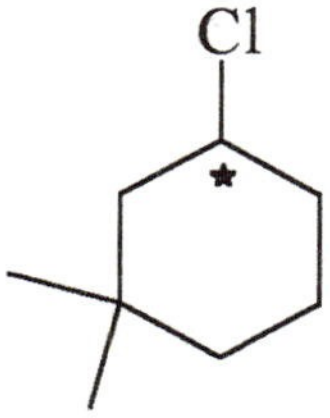

Cl

f) This molecule has no chiral center because identical substituents are encountered as one proceeds in either direction around the ring.

6.17 a) The right half of the face mirrors the left half , so there is a plane of symmetry.
b) A pencil can be split along its length into two identical pieces, so it has a plane of symmetry.
c) An ear does not have a plane of symmetry.
d) This molecule has a plane of symmetry.
e) This molecule does not have a plane of symmetry.
f) This molecule has a plane of symmetry that runs through the double bonded carbon and oxygen atoms.
g) This molecule has a plane of symmetry that passes through the methyl carbon, the central carbon and the hydrogen.
h) This molecule does not have a plane of symmetry.

6.18 Use the following steps to designate the configuration of the chiral centers:

Step 1: Assign priorities from 1 through 4 to the four groups bonded to the chiral center using the Cahn-Ingold-Prelog sequence rules outlined in Section 6.3 in the text (or solution 6.3 in this chapter). The group with the highest priority gets number 1 and the lowest priority group gets number 4.

Step 2: View the molecule at the chiral center with group number 4 pointed directly away from you. If the priority numbers of the remaining groups cycle in a clockwise direction (1 → 2 → 3→ 1), the chiral center has the ***R*** configuration. If the cycle is in a counterclockwise direction then the chiral center has the ***S*** configuration.

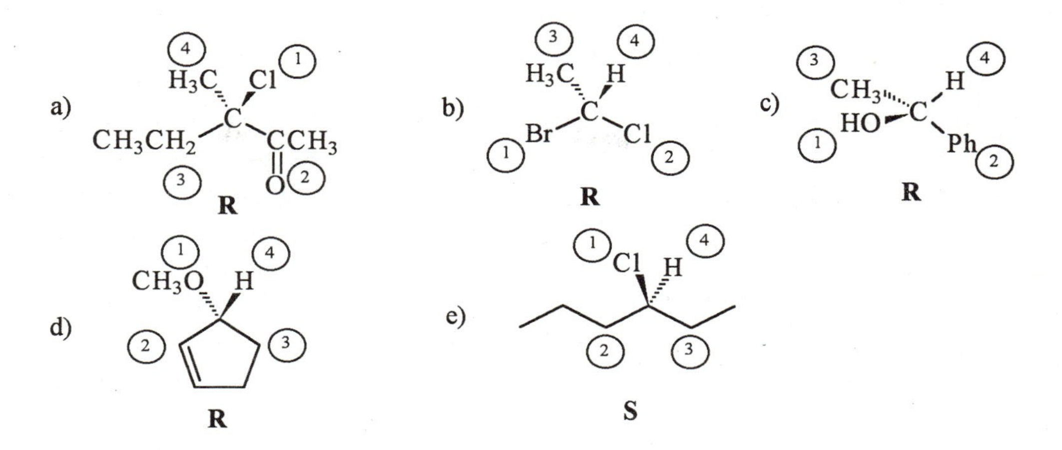

6.19 When drawing the structure of a chiral molecule from its name, follow the sequence outlined below:

1. Draw the structure without stereochemistry and label the chiral centers.
2. Assign priorities to the groups attached to the chiral center.
3. Draw a tetrahedral carbon and place the lowest priority group on the bond pointed away from you.
4. Place group number 1 on any one of the other three bonds.
5. If the configuration is ***R*** at the chiral center, then place groups 2 and 3 in a clockwise direction from group 1. If the configuration is **S**, then place groups 2 and 3 in a counterclockwise direction from group 1.

b) H CH_2CH_3

c) ① Br H ④ ③ ② $\Longrightarrow$ ① Br H ④ $\Longrightarrow$ ① Br H ④ ③ ②

R - clockwise

6.20 a) False. Enantiomeric molecules exhibit different properties only in a chiral environment.

b) False. Enantiomers have identical physical properties.

c) True. Water is not chiral.

d) Cannot be determined. The direction of rotation of plane polarized light by a chiral molecule has no relationship to the assigned configuration of the molecule and can only be determined by experiment.

e) True. The assignment of *d* to a chiral molecule denotes that the compound rotates plane polarized light in the clockwise or + direction.

6.21 Remember, in the Fisher projection a tetrahedral carbon is represented by a cross. The horizontal line of the cross represents bonds projecting above the plane and the vertical line represents bonds projecting behind the plane. Draw the horizontal line as wedged bonds and the vertical lines as dashed bonds and determine the configuration at the chiral center.

a)

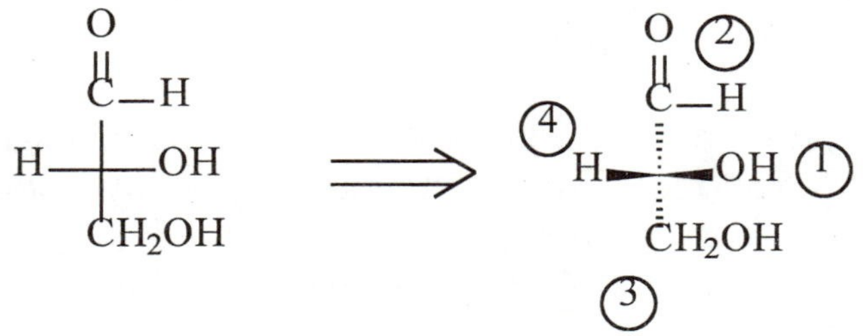

O
‖
C—H
H—+—OH
CH_2OH

⟹

O (2)
‖
C—H
(4) H—+—OH (1)
CH_2OH
(3)

R

b)

O
‖
C—OH
H_2N—+—H
CH_2OH

⟹

O (2)
‖
C—OH
(1) H_2N—+—H (4)
CH_2OH
(3)

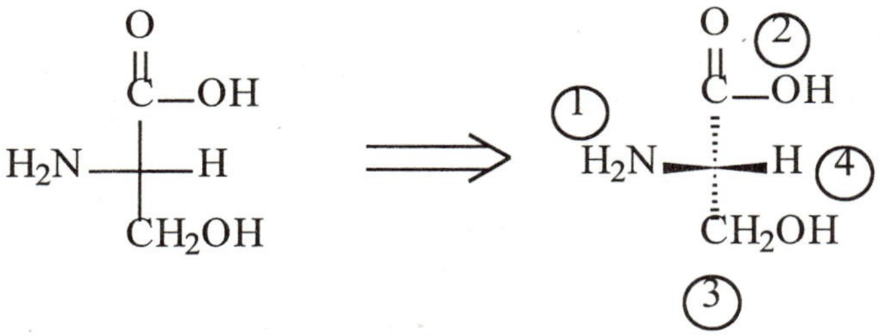

S

6.22 (2*R*,3*R*)-2,3,4-Trihydroxybutanal is D-erythrose; (2*S*,3*R*)-2,3,4-trihydroxybutanal is D-threose; L-erythrose is (2*S*,3*S*)-2,3,4-trihydroxybutanal.

6.23 Enantiomers are nonsuperimposable mirror images of a molecule. Diastereomers are non-mirror image stereoisomers of a molecule.

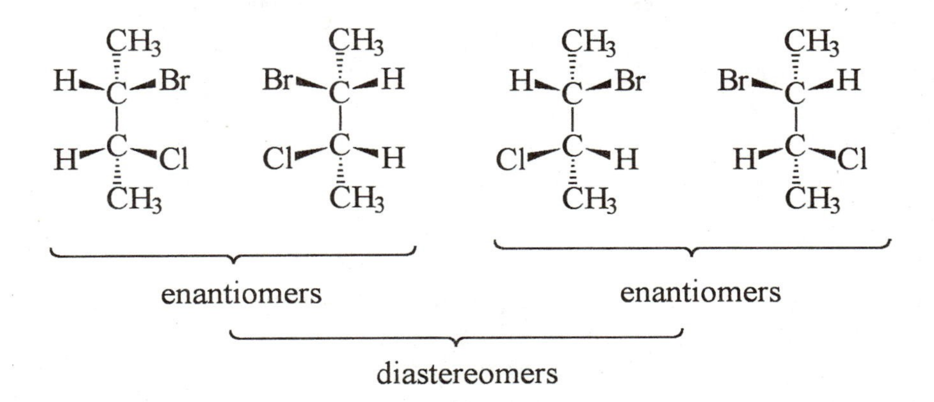

6.24

a) Cl OH — 4

b) 4

c) HO — 16

d) none

6.25

CH3 / H–C–Cl / H–C–Cl / CH3 — meso

CH3 / H–C–Cl / Cl–C–H / CH3 — rotates

CH3 / Cl–C–H / H–C–Cl / CH3 — rotates

has a plane of symmetry

6.26 The *cis*-diastereomer of 1,2-dimethylcyclopropane is *meso* and does not rotate plane polarized light because it has a plane of symmetry bisecting the ring. The *trans* isomer exists as a pair of enantiomers that do rotate plane polarized light.

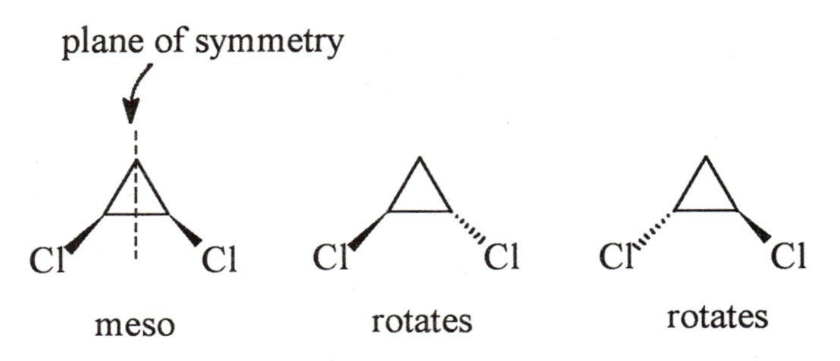

6.27 b) yes c) no, meso d) yes e) no, meso

6.28 To construct a Fisher projection, the molecule is first arranged with the horizontal bonds to its chiral center projecting out of the page and the vertical bonds projecting into the page. In the Fisher projection a tetrahedral carbon is represented by a cross. The horizontal line of the cross represent bonds projecting above the page and the vertical line represent the bonds projecting into the page.

a) CH_2OH (top); H (left), Cl (right); CH_3 (bottom)

b) CO_2H (top); HO (left), H (right); CH_3 (bottom)

c) CH=O (top); H_3C (left), H (right); CH_2CH_3 (bottom)

6.29 The approach to this problem is similar to that described previously in solution 6.20. First transform the Fisher projection into wedge and dash bond structure and then determine the configuration.

b) CO_2H (top); H (left), NH_2 (right); CH_3 (bottom) $\Longrightarrow$ ② CO_2H (top, dash); ④ H (left, wedge), NH_2 ① (right, wedge); CH_3 ③ (bottom, dash)

R

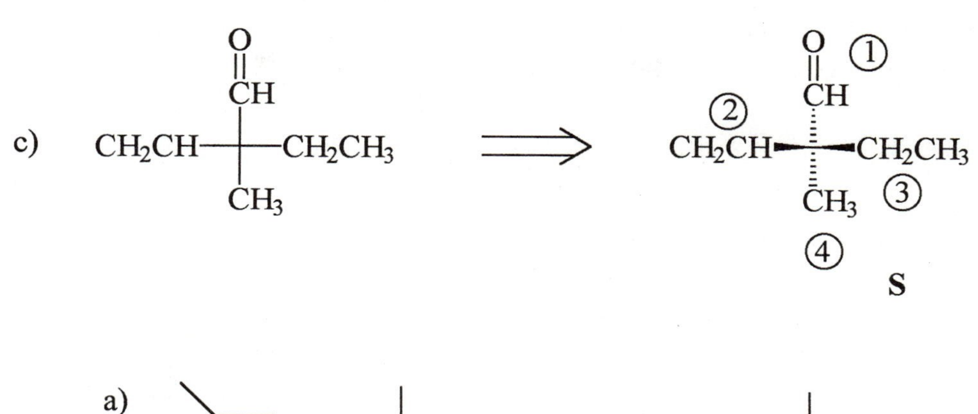

6.30

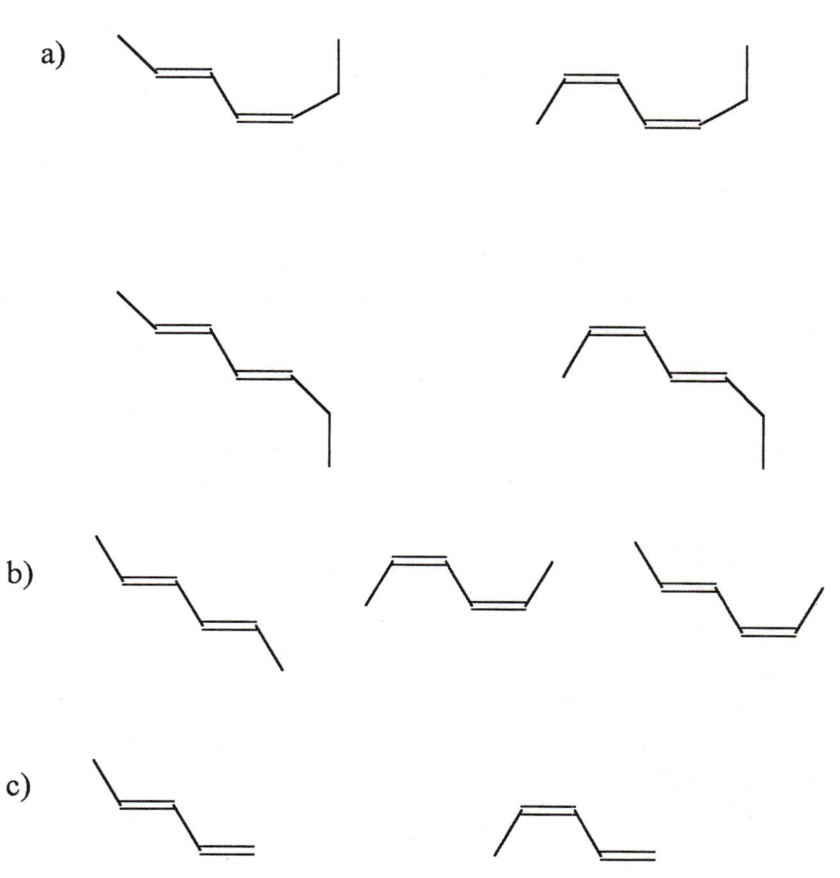

6.31

a) —(benzene ring) b) $-C\equiv N$ d) $-C(CH_3)_2-CH_3$ e) $-C\equiv CH$

6.32 a) *Z* b) *Z* c) *Z* d) *E*

6.33 The energy v. dihedral angle plot for 2-methylpropane is similar to that of propane shown in solution 6.5. However, the energy difference between the high and the low energy conformers of 2-methylpropane is larger than that for propane.

6.34

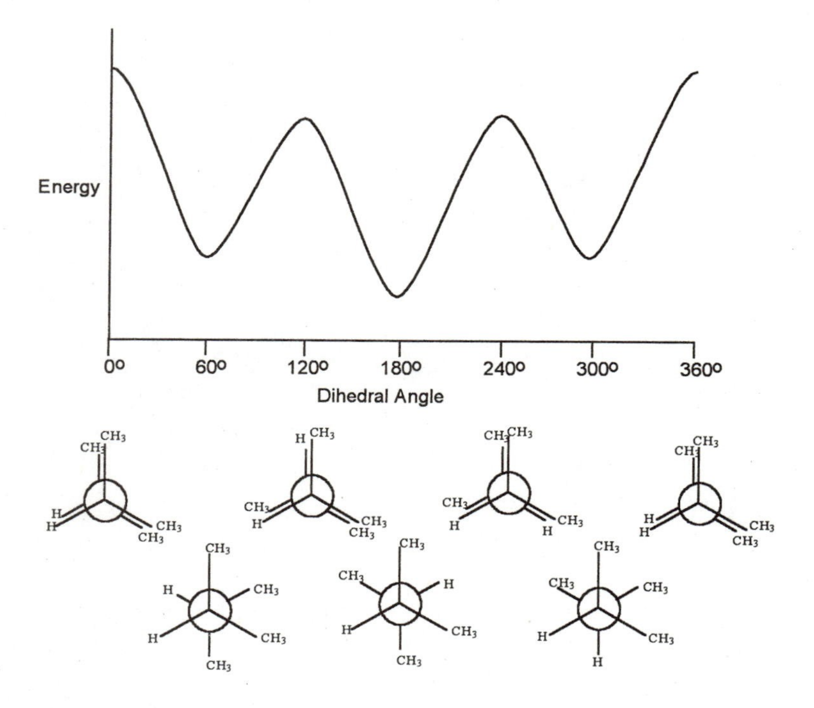

6.35 a) The three sp^3 carbons of cyclopropane form an equilateral triangle with an internuclear angle of 60°. This angle is much smaller than the tetrahedral angle of 109.5° required for the atomic orbitals of the sp^3 hybridized carbons, causing angle strain in the molecule. In addition to the angle strain, the rigid and flat cyclopropane ring also has a significant amount of torsional strain due to the eclipsed conformation about each C-C bond.

b) If the four carbons of the cyclobutane lie flat, this square molecule will have angle strain and a considerable amount of torsional strain due to eclipsed conformations about each C-C bond. As the ring distorts from planarity, the torsional strain decreases while the angle strain increases. However, the increase in angle strain is less than the decrease in torsional strain resulting in a net decrease in ring strain. Therefore, the lowest energy conformer of cyclobutane is slightly nonplanar.

c) The internuclear angle of planar cyclopentane ring is 108°, so it would have little or no angle strain. However, planar cyclopentane would have a considerable amount of torsional strain due to eclipsed conformations at all C-C bonds. By distorting from planarity the cyclopentane ring can relieve most of its torsional strain without increasing its angle strain significantly. Therefore the low energy conformation of cyclopentane is nonplanar and shaped somewhat like an envelope.

d) If cyclohexane were planar, its internuclear angle would be 120^0, which is larger than the tetrahedral angle of 109.5^0. Planar cyclohexane would also have a substantial amount of torsional strain. When the ring distorts from planarity its angle strain is decreased. There are two nonplanar conformations of cyclohexane, called the chair and the boat, that are free of angle strain. The chair conformer is the lowest energy conformation because it has neither torsional nor angle strain. The boat conformer has torsional and steric strain.

e) Cyclodecane is nonplanar. Despite its nonplanar geometry it has some angle strain and some torsional strain. In addition, cyclodecane has transannular strain that results from steric interactions of C-H bonds that point towards the center of the ring.

6.36 Both the -C≡N and -C≡CH groups are linear in geometry, and in the axial position both are aligned parallel to the ring axis, away from other axial substituents. The $-CH_3$ group has a tetrahedral geometry, and one hydrogen of the methyl group must be pointed over the ring to interact with the other axial substituents. Therefore, the axial strain energy is larger for the methyl group.

6.37 The chair conformation with two of the three methyl groups in the equatorial positions is more stable.

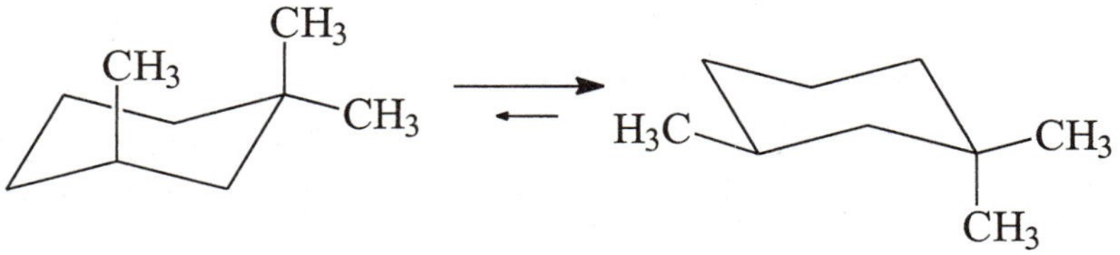

more stable

The axial strain energies listed in Table 6.2 of this chapter are 1,3-diaxial interactions between a group and two axial hydrogens on the same side of the ring. The 1,3-diaxial interactions caused between two bulky groups, such as two methyl groups in this case, will be considerably larger. Therefore, it is not possible to determine the exact energy difference between the conformations shown above using the axial strain values listed in this chapter.

6.38

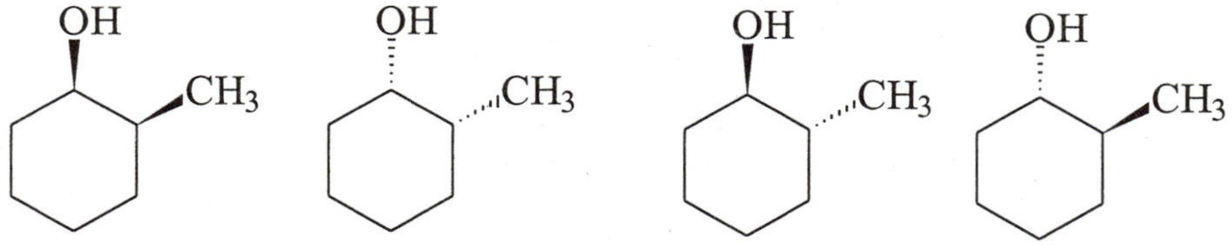

two enantiomers of the *cis*-diastereomer

two enantiomers of the *trans*-diastereomer

Both enantiomers of the *cis*-diastereomer are of equal stability as are both enantiomers of the *trans*-diastereomer. The *trans*-diastereomer of 2-methylcyclohexane is more stable than the *cis* -diastereomer because it has the lowest energy conformation with both substituents equatorial.

6.39

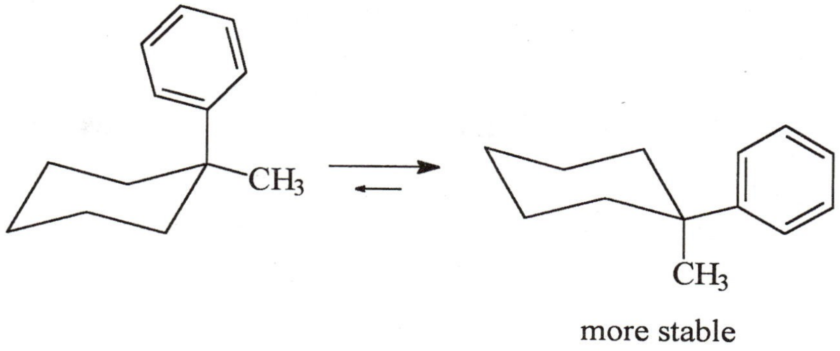

more stable

The axial destabilization energy of the phenyl group (2.9 kcal/mol) is larger than that of the methyl group (1.7 kcal/mol). Therefore the chair conformer with the phenyl group equatorial is more stable than the other conformer by 1.2 kcal/mol.

6.40 In general, bulky substituents prefer to be equatorial on a cyclohexane ring in order to minimize the axial destabilization energy. In this compound, one group is equatorial and the other group is axial. The axial destabilization energy of the bulky *t*-butyl group is so large that it is almost always equatorial. Therefore, the conformation with the methyl group axial will be more stable and will predominate at equilibrium.

6.41 The axial strain energy for chlorine is 0.5 kcal/mole; for methyl is 1.7 kcal/mol; and for isopropyl is 2.2 kcal/mol.

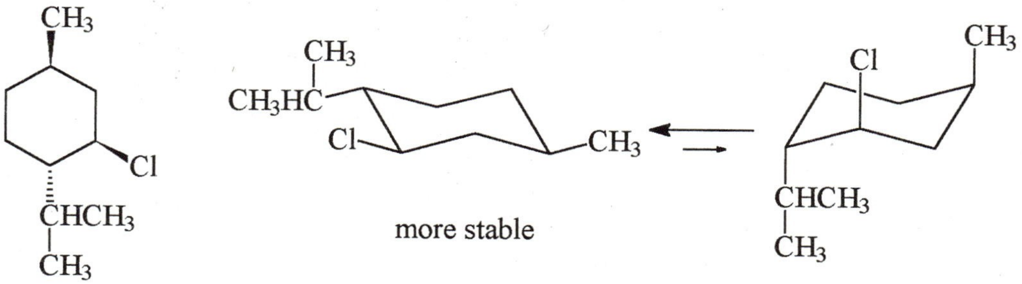

more stable

For this stereoisomer, the groups are either all axial or all equatorial. Obviously the equilibrium is greatly in favor of the conformer with all the groups equatorial. The axial destabilization energy for this conformer is zero.

more stable

For this stereoisomer, the conformer with both the alkyl groups equatorial is more stable. The axial destabilization energy of this conformation is 0.5 kcal/mol. Overall, the first stereoisomer is more stable than the second by 0.5 kcal/mol, due to the axial chlorine in the latter.

6.42 a) R b) R c) R d) R e) S
f) S g) R h) S

6.43

a) 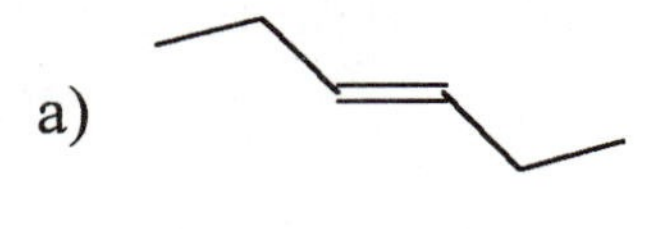

b)

c) 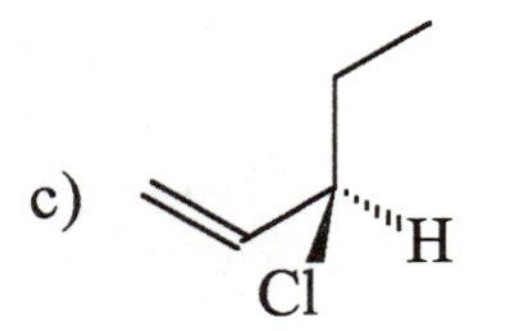

d)

6.44 a) 8 b) 3 c) 8 d) none

6.45

a)

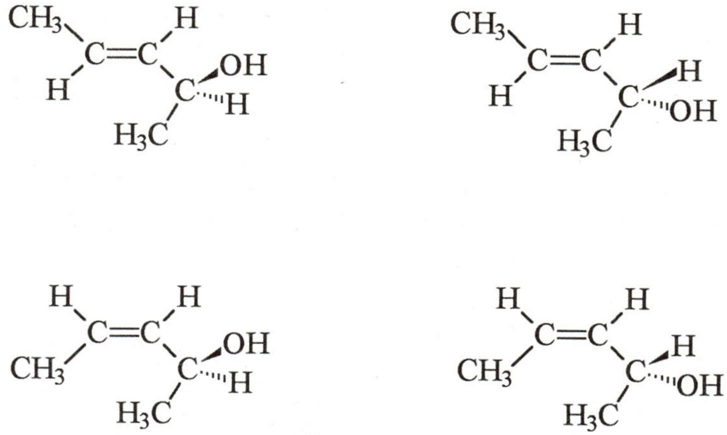

b)

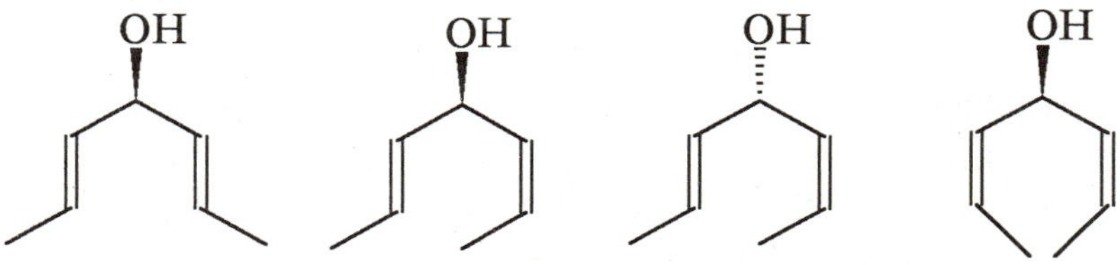

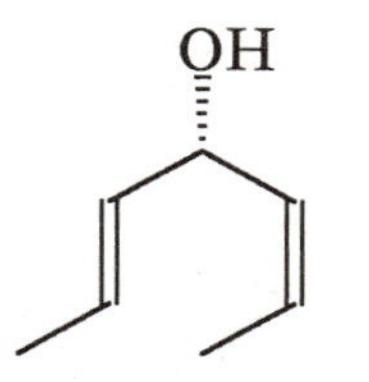

6.46 a) This compound has a plane of symmetry and is a meso compound, so it will not rotate plane polarized light.
b) This is a chiral compound, so it will rotate plane polarized light.
c) This is a meso compound, so it will not rotate plane polarized light.
d) This is a chiral compound, so it will rotate plane polarized light.
e) This has two planes of symmetry and is a meso compound, so it will not rotate plane polarized light.
f) This has a plane of symmetry, so it will not rotate plane polarized light.
g) This has no plane of symmetry so it will rotate plane polarized light
h) This has a plane of symmetry and is a meso compound, so it will not rotate plane polarized light
i) This does not have a plane of symmetry, so it will rotate plane polarized light.

6.47 Enantiomers exhibit different properties only when they are in a chiral environment.

a) True.
b) True.
c) True. Water is achiral.
d) True. They rotate plane polarized light in opposite directions but with the same magnitude.
e) False. Plane polarized light is chiral.
f) True.
g) True. Methanol is achiral.
h) True.
i) False. (S)-2-Butanol is chiral.
j) Cannot be determined. There is no relationship between the direction of rotation of plane polarized light and the absolute configuration.
k) False. The reagents are achiral, so a 50:50 mixture of enantiomers must be produced.

6.48 a) Identical b) Identical c) Identical
d) Enantiomers e) Diastereomers f) Diastereomers
g) Enantiomers h) Diastereomers i) Identical

6.49 a) Diastereomers b) Diastereomers c) Enantiomers
d) Enantiomers e) Identical f) Enantiomers

6.50 a) 66.5° b) +0.7°

c) The observed rotation is directly proportional to the concentration. For example, if the concentration of the solution is halved, the observed rotation will be half that of the rotation observed for the concentrated solution. The results can be distinguished for each case: half of +160° is + 80°; half of -200° is -100°; and half of +520° is + 260°.

6.51 The resolution of this amine using one enantiomer of a chiral carboxylic acid is similar to the scheme described in section 6.16 of the text.

The mixture of the *R* and *S* enantiomers of the amine is reacted with the *S* enantiomer of 2-chloropropanoic acid to produce a mixture of diastereomers of the salt.

Since diastereomers have different physical properties, the mixture can be separated by conventional separation techniques.

Each of the pure diastereomers of the salt is then treated with a strong base to regenerate the pure enantiomers of the amine.

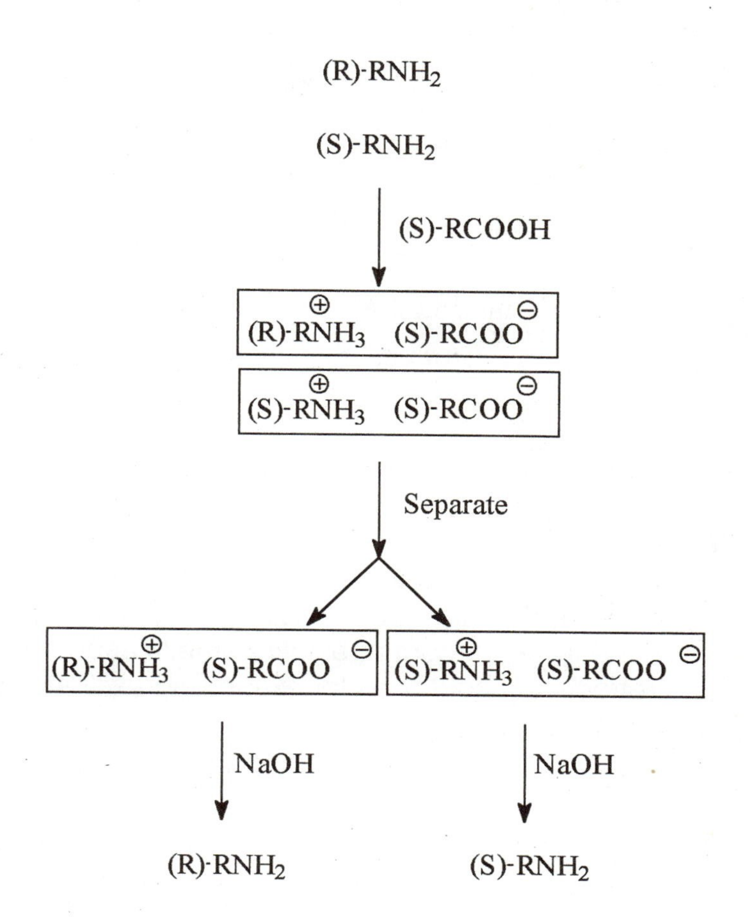

6.52 a) The DU of **X** = 1. Therefore **X** may have a ring or a pi bond.

b) Unknown **X** must have a double bond because it reacts with H_2 to form a saturated compound, **Y**.

c) Compound **X** must be chiral because it rotates plane polarized light. Compound **Y** is achiral.

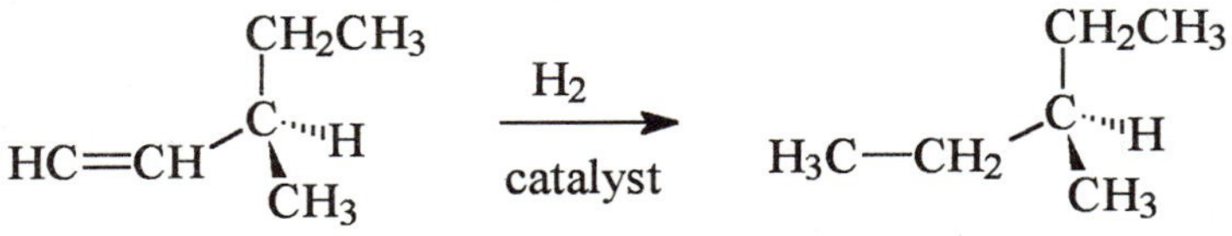

CH_2CH_3 | $HC{=}CH$–C···H, CH_3 $\xrightarrow[\text{catalyst}]{H_2}$ CH_2CH_3 | $H_3C{-}CH_2$–C···H, CH_3

compound **X** compound **Y**

(or enantiomer)

(or enantiomer)

6.53 In a substituted cyclohexane, for intramolecular hydrogen bonding to be effective the group with the polarized hydrogen and the group with the electronegative atom must be held in the proper geometry and proximity. The large amount of axial destabilization energy of a *tert*-butyl group will effectively hold a cyclohexane ring exclusively in the conformation with the *tert*-butyl group equatorial.

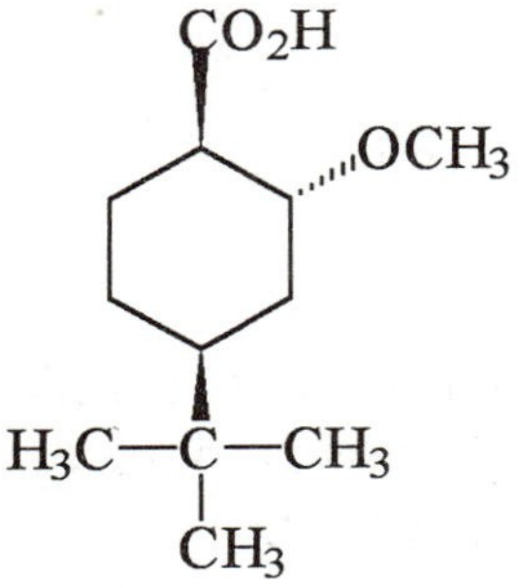

CO_2H OCH_3 $H_3C{-}C{-}CH_3$ CH_3

A

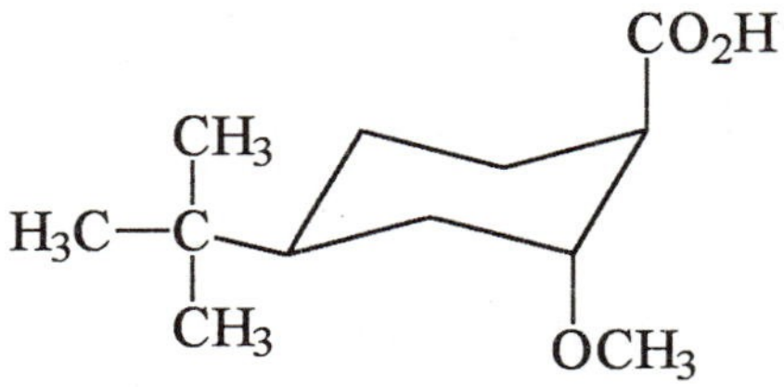

CO_2H CH_3 $H_3C{-}C$ CH_3 OCH_3

more stable conformation

In the more stable chair conformation of isomer **A**, the carboxylic acid group and the methoxy group are axial and on opposite sides of the ring. In this case the groups are too far apart to hydrogen bond.

B

more stable conformation

In isomer **B**, the stable conformation has all the groups equatorial. Therefore the proper geometry for intramolecular hydrogen bonding can be readily attained.

6.54 There are four stereoisomers for this compound.

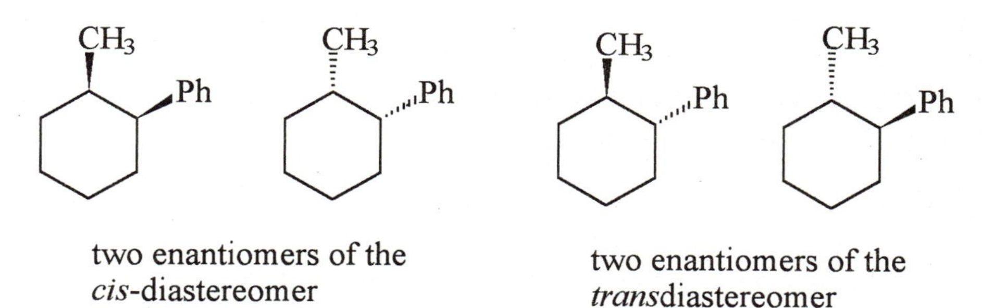

two enantiomers of the *cis*-diastereomer

two enantiomers of the *trans*diastereomer

Both enantiomers of the *cis*-diastereomer are of equal stability as are both enantiomers of the *trans*-diastereomer. The *trans*-diastereomer is more stable than the *cis* -diastereomer because it has the lowest energy conformation with both substituents equatorial. The methyl group is axial in the more stable conformer of the least stable diastereomer because the axial destabilization energy of the phenyl group is larger than that of the methyl group.

6.55

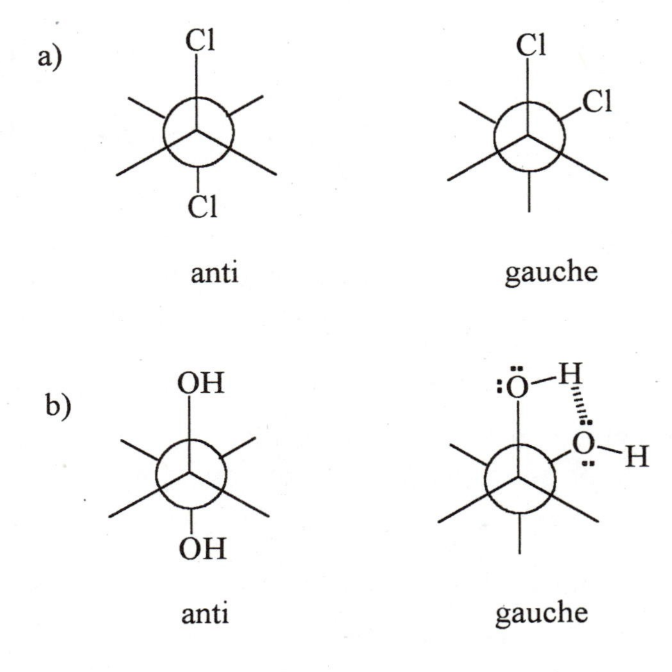

Besides steric and torsional strain, the stabilities of the conformations are influenced by dipole-dipole interactions (a and b) and hydrogen bonding (b).

6.56

CH_3 HO OH CH_3 HO OH CH_3 HO OH

chiral　　　　not chiral

6.57 *cis*-1,2-Dimethylcyclopropane has more steric strain than *trans*-1,2-dimethylcyclopropane. Therefore *trans*-1,2-dimethylcyclopropane is more stable.

6.58 This compound has a conformationally locked boat cyclohexane ring in which carbons 1 and 4 are connected by a CH_2 group (C-7). This compound has both angle strain and torsional strain.

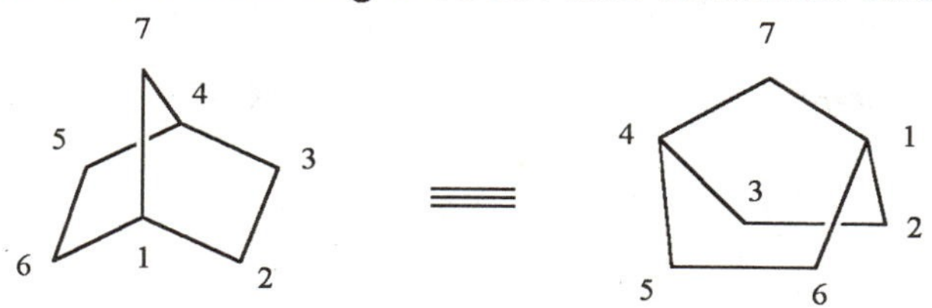

After completing this chapter, you should be able to:

1. Recognize compounds that exist as geometrical isomers and estimate the relative stabilities of these isomers.
   Problems 6.1, 6.2, 6.30.

2. Use the *Z* and *E* descriptors to designate the configurations of geometrical isomers.
   Problems 6.3, 6.4, 6.31, 6.32, 6.43.

3. Determine the conformations about a C-C single bond and estimate their relative energies.
   Problems 6.5, 6.6, 6.33, 6.34.

4. Determine the types and relative amounts of strain present in cyclic molecules.
   Problems 6.35, 6.57.

5. Draw the two chair conformations of cyclohexane derivatives and determine which is more stable.
   Problems 6.7, 6.9, 6.11, 6.36, 6.37, 6.41.

6. Use analysis of conformations to determine the relative stabilities of stereoisomeric cyclohexane derivatives.
   Problems 6.10, 6.12, 6.13, 6.14, 6.38, 6.39, 6.40, 6.54.

7. Identify chiral compounds, locate chiral centers, and determine how many stereoisomers exist for a particular compound.
   Problems 6.16, 6.23, 6.24, 6.44, 6.45, 6.48.

8. Locate any symmetry planes that are present in a molecule.
   Problem 6.17.

9. Designate the configuration of chiral centers as *R* or *S*.
   Problems 6.18, 6.19, 6.29, 6.42, 6.43.

10. Recognize the circumstances under which enantiomers have different properties.
    Problems 6.22, 6.47.

11. Understand when a compound, mixture or solution is optically active and what information this provides about the sample.
    Problems 6.22, 6.26, 6.27, 6.46.

12. Recognize *meso*-stereoisomers.
    Problems 6.25, 6.26, 6.27, 6.48, 6.56.

13. Understand the principles behind the process of separating enantiomers.
    Problem 6.52.

14. Be able to use Fischer projections properly.
    Problems 6.28, 6.29, 6.49.

Chapter 7
NUCLEOPHILIC SUBSTITUTION REACTIONS

7.1 S_N2 reactions always occur with inversion of configuration at the reaction center. The nucleophile approaches the carbon from the side opposite the leaving group.

a) 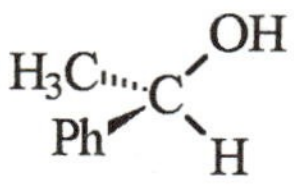 H_3C, Ph, C, OH, H   b) 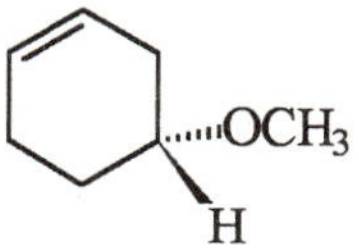 OCH_3, H   c) CH_3, CH_2CH_3, CH_2CH_2OH

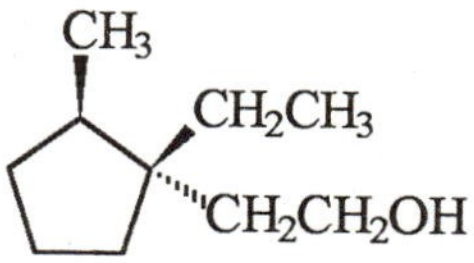

7.2 The rate of the S_N2 reaction is controlled by steric factors at the electrophilic carbon. The approach of the nucleophile is greatly impeded by bulky groups near the electrophilic carbon.

a) The right compound reacts faster because it has less steric hindrance.
b) The compound on the left has two bulky methyl groups on the carbon adjacent to the electrophilic carbon. Therefore, the right compound reacts faster because it has less steric hindrance.
c) A phenyl substituent on the reaction center provides added resonance stabilization to the transition state. Therefore, the left compound reacts faster.

7.3 In general the rate of S_N2 reaction depend on steric effects, and follows the trend: methyl > primary > secondary >> tertiary.

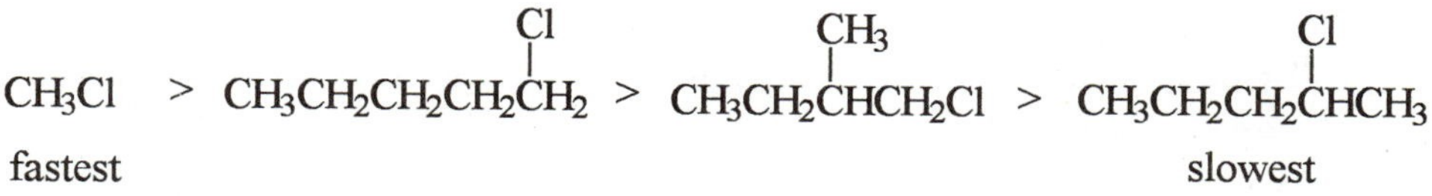

CH_3Cl > $CH_3CH_2CH_2CH_2CH_2Cl$ > $CH_3CH_2CH(CH_3)CH_2Cl$ > $CH_3CH_2CH_2CH(Cl)CH_3$

fastest ... slowest

7.4

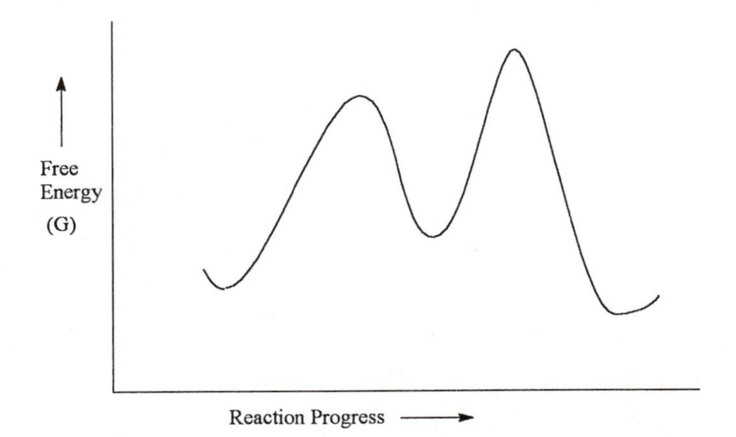

7.6 A common result in S_N1 reactions is racemization because the carbocation intermediate in the S_N1 reaction is sp^2 hybridized and has trigonal planar geometry. However, many S_N1 reactions result in an excess of the product of inversion due to the formation of an ion-pair.

a) Ph, CH_3, CH_3CH_2, OCH_3 on C: $(Ph)(CH_3)(CH_3CH_2)C{-}OCH_3$ + HCl
(racemic, perhaps with some excess inversion)

c) $CH_3CH_2{-}C(H)(H_3C){-}CH_2{-}C(CH_3)_2{-}O\overset{O}{\overset{\|}{C}}CH_3$ + HCl

7.7 For the S_N1 reaction, formation of the carbocation is the rate limiting step. Therefore, any effect that helps to stabilize the carbocation will increase the rate of S_N1 reaction.

a) The left compound is faster because it will produce a more stable tertiary carbocation.
b) The left compound has a faster rate because the carbocation intermediate will be resonance stabilized.
c) The left compound has a faster rate because the carbocation intermediate will be resonance stabilized.
d) The right compound has a faster rate because the methoxy group provides extra resonance stabilization of the carbocation.

7.8 The rate of S_N1 reactions depends on the degree of stabilization of the carbocation intermediate.

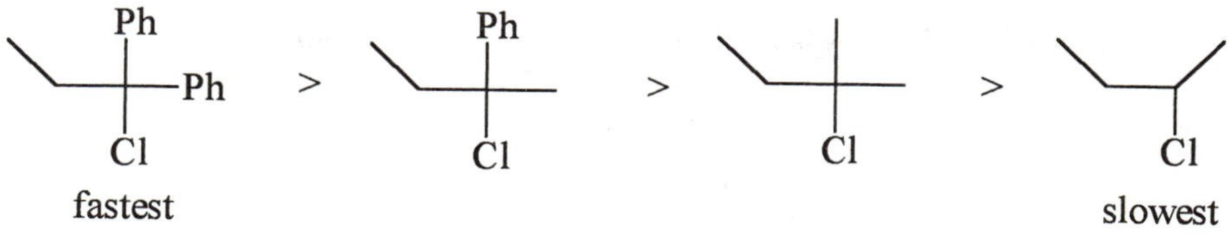

Ph, Ph, Cl (fastest) > Ph, Cl > Cl > Cl (slowest)

7.9 a) Primary substrates with a strong nucleophile (hydroxide ion) react by the S_N2 mechanism. The right reaction is faster because mesylate ion is a better leaving group than chloride ion.
b) Tertiary substrates react by the S_N1 mechanism. The right reaction is faster because iodide ion is a better leaving group than bromide ion.

7.10 a)

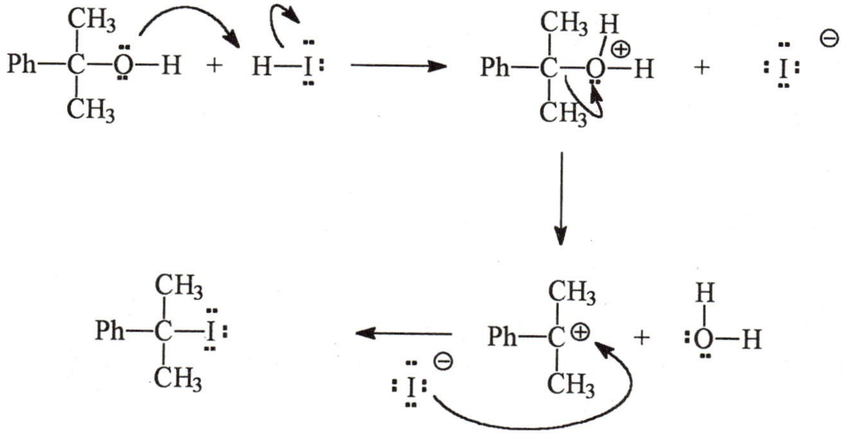

$$Ph-C(CH_3)_2-\ddot{O}-H + H-\ddot{I}: \longrightarrow Ph-C(CH_3)_2-\overset{\oplus}{O}H-H + :\ddot{I}:^{\ominus}$$

$$Ph-C(CH_3)_2^{\oplus} + H-\ddot{O}-H$$

$$:\ddot{I}:^{\ominus} + Ph-C(CH_3)_2^{\oplus} \longrightarrow Ph-C(CH_3)_2-\ddot{I}:$$

b)

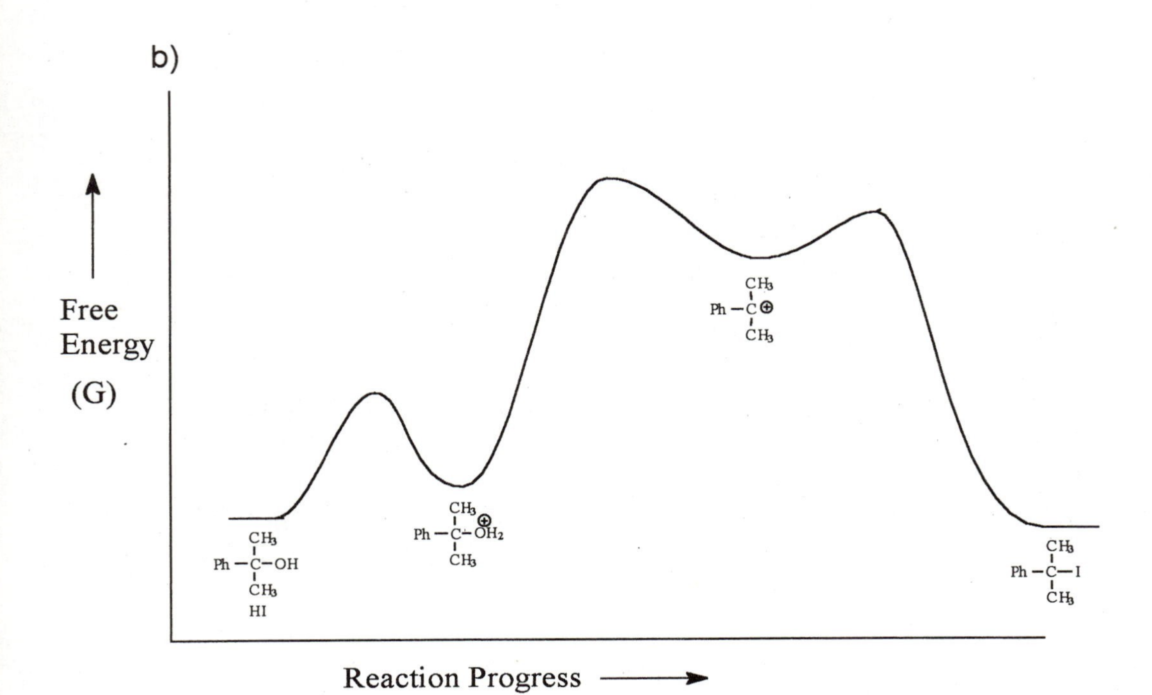

7.11 b) Because the leaving group is on a tertiary carbon, the reaction proceeds by an S_N1 mechanism. The rate determining step is the formation of the carbocation intermediate, so the reactivity of the nucleophile does not affect the rate of an S_N1 reaction. Therefore both reactions proceed at the same rate.

c) Because the leaving group is on a primary carbon, the reaction proceeds by an S_N2 mechanism. CH_3S^- is a stronger nucleophile than CH_3O^- (Nucleophilic strength increases down a column of the periodic table). Therefore, the right reaction is faster.

d) Because the leaving group is on a primary carbon, the reaction proceeds by an S_N2 mechanism. Here again, the only difference between the two reactions is the nucleophile. The left reaction is faster because the nucleophile is stronger. (It is a stronger base.)

7.12

a) S_N2 $CH_3CH_2\underset{}{C}HCH_3$ (with $:\ddot{Cl}:$ on C) $+ {}^{\ominus}:\ddot{S}H \longrightarrow CH_3CH_2CHCH_3$ (with $:\ddot{S}H$ on C) $+ :\ddot{Cl}:^{\ominus}$

b) S_N1 $CH_3CH_2C(CH_3)CH_3$ (with $:\ddot{Br}:$ on C) $\longrightarrow CH_3CH_2\overset{\oplus}{C}(CH_3)CH_3 + :\ddot{Br}:^{\ominus}$

$CH_3CH_2\overset{\oplus}{C}(CH_3)CH_3 + CH_3CH_2\ddot{O}H \longrightarrow CH_3CH_2C(CH_3)CH_3$ (with $H{-}\overset{\oplus}{\ddot{O}}{-}CH_2CH_3$ on C)

$CH_3CH_2C(CH_3)CH_3$ (with $H{-}\overset{\oplus}{\ddot{O}}{-}CH_2CH_3$ on C) $+ :\ddot{Br}:^{\ominus} \longrightarrow CH_3CH_2C(CH_3)CH_3$ (with $:\ddot{O}{-}CH_2CH_3$ on C) $+ H{-}\ddot{Br}:$

c) S_N2 $CH_3CH_2CH_2CH_2CH_2$ (with $:\ddot{O}Ts$ on C) $+ \ddot{N}H_2CH_3 \longrightarrow CH_3CH_2CH_2CH_2CH_2$ (with $\overset{\oplus}{N}H_2CH_3$ on C) $\;\; TsO:^{\ominus}$

d) S_N2 cyclopentyl $:\ddot{Cl}:$ $+ CH_3CH_2C(=O)\ddot{O}:^{\ominus} \longrightarrow$ cyclopentyl $:\ddot{O}C(=O)CH_2CH_3$ $+ :\ddot{Cl}:^{\ominus}$

7.13

$CH_3C(CH_3)CH_3$ (with $:\ddot{Cl}:$ on C) $\longrightarrow CH_3\overset{\oplus}{C}(CH_3)CH_3 + :\ddot{Cl}:^{\ominus}$

$CH_3\overset{\oplus}{C}(CH_3)CH_3 + CH_3\ddot{O}H \longrightarrow CH_3C(CH_3)CH_3$ (with $H{-}\overset{\oplus}{\ddot{O}}{-}CH_3$ on C)

$CH_3C(CH_3)CH_3$ (with $H{-}\overset{\oplus}{\ddot{O}}{-}CH_3$ on C) $+ :\ddot{Cl}:^{\ominus} \longrightarrow CH_3C(CH_3)CH_3$ (with $:\ddot{O}{-}CH_3$ on C) $+ H{-}\ddot{Cl}:$

7.14 The conditions of Figure 7.7 have a more polar solvent (water v. acetic acid) and a weaker nucleophile (water v. acetate anion) than the conditions of Figure 7.8 so the carbocation has a longer lifetime under the conditions of Figure 7.7.

7.15 Solvent polarity dramatically affects the reaction rate. To predict the effect of solvents on reaction rates, one should compare the polarity of the reactants with that of the transition state. If the transition state is more polar than the reactants, the transition state will be more stabilized than the reactants in a polar solvent. This will decrease $\Delta G^‡$, resulting in a faster reaction. On the other hand, if the reactants are more polar than the transition state, increasing solvent polarity will stabilize reactants more, resulting in an increase in the $\Delta G^‡$ and a decrease in the reaction rate. In general, rates of S_N1 reactions are faster in polar solvents while those of S_N2 reactions are faster in aprotic solvents.

b) This is a S_N1 reaction, so the transition state is more polar than the reactant. Therefore, it is faster in the more polar solvent, methanol.
c) This is a S_N2 reaction with a negative nucleophile, so the reactants are more polar than the transition state. Therefore, it is faster in the less polar solvent, pure methanol.
d) This is a S_N2 reaction with a negative nucleophile. Therefore, it is faster in the aprotic solvent, DMSO.

7.16 To determine whether a reaction will follow the S_N1 or the S_N2 mechanism, one should know the conditions that favor each one of these mechanisms.

The S_N1 mechanism is favored when:
the carbocation intermediate is stabilized; the solvent is polar; and only poor nucleophiles are present.

The S_N2 mechanism is favored when:
the electrophilic carbon center is not sterically hindered; the solvent is aprotic; and a good nucleophile present.

b) The substrate is a tertiary alkyl halide. Therefore, the reaction follows the S_N1 mechanism.
d) The leaving group is on a secondary carbon, so the mechanism depends on the nucleophile and the solvent. With a strong nucleophile and an aprotic solvent, the reaction follows the S_N2 mechanism.

e) The substrate is unhindered (less hindered than primary), so S_N2 is the preferred mechanism. It does not matter what the solvent or nucleophile is.

f) The allylic substrate can stabilize the carbocation intermediate. With a weak nucleophile and a polar solvent the reaction follows an S_N1 mechanism.

g) The allylic substrate can react by either mechanism. The nucleophile is strong and the solvent is aprotic, so an S_N2 mechanism is favored.

h) The secondary benzylic substrate provides resonance stabilization of the carbocation intermediate. The weak nucleophile and the polar solvent are ideally suited for the S_N1 mechanism.

7.17 The first step of both of these reactions are acid-base reactions. The resulting conjugate bases undergo intramolecular S_N2 reactions to form the products.

a) $Cl{-}CH_2CH_2CH_2COOH + {}^{\ominus}OH \longrightarrow Cl{-}CH_2CH_2CH_2CO_2^{\ominus} + H_2O$

$\downarrow$

(γ-butyrolactone) $+ Cl^{\ominus}$

b) $Br{-}CH_2CH_2CH_2CH_2CH_2\overset{\oplus}{N}H_3 + {}^{\ominus}OH \longrightarrow Br{-}CH_2CH_2CH_2CH_2CH_2\ddot{N}H_2 + H_2O$

$\downarrow$

(piperidinium, $H_2N^{\oplus}$) $Br^{\ominus}$

7.18 Elimination reactions compete with substitution reactions. The competition occurs because the nucleophile is also a base. When it reacts like a base, it abstracts a proton from the carbon adjacent to the leaving group, resulting in the formation of the elimination product.

a) 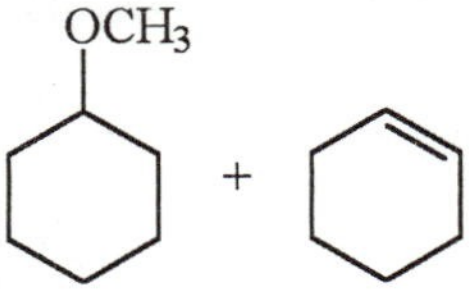 OCH_3 +

b) 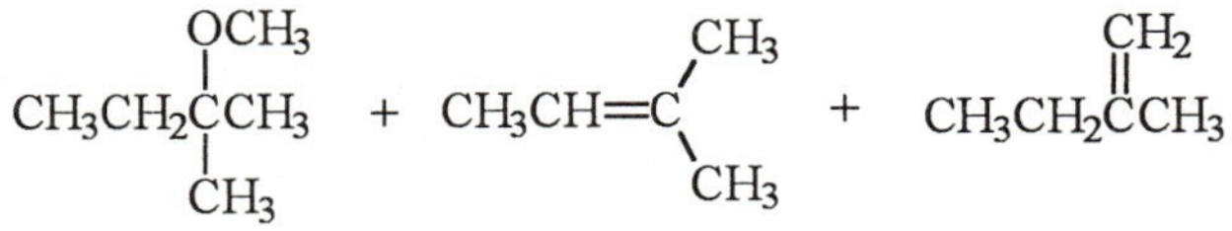 $CH_3CH_2\overset{OCH_3}{\underset{CH_3}{C}}CH_3$ + $CH_3CH{=}C(CH_3)_2$ + $CH_3CH_2\overset{CH_2}{\overset{\|}{C}}CH_3$

c) OH +

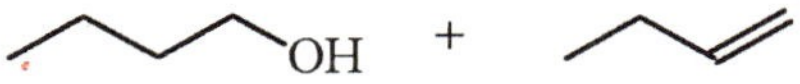

7.19 A carbocation may rearrange to form a more stable carbocation. Such rearrangement occurs by migration of a hydrogen or an alkyl group, with its bonding pair of electrons, from an adjacent carbon to the positively charged carbon.

a) 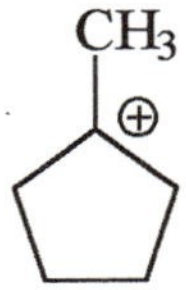 CH_3 ⊕

b) 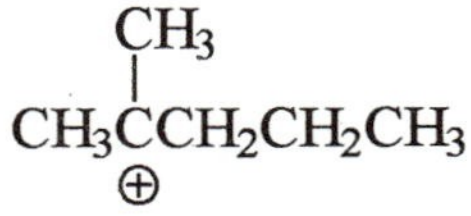 $CH_3\overset{CH_3}{\underset{\oplus}{C}}CH_2CH_2CH_3$

c) CH_3 ⊕

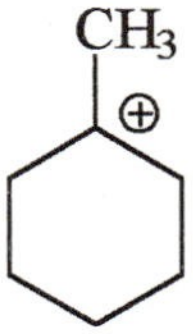

7.20 These are all S_N1 reactions. In (a) and (c) a secondary carbocation intermediate is formed in the initial step. This initially formed carbocation rearranges to form a more stable tertiary carbocation. Only a small amount of product is formed from the initial carbocation, so the major product is formed from the more stable cation. In (b) a resonance stabilized carbocation is formed. The nucleophile becomes bonded to either of the carbons that are positively charged in the resonance hybrid.

a) 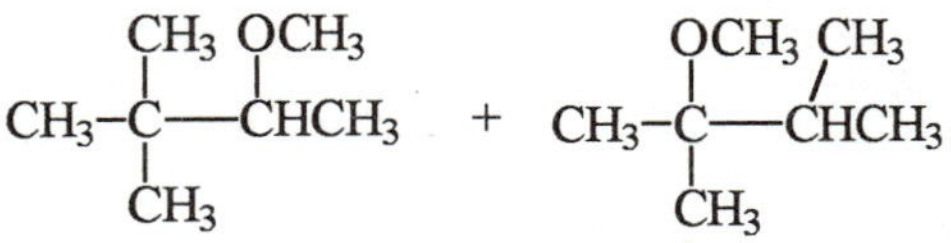 $CH_3{-}\overset{CH_3}{\underset{CH_3}{C}}{-}\overset{OCH_3}{C}HCH_3$ + $CH_3{-}\overset{OCH_3}{\underset{CH_3}{C}}{-}\overset{CH_3}{C}HCH_3$

b) OCH_2CH_3 + OCH_2CH_3

c) OH + OH

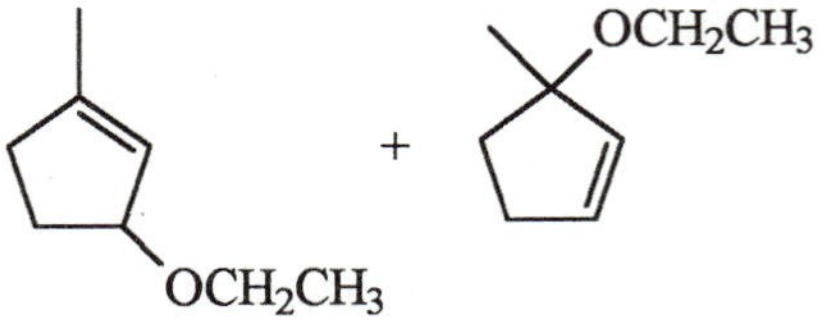

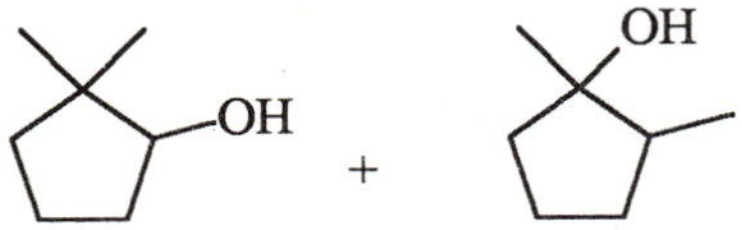

7.21 This reaction follows an S_N1 mechanism with rearrangement.

a)

H_3C, H, :Br:, H → :Br:$^{\ominus}$ + H_3C, H, $\oplus$, H → H_3C, $\oplus$, H, H

↓ $CH_3-\ddot{O}-H$

$H-\ddot{Br}:$ + H_3C, $\ddot{O}-CH_3$ ← :Br:$^{\ominus}$ + H_3C, H, $\oplus\ddot{O}-CH_3$

b)

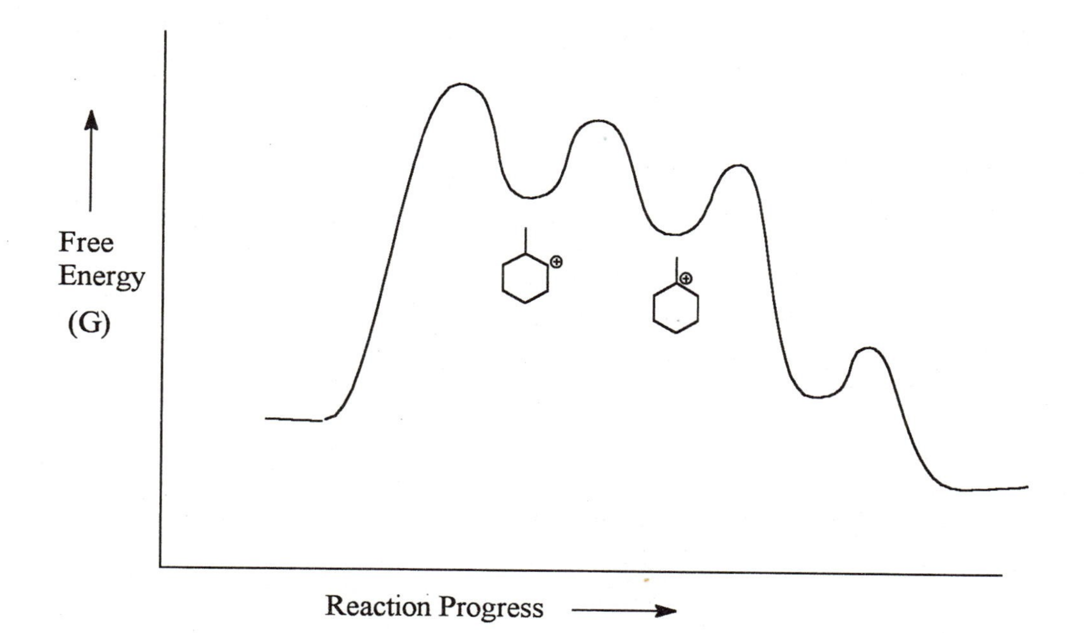

7.22 a) The right compound has a faster rate because iodide ion is a better leaving group than bromide ion.

b) The left compound has a faster rate because the transition state is resonance stabilized.

c) The left compound has a faster rate because the leaving group is on a less hindered primary carbon.

d) The right compound has a faster rate because the leaving group is on a primary carbon.

e) The left compound has a faster rate because the transition state is stabilized by resonance.

7.23 a) The right compound is faster because it will produce a more stable tertiary carbocation.

b) The right compound is faster because iodide ion is a better leaving group than chloride ion.

c) The left compound is faster because the carbocation intermediate will have greater resonance stabilization.

d) The left compound is faster because the carbocation intermediate is resonance stabilized.

e) The carbonyl group adjacent to the electrophilic carbon will destabilize the carbocation intermediate. Therefore, the right compound will react faster.

7.24

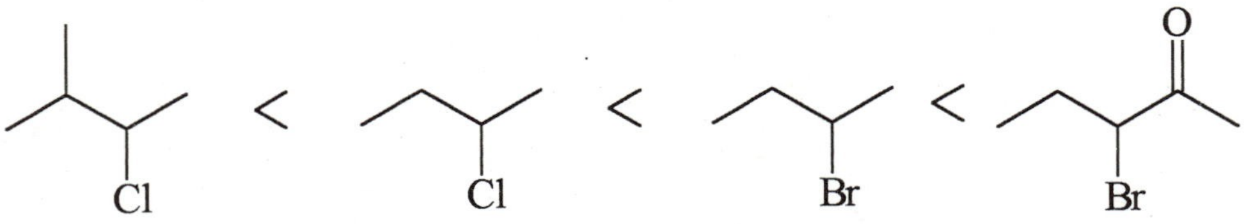

Cl < Cl < Br < O Br

7.25

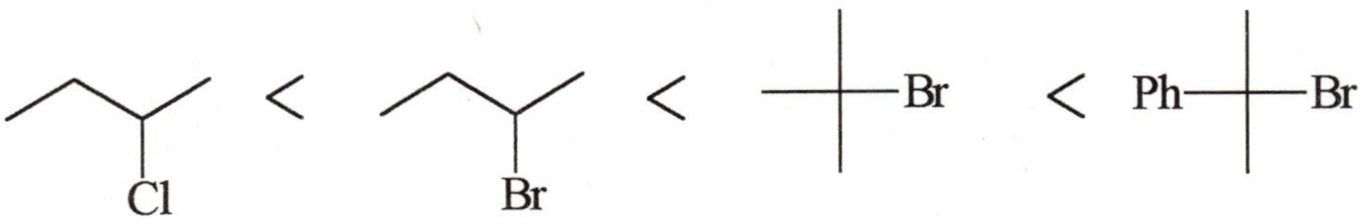

Cl < Br < Br < Ph Br

7.26 a) The substrate is a secondary alkyl halide so the mechanism depends on the nucleophile and the solvent. With a strong nucleophile and an aprotic solvent, the reaction follows the S_N2 mechanism.

OH (structure)

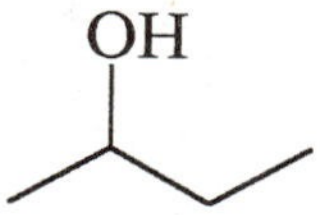

b) The substrate is a secondary alkyl halide so the mechanism depends on the nucleophile and the solvent. A poor nucleophile and a protic solvent favor the S_N1 mechanism.

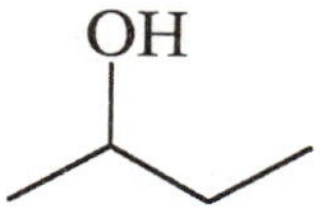

OH (structure)

c) The substrate is a tertiary alkyl halide. Because the S_N2 pathway does not occur at tertiary centers, the reaction follows an S_N1 mechanism.

SH (structure)

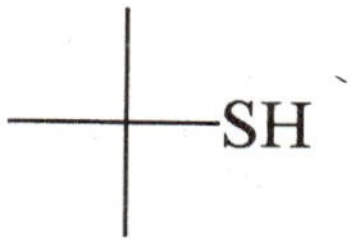

d) The leaving group is on a primary carbon, so S_N2 is the preferred mechanism. It does not matter what the solvent or nucleophile is.

OCH_3 (structure)

e) The allylic substrate can stabilize the carbocation intermediate. With a poor nucleophile and a polar solvent the reaction follows an S_N1 mechanism.

OH (structure)

f) The reaction follows the S_N2 mechanism because of the strong nucleophile in an aprotic solvent .

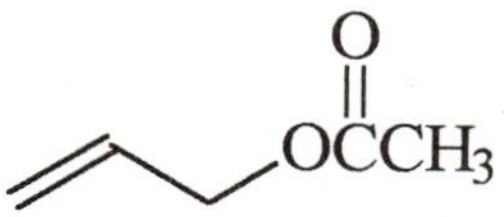

$OCCH_3$ with C=O (structure)

7.27

a) $CH_3C(=O)—O—C(H)(CH_2CH_3)(CH_2CH_2CH_3)$

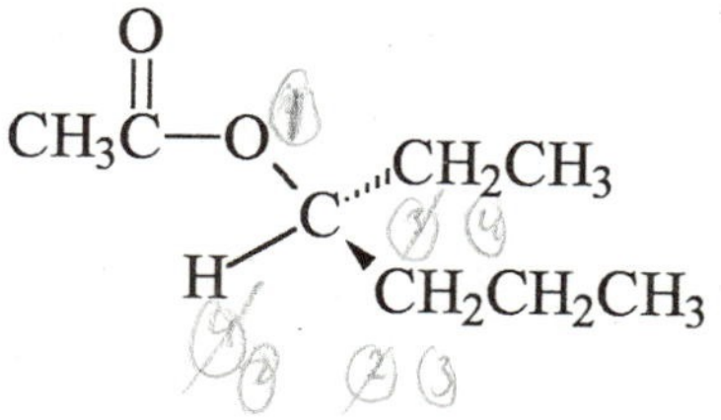

b) CH_3CH_2, CH_3, Ph, OCH_2CH_3 on C

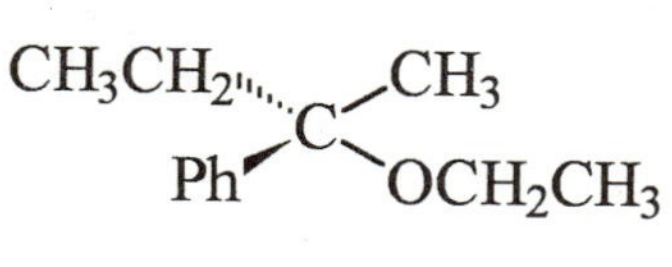

plus enantiomer

c) Ph, OH, CH_3 (cyclohexane) + HO, Ph, CH_3 (cyclohexane)

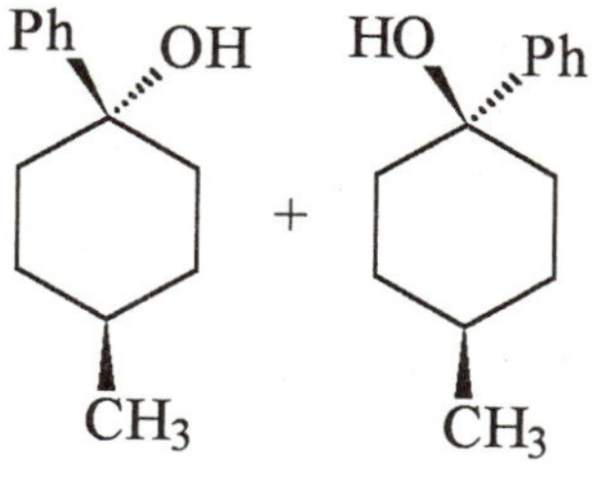

d) CN (structure)

e) 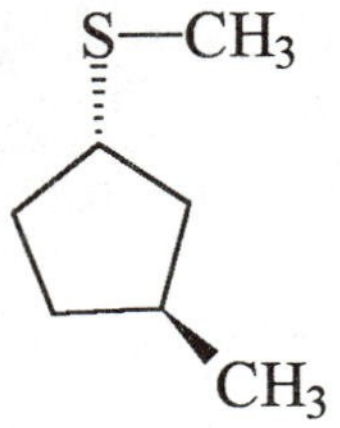 S—CH_3 / CH_3 (cyclopentane ring)

f)  CH_3O ... Cl

g) 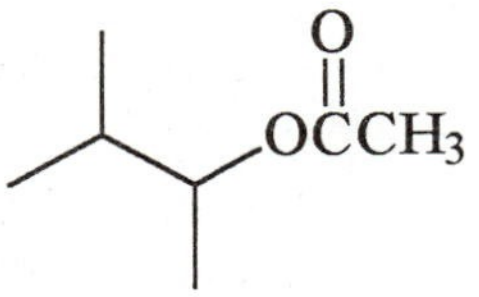 O || OCCH$_3$

h) 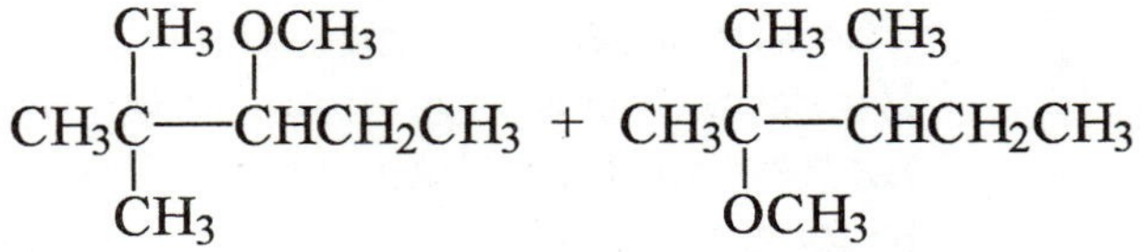 $CH_3C(CH_3)_2$—$CH(OCH_3)CH_2CH_3$ + $CH_3C(CH_3)(OCH_3)$—$CH(CH_3)CH_2CH_3$

major product

i) 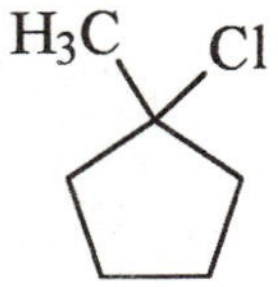 H_3C Cl

j)  $NH_3^{\oplus}$ $I^{\ominus}$

k) 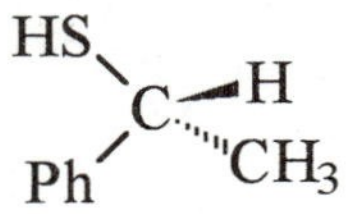 HS, C, H, Ph, CH_3

l) first step 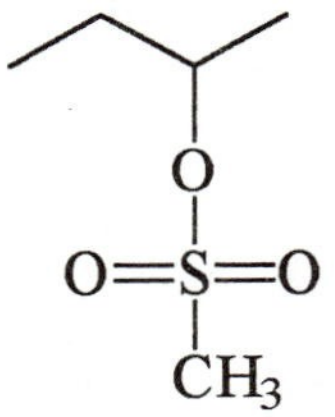 second step

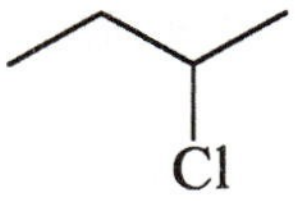

O

O=S=O

CH_3

Cl

m) $CH_3CH_2CH_2CH_2I$

7.28 a) The substrates are tertiary, so they follow the S_N1 mechanism. The right reaction is faster because bromide ion is a better leaving group than chloride ion.

b) The substrates are primary, so they follow the S_N2 mechanism. The left reaction is faster because the nucleophile is more reactive in an aprotic solvent.

c) A secondary substrate, poor nucleophile, resonance stabilization of the carbocation, and a polar solvent, favor the S_N1 mechanism . The right reaction is faster because the methyl group provides added stability to the resonance stabilized cation.

d) Primary substrates follow the S_N2 mechanism. The right reaction is faster because of less steric hindrance.

e) Secondary substrates with a poor nucleophile and a polar solvent favor the S_N1 mechanism . The left reaction is faster because the carbocation is resonance stabilized.

f) Primary substrates follow the S_N2 mechanism. The left reaction is faster because the nucleophile is stronger.

g) Tertiary substrates follow the S_N1 mechanism. The left reaction will be faster because methanol is more polar than ethanol.

h) The leaving group is bonded to a primary carbon, so these reactions follow the S_N2 mechanism. The left reaction is faster because intramolecular reactions are favored by entropy over intermolecular reactions.

i) Primary substrates follow the S_N2 mechanism. The left reaction is faster because the nucleophile is less hindered.

7.29 a)

$(CH_3)_3C-Br \longrightarrow (CH_3)_3C^{\oplus}\ \ :\ddot{O}(H)CH_2CH_3 \longrightarrow (CH_3)_3C-\overset{\oplus}{\ddot{O}}(H)CH_2CH_3\ \ :\ddot{O}(H)CH_2CH_3$

$\longrightarrow (CH_3)_3C-\ddot{O}CH_2CH_3 \quad H_2\overset{\oplus}{O}CH_2CH_3$

b)

$$CH_3CH_2CH_2CH_2{-}Cl + {:}\ddot{O}CH_3^{\ominus} \longrightarrow CH_3CH_2CH_2CH_2{-}\ddot{O}CH_3 + Cl^{\ominus}$$

c)

$$CH_3CH_2C(OH)(CH_3)CH_3 + H{-}\overset{\oplus}{O}H_2 \longrightarrow CH_3CH_2C(\overset{\oplus}{O}H_2)(CH_3)CH_3 \longrightarrow CH_3CH_2\overset{\oplus}{C}(CH_3)CH_3 + H_2O$$

$$CH_3CH_2\overset{\oplus}{C}(CH_3)CH_3 + {:}\ddot{Br}{:}^{\ominus} \longrightarrow CH_3CH_2C(Br)(CH_3)CH_3$$

7.30

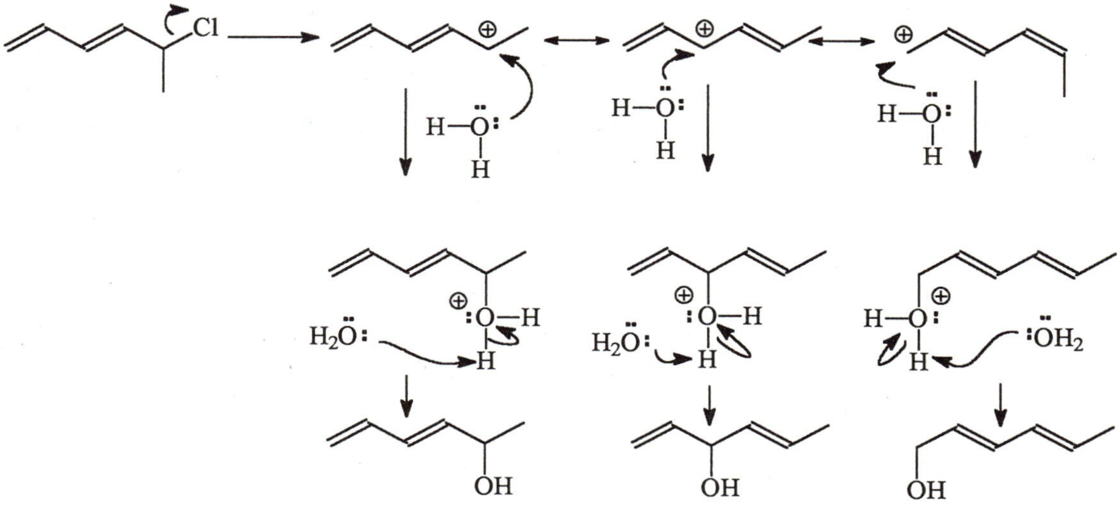

7.31 The trifluorosulfonate anion is a better leaving group because it is a weaker base due to the inductive stabilization effect of the CF_3 group.

7.32 The reaction follows the S_N1 mechanism, so the rate does not depend on the nucleophile. However, bromide ion is a better nucleophile than methanol. Therefore, when bromide ion is added to the reaction, the product is benzyl bromide rather than benzyl methyl ether.

7.33 Bromomethane is an unhindered substrate (less hindered than primary), so S_N2 is the preferred mechanism. This reaction is faster with hydroxide ion because it is a much stronger nucleophile than water. The tertiary substrate, 2-bromo-2-methylpropane, follows the S_N1 mechanism. The rate of a S_N1 reaction does not depend on the strength of the nucleophile.

7.34 The carbocation formed from this substrate is stabilized by resonance.

$$CH_3-\ddot{O}-\overset{\oplus}{C}H_2 \longleftrightarrow CH_3-\overset{\oplus}{O}=CH_2$$

7.35 The conditions for this reaction are favorable for a S_N1 mechanism. Cleavage of the ether occurs so as to form the more stable carbocation intermediate.

$$(CH_3)_3C{-}\ddot{O}{-}CH_3 + H{-}I \longrightarrow (CH_3)_3C{-}\overset{\oplus}{O}(H){-}CH_3 + :\ddot{I}:^{\ominus} \longrightarrow CH_3{-}\overset{\oplus}{C}(CH_3){-}CH_3 + H{-}\ddot{O}{-}CH_3$$

3° carbocation

$$\downarrow$$

$$CH_3{-}C(CH_3)_2{-}I$$

7.36 The reaction conditions, a weak nucleophile and a polar solvent, favor a S_N1 mechanism.

a) The left compound will give a precipitate more rapidly because the bromide ion is a better leaving group than chloride ion.
b) The left compound will give a precipitate more rapidly because the carbocation is resonance stabilized.
c) The right compound will give a precipitate more rapidly because a tertiary carbocation is more stable than a secondary carbocation.
d) The right compound will give a precipitate more rapidly because bromobenzene does not undergo nucleophilic substitution reactions.

7.37 The reaction conditions, a good nucleophile and an aprotic solvent, favor a S_N2 mechanism.

a) The right compound will give a precipitate more rapidly because the chlorine is attached to a less hindered secondary carbon.
b) The right compound will give a precipitate more rapidly because bromide ion is a better leaving group than chloride ion.
c) The right compound will give a precipitate more rapidly because it is a less hindered primary alkyl halide.
d) The left compound will give a precipitate more rapidly because the compound on the right has more steric hindrance due to the two methyl groups on the carbon adjacent to the electrophilic carbon.

7.38 The reaction conditions for Lucas test favor a S_N1 mechanism. Therefore the rate of the reaction depends on the stability of the carbocation intermediate.

a) right compound b) right compound
c) left compound

7.39 The chlorine on the carbon next to the carbon with the hydroxy group will destabilize, the carbocation intermediate by its inductive effect.

7.40 a) Propanol as nucleophile and as solvent (S_N1).
b) Acetate ion as nucleophile in an aprotic solvent (S_N2).
c) Acetate ion as nucleophile in a protic solvent such as acetic acid (S_N1).
d) Methyl amine as nucleophile (S_N2).
e) Hydrobromic acid as catalyst and nucleophile, water as solvent (S_N1).

7.41 This reaction occurs because the negative oxygen of the leaving group is stabilized by resonance with the benzene ring and the nitro group.

7.42 The substrate in both cases is a primary alkyl halide, so the S_N2 mechanism is favored. The right reaction is faster because the nucleophile is less sterically hindered.

7.43

$$CH_3CH_2-\ddot{O}-H \xrightarrow{H-A} CH_3CH_2-\overset{H}{\overset{|}{O^{\oplus}}}-H \xrightarrow{CH_3CH_2-\ddot{O}-H} CH_3CH_2-\underset{CH_2CH_3}{\overset{\oplus}{O}}-H \quad :A^{\ominus}$$

$$\downarrow$$

$$CH_3CH_2-O-CH_2CH_3$$

7.44

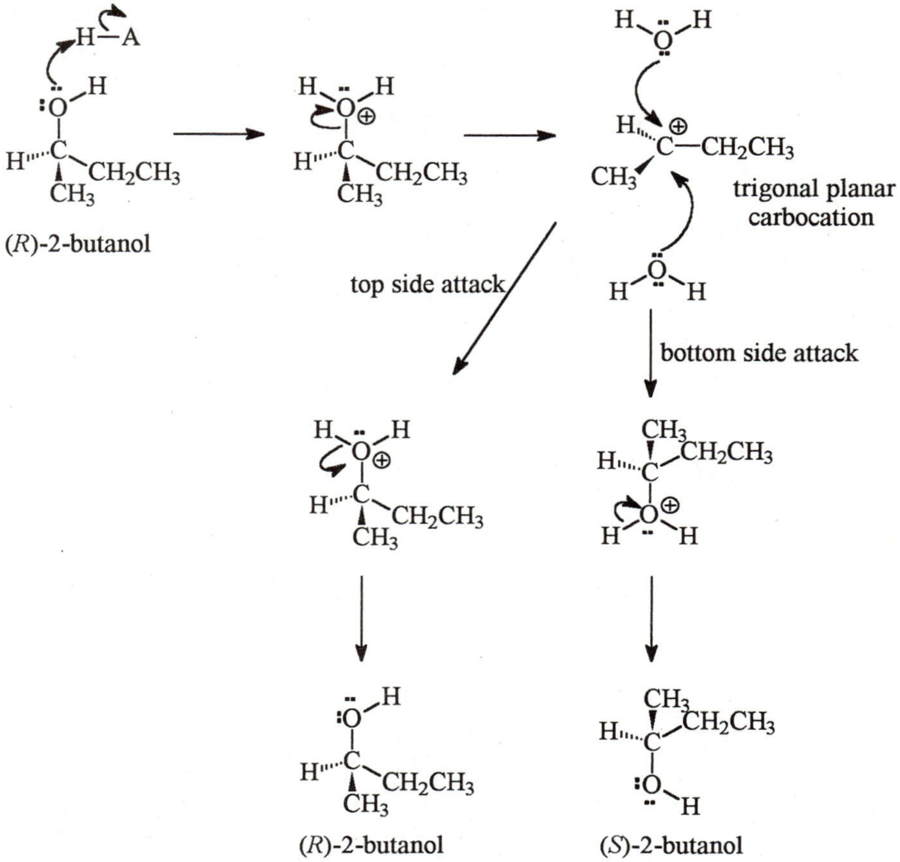

H—A

H_2O

trigonal planar carbocation

(*R*)-2-butanol

top side attack

bottom side attack

(*R*)-2-butanol

(*S*)-2-butanol

7.45

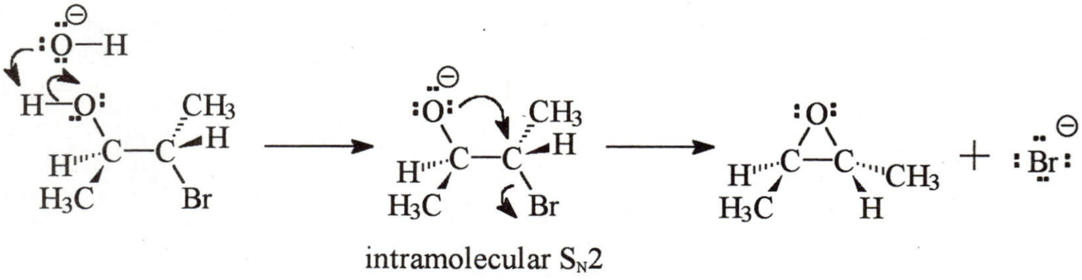

intramolecular S_N2

This is an intramolecular S_N2 reaction, where the nucleophile and the carbon attached to the leaving group must assume the proper configuration for **backside attack**. This requirement establishes the stereochemical outcome of this reaction.

7.46 substitution product elimination products

a)

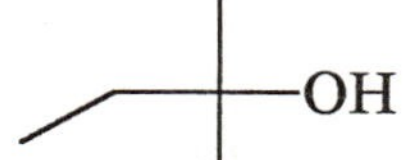

b) OCH$_3$

c)

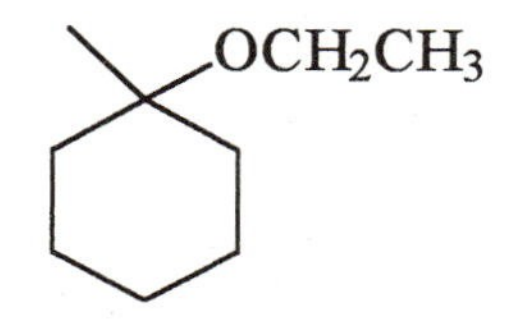

7.47

$CH_3C(CH_3)_2$—CHICH$_3$ ⟶ $CH_3C(CH_3)_2$—$\overset{\oplus}{C}H$—CH$_2$(H) :B ⟶ $CH_3C(CH_3)_2$—CH=CH$_2$

+ I$^{\ominus}$

$CH_3C(CH_3)_2$—$\overset{\oplus}{C}$HCH$_3$ ⟶ CH$_3\overset{\oplus}{C}$(CH$_2$—H)—C(CH$_3$)(H)CH$_3$ B: ⟶ CH$_3$C(=CH$_2$)—C(CH$_3$)(H)CH$_3$

methyl shift

⟶ CH$_3\overset{\oplus}{C}$(CH$_3$)—C(CH$_3$)(H)CH$_3$:B ⟶ CH$_3$C(CH$_3$)=C(CH$_3$)CH$_3$

The substitution products are:

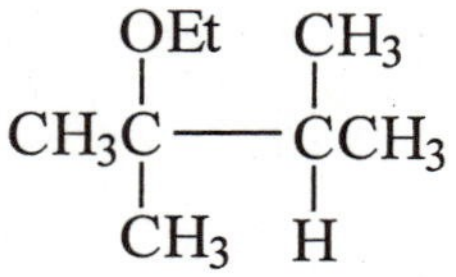

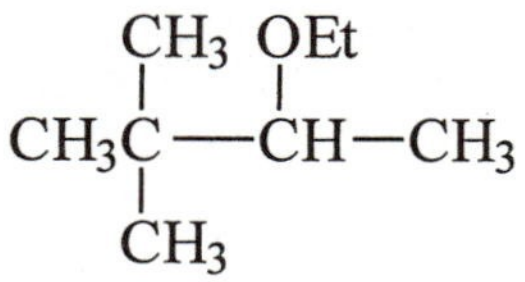

$$CH_3C(OEt)(CH_3){-}C(CH_3)(H)CH_3 \qquad CH_3C(CH_3)(CH_3){-}CH(OEt){-}CH_3$$

7.48

CH_3

⊕

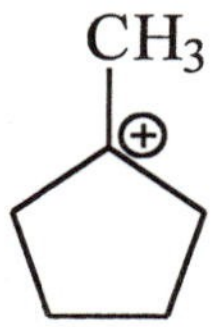

7.49 a) The reaction favors the S_N2 mechanism, so the rate of this reaction depends on the concentrations of both the alkyl chloride and the hydroxide ion. Therefore, the rate will double if the hydroxide ion concentration is doubled, and will decrease by a factor of 4 if the concentrations of both the alkyl chloride and the hydroxide ion are halved.

b) This reaction favors the S_N1 mechanism, so the rate of the reaction depends only on the concentration of the alkyl chloride. Therefore, the rate will be unaffected if the concentration of the hydroxide ion is doubled, and the rate will be halved if the concentrations of both the alkyl chloride and the hydroxide ion are halved.

7.50

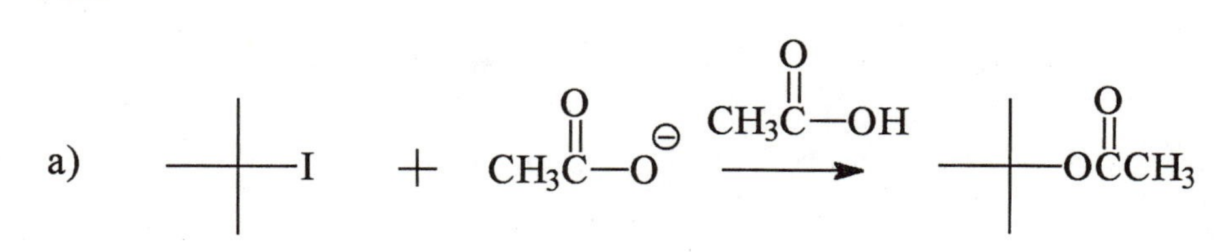

b) $CH_3CH(Br)CH_2CH_3 + {}^{\ominus}SCH_3 \xrightarrow{\text{DMSO}} CH_3CH(SCH_3)CH_2CH_3$

$CH_3CH(S^{\ominus})CH_2CH_3 + CH_3Cl \longrightarrow CH_3CH(SCH_3)CH_2CH_3$

c) $CH_3(CH_2)_5Br + CH_3O^{\ominus} \xrightarrow{\text{aprotic solvent}} CH_3(CH_2)_5OCH_3$

$CH_3(CH_2)_5O^{\ominus} + CH_3Cl \longrightarrow CH_3(CH_2)_5OCH_3$

7.51

a) OH-substituted methylcyclopentane $\xrightarrow[\text{TsCl}]{\text{pyridine}}$ OTs-substituted methylcyclopentane $+ \overset{\ominus}{S}H$ $\xrightarrow[S_N2]{\text{DMSO}}$ SH-substituted methylcyclopentane (inverted)

b) $CH_3CH_2C(Cl)(H)CH_3 + \overset{\ominus}{O}H \xrightarrow{\text{acetone}} CH_3CH_2C(H)(OH)CH_3$ (inverted)

$\xrightarrow[\text{pyridine}]{\text{TsCl}} CH_3CH_2C(H)(OTs)CH_3 \xrightarrow[CH_3\overset{\ominus}{S}]{\text{DMSO}} CH_3CH_2C(H)(SCH_3)CH_3$

7.52 The carbon of the cyanide anion is the stronger nucleophile because it is more basic.

7.53

(resonance structures: primary allylic carbocation $\longleftrightarrow$ tertiary carbocation)

The carbocation formed in this reaction has two resonance structures. The resonance structure to the right contributes more to the resonance hybrid because it is more stable. The nucleophile can bond to either carbon where some positive charge is located. Reaction at the tertiary carbon is favored because more positive charge is located there. Reaction at the primary carbon is favored because it is less hindered and allows the nucleophile to approach more readily.

7.54

$$CH_3CH_2\overset{\oplus}{C}H{-}CH_2 \ (\text{H on C2}) \xrightarrow[\text{shift}]{\text{hydride}} CH_3{-}CH_2\overset{\oplus}{C}HCH_3 \xrightarrow[\text{shift}]{\text{methyl}} \overset{\oplus}{C}H_2{-}\underset{CH_3}{\overset{H}{C}}{-}CH_3$$

$$\Big\downarrow \text{hydride shift}$$

$$CH_3{-}\underset{CH_3}{\overset{\oplus}{C}}{-}CH_3$$

After completing this chapter, you should be able to:

1. Write mechanisms for the S_N1 and S_N2 reactions.
   Problems 7.10, 7.12, 7.13, 7.21, 7.29, 7.35, 7.43, 7.44, 7.45.

2. Recognize the various nucleophiles and leaving groups and understand the factors that control their reactivities.
   Problems 7.11, 7.31, 7.32, 7.33, 7.41, 7.42, 7.52.

3. Understand the factors that control the rates of these reactions, such as steric effects, carbocation stabilities, the nucleophile, the leaving group, and solvent effects.
   Problems 7.2, 7.3, 7.7, 7.8, 7.15, 7.22, 7.23, 7.24, 7.25, 7.34, 7.39.

4. Be able to use these factors to predict whether a particular reaction will proceed by an S_N1 or an S_N2 mechanism and to predict what effect a change in reaction conditions will have on the reaction rate.
   Problems 7.9, 7.16, 7.28, 7.36, 7.37, 7.38.

5. Show the products of any substitution reaction.
   Problems 7.17, 7.18, 7.26, 7.27, 7.30, 7.40, 7.50, 7.51.

6. Show the stereochemistry of the products.
   Problems 7.1, 7.6, 7.27.

7. Recognize when a carbocation rearrangement is likely to occur and show the products expected from the rearrangement.
   Problems 7.19, 7.20, 7.47.

8. Show the structures of the products that result from the elimination reactions that compete with the substitution reactions.
   Problems 7.18, 7.46.

Chapter 8
ELIMINATION REACTIONS

8.1 An elimination reaction occurs when a proton and the leaving group are lost from adjacent carbons, resulting in the formation of a double bond.

a) $CH_3CH{=}CHCH_2CH_3$ b) c) +

8.2 This reaction follows the E2 mechanism, and anti elimination is preferred. Only the *(E)*-isomer of the alkene is produced from the (1*R*,2*S*)-diastereomer of the bromide.

H_3C Ph
C=C
Ph H

8.3 In E2 reactions, an anti-coplanar geometry of the leaving group and the hydrogen is required. Therefore, when trying to determine the elimination product, one should first draw the conformation with the hydrogen and the leaving group anti.

b) The hydrogen and the bromine are in the anti-coplanar geometry in the conformation shown. Anti elimination gives the *(Z)*-isomer of the product.

CH_3CH_2 Br
Ph–C–C–H
H Ph
⟶
CH_3CH_2 H
C=C
Ph Ph

c) In this case the a rotation of 60° about the C-C bond is needed to bring the hydrogen and the bromine to the required anti-coplanar geometry. Anti elimination then gives the (*E*)-isomer of the product.

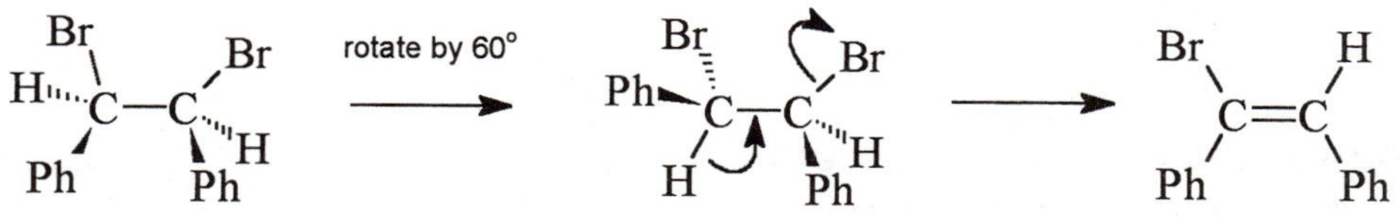

rotate by 60°

8.4 More of the *trans*-alkene is formed because the conformation that produces the *trans*-alkene is more stable because the bulky phenyl groups are anti.

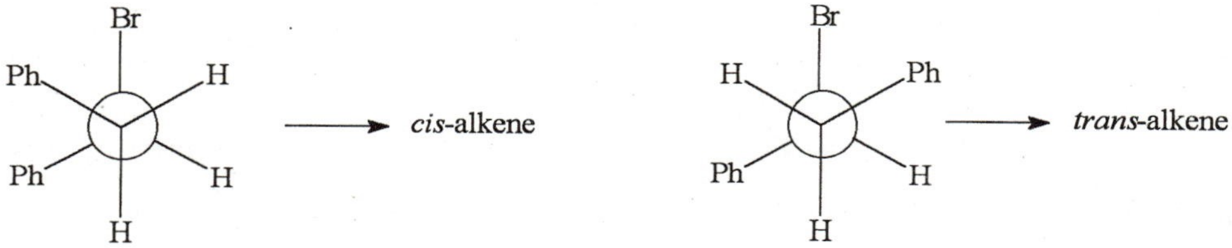

cis-alkene

trans-alkene

8.5 For anti elimination to occur in a cyclohexane ring, the hydrogen and the leaving group must be **trans** and both must be **axial**.

a) In this compound only one hydrogen (circled) is in the *trans*-diaxial arrangement with the chlorine that is required for elimination. This results in the production of a single alkene.

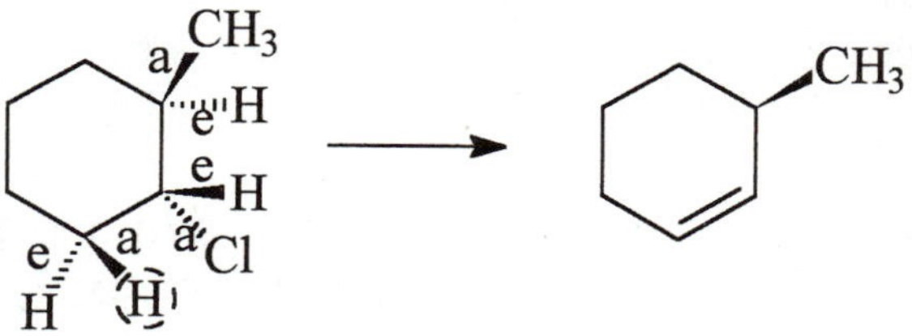

b) In this compound the bromine and two hydrogens (circled) are in a *trans*-diaxial arrangement. This results in the production of two isomeric alkenes.

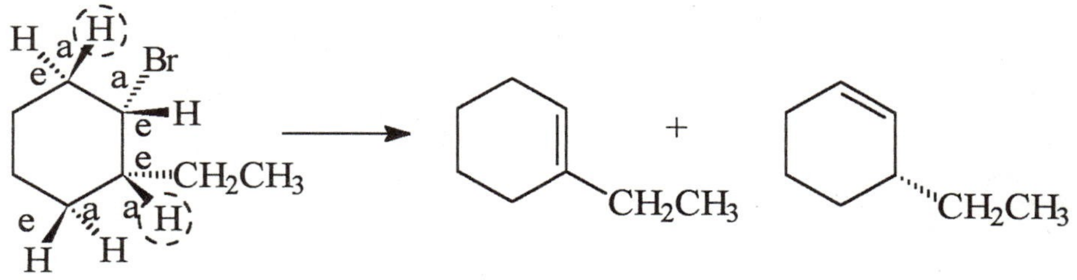

8.6 The bulky *t*-Bu group must be equatorial in both cases. For elimination to occur the leaving group, OTs, and a hydrogen on the adjacent carbon must be in a *trans*-diaxial arrangement. The isomer on the left has the OTs group axial, so it can readily undergo E2 elimination. The isomer on the right has the OTs group equatorial, so it cannot readily react.

8.7 The structural isomers that are often produced in elimination reactions have the double bond in different positions. The major product of these reactions can be predicted using Zaitsev's rule, which states that *the major product is the more highly substituted alkene.*

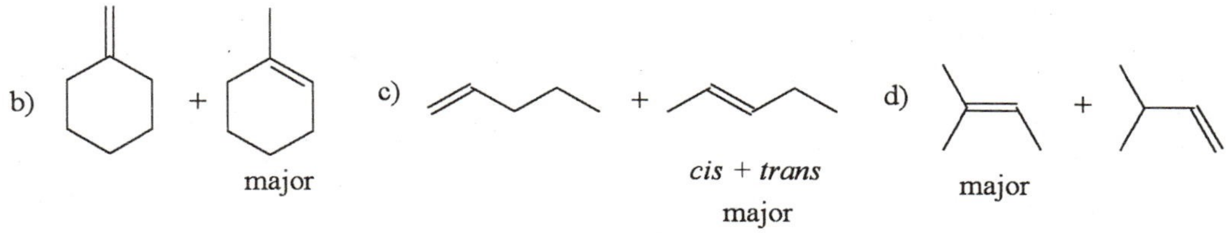

b) + major c) + *cis* + *trans* major d) + major

8.8 In this case (*E*)-2-butene is the major product because it is more stable and the conformation leading to it is more stable.

8.9 While most E2 reactions follow Zaitsev's rule, there are some exceptions. One such exception is observed with compounds that have a quaternary nitrogen atom, and is known as the Hofmann elimination. This type of elimination reaction follows Hofmann's rule, which states that *the major product is the least substituted alkene.*

b) $CH_3CH_2CH_2CH{=}CHCH_3$
+
$CH_3CH_2CH_2CH_2CH{=}CH_2$ major

c) 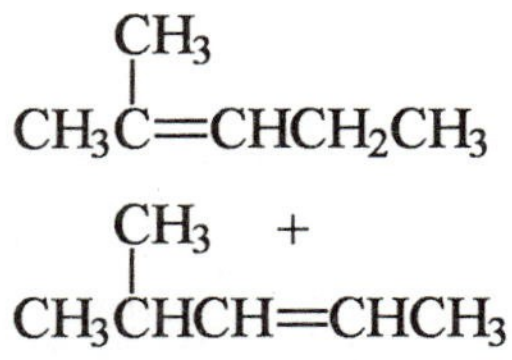 $CH_3C(CH_3){=}CHCH_2CH_3$
+
$CH_3CH(CH_3)CH{=}CHCH_3$ major

d) 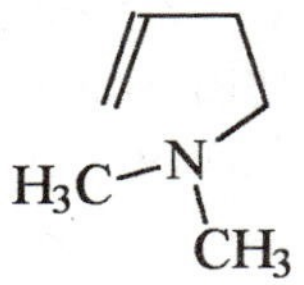 H₃C–N–CH₃ (ring structure)

e) 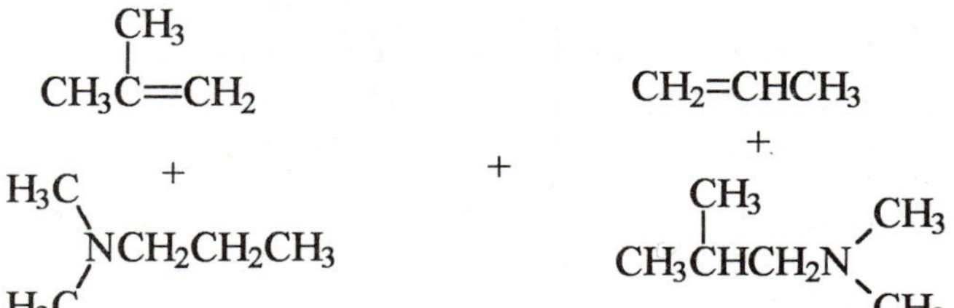 $CH_3C(CH_3){=}CH_2$
+
$(H_3C)_2NCH_2CH_2CH_3$
+
$CH_2{=}CHCH_3$
+
$CH_3CH(CH_3)CH_2N(CH_3)_2$
major

8.10

a)  Ph

b) 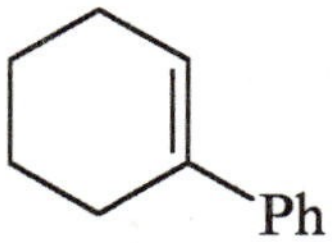 Ph

c) 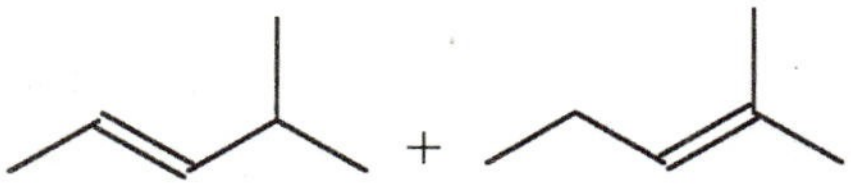 + major

d) 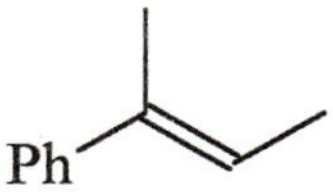 Ph

8.11

$$CH_3-\underset{CH_3}{\overset{CH_3}{C}}-\ddot{O}-H + H-\ddot{O}-SO_3H \longrightarrow CH_3-\underset{CH_3}{\overset{CH_3}{C}}-\overset{H}{\overset{\oplus}{O}}-H + :\overset{\ominus}{\ddot{O}}-SO_3H$$

$$\downarrow$$

$$H_3\overset{\oplus}{O}: + CH_2=C(CH_3)_2 \longleftarrow H_2\ddot{O}:\ H-CH_2-\underset{CH_3}{\overset{CH_3}{C}}\oplus + H_2\ddot{O}:$$

8.12 In this reaction a secondary carbocation is formed first. This carbocation then rearranges to form a more stable tertiary carbocation. Substitution and elimination products from both of these carbocations may also be formed.

$$CH_3-\underset{CH_3}{\overset{CH_3}{C}}-\overset{:\ddot{Br}:}{C}HCH_2CH_3 \longrightarrow CH_3-\underset{CH_3}{\overset{CH_3}{C}}-\overset{\oplus}{C}HCH_2CH_3 \longrightarrow CH_3-\overset{CH_3}{\underset{\oplus}{C}}-\overset{CH_3}{\underset{H}{C}}CH_2CH_3$$

$$CH_3\overset{\oplus}{\underset{..}{O}}H_2 + (H_3C)_2C=C(CH_3)CH_2CH_3 \longleftarrow CH_3\ddot{O}H$$

other products

$$CH_3-\underset{CH_3}{\overset{CH_3}{C}}-CH=CHCH_3 \quad CH_3-\underset{CH_3}{\overset{CH_3}{C}}-\overset{OCH_3}{C}HCH_2CH_3 \quad CH_3-\underset{OCH_3}{\overset{CH_3}{C}}-\overset{CH_3}{C}HCH_2CH_3 \quad CH_3-\overset{CH_2}{\overset{\|}{C}}-\overset{CH_3}{C}HCH_2CH_3$$

8.13 b) The bromine is bonded to a tertiary carbon and there is not a strong base present so the reaction proceeds by the $S_N1/E1$ mechanism. The S_N1 product should predominate. The E1 reaction will follow Zaitsev's rule.

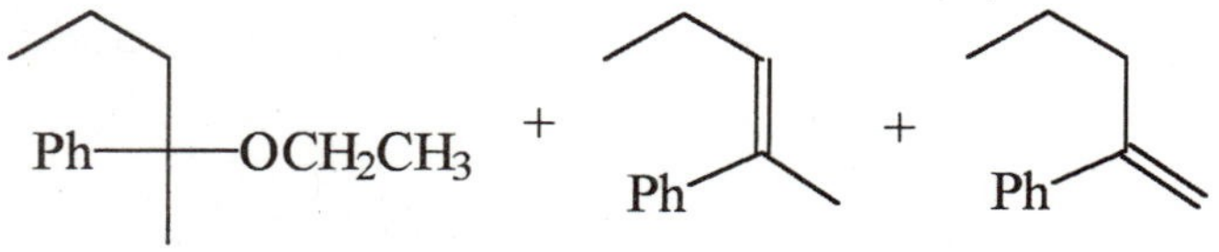

Ph—C(—OCH$_2$CH$_3$) + (Ph)C=C + (Ph)C=CH$_2$

major

c) A tertiary alkyl bromide and a weak base cause the reaction to proceed predominantly by the S_N1 mechanism. Products with both configurations at the reaction center will form. A minor amount of the elimination product, with the double bond in the more stable conjugated position, will also be formed.

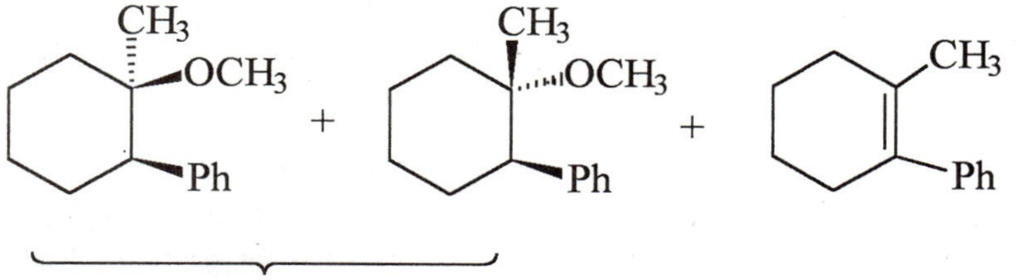

CH$_3$ OCH$_3$ Ph + CH$_3$ OCH$_3$ Ph + CH$_3$ Ph

major

d) same as (c)
e) A secondary alkyl bromide in the absence of a strong nucleophile or a strong base reacts by the S_N1/ E1 mechanism. Rearrangement of the initial carbocation will also occur. Substitution products from both water and ethanol will be formed.

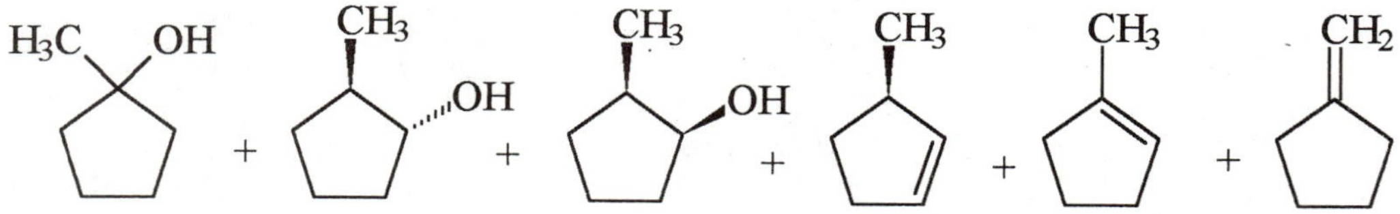

H$_3$C OH + CH$_3$ OH + CH$_3$ OH + CH$_3$ + CH$_3$ + CH$_2$

+ ethyl ethers corresponding to the alcohols

8.14 c) A secondary substrate with a weak base reacts predominately by the S_N2 mechanism. Minor amounts of E2 products may also be formed.

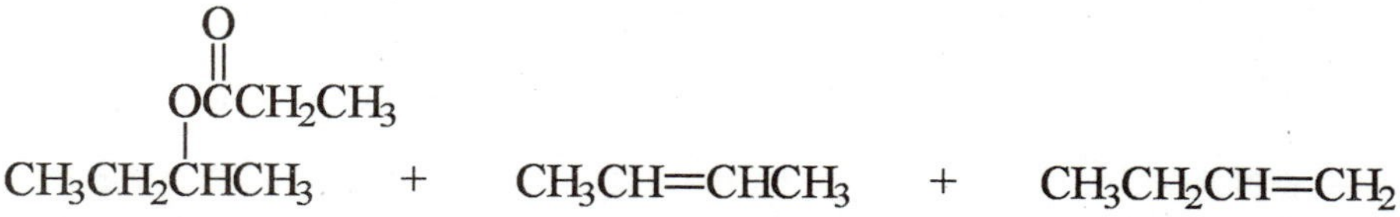

$CH_3CH_2CH(OC(=O)CH_2CH_3)CH_3$ + $CH_3CH{=}CHCH_3$ + $CH_3CH_2CH{=}CH_2$

d) A secondary substrate with a strong base reacts predominately by the E2 mechanism.

$$CH_3CH{=}CHCH_3 \quad + \quad CH_3CH_2CH{=}CH_2$$

major

e) A tertiary substrate in the absence of a strong base reacts by the S_N1 and E1 mechanisms.

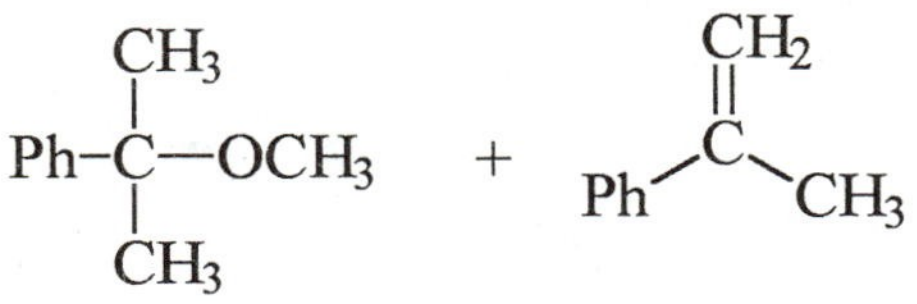

$$Ph{-}C(CH_3)_2{-}OCH_3 \quad + \quad Ph{-}C({=}CH_2){-}CH_3$$

major

8.15 a) The leaving group is on a primary carbon, so an S_N2 reaction occurs.

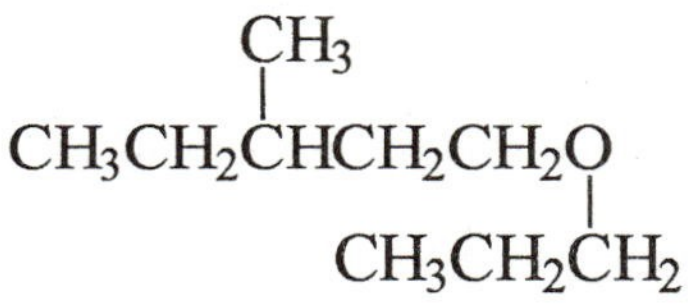

$$CH_3CH_2CH(CH_3)CH_2CH_2OCH_2CH_2CH_3$$

c) A secondary substrate with a strong base gives predominately E2 elimination.

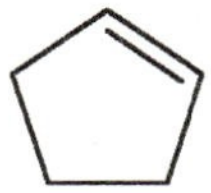

(cyclopentene)

d) A secondary substrate with a weak base that is moderately nucleophilic in an aprotic solvent favors a S_N2 reaction.

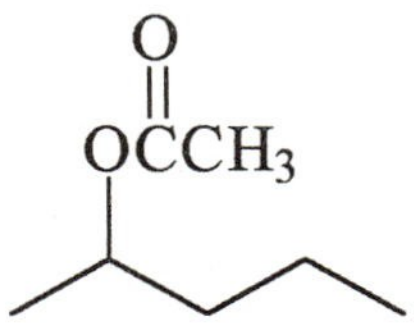

$$CH_3CH(OCCH_3)CH_2CH_3 \text{ (O=C, acetate ester)}$$

e) A tertiary substrate with a strong base favors E2 elimination.

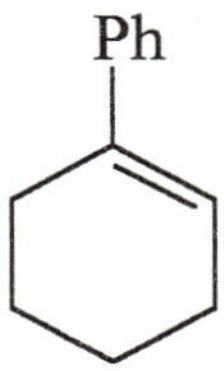

(1-phenylcyclohexene; Ph)

g) A secondary substrate in a polar solvent and in the absence of a strong base or nucleophile favors the $S_N1$and E1 reactions.

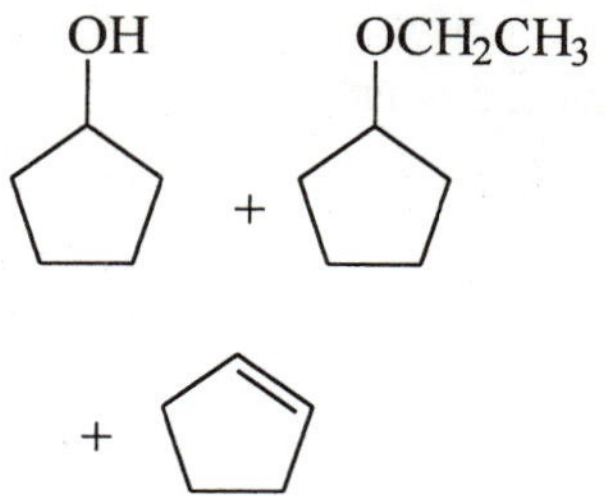

OH OCH_2CH_3

+

+

8.16

a)  + major

b) $CH_2{=}CH_2$ + $CH_2{=}CHCH_2CH_3$
major

c)

d)  + major

e) $Ph_2C{=}CH_2$

f) H_3C— + H_3C—
major

g)  +

major

h) 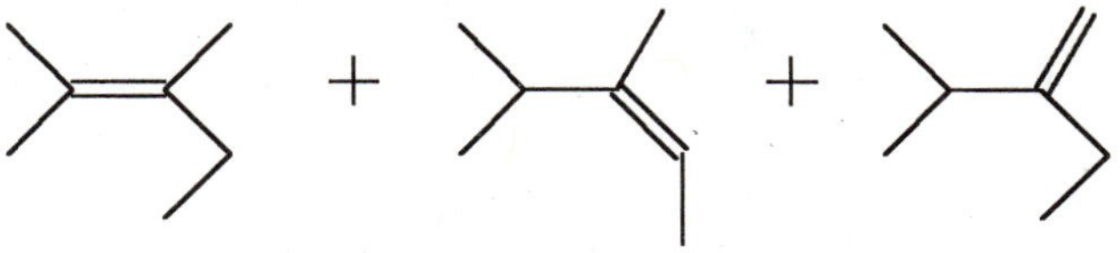 + +

major

i)

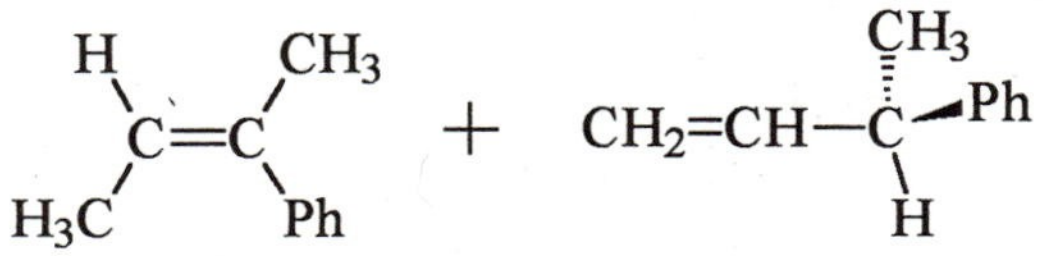

H CH3
C=C
H3C Ph

major

+ CH_2=CH—C(CH3)(H)—Ph

j) 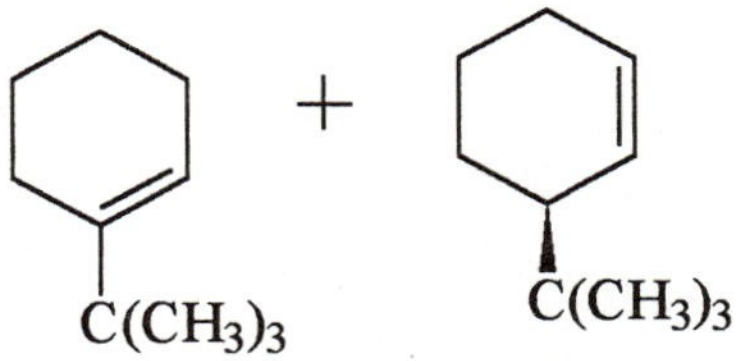 +

C(CH3)3
major

C(CH3)3

k)

CH3

l)  +

major

m) 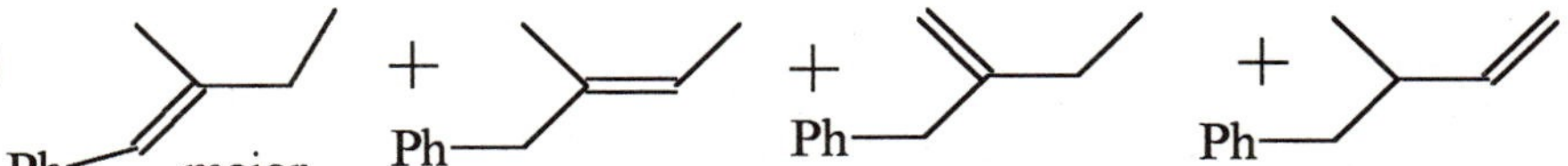 + + +

Ph major Ph Ph Ph

8.17 a) A secondary substrate with a moderate nucleophile and an aprotic solvent follows the S_N2 mechanism.

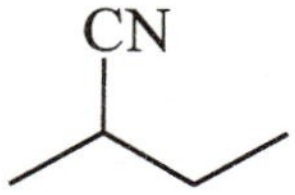

CN

b) A secondary substrate with a strong base reacts predominantly by the E2 mechanism.

major +

c) A secondary substrate with a weak nucleophile and polar solvent favors the S_N1 mechanism. A lesser amount of the E1 product may also be formed.

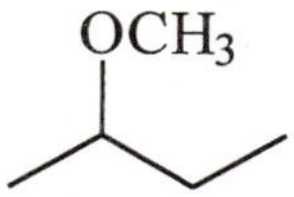

OCH_3

d) A primary substrate and a good nucleophile favor the S_N2 mechanism.

OEt

e) A hindered strong base and a primary substrate favor the E2 mechanism.

f) A tertiary substrate with a strong base favors the E2 mechanism.

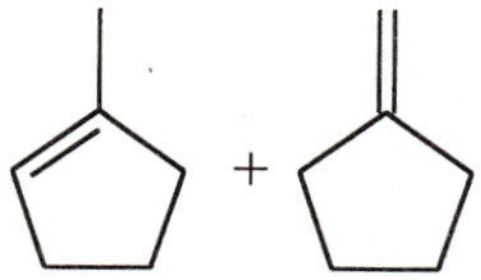

major +

g) A tertiary substrate in the absence of a strong base reacts by theS_N1 and E1 mechanisms.

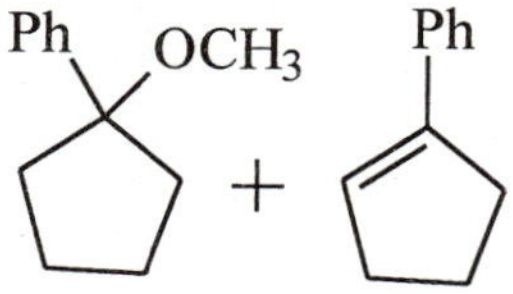

Ph OCH_3 + Ph

major

8.18

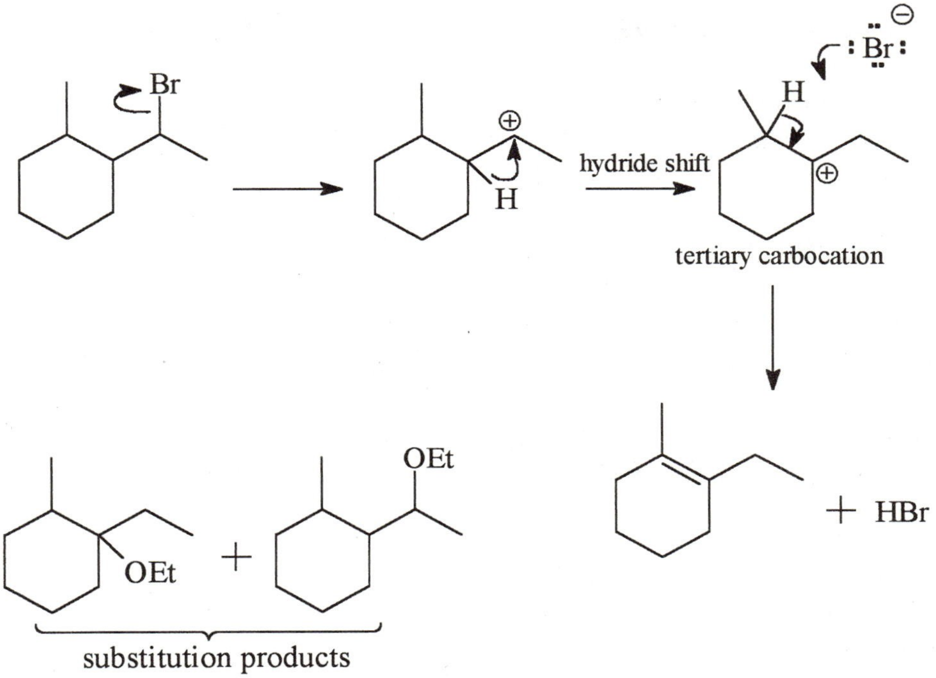

Br

hydride shift

H

H

:Br:

tertiary carbocation

+ HBr

OEt

OEt

substitution products

8.19 A tertiary substrate with a strong base favors E2 elimination. Normally the elimination products follow Zaitsev's rule with a less hindered base like ethoxide ion, producing more of the more highly substituted alkene.
When a sterically hindered, strong base, like *t*-BuO$^-$ ion, is used, the proportion of the less highly substituted alkene will increase.

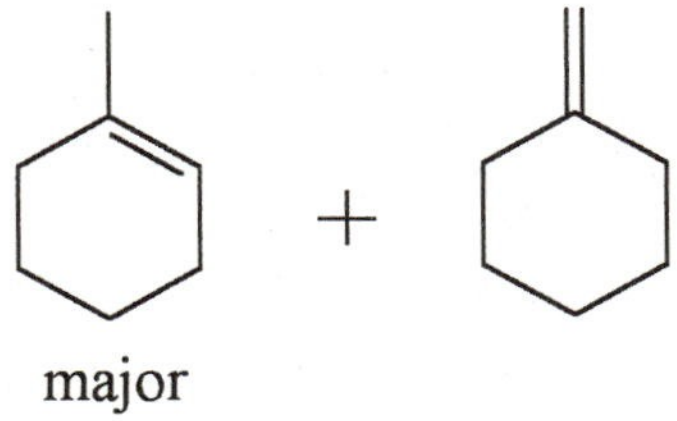

major

8.20 In the stereoisomer of 1,2,3,4,5,6-hexachlorocyclohexane shown, the Cl atoms are either all axial or all equatorial. Anti elimination cannot occur from the chair conformations of this stereoisomer because a Cl and a H are never in a *trans*-diaxial arrangement. All of the other stereoisomers of this compound will have at least one hydrogen anti to a chlorine in one of the conformations.

8.21

8.22 The most stable chair conformation of the *trans*-isomer has both the bromine and the *tert*-butyl group equatorial. However, elimination cannot occur from this conformation because bromine is not axial. Therefore, elimination occurs very slowly for the *trans*-isomer.

trans-isomer

more stable

For the *cis*-isomer, on the other hand, the most stable chair conformation is also the conformation needed for elimination, with the *tert*-butyl group equatorial and the bromine axial. The *cis*-isomer will react much faster than the *trans*-isomer because the conformation required for elimination is the more stable conformation.

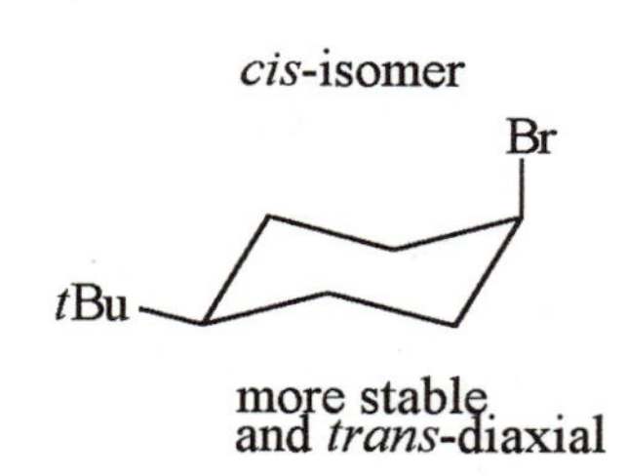

8.23 The *cis*-isomer has a faster rate of E2 elimination because the most stable chair conformation has the methyl equatorial and chlorine axial. Elimination can occur from this conformer. The stable conformation of the *trans*-isomer has both the chlorine and the methyl group equatorial. Elimination cannot occur from this conformation because chlorine is not axial.

8.24 The route using the conjugate base of cyclopentanol and methyl iodide will yield only the substitution product because methyl iodide cannot give an elimination product. In the other route, with bromocyclopentane (a secondary substrate) and methoxide anion (a strong base), E2 elimination will predominate. Therefore, the first route will give the highest yield of cyclopentyl methyl ether.

8.25 a) A E2 elimination product from a primary alkyl halide can be obtained by using a strong, hindered base. This reaction can be performed using potassium *tert*-butoxide in *tert*-butanol.

b) A S_N1 substitution product from a tertiary alkyl halide can be obtained by using a weak nucleophile in a polar solvent. This reaction can be carried out in methanol as the solvent and the nucleophile.
c) An alkene can be produced from a tertiary substrate by E2 elimination. This reaction can be carried out using a strong base such as sodium methoxide in methanol.
d) An alkene can be produced from a secondary substrate by E2 elimination using a strong base. This reaction can be accomplished with sodium ethoxide in ethanol.
e) A substitution product from a unhindered primary substrate can be obtained by using most nucleophiles. This reaction can be effected using sodium ethoxide in ethanol.
f) A substitution product from a secondary alkyl halide can be obtained by using a nucleophile that is not too basic. This reaction can be carried out by treating the alkyl bromide with methanol, an S_N1 reaction.

8.26 a) This terminal alkene can be prepared from a primary alkyl halide using a strong hindered base.

Br + *t*-BuOK $\xrightarrow{t\text{-BuOH}}$

b) This alkene can be prepared by reacting 4-bromoheptane with sodium methoxide in methanol. The alkyl halide is symmetrical so only a single elimination product is formed.

Br + CH_3ONa $\xrightarrow{CH_3OH}$

c) Cyclohexene can be prepared from chlorocyclohexane by E2 elimination using sodium methoxide in methanol.

Cl + CH_3ONa $\xrightarrow{CH_3OH}$

d) This alkene can be prepared from 4-bromo-1,2-dimethylcyclopentane by treating it with a strong base like sodium ethoxide in ethanol. Because the bromide is symmetrical, a single alkene is produced.

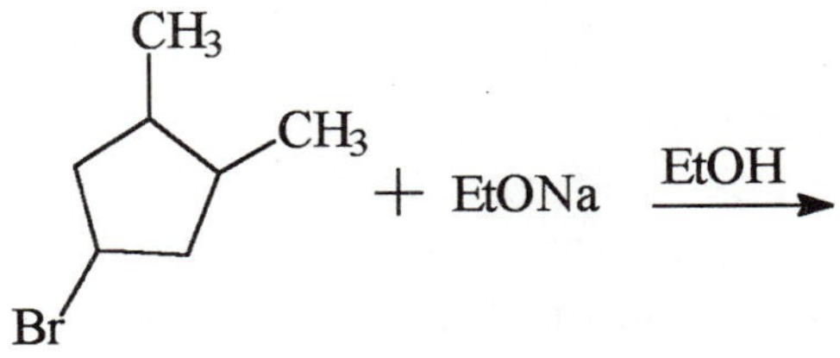

CH_3 CH_3 Br + EtONa $\xrightarrow{EtOH}$

8.27 The allylic substrate in ethanol follows the E1 and S_N1 mechanisms. The resonance structures have the positive charge at two different locations. Therefore two elimination and substitution products are possible.

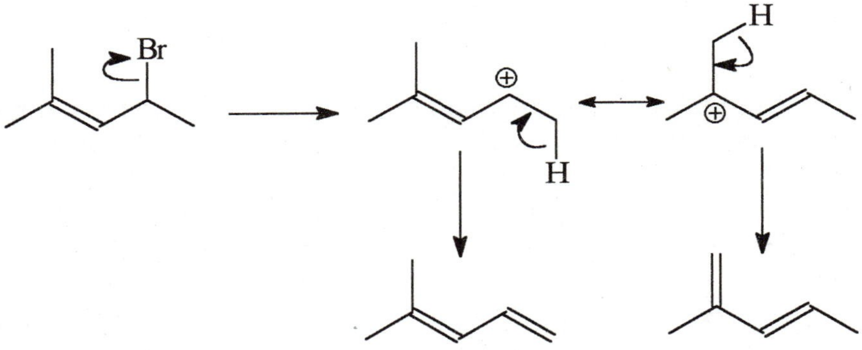

Br → ⊕ H ⟷ H ⊕

HOEt

H OEt

Br

OEt

HOEt

OEt

H

Br

OEt

8.28 The bond dissociation energy of a C-D bond is larger than that of a C-H bond by about 1.2 kcal/mol (5.0 kJ/mol). If a C-H bond is broken during the rate- determining step in a reaction, then replacing the hydrogen with deuterium results in a significant decrease in the rate of the reaction. In the E2 elimination involving cyclohexane rings, breaking of an C-H bond *trans* to the leaving group is required. In the chair conformation of the left compound only the deuteriums are *trans* to the bromine, but in the right compound the hydrogens are *trans* to the bromine. Therefore, the right compound will exhibit a faster rate of E2 elimination.

8.29

Ph—C(D)=CH_2 (major) + Ph—C(H)=CH_2

In this E2 elimination reaction, breaking of a C-H or a C-D bond is required in the rate determining step. The major product will be he one with the retention of the deuterium label because the C-H bond is easy to break than the C-D bond.

8.30 For syn elimination occur the hydrogen and the leaving group must be in a syn coplanar arrangement. This requires the molecule to be in a higher-energy eclipsed conformation. Although cyclopentane ring is nonplanar to relieve its torsional strain a fully staggered geometry is not easily attained. Therefore, the molecule can achieve the eclipsed geometry needed for syn elimination about as readily as the geometry needed for *anti* elimination.

8.31 a) The reaction follows the S_N1 and E1 mechanisms.

b)

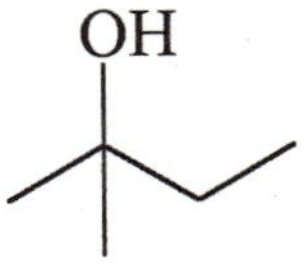

OH and OEt

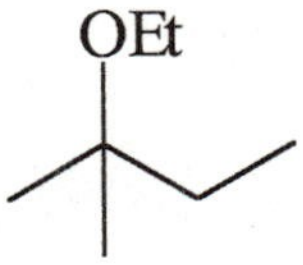

c)

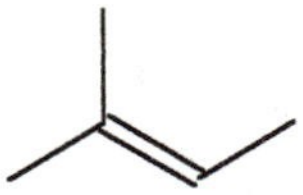

and

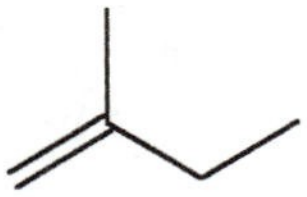

d) Iodide ion is a better leaving group than bromide ion, so the rate of reaction will be faster with 2-iodo-2-methylbutane. However, the amount of substitution and elimination products will not change because the leaving group is not involved in the second step of the mechanism.

8.32 Because of the relatively acidic hydrogen on the carbon adjacent to the carbonyl group, this compound reacts by the E1cb mechanism to produce this alkene.

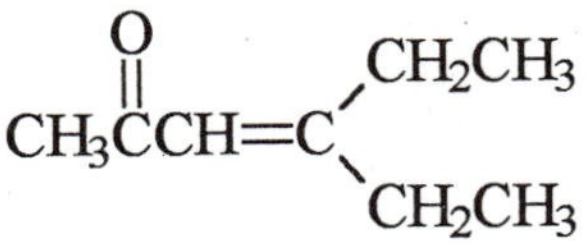

$CH_3C(=O)CH=C(CH_2CH_3)_2$

8.33 The ineffectiveness of DDT on some insects is due to the presence of an enzyme in these resistant insects that catalyzes the elimination of HCl to form DDE. This elimination reaction could be prevented by replacing the hydrogen with an alkyl group. Perhaps this would make the resulting compound an effective insecticide.

After completing this chapter, you should be able to:

1. Provide a detailed description, including stereochemistry, for the E2 mechanism and summarize the conditions that favor its occurrence.
   Problems 8.2, 8.3, 8.4, 8.6, 8.19, 8.20, 822, 8.23.

2. Provide similar information about the E1 mechanism.
   Problems 8.11, 8.12, 8.18, 8.27, 8.31.

3. Understand Zaitsev's rule and Hofmann's rule and when each applies.
   Problems 8.5, 8.7, 8.8, 8.9.

4. Show the expected products, including stereochemistry and regiochemistry, of any elimination reaction.
   Problems 8.1, 8.10, 8.13, 8.16, 8.21, 8.26.

5. Predict whether the major pathway that will be followed under a particular set of reaction conditions will be S_N1, S_N2, E1, or E2.
   Problems 8.14, 8.15, 8.17, 8.24, 8.25.

Chapter 9
SYNTHETIC USES OF SUBSTITUTION AND ELIMINATION REACTIONS

9.1

a) 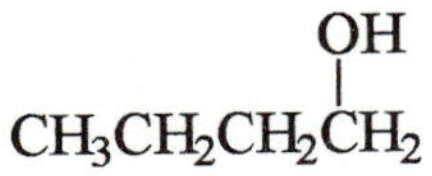 $CH_3CH_2CH_2CH_2OH$ (OH on the terminal CH_2)

b) 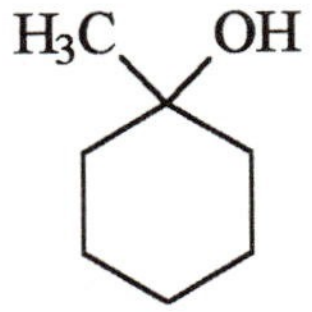 1-methylcyclohexanol (cyclohexane ring with H_3C and OH on the same carbon)

c) 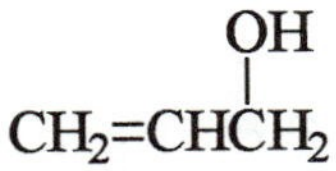 $CH_2{=}CHCH_2OH$ (OH on the terminal CH_2)

9.2 This is an S_N1 reaction, so only the chlorine bonded to the tertiary carbon is replaced.

9.3 Stereochemistry is retained because the C-O bond of the product is not broken in the reaction.

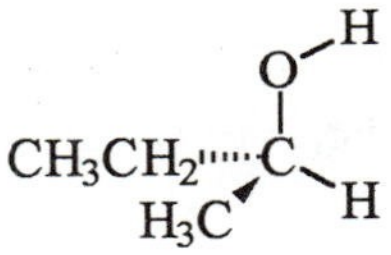

CH_3CH_2 (hashed bond), H_3C (wedged bond), H, and O–H bonded to C

9.4

a) 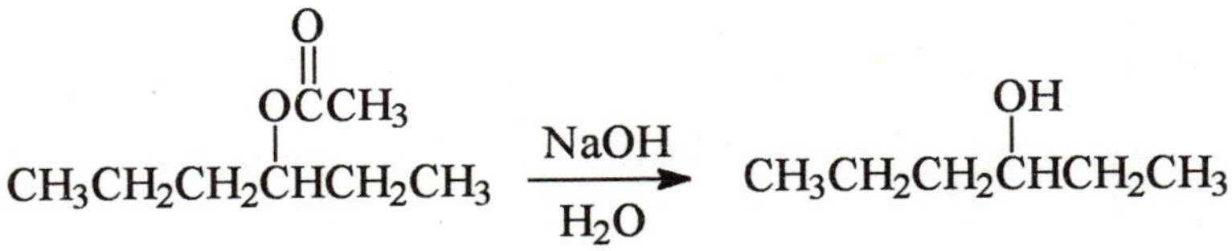 $CH_3CH_2CH_2CH(OCCH_3)CH_2CH_3$ (the OC(=O)CH₃ group is on the CH) $\xrightarrow[H_2O]{NaOH}$ $CH_3CH_2CH_2CH(OH)CH_2CH_3$

b) 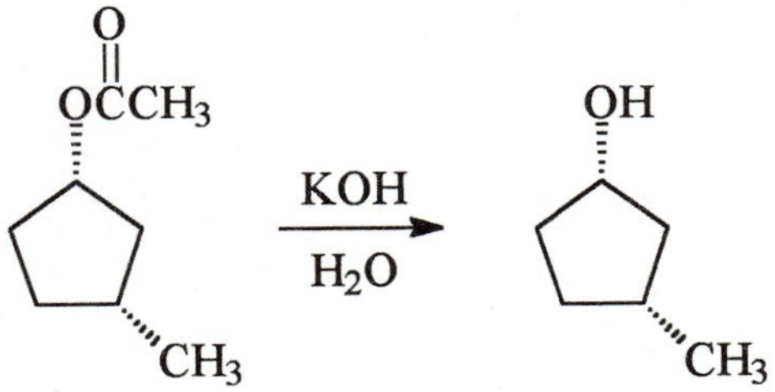 Cyclopentane with $OCCH_3$ (C=O; hashed bond) and CH_3 (hashed bond) $\xrightarrow[H_2O]{KOH}$ cyclopentane with OH (hashed bond) and CH_3 (hashed bond)

9.5

a) 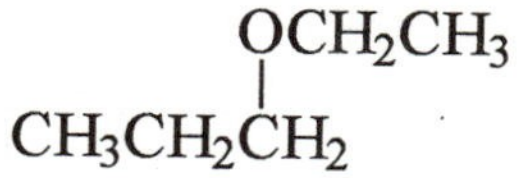 $CH_3CH_2CH_2$ with OCH_2CH_3 attached

b) 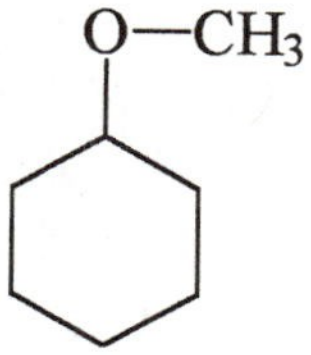 cyclohexane with $O{-}CH_3$

c) 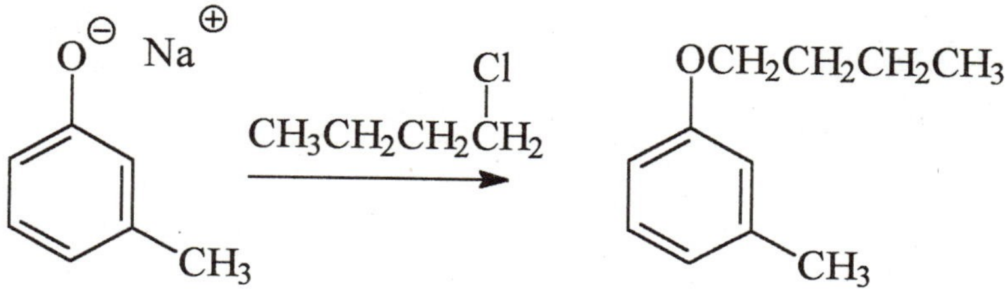 3-methylphenoxide ($O^{\ominus}$ $Na^{\oplus}$, CH_3) $\xrightarrow{CH_3CH_2CH_2CH_2Cl}$ 3-methylphenyl butyl ether ($OCH_2CH_2CH_2CH_3$, CH_3)

9.6 Although nitrogen is a better nucleophile than oxygen, the oxygen atom is less sterically hindered. The steric effect is winning.

9.7 a) Methyl iodide cannot undergo an elimination reaction, so its only reaction with *t*-BuO^- ion is substitution. The left route gives primarily E2 elimination. Therefore, the right route is better.

b) The right route is better because elimination cannot occur.

9.8 a) To minimize the competing E2 reaction, the conjugate base of the primary alcohol should be reacted with methyl iodide.

$$CH_3CH_2CH_2CH_2OH \xrightarrow[2)\ CH_3I]{1)\ Na} CH_3CH_2CH_2CH_2OCH_3$$

c) Since direct nucleophilic substitution cannot be effected on a benzene ring, the conjugate base of phenol must be reacted with the primary alkyl halide.

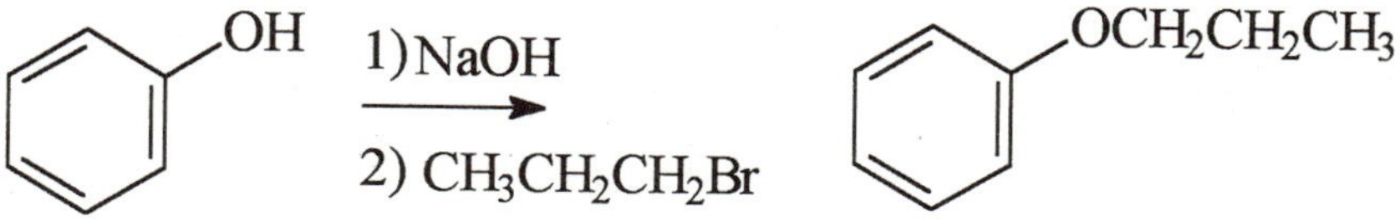

$$C_6H_5OH \xrightarrow[2)\ CH_3CH_2CH_2Br]{1)\ NaOH} C_6H_5OCH_2CH_2CH_3$$

d) To minimize E2 elimination, the conjugate base of the secondary alcohol must be reacted with the primary alkyl halide.

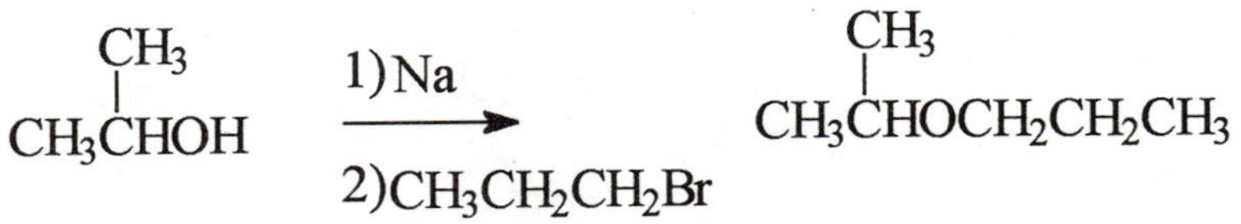

$$CH_3CH(CH_3)OH \xrightarrow[2)\,CH_3CH_2CH_2Br]{1)\,Na} CH_3CH(CH_3)OCH_2CH_2CH_3$$

9.9

a) 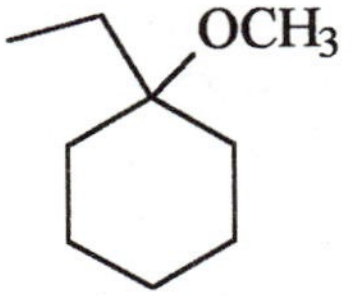 1-ethyl-1-methoxycyclohexane (structure: cyclohexane ring with ethyl and OCH_3 on the same carbon)

b) 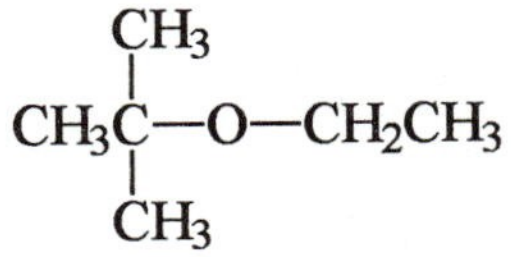 $CH_3C(CH_3)_2{-}O{-}CH_2CH_3$

c) $Ph_2CHOCH_2CH_2Cl$

9.12 The first step of this reaction is an acid-base reaction, involving OH^- and the hydrogen of the hydroxy group on the substrate. The conjugate base of the substrate then undergoes an intramolecular S_N2 reaction, displacing Br with inversion.

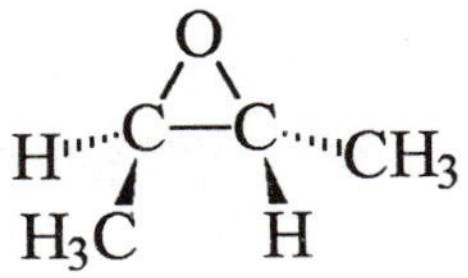

(Epoxide structure: H, H_3C on one ring carbon; CH_3, H on the other; O bridging)

9.13 Formation of a tertiary carbocation is faster in an S_N1 reaction.

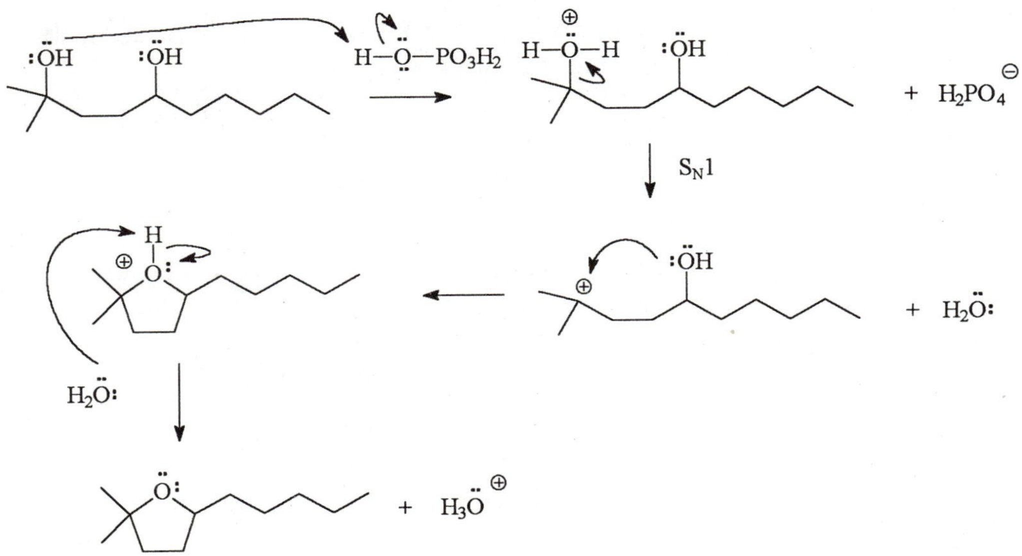

(Mechanism: diol protonated by $H{-}O{-}PO_3H_2$ at the tertiary OH, giving $+\ H_2PO_4^-$; loss of H_2O (S_N1) gives tertiary carbocation $+\ H_2\ddot{O}$:; intramolecular attack by the secondary :ÖH gives protonated cyclic ether; deprotonation by $H_2\ddot{O}$: gives the tetrahydrofuran product $+\ H_3O^+$)

9.14

a) 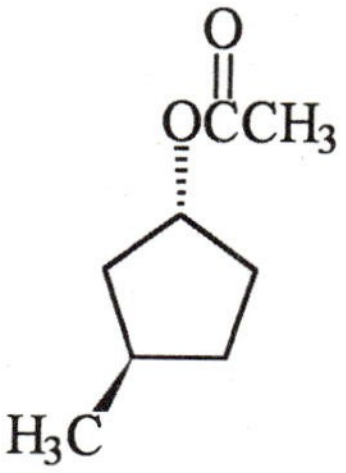 $OCCH_3$ (O) ; H_3C

b) 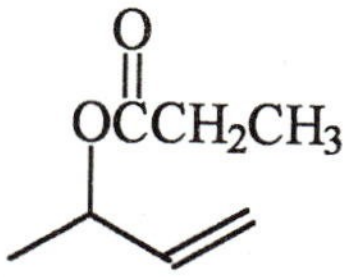 $OCCH_2CH_3$ (O)

c) 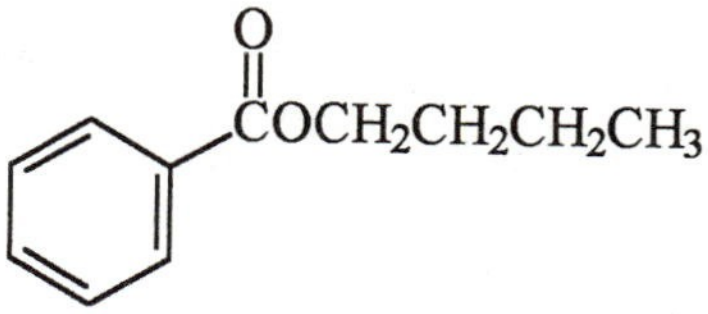 $COCH_2CH_2CH_2CH_3$ (O)

9.15

a) 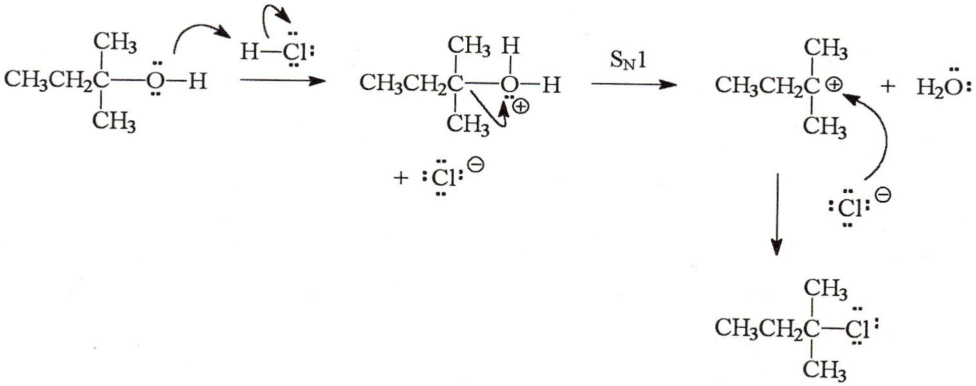 $CH_3CH_2C(CH_3)_2-\ddot{O}-H$ + $H-\ddot{Cl}:$ $\longrightarrow$ $CH_3CH_2C(CH_3)_2-\overset{+}{O}(H)-H$ + $:\ddot{Cl}:^{\ominus}$ $\xrightarrow{S_N1}$ $CH_3CH_2\overset{\oplus}{C}(CH_3)_2$ + $H_2\ddot{O}:$

$CH_3CH_2\overset{\oplus}{C}(CH_3)_2$ + $:\ddot{Cl}:^{\ominus}$ $\longrightarrow$ $CH_3CH_2C(CH_3)_2-\ddot{Cl}:$

9.16

a) 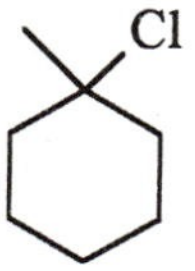 Cl

b)  Cl

c) 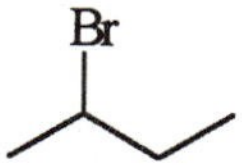 Br

d) $CH_3CH_2CH_2I$

e) 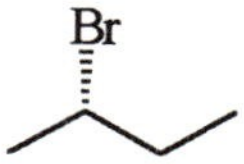 Br

f) 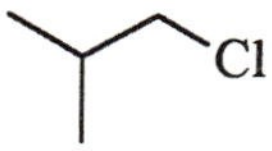 Cl

g)  Br

h) 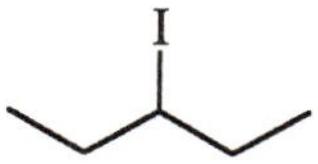 I

i) I

j) 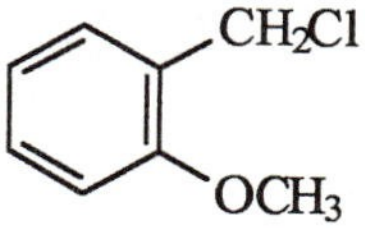 CH_2Cl ; OCH_3

9.17

a) 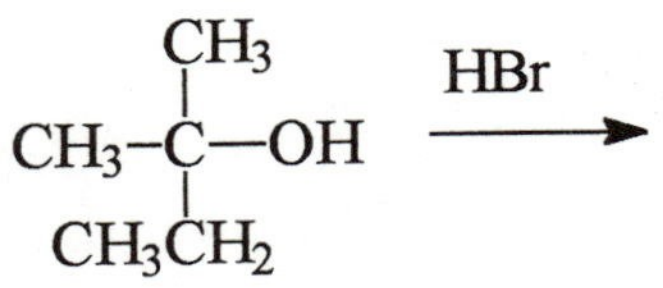 $CH_3-C(CH_3)(CH_2CH_3)-OH$ $\xrightarrow{HBr}$

b) $CH_3-C(CH_3)(Ph)-OH$ $\xrightarrow{HCl}$

c) OH (skeletal structure) $\xrightarrow[2)\ NaI]{1)\ TsCl,\ pyridine}$

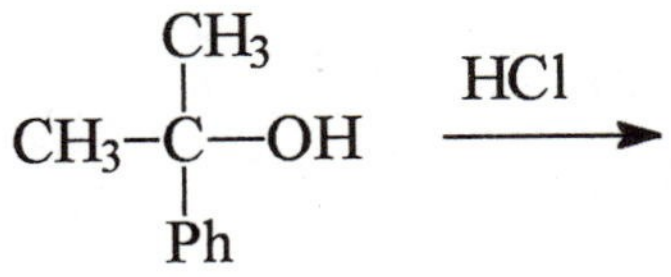

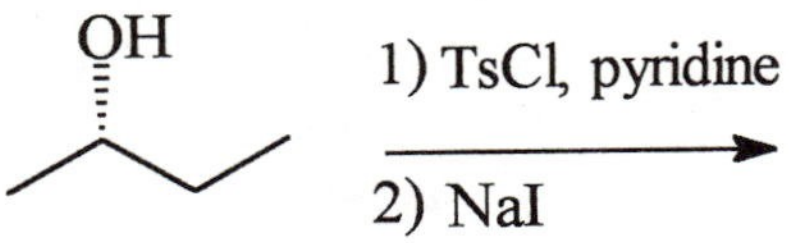

d) 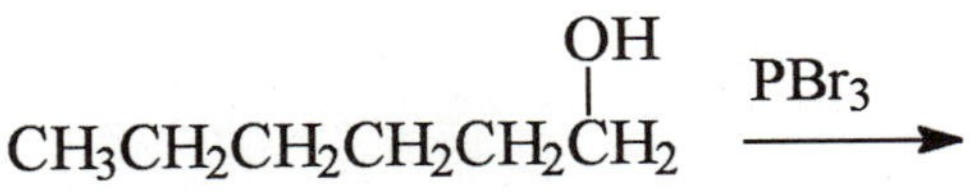 $CH_3CH_2CH_2CH_2CH_2CH_2OH$ $\xrightarrow{PBr_3}$

e)  OH (skeletal structure) $\xrightarrow{HBr}$

f) $CH_3CH_2CH_2CH_2OH$ $\xrightarrow{SOCl_2}$

9.18

a) 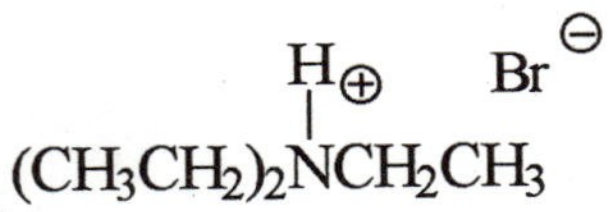 $(CH_3CH_2)_2\overset{\oplus}{N}HCH_2CH_3$ $Br^{\ominus}$

b) 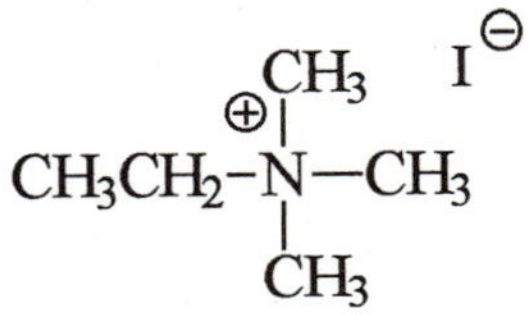 $CH_3CH_2-\overset{\oplus}{N}(CH_3)_2-CH_3$ $I^{\ominus}$

9.19

a) 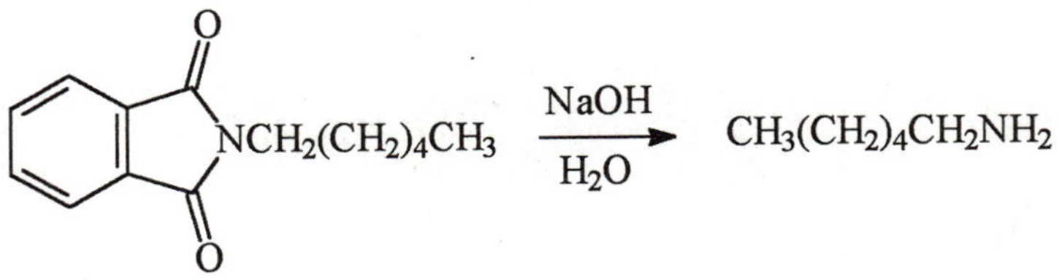 $NCH_2(CH_2)_4CH_3$ (phthalimide) $\xrightarrow[H_2O]{NaOH}$ $CH_3(CH_2)_4CH_2NH_2$

b) 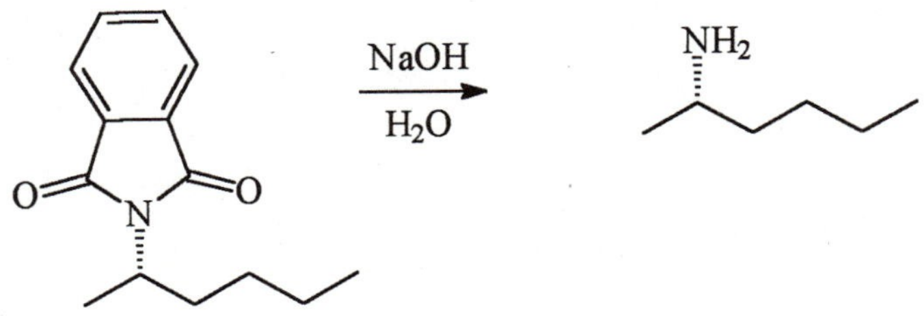 N (phthalimide, skeletal structure) $\xrightarrow[H_2O]{NaOH}$ NH_2 (skeletal structure)

9.20

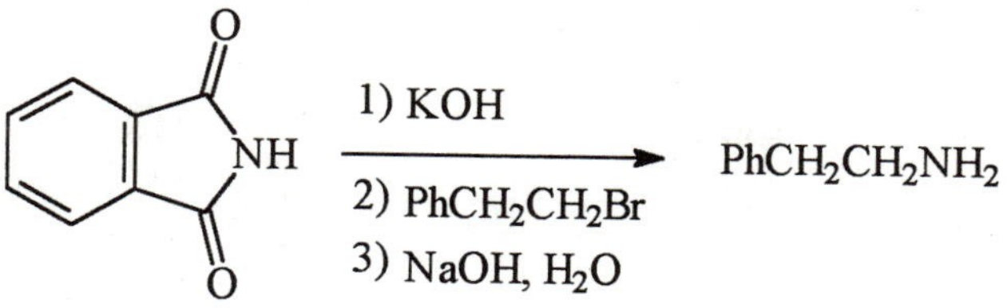

Phthalimide (NH) $\xrightarrow[\text{2) } PhCH_2CH_2Br \text{ 3) NaOH, } H_2O]{\text{1) KOH}}$ $PhCH_2CH_2NH_2$

9.21

a)  heptane (skeletal structure)

b) 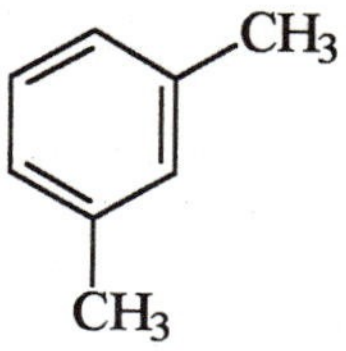 m-xylene: benzene ring with CH_3 and CH_3 in 1,3 positions

9.22

a) 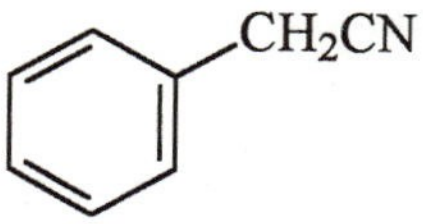 benzene ring–CH_2CN

b) $CH_3C{\equiv}C{-}CH_2CH_2CH_3$

c) 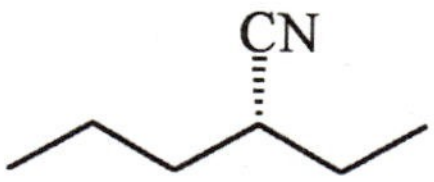 skeletal chain with CN on hashed wedge

d) $HC{\equiv}C{-}CH_2CH_3$ $\xrightarrow[\text{2) } CH_3I]{\text{1) } NaNH_2}$ $CH_3C{\equiv}C{-}CH_2CH_3$

e)  Cl–(chain)–CN

f) 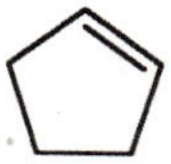 cyclopentene + $HC{\equiv}CH$

9.23

a) 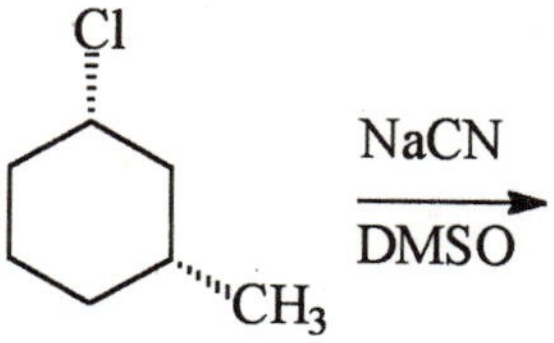 cyclohexane ring with Cl (hashed) and CH_3 (hashed) $\xrightarrow[\text{DMSO}]{\text{NaCN}}$

b) $Br{-}CH_2CH_2CH(CH_3)CH_3$ $\xrightarrow[NH_3\text{ (l)}]{H{-}C{\equiv}C{:}^{\ominus}\ Na^{\oplus}}$

c) $CH_3{-}C{\equiv}C{-}H$ $\xrightarrow[\text{2) } PhCH_2Br]{\text{1) } NaNH_2}$

9.24

a) 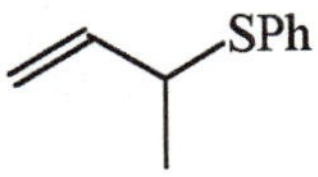 SPh b) $Ph_3\overset{\oplus}{P}—CH_2CH_2CH_3$ $Br^{\ominus}$ c) $CH_3CH_2CH_2CH_2SCH_3$ d) 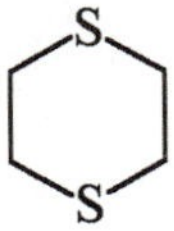 S S

9.25

a) 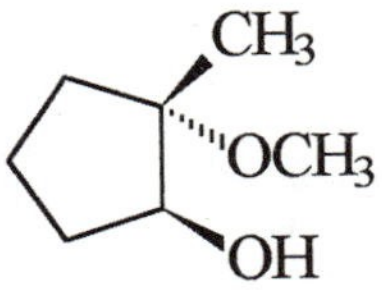 CH_3 OCH_3 OH b) 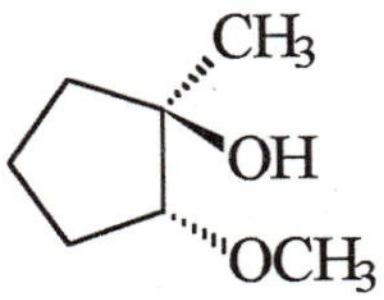 CH_3 OH OCH_3 c) 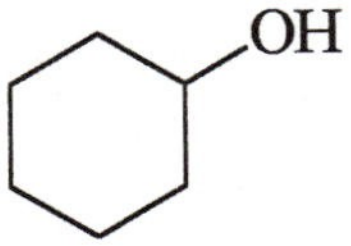 OH

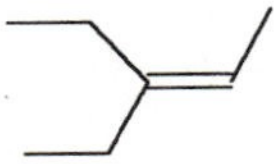

9.26

a) $CH_3CH_2CH{=}CH_2$ b) 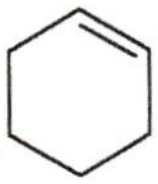 c)  major + minor d)

9.27 a) This is not a good way to prepare the desired alkene from the alkyl halide because the alkene shown would be a minor product according to Zaitsev's rule.

b) This is a good method to prepare the desired alkene. The alkene shown would be the major product because it is conjugated.

9.28 The top reaction is better because it can produce only the desired alkene. The bottom reaction would produce 1-pentene also.

9.29

a) 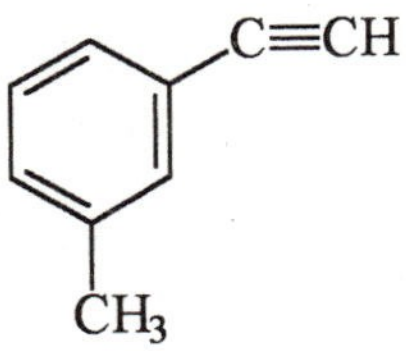 $C{\equiv}CH$ CH_3 b) 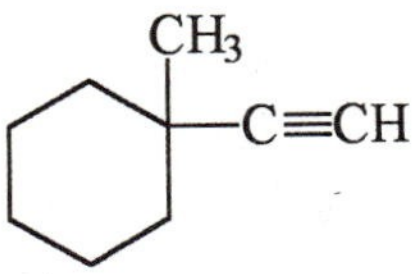 CH_3 $C{\equiv}CH$

9.30

a)  b) 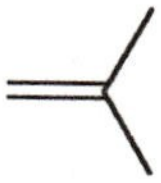 c) 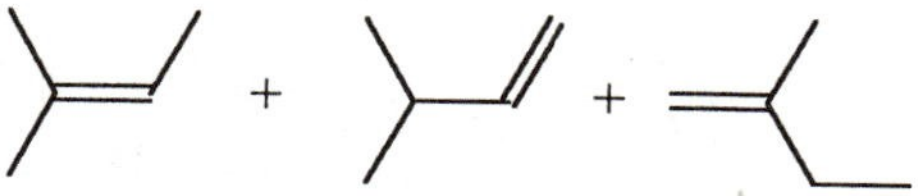 + +

major minor minor

9.31

a) 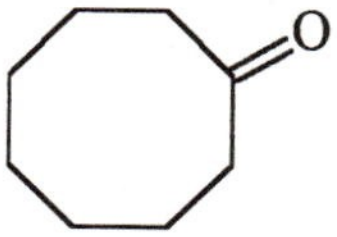 O

b) 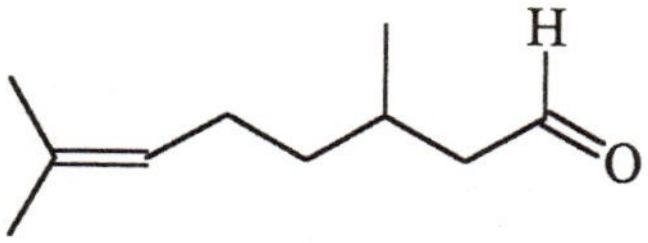 H O

c) 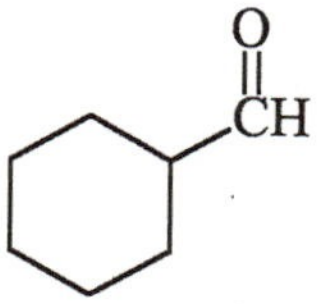 O CH

d) 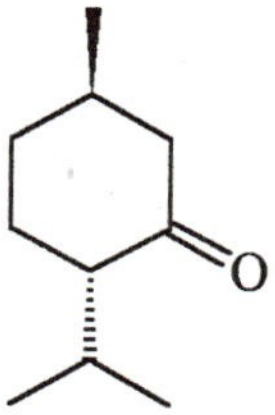 O

e) 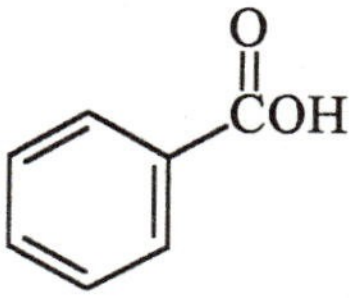 O COH

9.32

c) $CH_3CH_2CH_2CH_2Br$ $\xrightarrow[\text{2) KOH, } H_2O]{\text{1) NK}}$ $CH_3CH_2CH_2CH_2NH_2$ $\xrightarrow{\text{excess } CH_3I}$ $CH_3CH_2CH_2CH_2\overset{\oplus}{N}(CH_3)_3\ I^{\ominus}$

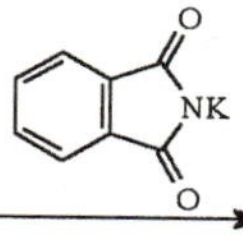

d) $CH_3CH_2\underset{}{C}HBr$ (CH_3) $\xrightarrow[\text{2) KOH, } H_2O]{\text{1) } CH_3CO_2^{\ominus}\text{ DMSO}}$ CH_3CH_2CHOH (CH_3) $\xrightarrow[\text{2) } CH_3CH_2I]{\text{1) Na}}$ $CH_3CH_2\text{–O–}CHCH_2CH_3$ (CH_3)

e) 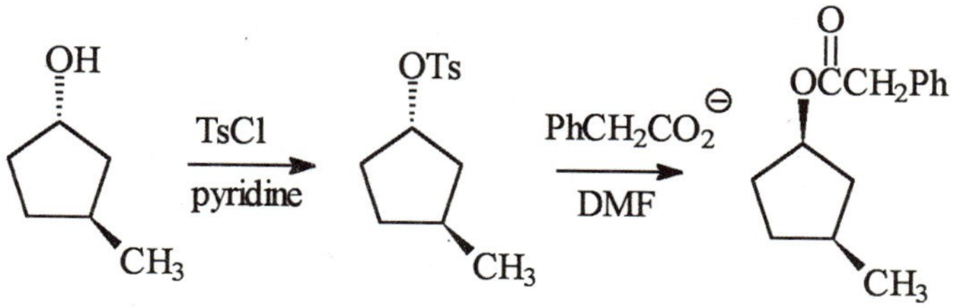 OH, CH_3 $\xrightarrow[\text{pyridine}]{\text{TsCl}}$ OTs, CH_3 $\xrightarrow[\text{DMF}]{PhCH_2CO_2^{\ominus}}$ O, $OCCH_2Ph$, CH_3

f) 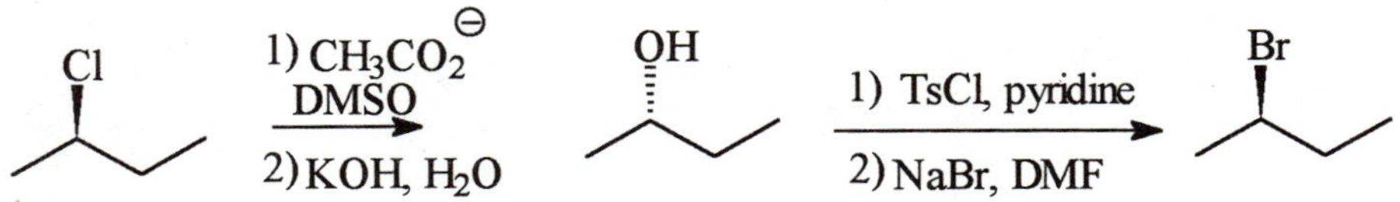 [2-chlorobutane] $\xrightarrow[\text{2) KOH, } H_2O]{\text{1) } CH_3CO_2^{\ominus}\text{, DMSO}}$ [2-butanol, OH] $\xrightarrow[\text{2) NaBr, DMF}]{\text{1) TsCl, pyridine}}$ [2-bromobutane, Br]

g) 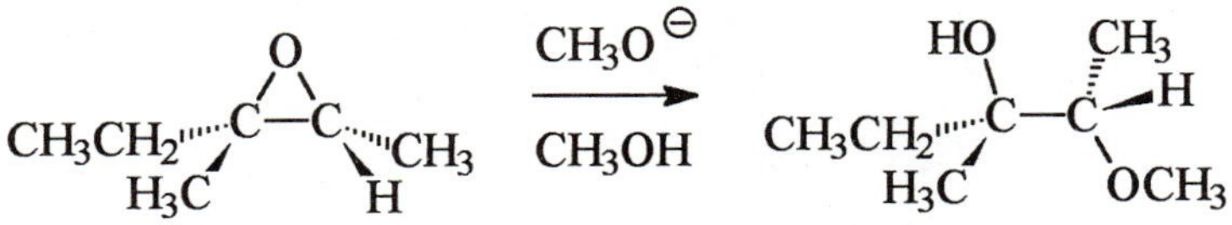 CH_3CH_2–C(O)(–H_3C)–C(–H)–CH_3 $\xrightarrow[CH_3OH]{CH_3O^{\ominus}}$ CH_3CH_2–C(HO)(H_3C)–C(CH_3)(H)(OCH_3)

h) $H-C\equiv C-H$ $\xrightarrow[\text{2) } PhCH_2Cl]{\text{1) } NaNH_2}$ $PhCH_2-C\equiv C-H$ $\xrightarrow[\text{2) } CH_3I]{\text{1) } NaNH_2}$ $PhCH_2-C\equiv C-CH_3$

9.33

d) 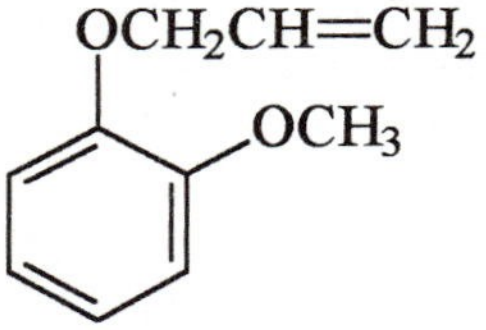 [benzene ring with $OCH_2CH=CH_2$ and OCH_3]

e)  [1,3-cyclohexadiene]

f) 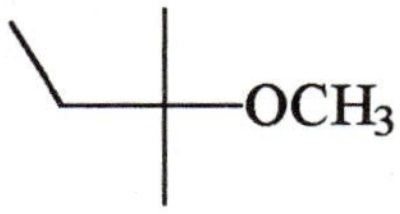 [tert-amyl methyl ether, $-OCH_3$]

g) 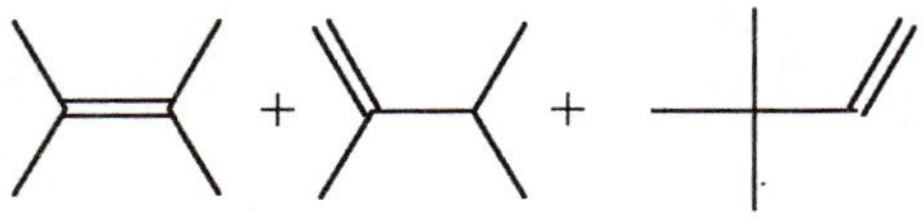 [2,3-dimethyl-2-butene] + [2,3-dimethyl-1-butene] + [3,3-dimethyl-1-butene]

major minor minor

h) 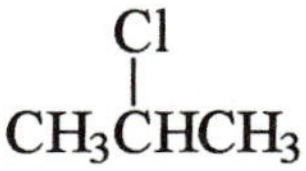 $CH_3CHClCH_3$ (Cl over $CH_3\overset{|}{C}HCH_3$)

i) $PhCH_2NH_2$

j) 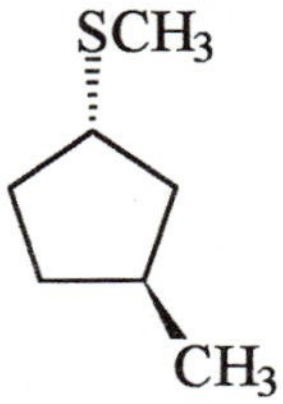 [cyclopentane with SCH_3 and CH_3]

k) 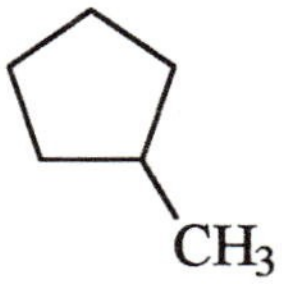 [methylcyclopentane, CH_3]

l) $CH_3CH_2CH_2CH_2CN$

m) $CH_3CH_2C{\equiv}C{-}CH_2CH_2CH_2Ph$ n) $CH_3CH(OH){-}CH_2SH$ o) cyclohexene

p) $Ph{-}C{\equiv}C{-}Ph$ q) 3,3-dimethylcyclopentanone r) $CH_3CH_2CH(CH_3)CH_2CHO$

9.34

a) butan-1-ol (OH) $\xrightarrow{SOCl_2}$ 1-chlorobutane (Cl)

b) 1-chlorobutane (Cl) $\xrightarrow[\text{2) KOH, } H_2O]{\text{1) potassium phthalimide (NK)}}$ butan-1-amine (NH_2)

c) butan-1-ol (OH) $\xrightarrow[H_2SO_4]{Na_2Cr_2O_7}$ butanoic acid (OH)

d) 1-chlorobutane (Cl) $\xrightarrow{CH_3CH_2S^{\ominus}}$ butyl ethyl sulfide (S)

e) $\ce{CH3CH2CH2CH2Cl}$ $\xrightarrow[t\text{-BuOH}]{t\text{-BuO}^{\ominus}}$ $\ce{CH3CH2CH=CH2}$

f) $\ce{CH3CH2CH2CH2Cl}$ $\xrightarrow[CH_3OH]{CH_3O^{\ominus}}$ $\ce{CH3CH2CH2CH2OCH3}$

g) $\ce{CH3CH2CH2CH2Cl}$ $\xrightarrow[DMSO]{NaCN}$ $\ce{CH3CH2CH2CH2CN}$

h) $\ce{CH3CH2CH2CH2Cl}$ $\xrightarrow{LiAlH_4}$ $\ce{CH3CH2CH2CH3}$

i) $\ce{CH3CH2CH2CH2Cl}$ $\xrightarrow[DMSO]{CH_3CH_2CO_2^{\ominus}}$ $\ce{CH3CH2CH2CH2OC(=O)CH2CH3}$

j) $\ce{CH3CH2CH2CH2OH}$ $\xrightarrow{PCC}$ $\ce{CH3CH2CH2C(=O)OH}$

9.35

a) 2,2-dimethyloxirane $\xrightarrow[2)\ H_2O]{1)\ LiAlH_4}$ $\ce{(CH3)3C-OH}$ $\xrightarrow{HBr}$ $\ce{(CH3)3C-Br}$

b) 2,2-dimethyloxirane $\xrightarrow[H_2SO_4]{CH_3OH}$ $\ce{CH3O-C(CH3)2-CH2OH}$

c) 2,2-dimethyloxirane $\xrightarrow[CH_3OH]{CH_3S^{\ominus}}$ $\ce{(CH3)2C(OH)-CH2SCH3}$

9.36

a) Br, CH_2OH

b) 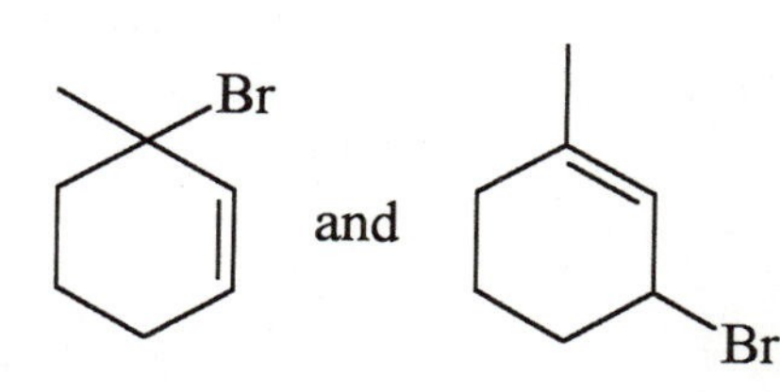

c) O

d) Br

e) Cl

f) NH_2

g) $CH_3CH_2\text{—}\overset{\oplus}{P}Ph_3$

h) OH, SCH_2CH_3

i) $PhCH{=}CH_2$

j) O

k) O

l) $PhCH_2O\overset{O}{\overset{\|}{C}}CH_2CH_3$

m) 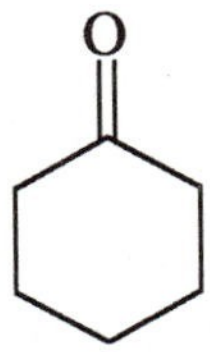 cyclohexanone (ring with C=O)

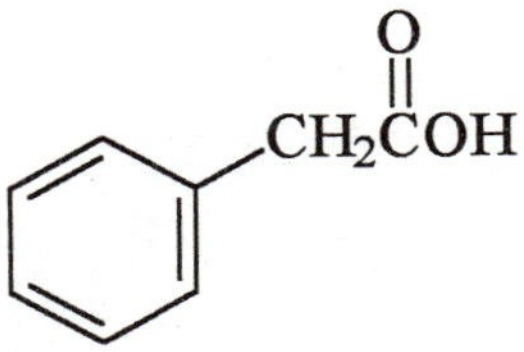

n) $C_6H_5CH_2COOH$ (benzene ring–$CH_2\overset{O}{\overset{\|}{C}}OH$)

o) CN (wedge-hashed substituent on a pentane chain)

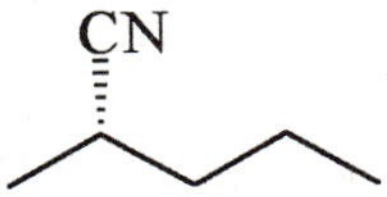

p) $CH_3CH_2CH_2C{\equiv}CCH_2CH_2CH_3$

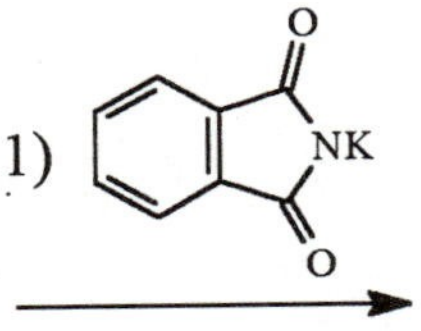

q) 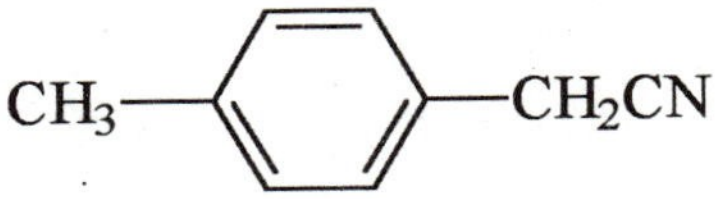 CH_3–(benzene ring)–CH_2CN

r) 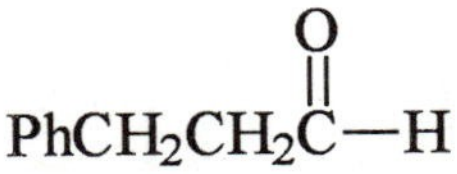 $PhCH_2CH_2\overset{O}{\overset{\|}{C}}{-}H$

s) 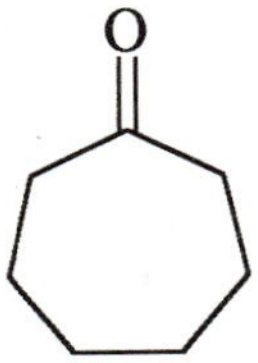 cycloheptanone (ring with C=O)

9.37

a) PBr_3

b) $\xrightarrow[\text{2) KOH, } H_2O]{\text{1) potassium phthalimide (NK)}}$

c) KCN, DMF

d) $\xrightarrow[\text{DMF}]{CH_3CH_2CO_2^{\ominus}}$

e) KOH, H_2O

f) H_2O

g) PPh_3

h) $\xrightarrow[CH_3OH]{CH_3O^{\ominus}}$

i) $\xrightarrow[\text{EtOH, } \Delta]{\text{KOH}}$

j) $\xrightarrow[CH_3OH]{CH_3O^{\ominus}}$ k) $CH_3CH_2CH_2OH$ l) $SOCl_2$

m) $\xrightarrow[heat]{H_2SO_4}$ n) $\xrightarrow[2)\ CH_3I]{1)\ Na}$ o) $\xrightarrow[acetone]{CrO_3\ H_2SO_4,\ H_2O}$

p) $\xrightarrow{CH_3C{\equiv}C^{\ominus}\ Na^{\oplus}}$ q) $PhS^{\ominus}$ r) $\xrightarrow[DMF]{CH_3CH_2O^{\ominus}}$

s) PCC t) Ag_2O, H_2O, THF

9.38

a) 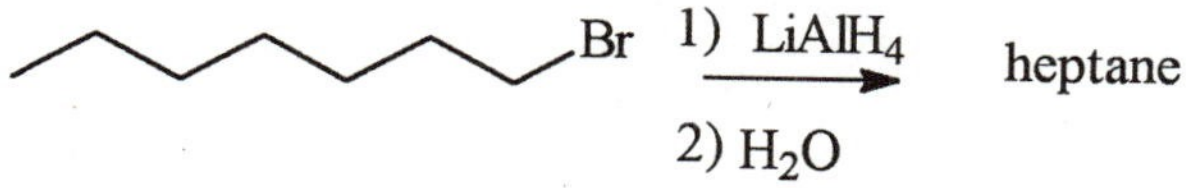 1-bromoheptane (Br) $\xrightarrow[2)\ H_2O]{1)\ LiAlH_4}$ heptane

b) 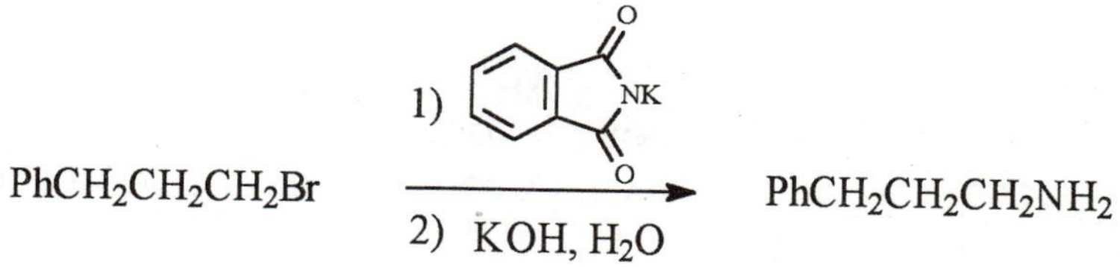 $PhCH_2CH_2CH_2Br$ $\xrightarrow[2)\ KOH,\ H_2O]{1)\ \text{potassium phthalimide (NK)}}$ $PhCH_2CH_2CH_2NH_2$

c) $PhCH_2Cl$ $\xrightarrow{HC{\equiv}C^{\ominus}\ Na^{\oplus}}$ $PhCH_2{-}C{\equiv}CH$

d) 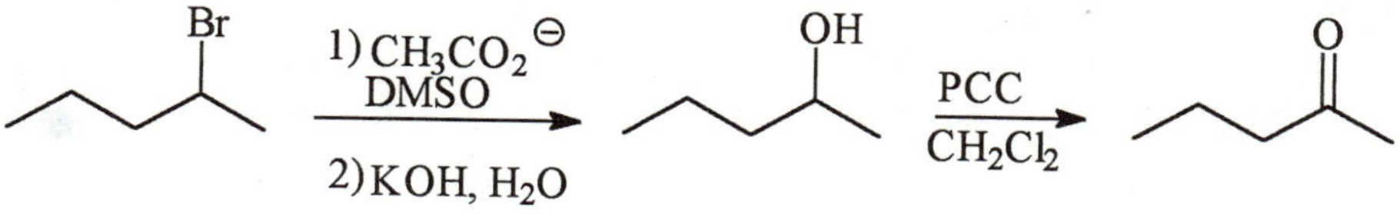 2-bromopentane (Br) $\xrightarrow[2)\ KOH,\ H_2O]{1)\ CH_3CO_2^{\ominus}\ DMSO}$ 2-pentanol (OH) $\xrightarrow[CH_2Cl_2]{PCC}$ 2-pentanone (O)

e) Br $\xrightarrow{KOH,\ H_2O}$ OH $\xrightarrow[H_2SO_4]{Na_2Cr_2O_7}$ OH (O)

f) Br + $^{\ominus}ONa^{\oplus}$ ⟶ O

g) Br, H_3C $\xrightarrow[2)\ KOH,\ H_2O]{1)\ CH_3CO_2^{\ominus}\ \ DMSO}$ OH, H_3C

9.39 Br, CH_3, OH $\xrightarrow[H_2O]{NaOH}$ CH_3, O $\xrightarrow[DMF]{KCN}$ CH_3, OH, CN

9.40 a) The product shown will only be a very minor product because a tertiary alkyl chloride with a strong base favors E2 elimination.

b) The product shown will not form because direct nucleophilic substitution on a benzene ring does not occur.

c) This product will not form because under basic conditions. The nucleophile will attack the less hindered carbon of the epoxide ring.

d) The reaction of primary alcohols with HCl requires the presence of the Lewis acid $ZnCl_2$.

9.41 a) The acetylide anion produced in the first step is a strong base. Therefore, the reaction will favor E2 elimination.

b) Multiple alkylation will compete, giving a low yield of the desired primary amine.

c) The ether will not be formed because the secondary alkyl chloride and the strong base will favor E2 elimination.

d) Nucleophilic substitution on a benzene ring does not occur.

e) This reaction follows the S_N1 mechanism. Therefore, the product will be a mixture of stereoisomers.

f) This is an E1 elimination reaction. The initially formed secondary carbocation can rearrange to form a more stable tertiary carbocation. Therefore, this reaction will produce a mixture of products in which the product shown will be only a minor component.

9.42 a)

$$H_3C{-}\overset{CH_3}{\underset{CH_3}{C}}{-}Br \longrightarrow H_3C{-}\overset{CH_3}{\underset{CH_3}{C^{\oplus}}} \quad H{-}\ddot{O}{-}CH_3 \longrightarrow H_3C{-}\overset{CH_3}{\underset{CH_3}{C}}{-}\overset{\oplus}{\ddot{O}}(H){-}CH_3 \quad :\ddot{Br}:^{\ominus}$$

$$\downarrow$$

$$HBr + H_3C{-}\overset{CH_3}{\underset{CH_3}{C}}{-}O{-}CH_3$$

b)

$$CH_3CH_2{-}\ddot{O}H + H{-}O{-}\overset{O}{\underset{O}{S}}{-}OH \longrightarrow CH_3CH_2{-}\overset{\oplus}{\ddot{O}}H_2 + {}^{\ominus}:\ddot{O}{-}\overset{O}{\underset{O}{S}}{-}OH$$

$$CH_3CH_2{-}\overset{\oplus}{\ddot{O}}H_2 + CH_3CH_2{-}\ddot{O}{-}H \longrightarrow CH_3CH_2{-}\overset{H}{\overset{\oplus}{\ddot{O}}}{-}CH_2CH_3 \quad H{-}\ddot{O}{-}H$$

$$\downarrow$$

$$CH_3CH_2OCH_2CH_3 + H_3O^{\oplus}$$

c)

$H-O-SO_2-OH$

$+\ {}^{\ominus}O-SO_2-OH$

$\ddot{O}CH_3$ H

OH

$\overset{\oplus}{O}CH_3$ H

$^{\ominus}O-SO_2-OH$

H_2SO_4 + OH, OCH_3

d)

H—I

$:\ddot{I}:^{\ominus}$

$\ddot{O}H$

$\overset{\oplus}{O}H_2$

$:\ddot{I}:^{\ominus}$

$\oplus$ + H_2O

I

e)

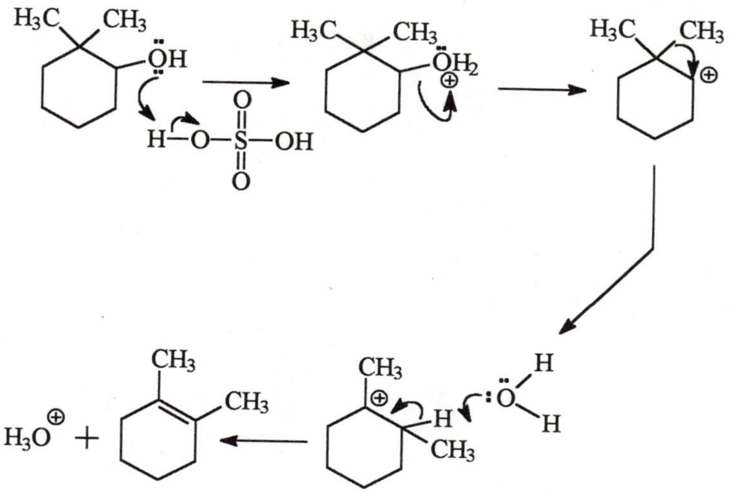

H_3C CH_3 $\ddot{O}H$ ⟶ H_3C CH_3 $\ddot{O}H_2^{\oplus}$ ⟶ H_3C CH_3 $\oplus$

H—O—S(=O)(=O)—OH

$H_3O^{\oplus}$ + CH_3 CH_3 ⟵ CH_3 $\oplus$ H CH_3 :Ö(H)H

9.43 Attack according to path a gives one enantiomer whereas attack according to path b gives the other.

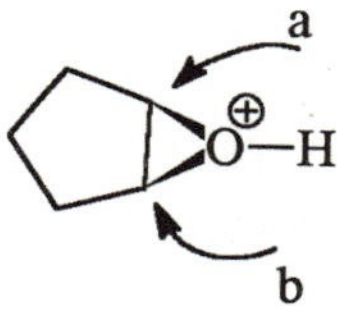

a, b, $\overset{\oplus}{O}$—H

9.44

CH_2–Br ⟶ $\overset{\oplus}{C}H_2$ ⟷ CH_2 $\oplus$ + :Br:$^{\ominus}$

H, H—Ö—Et

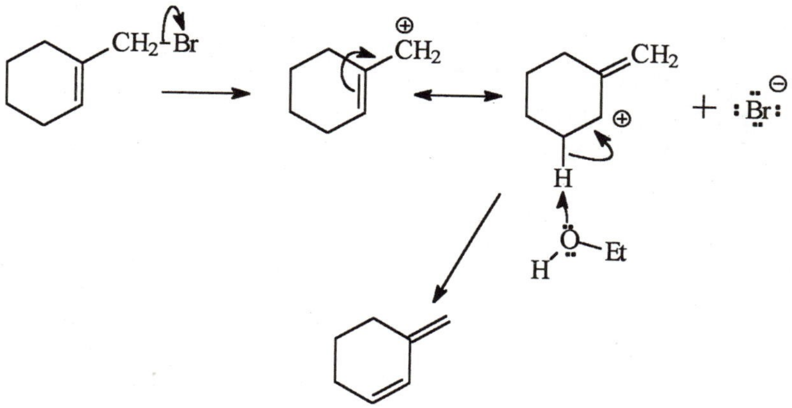

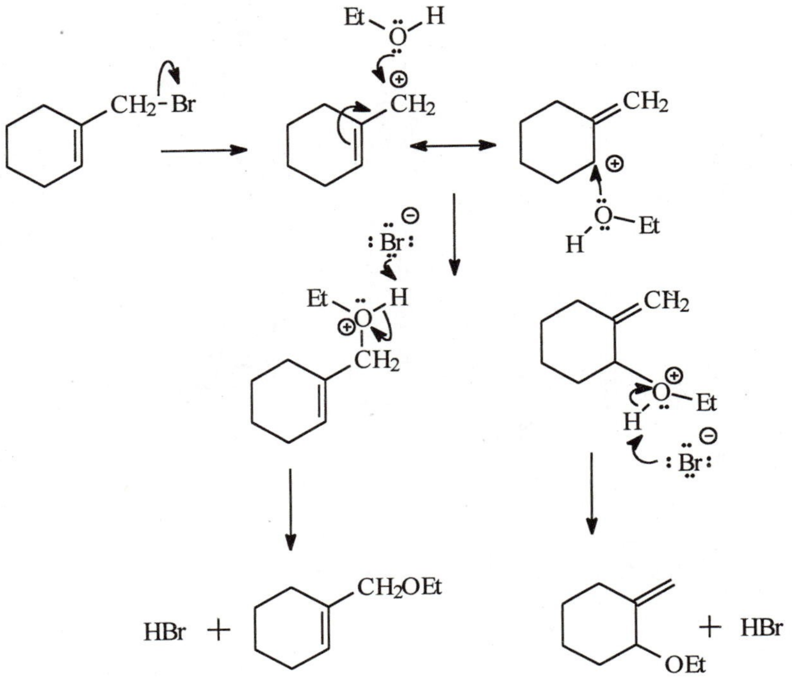

9.45 a) hydrogens removed so oxidation.
b) hydrogens added so reduction.
c) hydrogens removed so oxidation.
d) hydrogens added and oxygen removed so reduction.

9.46

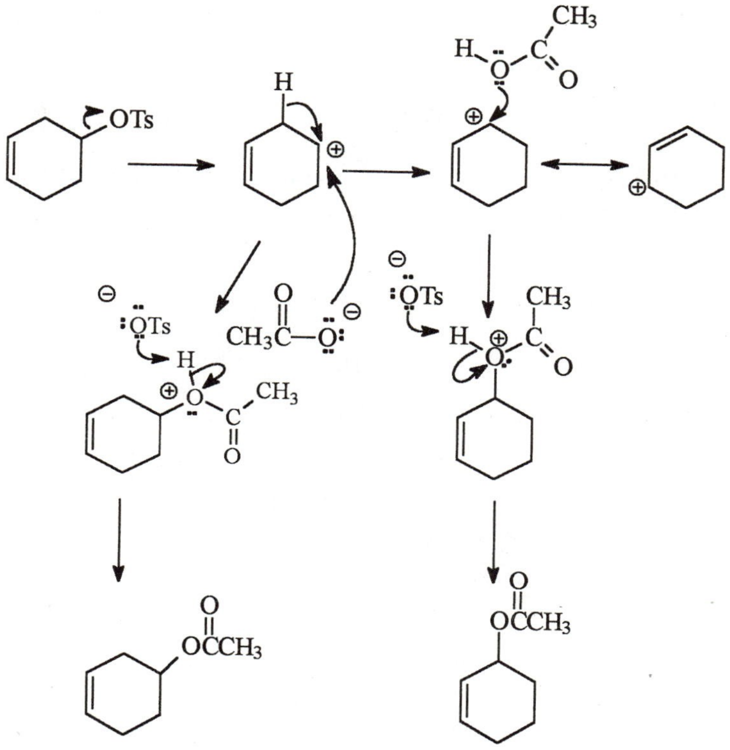

OTs

H

CH$_3$

H–Ö–C=O

$^{\ominus}$:ÖTs

CH$_3$C(=O)–Ö:$^{\ominus}$

$^{\ominus}$:ÖTs

H–O$^{\oplus}$–C(=O)CH$_3$

OCCH$_3$ (O)

OCCH$_3$ (O)

9.47

$$H_2C\overset{O}{—}CH_2 \xrightarrow[H_2O]{Na_2S} HO—CH_2CH_2—S—CH_2CH_2—OH$$

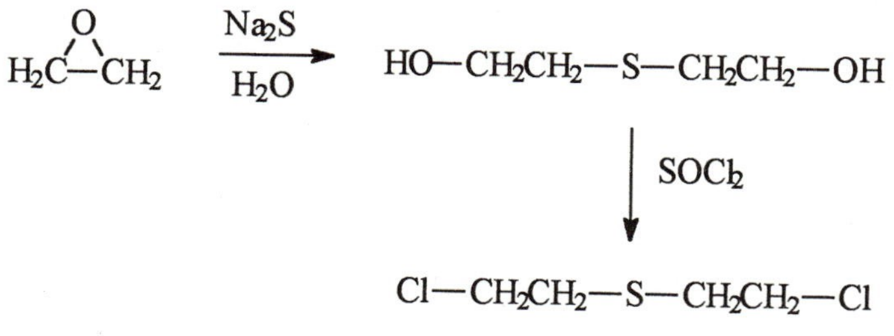

$$\downarrow SOCl_2$$

$$Cl—CH_2CH_2—S—CH_2CH_2—Cl$$

9.48 The conjugate base produced in the first step of the reaction acts as the nucleophile in the second step. The H on the oxygen attached directly to the benzene ring is more acidic, due to resonance stabilization of its conjugate base and is preferentially removed by the base.

9.49 A chlorine attached to an sp^2-hybridized carbon does not leave in S_N1 or S_N2 reactions.

9.50

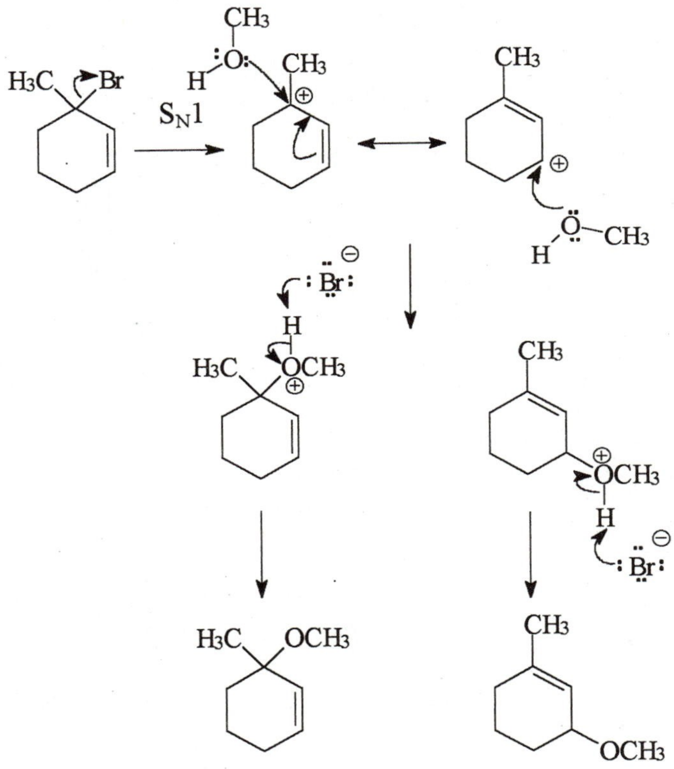

S_N1

9.51

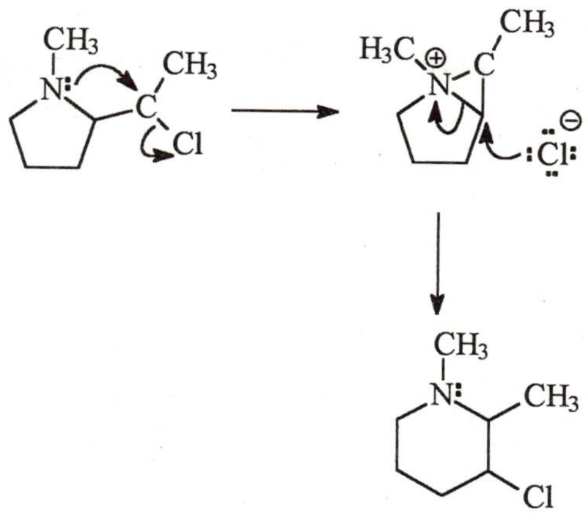

After completing this chapter, you should be able to:

1. Show the major product(s) of any of the reactions discussed in this chapter.
   Problems 9.1, 9.5, 9.9, 9.16, 9.18, 9.21, 9.24, 9.26, 9.29, 9.30, 9.36, 9.40.

2. Show the stereochemistry of the product(s).
   Problems 9.3, 9.4, 9.12, 9.14, 9.19, 9.22, 9.25, 9.31, 9.43.

3. Write the mechanism of the reaction.
   Problems 9.10, 9.11, 9.13, 9.15, 9.42, 9.44, 9.46, 9.50, 9.51.

4. Synthesize compounds using these reactions.
   Problems 9.7, 9.8, 9.17, 9.20, 9.23, 9.27, 9.28, 9.32, 9.33, 9.34, 9.35, 9.37, 9.38, 9.39, 9.41.

Chapter 10
ADDITIONS TO CARBON-CARBON DOUBLE AND TRIPLE BONDS

10.1 Like an S_N1 reaction, the first step is the rate determining step in this electrophilic addition reaction. Since the first step in both reaction mechanisms is the formation of a carbocation intermediate, structural features that stabilize the carbocation intermediate accelerate the electrophilic addition reaction.

a) $CH_2{=}CH_2$ < $CH_3CH_2CH{=}CH_2$ < $CH_3CH_2C(CH_3){=}CH_2$

slowest ... fastest

b) $CH_3CH{=}CH_2$ < $C_6H_5{-}CH{=}CH_2$ < $p\text{-}CH_3O{-}C_6H_4{-}CH{=}CH_2$

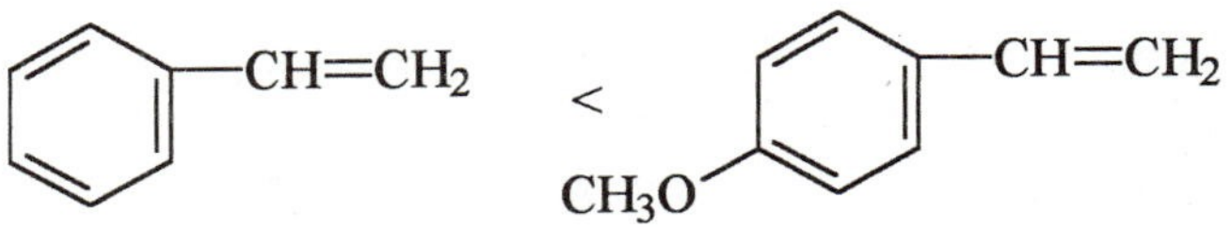

slowest ... fastest

10.2 The two secondary carbocations shown below, of approximately equal stabilities, can be formed, resulting in the formation of two products.

$$CH_3\overset{\oplus}{C}HCH_2CH_2CH_2CH_3 \quad \text{and} \quad CH_3CH_2\overset{\oplus}{C}HCH_2CH_2CH_3$$

10.3

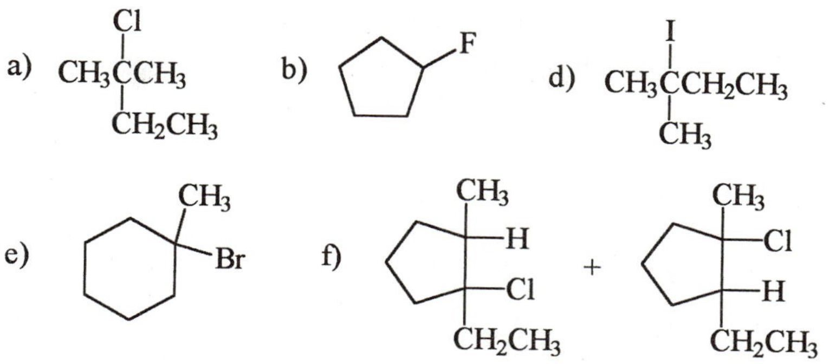

a) $CH_3C(Cl)(CH_2CH_3)CH_3$

b) Cyclopentyl–F

d) $CH_3C(I)(CH_3)CH_2CH_3$

e) 1-Bromo-1-methylcyclohexane: CH_3, Br

f) Cyclopentane with CH_3, H, Cl, CH_2CH_3 + cyclopentane with CH_3, Cl, H, CH_2CH_3

10.4

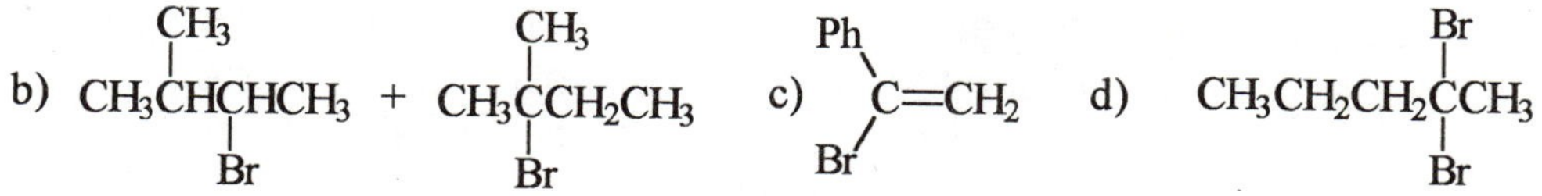

b) $CH_3CH(CH_3)CH(Br)CH_3$ + $CH_3C(CH_3)(Br)CH_2CH_3$

c) Ph(Br)C=CH_2

d) $CH_3CH_2CH_2C(Br)_2CH_3$

10.5

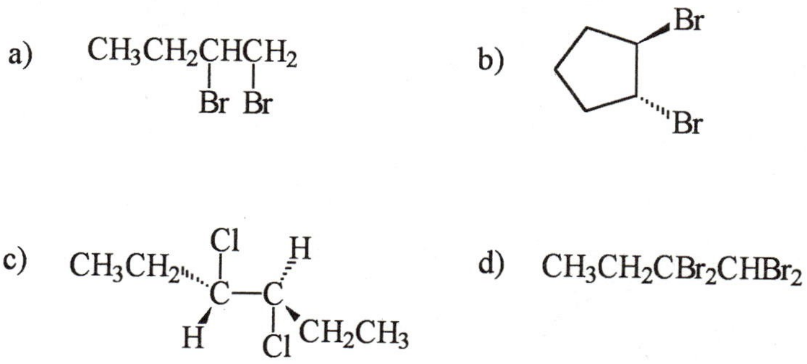

a) $CH_3CH_2CH(Br)CH_2Br$

b) trans-1,2-Dibromocyclopentane (Br, Br)

c) CH_3CH_2, Cl, H–C–C–H, Cl, CH_2CH_3

d) $CH_3CH_2CBr_2CHBr_2$

10.6

Cyclohexene + :Br–Br: → bromonium ion (Br:⊕) + ⊖ :Br: → trans-1,2-dibromocyclohexane (Br, Br) + trans-1,2-dibromocyclohexane (Br, Br)

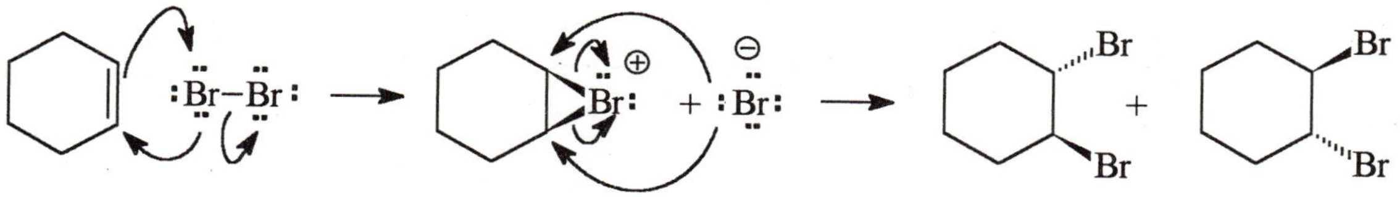

10.7 The right one reacts faster because the methyl group makes the double bond more nucleophilic.

10.8 A bromonium ion is more stable than a chloronium ion because bromine, being less electronegative than chlorine, is better able to accommodate a positive charge. Therefore, the left reaction should be more stereoselective than the right reaction.

10.9

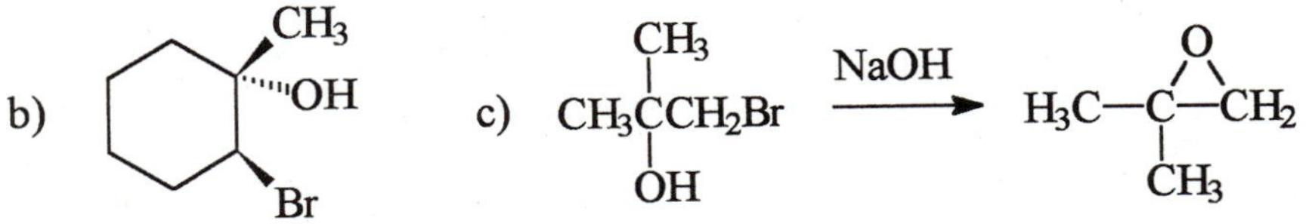

b) Cyclohexane with CH_3, OH and Br substituents

c) $CH_3C(CH_3)(OH)CH_2Br \xrightarrow{NaOH}$ epoxide: $H_3C-C(CH_3)(-O-)CH_2$

10.10 These are acid catalyzed hydrolysis reactions, where the electrophile is H^+ ion and the nucleophile is H_2O. Remember, the first step of electrophilic addition is the formation of a carbocation intermediate, so carbocation rearrangement is inevitable.

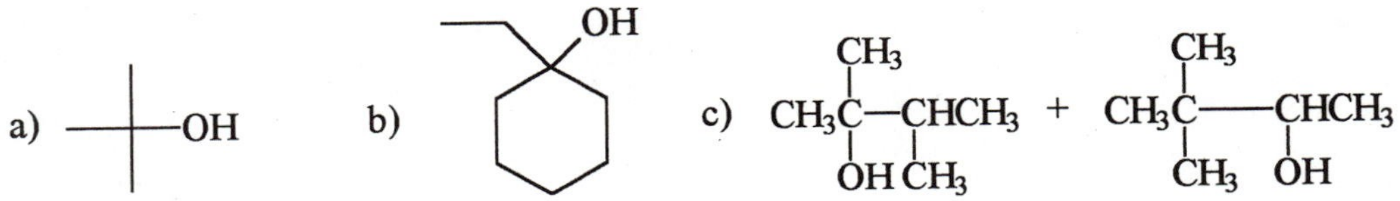

a) $(CH_3)_3C-OH$

b) 1-ethylcyclohexanol (OH)

c) $CH_3C(OH)(CH_3)-CHCH_3(CH_3)$ + $CH_3C(CH_3)(CH_3)-CHCH_3(OH)$

10.11

$CH_3CH{=}CH_2 + H{-}\ddot{O}^{\oplus}{-}H\,(H) \longrightarrow CH_3\overset{\oplus}{C}HCH_3 + H{-}\ddot{O}{:}\,(H)$

$CH_3\overset{\oplus}{C}HCH_3 + H_2\ddot{O}{:} \longrightarrow CH_3CH(\overset{\oplus}{O}H_2)CH_3$

$CH_3CH(\overset{\oplus}{O}H_2)CH_3 + H_2\ddot{O}{:} \longrightarrow CH_3CH(\ddot{O}H)CH_3 + H{-}\overset{\oplus}{\ddot{O}}{-}H\,(H)$

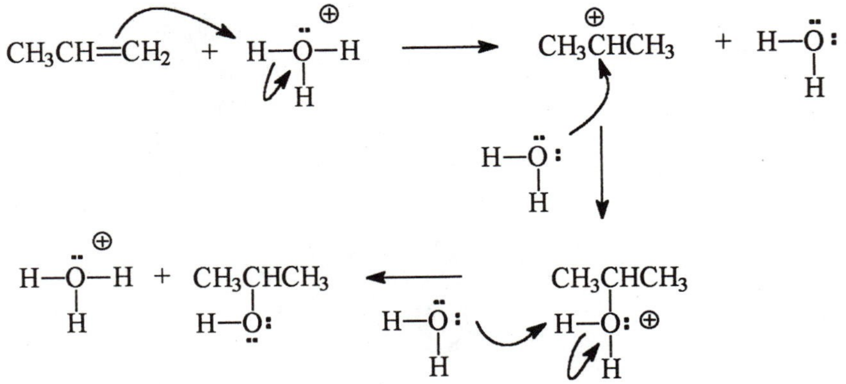

Phenylethene reacts faster because the intermediate carbocation formed from this alkene is stabilized by resonance.

10.12

$$CH_3C(CH_3){=}CH_2 + H{-}\overset{\oplus}{O}(H){-}CH_3 \longrightarrow CH_3\overset{\oplus}{C}(CH_3)CH_3 + H{-}\ddot{O}{:}{-}CH_3$$

$$CH_3\overset{\oplus}{C}(CH_3)CH_3 + H{-}\ddot{O}{:}{-}CH_3 \longrightarrow CH_3C(CH_3)(\overset{\oplus}{O}(H)CH_3)CH_3$$

$$CH_3C(CH_3)(\overset{\oplus}{O}(H)CH_3)CH_3 + H{-}\ddot{O}{:}{-}CH_3 \longrightarrow H_3C{-}\overset{\oplus}{O}(H){-}H + CH_3C(CH_3)(\ddot{O}{-}CH_3)CH_3$$

10.13

a) OH

b) OH

c) $CH_3CH_2\overset{O}{\overset{\|}{C}}CH_2CH_2CH_3$

d) $CH_3CH_2CH_2\overset{O}{\overset{\|}{C}}CH_3$

10.14 The left synthesis is better because only 3-hexanone, the desired ketone, is produced. The right reaction also produces 2-hexanone.

10.15

a) OH

b) HO, H, H, C, C, CH_3, H_3C, CH_2CH_3

c) OH, CH_2CH_3

10.16

$$CH_3C(CH_3)_2-CH(OH)CH_2CH_3 \quad + \quad CH_3C(CH_3)_2CH_2CH(OH)CH_3$$

The right alcohol is the major product because the boron prefers to bond to the less hindered carbon of the alkene.

10.17

b) (hex-3-ene) $\xrightarrow[\text{2) } NaBH_4\text{, } NaOH]{\text{1) } Hg(O_2CCH_3)_2\text{, } H_2O}$ (hexan-3-ol, OH)

c) (hex-1-ene) $\xrightarrow[\text{2) } H_2O_2\text{, } NaOH]{\text{1) } BH_3\text{, THF}}$ (hexan-1-ol, OH)

10.18

a) $CH_3CH_2CH_2C(=O)CH_2CH_2CH_2CH_3$ b) $PhCH_2CH_2CH_2CH(=O)$

10.19

b) $CH_3CH_2CH_2CH_2CH_2CH(=O) \xleftarrow[\text{2) } H_2O_2\text{, } NaOH]{\text{1) disiamylborane}} CH_3CH_2CH_2CH_2C{\equiv}CH$

$CH_3CH_2CH_2CH_2Br \xrightarrow{HC{\equiv}C{:}^{\ominus}\ Na^{\oplus}} CH_3CH_2CH_2CH_2C{\equiv}CH$

c) $CH_3(CH_2)_3C(=O)CH_2(CH_2)_3CH_3 \xleftarrow[H_2SO_4,\ Hg^{2\oplus}]{H_2O} CH_3(CH_2)_3C{\equiv}C(CH_2)_3CH_3$

$CH_3(CH_2)_3C{\equiv}CH \xrightarrow[\text{2) } CH_3CH_2CH_2CH_2Br]{\text{1) } NaNH_2} CH_3(CH_2)_3C{\equiv}C(CH_2)_3CH_3$

as part (b) $\longrightarrow$ $CH_3(CH_2)_3C{\equiv}CH$

10.20

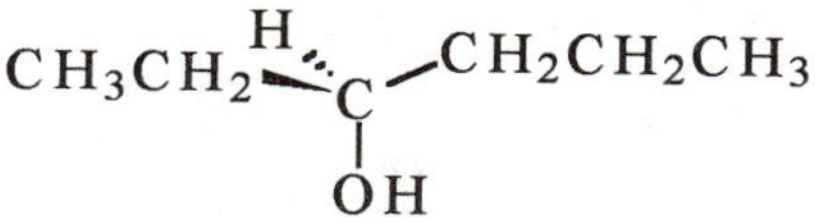

CH_3CH_2–C(H)(OH)–$CH_2CH_2CH_3$ (*S*)-enantiomer

10.21 (*R*)-2-Butanol can be prepared in good enantiomeric excess by using (*S,S*)-*trans*-2,5-dimethylborolane.

10.22

a) 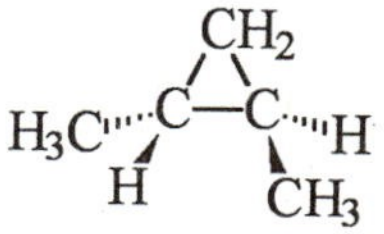 H_3C–C(H)–C(H)(CH_3) bridged by CH_2 b) H_3C–C(H)–C(H)–CH_3 bridged by CH_2 c) cyclopentane fused to CBr_2 d) 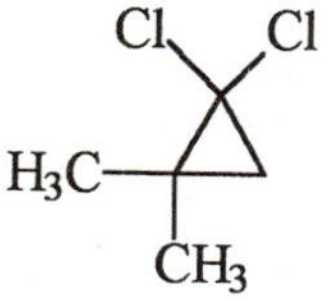 H_3C, CH_3, Cl, Cl cyclopropane

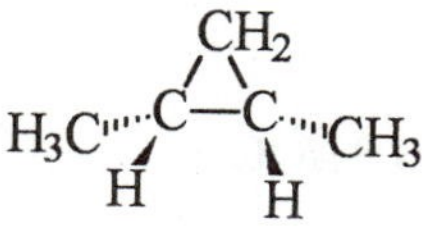

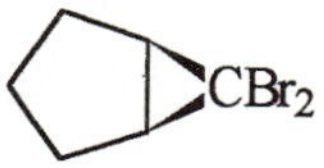

10.23 These are epoxidation reactions, in which an oxygen atom is added to the alkene to form a three membered ring containing an oxygen. Like the singlet carbene, the addition occurs with syn stereochemistry.

a) 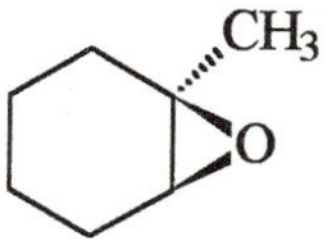 epoxycyclohexane with CH_3 b)  H_3C–C(H)–C(H)–CH_3 bridged by O ⟶ HO, CH_3, H_3C–C(H)–C(H)(OH)

10.24 The reagents OsO_4 and $KMnO_4$ add hydroxy groups to both carbons of the double bond. The addition occurs with a syn stereochemistry.

a) 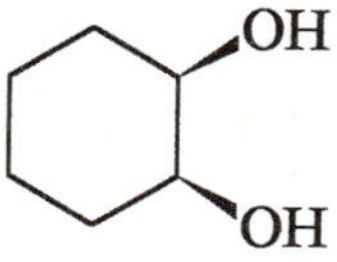 cyclohexane-1,2-diol (OH, OH) b) 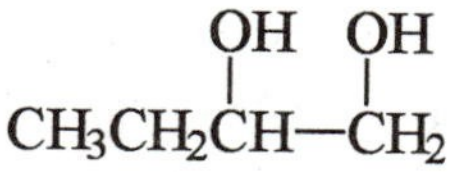 $CH_3CH_2CH(OH)-CH_2(OH)$ c) 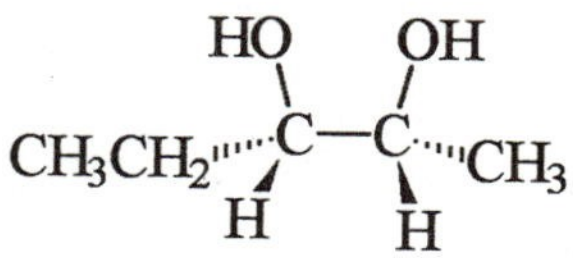 CH_3CH_2–C(H)(HO)–C(H)(OH)–CH_3

10.25

a) 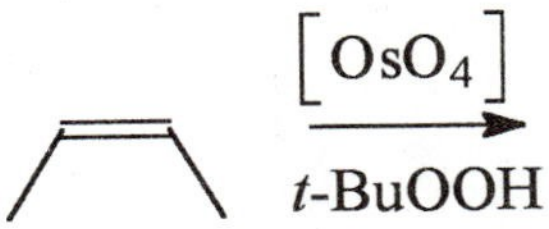 $\xrightarrow[t\text{-BuOOH}]{[OsO_4]}$ b) 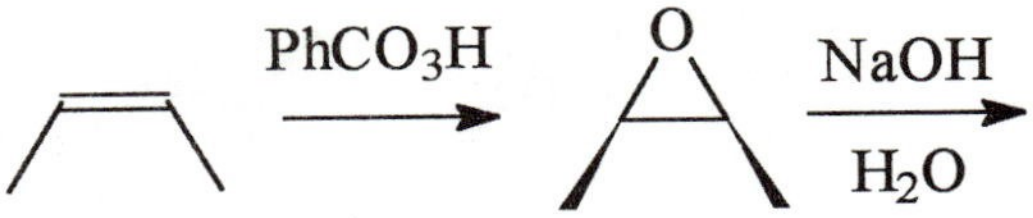 $\xrightarrow{PhCO_3H}$ $\xrightarrow[H_2O]{NaOH}$

10.26 Ozone reacts with carbon-carbon double bond to form the ozonide. The O-O bond of the ozonide is cleaved on treatment with a reducing agent like dimethyl sulfide, to form two carbonyl groups.

a) 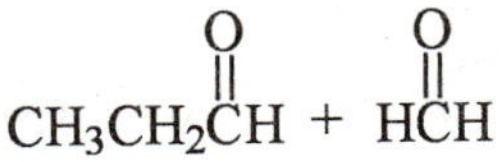 $CH_3CH_2\overset{O}{\overset{\|}{C}}H + H\overset{O}{\overset{\|}{C}}H$ b) 2 $CH_3\overset{O}{\overset{\|}{C}}CH_3$ c) O, CH_3, H, O

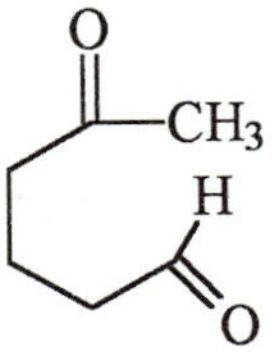

d) 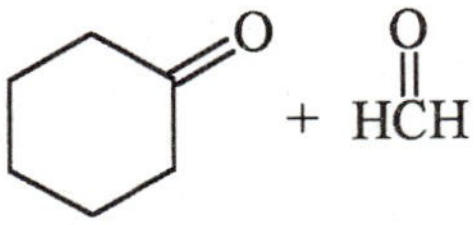 O $+ H\overset{O}{\overset{\|}{C}}H$ e) 2 H, O, H, O

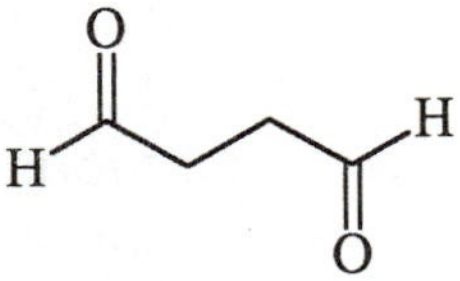

10.27

a) 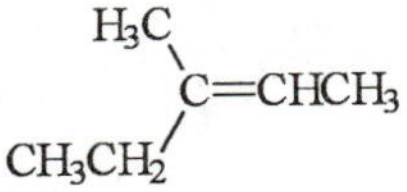 H_3C, $C{=}CHCH_3$, CH_3CH_2 b) 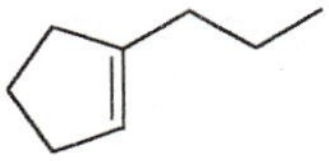 d) 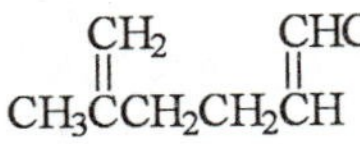 $CH_3\overset{CH_2}{\overset{\|}{C}}CH_2CH_2\overset{CHCH_3}{\overset{\|}{C}}H$ or $CH_3\overset{CHCH_3}{\overset{\|}{C}}CH_2CH_2\overset{CH_2}{\overset{\|}{C}}H$

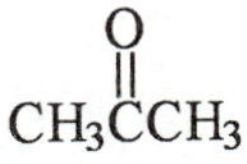

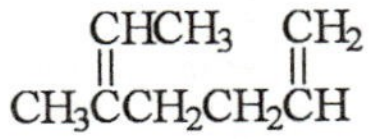

10.28

a) 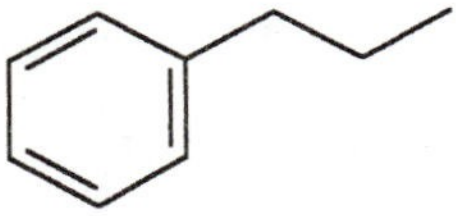 b) 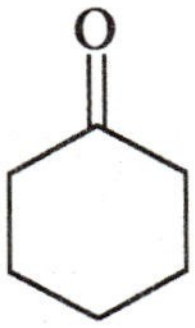 O c) $CH_3CH_2CH_2CH_2CH_3$

d) 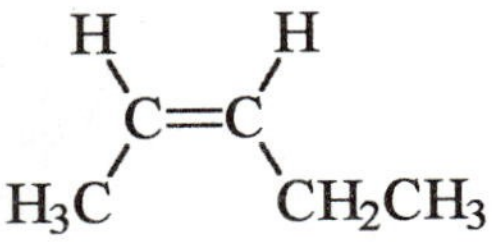 H, H, $C{=}C$, H_3C, CH_2CH_3 e)

10.29

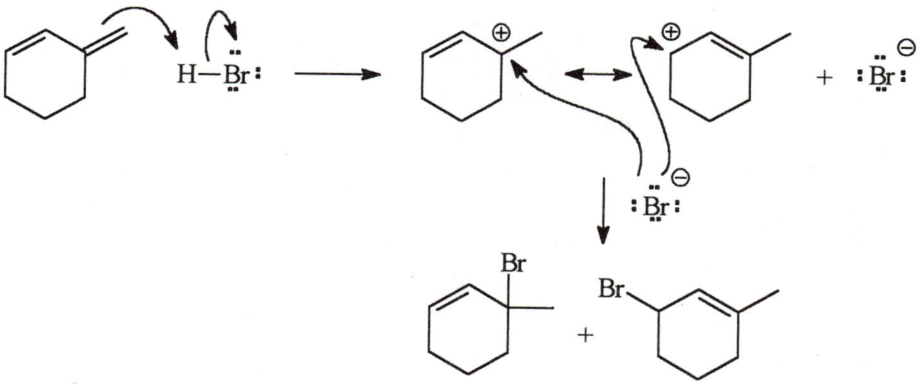

H—Br: → (carbocation) ↔ (carbocation) + :Br:$^{\ominus}$

:Br:$^{\ominus}$

↓

Br + Br

The H$^{\oplus}$ adds so as to give the most stable carbocation in the first step.

10.30

a) 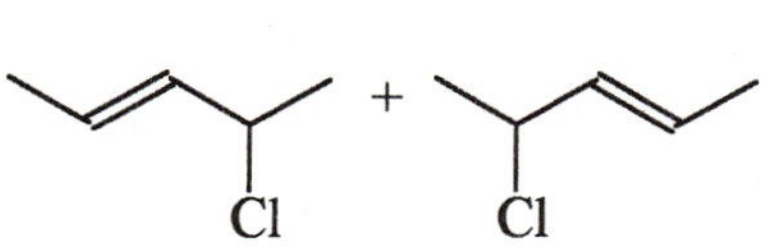 Cl + Cl

b) 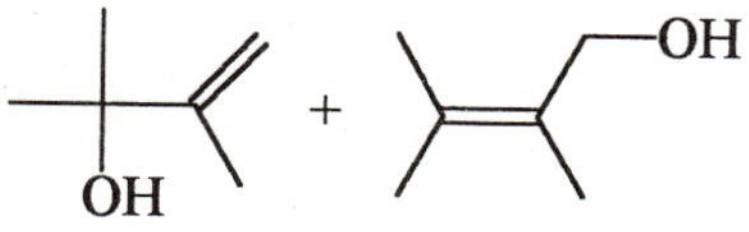 OH + OH

10.31

a) 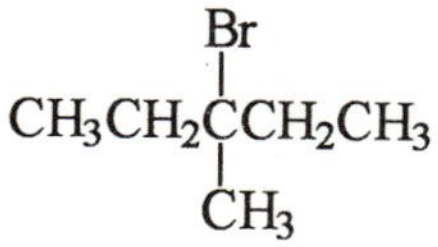 $CH_3CH_2C(Br)(CH_3)CH_2CH_3$

b)

c) 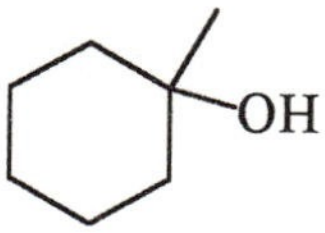 OH

d) 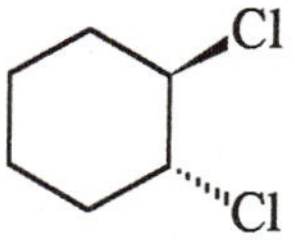 Cl, Cl

e) 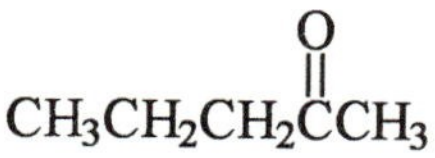 $CH_3CH_2CH_2\overset{O}{\overset{\|}{C}}CH_3$

f) 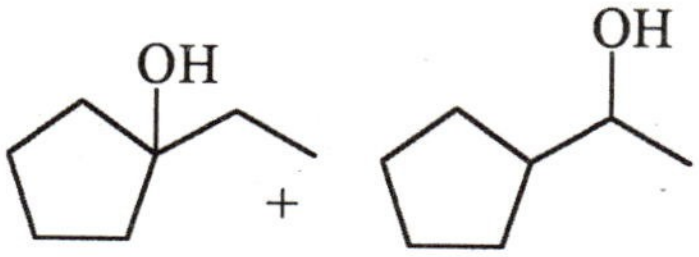 OH + OH

g) 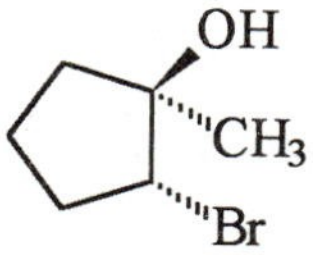 OH, CH_3, Br

h) 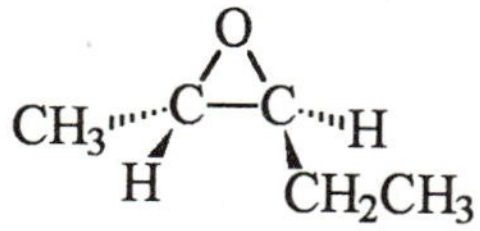 CH_3, O, H, H, CH_2CH_3

i) 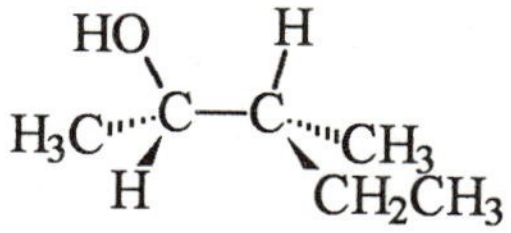 HO, H, H_3C, H, CH_3, CH_2CH_3

j) 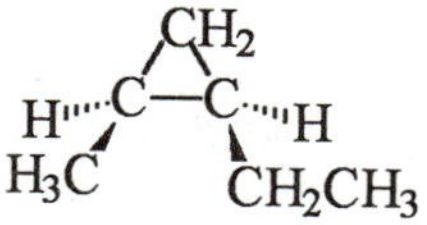 CH_2, H, H_3C, H, CH_2CH_3

k) 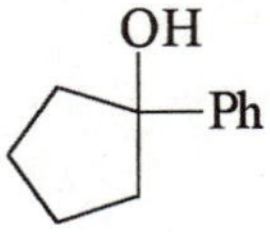 l) 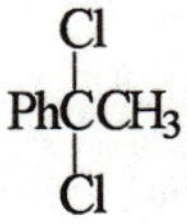 m) 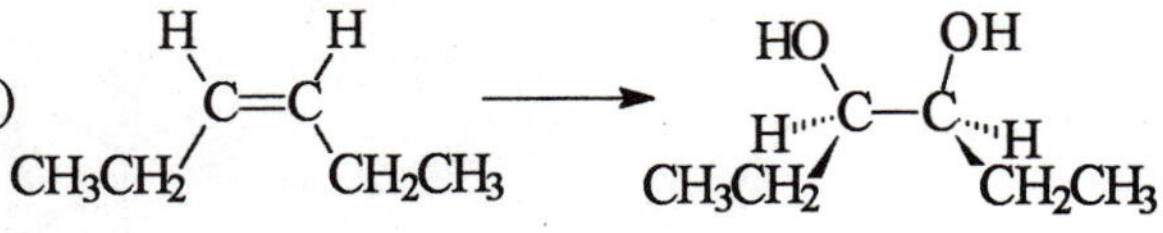

n) 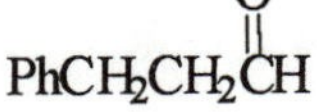

10.32

a) 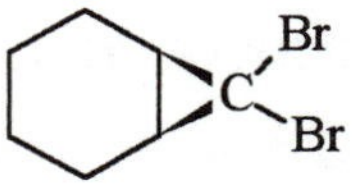 b) 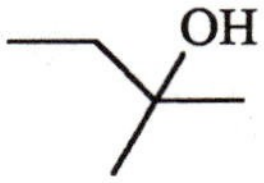 c) 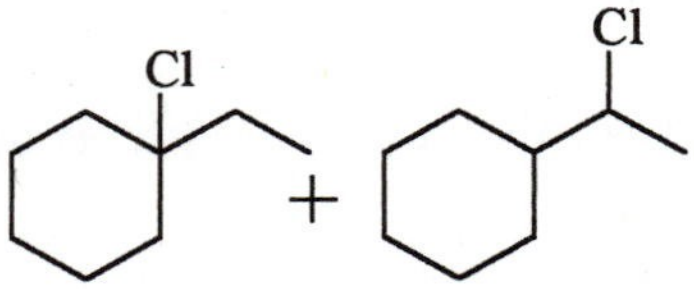

d) 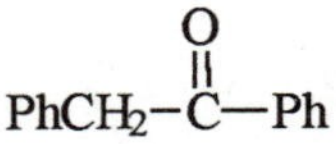 e) 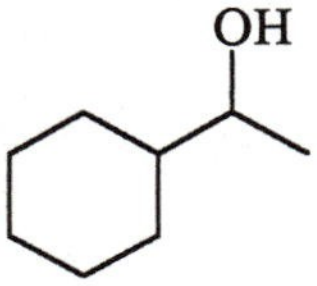 f) 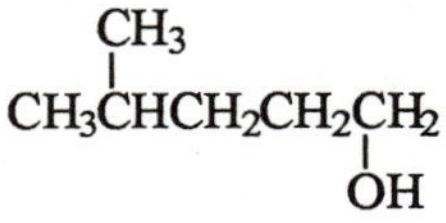

g) 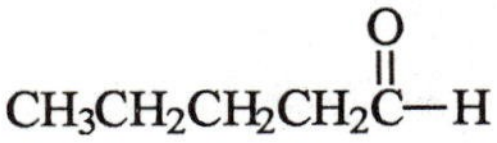 h) 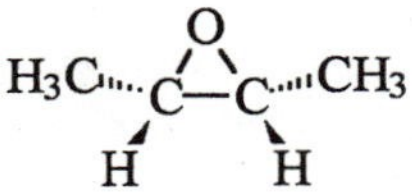

i) 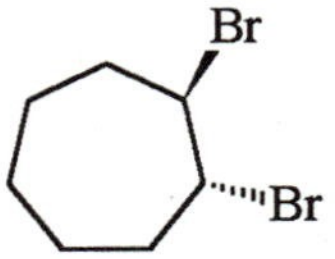 j) 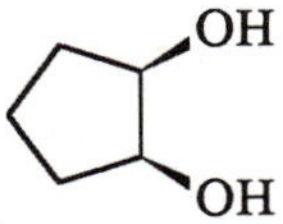 k) 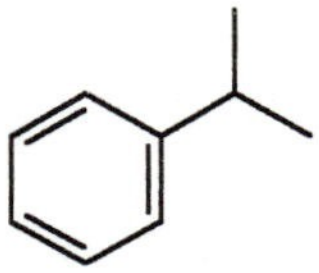

l) 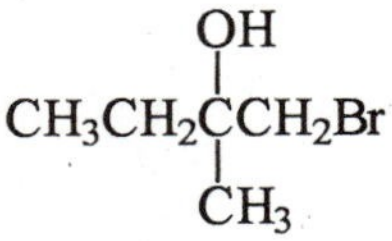 m) 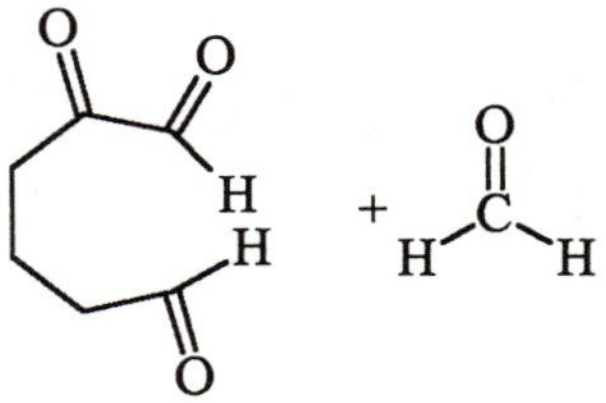

n) 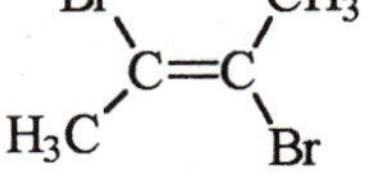

10.33

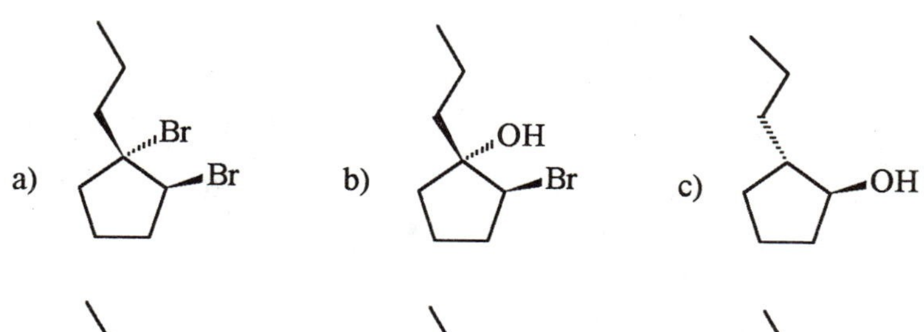

d) Br

e) OH

f) OH OH

g) O O H

10.34

a) H Ph C=C Ph H

b)

c)

d)

e) H H C=C CH_3CH_2 CH_3

f)

10.35 a)

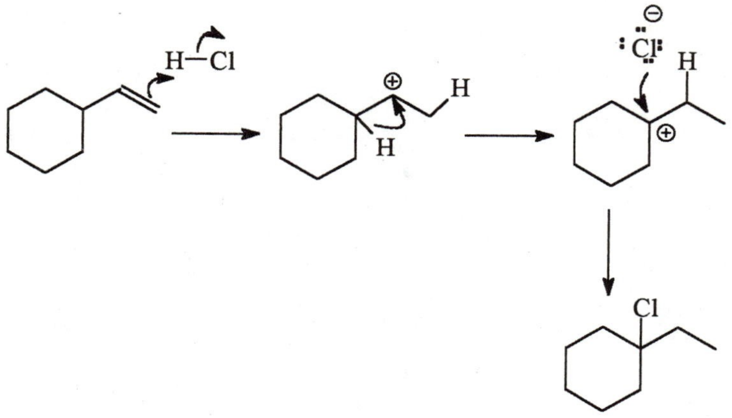

H—Cl
H
H
Cl
H
Cl

b)

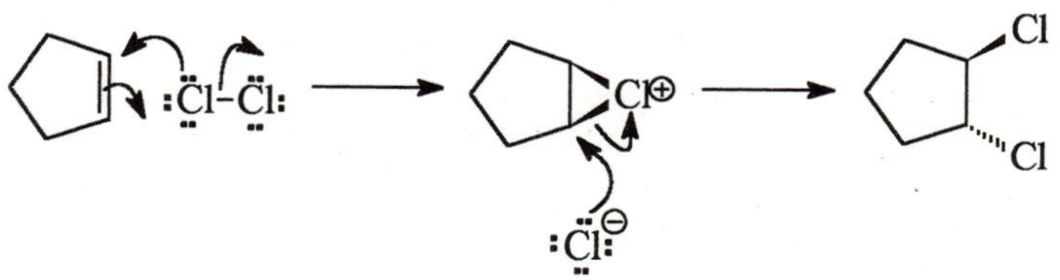

Cl–Cl
Cl
Cl
Cl
Cl

c)

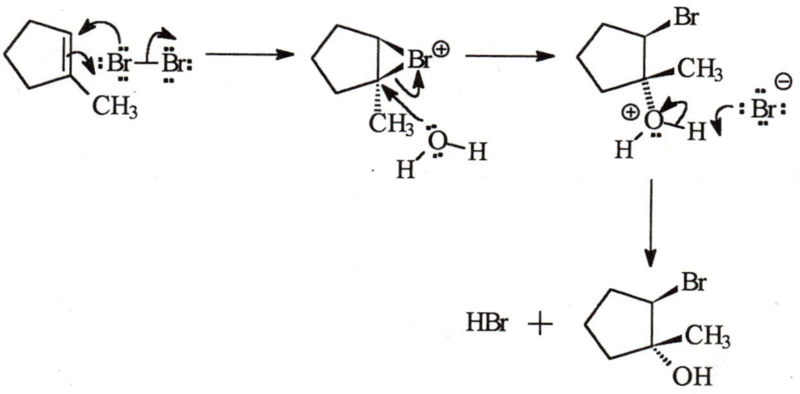

Br—Br
CH3
Br
CH3
O—H
H
Br
CH3
O
H
H
Br
HBr +
Br
CH3
OH

d)

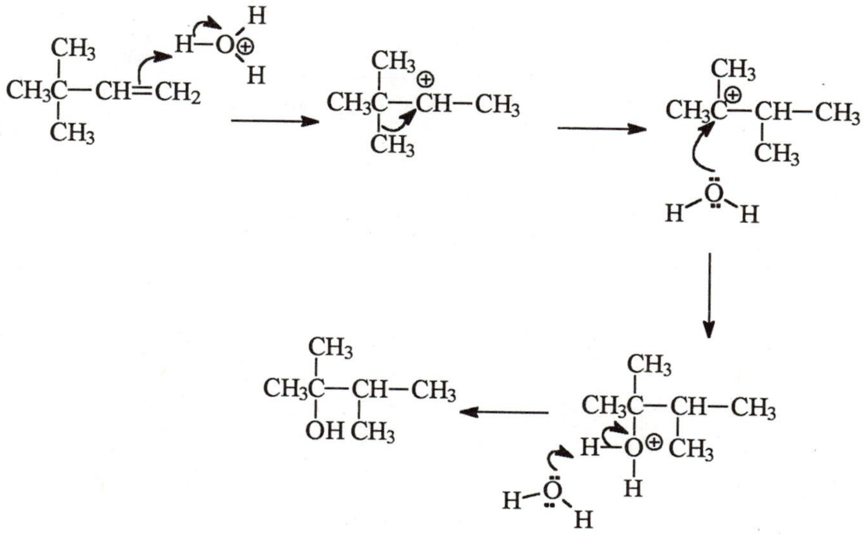

CH3 H—O⊕ H H
CH3C—CH=CH2
CH3
CH3 ⊕
CH3C—CH—CH3
CH3
CH3
CH3C⊕—CH—CH3
CH3
H—Ö—H

CH3
CH3C—CH—CH3
OH CH3
CH3
CH3C—CH—CH3
H—O⊕ CH3
H
H—Ö—H

e)

Cl
Cl—C—H ⊖:ÖH
Cl
Cl
Cl—C:⊖
Cl
+ H2O
Cl—C:
Cl
+ :Cl:⊖

Cl Cl
C
CH3C=CH
H CH3
Cl Cl
C
CH3C—CH
H CH3

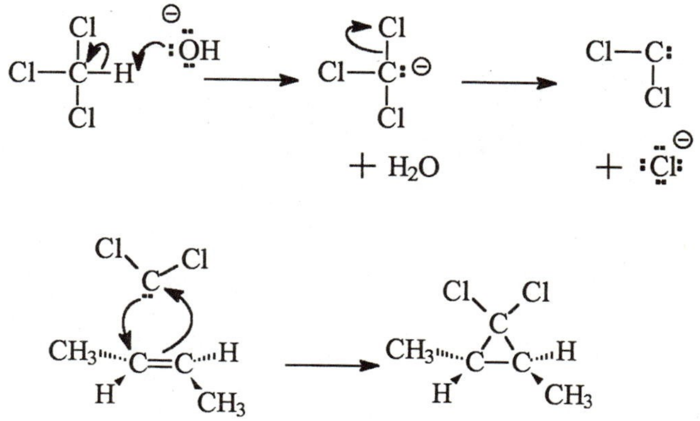

10.36

a) [1-pentene] $\xrightarrow[\text{2) } H_2O_2\text{, NaOH}]{\text{1) } BH_3\text{, THF}}$

b) [1-pentene] $\xrightarrow[\text{2) NaBH}_4]{\text{1) Hg(O}_2\text{CCH}_3)_2\text{, H}_2\text{O}}$

c) [1-pentene] $\xrightarrow[t\text{-BuOOH}]{OsO_4}$

d) [1-pentene] $\xrightarrow[H_2O]{Cl_2}$

e) [1-pentene] $\xrightarrow{HCl}$

f) [1-pentene] $\xrightarrow[\text{2) } H_2O_2\text{, NaOH}]{\text{1) } BH_3\text{, THF}}$ [pentan-1-ol] OH $\xrightarrow{PCC}$

g) [1-pentene] $\xrightarrow[Pt]{H_2}$

h) [1-pentene] $\xrightarrow{CH_3COOH \text{ (O above C)}}$

10.37

a) $CH_3C{\equiv}CH \xrightarrow[\text{2) } CH_3Br]{\text{1) } NaNH_2} CH_3C{\equiv}CCH_3 \xrightarrow[H_2SO_4]{HgSO_4,\ H_2O}$

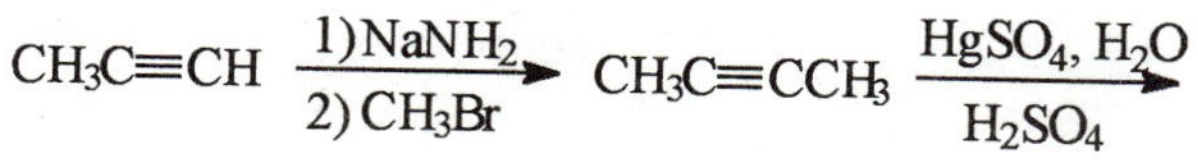

b) $CH_3C{\equiv}CH \xrightarrow[\text{2) } CH_3CH_2CH_2Br]{\text{1) } NaNH_2} CH_3C{\equiv}CCH_2CH_2CH_3$

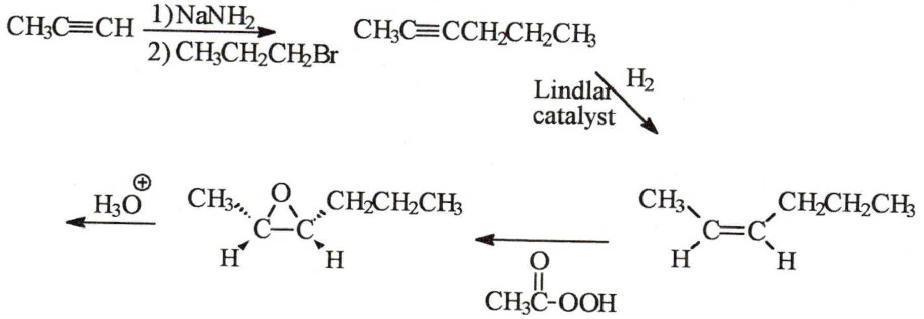

$\xrightarrow[\text{Lindlar catalyst}]{H_2}$ CH3/H–C=C–H/CH2CH2CH3 (cis alkene)

$\xrightarrow{CH_3\overset{O}{\overset{\|}{C}}\text{-OOH}}$ epoxide: CH3, H, O, CH2CH2CH3, H (C–C with O bridge)

$\xrightarrow{H_3O^{\oplus}}$

c) $CH_3C{\equiv}CH \xrightarrow[\text{2) } CH_3CH_2CH_2Br]{\text{1) } NaNH_2} CH_3C{\equiv}CCH_2CH_2CH_3$

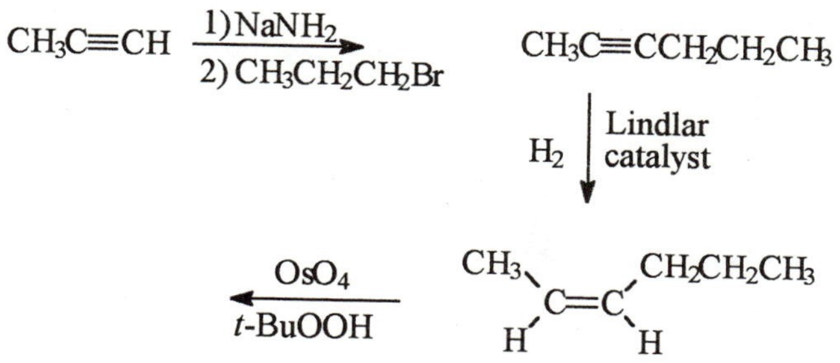

$\xrightarrow[\text{Lindlar catalyst}]{H_2}$ CH3/H–C=C–H/CH2CH2CH3 (cis alkene)

$\xrightarrow[t\text{-BuOOH}]{OsO_4}$

d) cyclohexanol (OH) $\xrightarrow[\text{heat}]{H_3PO_4}$ cyclohexene $\xrightarrow[H_2O]{Br_2}$

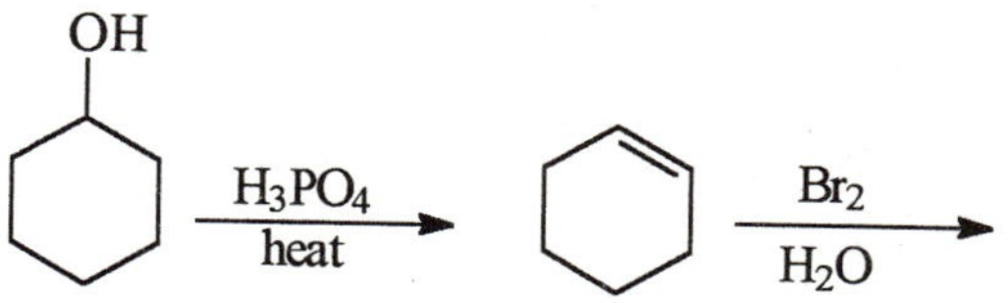

e) $HC{\equiv}CH$ $\xrightarrow[\text{2) } CH_3CH_2CH_2Br]{\text{1) } NaNH_2}$ $HC{\equiv}CCH_2CH_2CH_3$

H_2 $\downarrow$ Lindlar catalyst

$\xleftarrow[\text{2) } H_2O_2\text{, NaOH}]{\text{1) } BH_3\text{, THF}}$ H, H, C=C, $CH_2CH_2CH_3$, H

f) $CH_3C{\equiv}CH$ $\xrightarrow[\text{2) } CH_3Br]{\text{1) } NaNH_2}$ $CH_3C{\equiv}CCH_3$

H_2 $\downarrow$ Lindlar catalyst

$\xleftarrow[Zn(Cu)]{CH_2I_2}$ CH_3, H, C=C, CH_3, H

g) OH $\xrightarrow[\text{heat}]{H_3PO_4}$ $\xrightarrow{HCl}$

h) Cl, Ph $\xrightarrow{NaOH}$ Ph $\xrightarrow[\text{2. } H_2O_2\text{, NaOH}]{\text{1. } BH_3\text{, THF}}$

10.38

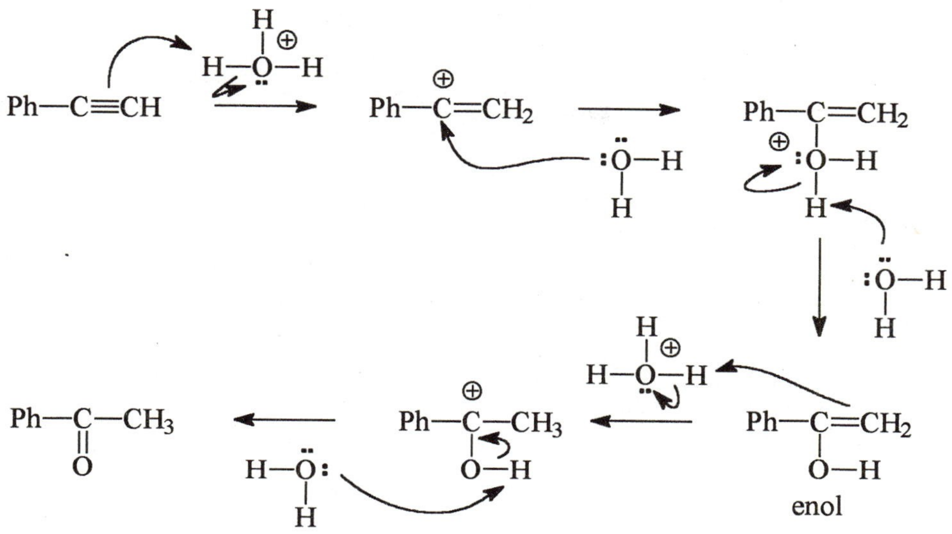

Ph—C≡CH → Ph—C(+)=CH₂ → Ph—C(=CH₂)—O(+)H₂ → Ph—C(=CH₂)—OH (enol) → Ph—C(+)(—CH₃)—OH → Ph—C(=O)—CH₃

enol

10.39 a) The left compound is has a faster rate because the carbocation intermediate is more stable.

b) The right compound has a faster rate because the nitro group destabilizes the carbocation intermediate.

c) The right compound has a faster rate because the carbocation intermediate is resonance stabilized.

10.40

a) The two chloronium ions that are formed in this reaction are non-superimposable mirror image structures.

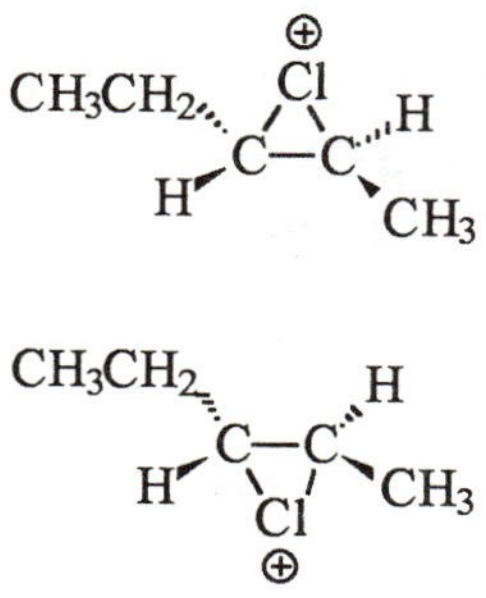

b) The products that are formed by attack of the chloride ion at each carbon of the two chloronium ions are enantiomers. The formation of the two enantiomers from the two chloronium ions is shown below.

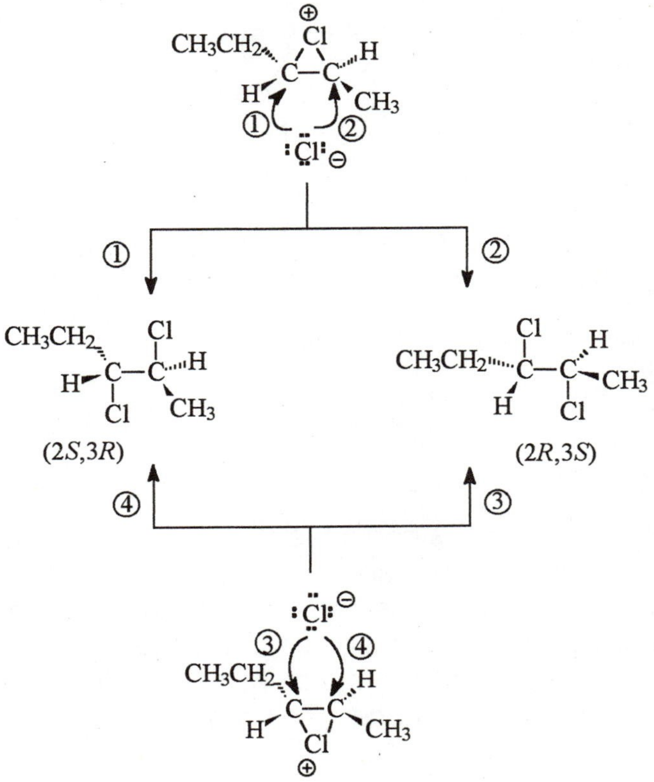

⊕
Cl
CH_3CH_2 C—C H
H CH_3
① ②
:Cl: ⊖

① ②

CH_3CH_2 Cl
H—C—C H
Cl CH_3
(2*S*,3*R*)

Cl H
CH_3CH_2—C—C—CH_3
H Cl
(2*R*,3*S*)

④ ③

:Cl: ⊖
③ ④
CH_3CH_2 H
C—C
H Cl CH_3
⊕

c) The two electrophilic carbons of the cyclic chloronium ion are not identical. Therefore, the percentages of nucleophile attack at these two carbon sites of one chloronium ion are not necessarily identical.

d) The nucleophile attack at the same carbons of the two chloronium ions will produce a racemic mixture. For example, in the above scheme in part (b), paths **1** and **3** are identical but the products are enantiomers; similarly paths **2** and **4** are identical but produce enantiomers.

10.41

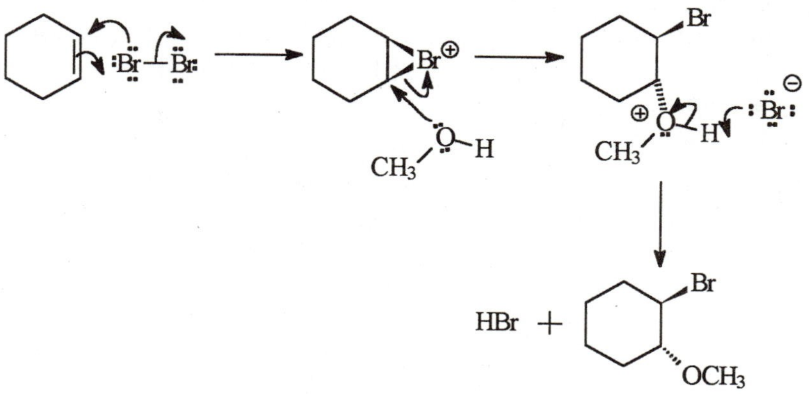

$$HBr + \text{trans-1-bromo-2-methoxycyclohexane}$$

10.42 In the first step of the oxymercuration reaction, the mercury electrophile adds to the double bond to form the cyclic mercurinium ion and then water reacts with the mercurinium ion as a nucleophile. If the reaction is run in methanol as solvent, the initially formed mercurinium ion will react with methanol in the same manner to give the following product.

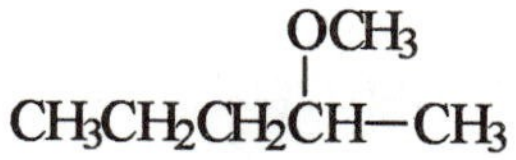

$$CH_3CH_2CH_2CH(OCH_3){-}CH_3$$

10.43

$$CH_2{=}C(OH){-}CH_3 \longrightarrow CH_2{=}C(O^-){-}CH_3 \longleftrightarrow {}^-CH_2{-}C(=O){-}CH_3$$

$$\downarrow$$

$$:\ddot{O}H^- + CH_3{-}C(=O){-}CH_3$$

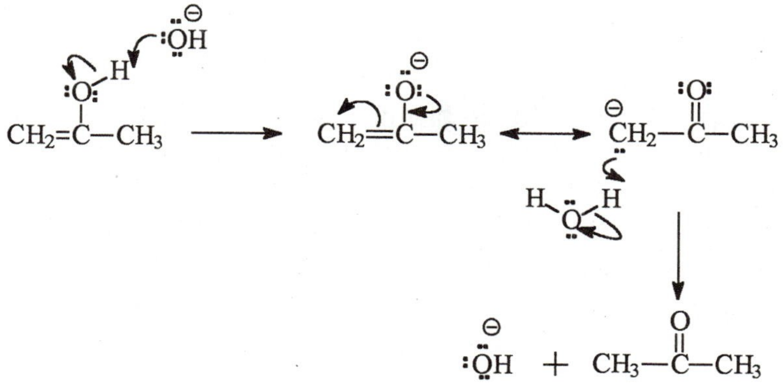

10.44 a) The DU for this compound is 2.

b) The disappearance of the bromine color is an indication of the presence of a carbon-carbon multiple bond(s) in the compound.

c) The catalytic hydrogenation of the unknown yields a product with two more hydrogens in its molecular formula than that of the unknown. This indicates that there is only one carbon-carbon double bond in the unknown. Therefore it must also have a ring.

d) The structure of the unknown is

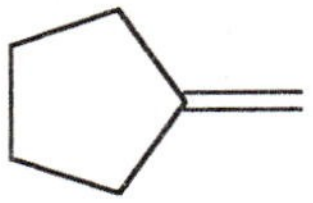

10.45 a) The DU of the compound is 2.

b) The result of the catalytic hydrogenation indicates that the unknown compound has two pi bonds.

c) The result of hydrogenation in the presence of Lindlar catalyst shows that the unknown has a carbon-carbon triple bond.

d) The structure of the unknown is

$$CH_3CH(CH_3){-}C{\equiv}C{-}CH_2CH_3$$

10.46 Addition reactions involving boranes are regioselective and are greatly influenced by sterics. Since the carbons of the double bond are both monosubstituted, BH_3 adds to both carbons giving slightly more of the product in which it has added to the less hindered carbon attached to the methyl group. The steric effect is more pronounced with disiamylborane because it has two bulky alkyl groups attached to the boron. Therefore, disiamylborane adds almost entirely to the less hindered carbon.

10.47 In this alkene substrate, the primary carbocation intermediate is more stable than the secondary carbocation because the strong inductive effect of the CF_3 group destabilizes the secondary carbocation. Therefore, the reaction occurs with anti-Markovnikov regiochemistry.

10.48 The hydration reaction of this alkene 10^{15} times faster than of ethene because the carbocation intermediate formed from this vinyl ethyl ether is resonance stabilized.

$$CH_3CH_2-\ddot{\underset{\cdot\cdot}{O}}-\overset{\oplus}{C}H-CH_3 \longleftrightarrow CH_3CH_2-\overset{\oplus}{\underset{\cdot\cdot}{O}}=CH-CH_3$$

10.49 Of the two vinyl cation intermediates that could be formed from the addition of HCl to propyne, the secondary vinyl cation is more stable.

$$CH_3-C\equiv CH \xrightarrow{HCl} \underset{\text{primary vinyl cation}}{CH_3-CH=\overset{\oplus}{C}H} \quad \underset{\text{secondary vinyl cation}}{CH_3-\overset{\oplus}{C}=CH_2}$$

10.50 Addition of chlorine or bromine to simple alkenes results in an anti addition because the halonium ion is more stable than the carbocation that might be formed instead. However, deviation from anti addition occurs as the carbocation becomes more stable due to resonance or inductive effects. The reaction that proceeds through a carbocation intermediate should result in a mixture of syn and anti additions because the planar carbocation can be attacked from either side by the incoming nucleophile.

The first reaction proceeds exclusively through a halonium ion with anti addition due to the destabilization of the carbocation by the two CF_3 groups on the benzene ring. The observation of both the syn and anti addition products in the next two reactions indicates that these reactions proceed at least partly via a carbocation rather than a halonium ion. Because the CH_3O group stabilizes the carbocation, the third reaction gives more syn addition than the second. The addition of Cl_2 to 1-phenylpropene produces more syn addition than the addition of Br_2 because a chloronium ion is less stable than a bromonium ion (Cl is more electronegative) and, thus, the addition of Cl_2 proceeds mainly or entirely via a carbocation intermediate.

10.51

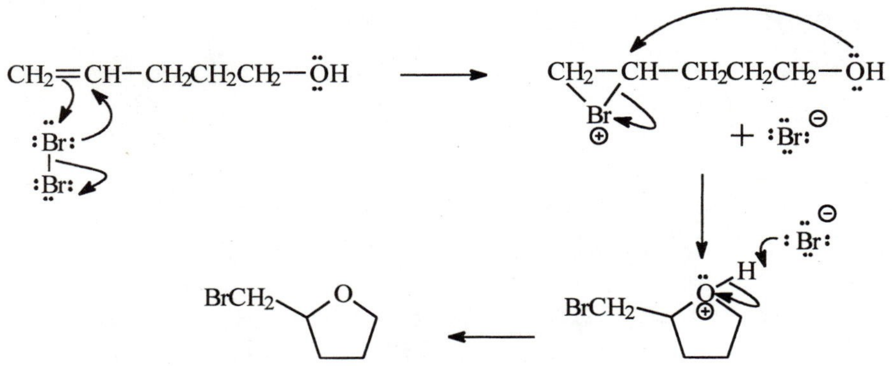

$$CH_2{=}CH{-}CH_2CH_2CH_2{-}\ddot{O}H \xrightarrow{Br_2} CH_2{-}CH{-}CH_2CH_2CH_2{-}\ddot{O}H \;(\text{bromonium } Br^{+}) + :\ddot{Br}:^{-}$$

$$\longrightarrow BrCH_2\text{-(oxolanium, } O^{+}\text{–H)} + :\ddot{Br}:^{-} \longrightarrow BrCH_2\text{-(tetrahydrofuran)}$$

BrCH₂ O BrCH₂

10.52 a) The DU of limonene is 3. The hydrogenation results indicate that limonene has only two pi bonds. The other DU must be due to a ring.

b) Possible structures for limonene are

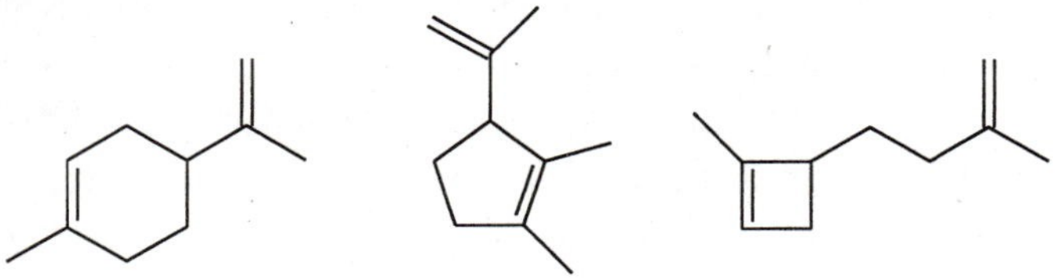

10.53

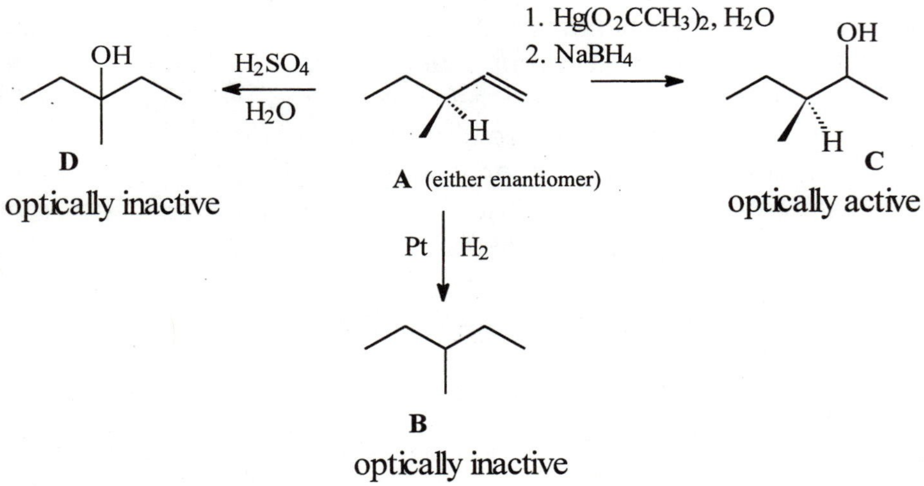

OH

H_2SO_4

H_2O

D

optically inactive

1. $Hg(O_2CCH_3)_2$, H_2O

2. $NaBH_4$

H

A (either enantiomer)

OH

H

C

optically active

Pt | H_2

B

optically inactive

After completing this chapter, you should be able to:

1. Show the products, including regiochemistry and stereochemistry, resulting from the addition to alkenes of all of the reagents discussed in this chapter.
   Problems 10.3, 10.4, 10.5, 10.9, 10.10, 10.13, 10.15, 10.16, 10.22, 10.23, 10.24, 10.26, 10.27, 10.28, 10.31, 10.32, 10.33, 10.34, 10.42.

2. Show the products, including regiochemistry and stereochemistry, resulting from the addition to alkynes of selected reagents as discussed in this chapter. These reagents include HX, X_2, $Hg^{2\oplus}/H^{\oplus}/H_2O$, disiamylborane, and H_2.
   Problems 10.4, 10.5, 10.13, 10.18, 10.28, 10.31, 10.32.

3. Show the mechanisms for any of these additions.
   Problems 10.2, 10.6, 10.11, 10.12, 10.29, 10.35, 10.38, 10.40, 10.41, 10.43.

4. Show rearranged products when they are likely to occur.
   Problems 10.4b, 10.10c, 10.31f, 10.32c.

5. Predict how the rate of addition varies with the structure of the alkene.
   Problems 10.1, 10.7, 10.39.

6. Predict the products from additions to conjugated dienes.
   Problem 10.30.

7. Use these reactions in synthesis.
   Problems 10.14, 10.17, 10.19, 10.25, 10.36, 10.37.

Chapter 11
FUNCTIONAL GROUPS AND NOMENCLATURE II

11.1 a) (1-methylpropyl)benzene b) o-chlorotoluene
c) 1-bromo-4-methoxybenzene d) 4-butyl-3-chlorotoluene
e) 3-ethyl-4-phenylcyclohexene f) 3-phenyl-1-butanol

11.2

a) 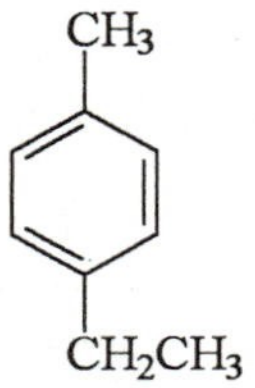 CH_3 CH_2CH_3 b) 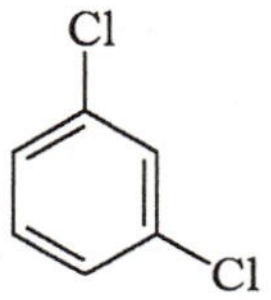 Cl Cl c) 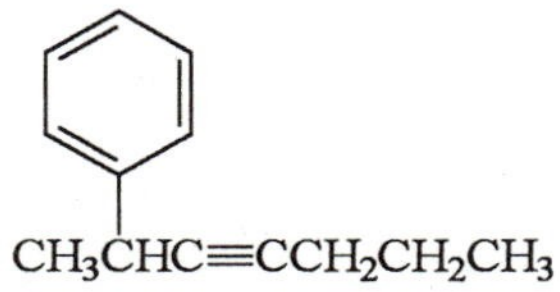 $CH_3CHC{\equiv}CCH_2CH_2CH_3$

d) 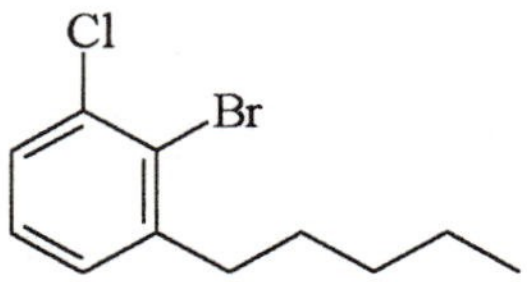 Cl Br e) 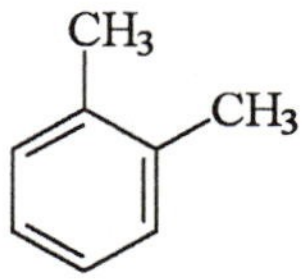 CH_3 CH_3 f) 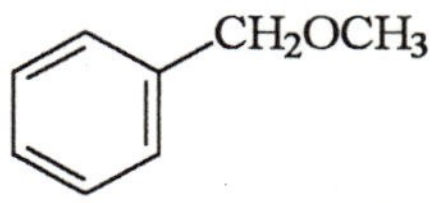 CH_2OCH_3

11.3 a) m-propylphenol b) 3,5-dibromophenol c) p-methoxyphenol

11.4

a) 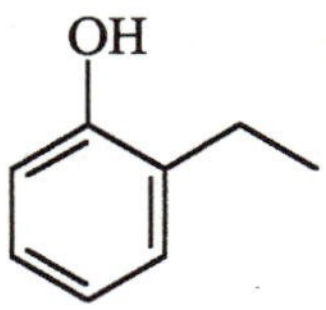 OH b) 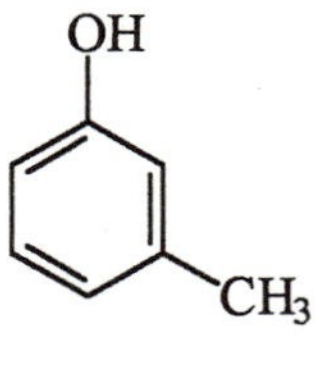 OH CH_3 c) 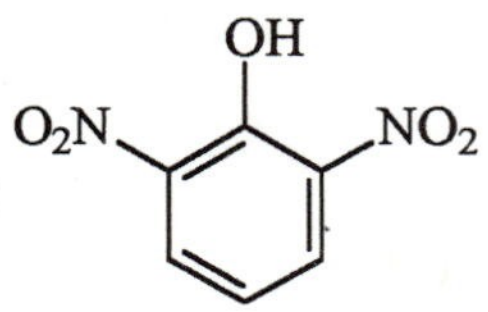 OH O_2N NO_2

11.5

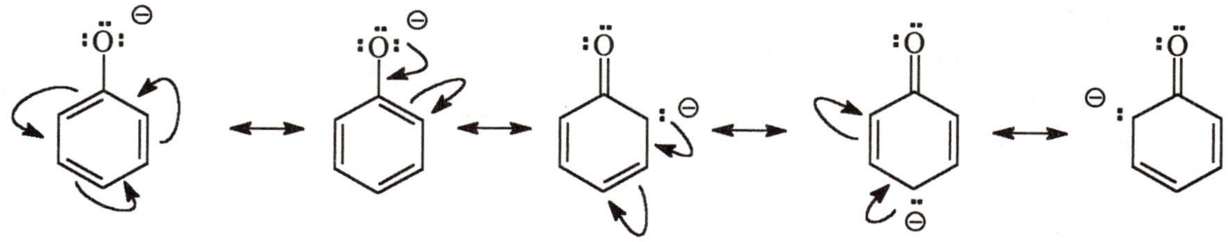

11.6 In acid-base equilibrium reactions, the equilibrium favors the formation of the weak acids and weak bases. As the strength of an acid increases, the pK_a decreases.

a) 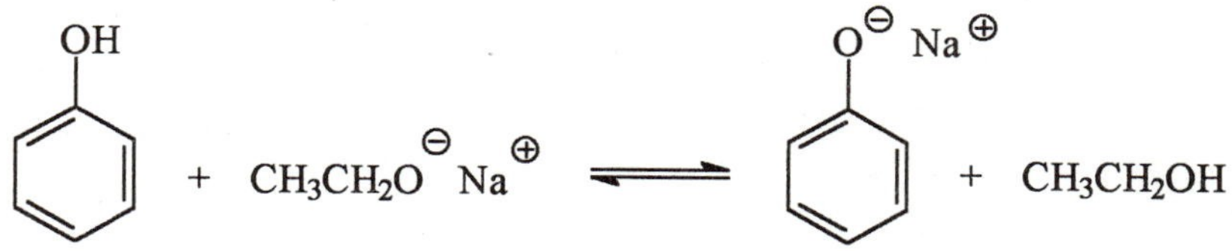 (phenol) + $CH_3CH_2O^{\ominus}\ Na^{\oplus}$ ⇌ (sodium phenoxide, $O^{\ominus}\ Na^{\oplus}$) + CH_3CH_2OH — products favored

b) 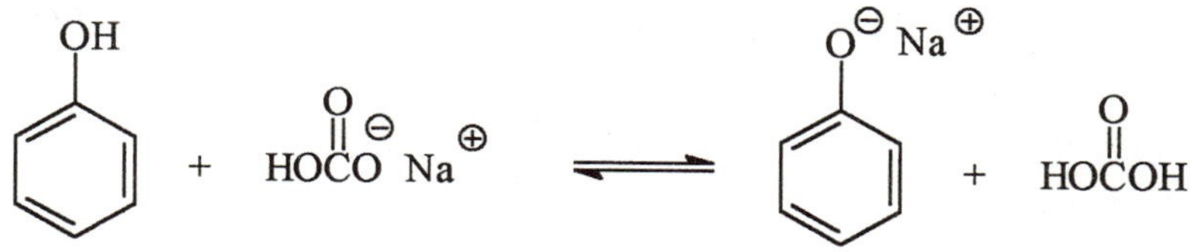 (phenol) + $HOCO_2^{\ominus}\ Na^{\oplus}$ (HOC(=O)O$^{\ominus}$ Na$^{\oplus}$) ⇌ (sodium phenoxide, $O^{\ominus}\ Na^{\oplus}$) + HOC(=O)OH — reactants favored

11.7 a) hexanal
b) 5-methyl-3-hexanone
c) (*Z*)-4-methyl-2-hexenal
d) p-chlorobenzaldehyde
e) 5-(2-methylpropyl)-3-cycloheptenone
f) 2,4-pentanedione
g) 3-phenylcyclopentanone
h) 4-methylpent-3-en-2-one

11.8

a)

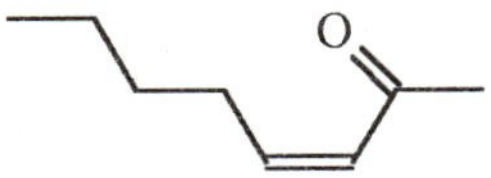

b)

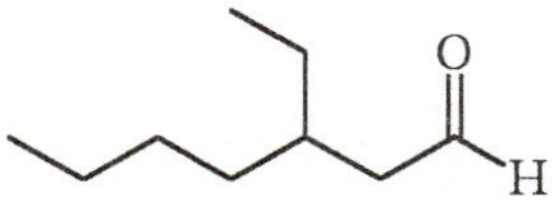

c) 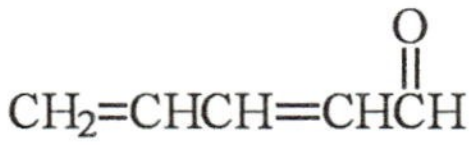 $CH_2{=}CHCH{=}CHCH$ (with C=O on the final carbon: $CH_2{=}CHCH{=}CHCHO$)

d) 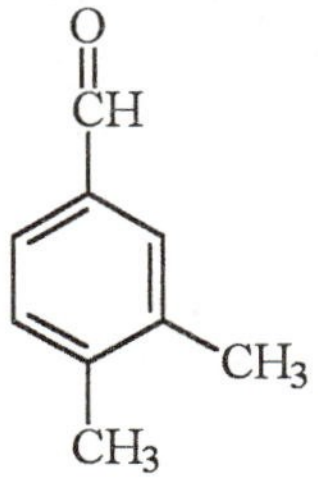 CH, CH_3, CH_3

e) 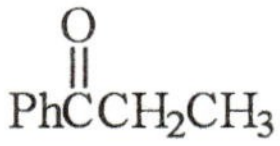 $PhCCH_2CH_3$ (with C=O: $PhCOCH_2CH_3$)

f)

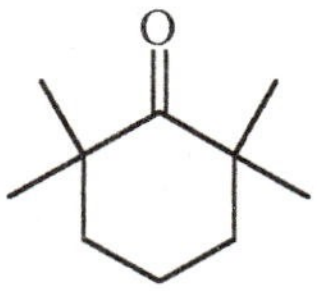

11.9 a) The circled hydrogens are more acidic because the conjugate base is stabilized by resonance.

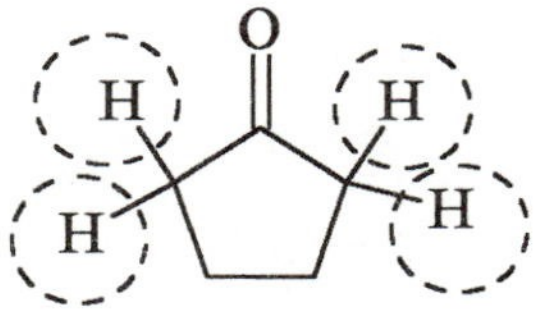

b) The circled hydrogens are more acidic because the conjugate base is stabilized by resonance with the carbonyl group and the benzene ring.

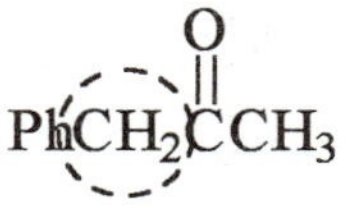

$PhCH_2CCH_3$ (O above C)

c) The circled hydrogens are more acidic because the conjugate base is stabilized by resonance with both of the adjacent carbonyl groups.

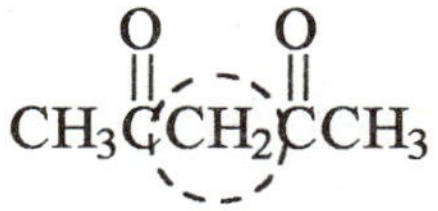

$CH_3CCH_2CCH_3$ (O above each C)

11.10 a) The equilibrium does not lie completely to the right because hydroxide ion is not a strong enough base. The conjugate acid of the base should have a larger pK_a than the substrate for complete removal of proton. The pK_a of acetone is about 20 whereas the pK_a of water (the conjugate acid of hydroxide ion)is only 15.7, so this equilibrium favors the reactants.

b) In order to completely remove a proton from acetone the conjugate acid of the base used must be a weaker acid than acetone.

LDA, $NaNH_2$, $CH_3\overset{O}{\overset{\|}{S}}\ddot{C}H_2^{\ominus}$, NaH

11.11 a) 3-methylpentanoic acid
b) p-bromobenzoic acid
c) (*Z*)-4,4-dichloro-2-pentenoic acid
d) 2-cyclohexene-1-carboxylic acid
e) 3-methyl-4-nitrobenzoic acid
f) 3-cyclopropylbutanoic acid

11.12

a) 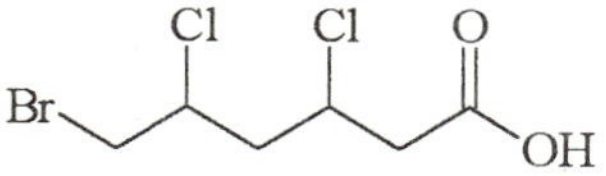 Br, Cl, Cl, O, OH b) 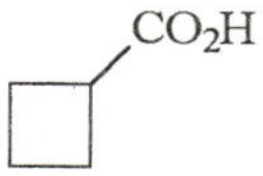 CO_2H c) 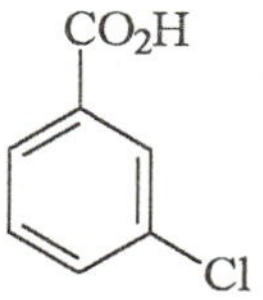 CO_2H, Cl d) 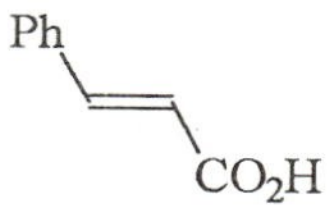 Ph, CO_2H

11.13

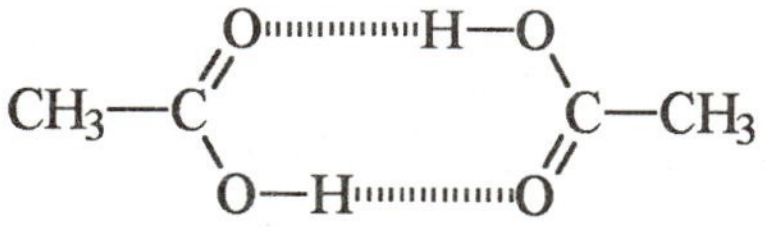

CH_3-C, O, H—O, O—H, O, $C-CH_3$

11.14 a) benzoyl chloride
b) propanoic anhydride
c) methyl propanoate
d) propyl 3-ethylbenzoate
e) cyclopentyl 4-methylpentanoate
f) *N*-methyl-3-pentynamide
g) ethanamide or acetamide
h) 2,3-dimethylbenzamide
i) isopropyl 3-cyanocyclopentanecarboxylate
j) pentanenitrile
k) potassium 2-methylpentanoate

11.15

a) $CH_3CH_2\overset{O}{\overset{||}{C}}Cl$ b) $CH_3\overset{O}{\overset{||}{C}}N(CH_3)_2$ c) $CH_3(CH_2)_3\overset{O}{\overset{||}{C}}O\overset{O}{\overset{||}{C}}(CH_2)_3CH_3$ d) $CO_2^{\ominus}$ $Na^{\oplus}$ NO_2

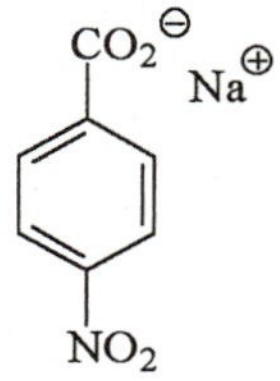

e) $CH_3(CH_2)_4\overset{O}{\overset{||}{C}}NH_2$ f) $CH_3\overset{O}{\overset{||}{C}}O\overset{CH_3}{\overset{|}{C}}HCH_3$ g) $Ph\overset{O}{\overset{||}{C}}OCH_2Ph$ h) $-\overset{O}{\overset{||}{C}}OCH_2CH_3$

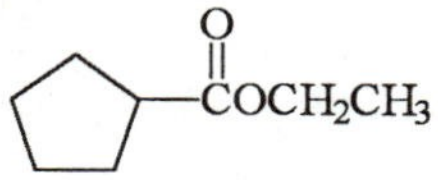

i) CN Cl

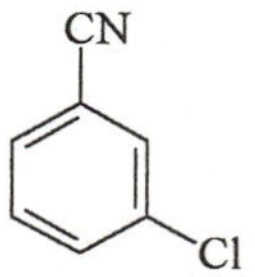

j) $CH_3CH_2CH_2CH_2\overset{CH_3}{\overset{|}{C}}HCH_2C{\equiv}N$

11.16

a) $CH_3CH_2CH_2\overset{O}{\overset{||}{C}}NH_2$ because it can form hydrogen bonds

b) $CH_3CH_2CH_2\overset{O}{\overset{||}{C}}OH$ because it can form hydrogen bonds

c) $\overset{O}{\overset{||}{C}}OH$ because it can form hydrogen bonds

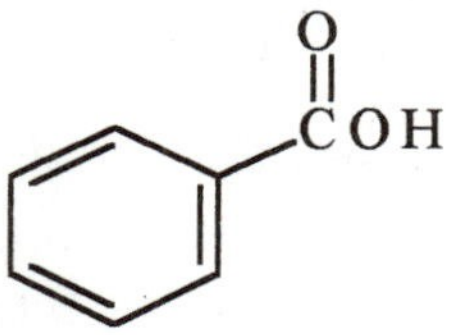

d) $CH_3CH_2CH_2\overset{O}{\overset{||}{C}}NH_2$ because it is more polar and forms stronger hydrogen bonds

11.17 a) cyclohexanethiol
b) dipropyl sulfide
c) 5-methyl-3-hexyne-1-thiol
d) benzenesulfonic acid
e) propyl p-toluenesulfonate

11.18

a) 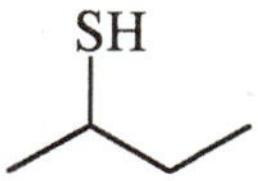 SH

b) 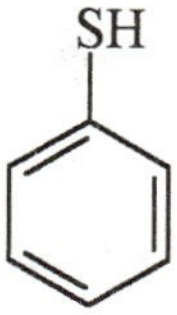 SH

c) 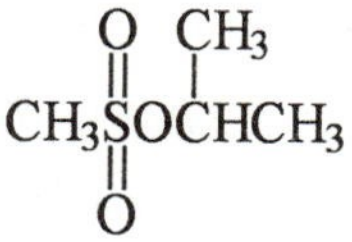 O CH_3 $CH_3SOCHCH_3$ O

d) 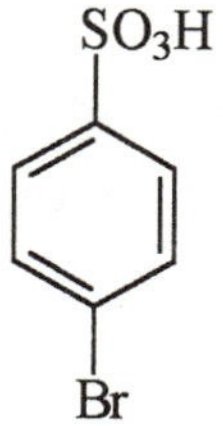 SO_3H Br

e) Ph—S—CCl_3

11.19 a) 4-hydroxy-2-cyclohexenone b) 2-amino-3-pentenoic acid
c) 1,4-butanedioic acid d) methyl 4-methyl-5-oxopentanoate
e) 3-cyanobenzaldehyde
f) 4-ethyl-3-hydroxy-*N*-methylhexanamide
g) 3-oxobutanenitrile h) isopropyl 3-oxo-4-phenylhexanoate

11.20

a) 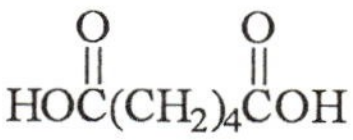 O O $HOC(CH_2)_4COH$

b) 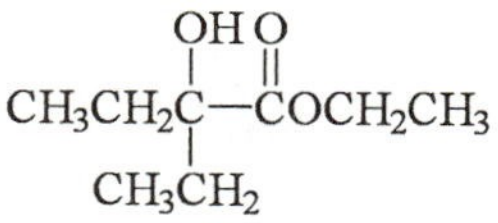 OH O $CH_3CH_2C—COCH_2CH_3$ CH_3CH_2

c) 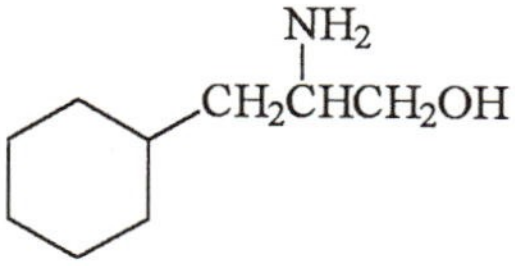 NH_2 CH_2CHCH_2OH

d) 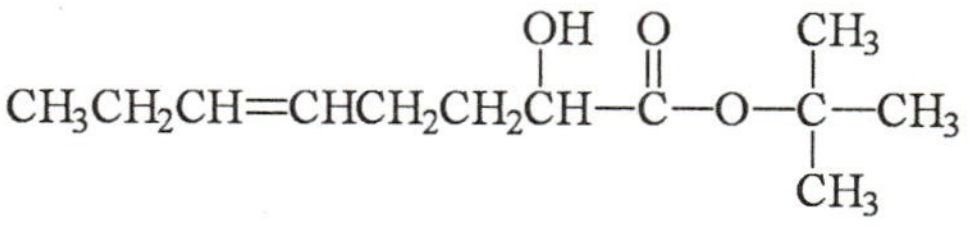 OH O CH_3 $CH_3CH_2CH{=}CHCH_2CH_2CH—C—O—C—CH_3$ CH_3

e) 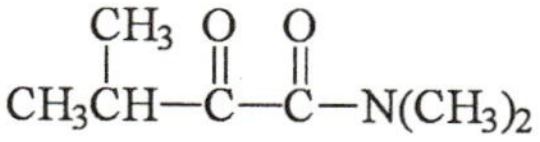 CH_3 O O $CH_3CH—C—C—N(CH_3)_2$

11.21 a) 1-bromo-4-chlorobenzene or p-bromochlorobenzene
b) 4,5-dimethyl-2-hexynal
c) 3-(1-methylpropyl)-3-hexenoic acid
d) 4-ethyl-6,6-dimethyloctanamide
e) 3,4-dimethylbenzonitrile
f) methyl acetate
g) 1-cyclobutyl-2-ethylbenzene or o-cyclobutylethylbenzene
h) 2-ethyl-3-methyl-2-cyclopentenone

i) 2-methylbenzoic acid or o-methylbenzoic acid
j) 4-methylpentanoyl chloride
k) 3-bromo-3-methylpentanenitrile
l) isopropyl 3-methyl-2-butenoate
m) 2-*t*-butyl-4-methylphenol
n) 5-phenyl-5-hexen-3-one
o) 2-methoxybenzaldehyde or o-methoxybenzaldehyde
p) sodium benzoate
q) *N*-methylcyclohexanecarboxamide
r) ethyl cyclopentanecarboxylate

11.22

a) 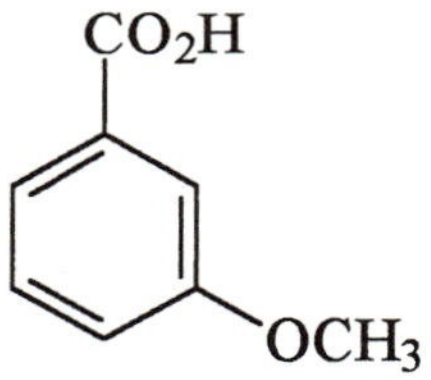 CO_2H / OCH_3

b) 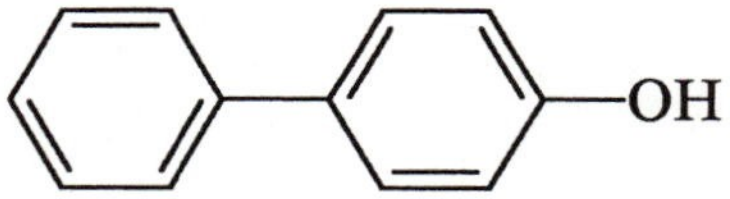 OH

c) 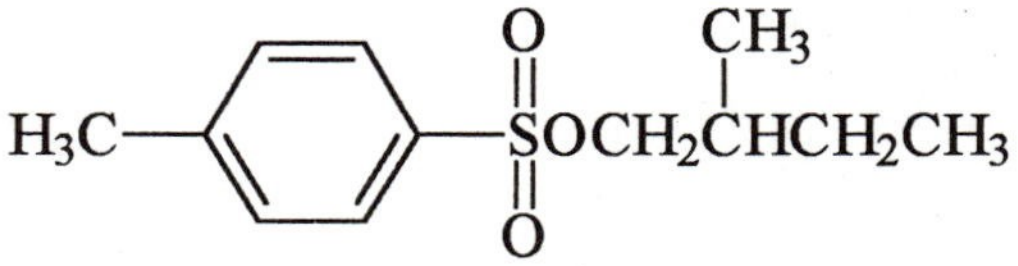 H_3C — $\overset{O}{\underset{O}{S}}OCH_2\overset{CH_3}{C}HCH_2CH_3$

d) 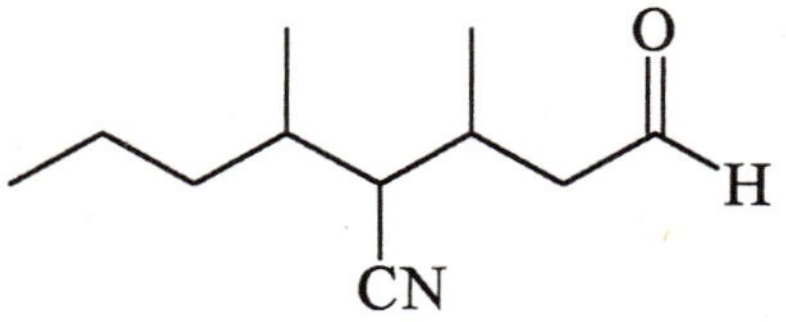 O / H / CN

e) 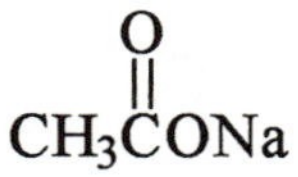 $CH_3\overset{O}{\overset{||}{C}}ONa$

f) 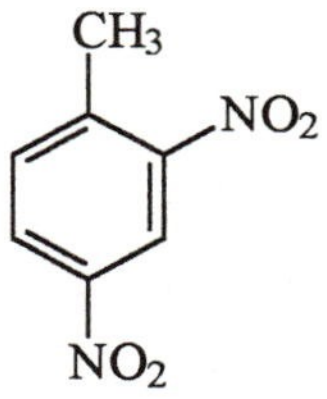 g) 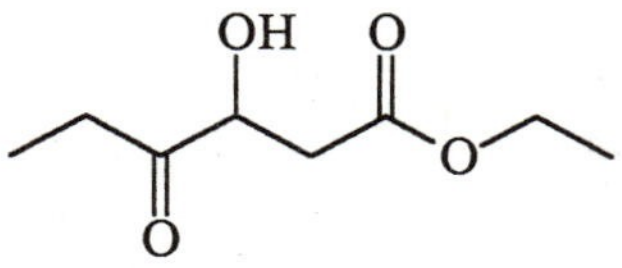

h) $CH_3\overset{O}{\overset{\|}{C}}Cl$ i) 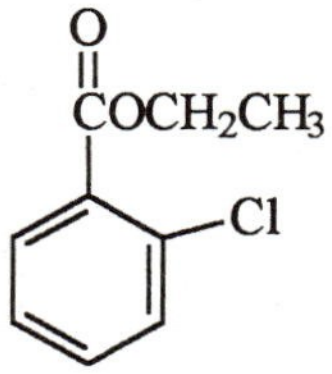

j) 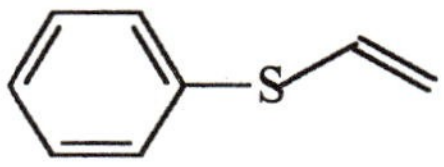 k) 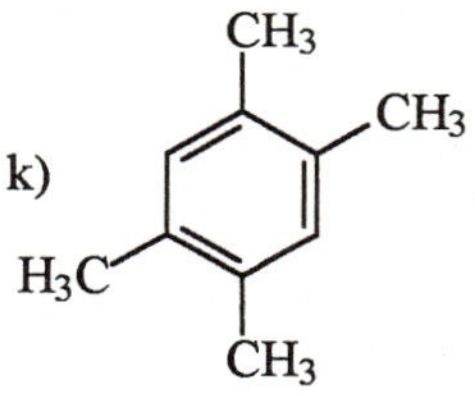

l) 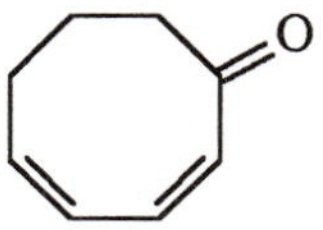 m) CH_3CN

n) 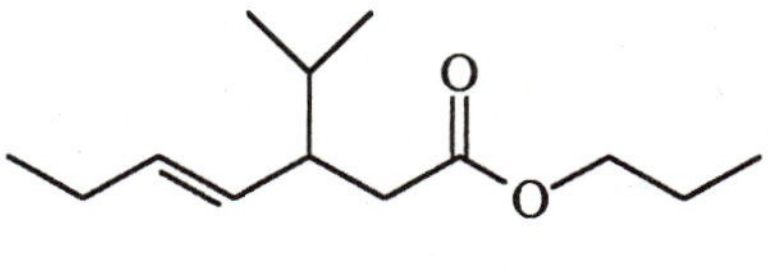

o) 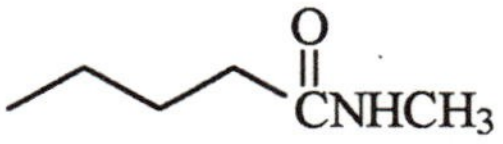 p) 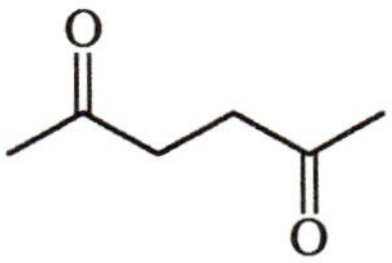

q)

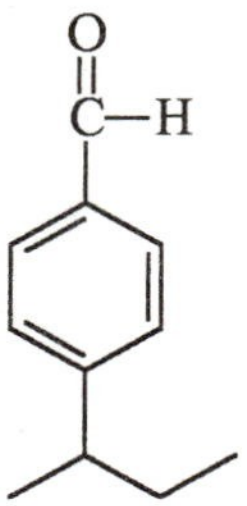

r)

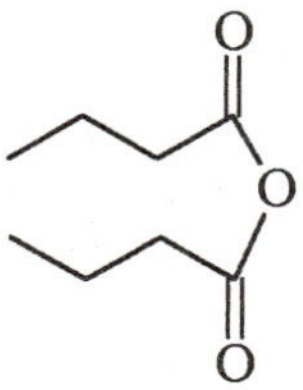

s)

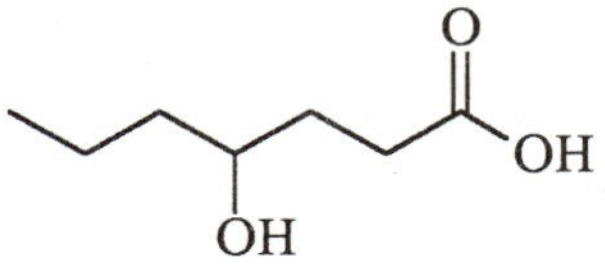

t)

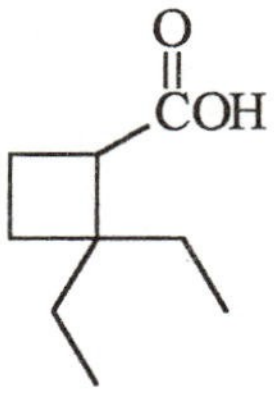

11.23 a) 3-oxo-4-hexenoic acid
b) 4-hydroxypentanal
c) 3-hydroxy-4-methoxybenzoic acid
d) 4-oxo-2-butenamide

11.24 a) The left compound has a higher melting point because it is more polar than heptane.
b) Cyclopentanol has a higher boiling point because it can form hydrogen bonds.
c) 2-Phenylethanoic acid is higher boiling because it is more polar and can form hydrogen bonds.

11.25 a) phosphine
b) thiol and alkene
c) sulfide and arene
d) sulfoxide and arene
e) sulfone
f) sulfonic acid

11.26 The most acidic hydrogens in these compounds are circled.

a) 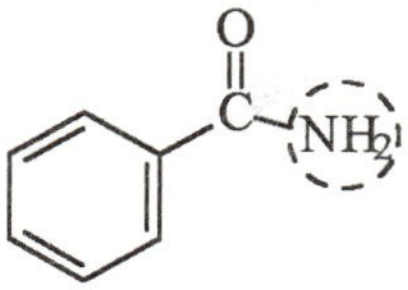 $C_6H_5-\overset{O}{\overset{\|}{C}}-NH_2$ (circled: NH_2)

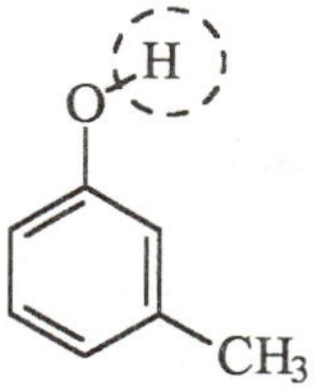

b) O–H on ring with CH_3 (circled: H)

c) $CH_3-\overset{O}{\overset{\|}{C}}-OCH_2CH_2CH_3$ (circled: CH_3)

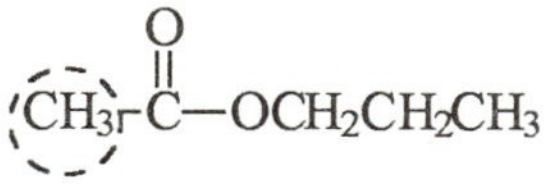

d) 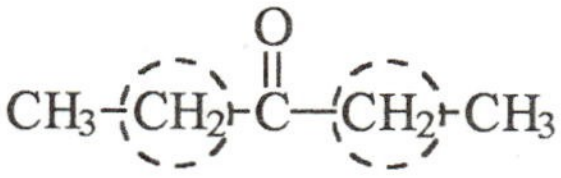 $CH_3-CH_2-\overset{O}{\overset{\|}{C}}-CH_2-CH_3$ (circled: CH_2, CH_2)

e) 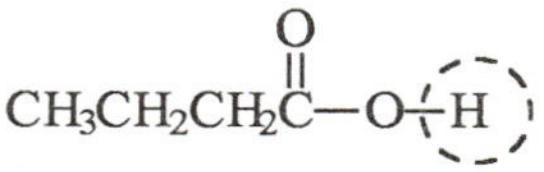 $CH_3CH_2CH_2\overset{O}{\overset{\|}{C}}-O-H$ (circled: H)

11.27 In the name ibuprofen, "ibu" , "pro", and "fen" are derived from the isobutyl, propanoic acid, and phenyl moieties in the structure.

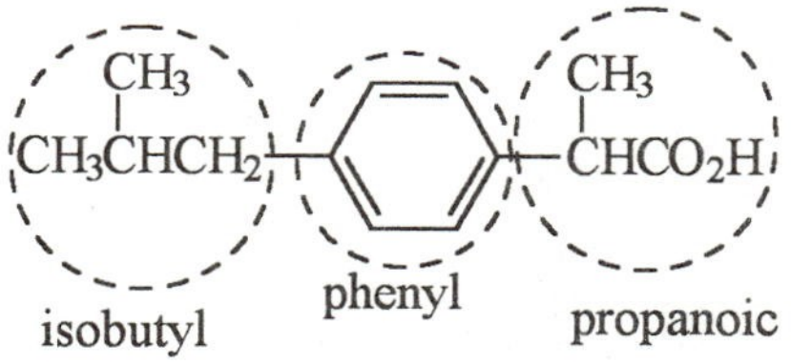

$$\underbrace{CH_3\overset{CH_3}{\overset{|}{C}}HCH_2}_{\text{isobutyl}}-\underbrace{C_6H_4}_{\text{phenyl}}-\underbrace{\overset{CH_3}{\overset{|}{C}}HCO_2H}_{\text{propanoic}}$$

11.28 Picric acid such a strong acid because the conjugate base is highly stabilized by the strong inductive and resonance stabilization effects of the three nitro groups.

11.29 Bicarbonate ion (the conjugate base of carbonic acid) is a strong enough base to deprotonate benzoic acid but not p-methylphenol. A typical scheme for the separation of these compounds from a mixture using bicarbonate as the base is:

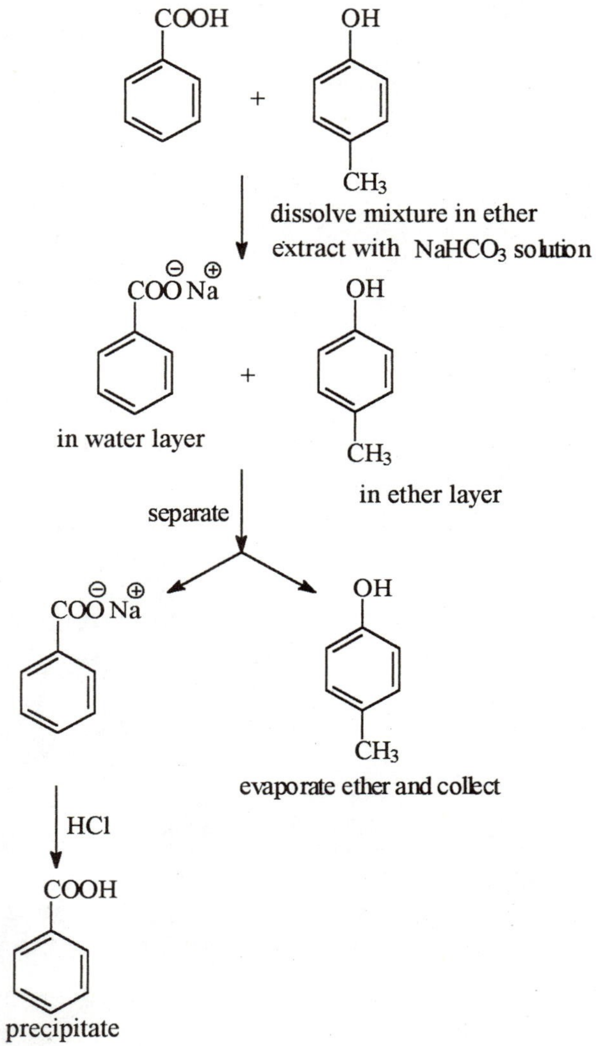

COOH + OH / CH_3

dissolve mixture in ether
extract with $NaHCO_3$ solution

$COO^{\ominus} Na^{\oplus}$ + OH / CH_3

in water layer

in ether layer

separate

$COO^{\ominus} Na^{\oplus}$

OH / CH_3

evaporate ether and collect

HCl

COOH

precipitate

11.30

$$\begin{array}{c} CH_3 \\ | \\ H_3C-C-H \\ | \\ CH_3 \end{array}$$

A

$$\begin{array}{c} CH_2Cl \\ | \\ H_3C-C-H \\ | \\ CH_3 \end{array}$$

B

$$\begin{array}{c} CH_3 \\ | \\ H_3C-C-Cl \\ | \\ CH_3 \end{array}$$

C

$$\begin{array}{c} CH_2Cl \\ | \\ H_3C-C-Cl \\ | \\ CH_3 \end{array}$$

D

$$\begin{array}{c} CH_2Cl \\ | \\ H_3C-C-H \\ | \\ CH_2Cl \end{array}$$

E or F

$$\begin{array}{c} CHCl_2 \\ | \\ H_3C-C-H \\ | \\ CH_3 \end{array}$$

E or F

$$\begin{array}{c} CH_2Cl \\ | \\ H_3C-C-Cl \\ | \\ CH_3 \end{array}$$

D

Based on these experiments it cannot be determined whether E is 1,3-dichloro-2-methylpropane and F is 1,1-dichloro-2-methylpropane or vice versa.

After completing this chapter, you should be able to:

1. Name an aromatic compound, a phenol, an aldehyde, a ketone, a carboxylic acid, an acid chloride, an anhydride, an ester, an amide, a nitrile, and a carboxylic acid salt.
   Problems 11.1, 11.3, 11.7, 11.11, 11.13, 11.21.

2. Draw the structure of a compound containing one of these functional groups when the name is provided.
   Problems 11.2, 11.4, 11.8, 11.12, 11.14, 11.22.

3. Recognize the common functional groups that contain sulfur or phosphorus.
   Problems 11.17, 11.18.

4. Name a compound containing more than one functional group or draw the structure of such a compound when the name is provided.
   Problems 11.19, 11.20, 11.23, 11.25.

5. Understand how the physical properties of these compounds depend on the functional group that is present.
   Problems 11.16, 11.24.

Chapter 12
STRUCTURE DETERMINATION BY SPECTROSCOPY, I INFRARED AND NUCLEAR MAGNETIC RESONANCE SPECTROSCOPY

12.1 a) 3.33×10^{-3} cm b) 9.68×10^{14} s^{-1} c) 0.858 kcal/mol (3.59 kJ/mol) d) 92.3 kcal/mol (387 kJ/mol)

12.2 This frequency of light falls in the infrared region of the electromagnetic spectrum.

12.3 The energy of light with this frequency is 2.86×10^{-5} kcal/mol (1.19×10^{-4} kJ/mol).

12.4 a) The C-H bond absorbs at a higher wavenumber because hydrogen is a lighter atom than deuterium.
b) The C≡C bond absorbs at a higher wavenumber because a triple bond is stronger than a double bond.
c) The C-Cl bond absorbs at a higher wavenumber because the chlorine is a lighter atom than iodine and the C-Cl bond is stronger than the C-I bond.

12.5 a) OH, 3000 cm^{-1}, very broad; =CH, 3100-3000 cm^{-1}; -CH, 3000-2850 cm^{-1}
b) =CH, 3100-3000 cm^{-1}; -CH, 3000-2850 cm^{-1}
c) NH_2, two bands, 3400-3250 cm^{-1}; =CH, 3100-3000 cm^{-1}
d) OH, 3550-3200 cm^{-1}, broad; =CH, 3100-3000 cm^{-1}; -CH, 3000-2850 cm^{-1}
e) NH, one band, 3400-3250 cm^{-1}; -CH, 3000-2850 cm^{-1}
f) -CH, 3000-2850 cm^{-1}; CHO, 2830-2700 cm^{-1}, two bands

12.6 The band in the triple bond region at 2150-2100 cm^{-1} is much stronger for 1-hexyne than it is for the more symmetrical 3-hexyne. In addition, the ≡CH band near 3300 cm^{-1} in the spectrum of 1-hexyne confirms the presence of the triple bond.

12.7 The 3000 - 2900 cm^{-1} range is good region in the IR spectrum to monitor hydrocarbon emissions because most hydrocarbons have -CH bonds that absorb in this region.

12.8 b) The cabonyl group is part of a five membered ring, therefore, the band will be shifted to a higher wavenumber from the base position for a ketone. The predicted position is 1745 cm^{-1}.

c) This is a conjugated aldehyde, so the carbonyl band will be shifted to a lower wavenumber by 20-40 cm^{-1} from the base position of 1730 cm^{-1}. The predicted position is 1710-1690 cm^{-1}.

d) This is a conjugated ketone, so the carbonyl band will be shifted to a lower wavenumber by 20-40 cm^{-1} from the base position of 1715 cm^{-1}. The predicted position is 1695-1675 cm^{-1}.

e) This is a conjugated ester. The ester carbonyl band (base position 1740 cm^{-1}) will be shifted to lower wavenumbers by 20-40 cm^{-1}. The predicted position is 1720-1700 cm^{-1}.

f) This is a conjugated carboxylic acid. The predicted position is 1710 cm^{-1} - (20-40 cm^{-1}) = 1690-1670 cm^{-1}.

12.9 a) The left compound, a ketone, has its carbonyl absorption near 1715 cm^{-1} whereas the right compound, an aldehyde, has its carbonyl peak near 1730 cm^{-1} and has two bands in the region of 2830-2700 cm^{-1}.

b) The left ester (non-conjugated) has its carbonyl absorption near 1740 cm^{-1} whereas the right ester (conjugated) has its carbonyl absorption near 1720-1700 cm^{-1}.

c) The right compound, a carboxylic acid has a very broad band near 3000 cm^{-1} whereas the left compound, a conjugated ester will not show this feature.

12.10 a) =CH, 3100-3000; -CH, 3000-2850; C=O, 1695-1675; C=C, 1660-1640.

b) $-NH_2$, 3400-3250 (two bands); =CH, 3100-3000; -CH, 3000-2850; aromatic ring, 1600-1450 (four bands), 900-675.

c) =CH, 3100-3000, -CH, 3000-2850; C=C, 1660-1640; C-O, 1300-1000.

d) OH, 3550-3200 (broad); -CH, 3000-2850; C-O, 1300-1000.

e) ≡CH, 3300; -C-H, 3000-2850; C≡C, 2150-2100; NO_2, 1550, 1380.

f) =CH, 3100-3000; -CH, 3000-2850; aldehyde CH, 2830-2700 (two bands); C=O, 1710-1690; aromatic ring, 1600-1450 (four bands), 900-675.

12.11 a) 1-Butyne has absorptions at 3300 cm^{-1} (≡CH) and 2150-2100 cm^{-1} (C≡C) that are not present in the spectrum of 1-butene.

b) Benzyl alcohol has absorptions at 3100-3000 (=CH), 1600-1450 (aromatic ring), and 900-675 cm^{-1} that are not present in the spectrum of *t*-butanol.

c) Benzaldehyde has two aldehydic CH bands at 2830-2700 cm^{-1} that are not present in the spectrum of the ketone, acetophenone.

d) The primary amine has two bands in the 3400-3250 cm^{-1} region whereas the secondary amine has only one.

12.12 b) The strong absorption at 1715 cm^{-1} is typical for a carbonyl group of a ketone. The absorption in the region of 3000 - 2850 cm^{-1} is indicative of H's bonded to sp^3 -hybridized C's. There is no evidence for the presence of a C=C bond.

c) The strong and broad OH absorption centered at 3000 cm^{-1} along with the carbonyl absorption at 1716 cm^{-1} are typical for a carboxylic acid. There is no evidence for the presence of a C=C bond.

d) The two bands in the region of 2830 - 2700 cm^{-1} and the carbonyl absorption at 1696 cm^{-1} provide evidence for the presence of an aldehyde. The observed shift of the carbonyl band to a lower wavenumber indicates that it is a conjugated aldehyde. The absorptions in the regions of 3100-3000 cm^{-1} are due sp^2 hybridized C-H bonds, and those in the region of 3000-2850 cm^{-1} are due to sp^3 hybridized C-H bonds. The appearance of four bands in the 1600-1450 cm^{-1} region is indicative the presence of an aromatic ring.

12.13 Because air is transparent in the region of 3000-2850 cm^{-1}, where methane has strong absorptions, methane has the potential to act as a greenhouse gas.

12.14 The position of absorption or chemical shift is given by:

$$\delta = \frac{\text{observed position of peak (Hz)}}{\text{operating frequency of instrument (Hz)}} \times 10^6$$

a) 7.4 δ

b) 1480 Hz on a 200 MHz and 2220 Hz on a 300 MHz instrument .

c) 7.4 δ

12.15 The number of absorption signals in the NMR spectrum is equal to the number of different types (chemically nonequivalent) hydrogens in a molecule.

b) We expect three NMR signals. The CH_3 groups are chemically nonequivalent. The hydrogens of the CH_2 group are enantiotopic, so they have the same chemical shift.

c) This molecule has 4 different types of hydrogens, so four signals will be found. These are the two different methyl groups, the two Hs ortho to the carbonyl group, and the two Hs meta to the carbonyl group.

e) The two CH_3 groups and the two vinyl hydrogens are different. Therefore expect 4 signals.

f) The two OH groups are identical and the two vinylic hydrogens are identical, so expect 2 signals.

12.16 It is possible to estimate the chemical shifts of most types of alkyl hydrogens using the values in Table 12.4. Start with the chemical shift for the particular type of hydrogen from the table. This value is then corrected according to the chemical environment of the hydrogens using the following rules:

If the group is not CH_3, then correct the value by adding 0.3 δ if it is CH_2 or 0.7 δ if it is CH.

If another electronegative group is attached to the same carbon (α-carbon) as the hydrogen under consideration, add the value from the table minus 0.9 δ. If another electronegative group is attached to the carbon adjacent to the carbon (β-carbon) bearing the hydrogen under consideration, add 0.3 δ to the Table value.

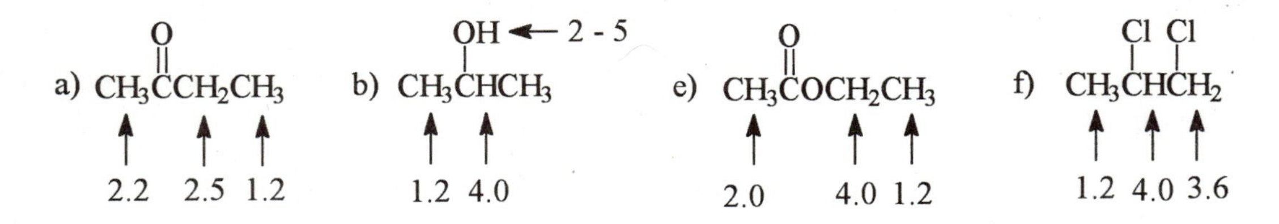

12.17 The signal from H_m will appear as a triplet (1:2:1) because the two inner lines overlap.

12.18

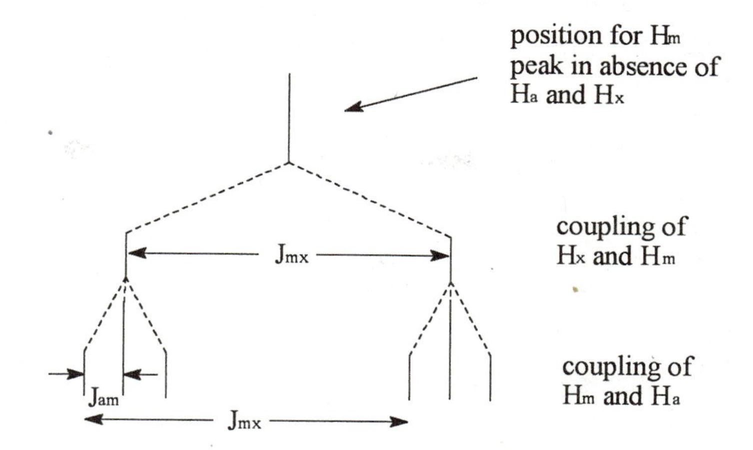

12.19 In general, the absorption peak from a hydrogen(s) that is coupled to *n* equivalent hydrogens is split into *n* + *1* peaks.

b) a = x = 3 c) a = 7, x = 2 d) a = 3, x = 2

e) a = 5, x = 2 f) a = 3, m = 6, x = 3

12.20

b) CH_3CH_2OH

CH_3	CH_2	OH	
1.2	3.6	2-5	chemical shift
3	4	1	number of lines
3	2	1	integral

c) $CH_3CH(Cl)CH_3$

CH_3	CH
1.2	3.7
2	7
6	1

d) $CH_3CH_2OCH_2CH_3$

CH_3	CH_2
1.2	3.6
3	4
3	2

e) $C_6H_5CH_2CH_2NO_2$

C_6H_5	CH_2	CH_2NO_2
7-8	2.9	4.7
1-m	3	3
5	2	2

f) $CH_3C(=O)OCH_2CH_3$

CH_3	CH_2	CH_3
2.0	4.0	1.2
1	4	3
3	2	3

g) $CH_3CH_2CH_2Cl$

CH_3	CH_2	CH_2Cl
0.9	1.5	3.3
3	6-m	3
3	2	2

h) CH_3CHCl_2

CH_3	$CHCl_2$
1.5	5.8
2	4
3	1

12.21

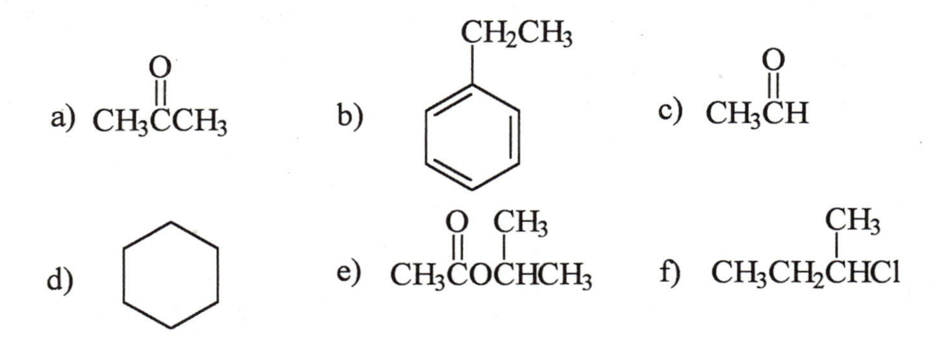

12.22 The number of absorption in the ^{13}C-NMR spectrum depends on the number of different types of carbons in a molecule.

b) 1 c) 3 d) 6 e) 3

12.23 The chemical shifts for carbons are approximately 20 times larger than the chemical shifts of the hydrogens attached to that carbon.

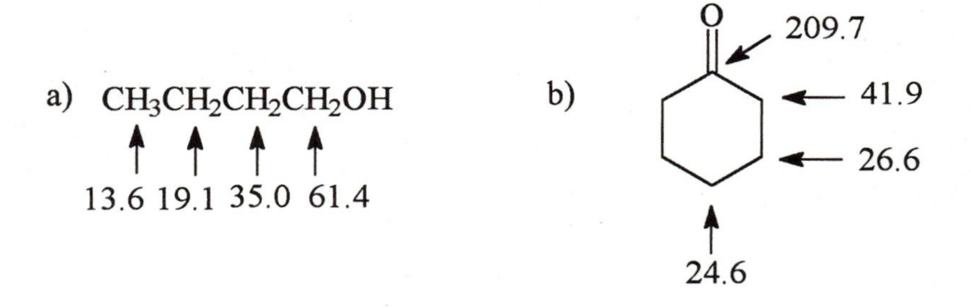

12.24 a) The DU for this compound is 5. There are 5 different types of carbons in this molecule. The signal around 190 δ is due to a carbonyl carbon of an aldehyde. The four signals in the 140 -120 δ region are due aromatic carbons.

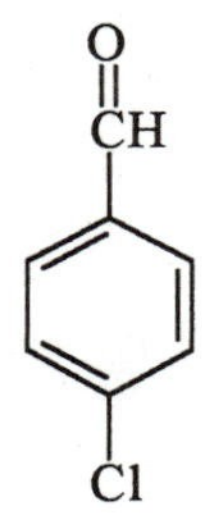

b) This compound has a DU of 1, and contains five different carbons. The signal at ~147 δ is an alkene type carbon that is not bonded to any hydrogens and the one at ~110 δ is also an alkene type carbon but bonded to two hydrogens. The CH_2 carbon at ~30 δ , and the CH_3 carbon at ~22 δ are bonded to an alkene type carbon. The signal at ~12 δ is due to a methyl group.

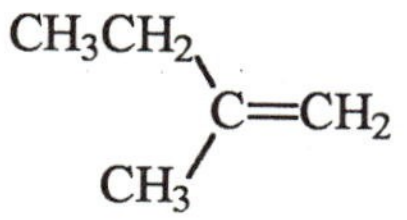

$CH_3CH_2(CH_3)C{=}CH_2$

c) The compound has a DU of 2. The compound must have symmetry, because the spectrum shows the presence of only three different types of carbons. There are probably two of each type of carbons. The signal at ~128 δ is due to an alkene type carbon that is bonded to one hydrogen.

12.25 a) This compound has a DU of 5. The two bands in the region of 2830 - 2700 cm^{-1} and the carbonyl absorption at 1706 cm^{-1} provide evidence for the presence of an aldehyde. The observed shift of the carbonyl band to a lower wavenumber indicates that it is a conjugated aldehyde. The absorptions in the regions of; 3100 - 3000 cm^{-1} are due sp^2 hybridized C-H bonds, and those at 3000 - 2850 cm^{-1} are due to sp^3 hybridized C-H bonds. The appearance of four bands in the 1600 - 1450 cm^{-1} region suggests the presence of an aromatic ring. The ^{1}H-NMR spectrum shows presence of five different types of hydrogens in this compound. The singlet at 9.8 δ is typical of an aldehydic hydrogen. The two doublets in the 8 - 7 δ region are due to the hydrogens of a para disubstituted benzene derivative. The 2 Hs that appear as quartet at 2.7 δ must be coupled to the 3 Hs appear as a triplet at 2.3 δ. All of the features combined with IR data and the DU value are consistent with 4-ethylbenzaldehyde.

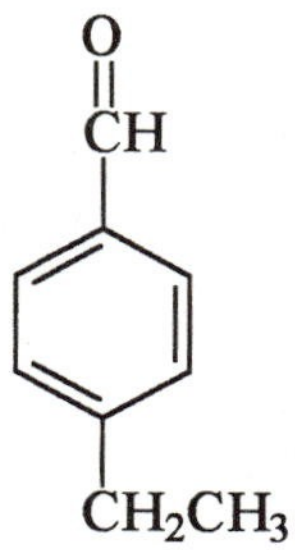

O
||
CH

CH_2CH_3

b) This compound has a DU of 1. The carbonyl absorption at 1741 cm^{-1} along with the strong absorptions in the 1300-1000 cm^{-1} region suggest the presence of an ester group. The bands in the region of 3000-2850 cm^{-1} are due to sp^3 hybridized C-H bonds. The ^1H-NMR spectrum shows the presence of four types of hydrogens and the integrals provide the actual number of hydrogens. The signal at 4.2 δ must be due to a CH_2 attached to the oxygen of the ester group. The CH_2 protons of the two triplets (somewhat distorted) at 3.7 δ and 2.9 δ must be coupled. Therefore the 2 Hs that appear as a quartet at 4.2 δ are coupled to the 3 Hs that appear as a triplet at 1.3 δ. The IR, NMR, and DU data fit ethyl 3-bromopropanoate.

$$BrCH_2CH_2\overset{\overset{O}{\|}}{C}OCH_2CH_3$$

c) This compound has a DU of 4. The strong and broad OH absorption in c)the 3500-3200 cm^{-1} region indicates that the compound is an alcohol. The absorptions in the regions of 3100-3000 cm^{-1} are due to sp^2 -hybridized C-H bonds, and those at 3000-2850 cm^{-1} are due to sp^3 hybridized C-H bonds. The appearance of four bands in the 1600-1450 cm^{-1} region and the strong bands in the 900-675 region suggest the presence of an aromatic ring. The NMR spectrum shows the presence of four different types of hydrogens and the integrals match the actual number of hydrogens. The signal at 7.3 δ is due to five aromatic hydrogens. The singlet at 2.7 δ is probably due to the hydroxy hydrogen, which is not coupled due to rapid chemical exchange. The 1H that appears as a quartet at 4.8 δ must be coupled to 3 Hs appearing as a doublet at 1.4 δ. These data fit 1-phenyl-1-ethanol.

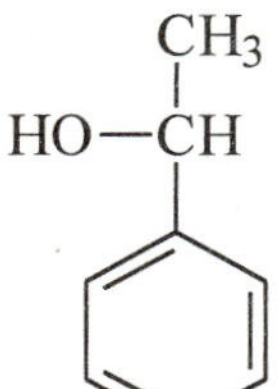

CH3
|
HO—CH
|
(phenyl ring)

12.26 The IR spectrum of this compound shows an OH band in the 3500-3200 cm^{-1} region and a strong absorption in the C-O region. The absorption in the region of 3000-2850 cm^{-1} is due to sp^3 hybridized C-H bonds. The IR data along with its DU of 0 indicate that this is a saturated alcohol. There are only four signals (all CH_2's) in the ^{13}C-NMR spectrum indicating that the compound has symmetry. The CH_2 group at 63 δ must be attached to an oxygen. The absence of any CH_3 or CH groups indicates that the OH groups must be on the ends of the chain of carbons.

$$HOCH_2CH_2CH_2CH_2CH_2CH_2CH_2OH$$

12.27 a) The DU is 0. The IR spectrum shows OH, sp^3 CH, and C-O, so the compound is an aliphatic alcohol. The ^{1}H-NMR is somewhat complex. The singlet at 2.6 δ is probably due to the hydrogen of the OH, which is not coupled due to rapid chemical exchange. The ^{13}C-NMR spectrum is more useful in this case. It shows the presence of five different carbons. The OH group must be linked to the CH_2 at 68 δ. There are two CH_3's, a CH_2, and a CH group that are not attached directly to any electronegative groups. Because its chemical shift is further down field the CH is probably closer to the OH. This allows the fragment $CHCH_2OH$ to be written. Attaching the remaining fragments, a CH_2 and two different CH_3 groups, results in the following structure.

$$CH_3CH_2\underset{}{\overset{CH_3}{\overset{|}{C}}}HCH_2OH$$

b) The DU is 2. The IR spectrum indicates the presence of a C=O group and sp^3 C-H bonds. The presence of only four signals in the ^{13}C-NMR spectrum indicates that the compound has symmetry. The signal at ~214 δ is due to a carbonyl carbon that is attached to no hydrogens, so the compound is a ketone. Since there is only one carbonyl group in the compound, there must be two of each type of CH_2 groups shown in the spectrum. The ^{1}H- NMR shows only two types of hydrogens. The distorted peaks indicate complex coupling of the different hydrogens. These data fit cycloheptanone.

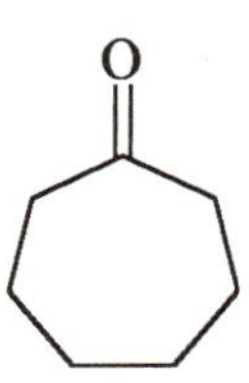

c) This compound has a DU of 5. The two bands in the region of 2830-2700 cm^{-1} and the carbonyl absorption near 1730 cm^{-1} indicate the presence of an aldehyde. The absorptions in the regions of; 3100 - 3000 cm^{-1} are due sp^2 hybridized C-H bonds, and those at 3000-2850 cm^{-1} are due to sp^3 hybridized C-H bonds. The appearance of four bands in the 1600-1450 cm^{-1} region suggests an aromatic ring. The ^{13}C-NMR shows the aldehyde carbon at 200 δ. The three types of aromatic CHs around 130-125 δ are an indication of a monosubstituted benzene ring. The C of the aromatic ring that is bonded to the substituent appears at 147 δ. The CH at ~53 δ is possibly attached to two electron withdrawing groups. The ^{1}H-NMR shows the presence of four types of hydrogens and the integrals match the actual number of hydrogens. The hydrogen of the aldehyde is at 9.7 δ. It is a doublet so the aldehyde group must be attached to a CH group. The five protons of the monosubstituted aromatic ring appear at 7.3 δ. The 1H that appears as a quadruplet must be coupled to the 3Hs appear as a doublet at 1.5 δ. This 1H must also be coupled to the aldehyde H, but the coupling is too small to be seen without expanding the peak. These data are consistent with the following compound.

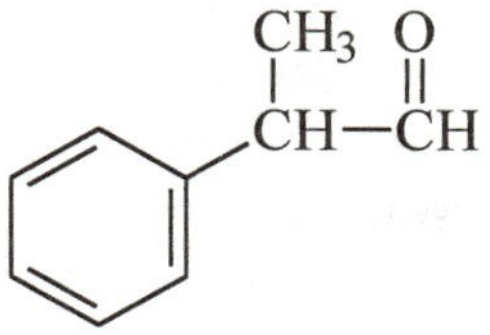

CH_3 O

CH—CH (attached to a benzene ring; C=O double bond)

12.28 a) NH_2, two bands, 3400-3250 cm^{-1}; ≡CH, 3300 cm^{-1}; -CH, 3000-2850 cm^{-1}; C≡C, 2150 - 2100 cm^{-1}.

b) OH, 3000 cm^{-1}, very broad; -CH, 3000-2850 cm^{-1}; -C≡N, 2260-2200 cm^{-1}, medium; C=O, 1710 cm^{-1}.

c) -CH, 3000-2850 cm^{-1}; C=O, 1740 cm^{-1}; C-O, 1300-1000 cm^{-1}.

d) =CH, 3100-3000 cm^{-1}; -CH, 3000-2850 cm^{-1}; C=O, 1695-1675 cm^{-1}; aromatic ring, 1600-1450 cm^{-1}, four bands, and 900-675 cm^{-1}.

12.29 a) The broad absorption centered at 3300 cm^{-1} indicates the presence of a hydroxy group. The absorption at 2900 cm^{-1} suggests the presence of hydrogens on sp^3 hybridized Cs. The bands in the 1300-1000 cm^{-1} region are possibly due to C-O bond. The compound is an alcohol.

b) The two bands in the 3400 - 3250 cm^{-1} region are typical of a NH_2 group. The absence of carbonyl band indicates that this is a primary amine. The absorption at 2900 cm^{-1} suggests the presence of hydrogens on sp^3 hybridized Cs.

c) The absorptions in the region of 3100 - 3000 cm^{-1} are due sp^2 hybridized C-H bonds, and those at 3000 - 2850 cm^{-1} are due to sp^3 hybridized C-H bonds. The appearance of four bands in the 1600-1450 cm^{-1} region and the strong band near 750 cm^{-1} are indicative of an aromatic ring. The strong absorption at 1683 cm^{-1} is due to a carbonyl group. The absence of bands for other carbonyl containing functional groups indicates that the compound is a ketone or an ester. It is often difficult to determine with certainty whether a compound is a ketone or an ester based solely on its IR spectrum. In this case there is a strong band near 1250 cm^{-1}, but it is not as broad as the C-O band of an ester normally appears (see spectrum d in this problem for an IR spectrum of an ester). Note that the carbonyl group of a conjugated ketone should appear at 1695-1675 cm^{-1} whereas that of a conjugated ester should appear at 1720-1700 cm^{-1}. This compound is actually a conjugated ketone.

d) The absorptions in the region of 3100-3000 cm^{-1} are due sp^2 hybridized C-H bonds, and those in the region of 3000-2850 cm^{-1} are due to sp^3 -hybridized C-H bonds. The appearance of four bands in the 1600 - 1450 cm^{-1} region is indicative of an aromatic ring. The strong absorption at 1715 cm^{-1} is due to a carbonyl group. The strong C-O band near 1200 cm^{-1} indicates that the compound is an ester. The observed shift of the carbonyl band to a lower wavenumber indicates that it is a conjugated ester.

12.30

a) 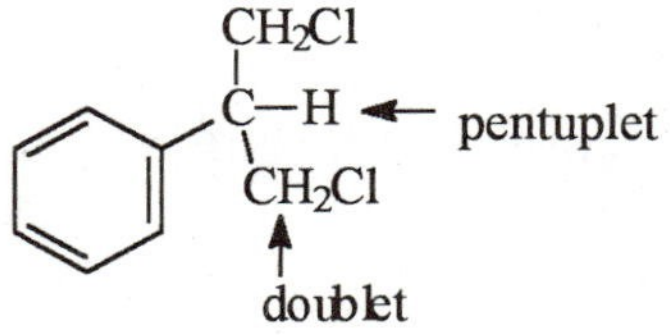 C_6H_5–CH(CH_2Cl)$_2$: C–H ← pentuplet; CH_2Cl ← doublet

b) 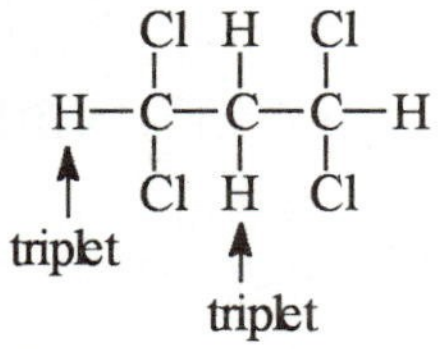 H–C(Cl)(Cl)–C(H)(H)–C(Cl)(Cl)–H: terminal H ← triplet; central H ← triplet

c) 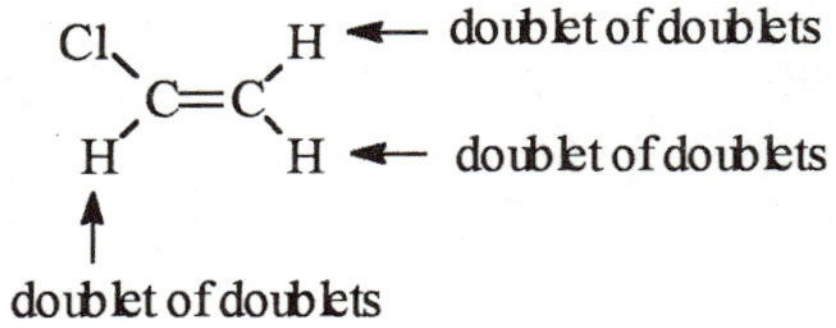 Cl(H)C=C(H)(H): H ← doublet of doublets; H ← doublet of doublets; H ← doublet of doublets (diastereotopic hydrogens)

12.31

a) CH_2 (diphenylmethane)

	a)	
chemical shift	7-8	(2.3+.3+2.3-.9) = 4.0
multiplicity	m	1
integral	5	1

b)

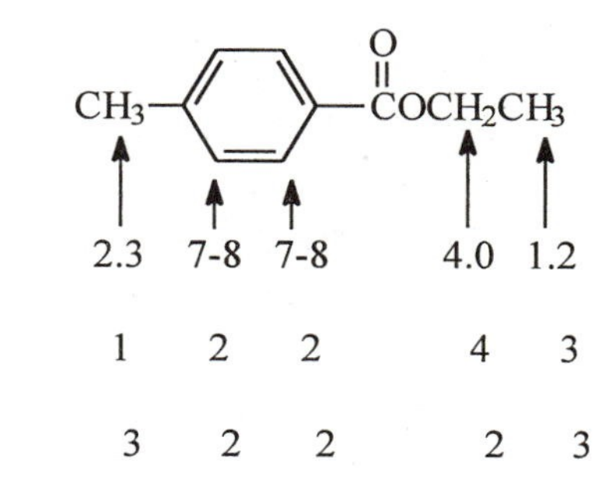

c)

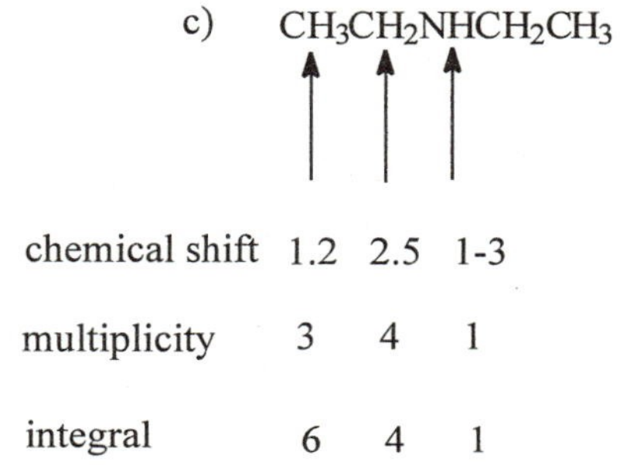

	c)		
chemical shift	1.2	2.5	1-3
multiplicity	3	4	1
integral	6	4	1

d)

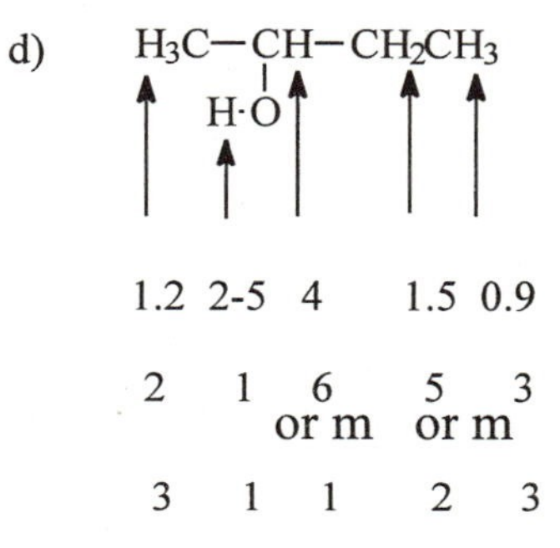

12.32

CH_2CH_3

120-160

CH_2CH_3

120-160 52 24

12.33 a) The absorption for the carbonyl group of the left compound should appear near 1715 cm^{-1} in its IR spectrum. The right compound is a conjugated ketone, so the carbonyl absorption will appear between 1695 and 1675 cm^{-1}. The ^{1}H-NMR spectrum of the left compound has two singlets in the upfield region whereas the right compound has a triplet and a quartet.

b) The right compound has two peaks in its ^{13}C-NMR spectrum whereas the left compound has 3 peaks.

c) The right compound is an aldehyde and the left is a ketone. These can be distinguished by IR spectroscopy. The aldehyde has two absorptions in the 2850-2700 cm^{-1} region and its carbonyl absorption is near 1730 cm^{-1}. The ketone has no absorption in the 2850-2700 cm^{-1} region and its carbonyl absorption appears near 1715 cm^{-1}. The ^{1}H-NMR spectrum of the aldehyde will show an peak around 10 δ, which will be absent in the ketone.

d) The left compound is a symmetrical ketone. The ^{1}H-NMR spectrum of this compound will show two signals, a triplet and a quartet. The right compound will show four signals.

12.34

a) 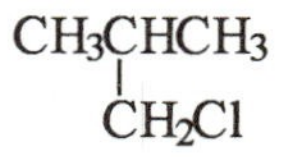 $CH_3CH(CH_2Cl)CH_3$

b) $Br-CH_2CH_2CH_2-Br$

c) 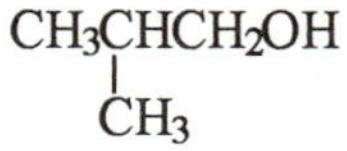 $CH_3CH(CH_3)CH_2OH$

d) 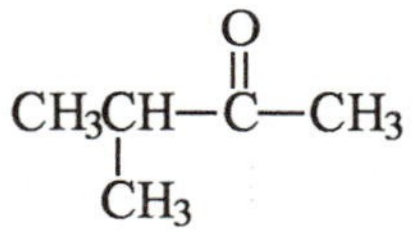 $CH_3CH(CH_3)-C(=O)-CH_3$

12.35 The CH at 190 δ indicates an aldehyde. The four signals between 140 - 120 δ are due to a monosubstituted benzene ring. These data fit benzaldehyde.

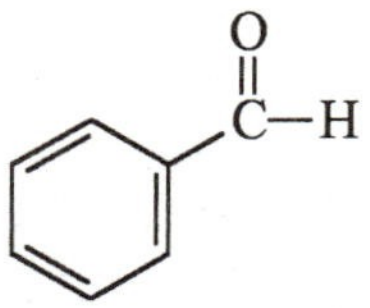

$C_6H_5-C(=O)-H$

12.36 The DU is 4. The IR spectrum shows the presence of =CH, 3100-3000 cm^{-1}; -CH, 3000-2850 cm^{-1}; and an aromatic ring, 1600-1450 cm^{-1} (four bands). If an aromatic ring is present , it accounts for the DU of 4. The ^{1}H-NMR spectrum shows the presence of three types of hydrogens. The multiplet for 5 Hs at 7.2 δ indicates a monosubstituted aromatic ring. The hydrogens of the two distorted triplets at 3.6 δ and 3.0 δ are coupled, indicating the structural feature $-CH_2CH_2-$. Examining the chemical shifts of these methylene hydrogens indicates that one CH_2 must be attached to

Br and other to the aromatic ring. The structure can be assigned as 1-bromo-2-phenylethane.

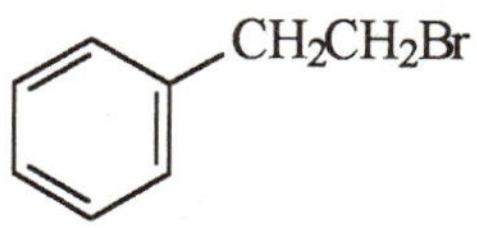

$$C_6H_5CH_2CH_2Br$$

12.37 The DU for this compound is 3. The IR spectrum indicates the presence of -CH, 3000-2850 cm^{-1}; C≡N, 2250 cm^{-1}; C=O of ester, 1740 cm^{-1}; and C-O, about 1200 cm^{-1}. The ^{1}H-NMR spectrum indicates the presence of three types of hydrogens. The integrals match the actual number of hydrogens. The 2 Hs appearing as the quartet at 4.2 δ must be coupled to the 3 Hs appearing as the triplet at 1.2 δ. The chemical shifts suggest that the CH_3CH_2 fragment is linked to the oxygen of an ester group. The singlet at 3.5 δ is consistent with a CH_2 group which is directly attached to a carbonyl and another electron withdrawing group. The structure can be assigned as ethyl 2-cyanoethanoate.

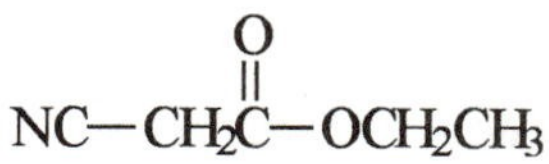

$$NC-CH_2\overset{O}{\overset{||}{C}}-OCH_2CH_3$$

12.38 The DU for this compound is 1. The IR spectrum indicates the presence of -CH, 3000-2850 cm^{-1}; C=O of ester, 1733 cm^{-1}; and C-O, 1200 cm^{-1}. The integrals in the ^{1}H-NMR spectrum suggest that actual number of hydrogens under each of the signals must be double that of the integral value. The quartet at 4.2 δ is due to hydrogens of a CH_2 that is directly linked to the oxygen of the ester group. This CH_2 group must be coupled to a CH_3 group which splits it into a quartet. The quartet at 2.3 δ is due to hydrogens of a CH_2 group which is directly linked to a carbonyl group. This CH_2 is also attached to a CH_3 group resulting in its signal appearing as a quartet. Careful examination of the signal at 1.2 δ reveals that the multiplet is actually two triplets. These are the two methyl groups, each split into a triplet by the CH_2 group to which they are bonded. The structure can be assigned as ethyl propanoate.

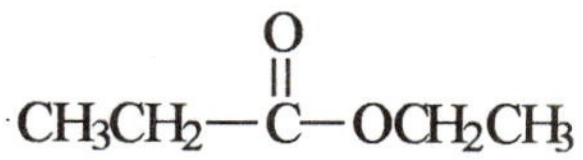

$$CH_3CH_2-\overset{O}{\overset{||}{C}}-OCH_2CH_3$$

12.39 The IR spectrum is typical of a carboxylic acid. The DU of 2 and the absence of alkene and aromatic bands (no C=C) suggest the presence of a ring (C=O and ring give DU = 2). The singlet at 12 δ in the ^{1}H-NMR spectrum is consistent with the hydrogen of a carboxylic acid. The presence of only four different types of carbons in the ^{13}C-NMR spectrum indicates some symmetry in the structure. Probably there are two of both kinds of the CH_2 groups to account for a total of 6 carbons and 10 hydrogens. The compound is cyclopentanecarboxylic acid.

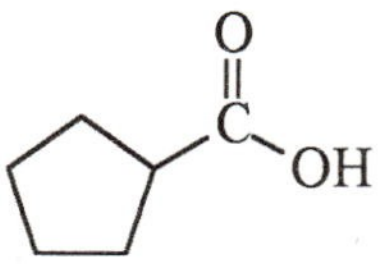

O
||
C
OH

12.40 The carbonyl band at 1741 cm^{-1} and the C-O absorption near 1200 cm^{-1} indicates that the unknown is an ester. This accounts for the DU of 1. The ^{1}H-NMR spectrum shows the presence of five types of hydrogens and the integrals match the actual numbers. The singlet at 2.1 δ results from a CH_3 group that is attached to the carbonyl group of the ester. The triplet at 4.2 δ results from a CH_2 attached to the oxygen of the ester group. This CH_2 must be bonded to a CH_2 because it appears as a triplet. The signal at 1.6 δ appears to contain five lines. This indicates a CH_2 group (integral =2) bonded to two CH_2 groups (at 4.2 and 1.4 δ) with similar coupling constants. The signal at 1.4 δ appears to contain six lines. This indicates a CH_2 group bonded to a CH_2 group (at 1.6 δ) and a CH_3 group (at 1.0 δ) with similar coupling constants. Overall, then, a butyl group appears to be present. The ^{13}C-NMR spectrum confirms these assignments. The compound is butyl ethanoate

$$CH_3\overset{\overset{\large O}{||}}{C}{-}OCH_2CH_2CH_2CH_3$$

12.41 The IR spectrum shows the presence of an OH band at 3300 cm^{-1} and -CH bands at 3000-2850 cm^{-1}. These IR bands and the DU of zero suggest that the compound is a saturated alcohol. The singlet at 2.6 δ in the ^{1}H-NMR spectrum is due to H of the hydroxy group. The triplet at 3.6 δ is due to hydrogens of a CH_2 group attached to the hydroxy group. The ^{13}C-NMR spectrum shows the presence of eight different types of carbons: seven CH_2 groups and one CH_3 group. The unknown is 1-octanol.

$$CH_3CH_2CH_2CH_2CH_2CH_2CH_2CH_2{-}OH$$

12.42 The IR spectrum shows the presence of a carbonyl group and sp^3 hybridized C-H bonds. The position of the carbonyl band (1721 cm^{-1}) is characteristic of a ketone. The absence of absorptions due to =C-H and C=C bands and the DU of 2 suggest that the compound is a cyclic ketone. The ^{1}H-NMR spectrum is complex. One useful piece of information is provided by the doublet near 1 δ with an integral of three. This indicates the presence of a methyl group attached to a carbon with one hydrogen, $CHCH_3$. The ^{13}C-NMR shows the presence of only five different carbons, indicating some symmetry in the structure. Since there is only one carbonyl carbon (near 210 δ) and one $CHCH_3$ (from the ^{1}H-NMR spectrum), there must be two of each of the other CH_2 groups to match the formula. The carbonyl carbon, the four CH_2 groups and the CH group must be part of the ring. The compound is 4-methylcyclohexanone.

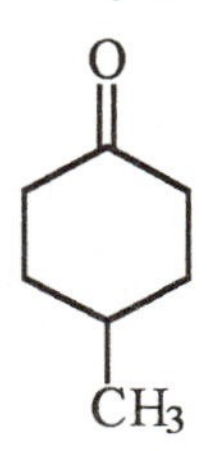

12.43

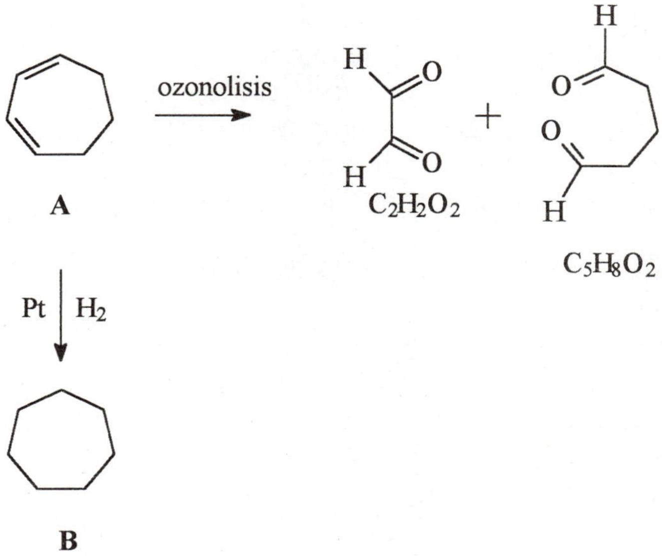

A ozonolisis → $C_2H_2O_2$ + $C_5H_8O_2$

A Pt H₂ → B

12.44

$$CH_3-\underset{OH}{\overset{CH_3}{C}}-CH_2CH_3$$

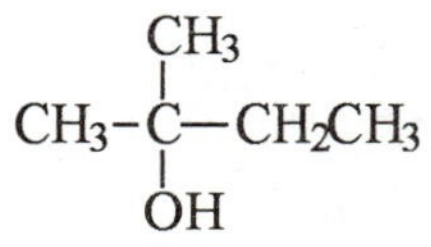

12.45 There will be three absorptions in the ^{1}H-NMR spectrum of this compound. The hydrogens on the carbons bonded to the chlorines are chemically equivalent and will produce one signal. The hydrogens of CH_2 group are diastereotopic, and will give two different signals.

After completing this chapter, you should be able to:

1. Predict the important absorption bands in the IR spectrum of a compound. Problems 12.4, 12.5, 12.6, 12.8, 12.9, 12.10, 12.11, 12.28.

2. Determine the functional group that is present in a molecule by examination of its infrared spectrum. Problems 12.12, 12.29.

3. Predict the chemical shifts, multiplicity, and integrals of peaks in the ^{1}H-NMR spectrum of a compound.
   Problems 12.15, 12.16, 12.17, 12.18, 12.19, 12.20, 12.30, 12.31.

4. Predict the number and general chemical shifts of peaks in the ^{13}C-NMR spectrum of a compound.
   Problem 12.22, 12.23, 12.32.

5. Determine the hydrocarbon skeleton of a compound by examination of its ^{1}H and/or ^{13}C-NMR spectrum.
   Problems 12.21, 12.24, 12.35.

6. Use a combination of these techniques to determine the structure of an unknown compound.
   12.25, 12.26, 12.27, 12.33, 12.34, 12.36, 12.37, 12.38, 12.39, 12.40, 12.41, 12.42.

Chapter 13
SPECTROSCOPY II
ULTRAVIOLET-VISIBLE SPECTROSCOPY AND MASS SPECTROMETRY

13.1 According to the Lambert-Beer law, **A = ε c l**,
where **A** is the absorbance of the solution,
ε is the molar extinction coefficient,
c is the concentration of the solution, and
l is the path length of the cell.

The absorbance of the solution of anthracene is 0.349.

13.2 The molar absorptivity of benzophenone is 1.97×10^4 $M^{-1}cm^{-1}$.

13.3 The concentration of the solution is 2.38×10^{-5} M.

13.4 The smaller molar absorptivity and the wavelength indicate that the transition is $n \rightarrow \pi^*$.

13.5 The absorption at 213 nm is due to a $\pi \rightarrow \pi^*$ transition (higher energy, larger ε) and that at 320 nm is due to a $n \rightarrow \pi^*$ transition (lower energy, smaller ε).

13.6 a) The left ketone is conjugated and should absorb at longer wavelength than the right ketone, which is not conjugated.
b) The right compound is conjugated and should absorb at longer wavelength than the left compound, which is not conjugated.
c) The left compound has three conjugated double bonds and should absorb at longer wavelength than the right compound, which has only two conjugated double bonds.
d) Cyclohexanone has an absorption for a $n \rightarrow \pi^*$ transition in the accessible UV region, whereas the alcohol shows no such absorption.

13.7 Butane = 58.0783; acetone = 58.0419. Since the molecular masses of these compounds differ by a few hundredths of a mass unit, they can be easily distinguished by a high resolution mass spectrometer.

13.8 The predicted intensity of the M+1 peak for butane is 4.4% and that for acetone is 3.3%. The various errors that occur in measuring these intensities may be larger than the difference in the intensities of these two peaks. Therefore, the intensity difference cannot be used to distinguish between these compounds.

13.9 b) M : M+2 : M+4 = 1 : 0.67 : 0.11 (9:6:1)
c) M : M+2 : M+4 = 1 : 1.33 : 0.33 (3:4:1)

13.10 a) The similar intensities of the $M^{+\cdot}$ (m/z 164) and the M+2 (m/z 166) peaks indicate that there is one Br is present in this compound.
b) The appearance of the molecular ion at odd m/z (73) indicates the presence of an odd number of nitrogens in the compound.

13.11

a) $CH_3\overset{\cdot\oplus}{N}HCH_3$ b) (cyclohexane radical cation: one C–C bond of the ring carrying • and ⊕, with H on each carbon)

13.12

a) $[CH_3CH_2CH_2 \wr CH_2CH_3]^{+\cdot} \longrightarrow CH_3CH_2\overset{\oplus}{C}H_2$ (m/z 43) $+ \cdot CH_2CH_3$

$\longrightarrow CH_3CH_2\overset{\cdot}{C}H_2 + \overset{\oplus}{C}H_2CH_3$ (m/z 29)

b) $[CH_3CH_2CH(CH_2CH_3) \wr CH_2CH_3]^{+\cdot} \longrightarrow CH_3CH_2\overset{\oplus}{C}H(CH_2CH_3)$ (m/z 71) $+ \cdot CH_2CH_3$

$CH_3CH_2\overset{\cdot}{C}H(CH_2CH_3) + \overset{\oplus}{C}H_2CH_3$ (m/z 29)

13.13

a) $\left[CH_3CH(CH_3)-CH(CH_2CH_3)CH_2CH_3 \right]^{+\cdot} \longrightarrow CH_3\overset{\oplus}{C}H(CH_3) + \cdot CH(CH_2CH_3)CH_2CH_3$

m/z 43

b) $\oplus CH(CH_2CH_3)CH_2CH_3$

m/z 71

13.14

b) $\left[C_6H_5CH_2-Cl \right]^{+\cdot} \longrightarrow C_6H_5\overset{\oplus}{C}H_2 + \cdot\ddot{Cl}:$

m/z 91

c) $CH_3CH_2CH_2-CH_2\overset{\cdot\oplus}{O}H \longrightarrow CH_3CH_2\dot{C}H_2 + \overset{\oplus}{:O}{=}CH_2 \;(-H)$

m/z 31

d) $CH_3CH_2-C(=\overset{\cdot\oplus}{O}:)-CH_2CH_2CH_2CH_3 \longrightarrow CH_3CH_2C\equiv\overset{\oplus}{O}: + \dot{C}H_2CH_2CH_2CH_3$

m/z 57

$\longrightarrow CH_3\dot{C}H_2 + :\overset{\oplus}{O}\equiv CCH_2CH_2CH_2CH_3$

m/z 85

$\longrightarrow CH_3CH_2C(\overset{\oplus\cdot}{O}-H)=CH_2 + CH_2=CHCH_3$

m/z 72 (McLafferty rearrangement)

13.15 The concentration should be 2.0×10^{-5} M.

13.16 a) $\pi \rightarrow \pi^*$ at 252 nm and $n \rightarrow \pi^*$ at 325 nm.
b) $\pi \rightarrow \pi^*$ at 235 nm.
c) $n \rightarrow \pi^*$ at 299 nm.
d) $\pi \rightarrow \pi^*$ at 227 nm.

13.17 a) 2-Butanone will show an absorption maximum for its $n \rightarrow \pi^*$ transition in this region.
b) The $\pi \rightarrow \pi^*$ transition of 1,3-pentadiene does not occur in the accessible UV region because it is not conjugated.
c) This compound has conjugated double bonds. Therefore its $\pi \rightarrow \pi^*$ transition will occur in this region.
d) This compound has an aromatic ring. Therefore a $\pi \rightarrow \pi^*$ transition will occur in this region.
e) This compound has two fused aromatic rings. Therefore a $\pi \rightarrow \pi^*$ transition will occur in this region.
f) This compound will not show an absorption in this region.
g) This compound will not show an absorption in this region.
h) This compound will show absorption maxima for $\pi \rightarrow \pi^*$ and $n \rightarrow \pi^*$ transitions in this region.

13.18

a) $[CH_3CH_2CH_2CH_2 \wr CH_2CH_3]^{\cdot+}$ (m/z 86) $\longrightarrow$ $CH_3CH_2CH_2\overset{\oplus}{C}H_2$ (m/z 57) + $\cdot CH_2CH_3$

$\longrightarrow$ $CH_3CH_2CH_2\dot{C}H_2$ + $\overset{\oplus}{C}H_2CH_3$ (m/z 29)

$\longrightarrow$ $CH_3CH_2\overset{\oplus}{C}H_2$ (m/z 43) + $\dot{C}H_2CH_2CH_3$

b) $\left[CH_3CH_2CH_2 \,\wr\, \underset{}{\overset{CH_3}{|}}{CH} \,\wr\, CH_2CH_3 \right]^{+\cdot}$ $\longrightarrow$ $CH_3CH_2CH_2\overset{CH_3}{\underset{\oplus}{CH}}$ + $\cdot CH_2CH_3$

m/z 100 → m/z 71

$CH_3CH_2CH_2\overset{CH_3}{\underset{\cdot}{CH}}$ + $\overset{\oplus}{C}H_2CH_3$

m/z 29

$\overset{\oplus}{C}H_3CH_2CH_2$ m/z 43 + $CH_3CH_2\overset{CH_3}{\underset{\oplus}{CH}}$ m/z 57

$CH_3CH_2\overset{CH_3}{\underset{\cdot}{CH}}$ $CH_3CH_2\dot{C}H_2$

c) $CH_3 \,\wr\, \overset{CH_3}{\underset{CH_3}{C}}-\overset{\cdot\oplus}{\underset{\cdot\cdot}{O}}H$ $\longrightarrow$ $\dot{C}H_3$ + $\oplus\overset{CH_3}{\underset{CH_3}{C}}-OH$

m/z 74 m/z 59

$\xrightarrow{-H_2O}$ $CH_3-\overset{\dot{C}H_2}{\underset{CH_3}{C}}\oplus$

m/z 56

d) $\left[\text{4-}CH_3C_6H_4-CH_2 \,\wr\, CH_2CH_3 \right]^{+\cdot}$ $\longrightarrow$ $\text{4-}CH_3C_6H_4-\overset{\oplus}{C}H_2$ + $\cdot CH_2CH_3$

m/z 134 m/z 105

e) $\left[CH_3CH_2 - C(CH_3)_2 - CH_2CH_2CH_3 \right]^{+\bullet}$ (m/z 114) $\longrightarrow$ $CH_3CH_2C^{\oplus}(CH_3)_2 + {}^{\bullet}CH_2CH_2CH_3$ (m/z 71)

$\longrightarrow$ $CH_3CH_2CH_2C^{\oplus}(CH_3)_2 + {}^{\bullet}CH_2CH_3$ (m/z 85)

$\longrightarrow$ $CH_3CH_2 - \overset{\oplus}{C}(CH_3) - CH_2CH_2CH_3 + {}^{\bullet}CH_3$ (m/z 99)

m/z 114

m/z 71

m/z 85

m/z 99

f) $CH_3 - C(\overset{\bullet\oplus}{:OH})(CH_3) - CH_2CH_3$ (m/z 88) $\longrightarrow$ $CH_3 - C^{\oplus}(OH)(CH_3) + {}^{\bullet}CH_2CH_3$ (m/z 59)

$\longrightarrow$ $^{\oplus}C(OH)(CH_3) - CH_2CH_3 + {}^{\bullet}CH_3$ (m/z 73)

$\downarrow$ - H_2O

$\left[CH_3 - C(CH_3) = CHCH_3 \right]^{+\bullet}$ (m/z 70)

m/z 88

m/z 59

m/z 70

m/z 73

g) $\left[CH_3 \wr C(=O) \wr CH_2CH_2CH_2CH_3\right]^{\dot{+}} \longrightarrow CH_3C{\equiv}\overset{\oplus}{O}: + CH_3CH_2CH_2\dot{C}H_2$

m/z 100 → m/z 43

$\longrightarrow CH_3CH_2CH_2CH_2C{\equiv}\overset{\oplus}{O}: + {}^{\bullet}CH_3$

m/z 85

McLafferty Rearrangement

$CH_3{-}C(\overset{\oplus}{\dot{O}}H){=}CH_2 + CH_2{=}CHCH_3$

m/z 58

h) $CH_3CH_2CH_2 \wr CH(OH) \wr CH_3 \longrightarrow CH_3CH_2CH_2CH{=}\overset{\oplus}{O}H + HC(=\overset{\oplus}{O}H)CH_3 + \text{m/z } 70$

m/z 88 — m/z 73 — m/z 45 — (loss of H_2O)

i) $\left[CH_3CH{=}CHCH_2 \wr CH_2CH_3\right]^{\dot{+}} \longrightarrow CH_3CH{=}CH\overset{\oplus}{C}H_2 + {}^{\bullet}CH_2CH_3$

m/z 84 — m/z 55

13.19 a) The major fragment in the mass spectrum of butane is an ethyl cation at m/z 29. For 2-methylpropane, the major fragment should be a 2-propyl cation at m/z 43.

b) Both ketones will produce fragment ions with the same m/z value due to the cleavage of the bonds to the carbonyl carbon. However, the fragmentation ions of the McLafferty rearrangement in these two ketones are different. The left ketone will show a peak at m/z 58 while the right compound will show a peak at m/z 72.

c) The left ketone will show peaks at m/z 43 and 71 due to the cleavage of the bonds to the carbonyl carbon and one at m/z 58 due to the McLafferty rearrangement. The right ketone will show a peak at m/z 57 due to the cleavage of the bonds to the carbonyl carbon. This ketone cannot undergo a McLafferty rearrangement.

13.20 Neopentane does not show a molecular ion peak in its mass spectrum because fragmentation to produce the tertiary carbocation occurs very readily. Therefore, the base ion in the mass spectrum of neopentane has m/z 57.

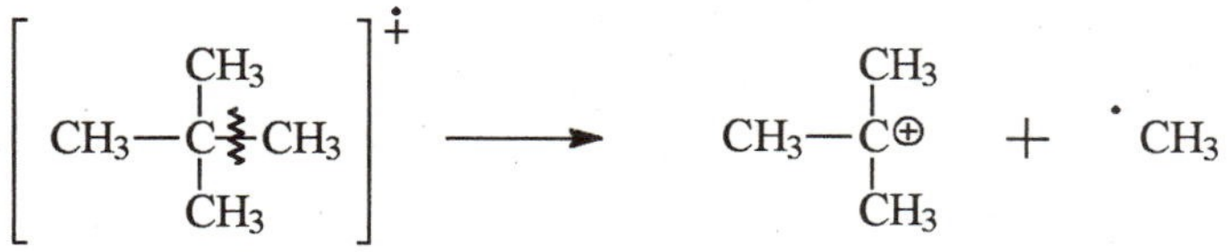

$$\left[CH_3-\underset{CH_3}{\overset{CH_3}{C}} \wr CH_3 \right]^{\dot{+}} \longrightarrow CH_3-\underset{CH_3}{\overset{CH_3}{C^{\oplus}}} + \cdot CH_3$$

m/z 72 molecular ion → m/z 57 base ion

13.21 The peaks at m/z 115, 101, and 73 are due to fragment ions produced by the cleavage of one of the bonds between the hydroxy carbon and an adjacent carbon in 3-methyl-3-heptanol. The resulting cations are relatively stable because they satisfy the octet rule at all of the atoms.

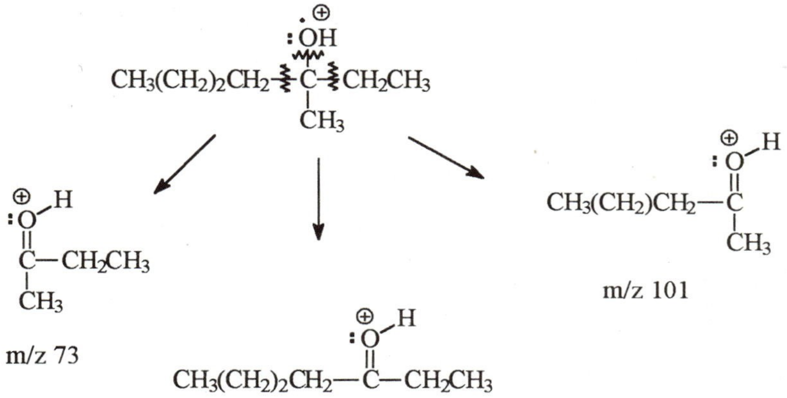

$$CH_3(CH_2)_2CH_2 \wr \underset{CH_3}{\overset{:\dot{O}H^{\oplus}}{C}} \wr CH_2CH_3$$

$$\underset{CH_3}{\overset{:\overset{\oplus}{O}-H}{\overset{\|}{C}}}-CH_2CH_3$$

m/z 73

$$CH_3(CH_2)_2CH_2-\overset{:\overset{\oplus}{O}-H}{\overset{\|}{C}}-CH_2CH_3$$

m/z 115

$$CH_3(CH_2)CH_2-\underset{CH_3}{\overset{:\overset{\oplus}{O}-H}{\overset{\|}{C}}}$$

m/z 101

13.22 Both the ketones will produce fragment ions at m/z 43 and 85 due to the cleavage of the bonds to the carbonyl carbon. However, the fragment ions resulting from the McLafferty rearrangement will be different for each of these ketones. The McLafferty rearrangement of 3-methyl2-pentanone will produce the ion with m/z 72, while that of 4-methyl-2-pentanone will give m/z 58. Therefore, the upper spectrum belongs to 4-methyl-2-pentanone and the lower spectrum belongs to 3-methyl-2-pentanone.

The fragment ion responsible for the peak at m/z 43 in both ketones arises from the same cleavage product:

$$CH_3-\overset{\oplus}{C}\equiv O:$$

m/z 43

13.23 The peaks at m/z 186, 188 and 190 in a 1:2:1 ratio suggest the presence of two bromine atoms. Subtracting two ^{79}Br from 186 leaves a mass of 28, corresponding to C_2H_4. Therefore the formula is with $C_2H_4Br_2$. The unknown is one of the two isomers with this formula, 1,1-dibromoethane or 1,2-dibromoethane. Determining which of these two isomers is correct is more difficult. Both are expected to lose a bromine atom to give C_2H_4Br at m/z 107 and 109, the base ions in the spectrum. However, only 1,2-dibromoethane can fragment to give CH_2Br with m/z 93 and 95. The presence of these peaks in the spectrum indicate that the unknown is 1,2-dibromoethane.

$$\left[\underset{\text{Br}}{CH_2} - \underset{\text{Br}}{CH_2} \right]^{\dot{+}} \longrightarrow \overset{\oplus}{C}H_2Br + \dot{C}H_2Br$$

m/z 186, 188, 190 → m/z 93, 95

13.24 The DU for both isomers is 1. Based on their IR spectra, this must result from a carbonyl group in each. Therefore one must be propanal and other must be 2-propanone. Based on the position of the carbonyl absorption in the IR spectra **A** must be propanal and **B** must be 2-propanone. This confirmed by their mass spectra. The major fragmentation pathway in aldehyde and ketone is cleavage of the bond to the carbonyl carbon.

$$\underset{\textbf{A}}{CH_3CH_2-\overset{O}{\overset{\|}{C}}-H} \longrightarrow \underset{m/z\ 57}{CH_3CH_2\overset{\oplus}{C}=O} + \underset{m/z\ 29}{H\overset{\oplus}{C}=O}$$

$$\underset{\textbf{B}}{CH_3-\overset{O}{\overset{\|}{C}}-CH_3} \longrightarrow \underset{m/z\ 43}{CH_3\overset{\oplus}{C}=O}$$

13.25 The ^{1}H- NMR peak near 7.25 δ and the DU of 4 are consistent with a benzene ring in the structures of the both isomers. The fact that the area under the NMR peak at 7.25 δ is 5 indicates that the benzene ring is monosubstituted in each case. Both compounds are therefore monosubstituted benzene derivatives. Subtracting C_6H_5 from the formula leaves C_3H_7, so the remaining fragment must be either a propyl or a isopropyl group. The base peak at m/z 91 in the mass spectra of isomer **D** is typical of a benzyl ion fragment, indicating that the structure is propylbenzene. The peak at m/z 105 in mass spectrum of **C** is consistent with the structure isopropylbenzene.

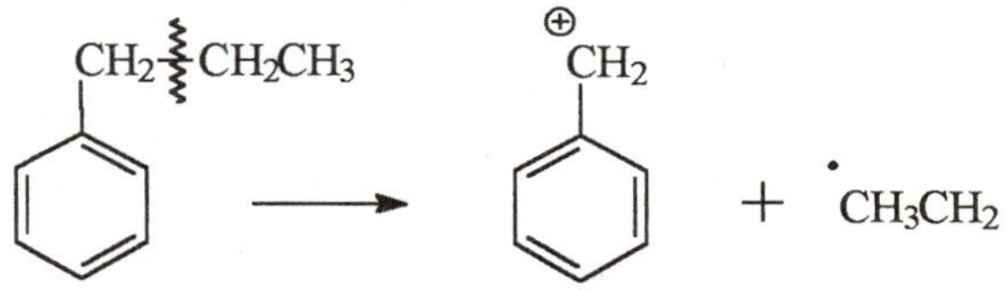

CH_2–CH_2CH_3 → $\overset{\oplus}{C}H_2$ + $\dot{C}H_3CH_2$

D

m/z 91

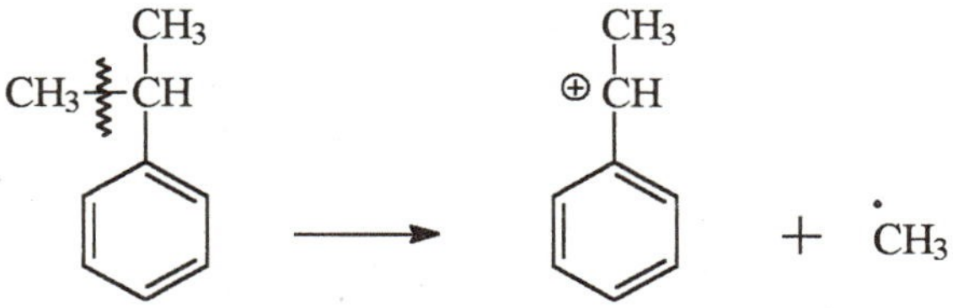

CH_3

CH_3–CH → $\oplus$ CH (CH_3) + $\dot{C}H_3$

C

m/z 105

13.26 The DU of **E** and **F** is 3. Both produce **G**, with DU of 1, upon catalytic hydrogenation. **G** cannot have a pi bond because it is produced by hydrogenation so its DU must be due to a ring. Therefore **E** and **F** must have two pi bonds and a ring. Since **G** shows only a single peak in its ^{13}C-NMR spectrum, it must be cyclohexane. Thus **E** and **F** are isomers of cyclohexadiene. The λ_{max} at 259 nm in the UV spectrum of compound **F**, suggests that the two double bonds are conjugated. The absence of any absorption maximum above 200 nm in the UV spectrum of **E** is consistent with non-conjugated double bonds.

E

F

13.27 The ^{1}H-NMR spectrum shows the presence of three different hydrogens in 2:2:3 ratio. The two hydrogens at 3.5 δ are coupled to the two hydrogens appearing at 1.8 δ as are the three hydrogens appearing at 1.0 δ. This suggests the presence of a propyl group. The peaks of equal intensity at m/z 122 and 124 in the mass spectrum indicate the presence of a bromine. The unknown is 1-bromopropane.

$$CH_3CH_2CH_2Br$$

After completing this chapter, you should be able to:

1. Determine whether a compound will absorb light in the ultraviolet or visible region.
   Problems 13.6, 13.17.

2. Identify the chromophore and type of transition responsible for absorption of UV-visible radiation.
   Problems 13.16.

3. Use a high-resolution mass spectrum to determine the formula of a compound.
   Problems 13.7.

4. Determine whether sulfur, chlorine, bromine, or nitrogen is present in a compound by examination of the M, M + 1, and M + 2 peaks in its mass spectrum.
   Problems 13.9, 13.10.

5. Explain the major fragmentation pathways for compounds containing some of the simple functional groups.
   Problems 13.12, 13.13, 13.14, 13.18, 13.19, 13.20, 13.21, 13.22, 13.23, 13.24, 13.25.

Chapter 14
ADDITIONS TO THE CARBONYL GROUP

14.1

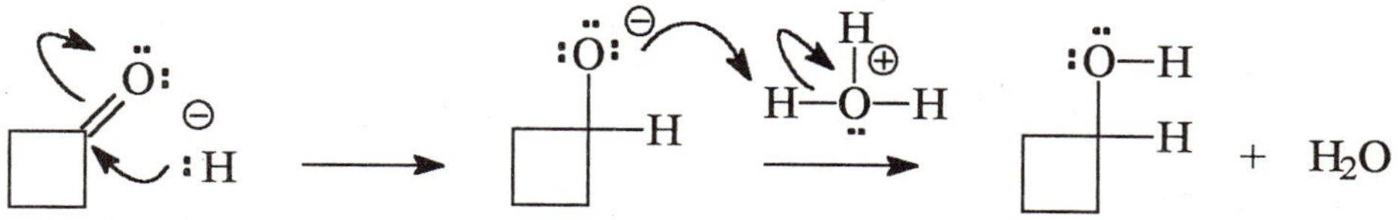

:Ö: ⊖ :H ⟶ :Ö: ⊖ —H ; H—Ö—H (H, ⊕) ⟶ :Ö—H, —H + H_2O

14.2

a) 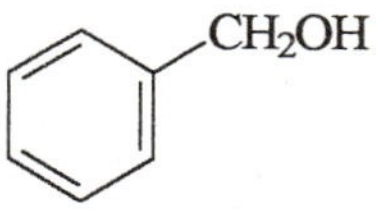 $C_6H_5CH_2OH$ (CH₂OH on benzene ring) b) OH, $CH_3CH_2CHCH_3$ c) OH, $CH_3CH{=}CHCH_2CH_2CH_2$ d) 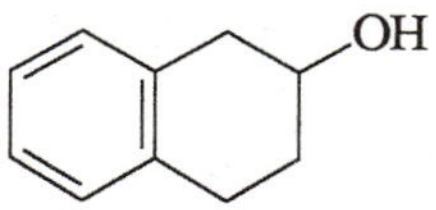 OH

14.3

H—O:⊕ (H, H)

:O, $H_3C—C—CH_3$ ⇌ Ö—H ⊕, $H_3C—C—CH_3$, :Ö[18]—H, H ⇌ :Ö—H, $H_3C—C—CH_3$, ⊕ :O[18]—H, H ; :Ö—H, H ⇌ :Ö—H, $H_3C—C—CH_3$, :O[18]—H

H—O:⊕ (H, H)

:Ö[18], $H_3C—C—CH_3$ ⇌ H—Ö:, H ; $H_3C—C—CH_3$, ⊕ O[18]—H ⇌ H—Ö—H ⊕, $H_3C—C—CH_3$, :O[18]—H

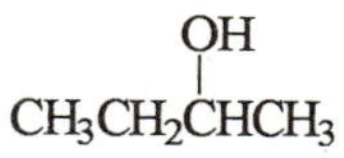

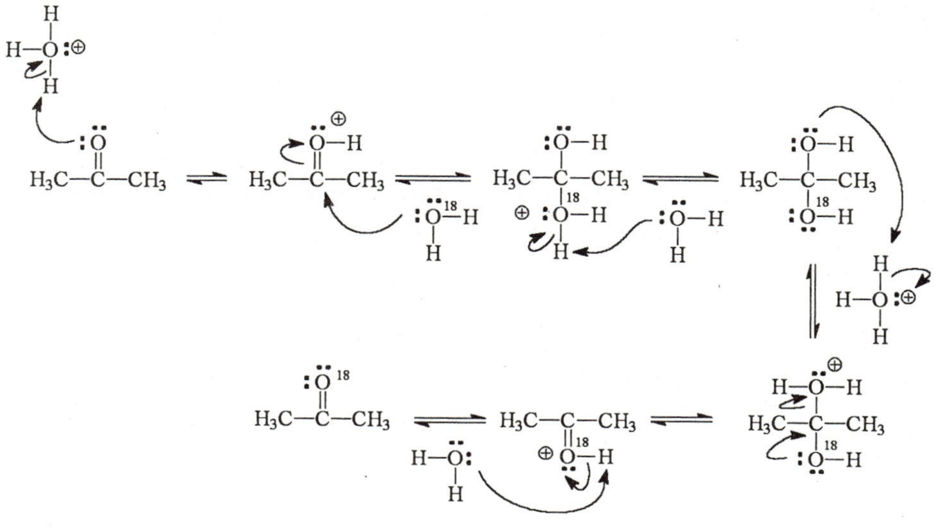

14.4 a) right compound due to inductive electron withdrawing effect of F
b) right compound due to inductive and steric effects
c) left compound due to less steric hindrance
d) right compound due to inductive electron withdrawing effect of Cl
e) not much difference because the steric effect is too far from the reacting carbonyl carbon

14.5

$$CH_3CH_2\overset{O}{\overset{\|}{C}}H < \overset{Cl}{\overset{|}{C}}H_2CH_2\overset{O}{\overset{\|}{C}}H < CH_3\overset{Cl}{\overset{|}{C}}H\overset{O}{\overset{\|}{C}}H < CH_3CF_2\overset{O}{\overset{\|}{C}}H$$

smallest K largest K

14.6

a) HO, CN, C, H, H_3C

b) $CH_3CH_2\overset{OH}{\overset{|}{C}}(CN)CH_3$

c) $CH_3\overset{CH_3}{\overset{|}{C}}HCH_2\overset{OH}{\overset{|}{C}}H(CN)$

14.7 a) right compound because aldehydes are less sterically hindered than ketones
b) left compound because resonance makes the carbonyl carbon of the right compound less electrophilic
d) right compound because the electron withdrawing nitro group makes its carbonyl carbon more electrophilic

14.8

a) $CH_3CH_2CH_2CH_3$ b) D c) $CH_3CH_2C{\equiv}CMgBr$ + CH_3CH_3

14.9

a) $C_6H_5C(OH)(CH_3)CH_2CH_3$

b) $CH_3CH_2CH_2CH_2CH_2OH$

c) $CH_3CH_2CH_2CH(OH)CH_2CH_2CH_3$

d) $CH_3CH_2C(C_6H_5){=}CHCH_3$

e) 1-methylcyclohexanol (HO, CH_3 on the same ring carbon)

f) $CH_3CH_2CH_2CH_2CH_2CO_2H$

g) $CH_3C{\equiv}CCH(OH)CH_3$

14.10

a) CH_3CH_2CHO $\xrightarrow[2)\ H_3O^{\oplus}]{1)\ CH_3MgI}$ or CH_3CHO $\xrightarrow[2)\ H_3O^{\oplus}]{1)\ CH_3CH_2MgI}$

b) $C_6H_5CH_2Br$ $\xrightarrow{1)\ Mg,\ ether}$ 2) HCHO 3) $H_3O^{\oplus}$

c) bromocyclohexane $\xrightarrow{1)\ Mg,\ ether}$ 2) CO_2 3) $H_3O^{\oplus}$

d) $PhCOPh$ $\xrightarrow[2)\ H_3O^{\oplus}]{1)\ PhMgBr}$

e) $CH_3CH_2COCH_2CH_3$ $\xrightarrow[2)\ NH_4Cl,\ H_2O]{1)\ CH_3Li}$ or $CH_3CH_2COCH_3$ $\xrightarrow[2)\ NH_4Cl,\ H_2O]{1)\ CH_3CH_2MgBr}$

f) cyclohexanone $\xrightarrow[2)\ NH_4Cl,\ H_2O]{1)\ HC{\equiv}CMgBr}$

g) cyclopentyl–$COCH_2CH_3$ $\xrightarrow[2)\ NH_4Cl,\ H_2O]{1)\ CH_3Li}$ or

cyclopentyl–$COCH_3$ $\xrightarrow[2)\ NH_4Cl,\ H_2O]{1)\ CH_3CH_2MgBr}$ or $CH_3COCH_2CH_3$ $\xrightarrow[2)\ NH_4Cl,\ H_2O]{1)\ \text{cyclopentyl–MgBr}}$

14.11 An acid-base reaction between the Grignard reagent and the hydroxy group on the aldehyde destroys the Grignard reagent.

14.12

$$Ph_3P + PhCH_2Cl \xrightarrow{S_N2} Ph_3\overset{\oplus}{P}-CH_2Ph \; Cl^{\ominus}$$

14.13

a) $PhCH{=}CHCH_2CH_3$

b) 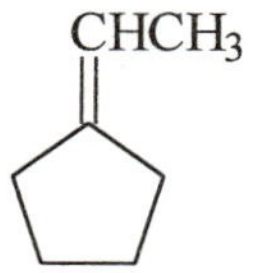 (cyclopentane ring with exocyclic $=CHCH_3$)

c) $HC{\equiv}CCH{=}CHCH_2CH_2CH_3$

d) 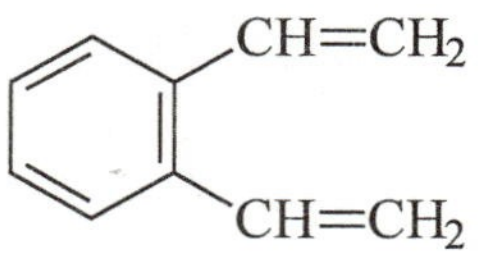 (benzene ring with two ortho $CH{=}CH_2$ groups)

e) $PhCH{=}CH\overset{O}{\overset{\|}{C}}OEt$

14.14 The conjugate base (the ylide) is stabilized by resonance by the phenyl group.

14.15

a) 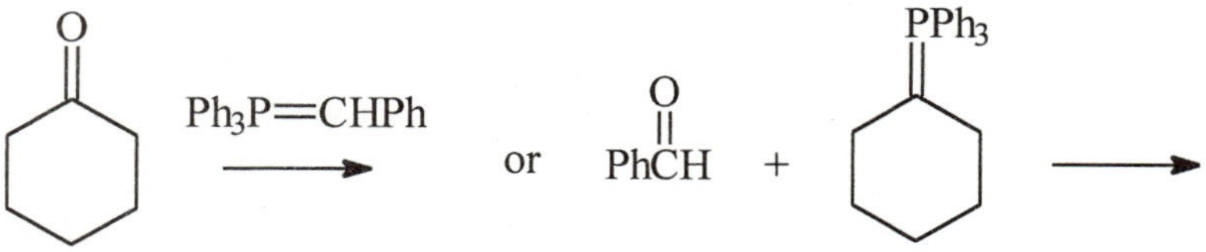 cyclohexanone $\xrightarrow{Ph_3P{=}CHPh}$ or $Ph\overset{O}{\overset{\|}{C}}H$ + (cyclohexylidene)$=PPh_3$ $\longrightarrow$

b) $PhCH_2\overset{O}{\overset{\|}{C}}H \xrightarrow{Ph_3P{=}CHCH_2CH_3}$ or $CH_3CH_2\overset{O}{\overset{\|}{C}}H \xrightarrow{Ph_3P{=}CHCH_2Ph}$

c) $CH_3CH_2\overset{O}{\overset{\|}{C}}H \xrightarrow{Ph_3P{=}CH\overset{O}{\overset{\|}{C}}CH_2CH_3}$

d) 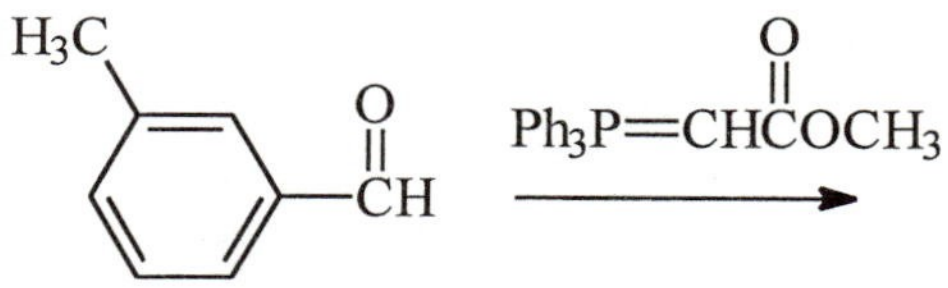 (H_3C-substituted benzene ring with $-\overset{O}{\overset{\|}{C}}H$, meta) $\xrightarrow{Ph_3P{=}CH\overset{O}{\overset{\|}{C}}OCH_3}$

14.16

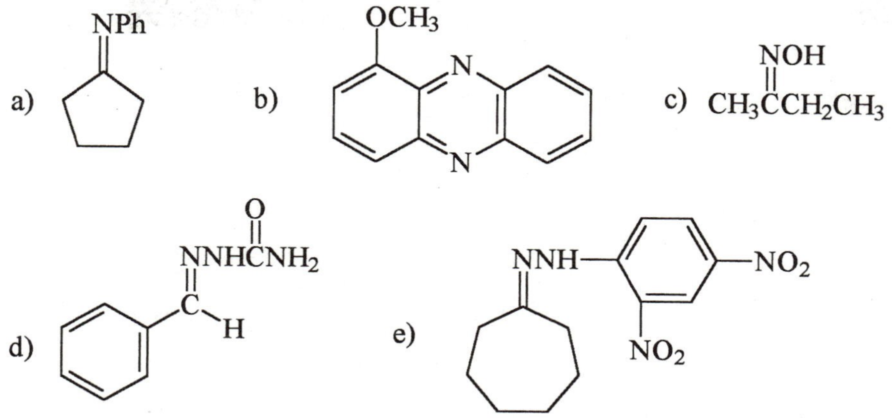

a) NPh

b) OCH$_3$ N N

c) NOH $CH_3CCH_2CH_3$

d) O NNHCNH$_2$ C H

e) NNH— —NO$_2$ NO$_2$

14.17

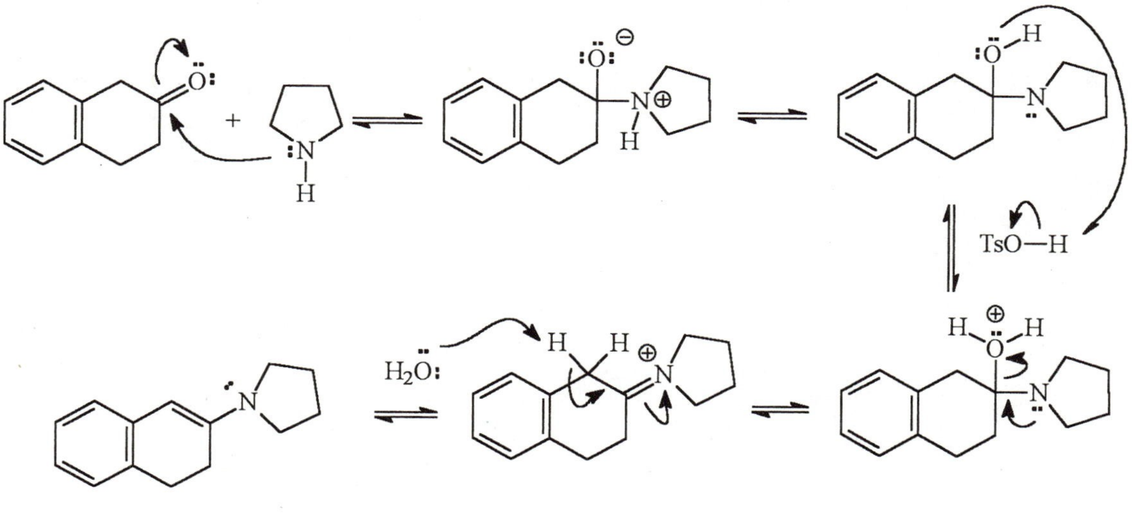

+ N H

O⊖ N⊕ H

O H N

TsO—H

H ⊕O H N

H$_2$O: H H ⊕N

N

14.18 The enamine in produced in problem 14.7 is conjugated whereas the enamine shown in this problem is not. Formation of the more stable conjugated enamine is preferred.

14.19

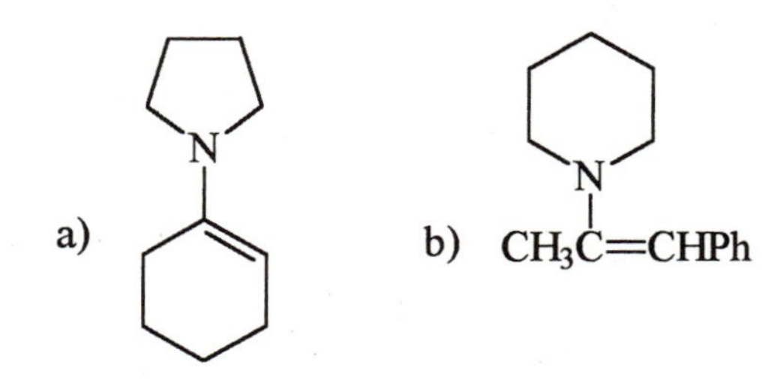

14.20

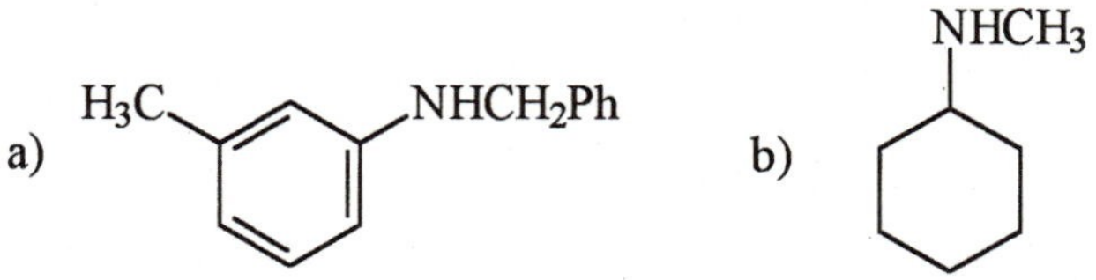

a) H_3C, $NHCH_2Ph$

b) $NHCH_3$

14.21 The right compound gives more cyclic hemiacetal at equilibrium because a five-membered ring is more stable than a four-membered ring.

14.22

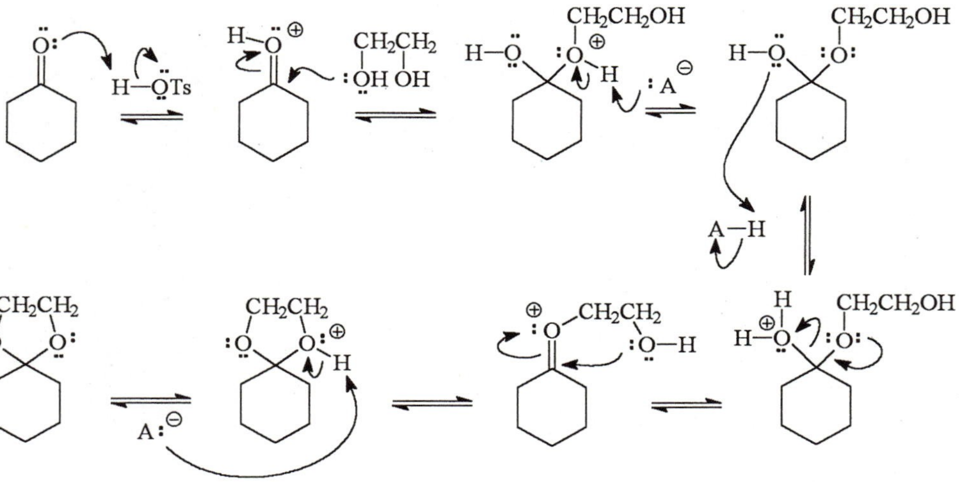

H—OTs; CH_2CH_2 :OH OH; CH_2CH_2OH; H—O; :A; A—H; CH_2CH_2; :O—H; A:

14.23

a) 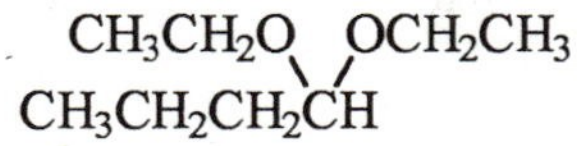 CH_3CH_2O OCH_2CH_3
$CH_3CH_2CH_2CH$

b) 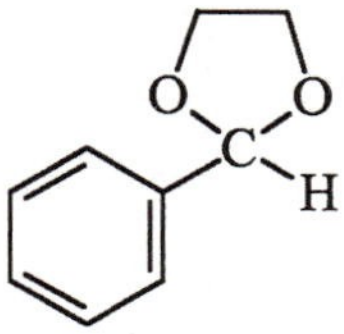 O O C H

c) S S C H

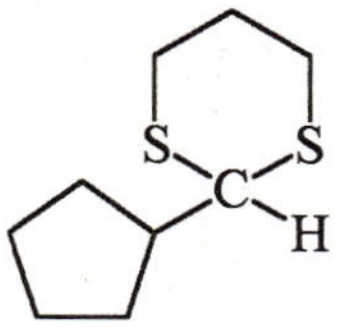

14.24

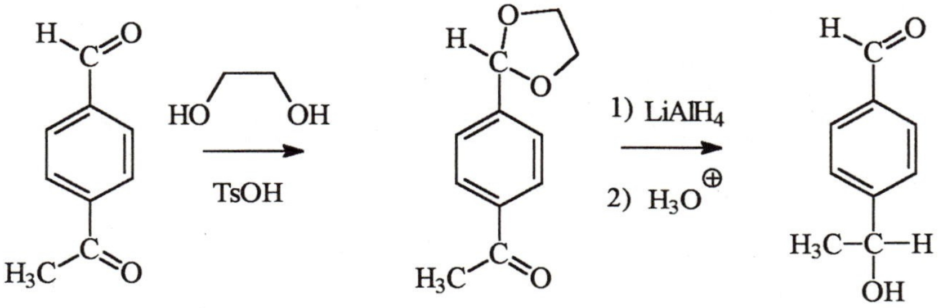

H C=O

H_3C C=O

HO OH

TsOH

H C O O

H_3C C=O

1) $LiAlH_4$

2) $H_3O^{\oplus}$

H C=O

H_3C-C-H

OH

14.25

a) 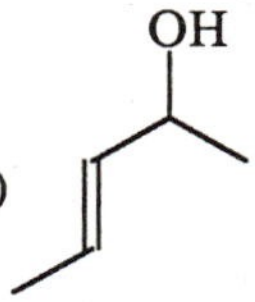 OH

b) 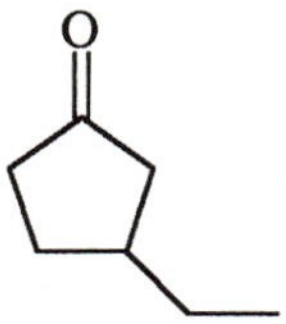 O

c) 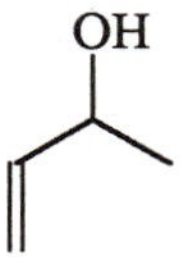 OH

d) 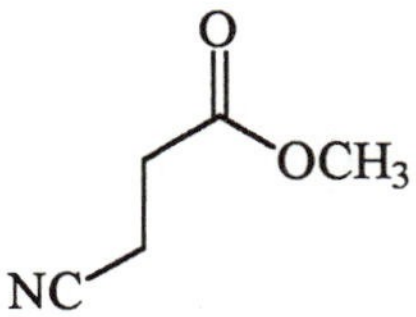 O OCH_3 NC

e) 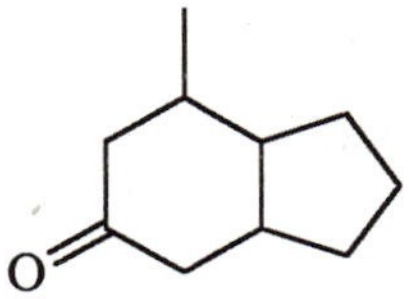 O

14.26

a) $PhCH_2CO_2H$

b) 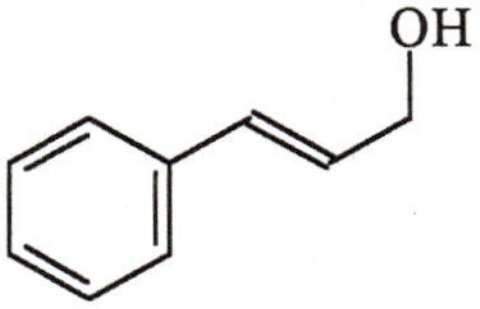 OH

c) 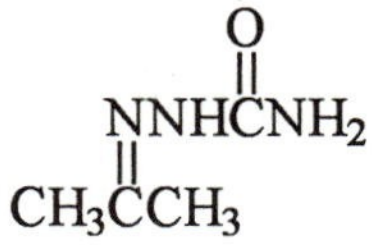 $NNH\overset{O}{\overset{\|}{C}}NH_2$
$CH_3\overset{\|}{C}CH_3$

d) 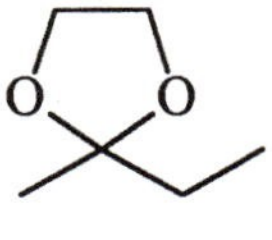 O O

e) 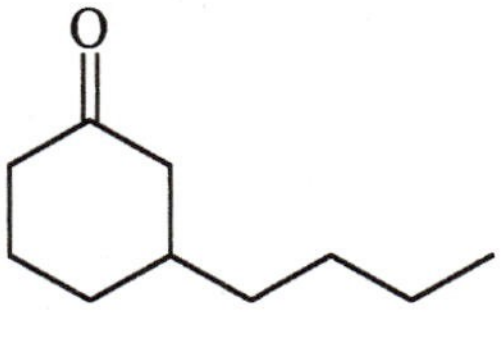 O

f) 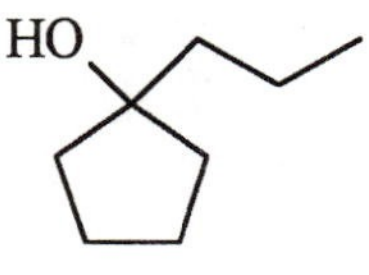 HO

g) 4-(1-cyclohexenyl)morpholine structure

h) $m\text{-}O_2NC_6H_4CH(OCH_3)_2$

i) $Ph-CH(NHCH_3)-CH(CH_3)CH_3$

j) $PhCH{=}CPh_2$ k) $CH_3CH_2CH_2CH_2CH(OH)CN$ l) $CH_3CH_2CH_2CH_2CH_2OH$

m) CHPh (=CHPh on cyclohexane)   n) $PhC(=NOH)Ph$ o) $PhCH(CN)-CH_2C(=O)Ph$

p) $C_6H_5NHCH_2CH_2Ph$ q) $CH_3CH_2CH_2CH_2CH{=}CHC(=O)OCH_3$ r) $m\text{-}CH_3C_6H_4CH(OH)CH_3$

14.27

a) $C_6H_5CH_2OH$ b) $C_6H_5CH(OH)C_6H_{11}$ c) $C_6H_5CH(CN)OH$

d) $C_6H_5CH(OCH_3)_2$ e) $C_6H_5CH{=}NNHC(=O)NH_2$ f) $C_6H_5CH{=}CHCH_2CH_2CH_3$

14.28

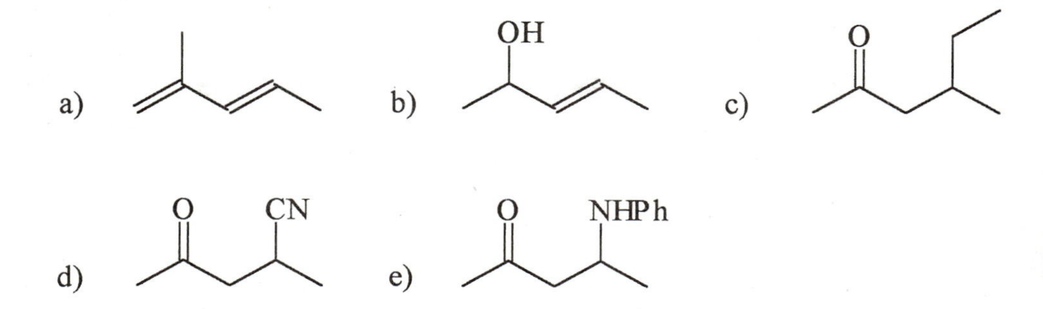

14.29 The reaction gives poor a yield because the carbonyl carbon is highly sterically hindered.

14.30

(2,6-dimethylphenyl)CCH_2CH_3 (C=O) < $C_6H_5CCH_2CH_3$ (C=O) < C_6H_5CH (C=O) < CH_3CH_2CH (C=O) < CH_3CF_2CH (C=O)

smallest K — largest K

14.31

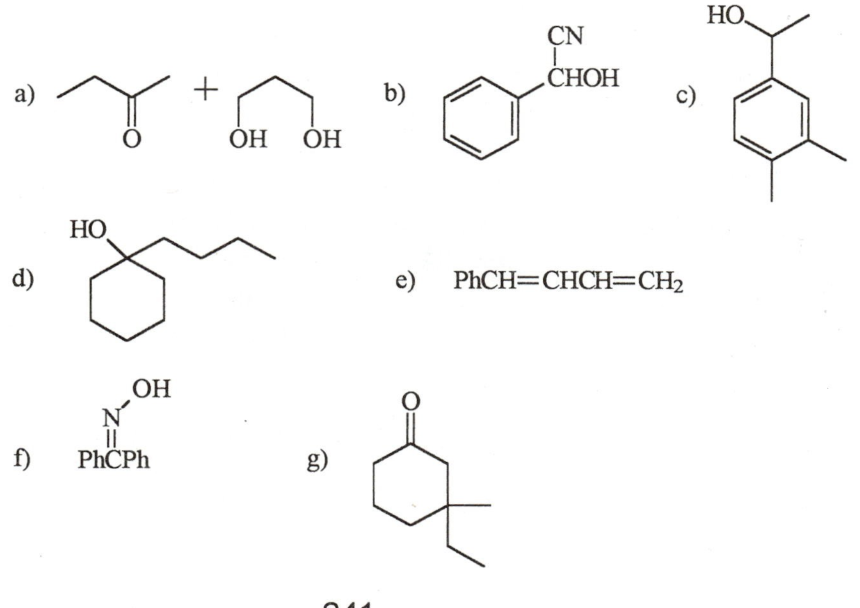

14.32

a) 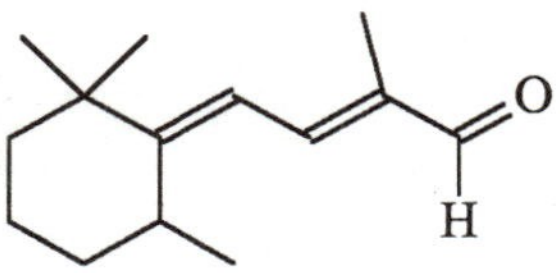 O H

b) 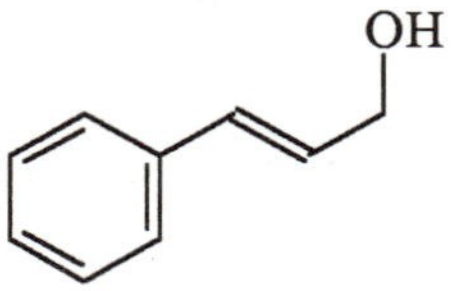 OH

c) 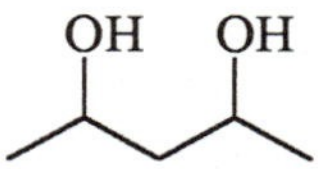 OH OH

d)  $CH_3CH(COOH)CH_2CH_3$

e) 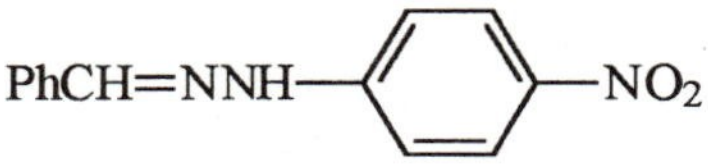 PhCH=NNH—C_6H_4—NO_2

f) 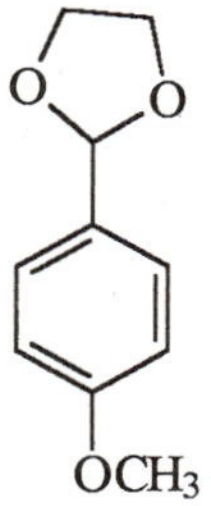 O O, OCH_3

g) $PhNHCH_2CH_2—C{\equiv}N$

14.33 The amino group is strongly electron donating by resonance while the nitro group is strongly electron withdrawing by both its resonance and inductive effects. An electron withdrawing group in the para position makes the carbonyl carbon of the aldehyde group more reactive toward nucleophiles. Therefore, the right compound has a higher equilibrium constant because its carbonyl carbon is more electrophilic.

14.34 a) The left compound has a larger equilibrium constant because the carbonyl carbon is less sterically hindered than in the right compound.

b) The left compound has a larger equilibrium constant because its carbonyl group is less stabilized by resonance.

c) The right compound has a larger equilibrium constant because the carbonyl group is more electrophilic due to the electron withdrawing CCl_3 group.

14.35

a) 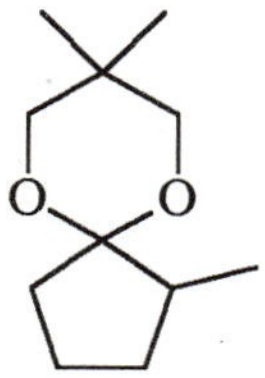 O O

b)

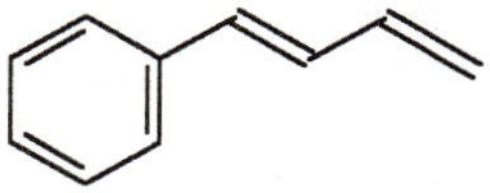

c) 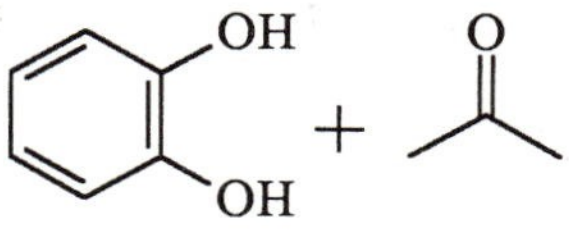 OH OH + O

d) 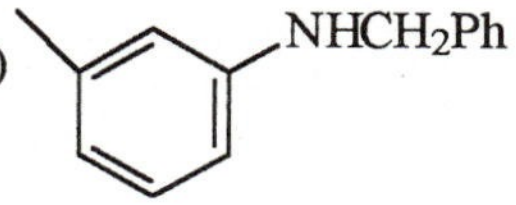 $NHCH_2Ph$ e) $CH_3CH_2CH_2CH{=}NPh$ f) 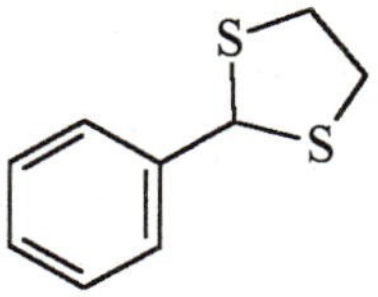 S S

g) 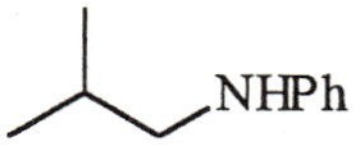 NHPh h) 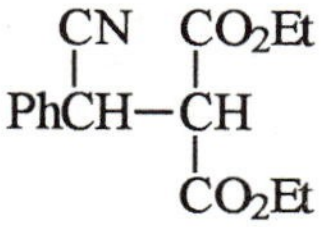 $PhCH(CN)-CH(CO_2Et)_2$

CN CO_2Et
PhCH—CH
CO_2Et

14.36

a) 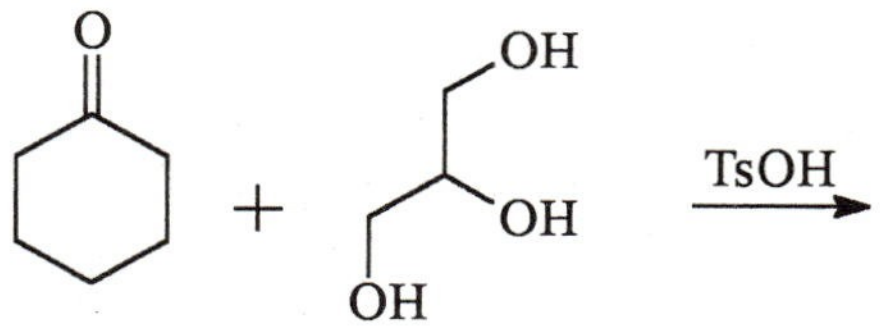 O + OH OH OH $\xrightarrow{TsOH}$

b) 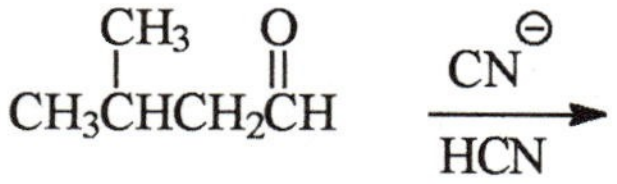 $CH_3CH(CH_3)CH_2CH{=}O$ $\xrightarrow[HCN]{CN^{\ominus}}$

c) 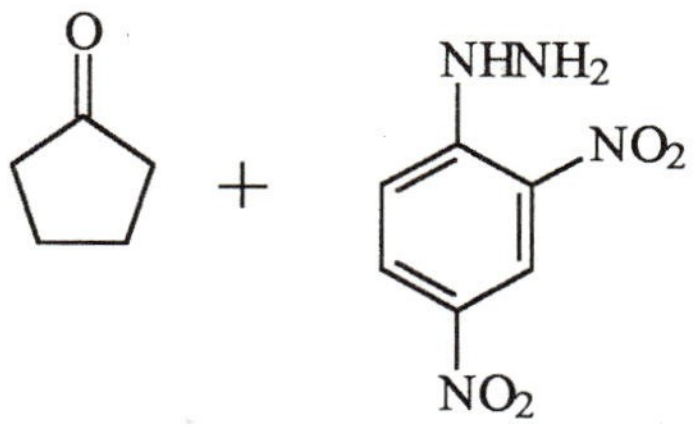 O + $NHNH_2$ NO_2 NO_2

d) NaBH$_4$ / CH$_3$OH →

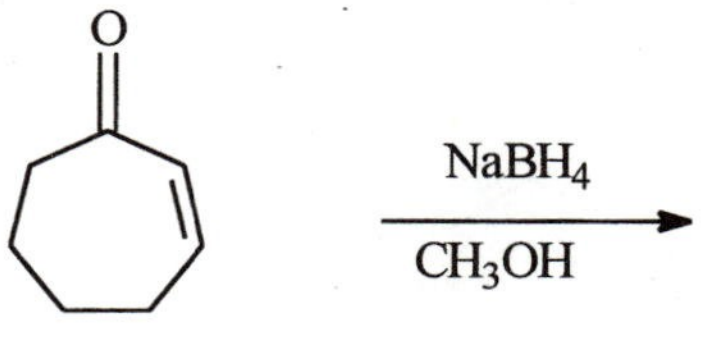

e) PPh$_3$ →

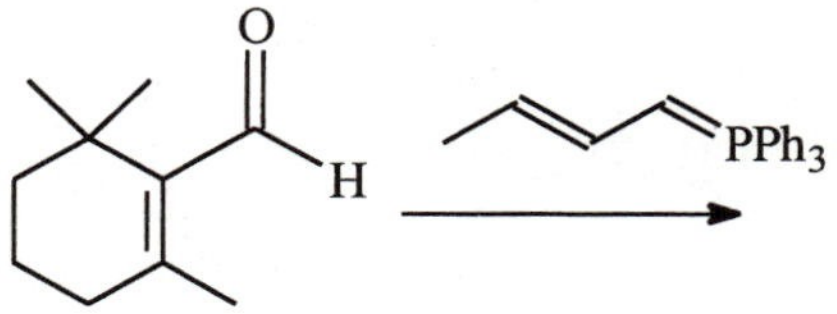

f) 1) MgBr 2) $H_3O^{\oplus}$ →

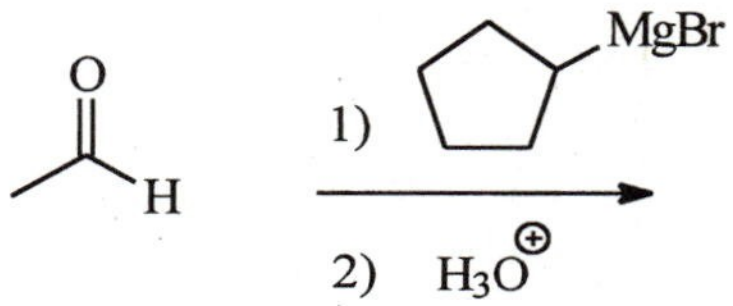

14.37

a) $CH_3\overset{O}{\overset{||}{C}}H$ 1) CH_3CH_2MgBr 2) $H_3O^{\oplus}$ →

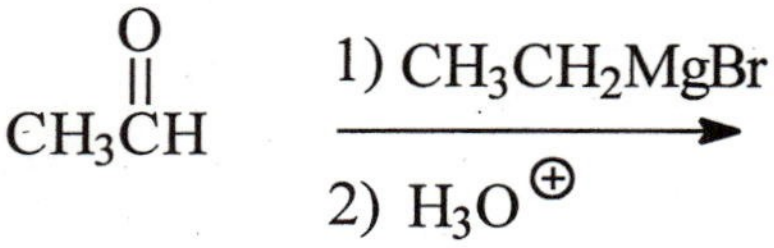

b) Br Mg / ether → MgBr 1) 2) NH_4Cl, H_2O →

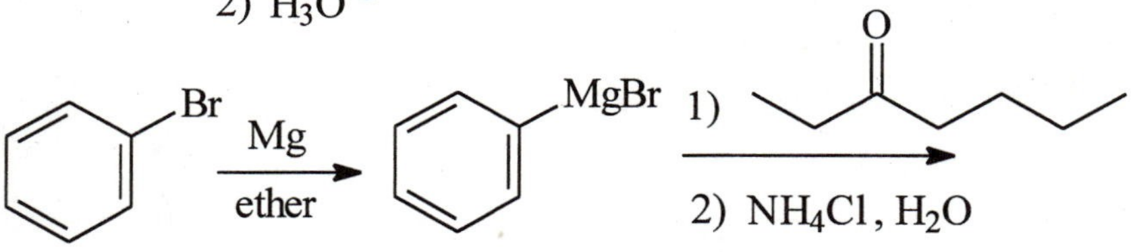

c) $Ph_3P{=}CH_2$ → or 1) CH_3MgBr 2) $H_3O^{\oplus}$ →

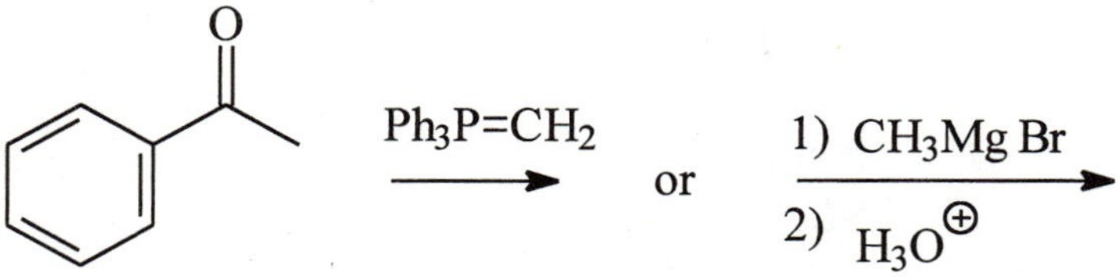

d) $PhCH(=O)$ + cyclopentylamine $\xrightarrow{NaBH_3CN}$

e) (E)-pent-3-en-2-one $\xrightarrow[2)\ NH_4Cl,\ H_2O]{1)\ CH_3MgBr}$

f) (E)-pent-3-en-2-one $\xrightarrow[2)\ H_3O^{\oplus}]{1)\ (CH_3)_2CuLi}$

g) 2-bromopropane $\xrightarrow[\text{ether}]{Mg}$ isopropyl-MgBr $\xrightarrow[2)\ H_3O^{\oplus}]{1)\ CH_3CHO}$

14.38

a) $CH_3CH{=}O$ + $H{-}OH_2^{\oplus}$ ⇌ $CH_3CH{=}OH^{\oplus}$ + $H_2\ddot{O}$: ⇌ $CH_3CH(OH){-}OH_2^{\oplus}$ + $H_2\ddot{O}$: ⇌ $CH_3CH(OH)_2 + H_3O^{\oplus}$

b) $CH_3CH{=}O$ + $^{\ominus}$:CN ⇌ $CH_3CH(O^{\ominus})CN$ + $H{-}CN$ ⇌ $CH_3CH(OH)CN$

c) $C_6H_5{-}\overset{\ddot{O}:}{\overset{\|}{C}}H$ + $\ominus$ H: ⟶ $C_6H_5{-}CH(H){-}\ddot{O}:^{\ominus}$ + $H{-}\ddot{O}CH_3$ ⟶ $C_6H_5{-}CH_2{-}OH$

Labels as printed: :Ö: , CH, ⊖ H: ; ⊖ :Ö: , H—ÖCH$_3$, CH, H ; OH, CH$_2$

d) $CH_3CH{=}\ddot{O}$ + $Ph{-}MgBr$ ⟶ $CH_3CH(Ph){-}\ddot{O}:^{\ominus}$ + $H{-}OH_2^{\oplus}$ ⟶ $CH_3CH(Ph){-}OH$ + $H_2\ddot{O}:$

Labels as printed: :Ö: , CH$_3$CH , Ph—MgBr ; ⊕ H—OH$_2$, ⊖ :Ö: , CH$_3$CH , Ph ; OH , CH$_3$CH , Ph + H$_2$Ö:

e) $CH_3CH_2CH{=}\ddot{O}$ + $\ominus\,\ddot{C}H_2{-}\overset{\oplus}{P}Ph_3$ ⟶ $CH_3CH_2C(H)(CH_2{-}\overset{\oplus}{P}Ph_3){-}\ddot{O}:^{\ominus}$ ⟶ $CH_3CH_2C(H){-}\ddot{O}:$ / $CH_2{-}PPh_3$ (ring) ⟶ $CH_3CH_2CH{=}CH_2$ + $Ph_3P{=}O$

Labels as printed: :Ö: , CH$_3$CH$_2$CH , ⊖ ¨CH$_2$—PPh$_3$ ⊕ ; H ⊖ , CH$_3$CH$_2$C—Ö: , CH$_2$—PPh$_3$ ⊕ ; H , CH$_3$CH$_2$C——Ö: , CH$_2$—PPh$_3$

CH$_3$CH$_2$CH=CH$_2$ + Ph$_3$P=O

f)

:Ö: CH H₂N̈OH ⇌ ⊖ :Ö: CH ⊕ NHOH H ⇌ :ÖH CH NHOH H—ÖH₂ ⊕

⇌ H ⊕ :O—H CH :NHOH ⇌ CH ⊕ NOH H H₂Ö: ⇌ CH NOH

g)

H—A :O: CH_3CCH_3 ⇌ ⊕ H :O CH_3CCH_3 EtÖH ⇌ :ÖH CH_3CCH_3 O ⊕ H Et ⊖A: ⇌ H—A :ÖH CH_3CCH_3 :ÖEt

⇌ H ⊕ H :O CH_3CCH_3 : ÖEt ⇌ EtÖH CH_3CCH_3 ÖEt ⊕ ⇌ Ä⊖ H ⊕ Et O: CH_3CCH_3 :ÖEt ⇌ :ÖEt CH_3CCH_3 :ÖEt

14.39 A conjugate addition reaction does not occur unless there is a group attached to the double bond that can help stabilize, by resonance, the carbanion intermediate. The top reaction does not occur because the methyl group will not stabilize the carbanion intermediate. The bottom reaction occurs because the nitro group can help stabilize the carbanion intermediate.

14.40

$CH_3C(=O)CH_3$ + C_6H_5MgBr ⟶ $CH_3C(O^-)(C_6H_5)CH_3$ $\xrightarrow{H-OH_2^+}$ $CH_3C(OH)(C_6H_5)CH_3$ $\xrightarrow{H-OH_2^+}$ $CH_3C(OH_2^+)(C_6H_5)CH_3$ ⟶ $H_3C-C^+(C_6H_5)-CH_2-H$ $\xrightarrow{H_2O:}$ $H_3C(C_6H_5)C{=}CH_2$

14.41

a) $CH_3CH_2CH_2CH{=}O$ $\xrightarrow[2)\ H_3O^{\oplus}]{1)\ CH_3MgBr}$ $CH_3CH_2CH_2CH(OH)CH_3$ $\xrightarrow{PCC}$ $CH_3CH_2CH_2C(=O)CH_3$ $\xrightarrow[2)\ NH_4Cl,\ H_2O]{1)\ CH_3CH_2MgBr}$ $CH_3CH_2CH_2C(OH)(CH_2CH_3)CH_3$

b) Cyclohexanecarbaldehyde ($\overset{O}{\parallel}$CH on cyclohexane) $\xrightarrow[\text{2) } H_3O^{\oplus}]{\text{1) } CH_3CH_2MgBr}$ cyclohexyl–$CH(OH)CH_2CH_3$ (HO–$CHCH_2CH_3$) $\xrightarrow{PCC}$ cyclohexyl–$C(=O)CH_2CH_3$ (O=CCH_2CH_3)

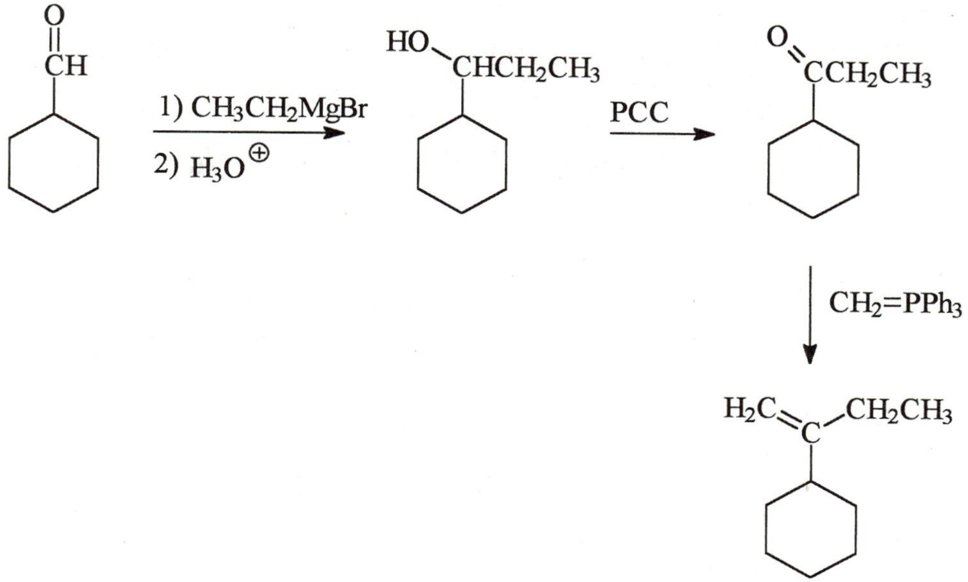

$\downarrow$ $CH_2{=}PPh_3$

cyclohexyl–$C(=CH_2)CH_2CH_3$ ($H_2C{=}C$–CH_2CH_3)

c) 2-cyclopentenone (O) $\xrightarrow{(CH_3)_2CuLi}$ 3-methylcyclopentanone (O, CH_3) $\xrightarrow{CH_2=PPh_3}$ 1-methylene-3-methylcyclopentane (CH_2, CH_3)

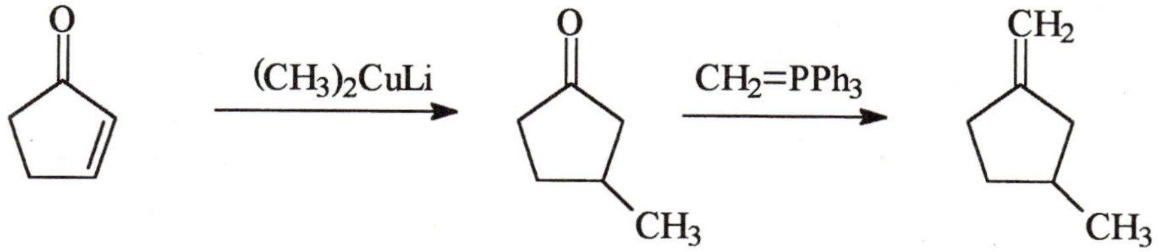

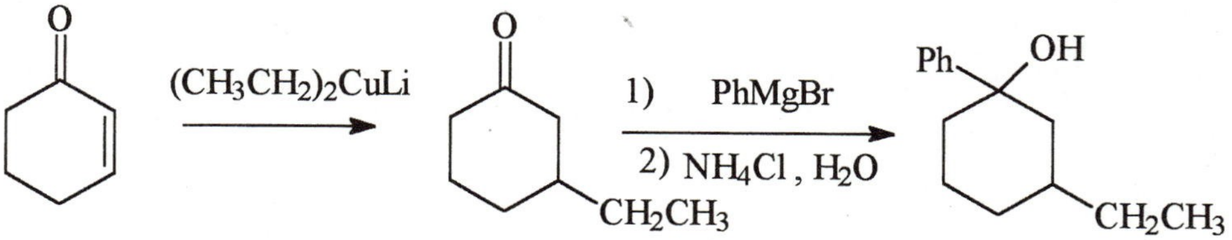

d) 2-cyclohexenone (O) $\xrightarrow{(CH_3CH_2)_2CuLi}$ 3-ethylcyclohexanone (O, CH_2CH_3) $\xrightarrow[\text{2) } NH_4Cl\text{ , } H_2O]{\text{1) PhMgBr}}$ 1-phenyl-3-ethylcyclohexanol (Ph, OH, CH_2CH_3)

e) Cyclohexanecarbaldehyde (O=CH) $\xrightarrow{CH_2{=}PPh_3}$ vinylcyclohexane ($CH_2{=}CH$) $\xrightarrow[\text{2) } H_2O_2\text{, NaOH}]{\text{1) } BH_3\text{, THF}}$ 2-cyclohexylethanol (CH_2OH, CH_2)

f) Cyclopentenone $\xrightarrow[\text{2) } H_3O^{\oplus}]{\text{1) } Ph_2CuLi}$ 3-phenylcyclopentanone (O, Ph) $\xrightarrow[\text{2) } H_3O^{\oplus}]{\text{1) } LiAlH_4}$ 3-phenylcyclopentanol (OH, Ph)

14.42

a) $BrCH_2CH_2CH_2\overset{O}{\overset{\|}{C}}H$ $\xrightarrow[TsOH]{HO\text{–}CH_2CH_2\text{–}OH}$ 2-(3-bromopropyl)-1,3-dioxolane (O, O, Br) $\xrightarrow[\text{2) } H_3O^{\oplus}]{\text{1) } HC{\equiv}C\text{:}^{\ominus}}$

b) $H\overset{O}{\overset{\|}{C}}\text{–}C_6H_4\text{–}\overset{O}{\overset{\|}{C}}Ph$ $\xrightarrow[TsOH]{HO\text{–}CH_2CH_2\text{–}OH}$ (1,3-dioxolan-2-yl)–C_6H_4–$\overset{O}{\overset{\|}{C}}Ph$ $\xrightarrow[\text{2) } NH_4Cl\text{, } H_2O]{\text{1) } CH_3MgBr}$

c) [2-cyclohexenone] $\xrightarrow[\text{Pt}]{H_2}$ [cyclohexanone] $\xrightarrow[\text{2) } NH_4Cl, \ H_2O]{\text{1) } CH_3CH_2MgBr}$

d) $CH_3\overset{O}{\overset{\|}{C}}H$ $\xrightarrow[\text{2) } H_3O^{\oplus}]{\text{1) } CH_3CH_2CH_2MgBr}$ [2-pentanol, OH] $\xrightarrow{PCC}$ [2-pentanone, O] $\xrightarrow{CH_2{=}PPh_3}$

e) $BrCH_2\overset{O}{\overset{\|}{C}}CH_3$ $\xrightarrow[\text{2) BuLi}]{\text{1) } PPh_3}$ $CH_3\overset{O}{\overset{\|}{C}}C{=}PPh_3$ $\xrightarrow{Ph\overset{O}{\overset{\|}{C}}H}$

f) $CH_3CH_2\overset{O}{\overset{\|}{C}}CH_3$ $\xrightarrow[\text{2) } NH_4Cl, \ H_2O]{\text{1) PhMgBr}}$ [2-phenyl-2-butanol; Ph, OH] $\xrightarrow{SOCl_2}$

g) $PhCH_2\overset{O}{\overset{\|}{C}}H$ $\xrightarrow[\text{2) } H_3O^{\oplus}]{\text{1) } CH_3MgBr}$ [Ph, OH] $\xrightarrow{PCC}$ [Ph, O]

$\downarrow$ $CH_3CH_2NH_2$, $NaBH_3CN$

14.43 The deprotonation of the phosphonium salt can be accomplished with NaOH because the electron pair of the conjugate base (carbanion of the ylide) is stabilized by resonance with the carbonyl group.

14.44 In this reaction the oxygen of the hydroxy group acts as an intramolecular nucleophile to first form a hemiacetal. Remember that intramolecular reactions are favored by entropy. The hemiacetal then reacts with methanol to form the acetal.

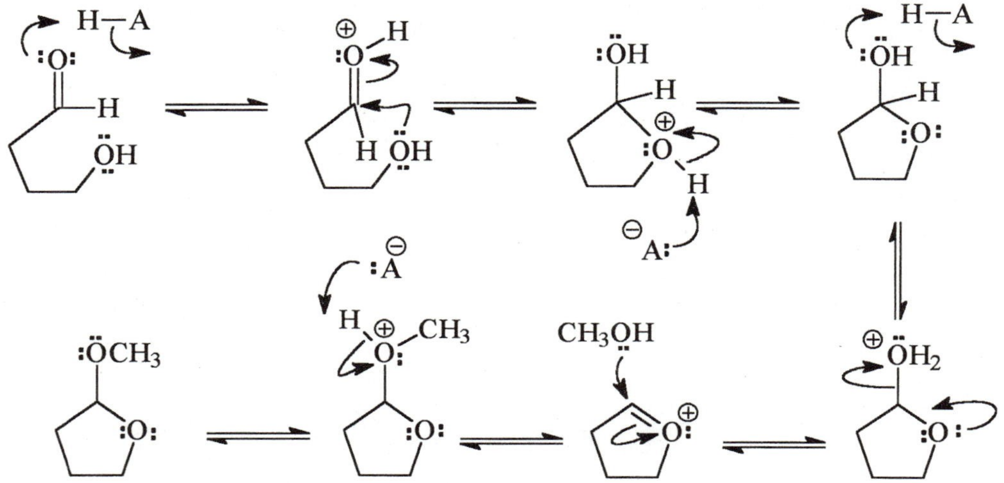

H–A :O: H ÖH ⊕ :O H H ÖH :ÖH H ⊕ :O H ⊖ A: H–A :ÖH H :O:

:ÖCH₃ :O: ⊖ :A H ⊕ CH₃ O: :O: CH₃ÖH ⊕ O: ⊕ ÖH₂ :O:

14.45

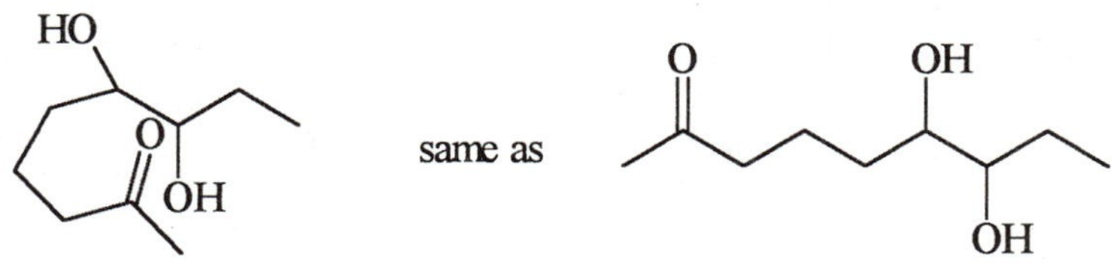

HO O OH same as O OH OH

14.46

CH₃ :N CH PhCH₂–MgCl ⊖ :NCH₃ H–ÖH CH CH₂Ph :NHCH₃ CHCH₂Ph

A B

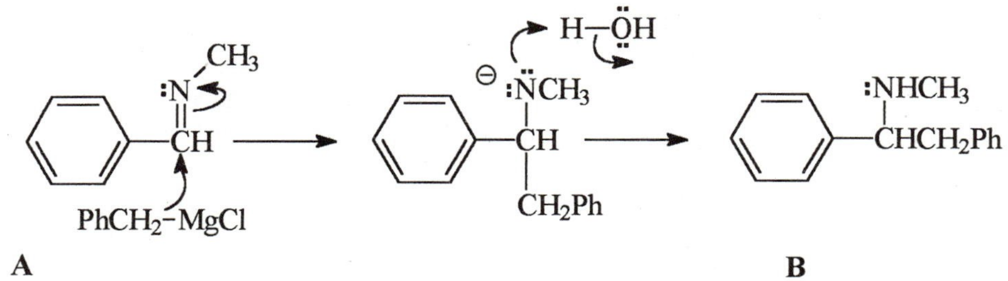

14.47 This is a reductive amination reaction. First one equivalent of ammonia reacts with one equivalent of benzaldehyde to form an imine which in turn is reduced to a primary amine. The primary amine formed then reacts with another equivalent of benzaldehyde to form another imine. The resulting imine is then reduced to a secondary amine.

$$\text{PhCH=O} \xrightarrow{NH_3} \text{PhCH=NH} \xrightarrow[Pt]{H_2} \text{PhCH}_2\text{NH}_2 \xrightarrow{\text{PhCH=O}} \text{PhCH=NCH}_2\text{Ph} \xrightarrow[Pt]{H_2} \text{PhCH}_2\text{NHCH}_2\text{Ph}$$

14.48

14.49 This compound can form a cyclic hemiacetal because it has an alcohol nucleophile and an aldehyde group in the same molecule. Although hemiacetals are normally not favored at equilibrium, formation of a six-membered cyclic hemiacetal has a larger equilibrium constant than a comparable intermolecular reaction. In this synthesis the equilibrium favors the cyclic hemiacetal as indicated by the absence of the carbonyl band in the IR spectrum.

14.50

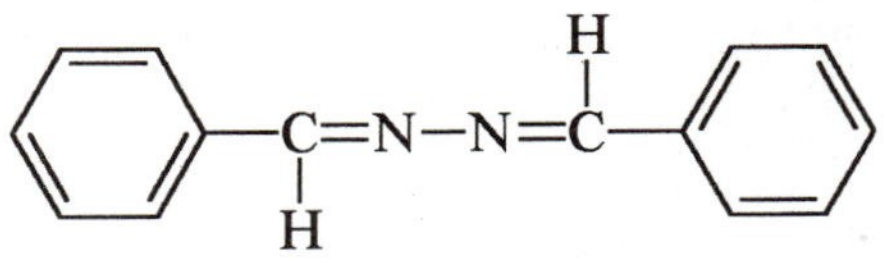

$$C_6H_5-\underset{H}{C}=N-N=\overset{H}{C}-C_6H_5$$

14.51 Grignard reactions with ketones give tertiary alcohols. These alcohols are very prone to the E1 elimination reaction. In this reaction the initially formed alcohol undergoes an acid catalyzed elimination reaction to produce a stable conjugated alkene. This is the reason for not observing the OH absorption in the 3500 - 3200 cm^{-1} region in the IR spectrum.

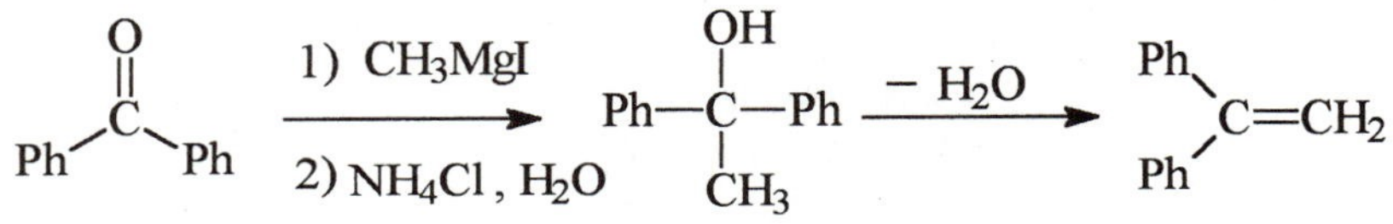

$$Ph-\overset{O}{\overset{||}{C}}-Ph \xrightarrow[2)\ NH_4Cl,\ H_2O]{1)\ CH_3MgI} Ph-\overset{OH}{\underset{CH_3}{C}}-Ph \xrightarrow{-\ H_2O} (Ph)_2C=CH_2$$

14.52 The NMR spectrum suggests that the product is 2-pentanone, the result of a conjugate addition reaction rather than a normal Grignard reaction.

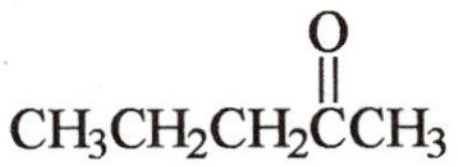

$$CH_3CH_2CH_2\overset{O}{\overset{||}{C}}CH_3$$

After completing this chapter, you should be able to:

1. Show the products resulting from the addition to aldehydes and ketones of all of the reagents discussed in this chapter.
   Problems 14.2, 14.6, 14.8, 14.9, 14.13, 14.16, 14.19, 14.20, 14.23, 14.26, 14.27, 14.31, 14.32, 14.35.

2. Show the products resulting from the addition of certain of these reagents to α,β-unsaturated compounds, noting whether 1,2- or 1,4-addition predominates.
   Problems 14.25, 14.26, 14.28, 14.31, 14.32, 14.35.

3. Show the mechanisms for any of these additions.
   Problems 14.1, 14.3, 14.17, 14.22, 14.38, 14.40.

4. Predict the effect of the structure of the aldehyde or ketone on the position of the equilibrium for these reactions.
   Problems 14.4, 14.5, 14.7, 14.21, 14.29, 14.30, 14.33, 14.34.

5. Use these reactions, in combination with the reactions from previous chapters, to synthesize compounds.
   Problems 14.10, 14.12, 14.15, 14.36, 14.37, 14.41, 14.42.

6. Use acetals as protective groups in syntheses.
   Problems 14.24, 14.42a, 14.42b.

Chapter 15
SUBSTITUTION AT THE CARBONYL GROUP

15.1 a) The left ester due to less steric hindrance.
c) The left compound due to a more electrophilic carbonyl carbon because of the electron withdrawing F.
d) The left compound due to a more electrophilic carbonyl carbon because of the electron withdrawing nitro group.

15.2 a) The equilibrium favors the products because acetic anhydride is more reactive than acetic acid or cyclopentyl acetate.
b) The equilibrium favors the reactants because an amide is less reactive than an ester.
c) The equilibrium favors the products because an acid chloride is more reactive than an amide.

15.3 The reaction of the ester is faster because an ester is more reactive than an amide.

15.4

$$CH_3CH_2CH_2\overset{\overset{O}{\|}}{C}Cl + NH_2CH_2CH_2CH_3 \longrightarrow$$

15.5

a) 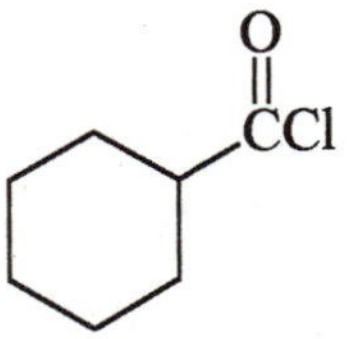 (cyclohexyl)$\overset{\overset{O}{\|}}{C}Cl$ b) 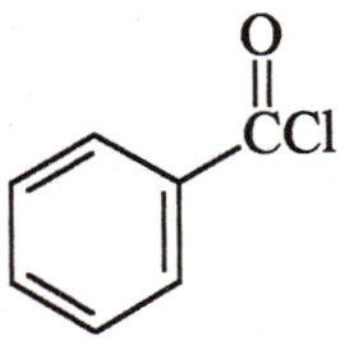 (phenyl)$\overset{\overset{O}{\|}}{C}Cl$ c) $CH_3CH_2CH_2\overset{\overset{O}{\|}}{C}Cl$

15.6

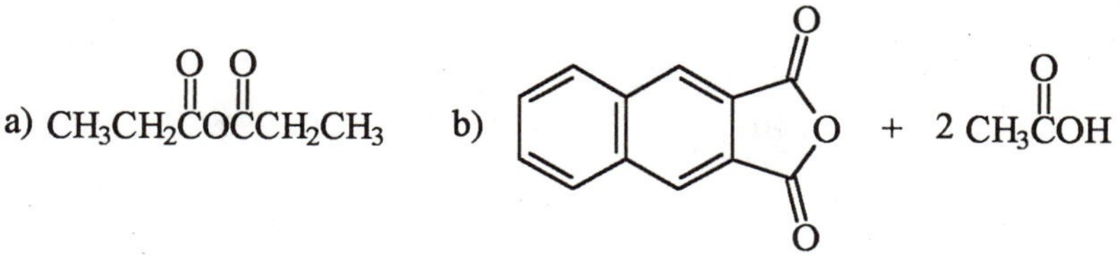

a) $CH_3CH_2\overset{O}{\overset{\|}{C}}O\overset{O}{\overset{\|}{C}}CH_2CH_3$ b) O O O + 2 $CH_3\overset{O}{\overset{\|}{C}}OH$

15.7

$CH_3{-}\overset{\ddot{O}:}{\overset{\|}{C}}{-}\ddot{Cl}:$ ⇌ $CH_3{-}C{-}\ddot{Cl}:$ ($:\ddot{O}^{\ominus}$ above, $\oplus\,:O{-}CH_3$ / H below) ⇌ $CH_3{-}\overset{\ddot{O}:}{\overset{\|}{C}}{-}\ddot{O}^{\oplus}{-}CH_3$ (H on O)

$:\ddot{O}{-}CH_3$ / H $:\ddot{Cl}:^{\ominus}$

$:\ddot{Cl}{-}H$ + $CH_3{-}\overset{\ddot{O}:}{\overset{\|}{C}}{-}\ddot{O}{-}CH_3$

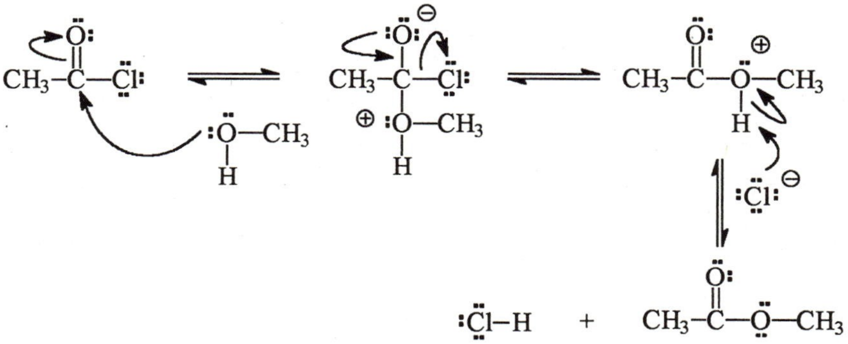

15.8 The reaction produces aspirin, an ester, rather than the more reactive anhydride.

15.9

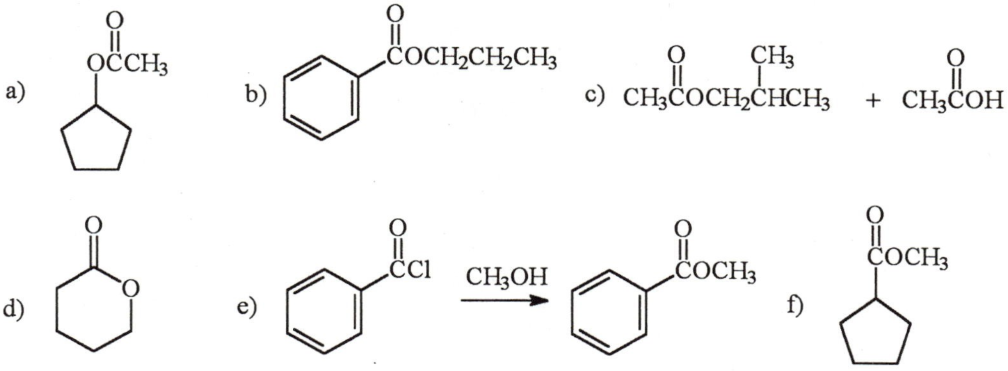

a) $O\overset{O}{\overset{\|}{C}}CH_3$ b) $\overset{O}{\overset{\|}{C}}OCH_2CH_2CH_3$ c) $CH_3\overset{O}{\overset{\|}{C}}OCH_2\overset{CH_3}{\overset{|}{C}}HCH_3$ + $CH_3\overset{O}{\overset{\|}{C}}OH$

d) O O e) $\overset{O}{\overset{\|}{C}}Cl$ $\xrightarrow{CH_3OH}$ $\overset{O}{\overset{\|}{C}}OCH_3$ f) $\overset{O}{\overset{\|}{C}}OCH_3$

15.10

$$CH_3-\overset{O}{\overset{\|}{C}}-OCH_2CH_3 + H-\overset{\oplus}{O}(H)-H \rightleftharpoons CH_3-\overset{\oplus OH}{\overset{\|}{C}}-OCH_2CH_3$$

$$CH_3-\overset{\oplus OH}{\overset{\|}{C}}-OCH_2CH_3 + H_2O \rightleftharpoons CH_3-C(OH)(OCH_2CH_3)-\overset{\oplus}{O}H_2$$

$$CH_3-C(OH)(OCH_2CH_3)-\overset{\oplus}{O}H_2 + H_2O \rightleftharpoons CH_3-C(OH)(OCH_2CH_3)-OH + H_3O^{\oplus}$$

$$CH_3-C(OH)_2-OCH_2CH_3 + H-\overset{\oplus}{O}(H)-H \rightleftharpoons CH_3-C(OH)_2-\overset{\oplus}{O}(H)CH_2CH_3$$

$$CH_3-C(OH)_2-\overset{\oplus}{O}(H)CH_2CH_3 \rightleftharpoons CH_3-\overset{\oplus OH}{\overset{\|}{C}}-OH + CH_3CH_2OH$$

$$CH_3-\overset{\oplus OH}{\overset{\|}{C}}-OH + H_2O \rightleftharpoons CH_3-\overset{O}{\overset{\|}{C}}-OH$$

15.11

$$\begin{array}{l} CH_2OH \\ | \\ CHOH \\ | \\ CH_2OH \end{array} \quad + \quad \begin{array}{c} {}^{\ominus}O\overset{O}{\overset{\|}{C}}(CH_2)_7CH{=}CH(CH_2)_7CH_3 \\ + \\ {}^{\ominus}O\overset{O}{\overset{\|}{C}}(CH_2)_{14}CH_3 \\ + \\ {}^{\ominus}O\overset{O}{\overset{\|}{C}}(CH_2)_{12}CH_3 \end{array}$$

15.13

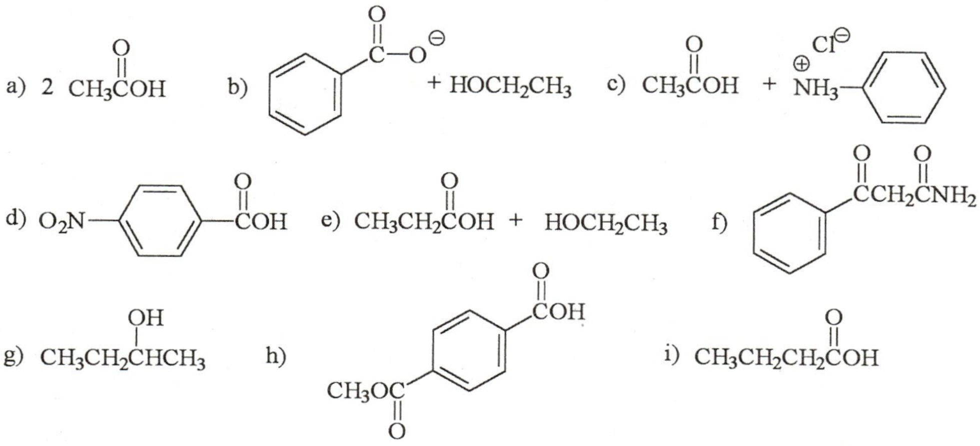

a) 2 CH_3COH (C=O)   b) $C_6H_5C(=O)O^{\ominus}$ + $HOCH_2CH_3$ c) CH_3COH (C=O) + $Cl^{\ominus}$ $\overset{\oplus}{N}H_3-C_6H_5$

d) $O_2N-C_6H_4-COH$ (C=O)   e) CH_3CH_2COH (C=O) + $HOCH_2CH_3$ f) $C_6H_5-CCH_2CNH_2$ (two C=O)

g) $CH_3CH_2CHCH_3$ (OH)   h) $CH_3OC(=O)-C_6H_4-COH$ (C=O)   i) $CH_3CH_2CH_2COH$ (C=O)

15.14

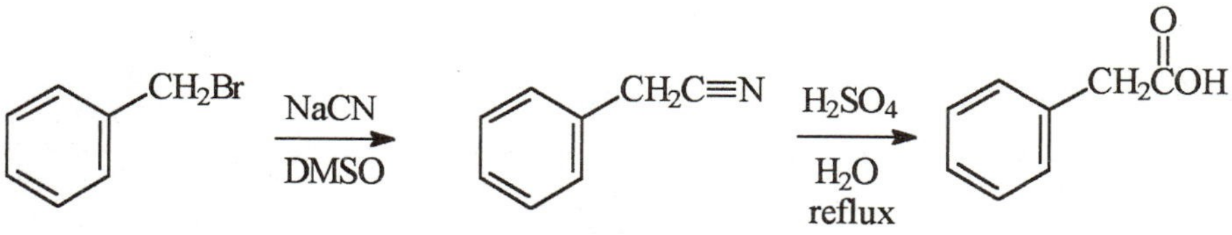

$$C_6H_5CH_2Br \xrightarrow[\text{DMSO}]{\text{NaCN}} C_6H_5CH_2C{\equiv}N \xrightarrow[\text{H}_2\text{O, reflux}]{\text{H}_2\text{SO}_4} C_6H_5CH_2COH \ (C=O)$$

15.15

a) $CH_3O-C_6H_4-NHCCH_3$ (C=O; meta)   b) $CH_3CH_2CH_2CN(CH_2CH_3)_2$ (C=O)

c) $C_5H_9-CNHPh$ (cyclopentyl, C=O)   d) $C_6H_5CNHCH_3$ (C=O) + CH_3OH

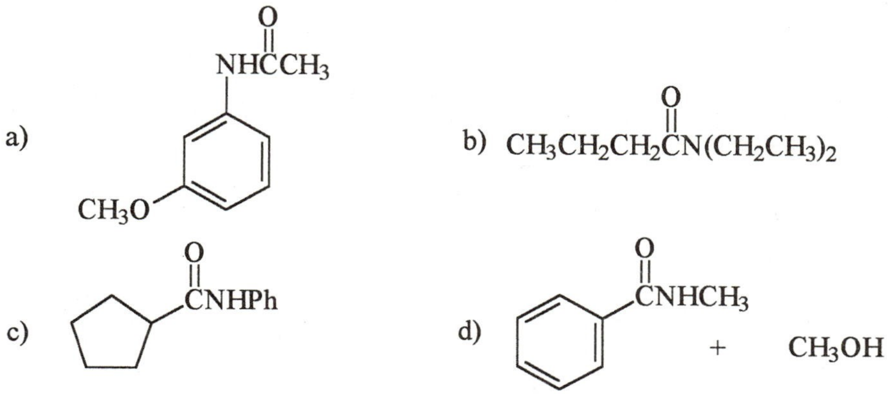

15.16

$$PhCH_2\ddot{N}H_2 + :\ddot{C}l-\overset{\overset{..}{O}:}{\overset{||}{C}}-OCH_2CH_3 \underset{}{\overset{\text{step one}}{\rightleftharpoons}} :\ddot{C}l-\overset{:\ddot{O}:^{\ominus}}{\overset{|}{C}}(-OCH_2CH_3)-\underset{\oplus}{N}H_2CH_2Ph$$

step two

$$PhCH_2\ddot{N}H-\overset{O}{\overset{||}{C}}-OCH_2CH_3 \rightleftharpoons PhCH_2\overset{\oplus}{N}H(H)-\overset{\ddot{O}:}{\overset{||}{C}}-OCH_2CH_3 + :\ddot{C}l:^{\ominus}$$

B:

In step two, chloride ion leaves because it is a weaker base and a better leaving group than ethoxide ion.

15.17

a) 1) $SOCl_2$; 2) cyclohexanol (cyclohexyl–OH)

b) 1) $SOCl_2$; 2) excess $CH_3CH_2CH_2NH_2$

d) 1) $H_3O^{\oplus}$, Δ; 2) $SOCl_2$; 3) CH_3CH_2OH

15.18

a) $CH_3CH_2CH_2CH_2OH$ + CH_3OH

b) $PhCH_2NHCH_2CH_3$

c) cyclohex-2-en-1-ol (ring with OH)

d) benzene ring with $\overset{O}{\overset{||}{C}}OCH_2CH_3$ and, meta, $\underset{OH}{\underset{|}{C}}HCH_3$

e) $CH_3CH_2CH_2CH_2CH_2NH_2$

15.19 If acid were used, the product, an amine, would also be protonated.

15.20

a) $CH_3CH(CH_3)-CHO$ (structure: $CH_3\overset{CH_3}{\overset{|}{C}}H-\overset{O}{\overset{||}{C}}H$)

b) m-methoxybenzaldehyde (benzene ring with $\overset{O}{\overset{||}{C}}H$ and OCH_3)

c) benzene ring $-CH_2\overset{O}{\overset{||}{C}}H$

15.21

a) 1) $SOCl_2$; 2) $LiAlH(Ot\text{-}Bu)_3$, -78° C

b) 1) $SOCl_2$; 2) $PhNH_2$; 3) $LiAlH_4$; 4) H_2O

c) 1) $H_3O^{\oplus}$, Δ; 2) $SOCl_2$; 3) $LiAlH(Ot\text{-}Bu)_3$, -78 °C

d) 1) $LiAlH_4$; 2) H_2O; 3) $CH_3\overset{O}{\overset{||}{C}}Cl$; 4) $LiAlH_4$; 5) H_2O

15.22

a) cyclohexyl$-\overset{OH}{\underset{Ph}{C}}-Ph$

b) $CH_3CH_2CH_2\overset{OH}{\underset{CH_3}{C}}CH_3$

c) cyclopentyl$-\overset{OH}{CH}-$cyclopentyl

15.23

$Ph\text{-}\overset{:\ddot{O}}{\overset{||}{C}}-\ddot{O}CH_2CH_3$ + Li—CH_3 → $Ph\text{-}\overset{:\ddot{O}:^{\ominus}}{\underset{CH_3}{C}}-\ddot{O}CH_2CH_3$ → $Ph\text{-}\overset{:\ddot{O}}{\overset{||}{C}}-CH_3$ + $:\ddot{O}CH_2CH_3^{\ominus}$

$Ph\text{-}\overset{:\ddot{O}}{\overset{||}{C}}-CH_3$ + Li—CH_3 → $Ph\text{-}\overset{:\ddot{O}:^{\ominus}}{\underset{CH_3}{C}}-CH_3$

$Ph\text{-}\overset{:\ddot{O}:^{\ominus}}{\underset{CH_3}{C}}-CH_3$ + $NH_4^{\oplus}$ (H—N—H with H above and below) → $\ddot{N}H_3$ + $Ph\text{-}\overset{:\ddot{O}-H}{\underset{CH_3}{C}}-CH_3$

15.24

a) 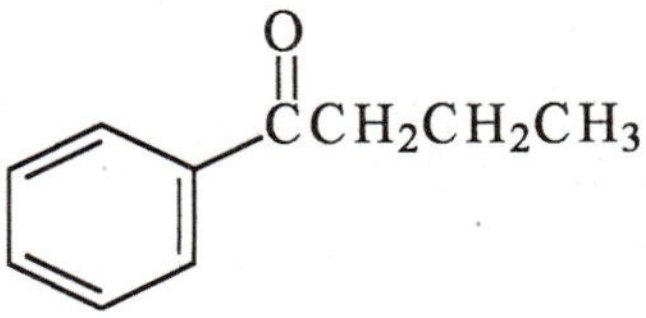 $C_6H_5C(=O)CH_2CH_2CH_3$ b) same as (a)

15.25

a) 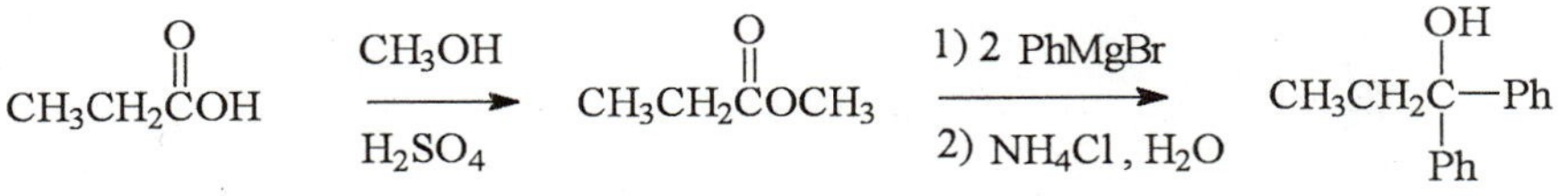 $CH_3CH_2COOH \xrightarrow[H_2SO_4]{CH_3OH} CH_3CH_2COOCH_3 \xrightarrow[2)\ NH_4Cl,\ H_2O]{1)\ 2\ PhMgBr} CH_3CH_2C(OH)(Ph)-Ph$

b) 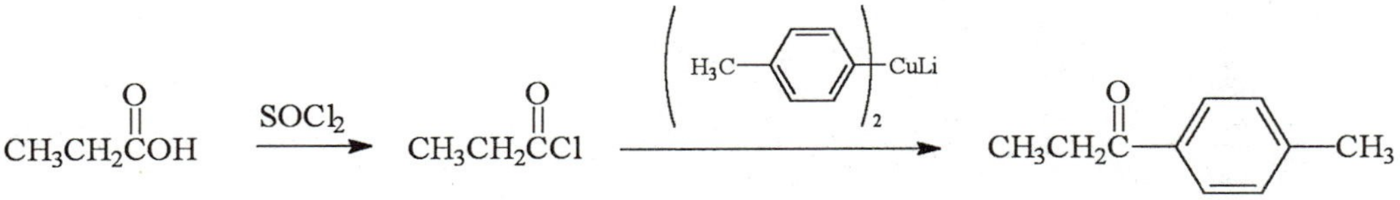 $CH_3CH_2COOH \xrightarrow{SOCl_2} CH_3CH_2COCl \xrightarrow{(H_3C-C_6H_4)_2CuLi} CH_3CH_2C(=O)-C_6H_4-CH_3$

c) $C_6H_5Br \xrightarrow[ether]{Mg} C_6H_5MgBr \xrightarrow[2)\ H_3O^{\oplus},\ \Delta]{1)\ C_6H_{11}C\equiv N} C_6H_5C(=O)C_6H_{11}$

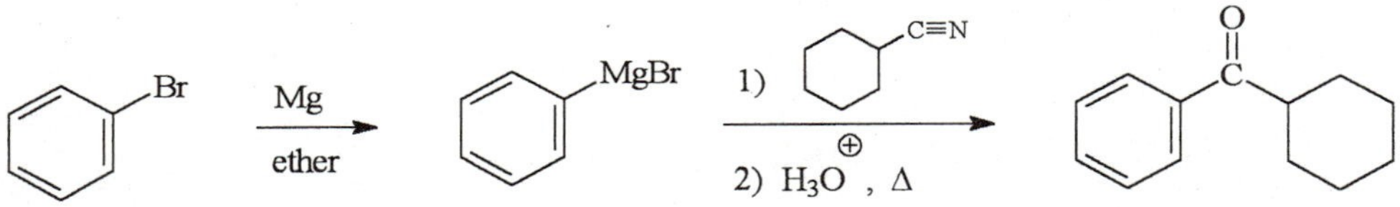

d) 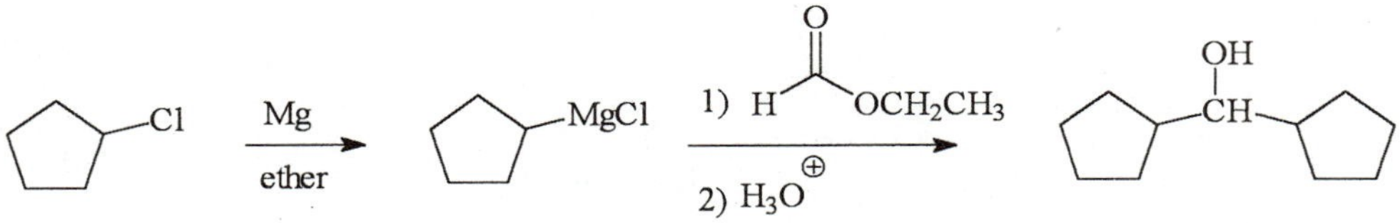 $C_5H_9Cl \xrightarrow[ether]{Mg} C_5H_9MgCl \xrightarrow[2)\ H_3O^{\oplus}]{1)\ HC(=O)OCH_2CH_3} C_5H_9-CH(OH)-C_5H_9$

15.26

a) 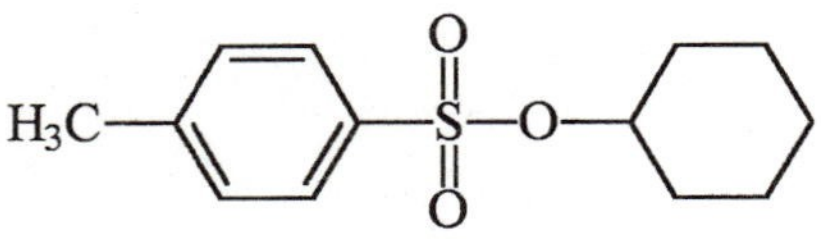 $H_3C-C_6H_4-SO_2-O-C_6H_{11}$ b) $C_6H_5NH-SO_2-CH_3$ c) $CH_3CH_2OP(=O)(OCH_2CH_3)OCH_2CH_3$

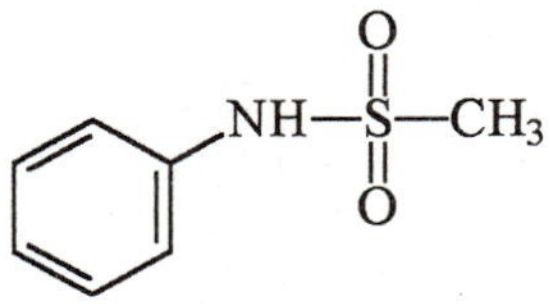

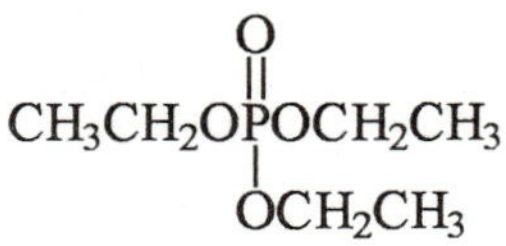

15.27

a) $CH_3CH_2O-\overset{O}{\overset{\|}{C}}-OCH_2CH_3$ b) $CH_3CH_2\overset{O}{\overset{\|}{C}}Cl$ c) $CH_3CH_2-C_6H_4-NH\overset{O}{\overset{\|}{C}}CH_3$ (para)

d) $PhCH_2\overset{OH}{\overset{|}{C}}(CH_2CH_3)CH_2CH_3$ e) $PhCH_2\overset{O}{\overset{\|}{C}}Cl \xrightarrow{PhCH_2\overset{O}{\overset{\|}{C}}O^{\ominus}} PhCH_2\overset{O}{\overset{\|}{C}}O\overset{O}{\overset{\|}{C}}CH_2Ph$ f) $C_6H_5CO_2CH_2CH_2CH_2CH_3$

g) $C_6H_4(CH_2OH)(CH_3)$ (ortho) h) $CH_3CH_2CH_2CH_2\overset{O}{\overset{\|}{C}}CH_2CH_2CH_2CH_3$ i) $CH_3CH_2CH_2CH_2NH_2$

j) $CH_2{=}CH\overset{O}{\overset{\|}{C}}OCH_2CH_3$ k) $CH_3\overset{CH_3}{\overset{|}{C}}HCOCl \xrightarrow[H_2O]{NH_3} CH_3\overset{CH_3}{\overset{|}{C}}HCONH_2$ l) $CH_3(CH_2)_{10}CH_2OSO_2-C_6H_4-CH_3$ (para)

m) $HO\overset{O}{\overset{\|}{C}}CH_2CH_2CH_2\overset{O}{\overset{\|}{C}}OH$ n) $H_3C-C_6H_4-CH_2CH_2OH$ (meta) o) $\overset{O}{\overset{\|}{C}}H-C_6H_4-OCH_3$ (para)

p) 1-naphthyl$-\overset{O}{\overset{\|}{C}}-CH_3$ q) $PhCH(O\overset{O}{\overset{\|}{C}}CH_3)-\overset{O}{\overset{\|}{C}}Ph$ r) $CH_3-C_6H_4-NH_2$ (para)

s) (2-methylphenyl)methanol: benzene ring with CH_2OH and CH_3 on adjacent carbons

t) $CH_3O\overset{O}{\overset{\|}{C}}-\overset{O}{\overset{\|}{C}}OCH_3$

u) cyclohexane ring bearing $CH_2N(CH_3)_2$

v) $H\overset{O}{\overset{\|}{C}}-C_6H_4-\overset{O}{\overset{\|}{C}}H$ (para)

15.28

a) cyclopropyl$-\overset{O}{\overset{\|}{C}}OCH_2CH_3$

b) cyclohexyl$-O\overset{O}{\overset{\|}{C}}CH_3$

c) benzene ring with CH_3 and $COOH$ on adjacent carbons

d) $C_6H_5CH_2CH_2OH$ + CH_3OH

e) skeletal structure: a carbon bearing OH, an ethyl group and two butyl groups

15.29

a) benzene ring with CH_3 and $NH\overset{O}{\overset{\|}{C}}CH_3$ on adjacent carbons + CH_3COOH

b) benzene ring with CH_3 and $\overset{O}{\overset{\|}{C}}O^{\ominus}Na^{\oplus}$ meta to each other + CH_3OH

c) $CH_3\underset{CH_3}{\underset{|}{C}}H\overset{O}{\overset{\|}{C}}NH_2$

d) benzene ring with CH_3 and CH_2NH_2 on adjacent carbons

e) cyclopentene ring bearing $\overset{O}{\overset{\|}{C}}-CH_3$

15.30

a) 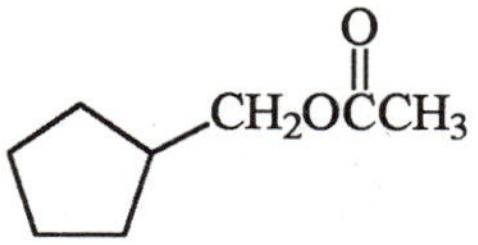 cyclopentyl–CH_2OCCH_3 (C=O)

b) 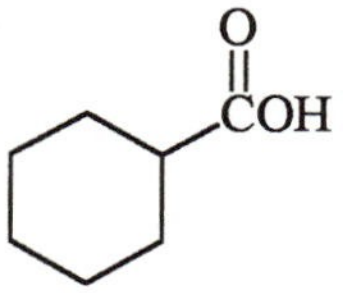 cyclohexyl–COH (C=O)

c) $CH_3CH_2CH_2NH_2$

d) 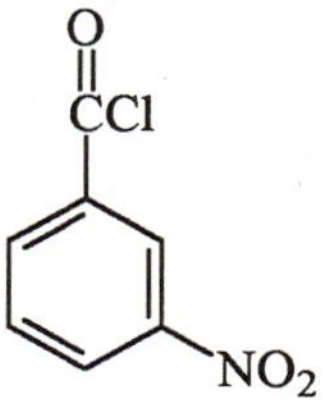 3-nitrophenyl–CCl (C=O); NO_2

e)  $CH_3\overset{CH_3}{C}=CHCH_2CH_2CH$ (C=O)

f) 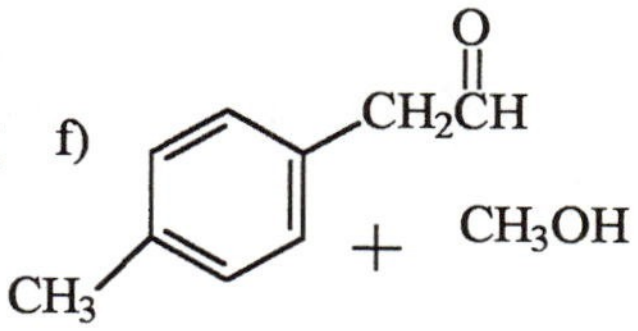 4-methylphenyl–CH_2CH (C=O) + CH_3OH

15.31

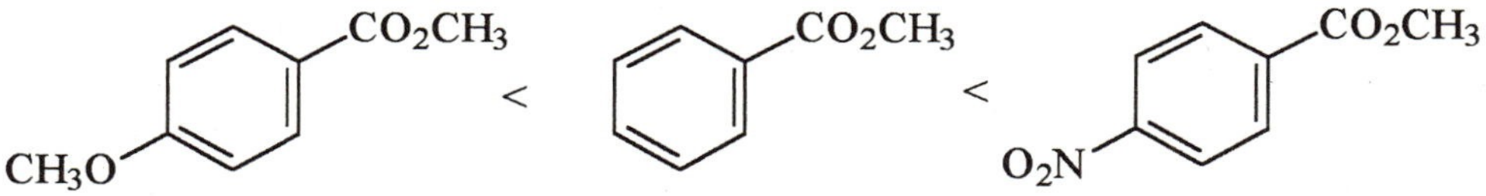

CH_3O–C_6H_4–CO_2CH_3 < C_6H_5–CO_2CH_3 < O_2N–C_6H_4–CO_2CH_3

slowest ... fastest

15.32

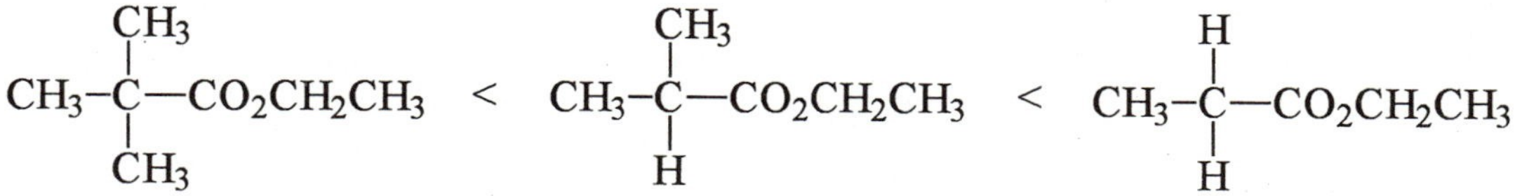

$$CH_3-\underset{CH_3}{\overset{CH_3}{C}}-CO_2CH_2CH_3 \; < \; CH_3-\underset{H}{\overset{CH_3}{C}}-CO_2CH_2CH_3 \; < \; CH_3-\underset{H}{\overset{H}{C}}-CO_2CH_2CH_3$$

15.33

a) 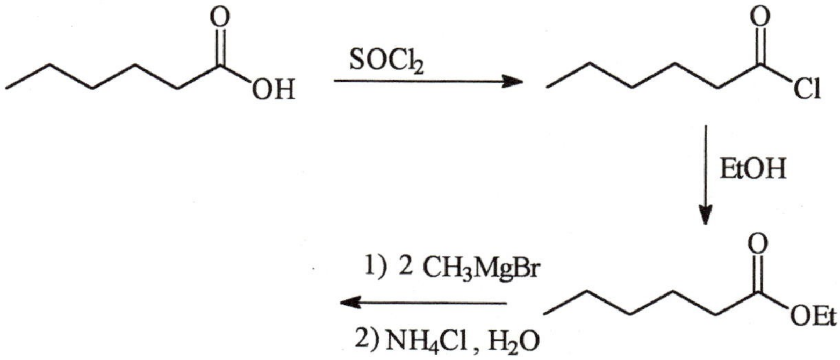 hexanoic acid (–C(=O)OH) $\xrightarrow{SOCl_2}$ hexanoyl chloride (–C(=O)Cl) $\xrightarrow{EtOH}$ ethyl hexanoate (–C(=O)OEt) $\xrightarrow[\text{2) } NH_4Cl,\ H_2O]{\text{1) } 2\ CH_3MgBr}$

b) Hexanoic acid ($CH_3CH_2CH_2CH_2CH_2COOH$) $\xrightarrow[\text{2) } H_3O^{\oplus}]{\text{1) } LiAlH_4}$

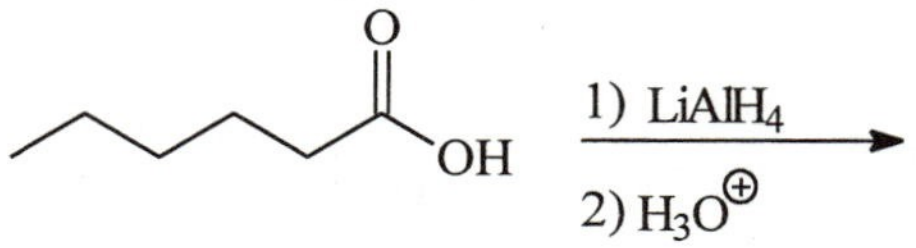

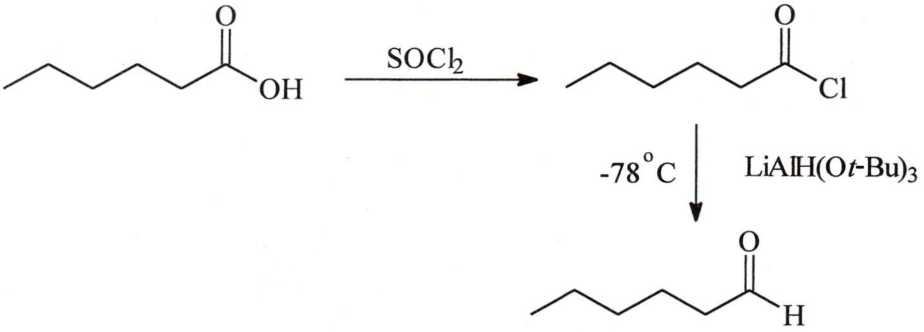

c) Hexanoic acid (OH) $\xrightarrow{SOCl_2}$ hexanoyl chloride (Cl)

-78 °C, $LiAlH(Ot\text{-}Bu)_3$ ↓

hexanal (H)

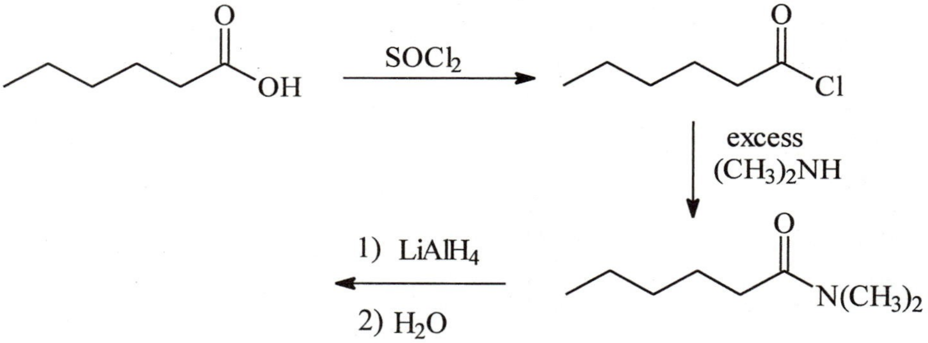

d) Hexanoic acid (OH) $\xrightarrow{SOCl_2}$ hexanoyl chloride (Cl)

excess $(CH_3)_2NH$ ↓

N,N-dimethylhexanamide ($N(CH_3)_2$)

$\xleftarrow[\text{2) } H_2O]{\text{1) } LiAlH_4}$

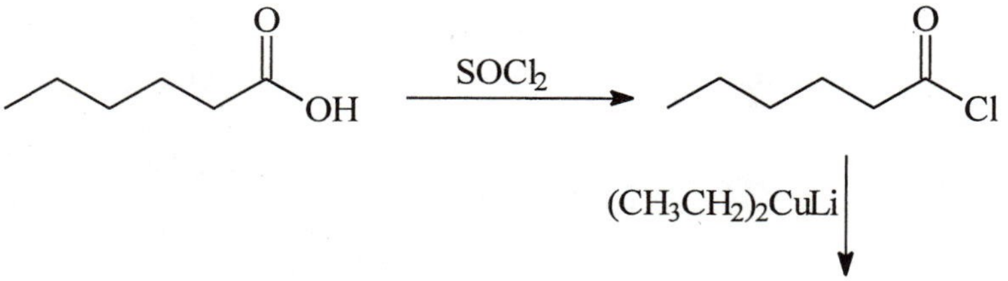

e) Hexanoic acid (OH) $\xrightarrow{SOCl_2}$ hexanoyl chloride (Cl)

$(CH_3CH_2)_2CuLi$ ↓

f) 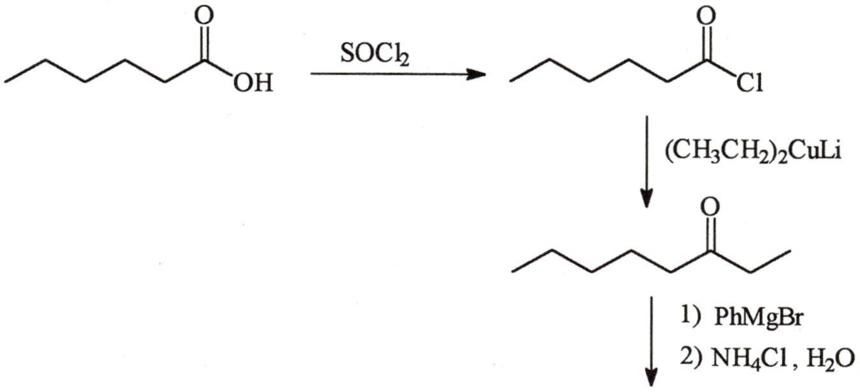 Hexanoic acid $\xrightarrow{SOCl_2}$ hexanoyl chloride (Cl) $\xrightarrow{(CH_3CH_2)_2CuLi}$ octan-3-one $\xrightarrow{\text{1) PhMgBr; 2) } NH_4Cl,\ H_2O}$

15.34

a) 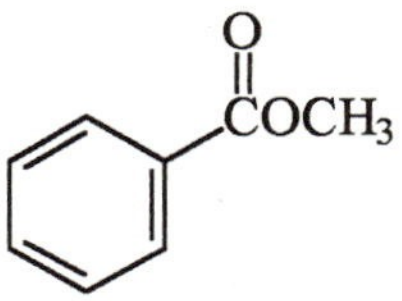 $C_6H_5COCH_3$ (C=O) b) $C_6H_5COCH(CH_3)_2$ (C=O) c) $C_6H_5CH_2OH$

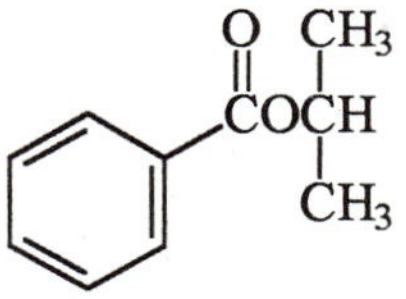

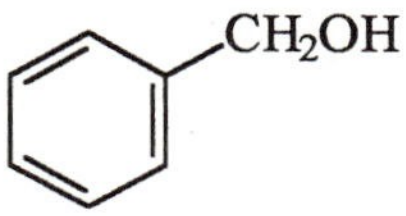

d) 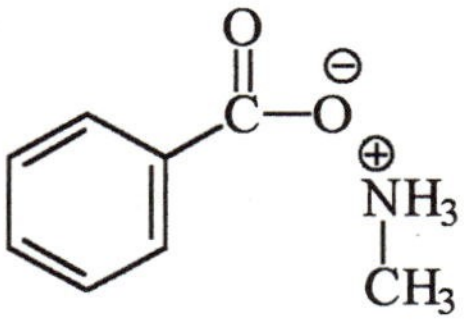 $C_6H_5C(=O)\text{–}O^{\ominus}\ \ {}^{\oplus}NH_3CH_3$ e) 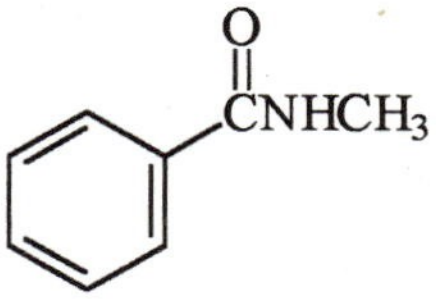 $C_6H_5CNHCH_3$ (C=O)

15.35

a) 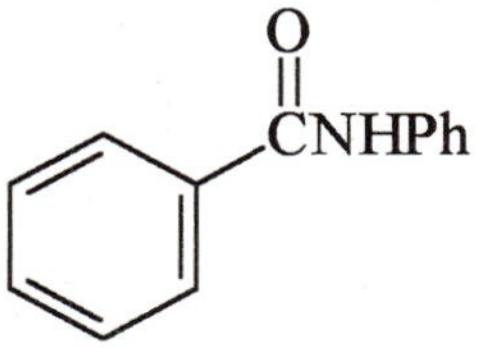 C_6H_5CNHPh (C=O) b) $C_6H_5COCH(CH_3)_2$ (C=O) c) $C_6H_5COCC_6H_5$ (both C=O)

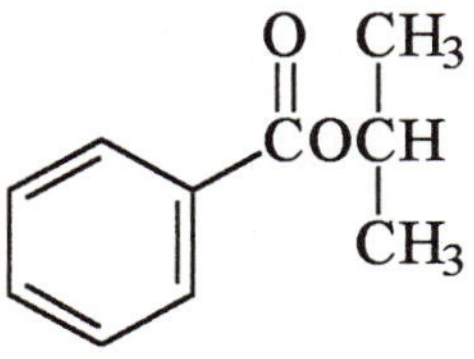

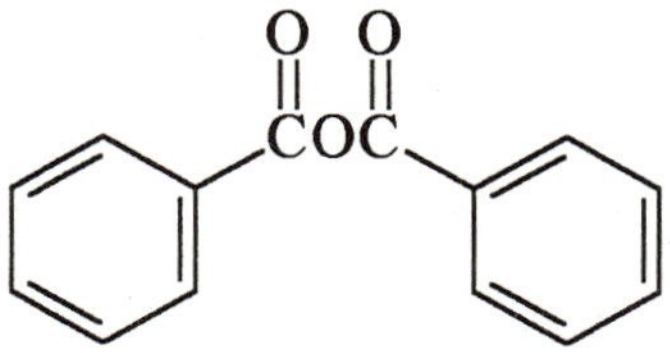

d) 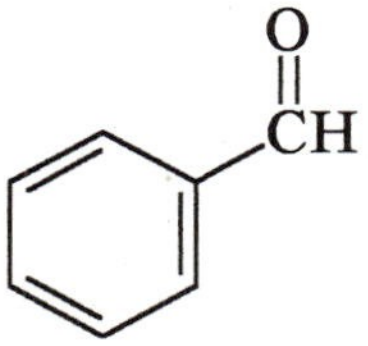 C_6H_5CH (C=O) e) 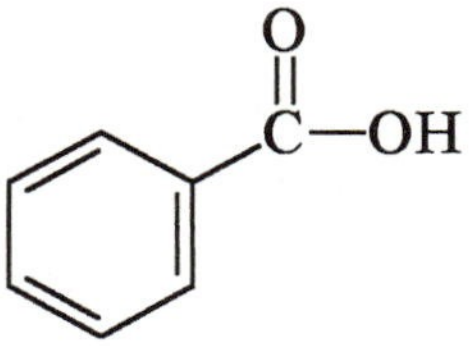 $C_6H_5C\text{—}OH$ (C=O) f) 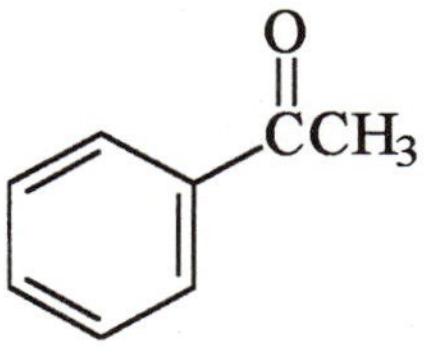 $C_6H_5CCH_3$ (C=O)

15.36

a) 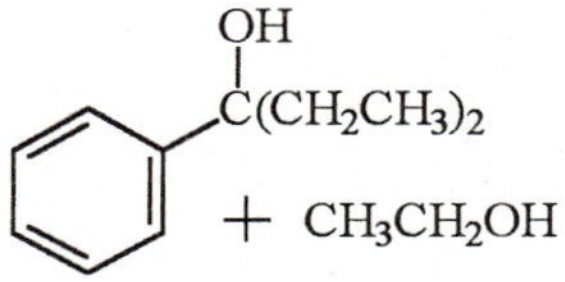 $C_6H_5C(OH)(CH_2CH_3)_2$ + CH_3CH_2OH

b) $C_6H_5CH{=}O$ + CH_3CH_2OH

c) $C_6H_5C(=O)O^{\ominus}Na^{\oplus}$ + CH_3CH_2OH

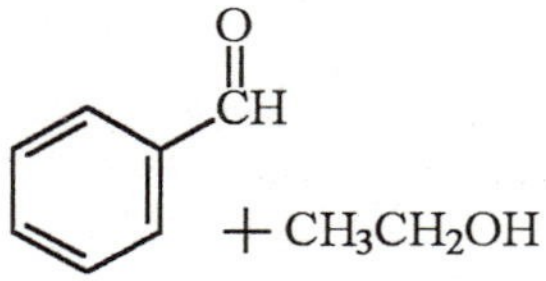

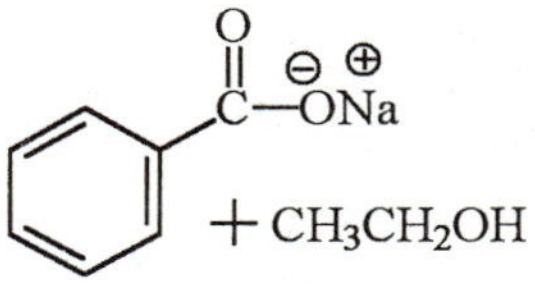

d) 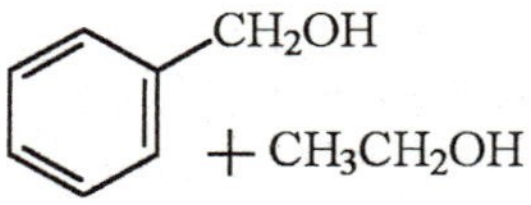 $C_6H_5CH_2OH$ + CH_3CH_2OH

e) 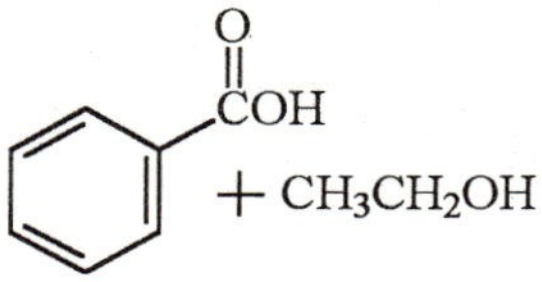 $C_6H_5C(=O)OH$ + CH_3CH_2OH

15.37

a) 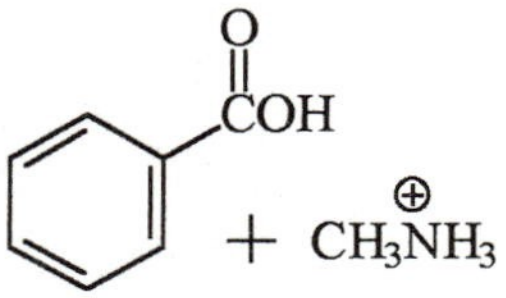 $C_6H_5C(=O)OH$ + $CH_3NH_3^{\oplus}$

b) 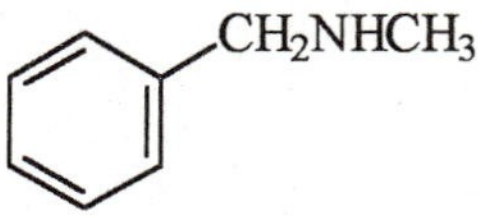 $C_6H_5CH_2NHCH_3$

15.38 a)

$CH_3{-}C{\equiv}N$ + $H_3O^{\oplus}$ ⇌ $CH_3{-}C{\equiv}N^{\oplus}{-}H$ (+ H_2O) ⇌ $CH_3{-}C(=NH){-}O^{\oplus}H_2$ (+ H_2O) ⇌ $CH_3{-}C(OH){=}N{-}H$ (+ $H_3O^{\oplus}$)

⇌ $CH_3{-}C(OH){=}N^{\oplus}H_2$ ⟷ $CH_3{-}C(NH_2){=}O^{\oplus}{-}H$ (+ H_2O) ⇌ $CH_3{-}C(=O){-}NH_2$

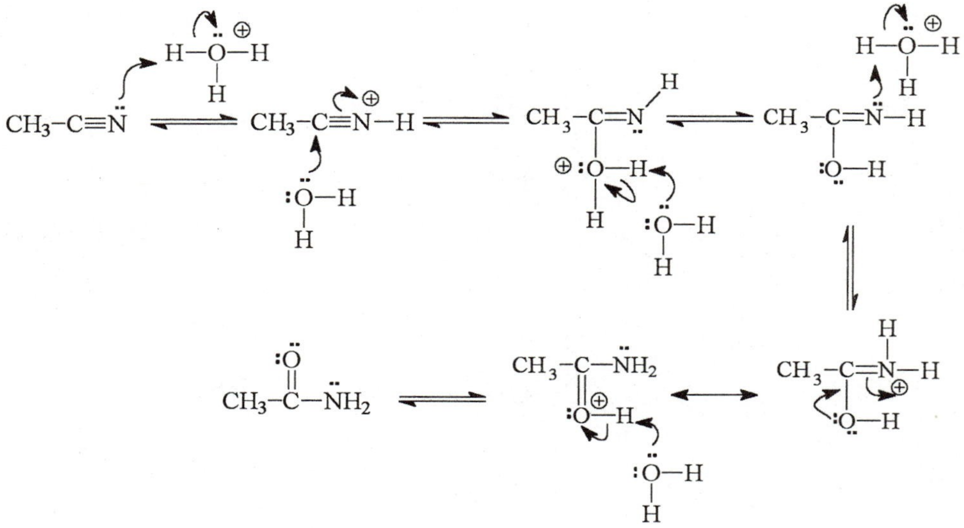

b)

$CH_3CH_2-C(=O)-OCH_3 + OH^- \rightleftharpoons CH_3CH_2-C(O^-)(OH)-OCH_3 \rightleftharpoons CH_3CH_2-C(=O)-O-H + CH_3O^- \rightleftharpoons CH_3CH_2-C(=O)-O^- + CH_3OH$

c)

$CH_3CH_2-C(=O)-OCH_3 + H_3O^+ \rightleftharpoons CH_3CH_2-C(=O^+H)-OCH_3 \rightleftharpoons CH_3CH_2-C(OH)(O^+H_2)-OCH_3 \rightleftharpoons CH_3CH_2-C(OH)(OH)-OCH_3 \rightleftharpoons CH_3CH_2-C(OH)(OH)-O^+(H)CH_3 \rightleftharpoons CH_3CH_2-C(=O^+H)-OH + CH_3OH \rightleftharpoons CH_3CH_2-C(=O)-OH$

d)

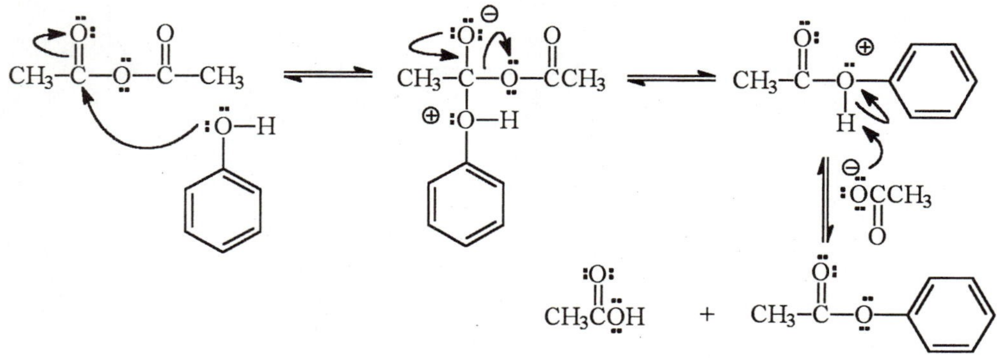

$CH_3-\overset{\overset{\displaystyle :\ddot{O}:}{\|}}{C}-\ddot{O}-\overset{\overset{\displaystyle O}{\|}}{C}-CH_3$ + $:\ddot{O}-H$ (Ph) ⇌ $CH_3-C(:\ddot{O}:^{\ominus})(\oplus:O-H)-\ddot{O}-\overset{O}{\|}CCH_3$ (Ph) ⇌ $CH_3-\overset{:\ddot{O}:}{\|}C-\ddot{O}^{\oplus}(H)$-Ph

$\ominus$ $:\ddot{O}CCH_3$, O

⇅

$:O:$ $CH_3C\ddot{O}H$ (C=O) + $CH_3-\overset{\ddot{O}:}{\|}C-\ddot{O}$-Ph

e)

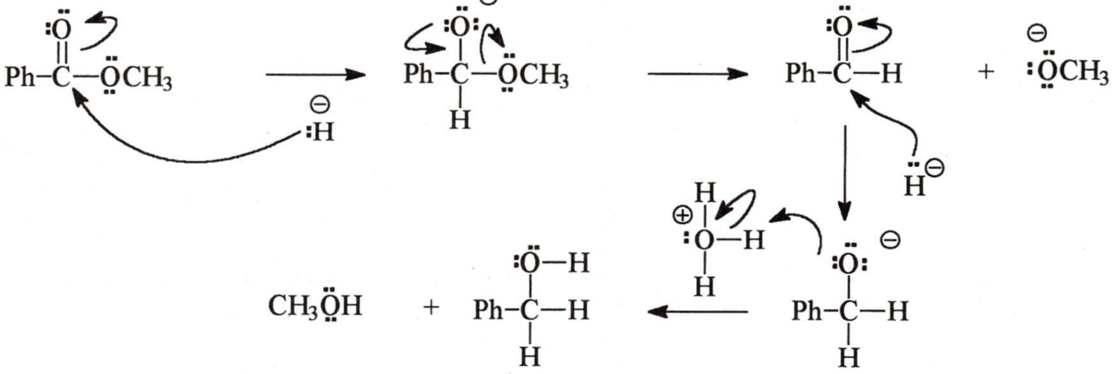

$Ph-\overset{:\ddot{O}}{\|}C-\ddot{O}CH_3$ + $\ominus$ $:H$ → $Ph-C(:\ddot{O}:^{\ominus})(H)-\ddot{O}CH_3$ → $Ph-\overset{:\ddot{O}}{\|}C-H$ + $\ominus$ $:\ddot{O}CH_3$

$Ph-\overset{:\ddot{O}}{\|}C-H$ + $\ddot{H}^{\ominus}$ ↓

$Ph-C(:\ddot{O}:^{\ominus})(H)-H$ + $\oplus$ $:O(H)(H)-H$

$CH_3\ddot{O}H$ + $Ph-C(:\ddot{O}-H)(H)-H$ ←

f)

$CH_3CH_2CH_2C{\equiv}N:$ + $\ominus$ $:H$ → $CH_3CH_2CH_2\overset{H}{C}{=}\ddot{N}:^{\ominus}$ + $\ddot{H}^{\ominus}$ → $CH_3CH_2CH_2C(H)(H)-\ddot{N}:^{2\ominus}$

↓ H $:\ddot{O}-H$

$CH_3CH_2CH_2CH_2NH_2$ ← $CH_3CH_2CH_2CH_2\ddot{N}H^{\ominus}$ + H $:\ddot{O}-H$

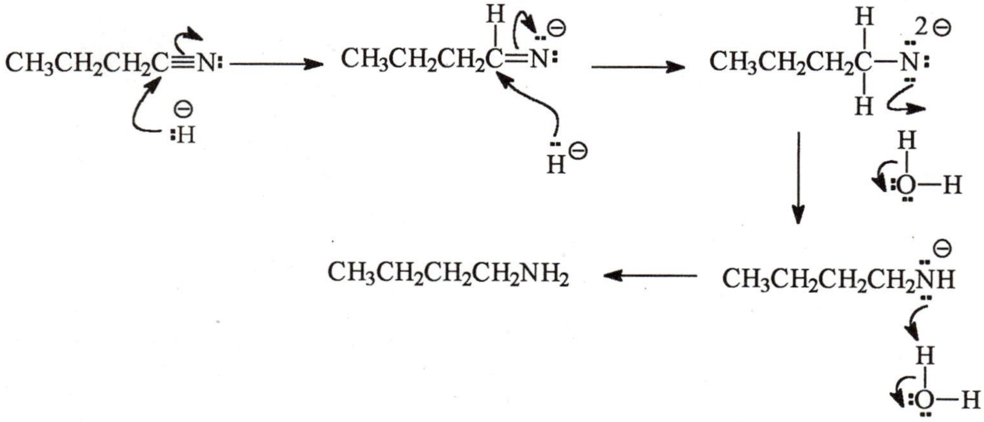

g)

$CH_3CH_2-C(=\ddot{O}:)-\ddot{O}CH_2CH_3$ ⟶ $CH_3CH_2-C(-\ddot{O}:^{\ominus})(CH_3)-\ddot{O}CH_2CH_3$ ⟶ $:\ddot{O}CH_2CH_3^{\ominus}$ + $CH_3CH_2-C(=\ddot{O}:)-CH_3$

Li—CH_3

$CH_3CH_2-C(=\ddot{O}:)-CH_3$ + Li—CH_3 ⟶ $CH_3CH_2-C(-\ddot{O}:^{\ominus})(CH_3)-CH_3$

$CH_3CH_2-C(-\ddot{O}:^{\ominus})(CH_3)-CH_3$ + $H-O^{\oplus}(H)-H$ ⟶ $CH_3CH_2\ddot{O}H$ + $CH_3CH_2-C(:\ddot{O}-H)(CH_3)-CH_3$

h)

$Ph-C\equiv N:$ + IMg—CH_3 ⟶ $Ph-C(CH_3)=\ddot{N}:^{\ominus}$ + $H-:\!O^{\oplus}(H)-H$ ⟶ $Ph-C(CH_3)=\ddot{N}H$ + $H-:\!O^{\oplus}(H)-H$ ⟶ $Ph-C(CH_3)=NH_2^{\oplus}$ + $:\ddot{O}-H$ (H)

⟶ $Ph-C(CH_3)(-\ddot{N}H_2)-O^{\oplus}(H)H$

$Ph-C(CH_3)(-\ddot{N}H_2)-O^{\oplus}(H)H$ ⇌ $Ph-C(CH_3)(-NH_3^{\oplus})-\ddot{O}-H$ ⇌ $Ph-C(CH_3)=:\!O^{\oplus}-H$ + $\ddot{N}H_3$ ⇌ $Ph-C(=:\!O:)-CH_3$ + $NH_4^{\oplus}$

i)

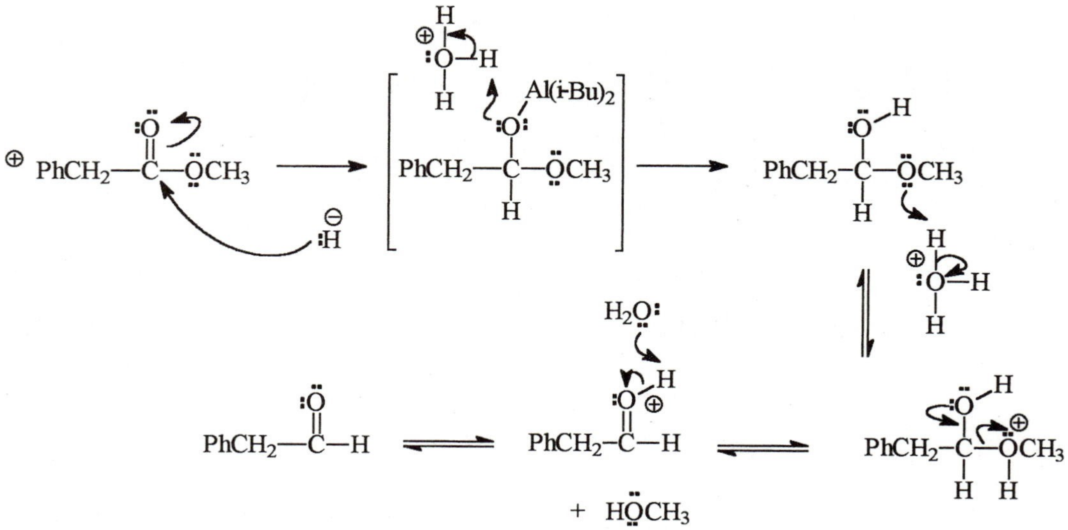

H
⊕ :O—H
H
Al(i-Bu)₂
:O:
:O:
⊕ PhCH₂—C—OCH₃ → [PhCH₂—C—OCH₃] → PhCH₂-C-OCH₃
H
:H ⊖
:O—H
H
H
H
⊕ :O—H
H

$PhCH_2-C(=O)-OCH_3 \xrightarrow{H^-} [PhCH_2-CH(OAl(i\text{-}Bu)_2)-OCH_3] \xrightarrow{H_3O^+} PhCH_2-CH(OH)-OCH_3$

H₂O:
:O: H
:O ⊕ H
PhCH₂—C—H ⇌ PhCH₂-C—H ⇌ PhCH₂-C-OCH₃ ⊕
H H
+ HOCH₃

$PhCH_2-CH(OH)-OCH_3 + H_3O^+ \rightleftharpoons PhCH_2-CH(OH)-O^+(H)CH_3 \rightleftharpoons PhCH_2-CH=O^+H + HOCH_3$

$PhCH_2-CH=O^+H + H_2O \rightleftharpoons PhCH_2-C(=O)-H$

j)

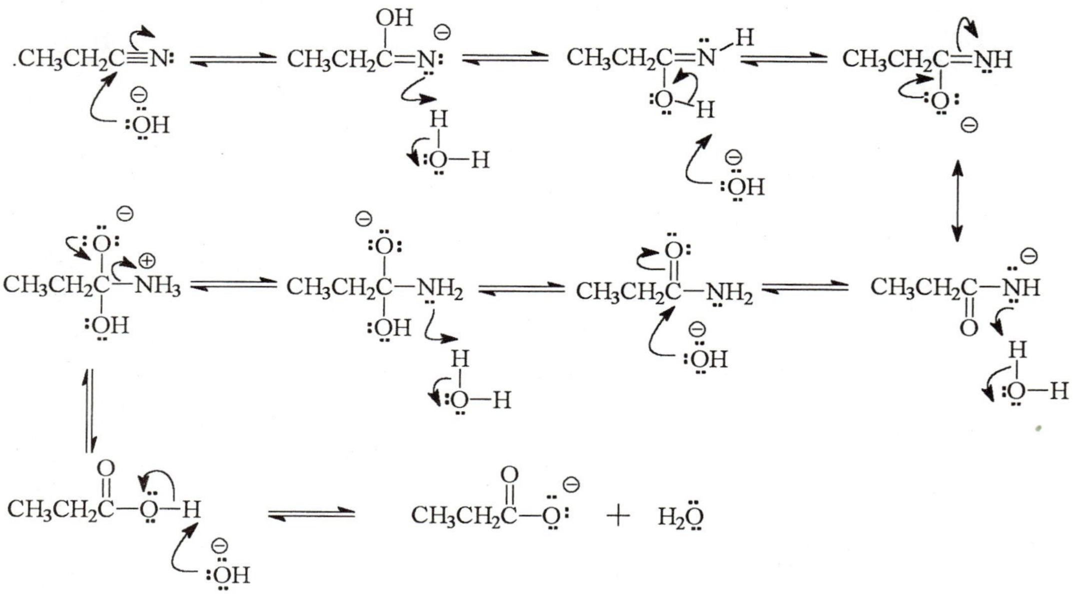

OH ⊖
H
CH₃CH₂C≡N: ⇌ CH₃CH₂C=N: ⇌ CH₃CH₂C=N ⇌ CH₃CH₂C=NH
⊖ :OH
H
:O—H
:O H
⊖ :OH
:O: ⊖

$CH_3CH_2C\equiv N + {}^-OH \rightleftharpoons CH_3CH_2C(OH)=N^- \xrightleftharpoons{H_2O} CH_3CH_2C(OH)=NH \xrightleftharpoons{^-OH} CH_3CH_2C(O^-)=NH \leftrightarrow CH_3CH_2C(=O)-NH^-$

⊖ :O: ⊕
⊖ :O:
:O:
⊖
CH₃CH₂C—NH₃ ⇌ CH₃CH₂C—NH₂ ⇌ CH₃CH₂C—NH₂ ⇌ CH₃CH₂C—NH
:OH
:OH
H
:O—H
⊖ :OH
O
H
:O—H

$CH_3CH_2C(=O)-NH^- + H_2O \rightleftharpoons CH_3CH_2C(=O)-NH_2 \xrightleftharpoons{^-OH} CH_3CH_2C(O^-)(OH)-NH_2 \xrightleftharpoons{H_2O} CH_3CH_2C(O^-)(OH)-NH_3^+$

O
O ⊖
CH₃CH₂C—O—H ⇌ CH₃CH₂C—O: + H₂O
⊖ :OH

$CH_3CH_2C(=O)-O-H + {}^-OH \rightleftharpoons CH_3CH_2C(=O)-O^- + H_2O$

15.39

a) $CH_3CH_2CH_2CH_2COOH \xrightarrow{SOCl_2} CH_3CH_2CH_2CH_2COCl \xrightarrow{(CH_3CH_2CH_2CH_2)_2CuLi} CH_3CH_2CH_2CH_2COCH_2CH_2CH_2CH_3 \xrightarrow{Ph_3P{=}CHCH_3}$

b) $C_6H_5COOH \xrightarrow{SOCl_2} C_6H_5COCl \xrightarrow{(CH_3)_2CuLi} C_6H_5COCH_3 \xrightarrow{Ph_3P{=}CH_2} C_6H_5C(CH_3){=}CH_2$

c) $C_6H_5CH_2COOH \xrightarrow{SOCl_2} C_6H_5CH_2CCl$ (C=O) $\xrightarrow{(CH_3CH_2)_2CuLi} C_6H_5CH_2CCH_2CH_3$ (C=O)

$\xrightarrow{\text{1) } CH_3MgBr \text{ 2) } NH_4Cl,\ H_2O} C_6H_5CH_2C(OH)(CH_3)CH_2CH_3 \xrightarrow{SOCl_2} C_6H_5CH_2C(Cl)(CH_3)CH_2CH_3$

d) Cyclopentane-COOH $\xrightarrow[\text{2) } H_3O^{\oplus}]{\text{1) } LiAlH_4}$ Cyclopentane-CH_2OH $\xrightarrow{SOCl_2}$ Cyclopentane-CH_2Cl

$\downarrow CH_3C{\equiv}C^{\ominus}$

Cyclopentane-$CH_2C{\equiv}CCH_3$

e) $CH_3CH_2CH_2CH_2Br$ $\xrightarrow{NaCN}$ $CH_3CH_2CH_2CH_2CN$ $\xrightarrow[\text{reflux}]{H_2SO_4,\ H_2O}$ $CH_3CH_2CH_2CH_2COOH$

$\downarrow CH_3OH,\ TsOH$

$CH_3CH_2CH_2CH_2COOCH_3$

15.40

a) $C_6H_5CH_2Br$ $\xrightarrow{NaCN}$ $C_6H_5CH_2CN$ $\xrightarrow[\text{2) } H_2O]{\text{1) } LiAlH_4}$ $C_6H_5CH_2CH_2NH_2$

b) Cyclohexyl-$\overset{O}{\overset{\|}{C}}OH$ $\xrightarrow{SOCl_2}$ cyclohexyl-$\overset{O}{\overset{\|}{C}}Cl$ $\xrightarrow{CH_3CH_2CH_2CH_2NH_2}$ cyclohexyl-C(=O)-NH-$CH_2CH_2CH_2CH_3$

$\downarrow$ 1) $LiAlH_4$ 2) H_2O

Cyclohexyl-CH_2-NH-$CH_2CH_2CH_2CH_3$

c) Phenyl-$\overset{O}{\overset{\|}{C}}OH$ $\xrightarrow[TsOH]{CH_3OH}$ phenyl-$\overset{O}{\overset{\|}{C}}OCH_3$ $\xrightarrow[2)\ H_3O^{\oplus}]{1)\ 2\ CH_3MgBr}$ phenyl-$\overset{CH_3}{\overset{|}{C}}=CH_2$

$\downarrow$ 1) BH_3 , THF 2) H_2O_2 , $\overset{\ominus}{O}H$

Phenyl-$\overset{CH_3}{\overset{|}{C}}HCH_2OH$

d) Cyclopentenyl-$\overset{O}{\overset{\|}{C}}OH$ $\xrightarrow[2)\ DIBAH\quad 3)\ H_3O^{\oplus}]{1)\ CH_3OH,\ H_2SO_4}$ cyclopentenyl-$\overset{O}{\overset{\|}{C}}H$ $\xrightarrow[2)\ H_3O^{\oplus}]{1)\ (CH_3)_2CuLi}$ 2-methylcyclopentyl-$\overset{O}{\overset{\|}{C}}H$

15.41 The equilibrium favors the reactants because phenoxide ion is a weaker base (resonance stabilized) than methoxide ion.

15.42

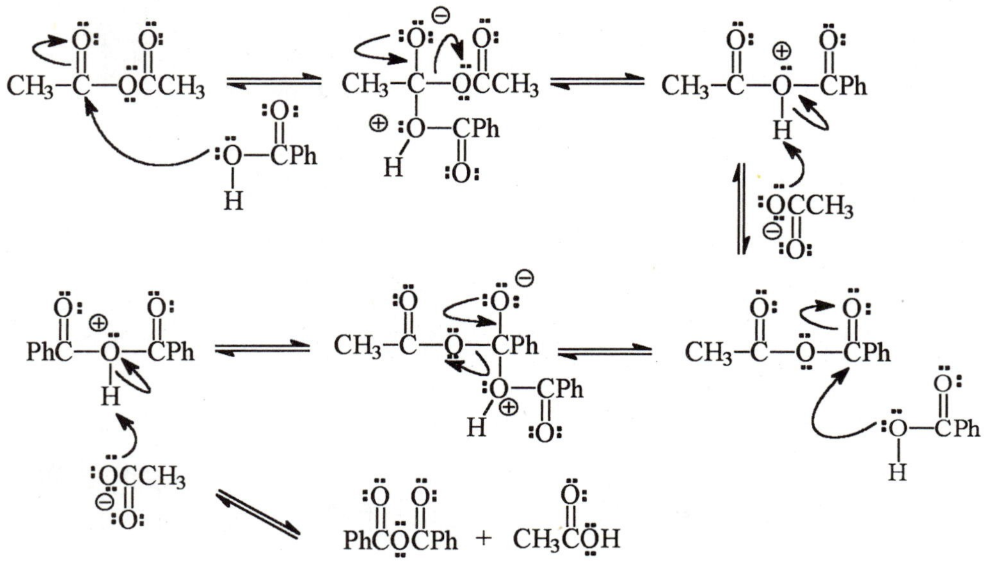

CH_3COCCH_3 + $PhCOOH$ ⇌ tetrahedral intermediate ⇌ $CH_3C(O)O^{+}(H)CPh$ ⇌ ($CH_3CO_2^{-}$) ⇌ $CH_3C(O)OC(O)Ph$ ⇌ tetrahedral intermediate ⇌ $PhC(O)O^{+}(H)C(O)Ph$ ⇌

$PhCOCPh$ + CH_3COH

15.43

$CH_3CO_2^{-}$ + $H^{18}OH$ ⇌ CH_3COOH + $^{18}OH^{-}$ ⇌ $CH_3C(O^{-})(OH)(^{18}OH)$ ⇌ $CH_3C(O)^{18}OH$ + ^{-}OH ⇌ $CH_3C(O)^{18}O^{-}$ ⟷ $CH_3C(O^{-})={}^{18}O$

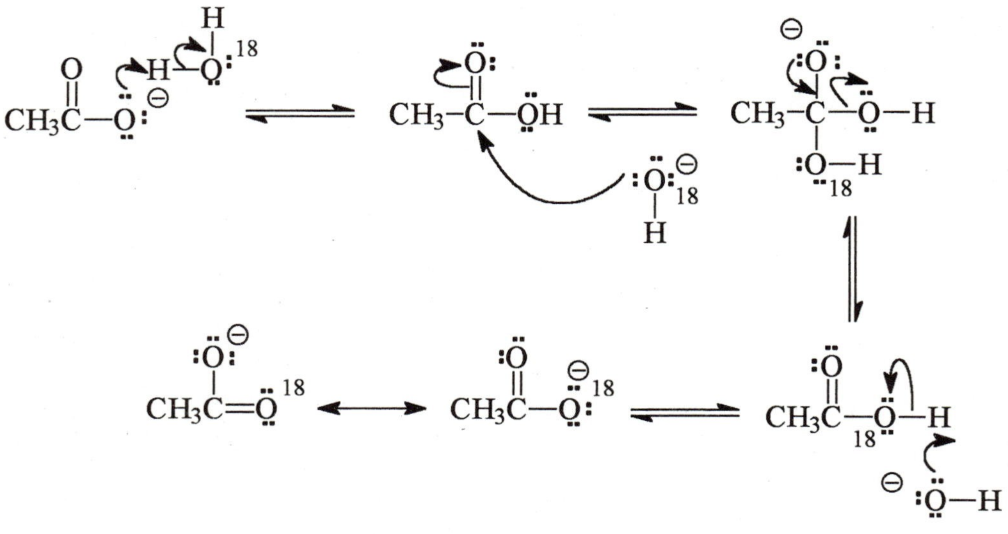

15.44

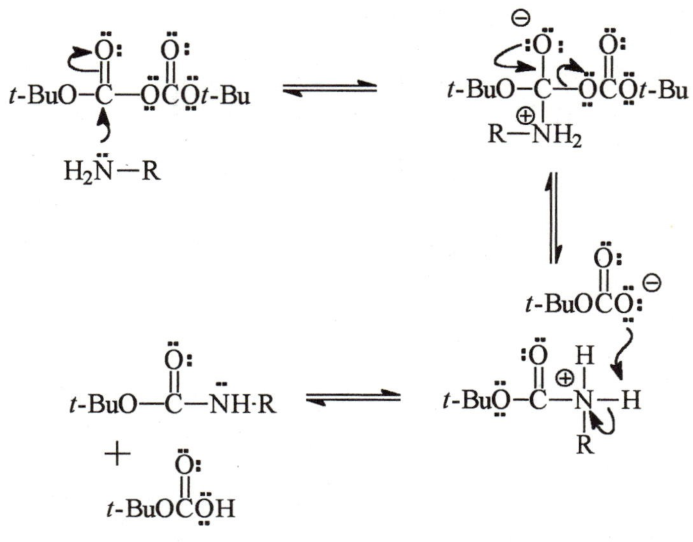

The products of this reaction are determined by the stability of the leaving group. The *t*-BuOCOO⁻ group can support a negative charge better than the *t*-BuO⁻ group.

15.45

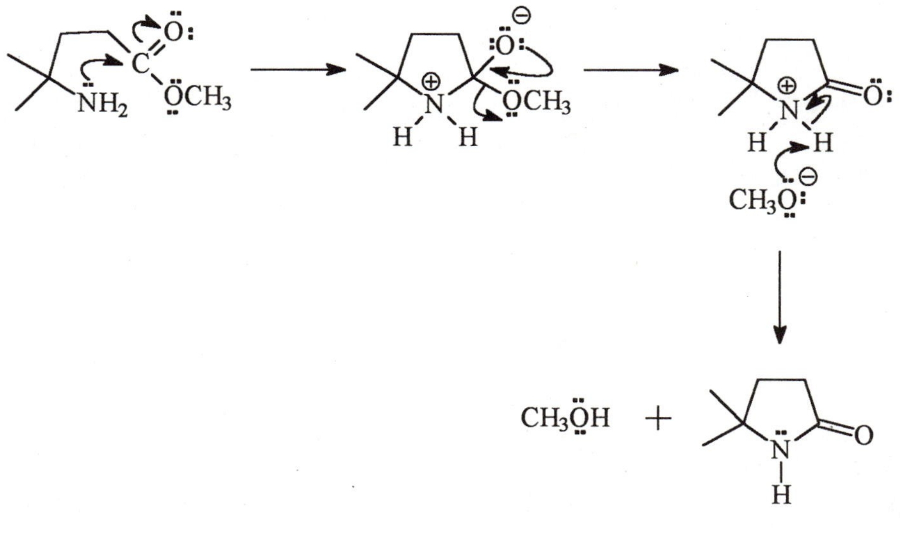

15.46

a)

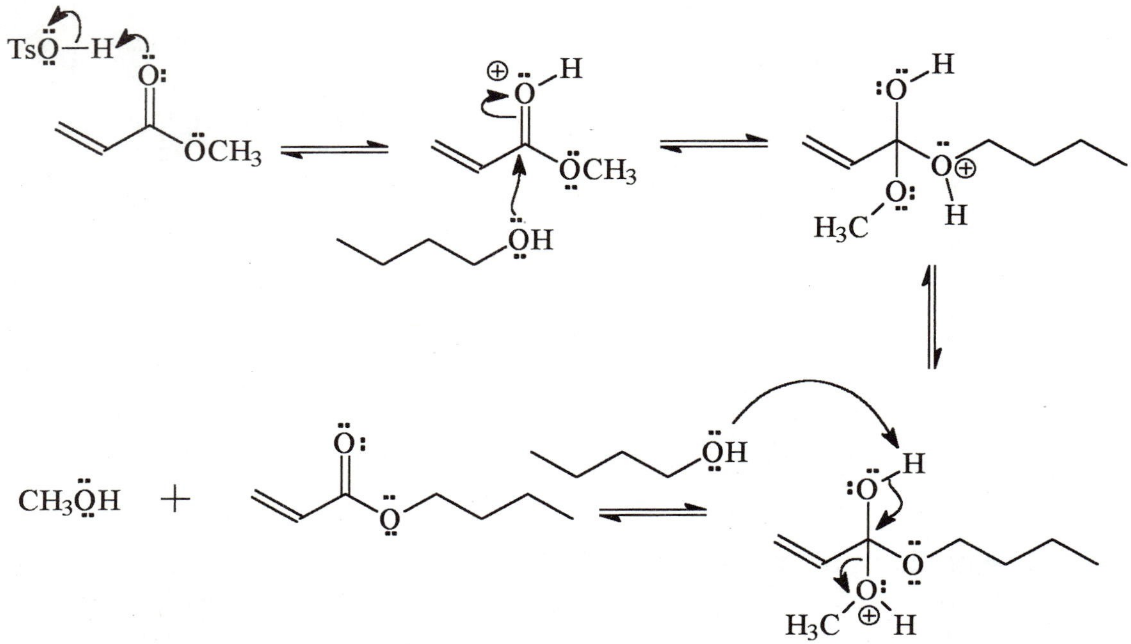

b) The equilibrium constant for this reaction should be about one because the stability of the reactants and the products are approximately the same.

c) The equilibrium can be driven toward the product either by using an excess of butanol in the reaction or by removing methanol by distillation during the reaction.

15.47

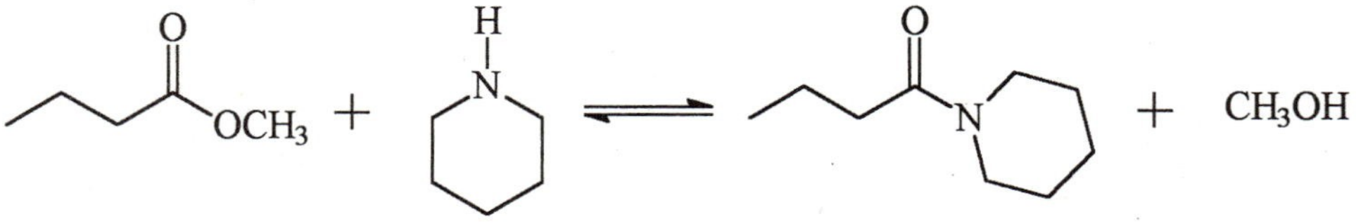

The products in this reaction are more stable than the reactants, so therefore it is not necessary to drive the equilibrium towards the products.

15.48 The equilibrium favors the product because the ring strain of the reactant is relieved during this reaction.

15.49

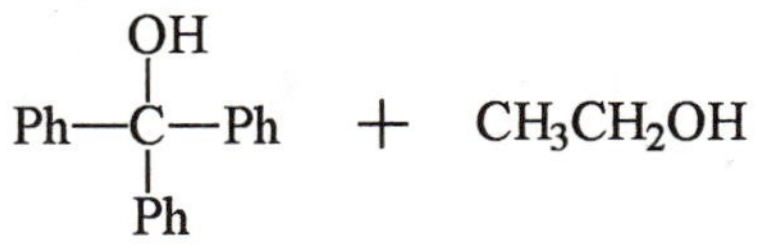

$$Ph-\underset{Ph}{\overset{OH}{C}}-Ph \quad + \quad CH_3CH_2OH$$

15.50

a) 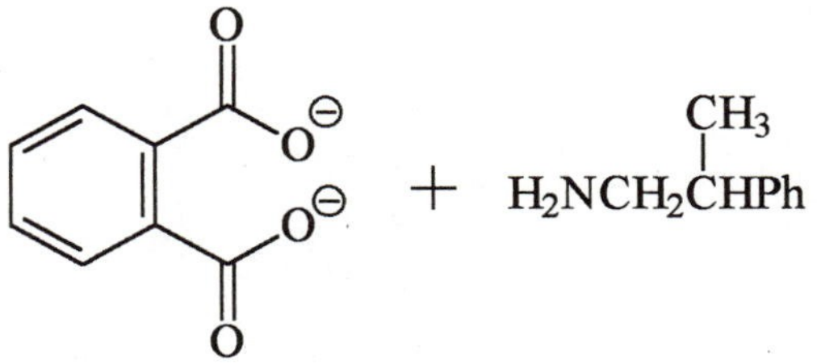 phthalate dianion (two $CO_2^{\ominus}$ groups on a benzene ring) $+ \; H_2NCH_2\overset{CH_3}{C}HPh$

b) $CH_2=CHCO_2CH_2CH_3$

c) $PhCH_2C(=O)NHCH_2CH_2CH_3$

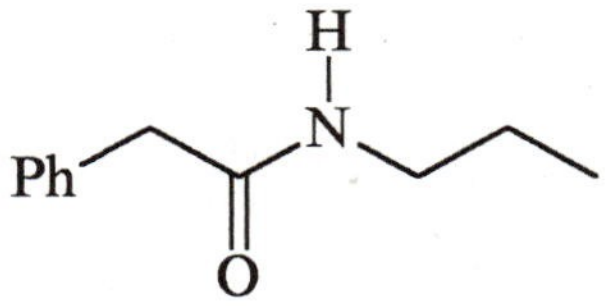

d) 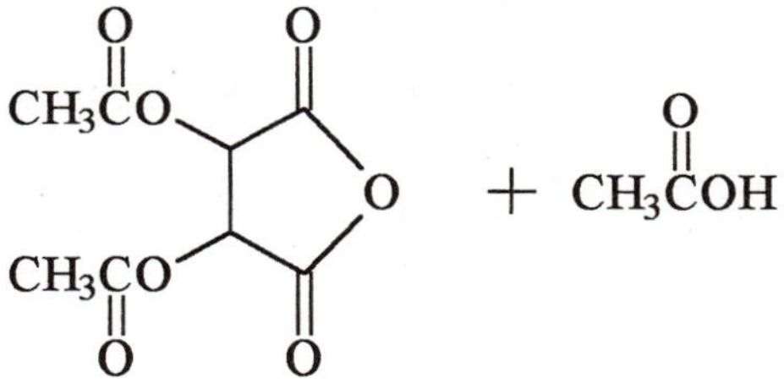 $CH_3\overset{O}{\overset{\|}{C}}O$, $CH_3\overset{O}{\overset{\|}{C}}O$ (diacetoxy cyclic anhydride, O in ring) $+ \; CH_3\overset{O}{\overset{\|}{C}}OH$

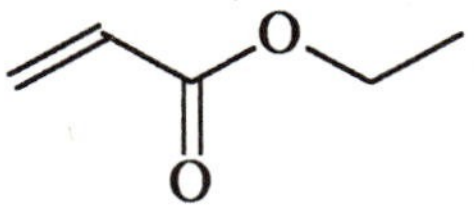

e) 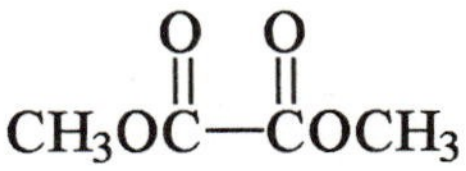 $CH_3O\overset{O}{\overset{\|}{C}}-\overset{O}{\overset{\|}{C}}OCH_3$

f) 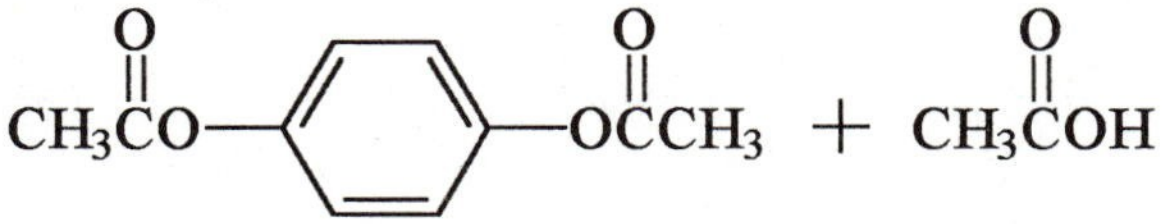 $CH_3\overset{O}{\overset{\|}{C}}O-C_6H_4-O\overset{O}{\overset{\|}{C}}CH_3 \; + \; CH_3\overset{O}{\overset{\|}{C}}OH$

g) 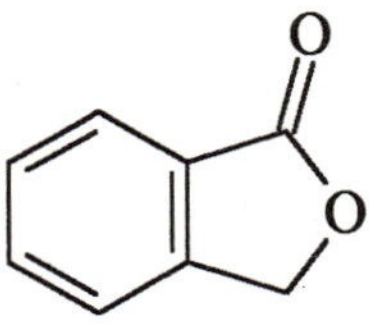 phthalide (benzo-fused lactone: C=O, O in ring)

15.51

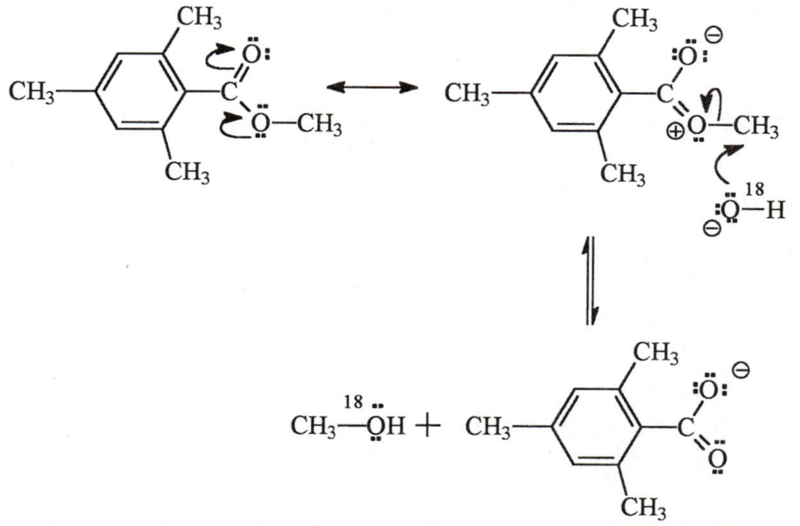

CH_3 CH_3 CH_3 C O O—CH_3 ⟷ CH_3 CH_3 CH_3 C $O^{\ominus}$ $^{\oplus}O$—CH_3 ^{18}O—H $\ominus$

⇌

CH_3—^{18}OH + CH_3 CH_3 CH_3 C $O^{\ominus}$ O

The regular saponification mechanism is not observed in this case because the carbonyl carbon is not accessible by the nucleophile due to steric hindrance by the two ortho methyl groups. Therefore the reaction follows a S_N2 mechanism.

15.52

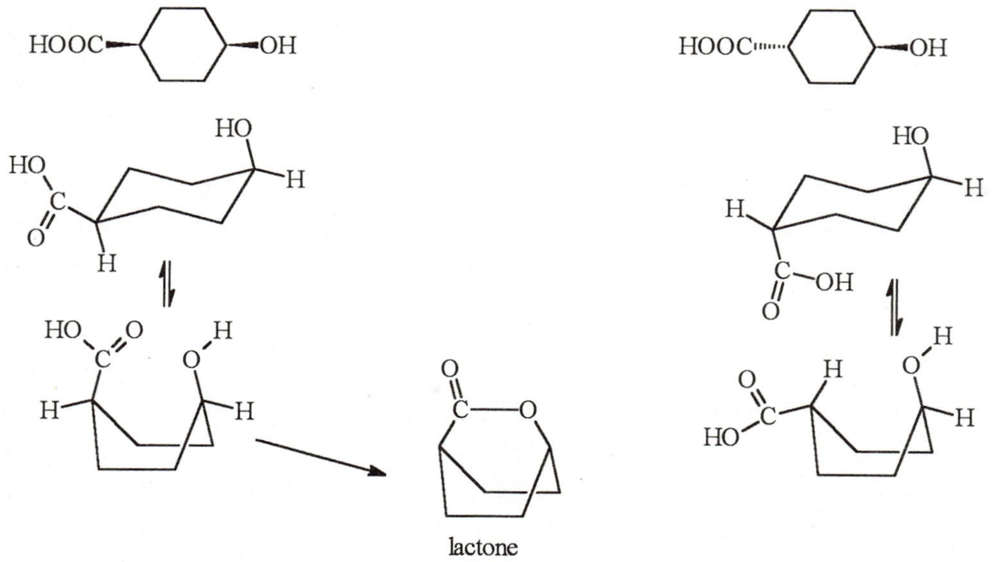

HOOC OH HOOC OH

HO HO C O H H HO H H C OH O H

⇌ ⇌

HO O C O H H H H

O C O

HO C O H O H H

lactone

Only the cis isomer forms the lactone because the boat conformation of the cis isomer provides the required geometry for intramolecular reaction to produce the lactone.

15.53 The lone pair of electrons on the nitrogen of an amide is delocalized by resonance and is less available for bonding. The proton becomes bonded to the oxygen in an amide because the electron density is higher at the oxygen than the nitrogen.

$$CH_3-\overset{\overset{\displaystyle :\ddot{O}:}{\|}}{C}-\ddot{N}H_2 \longleftrightarrow CH_3-\overset{\overset{\displaystyle :\ddot{O}:^{\ominus}}{|}}{C}=\overset{\oplus}{N}H_2$$

15.54

$$CH_3\overset{O}{\overset{\|}{C}}NH-C_6H_4-\overset{O}{\overset{\|}{\underset{\underset{O}{\|}}{S}}}-Cl + \text{2-aminothiazole} \longrightarrow CH_3\overset{O}{\overset{\|}{C}}NH-C_6H_4-\overset{O}{\overset{\|}{\underset{\underset{O}{\|}}{S}}}-\overset{H}{\overset{|}{N}}-\text{(2-thiazolyl)}$$

1) HCl, H_2O

2) $NaHCO_3$

$$H_2N-C_6H_4-\overset{O}{\overset{\|}{\underset{\underset{O}{\|}}{S}}}-\overset{H}{\overset{|}{N}}-\text{(2-thiazolyl)}$$

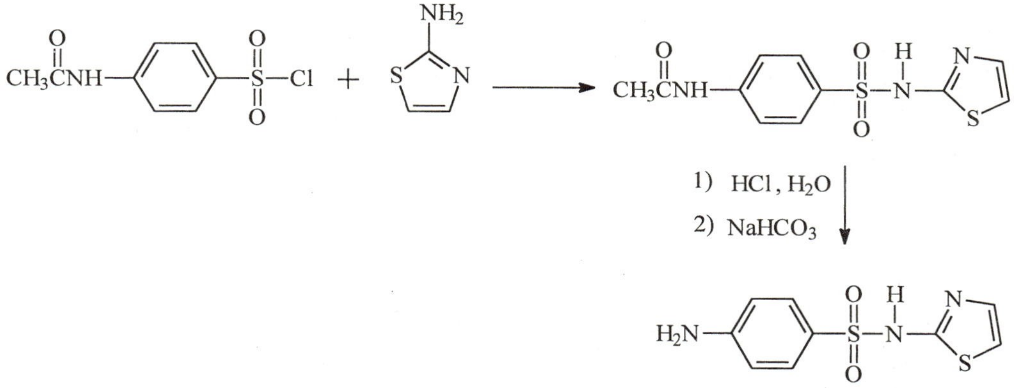

15.55 The carbonyl carbon of the reactant in the reaction shown is more electrophilic than that of an amide. Also in this reaction the leaving group is a more stable molecule (trimethyl amine) whereas in the hydrolysis of an amide the leaving group is an unstable anion (a strong base).

15.56 When succinic anhydride is heated in methanol it produces a molecule with an ester and a carboxyl group. This is apparent from the presence of -OH band in the IR spectrum and the signal at 10.25 δ in the NMR spectrum of the product. However, when this reaction is carried out in the presence of a catalytic amount of sulfuric acid, the carboxyl group reacts with methanol to form a diester. This is consistent with the IR spectrum of the product and the fact that it shows only three absorptions in its ^{13}C-NMR spectrum.

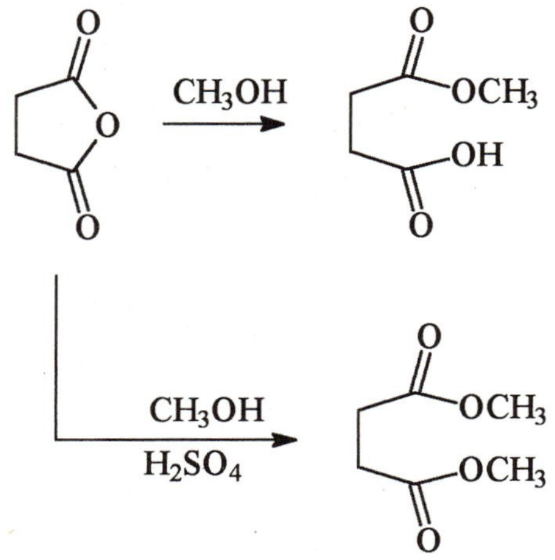

$$\text{succinic anhydride} \xrightarrow{CH_3OH} HO_2C\text{-}CH_2CH_2\text{-}CO_2CH_3$$

$$\text{succinic anhydride} \xrightarrow[H_2SO_4]{CH_3OH} CH_3O_2C\text{-}CH_2CH_2\text{-}CO_2CH_3$$

15.57

$$OHC\text{-}CH_2CH_2\text{-}CO_2CH_3 \xrightarrow[\text{2) } H_3O^{\oplus} \text{ heat}]{\text{1) } NaBH_4} \left[HOCH_2CH_2CH_2\text{-}CO_2CH_3 \right] \longrightarrow \text{lactone (five-membered ring)}$$

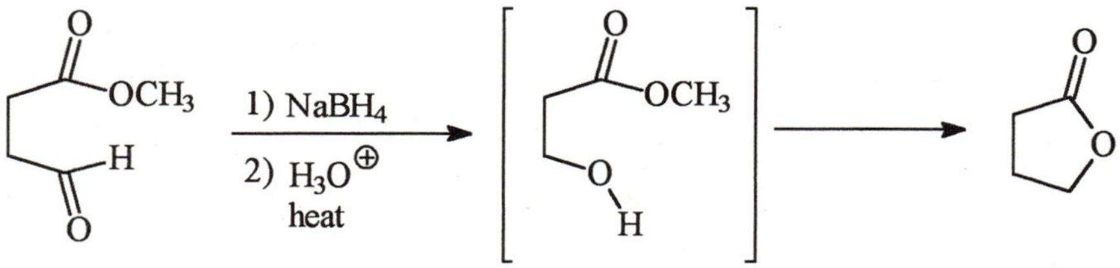

Sodium borohydride selectively reduces the aldehyde group of this compound. An alcohol is normally produced on treatment with an acid. However, in this reaction the alcohol formed during the work up step with acid undergoes an intramolecular esterification reaction to form a lactone as shown above. This is apparent from the absence of a -OH band in the IR spectrum of the product, and is consistent with the NMR spectrum. The carbonyl absorption occurs at higher wavenumber because it is part of a five membered ring ($1740 + 30 = 1780\ cm^{-1}$).

After completing this chapter, you should be able to:

1. Show the products of any of the reactions discussed in this chapter. Problems 15.5, 15.6, 15.9, 15.13, 15.15, 15.18, 15.20, 15.22, 15.24, 15.26, 15.27, 15.28, 15.29, 15.33, 15.34, 15.35, 15.36, 15.49.

2. Show the mechanism for these reactions.

Problems 15.7, 15.10, 15.12, 15.16, 15.23, 15.37, 15.41, 15.42, 15.43, 15.44, 15.45.

3. Predict the effect of a change in structure on the rate and equilibrium of a reaction.
   Problems 15.1, 15.2, 15.3, 15.8, 15.30, 15.31, 15.40.

4. Use these reactions to interconvert any of the carboxylic acid derivatives and to prepare aldehydes and ketones.
   Problems 15.4, 15.14, 15.17, 15.21, 15.25, 15.32, 15.46, 15.53.

5. Use these reactions in combination with reactions from previous chapters to synthesize compounds.
   Problems 15.38, 15.39.

Chapter 16
ENOLATE AND OTHER CARBON NUCLEOPHILES

16.1

a) $PhCH{=}C(OH){-}CH_3$ and $PhCH_2C(OH){=}CH_2$

This one is more stable because it is conjugated.

b) $CH_3CH_2C(OH){=}CHCH_3$

c) 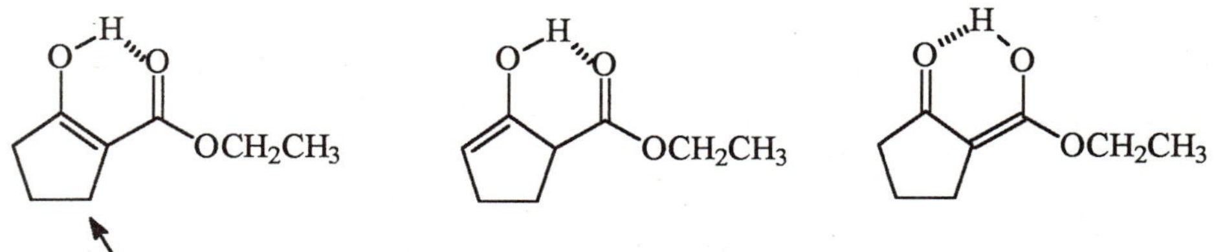 OCH₂CH₃ OCH₂CH₃ OCH₂CH₃

This one is more stable due to conjugation, hydrogen bonding and loss of the ketone rather than the ester carbonyl group.

16.2

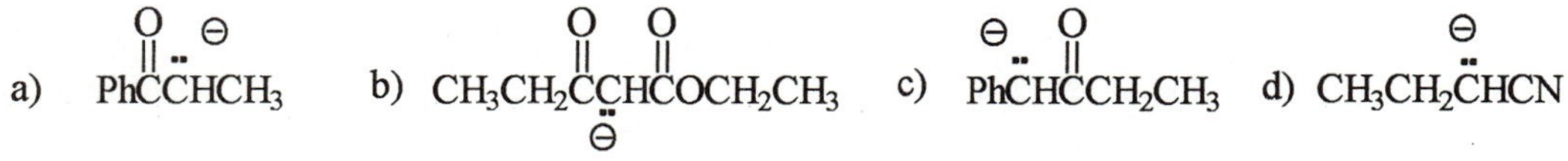

a) $PhC(=O)\ddot{C}^{\ominus}HCH_3$ b) $CH_3CH_2C(=O)\ddot{C}^{\ominus}HC(=O)OCH_2CH_3$ c) $Ph\ddot{C}^{\ominus}HC(=O)CH_2CH_3$ d) $CH_3CH_2\ddot{C}^{\ominus}HCN$

16.3

$Li{-}CH_2CH_2CH_2CH_3 \longrightarrow Li^{\oplus}\ {}^{\ominus}\!:\!\ddot{O}\!:\ CH_2CH_2CH_2CH_3$

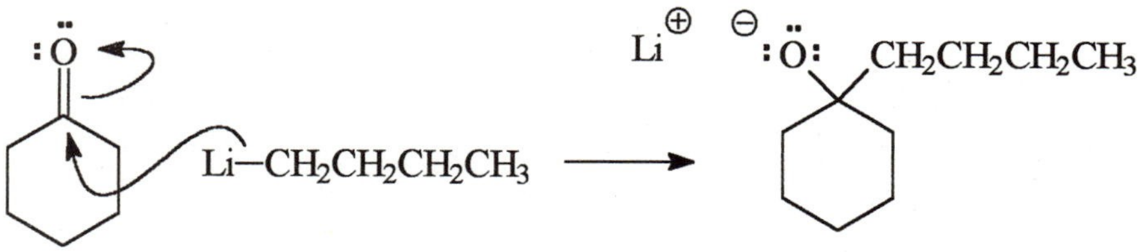

16.4

b) Cyclohexyl–CH(CN)Ph: cyclohexane ring bonded to $\overset{CN}{|}CHPh$

c) $CH_3CH_2CH(CH_2CH_2CH_2CH_3)\overset{O}{\overset{\|}{C}}OEt$

d) $Ph\overset{O}{\overset{\|}{C}}CH(CH_2CH_3)CH_2CH_3$

e) cyclopropane–CN

f) $CH_3CH_2CH_2CH(CH_2{-}CH{=}CH_2)\overset{O}{\overset{\|}{C}}OCH_3$

16.5

a) $CH_3\overset{O}{\overset{\|}{C}}CH(CH_2Ph)\overset{O}{\overset{\|}{C}}OEt \xrightarrow[2)\ H_3O^{\oplus},\ \Delta]{1)\ NaOH,\ H_2O} CH_3\overset{O}{\overset{\|}{C}}CH_2CH_2Ph$

b) $EtO\overset{O}{\overset{\|}{C}}CH(CH_2CH_2CH_3)\overset{O}{\overset{\|}{C}}OEt \xrightarrow[2)\ H_3O^{\oplus},\ \Delta]{1)\ NaOH,\ H_2O} CH_3CH_2CH_2CH_2\overset{O}{\overset{\|}{C}}OH$

c) 2-oxocyclohexane ring bearing an ethyl group and $\overset{O}{\overset{\|}{C}}OEt$ on C-1

d) $CH_3\overset{O}{\overset{\|}{C}}C(CH_3)(CH_2CH_2CH_2CH_3)\overset{O}{\overset{\|}{C}}OEt \xrightarrow[2)\ H_3O^{\oplus},\ \Delta]{1)\ NaOH,\ H_2O} CH_3\overset{O}{\overset{\|}{C}}{-}\overset{CH_3}{\overset{|}{C}}HCH_2CH_2CH_2CH_3$

e) $EtO\overset{O}{\overset{\|}{C}}CH\overset{O}{\overset{\|}{C}}OEt$ with cyclopentenyl substituent on CH

f) 2-methyl-2-(ethoxycarbonyl)cyclopentanone $\xrightarrow[2)\ H_3O^{\oplus},\ \Delta]{1)\ NaOH,\ H_2O}$ 2-methylcyclopentanone

g) $N{\equiv}CCH(CH_2CH_2CH_3)\overset{O}{\overset{\|}{C}}CH_3$

16.6

b) $EtOC(=O)CH_2C(=O)OEt$ $\xrightarrow[\text{2) } CH_3CH(Br)CH_3]{\text{1) NaOEt, EtOH}}$ $EtOC(=O)CH(CH(CH_3)CH_3)C(=O)OEt$ $\xrightarrow[\text{2) } H_3O^{\oplus},\ \Delta]{\text{1) NaOH, } H_2O}$ $CH_3CH(CH_3)CH_2CO_2H$

c) $CH_3C(=O)CH_2C(=O)OEt$ $\xrightarrow[\text{2) } PhCH_2Br\text{; 3) NaOEt, EtOH; 4) } CH_3I]{\text{1) NaOEt, EtOH}}$ $CH_3C(=O)C(CH_3)(CH_2Ph)C(=O)OEt$ $\xrightarrow[\text{2) } H_3O^{\oplus},\ \Delta]{\text{1) NaOH, } H_2O}$ $PhCH_2CH(CH_3)-C(=O)CH_3$

d) $EtOC(=O)CH_2C(=O)OEt$ $\xrightarrow[\text{2 } CH_3CH_2Br]{\text{2 NaOEt, EtOH}}$ $EtOC(=O)C(CH_2CH_3)(CH_2CH_3)C(=O)OEt$ $\xrightarrow[\text{2) } H_3O^{\oplus},\ \Delta]{\text{1) NaOH, } H_2O}$ $CH_3CH_2CH(CH_2CH_3)-CO_2H$

e) $EtOC(=O)CH_2C(=O)OEt$ $\xrightarrow[\text{2) cyclopentyl}-CH_2Br]{\text{1) NaOEt, EtOH}}$ $EtOC(=O)CH(CH_2\text{-cyclopentyl})C(=O)OEt$ $\xrightarrow[\text{2) } H_3O^{\oplus},\ \Delta]{\text{1) NaOH, } H_2O}$ cyclopentyl$-CH_2CH_2CO_2H$

f) $CH_3C(=O)CH_2C(=O)OEt$ $\xrightarrow[BrCH_2CH_2CH_2CH_2Br]{\text{2 NaOEt, EtOH}}$ 1-acetyl-1-(ethoxycarbonyl)cyclopentane ($CH_3C(=O)$, $C(=O)OEt$) $\xrightarrow[\text{2) } H_3O^{\oplus},\ \Delta]{\text{1) NaOH, } H_2O}$ cyclopentyl$-C(=O)CH_3$

16.7

a) $CH_3CH_2C(=O)CH_2C(=O)CH_2CH_3$ $\xrightarrow[\text{2) } CH_3CH_2CH_2Br]{\text{1) NaOEt, EtOH}}$

b) 2-oxocyclopentane-CO_2Et $\xrightarrow[\text{2) } PhCH_2Br]{\text{1) NaOEt, EtOH}}$

c) $NCCH_2CN$ $\xrightarrow[\text{2) } CH_3CH(CH_3)CH_2CH_2Br]{\text{1) NaH}}$

d) $CH_3CH(CH_3)CH_2CH_2CH(CN)CN$ $\xrightarrow[\text{2) } CH_3I]{\text{1) NaOEt, EtOH}}$

16.8

b) 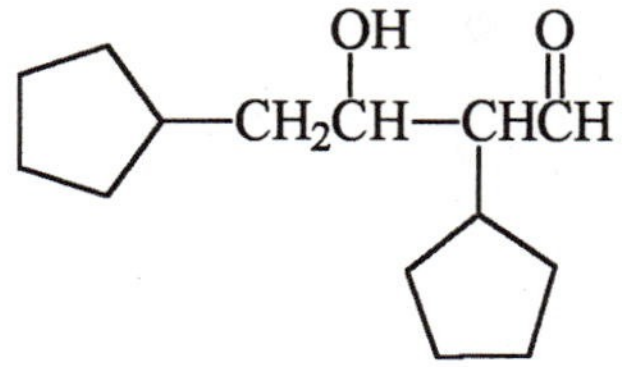 (cyclopentyl)$-CH_2\overset{OH}{CH}-\underset{\text{cyclopentyl}}{CH}\overset{O}{\overset{\|}{C}}H$ c) $CH_3\overset{CH_3}{CH}CH_2CH_2CH{=}\overset{O}{\overset{\|}{C}}H$ with $CH_3\underset{CH_3}{CH}CH_2$ substituent d) 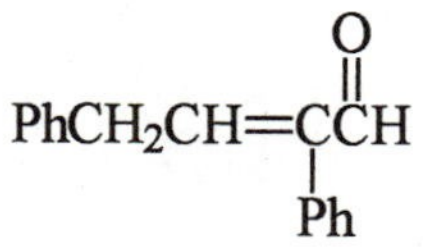 $PhCH_2CH{=}\underset{Ph}{C}\overset{O}{\overset{\|}{C}}H$

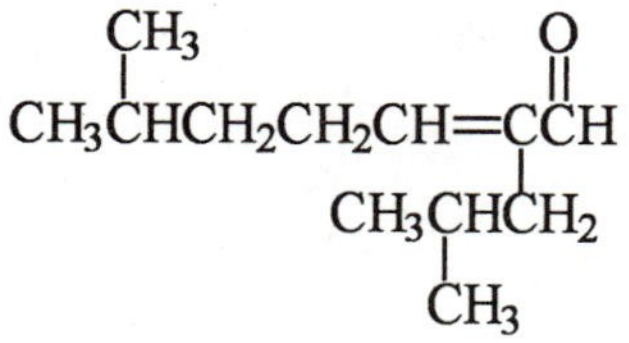

16.9

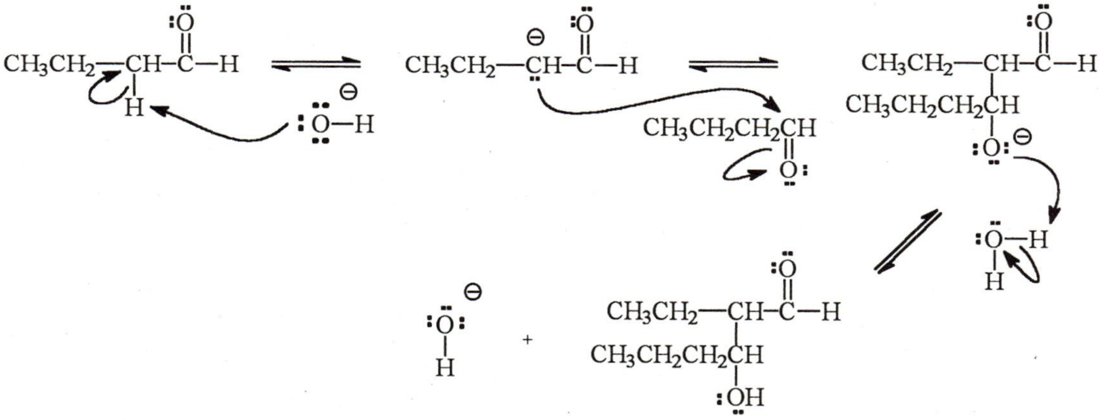

$CH_3CH_2-CH(H)-C(=\ddot{O})-H$ + $:\ddot{O}-H^{\ominus}$ ⇌ $CH_3CH_2-\overset{\ominus}{\ddot{C}}H-C(=\ddot{O})-H$ + $CH_3CH_2CH_2CH{=}\ddot{O}:$ ⇌ $CH_3CH_2-CH(-C(=\ddot{O})-H)-CH(CH_2CH_2CH_3)-\ddot{O}:^{\ominus}$ + $H-\ddot{O}-H$ ⇌ $:\ddot{O}:^{\ominus}-H$ + $CH_3CH_2-CH(-C(=\ddot{O})-H)-CH(CH_2CH_2CH_3)-\ddot{O}H$

16.10

b) 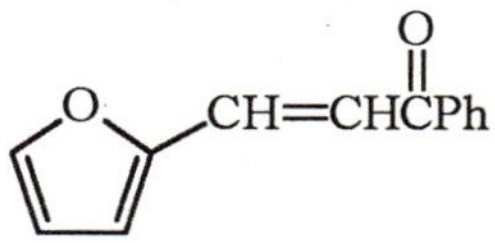 O CH=CH$\overset{O}{\overset{\|}{C}}$Ph c) H_3C CH=C(CH_3)–C(O) d) 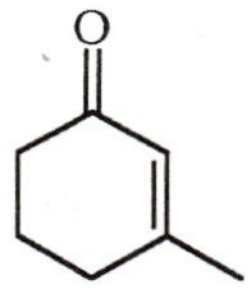 e) 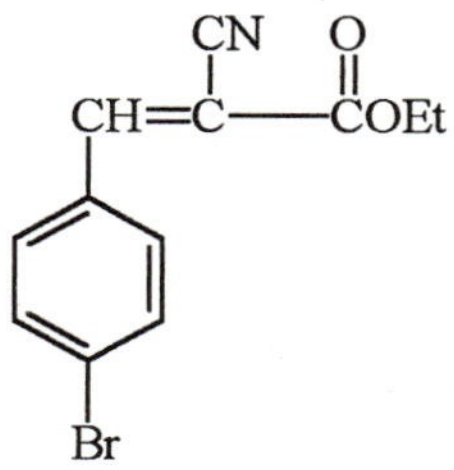 CH=C(CN)–$\overset{O}{\overset{\|}{C}}$OEt, Br f) 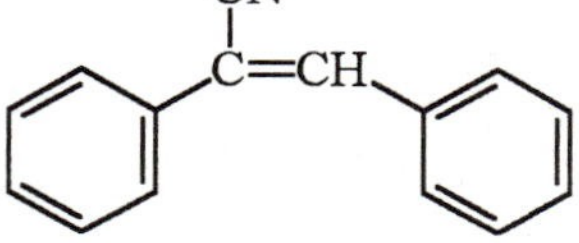 C(CN)=CH

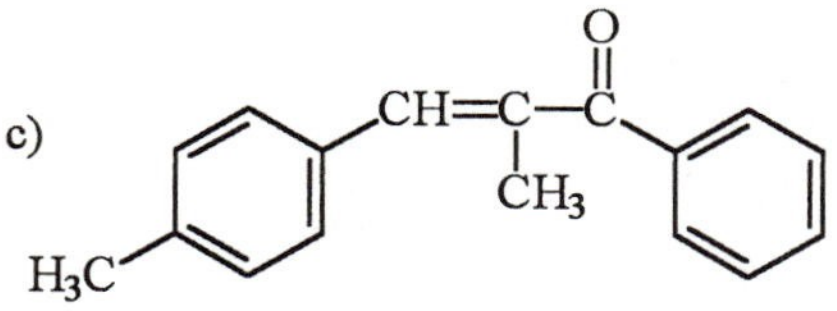

16.11

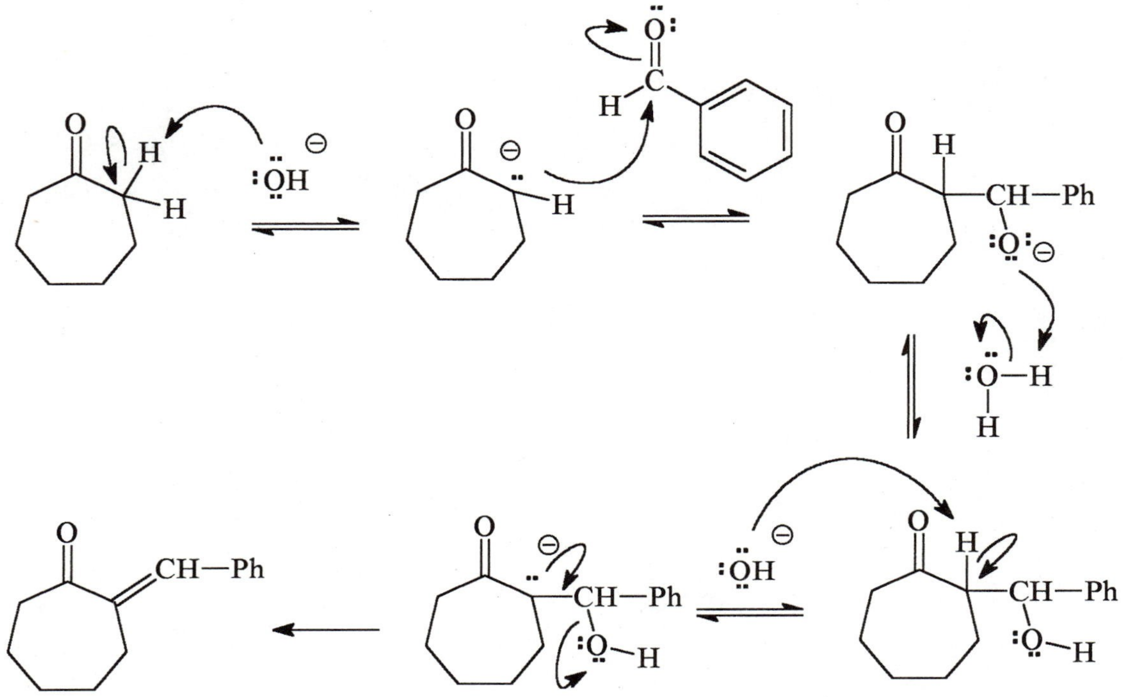

16.12

b) 2 CH_3CHCH_2CH (CH$_3$ on C-2; C=O on C-1) $\xrightarrow{NaOH}$

c) 2 (4-methylbenzaldehyde) + CH_3CCH_3 (C=O) $\xrightarrow[\Delta]{NaOH}$

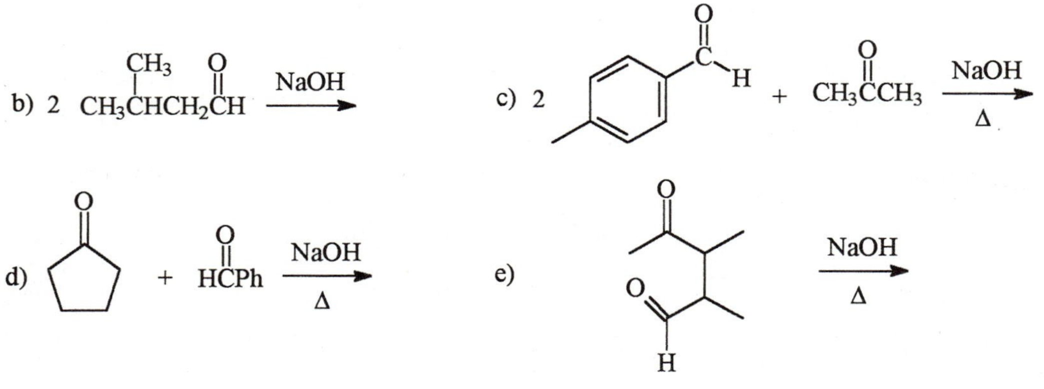

d) (cyclopentanone) + HCPh (C=O) $\xrightarrow[\Delta]{NaOH}$

e) $\xrightarrow[\Delta]{NaOH}$

16.13

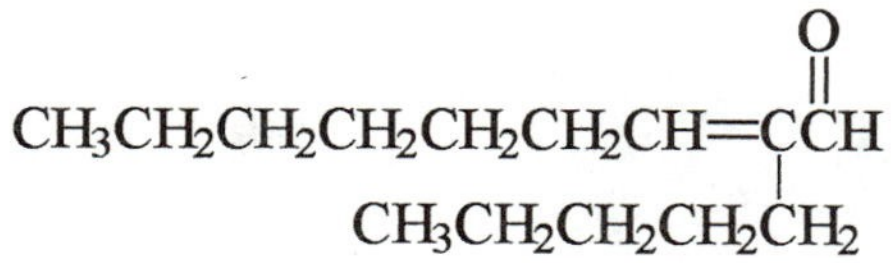

$$CH_3CH_2CH_2CH_2CH_2CH_2CH{=}\underset{\displaystyle CH_2CH_2CH_2CH_2CH_3}{\overset{\displaystyle O \atop \|}{C}}H$$

Use excess benzaldehyde to minimize the amount of this product.

16.14

a) 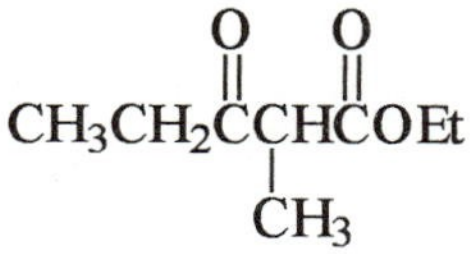 $CH_3CH_2\overset{O}{\overset{\|}{C}}\underset{CH_3}{CH}\overset{O}{\overset{\|}{C}}OEt$

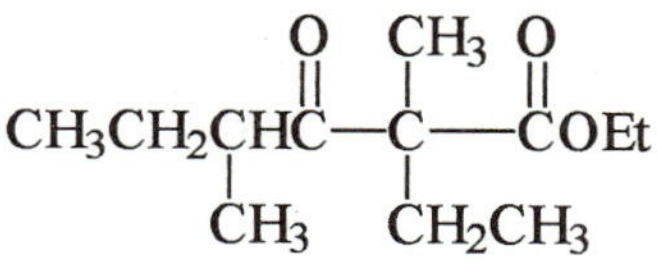

b) $CH_3CH_2\underset{CH_3}{CH}\overset{O}{\overset{\|}{C}}-\overset{CH_3}{\underset{CH_2CH_3}{C}}-\overset{O}{\overset{\|}{C}}OEt$

c) $C_6H_5CH_2\overset{O}{\overset{\|}{C}}-\underset{C_6H_5}{CH}\overset{O}{\overset{\|}{C}}OEt$

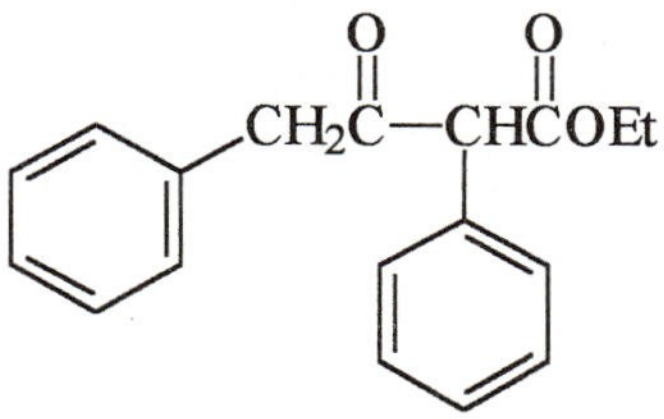

d) 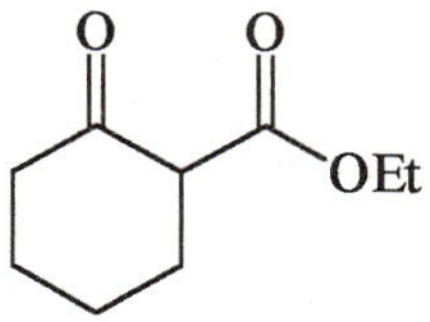 O O OEt

16.15 This product does not form because there is no acidic H between the two carbonyl groups so the equilibrium driving step cannot occur.

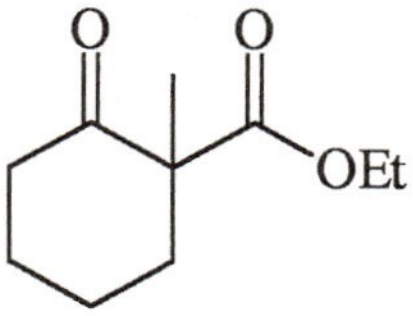

O O OEt

16.16

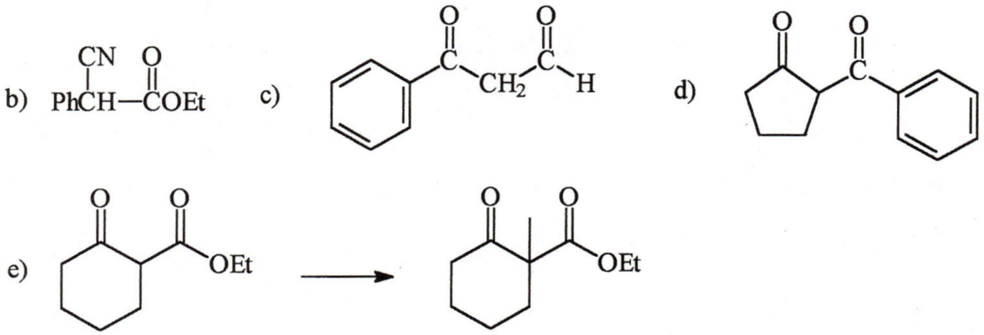

b) $PhCH(CN)-C(=O)OEt$

c) $PhC(=O)-CH_2-C(=O)H$

d)

e) → OEt

16.17

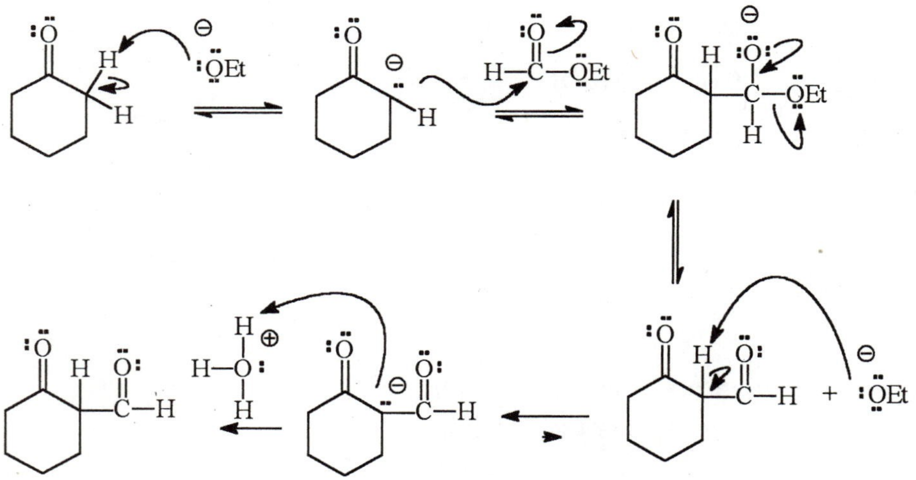

:Ö: H :ÖEt ⇌ :Ö: H ⇌ H–C(=O)–ÖEt ⇌ :Ö: H :Ö: C–ÖEt H

⇌

:Ö: H Ö: C–H H–O: H ⊕ H ← :Ö: Ö: C–H ⇌ :Ö: H Ö: C–H + :ÖEt

16.18

a) 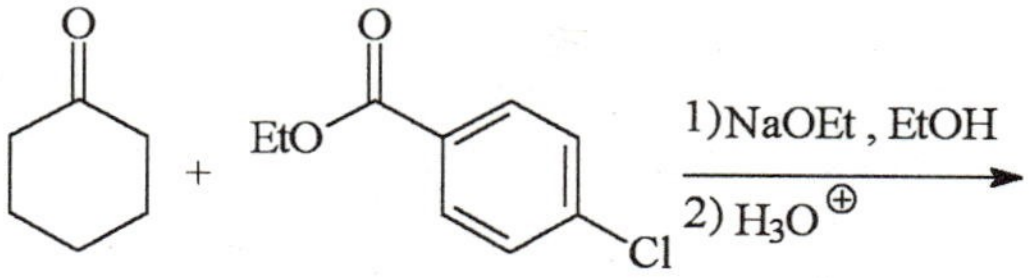 cyclohexanone + EtO–C(=O)–C$_6$H$_4$–Cl $\xrightarrow[2)\ H_3O^{\oplus}]{1)\ NaOEt,\ EtOH}$

b) CH_3COCH_3 + EtO–C(=O)–Ph $\xrightarrow[2)\ H_3O^{\oplus}]{1)\ NaOEt,\ EtOH}$

c) 2-methylacetophenone (CH_3, O) + EtO–C(=O)–H $\xrightarrow[2)\ H_3O^{\oplus}]{1)\ NaOEt,\ EtOH}$

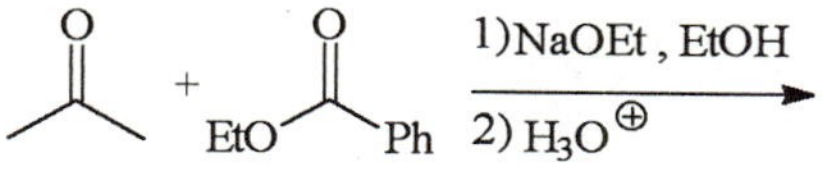

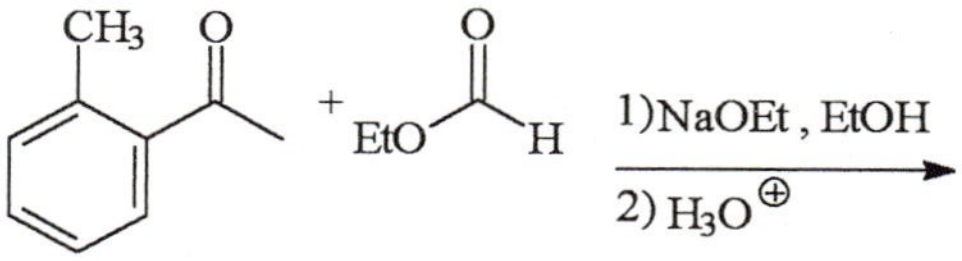

d) 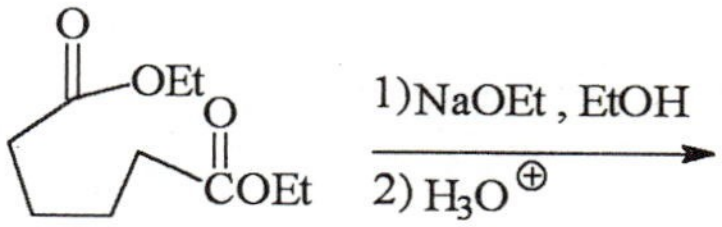 diester (OEt, O, COEt) $\xrightarrow[2)\ H_3O^{\oplus}]{1)\ NaOEt,\ EtOH}$

e) 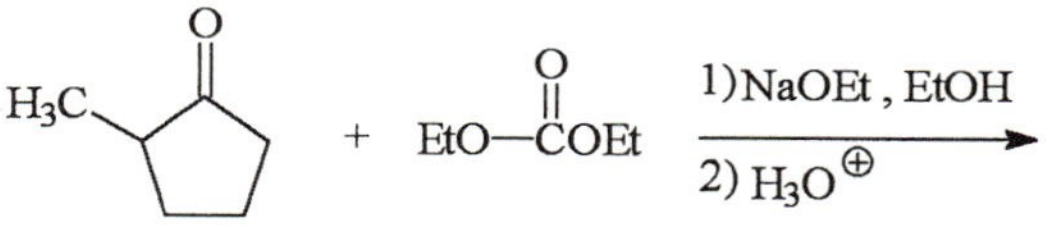 H_3C–cyclopentanone + $EtO{-}\overset{O}{\overset{\|}{C}}OEt$ $\xrightarrow[2)\ H_3O^{\oplus}]{1)\ NaOEt,\ EtOH}$

16.19

a) 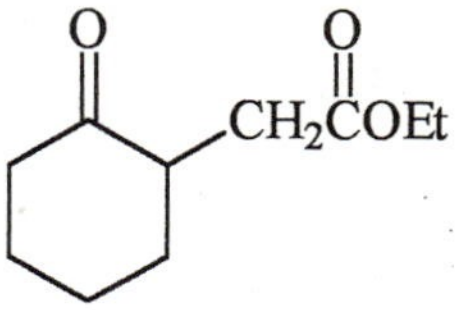 2-($CH_2\overset{O}{\overset{\|}{C}}OEt$)cyclohexanone

b) 2-allyl-1-tetralone

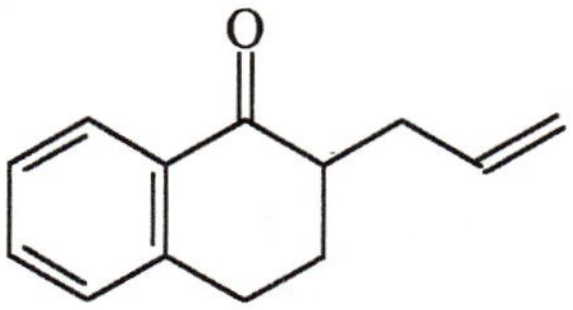

c) 2-propanoylcyclopentanone

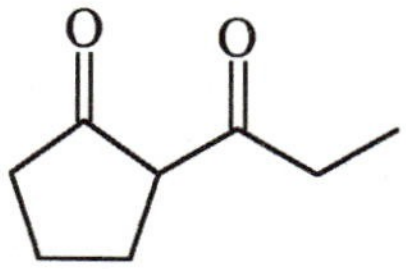

16.20

a) 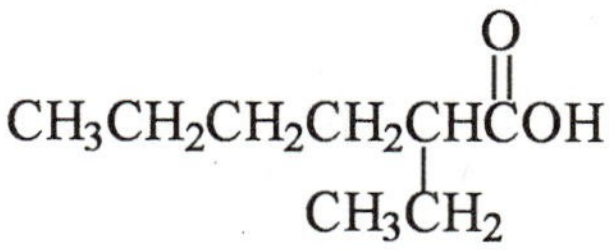 $CH_3CH_2CH_2CH_2\underset{CH_3CH_2}{\underset{|}{C}}H\overset{O}{\overset{\|}{C}}OH$

b) $CH_3CH_2CH_2CH_2\overset{O}{\overset{\|}{C}}\underset{CH_3}{\underset{|}{C}}HCH_3$

c) $CH_3CH_2CH_2\overset{O}{\overset{\|}{C}}CH_2\overset{O}{\overset{\|}{C}}OCH_3$

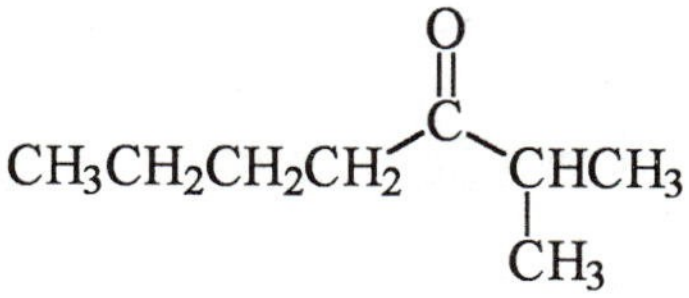

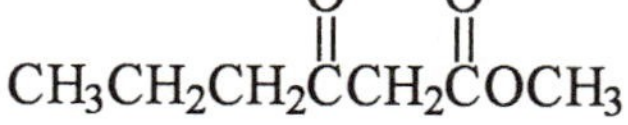

d) 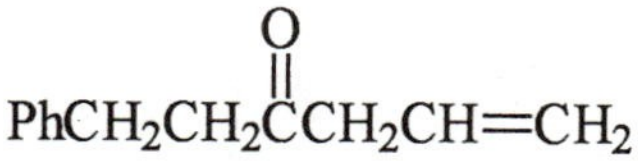 $PhCH_2CH_2\overset{O}{\overset{\|}{C}}CH_2CH{=}CH_2$

16.21

a) 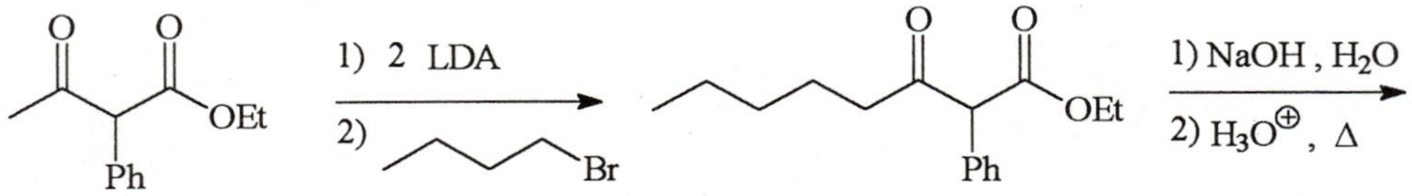 [ethyl 2-phenyl-3-oxobutanoate] $\xrightarrow[\text{2) } CH_3CH_2CH_2CH_2Br]{\text{1) 2 LDA}}$ [ethyl 3-oxo-2-phenyloctanoate] $\xrightarrow[\text{2) } H_3O^{\oplus},\ \Delta]{\text{1) NaOH, } H_2O}$

b) 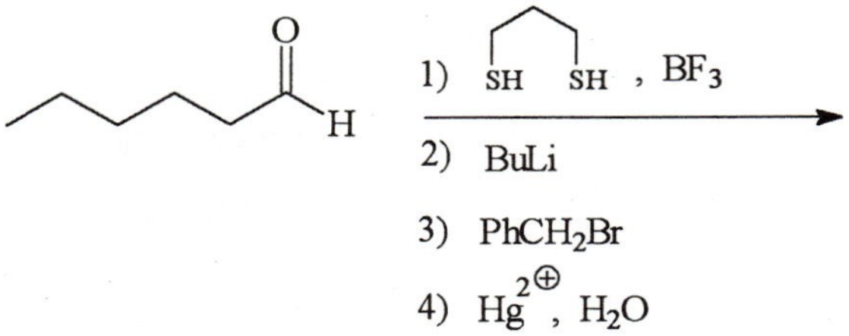 [hexanal] $\xrightarrow{\text{1) } HS(CH_2)_3SH,\ BF_3 \quad \text{2) BuLi} \quad \text{3) } PhCH_2Br \quad \text{4) } Hg^{2\oplus},\ H_2O}$

c) $PhCH_2CO_2H \xrightarrow{\text{1) 2 LDA} \quad \text{2) } CH_3I \quad \text{3) } H_3O^{\oplus}}$

16.22

b) 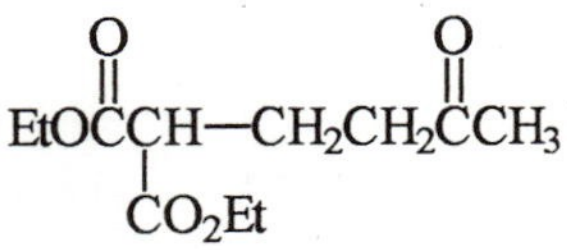 $EtO\overset{O}{\overset{\|}{C}}\underset{CO_2Et}{\underset{|}{C}}H{-}CH_2CH_2\overset{O}{\overset{\|}{C}}CH_3$

c) $Ph\underset{EtO_2C\underset{|}{C}HCN}{C}HCH_2\overset{O}{\overset{\|}{C}}Ph$

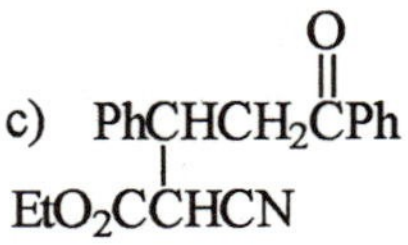

d) 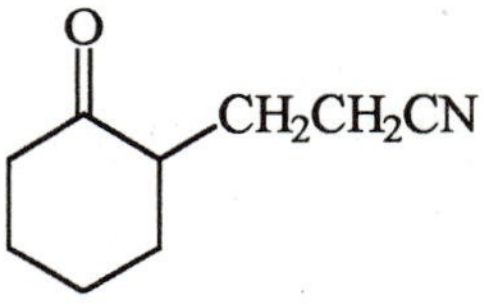 2-(CH_2CH_2CN)cyclohexanone

16.23 The hydrogen on the α-carbon bonded to the phenyl group is more acidic because the resulting enolate anion is stabilized by resonance with the phenyl group.

16.24

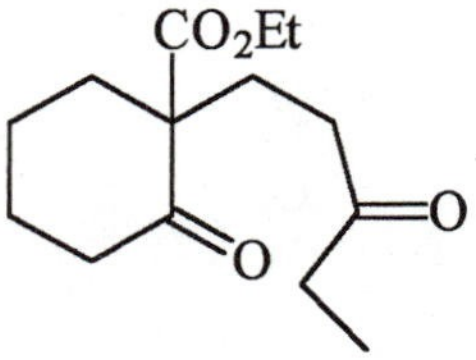

[2-oxocyclohexane with CO_2Et and $CH_2CH_2C(=O)CH_2CH_3$ on C-1]

16.25

b) 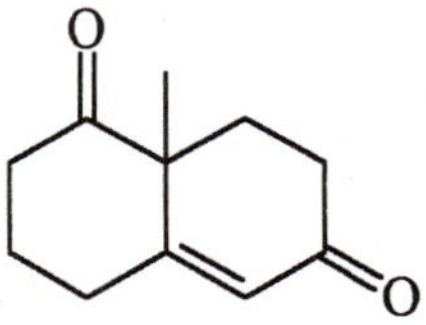 [bicyclic enedione structure]

16.26

a) $CH_3\overset{O}{\overset{\|}{C}}CH_2\overset{O}{\overset{\|}{C}}OEt \xrightarrow[2)\ CH_3CH_2CH_2Br]{1)\ NaOEt,\ EtOH} CH_3\overset{O}{\overset{\|}{C}}\underset{CH_2CH_2CH_3}{\underset{|}{C}H}\overset{O}{\overset{\|}{C}}OEt$

1) NaOEt, EtOH
2) $CH_2{=}CHCH_2Br$

$CH_3\overset{O}{\overset{\|}{C}}C(CH_2{=}CHCH_2)(CH_2CH_2CH_3)\overset{O}{\overset{\|}{C}}OEt$

1) NaOH, H_2O
2) $H_3O^{\oplus}$, Δ

$CH_3CH_2CH_2\underset{}{C}H(CH_2{=}CHCH_2)-\overset{O}{\overset{\|}{C}}CH_3$

b) 2 butanal (H) → NaOH, Δ → 2-ethylhex-2-enal (H) → H_2, Pt → 2-ethylhexanal (H)

c) ethyl butanoate (OEt) → 1) LDA; 2) $PhCH_2Br$ → ethyl 2-benzylbutanoate (Ph, OEt) → $H_3O^{\oplus}$, Δ → 2-benzylbutanoic acid (Ph, OH)

d) butanal (H) →
1) $HS(CH_2)_3SH$ (SH SH), BF_3
2) BuLi
3) 1-bromo-2-methylpropane (Br)
4) $Hg^{2\oplus}$, H_2O

e) O (cyclopentanone) — 1) CH_3OCOCH_3 (C=O), $NaOCH_3$, CH_3OH; 2) $H_3O^{\oplus}$ → O, O, C—OCH_3 — 1) 2 LDA; 2) CH_3I; 3) $H_3O^{\oplus}$; 4) $NaOCH_3$; 5) CH_3I → O, O, C—OCH_3

f) O (cyclohexanone) + 2 PhCH (O) — NaOH, Δ → PhCH, O, CHPh

g) EtOCCH$_2$COEt (O, O) — 1) NaOEt, EtOH; 2) $CH_3CH_2CH_2Br$; 3) NaOEt, EtOH; 4) $PhCH_2Cl$ → EtOC—C—COEt (O, O), $PhCH_2$, $CH_2CH_2CH_3$

↓ 1) NaOH, H_2O; 2) $H_3O^{\oplus}$, Δ

$CH_3CH_2CH_2CHCOH$ (O), $PhCH_2$ — 1) $SOCl_2$; 2) excess NH_3 → $CH_3CH_2CH_2CHCNH_2$ (O), $PhCH_2$

h) O (cyclopentanone) + CH_3OCOCH_3 (O) — 1) $NaOCH_3$, CH_3OH; 2) $H_3O^{\oplus}$ → O, O, C—OCH_3

↓ 1) 2 LDA; 2) CH_3I; 3) $H_3O^{\oplus}$

O, O, C—OCH_3

16.27

a) 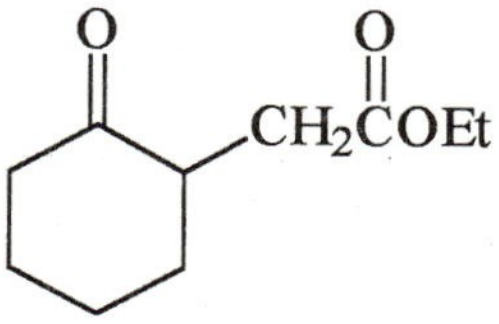 2-(CH_2COOEt)cyclohexanone: $CH_2\overset{O}{\overset{\|}{C}}OEt$

b) 2-ethyl-2-($\overset{O}{\overset{\|}{C}}OEt$)cyclopentanone: CH_3, CH_2, $\overset{O}{\overset{\|}{C}}OEt$

c) cyclopentyl–$\overset{O}{\overset{\|}{C}}OEt$

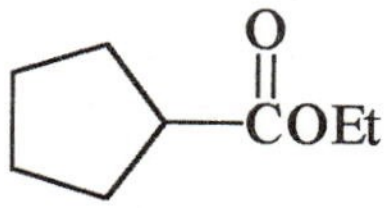

d) 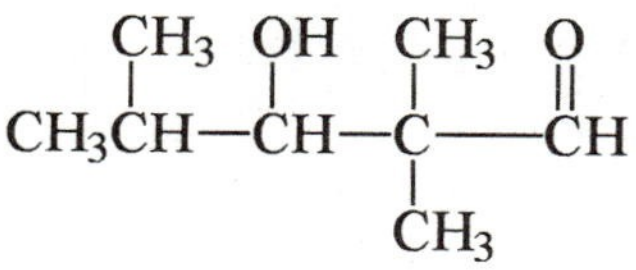 $CH_3\overset{CH_3}{\overset{|}{C}H}-\overset{OH}{\overset{|}{C}H}-\underset{CH_3}{\underset{|}{\overset{CH_3}{\overset{|}{C}}}}-\overset{O}{\overset{\|}{C}H}$

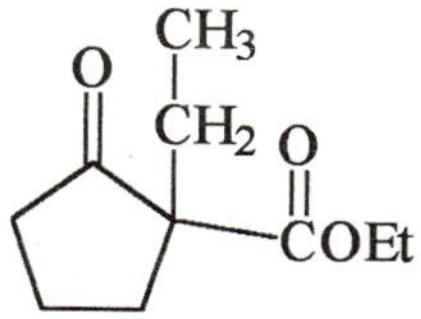

e) 2-methyl-5-($\overset{O}{\overset{\|}{C}}OEt$)cyclopentanone

f) $CH_3CH_2\overset{O}{\overset{\|}{C}}CH_2CH_2CH_2Ph$

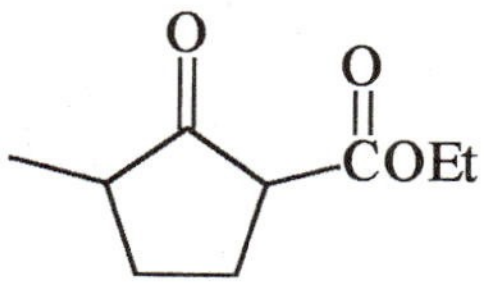

g) $Ph\underset{PhCH_2CH_2}{\underset{|}{C}H}\overset{O}{\overset{\|}{C}}OEt$

h) 1-acetylcyclopentene

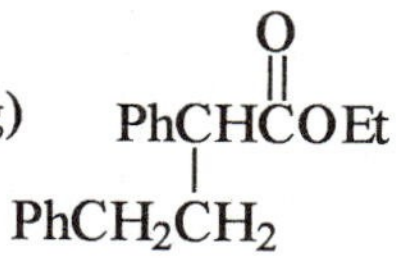

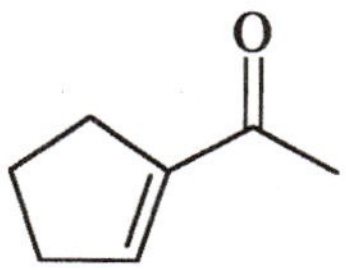

i) 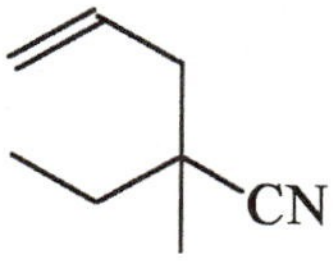 CN

j) 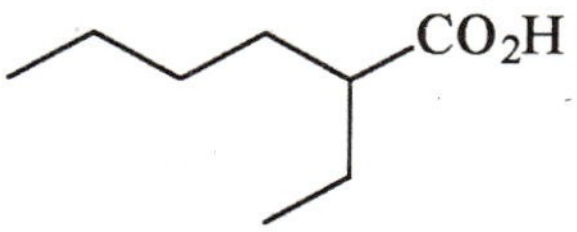 CO_2H

k) 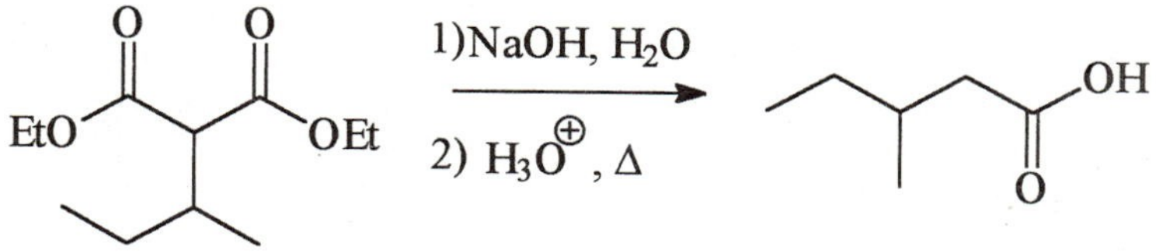 EtO, OEt $\xrightarrow[\text{2) } H_3O^{\oplus}, \Delta]{\text{1)NaOH, } H_2O}$ OH

l) 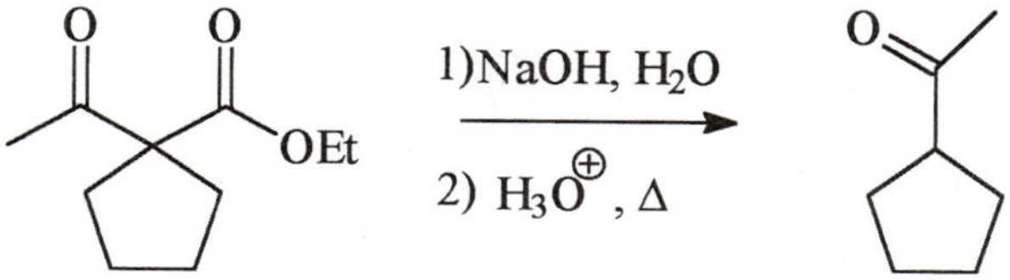 OEt $\xrightarrow[\text{2) } H_3O^{\oplus}, \Delta]{\text{1)NaOH, } H_2O}$ acetylcyclopentane

m)

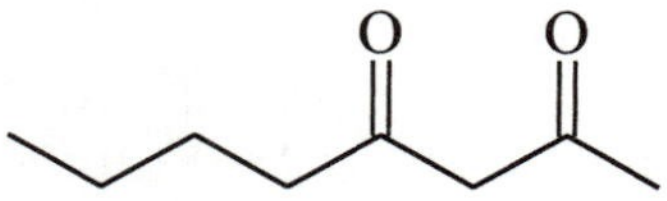

n)

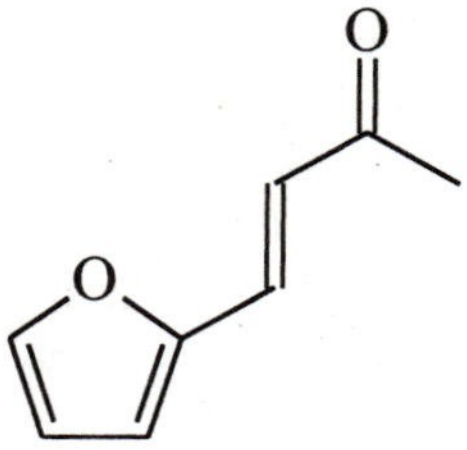

o) 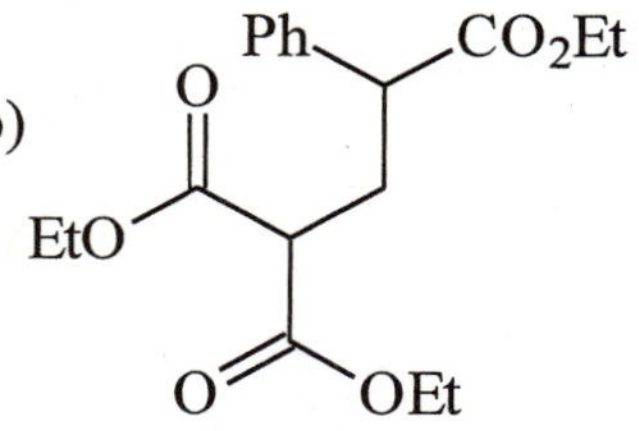 Ph, CO_2Et, EtO, O, O, OEt

p) 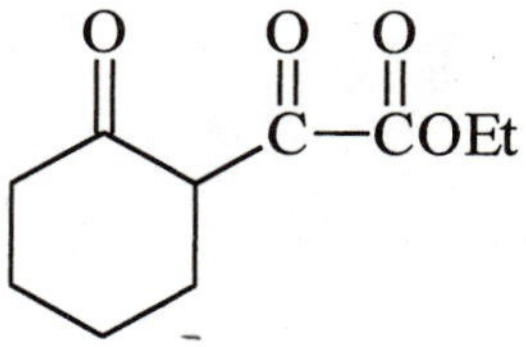 O, O, O, $\overset{\|}{C}$—$\overset{\|}{C}$OEt

16.28

a) 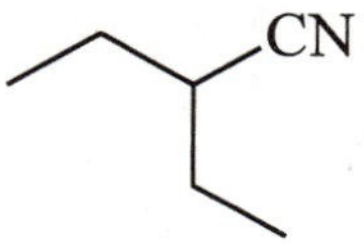 CN

b) 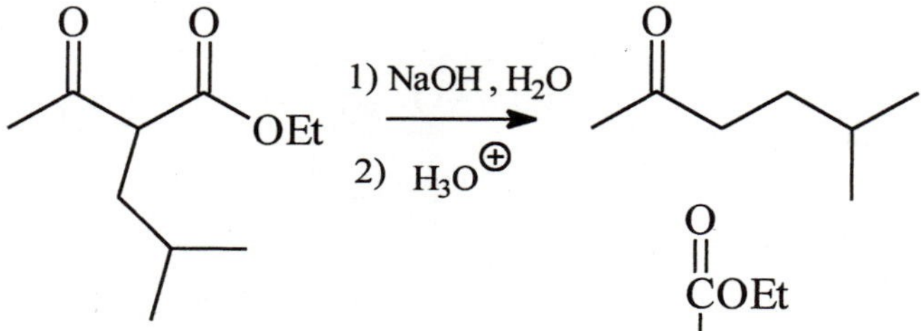 O, O, OEt

1) NaOH , H_2O

2) $H_3O^{\oplus}$

c) 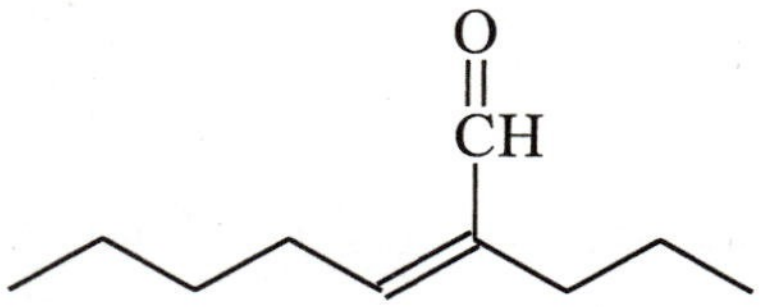 O, CH

d) 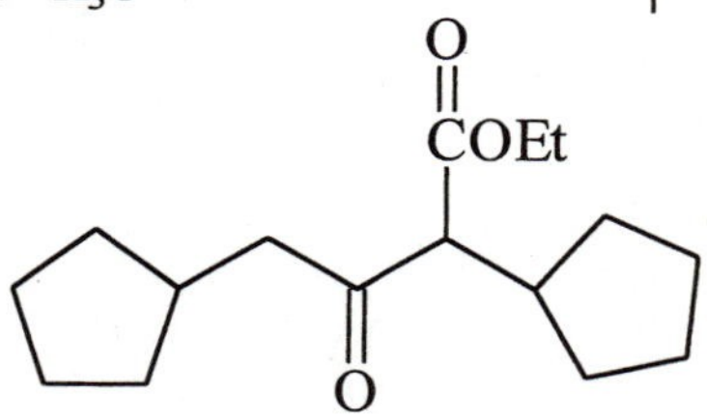 O, COEt, O

e) 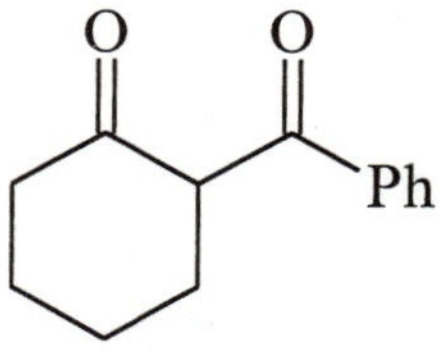 O, O, Ph

f) 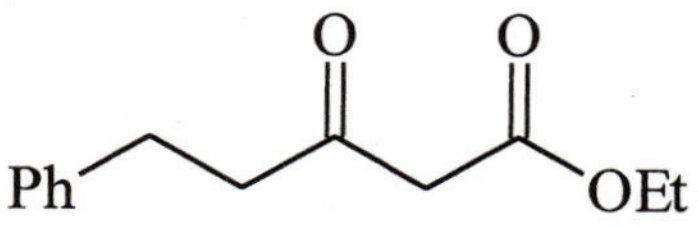 Ph, O, O, OEt

g) 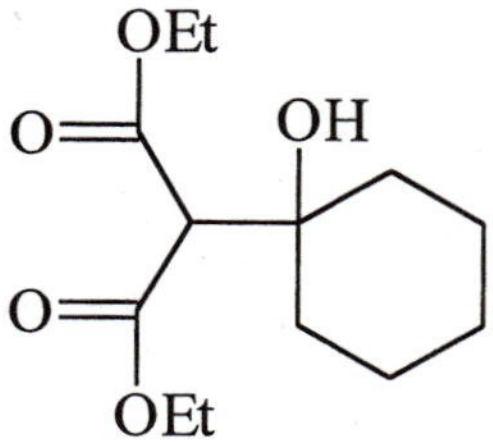 OEt, O, OH, O, OEt

16.29

a) 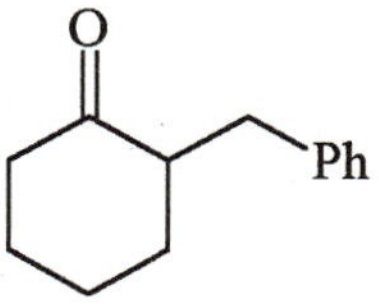 Ph

b) 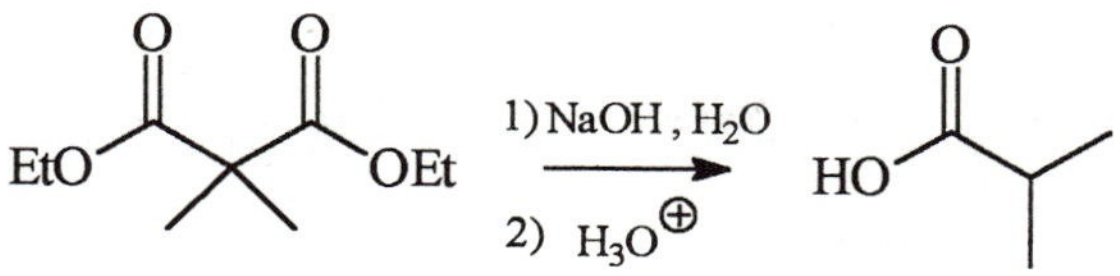 EtO OEt → 1) NaOH, H_2O 2) $H_3O^{\oplus}$ → HO

c)

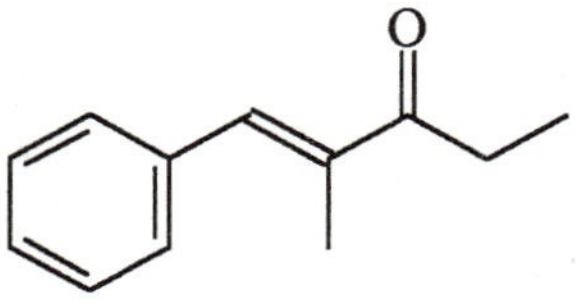

d) 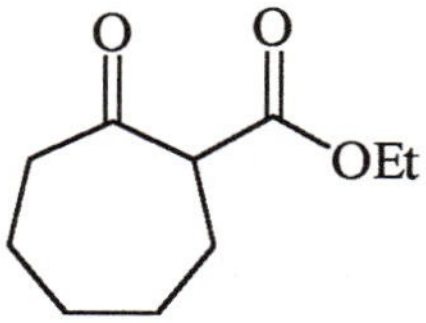 OEt

e) 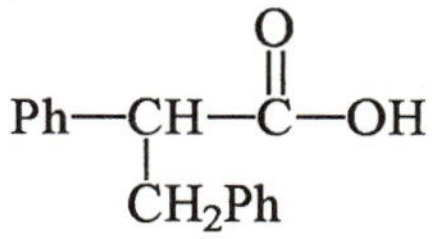 $Ph—CH(CH_2Ph)—C(=O)—OH$

f) 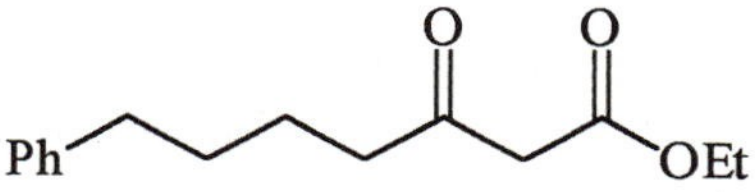 Ph OEt

16.30

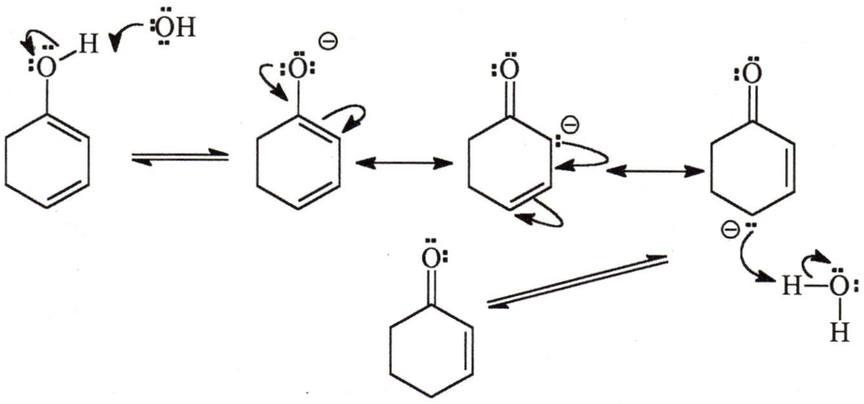

16.31

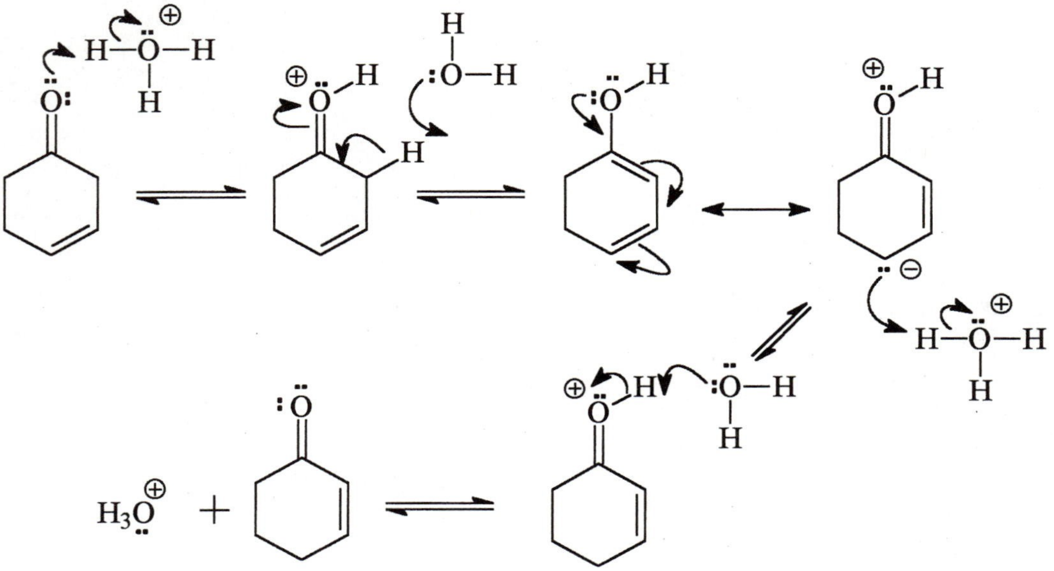

H–Ö–H (⊕), H
:Ö:
⊕Ö–H, H, :Ö–H, H
:Ö–H
⊕Ö–H
⊖
H–Ö–H (⊕), H
⊕Ö, H, :Ö–H, H
:Ö
$H_3O^{\oplus}$ +

16.32

:Ö
CH_3CH_2–CH–C–ÖEt, H, ⊖ :Ö–Et
⊖ :Ö
CH_3CH_2–CH–C–ÖEt
$CH_3CH_2CH_2C$–OEt, Ö:
:Ö
CH_3CH_2–CH–C–ÖEt
$CH_3CH_2CH_2C$–ÖEt
:Ö:⊖

H :Ö
CH_3CH_2–C–C–ÖEt
$CH_3CH_2CH_2C$
:Ö

$H_3O^{\oplus}$

⊖ :Ö
CH_3CH_2–C–C–ÖEt
$CH_3CH_2CH_2C$
:Ö

⊖ :Ö:
Et
+

H :Ö
CH_3CH_2–C–C–ÖEt
$CH_3CH_2CH_2C$
:Ö

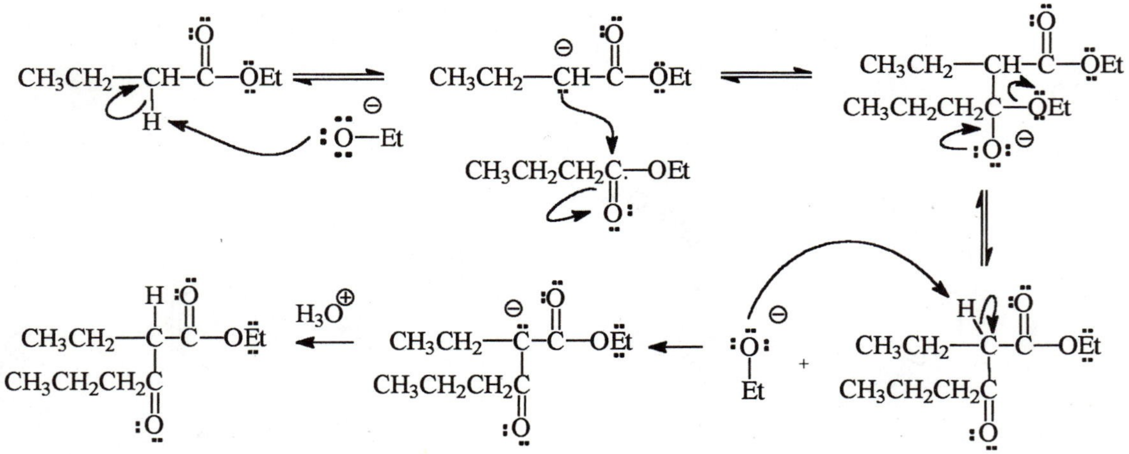

16.33

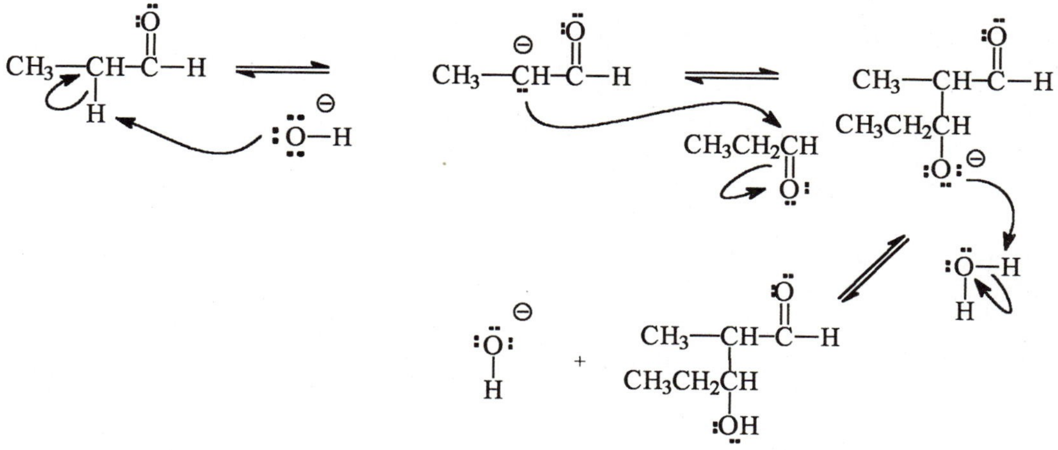

CH₃—CH—C(=O)—H ⇌ CH₃—CH⁻—C(=O)—H ⇌ CH₃—CH(CH(O⁻)CH₂CH₃)—C(=O)—H ⇌ OH^- + CH₃—CH(CH(OH)CH₂CH₃)—C(=O)—H

CH_3CH_2CH

$:\ddot{O}-H$

H

16.34

a) 2 CH₃CH₂CHO $\xrightarrow[H_2O,\ \Delta]{NaOH}$ CH₃CH₂CH=C(CH₃)CHO $\xrightarrow[Pt]{H_2}$

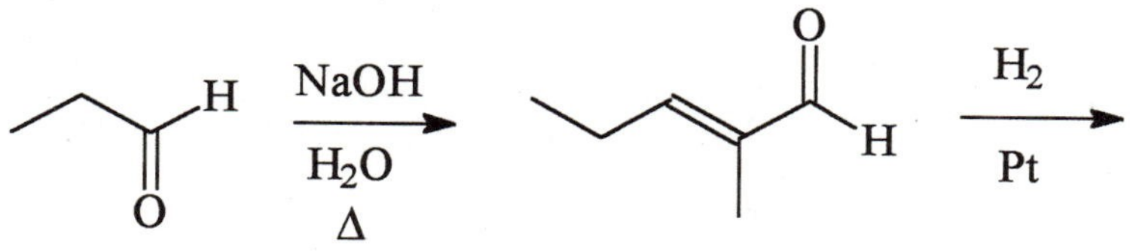

b) as (a) → CH₃CH₂CH=C(CH₃)CHO $\xrightarrow[2)\ H_3O^{\oplus}]{1)\ NaBH_4}$

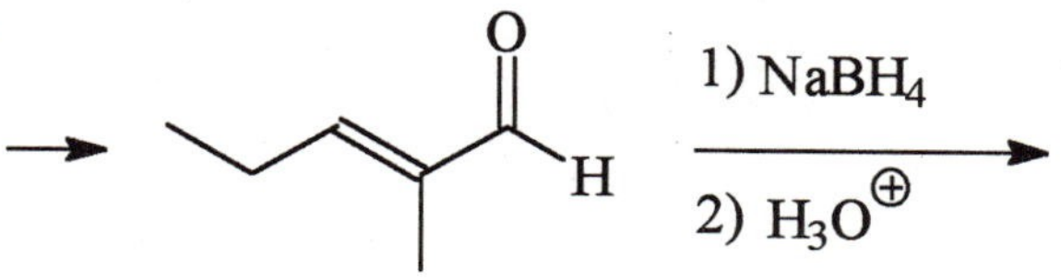

c) as (a) → CH₃CH₂CH₂CH(CH₃)CHO $\xrightarrow[2)\ H_3O^{\oplus}]{1)\ NaBH_4}$

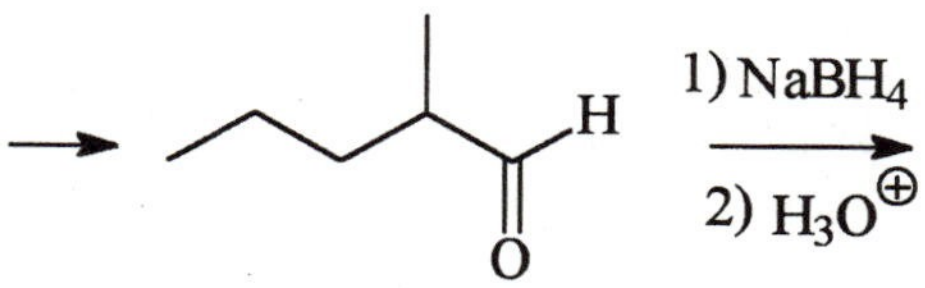

d) PhCHO + CH₃CH₂CHO $\xrightarrow[H_2O,\ \Delta]{NaOH}$ PhCH=C(CH₃)CHO $\xrightarrow[Pd]{H_2}$ PhCH₂CH(CH₃)CHO

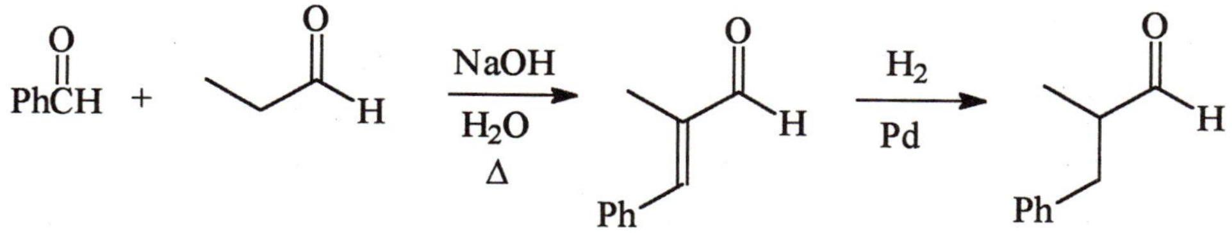

16.35

a) 2

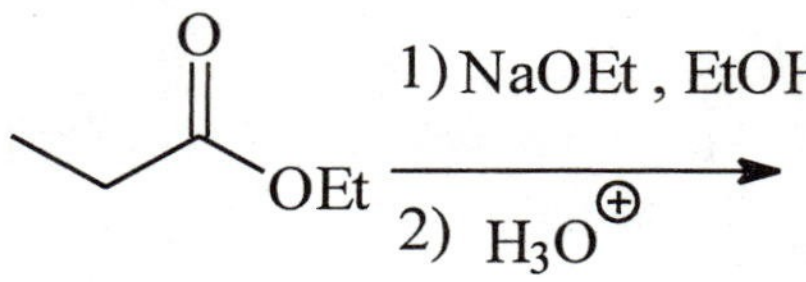

O

OEt

1) NaOEt, EtOH

2) $H_3O^{\oplus}$

b) as (a) →

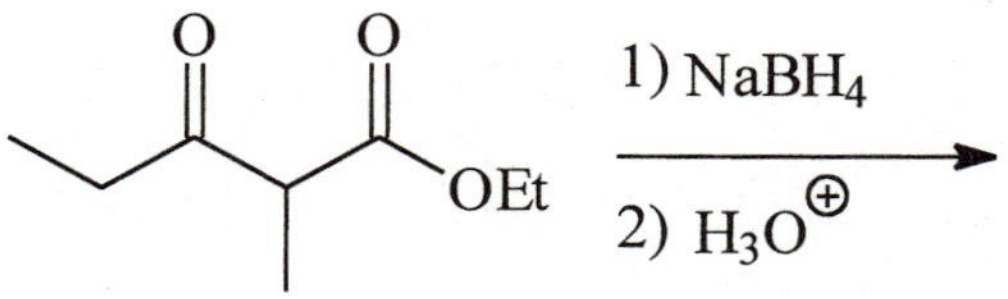

O O

OEt

1) $NaBH_4$

2) $H_3O^{\oplus}$

c) as (a) →

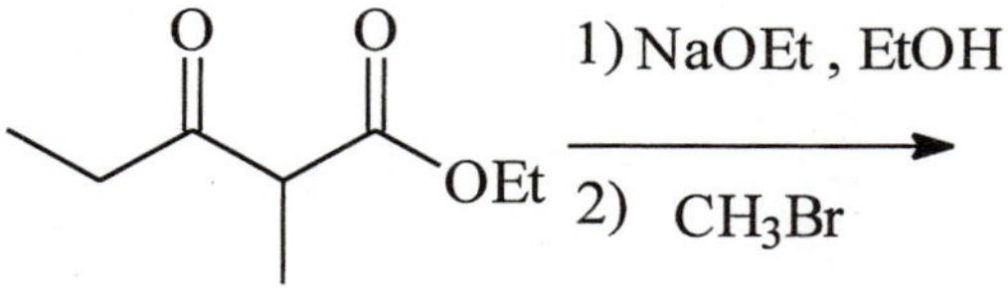

O O

OEt

1) NaOEt, EtOH

2) CH_3Br

d) as (a) →

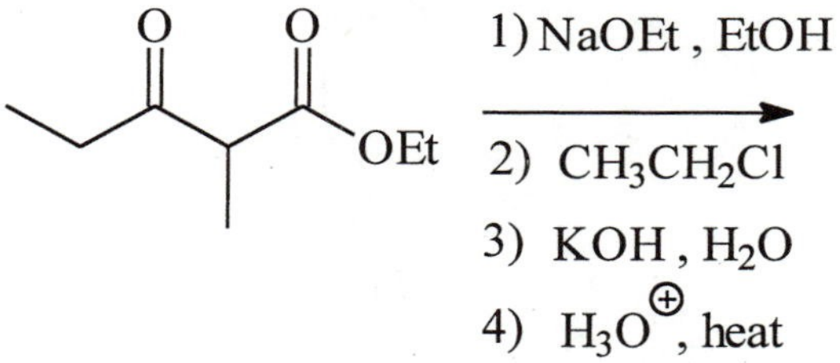

O O

OEt

1) NaOEt, EtOH

2) CH_3CH_2Cl

3) KOH, H_2O

4) $H_3O^{\oplus}$, heat

e) as (a) →

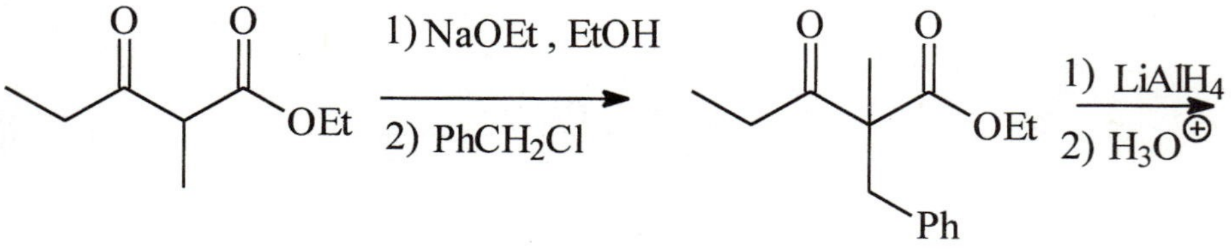

O O

OEt

1) NaOEt, EtOH

2) $PhCH_2Cl$

O O

OEt

Ph

1) $LiAlH_4$

2) $H_3O^{\oplus}$

f) as (a) →

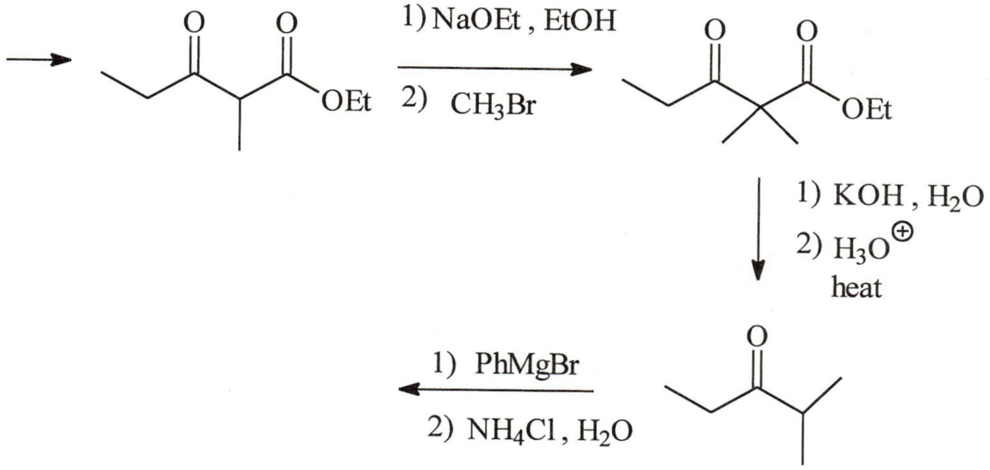

O O
OEt

1) NaOEt, EtOH
2) CH_3Br

O O
OEt

1) KOH, H_2O
2) $H_3O^{\oplus}$
heat

O

1) PhMgBr
2) NH_4Cl, H_2O

g)

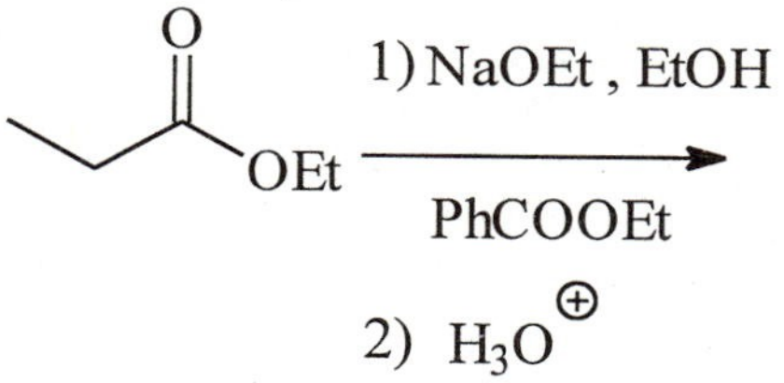

O
OEt

1) NaOEt, EtOH
PhCOOEt
2) $H_3O^{\oplus}$

16.36

a)

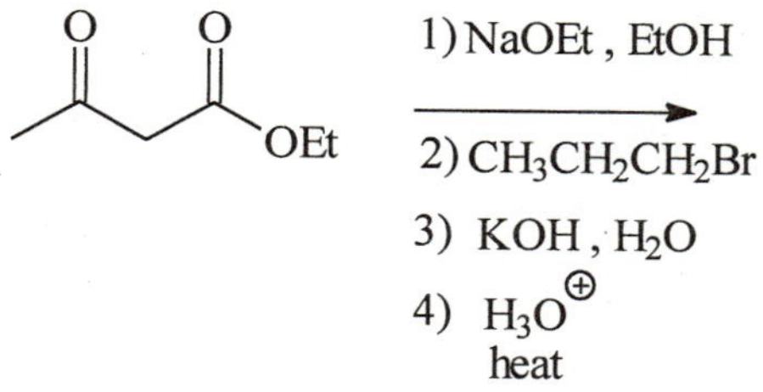

O O
OEt

1) NaOEt, EtOH
2) $CH_3CH_2CH_2Br$
3) KOH, H_2O
4) $H_3O^{\oplus}$
heat

b)

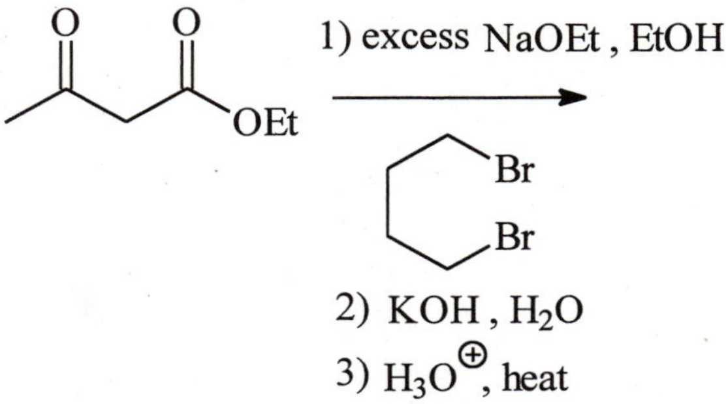

1) excess NaOEt , EtOH

Br(CH$_2$)$_4$Br

2) KOH , H_2O

3) $H_3O^{\oplus}$, heat

c)

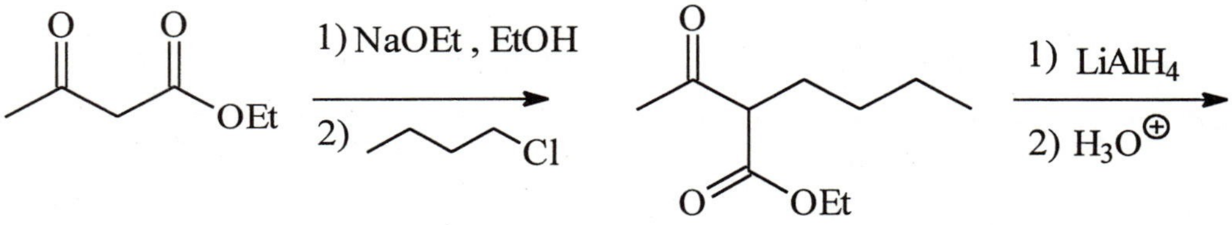

1) NaOEt , EtOH

2) Cl (1-chlorobutane)

1) $LiAlH_4$

2) $H_3O^{\oplus}$

d)

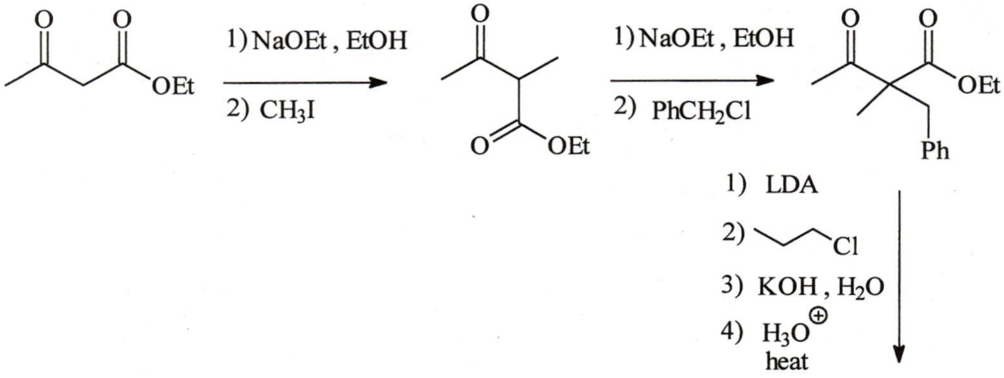

1) NaOEt , EtOH

2) CH_3I

1) NaOEt , EtOH

2) $PhCH_2Cl$

Ph

1) LDA

2) Cl (1-chloropropane)

3) KOH , H_2O

4) $H_3O^{\oplus}$
heat

16.37

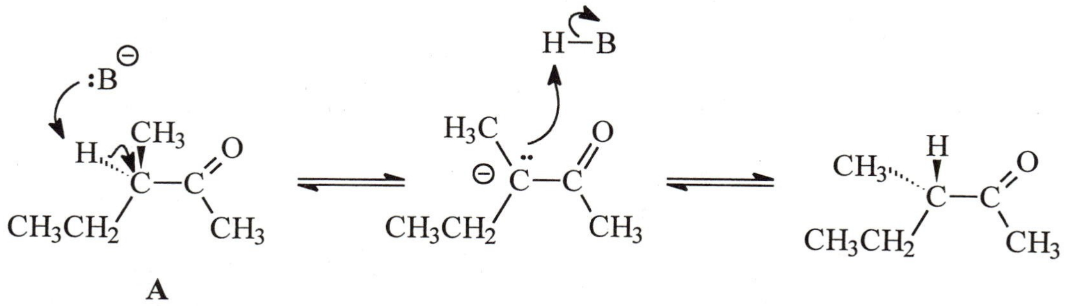

:B⊖ H CH$_3$ C—C O CH$_3$CH$_2$ CH$_3$ ⇌ H—B H$_3$C ⊖ C—C O CH$_3$CH$_2$ CH$_3$ ⇌ CH$_3$ H C—C O CH$_3$CH$_2$ CH$_3$

A

The planar enolate ion can be protonated from either side. **B** does not racemize because the chiral center is not the α-carbon.

16.38

a) 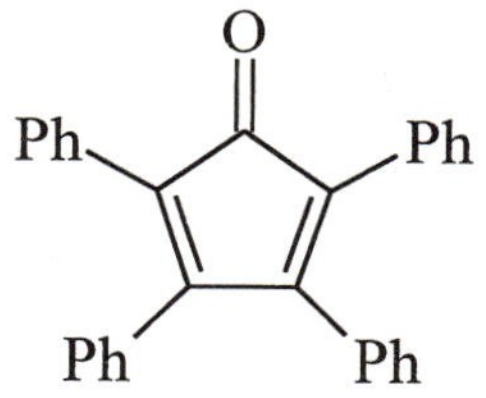 Ph Ph Ph Ph O

b) 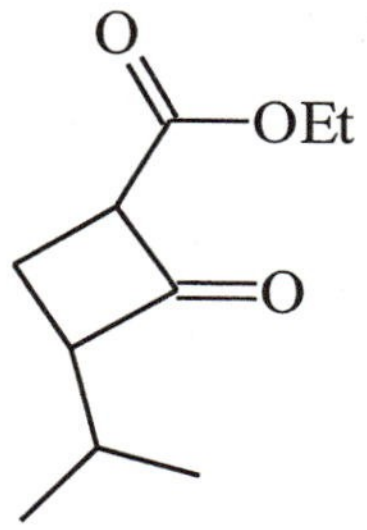 O OEt O

c) 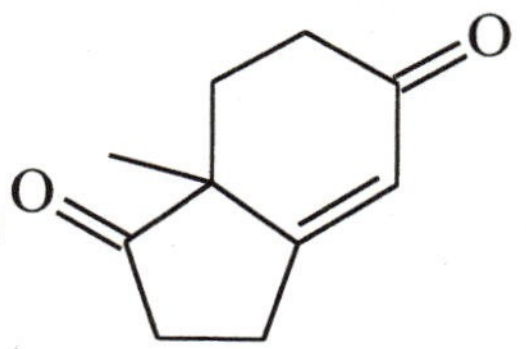 O O

d) 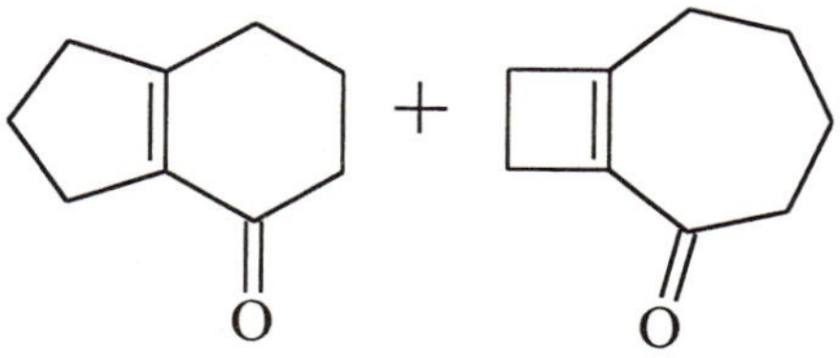 O + O

e) 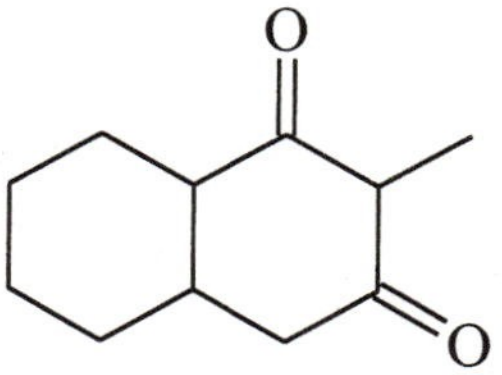 O O

f) 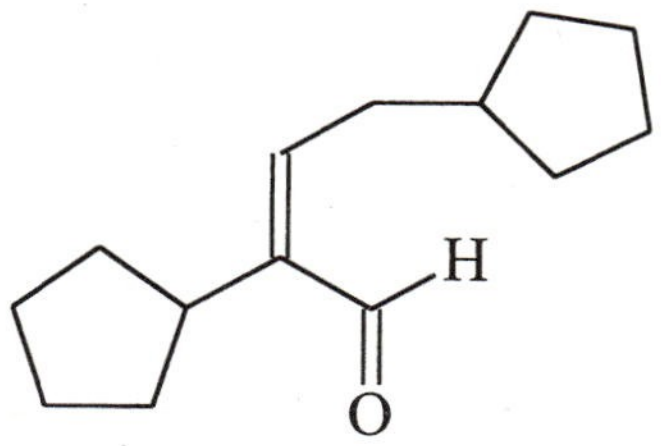 H O

g) 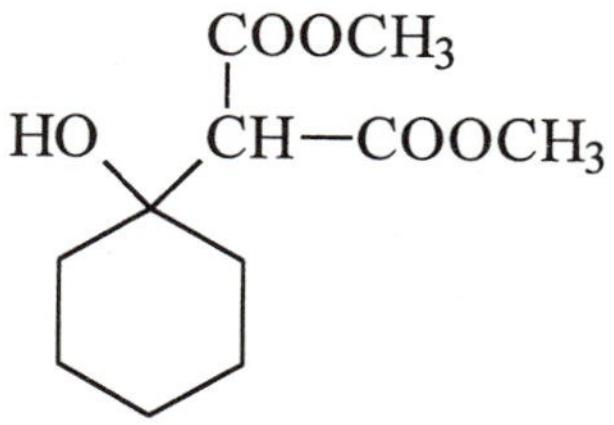 COOCH$_3$ HO CH—COOCH$_3$

16.39

a)

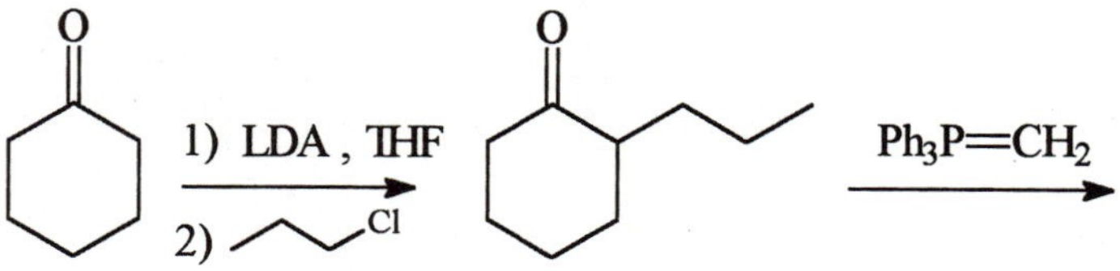

1) LDA, THF

2) Cl

$Ph_3P{=}CH_2$

b)

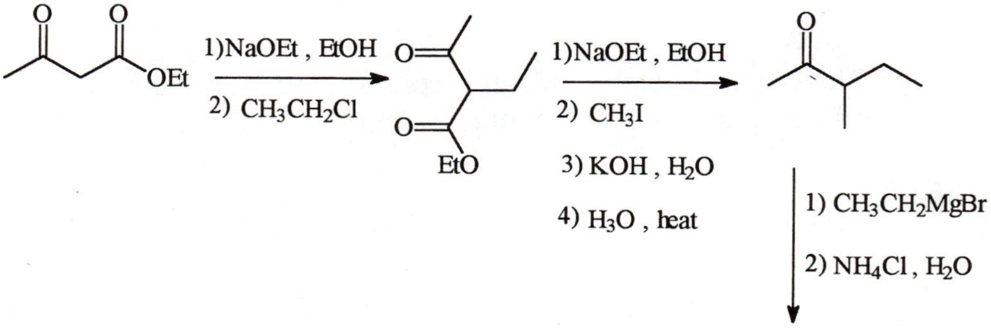

OEt

1) NaOEt, EtOH

2) CH_3CH_2Cl

EtO

1) NaOEt, EtOH

2) CH_3I

3) KOH, H_2O

4) H_3O, heat

1) CH_3CH_2MgBr

2) NH_4Cl, H_2O

c)

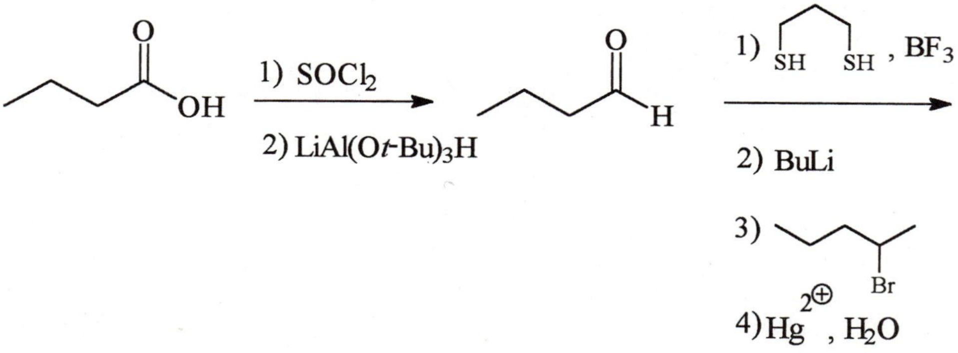

OH

1) $SOCl_2$

2) $LiAl(O\text{-}t\text{-}Bu)_3H$

H

1) SH SH, BF_3

2) BuLi

3) Br

4) $Hg^{2\oplus}$, H_2O

d)

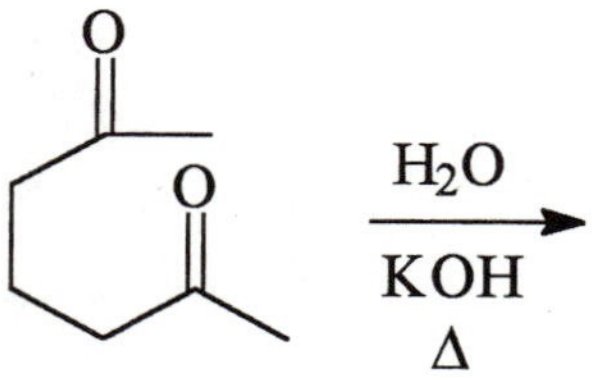

H_2O

KOH

Δ

e) O O → 1) NaOEt 2) Cl → O O → 1) 2 LDA, THF 2) $PhCH_2Cl$ 3) $H_3O^{\oplus}$

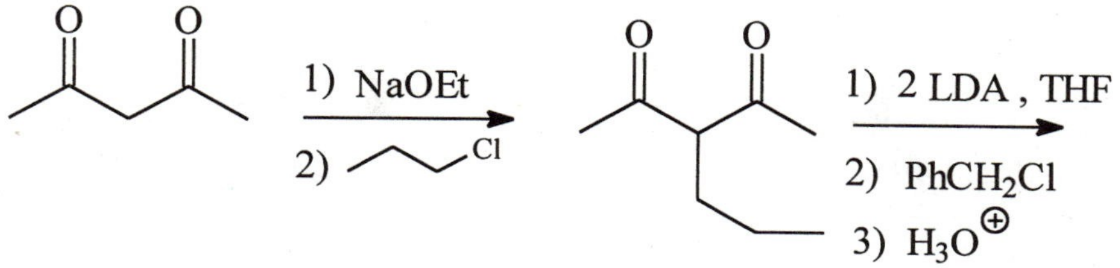

f) OH O → 1) $SOCl_2$ 2) EtOH → OEt O → 1) LDA 2) CH_3CH_2Cl → OEt O

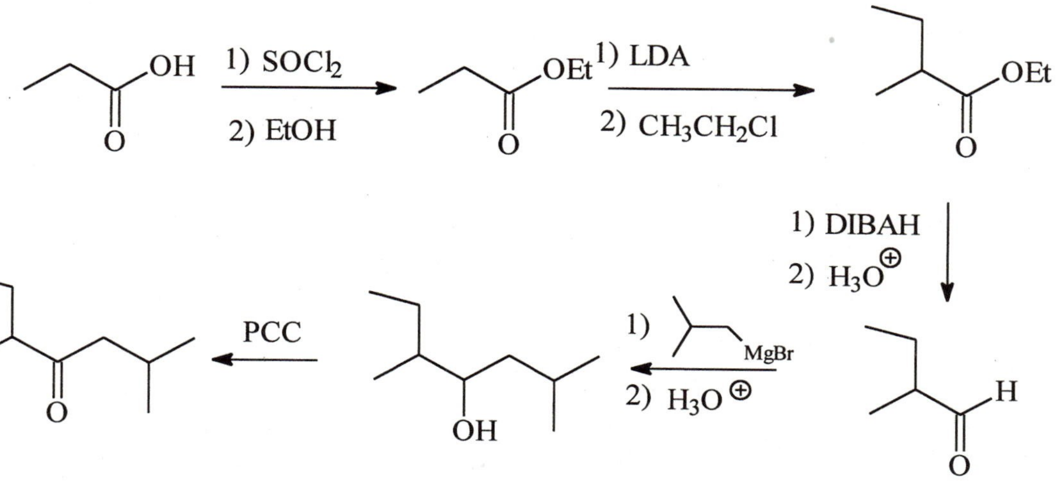

1) DIBAH 2) $H_3O^{\oplus}$ → H O

1) MgBr 2) $H_3O^{\oplus}$ → OH

PCC → O

g) EtO O O OEt → 1)NaOEt, EtOH 2) O Ph → OEt O O OEt O Ph → 1)NaOEt, EtOH 2) CH_3I → OEt O O OEt O Ph

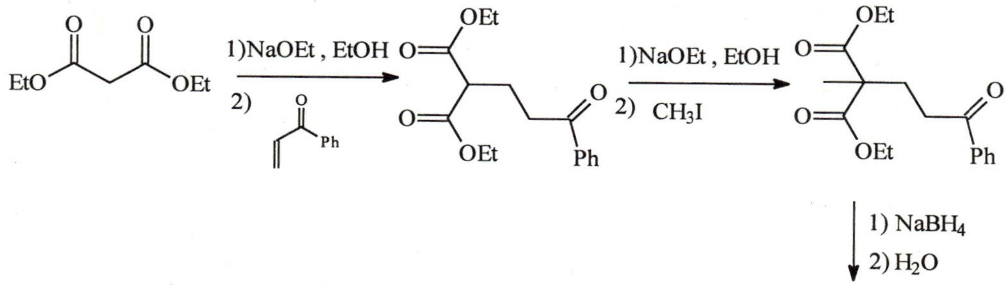

1) $NaBH_4$ 2) H_2O

16.40

a) EtOC(=O)–$CH_2CH_2CH_2$–C(=O)OEt $\xrightarrow{\text{1) NaOEt, EtOH; 2) } H_3O^{\oplus}}$ 2-(ethoxycarbonyl)cyclopentanone (O, O, OEt) $\xrightarrow{\text{1) NaOEt; 2) allyl chloride (Cl); 3) NaOH, } H_2O\text{; 4) } H_3O^{\oplus}, \Delta}$

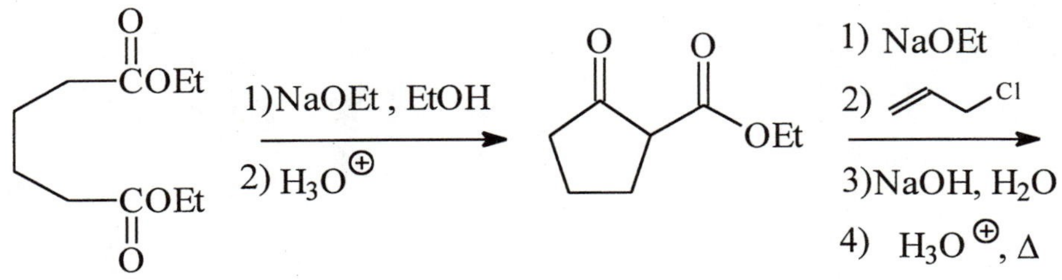

1)NaOEt , EtOH

2) $H_3O^{\oplus}$

1) NaOEt

2) Cl

3)NaOH, H_2O

4) $H_3O^{\oplus}$, Δ

b) cyclopentanone (O)

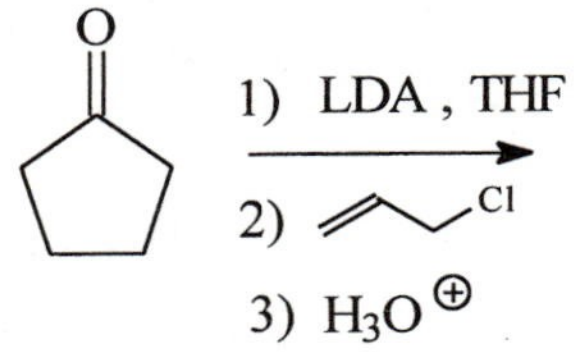

1) LDA , THF

2) Cl

3) $H_3O^{\oplus}$

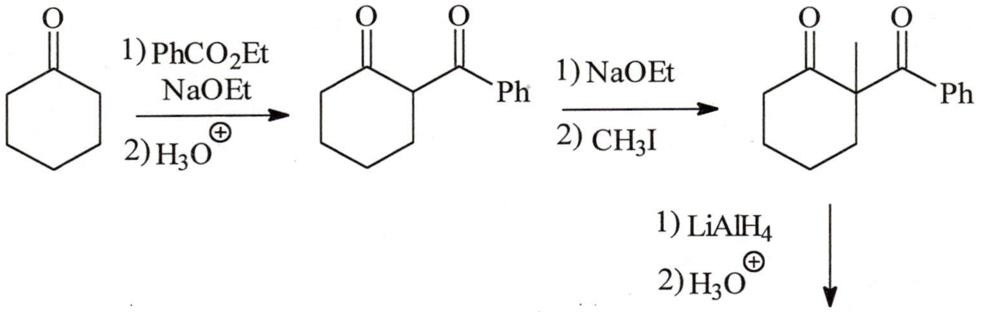

c) cyclohexanone (O)

1) $PhCO_2Et$
NaOEt

2) $H_3O^{\oplus}$

2-benzoylcyclohexanone (O, O, Ph)

1) NaOEt

2) CH_3I

2-benzoyl-2-methylcyclohexanone (O, O, Ph)

1) $LiAlH_4$

2) $H_3O^{\oplus}$

d) Benzaldehyde → 1) $CH_3CH_2CH_2MgBr$, 2) $H_3O^{\oplus}$ → 1-phenyl-1-butanol → PCC → 1-phenyl-1-butanone → (benzaldehyde), NaOH, H_2O, Δ → 2-ethyl-1,3-diphenyl-2-propen-1-one → H_2, Pt →

e) $CH_3\overset{O}{\overset{\|}{C}}CH_3$ $\xrightarrow[EtO\overset{O}{\overset{\|}{C}}OEt]{NaOEt}$ $CH_3\overset{O}{\overset{\|}{C}}CH_2\overset{O}{\overset{\|}{C}}OEt$ $\xrightarrow[PhCH=CH\overset{O}{\overset{\|}{C}}CH_3]{NaOEt}$

16.41

:Ö: :Ö: C—ÖEt :Ö:⊖ H

⇌

:Ö: ⊖ :Ö: C—OEt :Ö—H

⇌

EtÖ:⊖ :Ö: :Ö: C—Ö—H

↓

:Ö: :Ö: C—Ö:⊖ H—ÖH$_2$ ⊕

←

H :Ö: :Ö: C=Ö

←

:Ö—H

$+ CO_2$

←

:Ö:

16.42

:Ö: NC—CH—C—ÖEt H ⊖ :Ö—Et

⇌

:Ö: ⊖ NC—CH—C—ÖEt CH$_2$=CH—C—OEt Ö:

⇌

:Ö: NC—CH—C—ÖEt CH$_2$—CH—C—ÖEt ⊖ :Ö: H—ÖEt

⇌

:Ö: NC—CH—C—ÖEt CH$_2$—CH$_2$—C—ÖEt :Ö

16.43

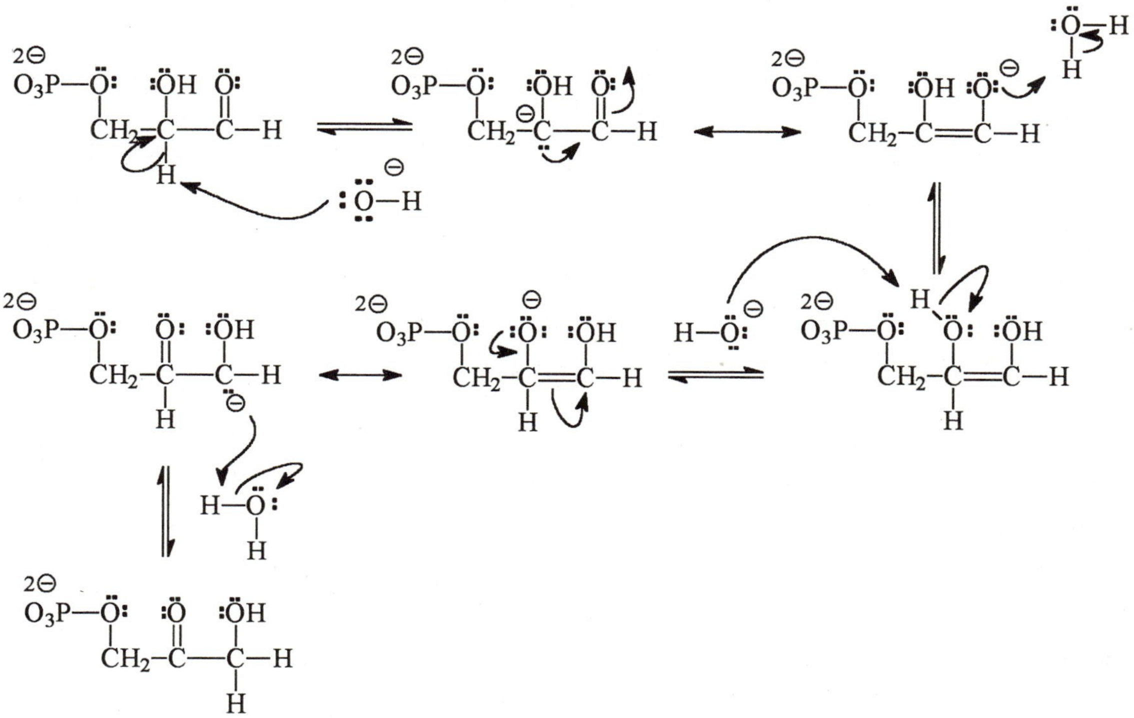

16.44

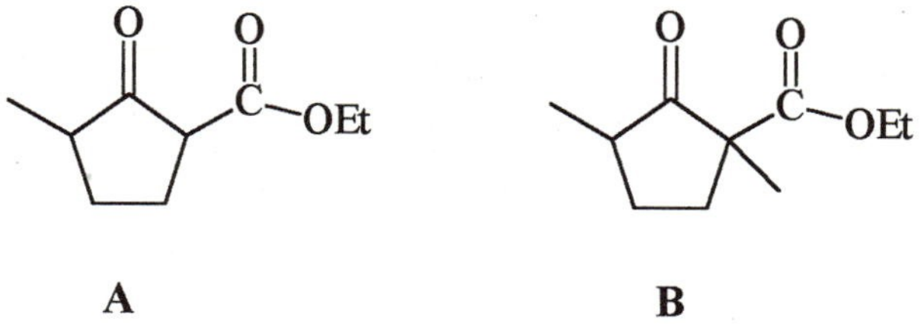

A **B**

The ester condensation in the first step cannot occur at the position occupied by the methyl group because the product does not have a H on the carbon between the two carbonyl groups and the equilibrium driving step cannot occur. The alkylation in the second step occurs at the carbon between the two carbonyl groups because the most acidic H is located there.

16.45

a)

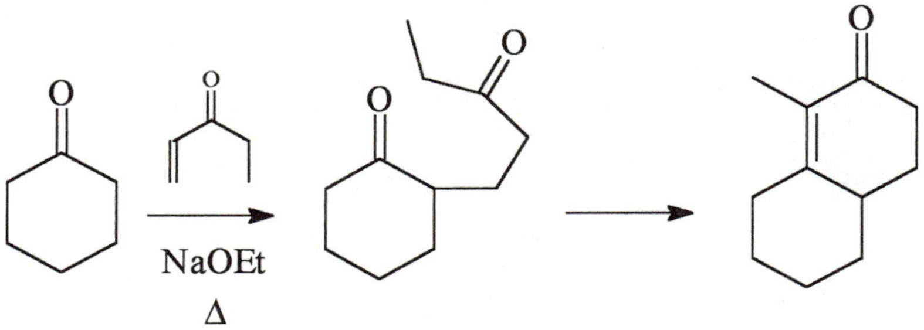

NaOEt

Δ

b)

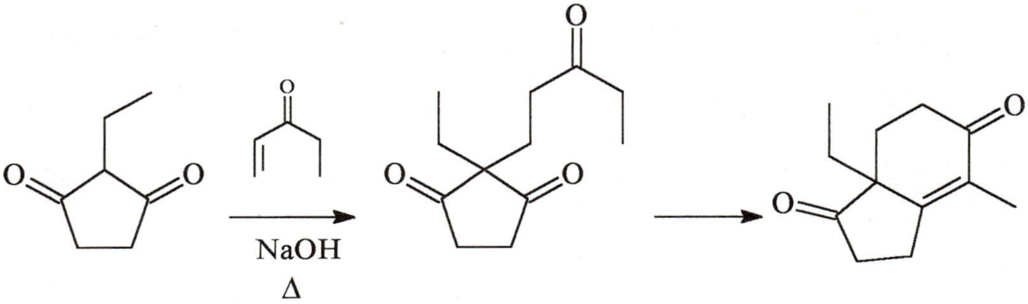

NaOH

Δ

16.46

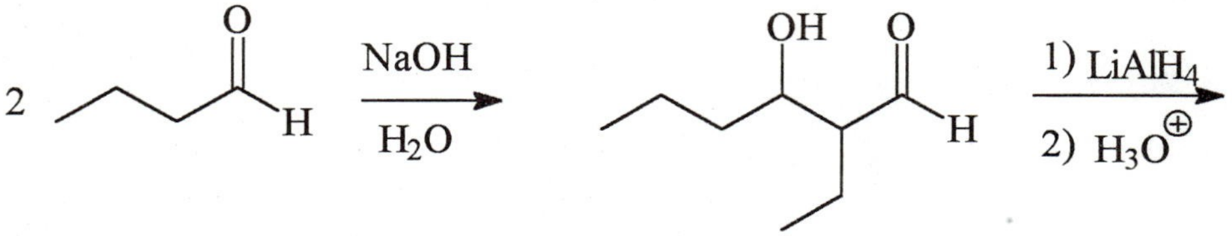

2

NaOH

H_2O

OH

1) $LiAlH_4$

2) $H_3O^{\oplus}$

16.47

2

NaOH

H_2O

Δ

1) $LiAlH_4$

2) $H_3O^{\oplus}$

3) H_2, Pt

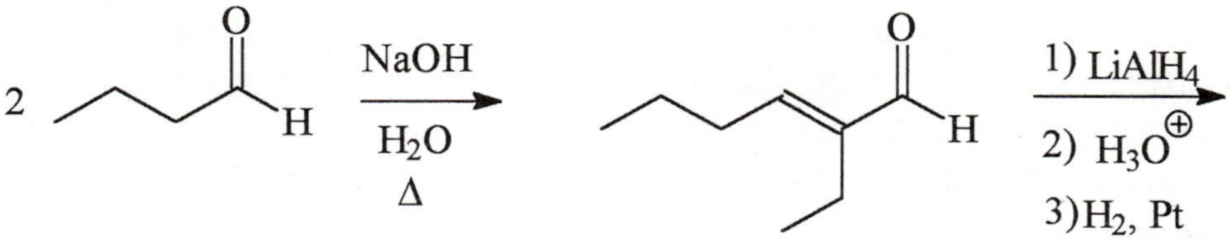

16.48

$CH_3{-}C(=\ddot{O}){-}C(CH_3)_2{-}C(=\ddot{O}){-}OEt + {}^{-}:\ddot{O}{-}Et \rightleftharpoons CH_3{-}C(:\ddot{O}:^{-})(EtO){-}C(CH_3)_2{-}C(=\ddot{O}){-}OEt \rightleftharpoons CH_3\overset{O}{\overset{\|}{C}}OEt + {}^{-}:C(CH_3)_2{-}C(=O){-}OEt$

${}^{-}:C(CH_3)_2{-}C(=O){-}OEt + H{-}\ddot{O}Et \longrightarrow H{-}C(CH_3)_2{-}C(=O){-}OEt$

16.49

$CH_3CH_2C(=\ddot{O}:){-}CH_2CH_3 + D{-}\ddot{O}D_2^{\oplus} \rightleftharpoons CH_3CH_2C(=O^{\oplus}{-}D){-}CH(H)CH_3 + :\ddot{O}D_2 \rightleftharpoons CH_3CH_2C(:\ddot{O}{-}D){=}CHCH_3$

$CH_3CH_2C(:\ddot{O}{-}D){=}CHCH_3 + D{-}\ddot{O}D_2^{\oplus} \rightleftharpoons CH_3CH_2C(=\ddot{O}^{\oplus}{-}D){-}CH(D)CH_3$

$CH_3CH_2C(=\ddot{O}^{\oplus}{-}D){-}CH(D)CH_3 + D_2\ddot{O}: \rightleftharpoons CH_3CH_2C(=\ddot{O}:){-}CH(D)CH_3$

If this reaction were continued, all four hydrogens on the carbons adjacent to the carbonyl group would be exchanged for deuterium.

16.50

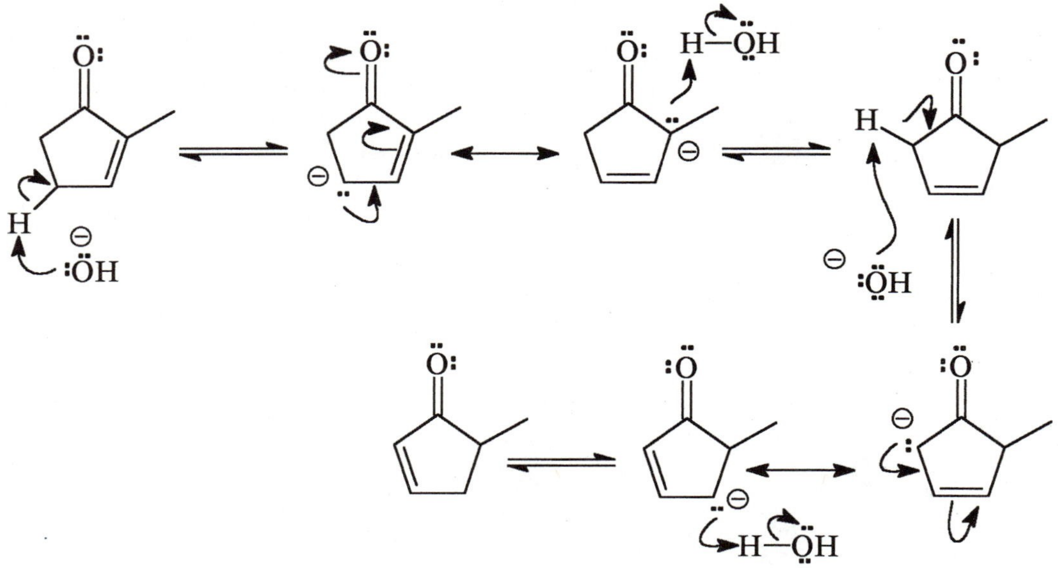

16.51

a)

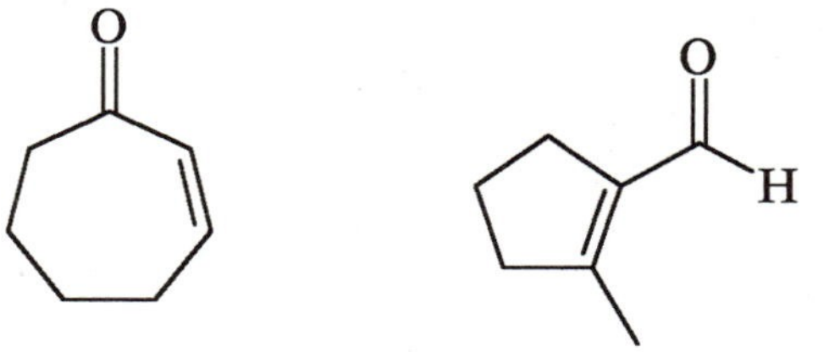

2-Cycloheptenone is not formed because a five-membered ring is favored by entropy. The second product is not formed because the aldehyde carbon is more electrophilic than the ketone carbon.

b)

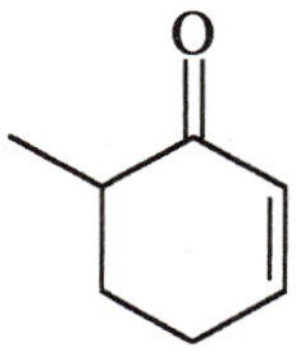

16.52

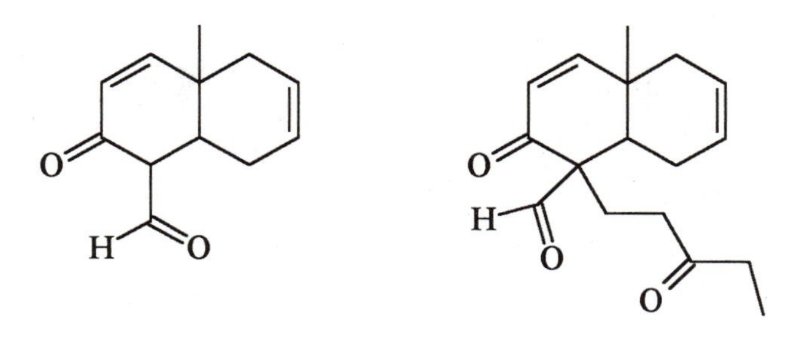

16.53

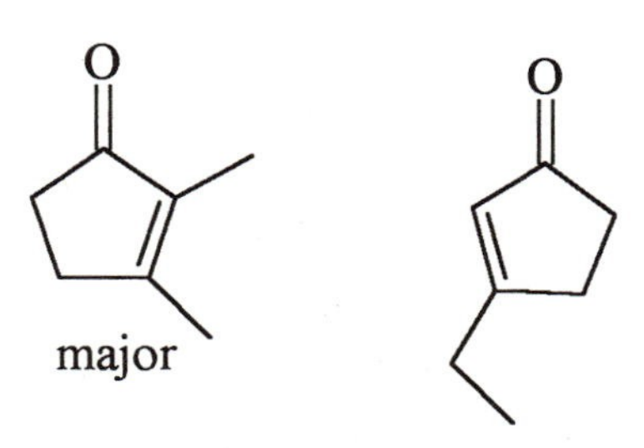

16.54

O
Ph

The H on C-1 is more acidic because the conjugate base is stabilized by resonance with the carbonyl group and the phenyl group.

16.55

O
CH=CH—C—OH

After completing this chapter, you should be able to:

1. Show the products of any of the reactions discussed in this chapter.
   Problems 16.1, 16.2, 16.4, 16.5, 16.8, 16.10, 16.14, 16.16, 16.19, 16.20, 16.22, 16.24, 16.25, 16.27, 16.28, 16.37, 16.43.

2. Show the mechanism for any of these reactions.
   Problems 16.9, 16.11, 16.17, 16.29, 16.30, 16.31, 16.32, 16.36, 16.40, 16.41, 16.47, 16.48, 16.49.

3. Use these reactions in combination with reactions from previous chapters to synthesize compounds.
   Problems 16.6, 16.7, 16.12, 16.18, 16.21, 16.26, 16.33, 16.34, 16.35, 16.38, 16.39, 16.44, 16.45, 16.46.

Chapter 17
BENZENE AND AROMATIC COMPOUNDS

17.1

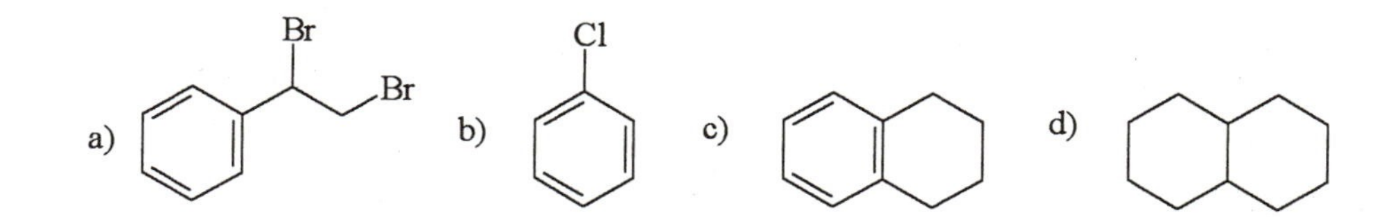

17.2 (5)(28.6 kcal/mol) - 80 kcal/mol = 63 kcal/mol.
(5)(120 kJ/mol) - 335 kJ/mol = 265 kJ/mol.
Yes, naphthalene should be termed aromatic.

17.3

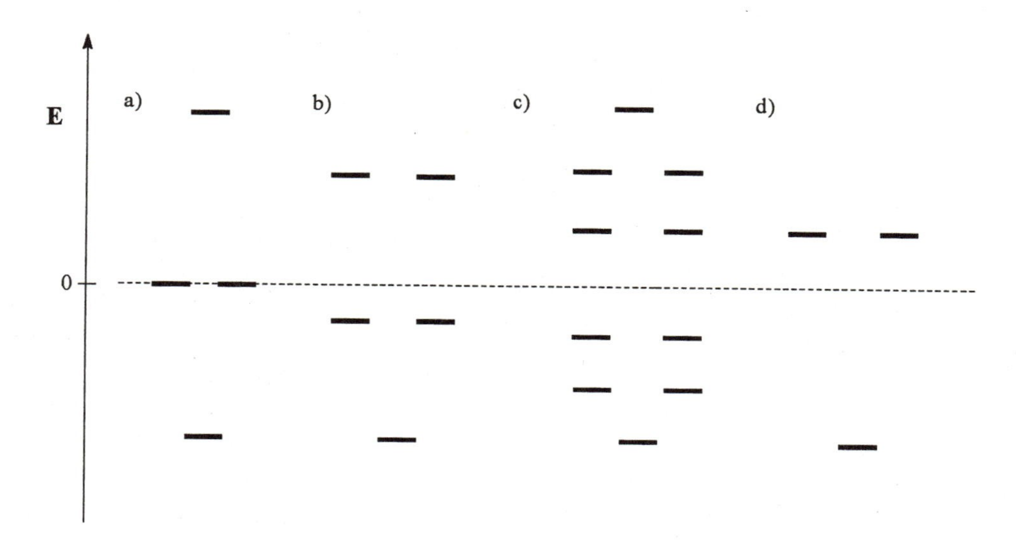

17.4

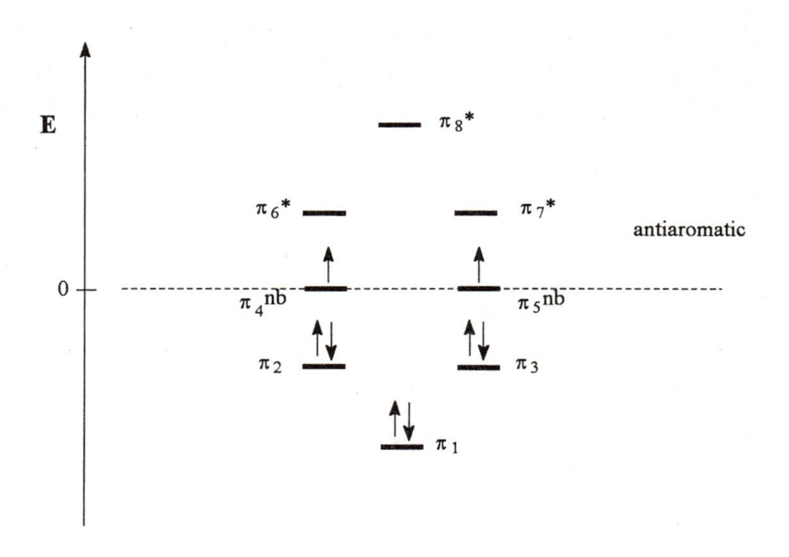

The HOMO of this compound is half-filled, so it is antiaromatic.

17.5

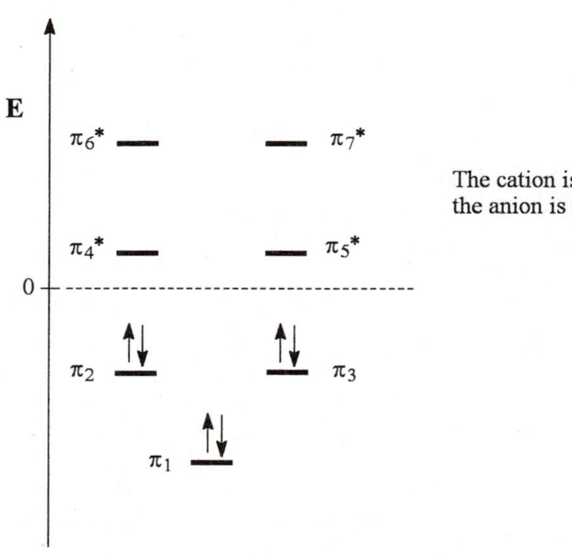

The cation is aromatic;
the anion is antiaromatic.

The MO energy diagram of the cycloheptatrienyl cation above has completely filled HOMOs. Therefore it is aromatic. The anion, on the other hand, has half-filled HOMOs. Therefore it is antiaromatic.

17.6

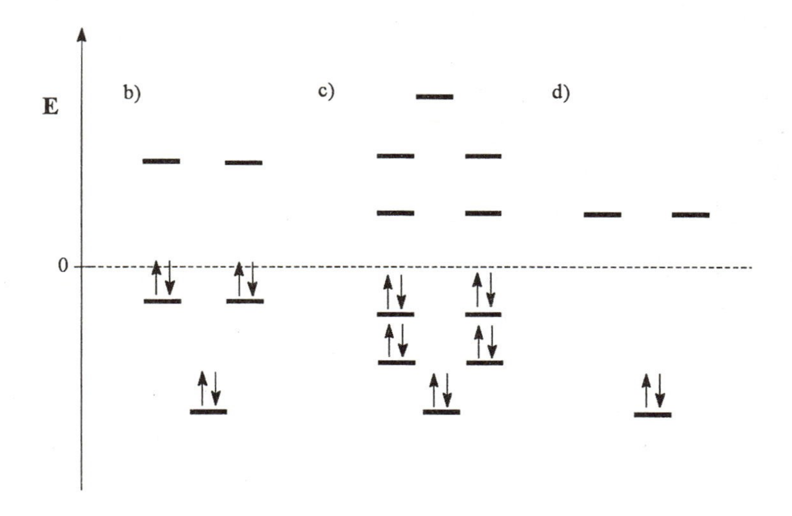

All of these compounds have fully filled HOMOs. Therefore they are all aromatic.

17.7 According to Huckel's rule, aromatic compounds are cyclic, fully conjugated, planar, and have (4n+2) π electrons whereas antiaromatic compounds are cyclic, fully conjugated, planar, and have 4n π electrons.
a) 4 π electrons, so antiaromatic.
b) 6 π electrons, so aromatic.
c) 10 π electrons, so aromatic, if planar.
d) 2 π electrons, so aromatic.

17.9 a) 6 π electrons, so aromatic (do not count unshared electron on Ns).
b) 6 π electrons, so aromatic (count one pair on O).
c) 6 π electrons, so aromatic (count one pair on S).
d) 6 π electrons, so aromatic (count one pair on O).
e) 6 π electrons, so aromatic (do not count unshared pairs on Ns).
f) 8 π electrons, so antiaromatic (count pair on N).

17.10

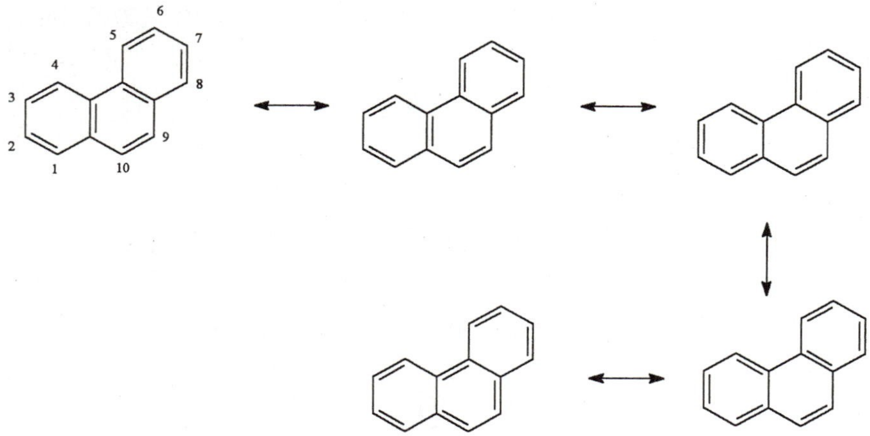

The bond between C-9 and C-10 should be the shortest because it is a double bond in four of the five resonance structures.

17.11

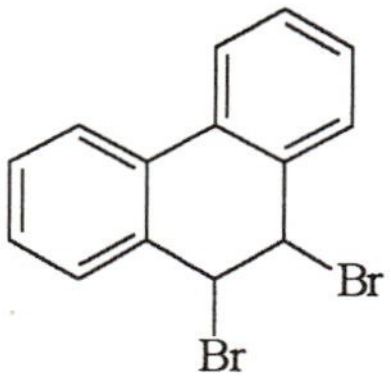

The calculated resonance energy for phenantherene is 92 kcal/mol (385 kJ/mol). The product has two benzene rings so it is approximate resonance energy is 2(36) = 72 kcal/mol (301 kJ/mol). Therefore the resonance energy lost is equal to (92 - 72) = 20 kcal/mol (83.6 kJ/mol).

17.12 This compound is aromatic. The diamagnetic ring current induces a magnetic field that is parallel to the external magnetic field on the outside of the ring and opposed in the center of the ring. This causes the NMR signals of hydrogens on the outside of the ring to appear downfield and the hydrogens inside the ring to appear upfield from the position of normal alkene hydrogens.

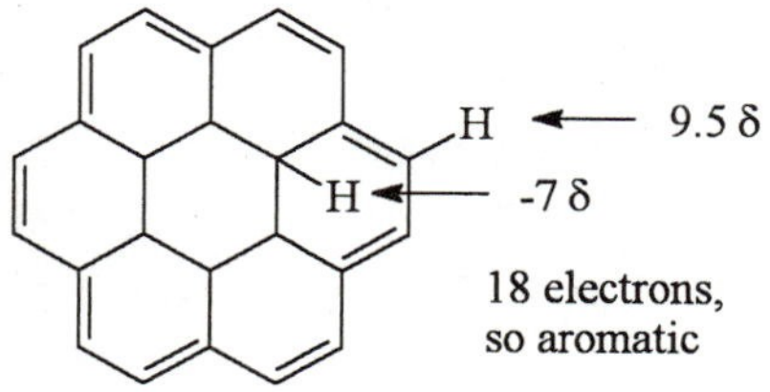

H ← 9.5 δ

H ← -7 δ

18 electrons, so aromatic

17.13 a) The left compound is a stronger acid because its conjugate base is aromatic.

b) The right compound is a stronger acid because conjugate base of the left compound is antiaromatic.

c) The right compound is a stronger acid because its conjugate base is aromatic.

17.14 The left compound has the faster rate of substitution by the S_N1 mechanism because the carbocation that is formed from it is aromatic.

17.15

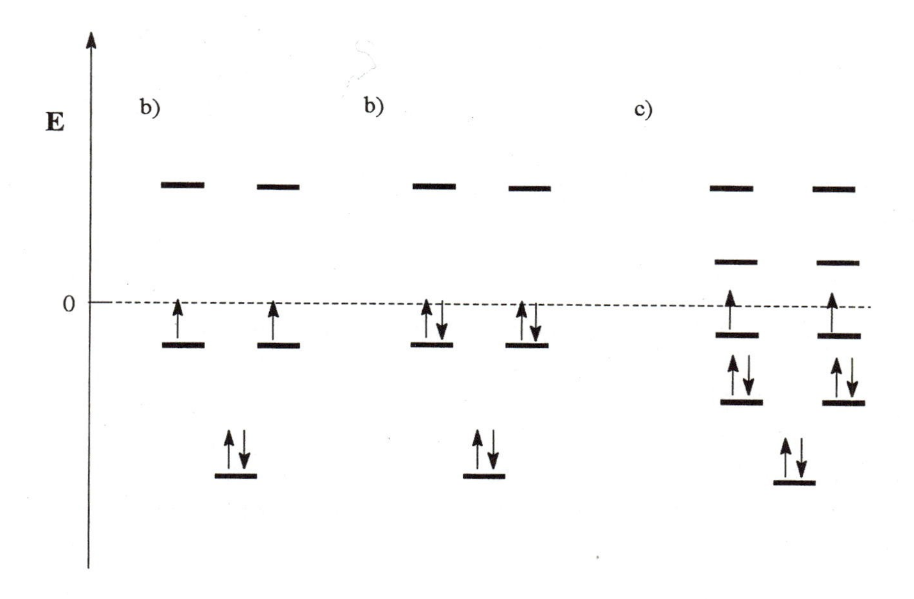

17.16 a) 8 π electrons, so antiaromatic.

b) 6 π electrons, so aromatic (one lone pair of electrons on the O is part of the cycle of π electrons).

c) The cycle of p orbitals is interrupted by a sp^3 hybridized carbon, so neither.

d) 6 π electrons, so aromatic (count one pair on the O).

e) The cycle of p orbitals is interrupted by a sp^3 hybridized nitrogen, so neither.

f) 10 π electrons, so aromatic, if it is planar.

17.17 The dianion of cyclooctatetraene is more stable than expected because it is aromatic (it has 10 π electrons) if it assumes a planar geometry.

17.18

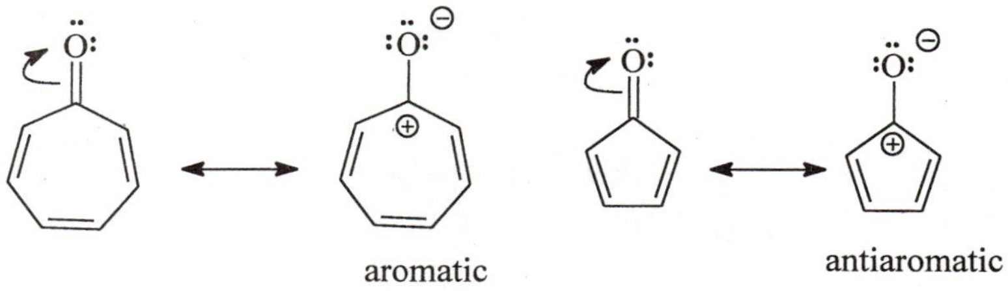

aromatic

antiaromatic

The electronegativity of the oxygen attracts the π electrons of the double bond of the carbonyl group. In the resonance structures of these compounds both carbonyl pi electrons are located on the oxygen. This resonance structure of 2,4,6-cycloheptatrienone is aromatic whereas the similar resonance structure of 2,4-cyclopentadienone is antiaromatic. Therefore 2,4,6-cycloheptatrienone is more stable than 2,4-cyclopentadienone.

17.19 The resonance structure of cyclopropenone is aromatic (see solution 17.18). Therefore cyclopropenone is less reactive than cyclopropanone despite its high angle strain.

17.20 The compound on the left and the one in the middle have an even number of pairs of π electrons, so they are antiaromatic and are unstable. The right compound is very stable because it is aromatic (odd number of pairs of π electrons).

17.21

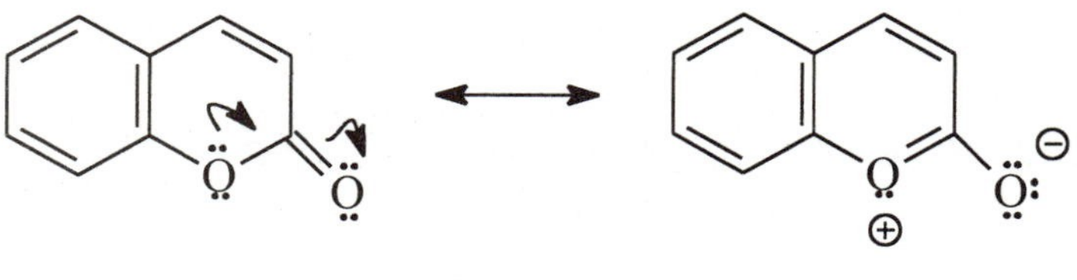

In the resonance structure of coumarin shown above, the oxygen containing ring has 6 π electrons, giving it aromatic character.

17.22 A resonance structure of the conjugate acid of this ketone is aromatic.

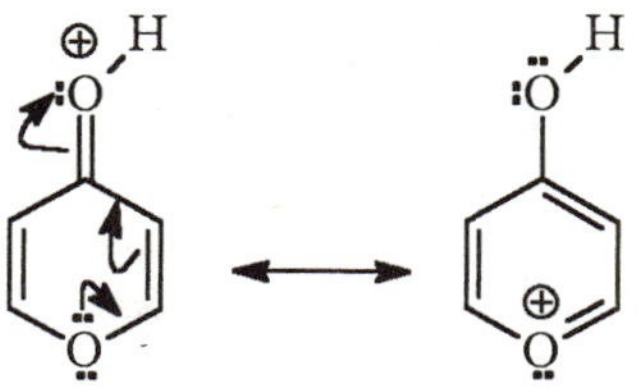

17.23 Benzocyclobutadiene is very unstable because it has an antiaromatic ring in addition to a considerable amount of angle strain.

17.24 a) There are 10 π electrons along the periphery of the ring ,so this compound is aromatic.

b) This compound has 14 π electrons along the periphery of the ring, so it is aromatic.

17.25 This compound has 14 π electrons (only one pair of the triple bond is part of the π system) so it is aromatic. The hydrogens on the periphery of the ring should appear in the region of 7 - 8 δ due to the diamagnetic ring current. The hydrogens on the inside of the ring should appear further upfield.

17.26 The nitrogen with no hydrogen attached is more basic because the lone pair of electrons is in an sp^2 atomic orbital and is not part of the aromatic cycle of p orbitals.

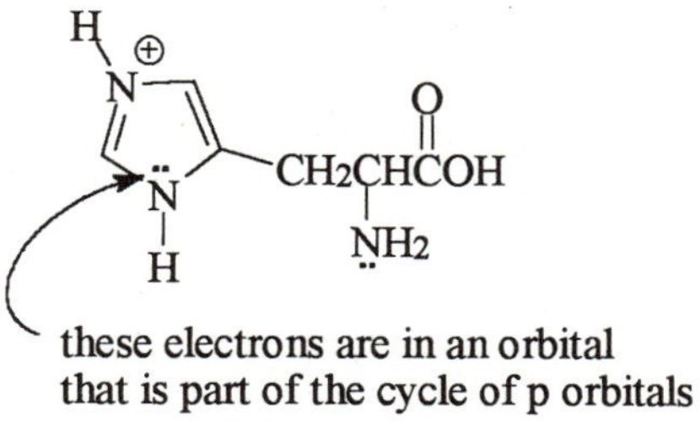

these electrons are in an orbital that is part of the cycle of p orbitals

17.27 The acidity of this compound is due to the extreme stability of its conjugate base. The conjugate base is aromatic and is also stabilized by the electron withdrawing inductive effects of the five CF_3 groups.

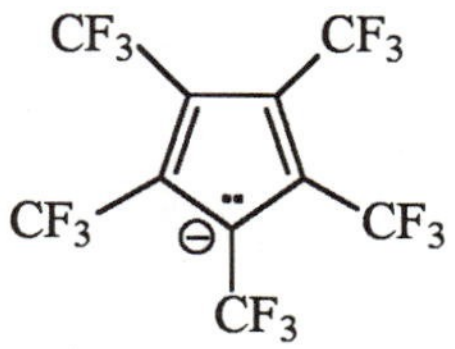

17.28 The resonance stabilization energy of pyrrole is 21.6 kcal/mole (91kJ/mol).

$$2\ (26.6\ \text{kcal/mol}) - (31.6\ \text{kcal/mol}) = 21.6\ \text{kcal/mol}$$

17.29 The dianion of this compound formed readily because it is aromatic (10 π electrons).

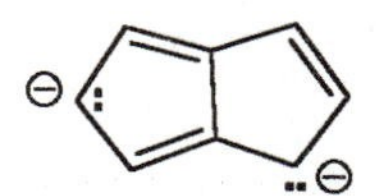

17.30 a) Indole can be viewed as resulting from the fusion of a benzene ring and a heterocyclic pyrrole ring, so its heterocyclic ring is aromatic.
b) The fused heterocyclic ring of benzimidazole is imidazole (see solution 17.8 in the text) which is aromatic.
c) The fused heterocyclic ring of quinoline is pyridine, which is aromatic.

17.31 Adenine can be viewed as resulting from the fusion of the two aromatic rings below, so it is aromatic.

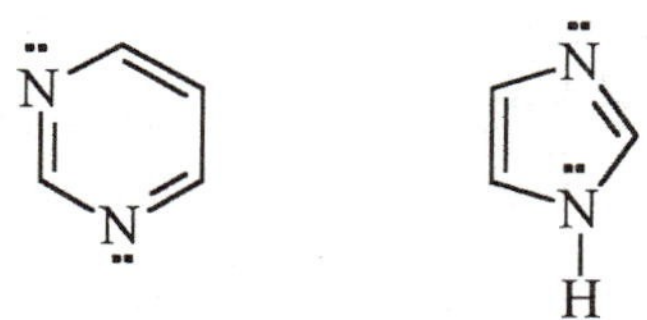

17.32 In resonance structure **B** both rings are aromatic. In this compound the electrons are drawn from the seven membered rig to the five membered ring creating a large dipole moment in the molecule.

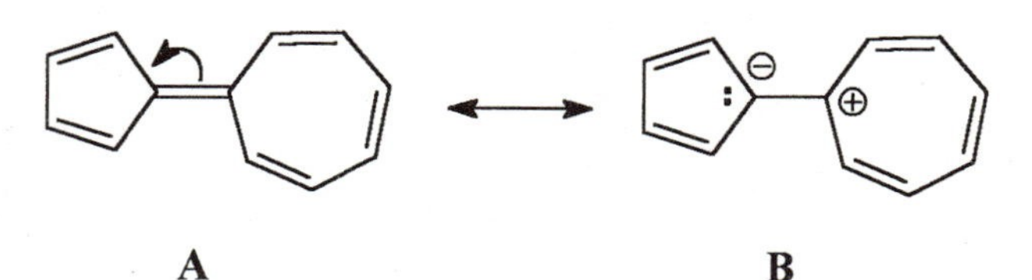

17.33 The cation formed in this reaction fulfills the criteria of aromaticity. All the hydrogens on the cation are on the periphery the ring and are magnetically equivalent, so they appear as a singlet. The signal is shifted downfield

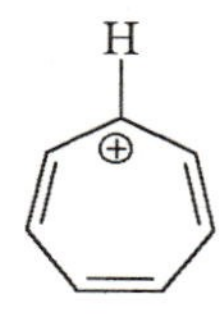

from the normal aromatic position because the lower electron density of the carbocation results in additional deshielding.

17.34 The dication species produced in this reaction is aromatic. The carbons of the methyl groups on the ring are all magnetically equivalent as are the four ring carbons. The ring carbons are very far downfield because they are part of an aromatic ring. They are deshielded even more due to the low electron density in the ring.

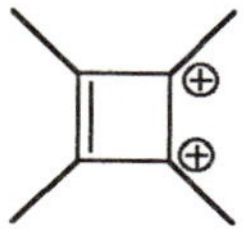

17.35 The benzylic cation rearranges to form cycloheptatrienyl cation, which is aromatic.

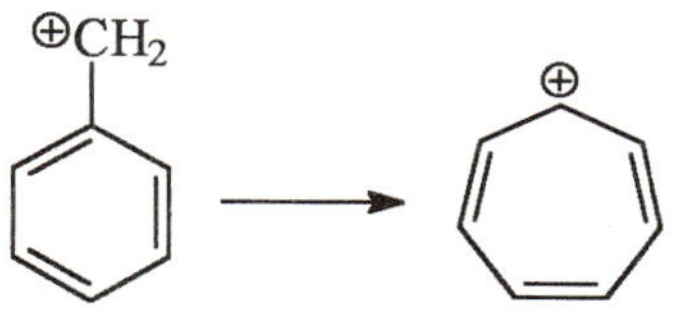

After completing this chapter, you should be able to:

1. Show the MO energy levels for planar, cyclic, conjugated compounds.
   Problems 17.3, 17.4, 17.5, 17.15.

2. Apply Hückel's rule and recognize whether a particular compound is aromatic, antiaromatic, or neither.
   Problems 17.6, 17.7, 17.8, 17.9, 17.16, 17.20, 17.21, 17.24, 17.30, 17.31.

3. Understand how aromaticity and antiaromaticity affect the chemistry (and NMR spectra) of compounds.
   17.1, 17.2, 17.12, 17.13, 17.14, 17.17, 17.18, 17.19, 17.22, 17.23, 17.25, 17.26, 17.27, 17.28, 17.29, 17.32.

Chapter 18
AROMATIC SUBSTITUTION REACTIONS

18.1

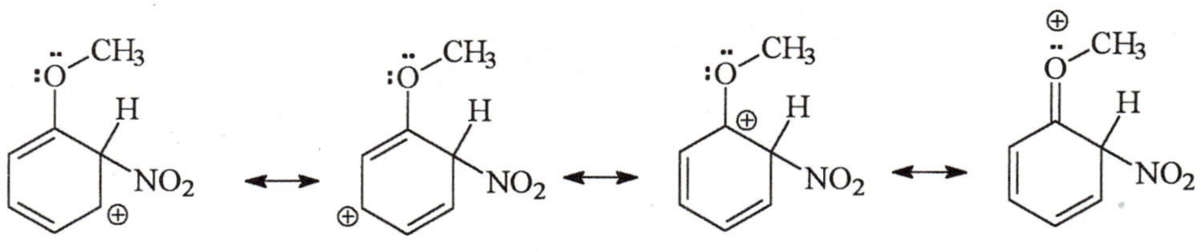

$:\ddot{O}-CH_3$, H, NO_2, ⊕ ⟷ $:\ddot{O}-CH_3$, H, NO_2, ⊕ ⟷ $:\ddot{O}-CH_3$, ⊕, H, NO_2 ⟷ ⊕ $\ddot{O}-CH_3$, H, NO_2

The last structure is especially stable because the octet rule is satisfied at all atoms.

18.2

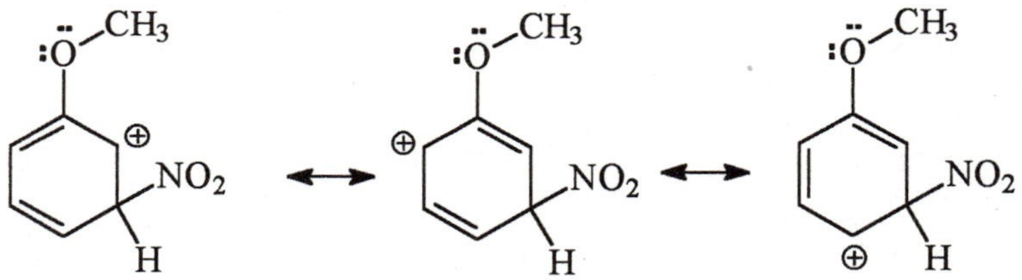

$:\ddot{O}-CH_3$, ⊕, NO_2, H ⟷ ⊕, $:\ddot{O}-CH_3$, NO_2, H ⟷ $:\ddot{O}-CH_3$, NO_2, ⊕, H

No especially stable resonance structure is formed because the positive charge is never located on the carbon that is bonded to the methoxy group.

18.3 a) The electron pair on the N can stabilize a positive charge on the adjacent carbon by resonance.

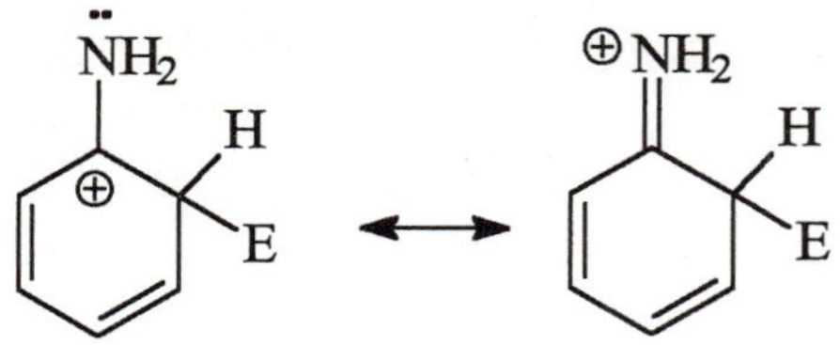

$\ddot{N}H_2$, H, ⊕, E ⟷ ⊕NH_2, H, E

b) The ethyl group stabilizes a positive charge on the adjacent carbon by hyperconjugation.

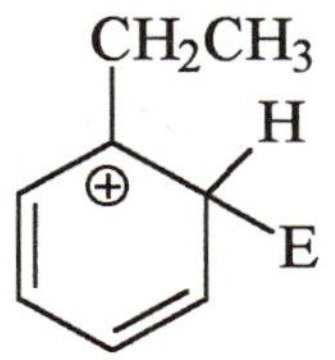

c) The phenyl group can stabilize a positive charge on the adjacent carbon by resonance.

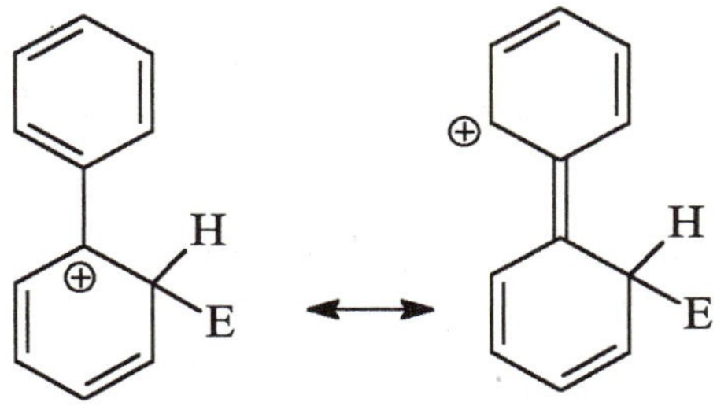

18.4 a) The electronegative fluorines make the CF_3 group electron withdrawing.
b) The positive nitrogen is an inductive electron withdrawing group.

18.5 a) Because of the unshared electrons on the S, this is an activating group and it directs ortho and para.
b) The S is electron deficient, much like the carbon of a carbonyl group, so this is a deactivating group and it directs meta.
c) The carbon is electron deficient, so this is a deactivating group and it directs meta.

18.6

b) 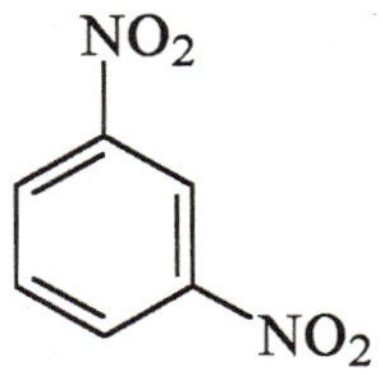 NO_2 NO_2

c) 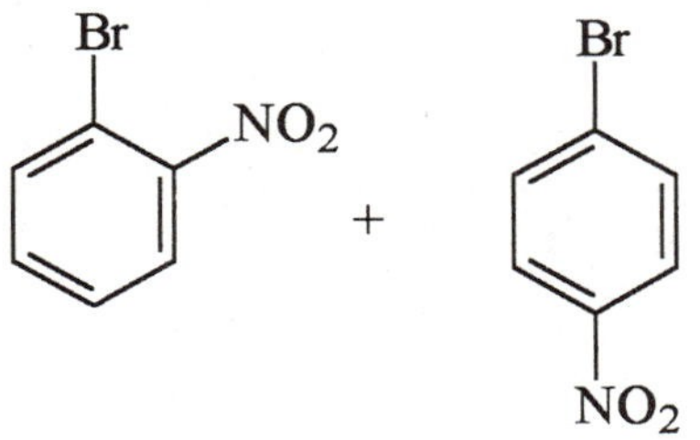 Br NO_2 + Br NO_2

18.7 a) Both groups are directing to the same position, ortho to the ethyl group and meta to the nitro group.

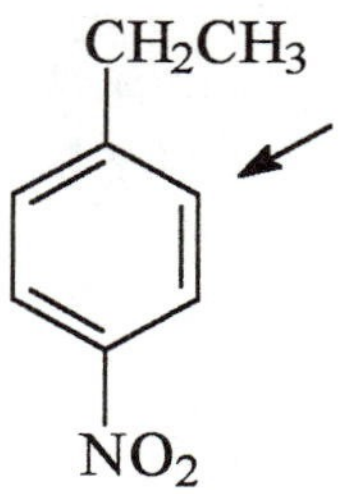

b) The groups are directing to different positions, so the more activating group will control the regiochemistry. The nitrogen attached to the ring is activating, whereas the carbonyl group is deactivating. Therefore the substitution occurs ortho and para to the nitrogen group, but not between the two groups.

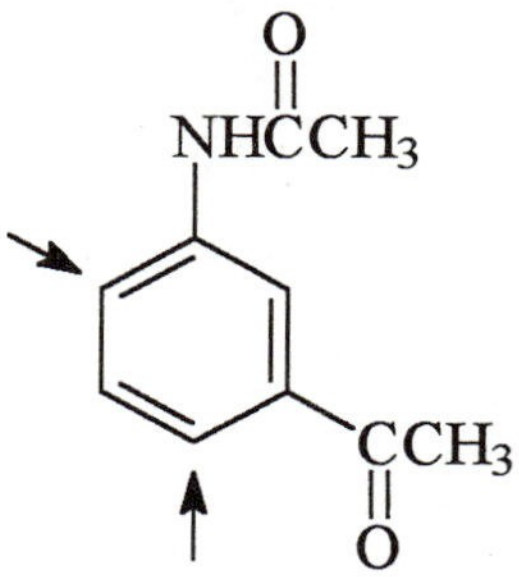

c) The methoxy group is a stronger activating group than the methyl group, so the substitution occurs ortho to the methoxy group.

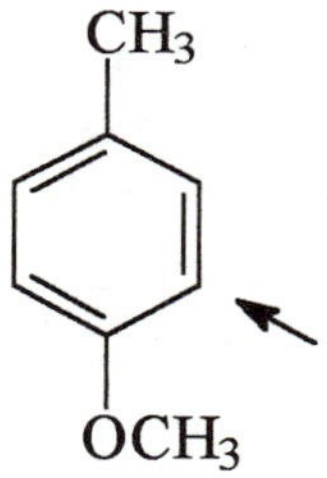

d) The groups are directing to the same positions, so the substitution occurs ortho to one and para to the other, but not at the position ortho to both because this position is too sterically hindered.

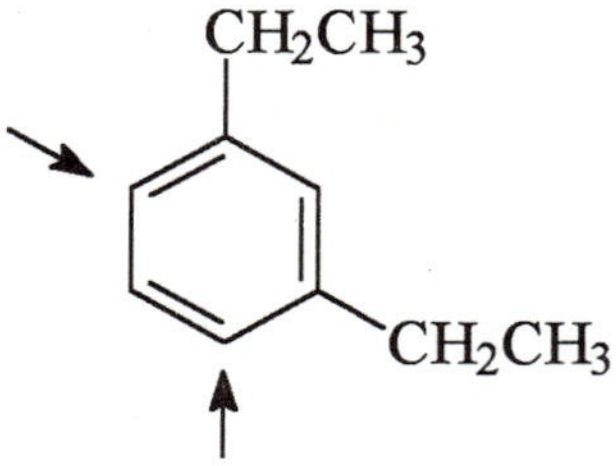

e) Both groups are directing to the same positions.

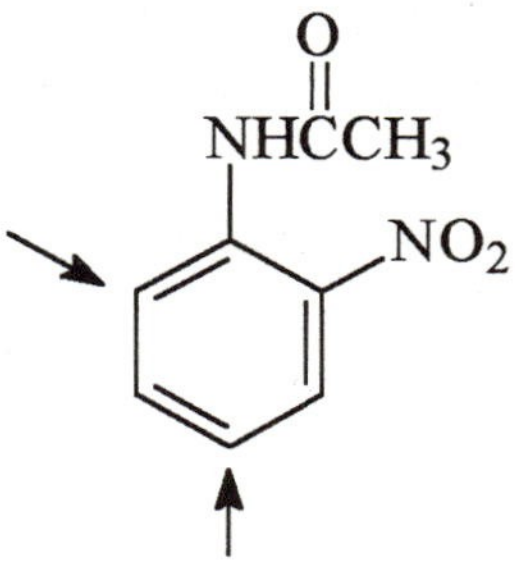

$NHCCH_3$ (C=O), NO_2

f) Substitution occurs ortho and para to the hydroxy group because it is more activating than chlorine.

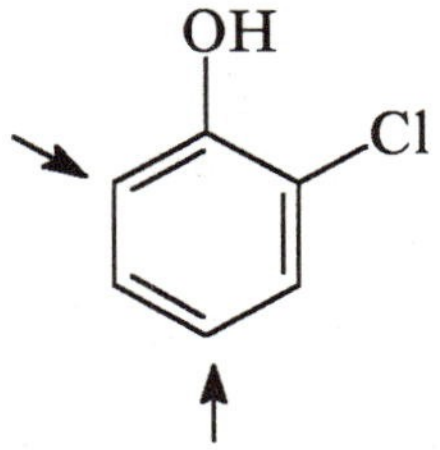

OH, Cl

18.8

a) 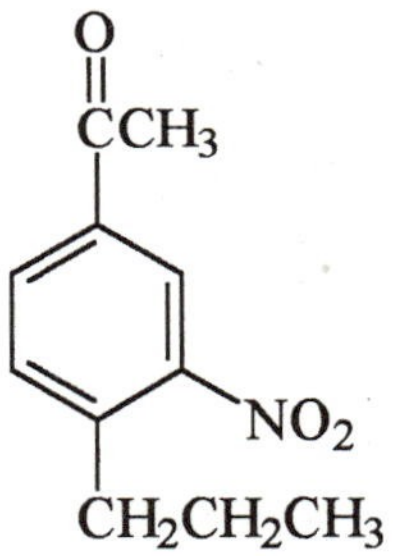 CCH_3 (C=O), NO_2, $CH_2CH_2CH_3$

b) 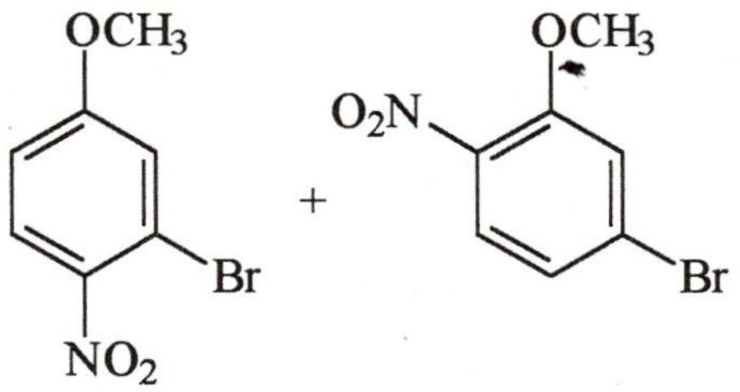 OCH_3, Br, NO_2 + O_2N, OCH_3, Br

18.9

a) 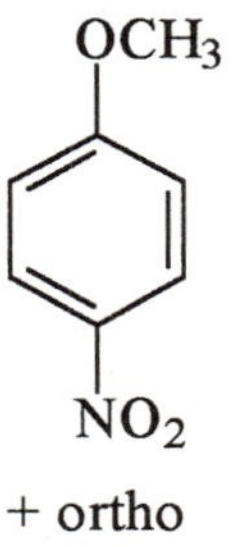 OCH_3, NO_2 + ortho

b) 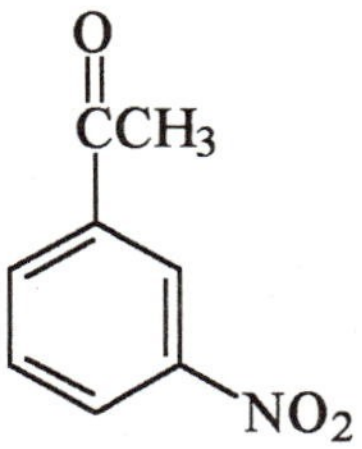 CCH_3 (C=O), NO_2

c) $NHCCH_3$ (C=O) $\xrightarrow[Ac_2O]{HNO_3}$ $NHCCH_3$ (C=O), NO_2 + ortho $\xrightarrow[H_2O]{KOH}$ NH_2, NO_2 + ortho

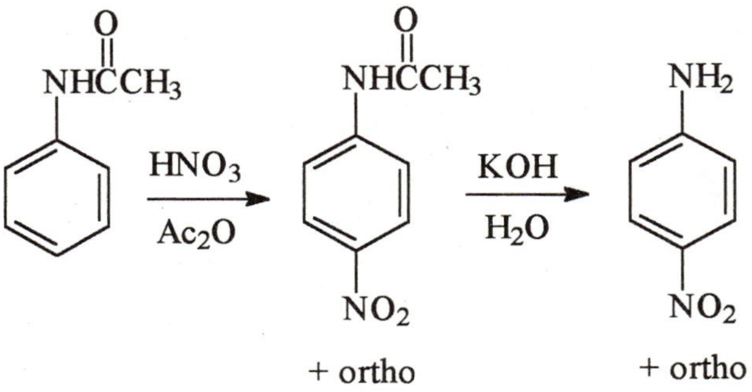

18.10

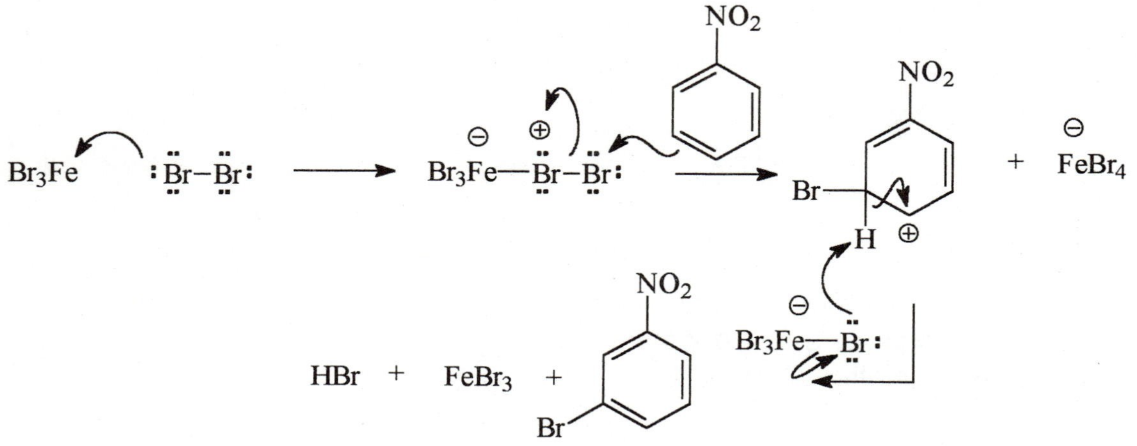

Br_3Fe :Br–Br: ⟶ $\overset{\ominus}{Br_3Fe}$–$\overset{\oplus}{Br}$–Br: ⟶ (Br, H, ⊕, NO_2 arenium ion) + $\overset{\ominus}{FeBr_4}$

NO_2

$\overset{\ominus}{Br_3Fe}$–Br:

HBr + $FeBr_3$ + (NO_2, Br)

18.11

a) 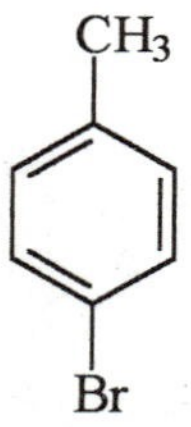 CH_3, Br + ortho

b) 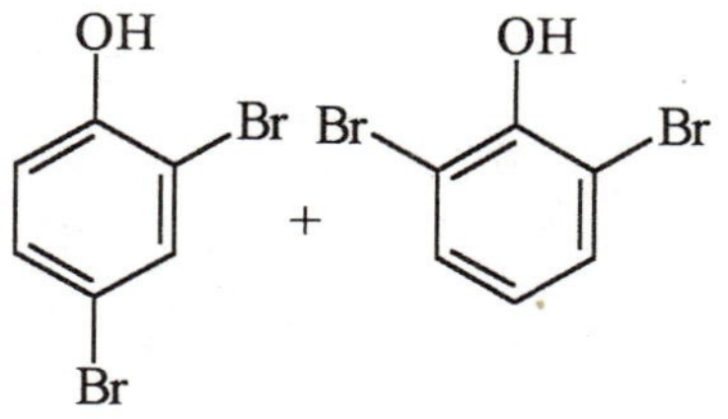 OH, Br, Br + OH, Br, Br

c) 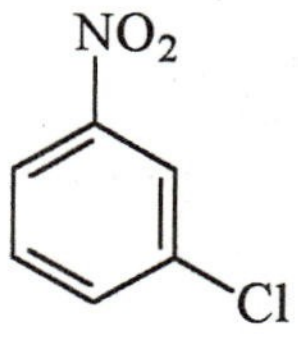 NO_2, Cl

d)  F, Cl + ortho

e) 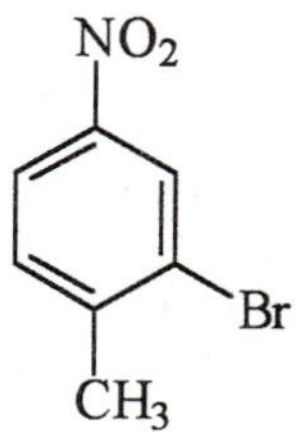 NO_2, Br, CH_3

f) 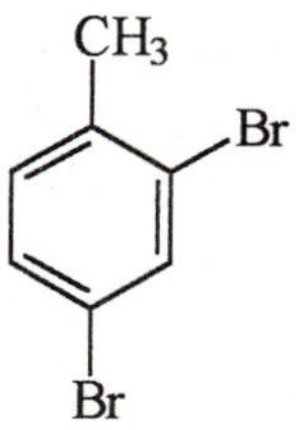 CH_3, Br, Br

18.12

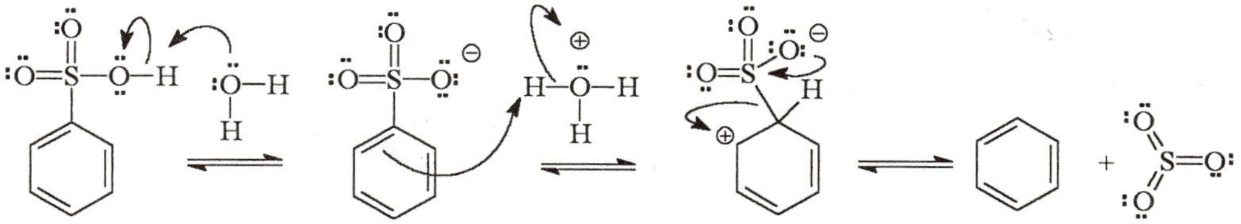

$:\ddot{O}=S(=\ddot{O}:)-\ddot{O}-H$ (benzenesulfonic acid) + $:\ddot{O}-H$ / H $\rightleftharpoons$ $:\ddot{O}=S(=\ddot{O}:)-\ddot{O}:^{\ominus}$ (benzenesulfonate) + $H-\ddot{O}^{\oplus}-H$ / H $\rightleftharpoons$ arenium ion ($:\ddot{O}=S(=\ddot{O}:)-\ddot{O}:^{\ominus}$, H, $\oplus$) $\rightleftharpoons$ benzene + $:\ddot{O}=S(=\ddot{O}:)=\ddot{O}:$

18.13

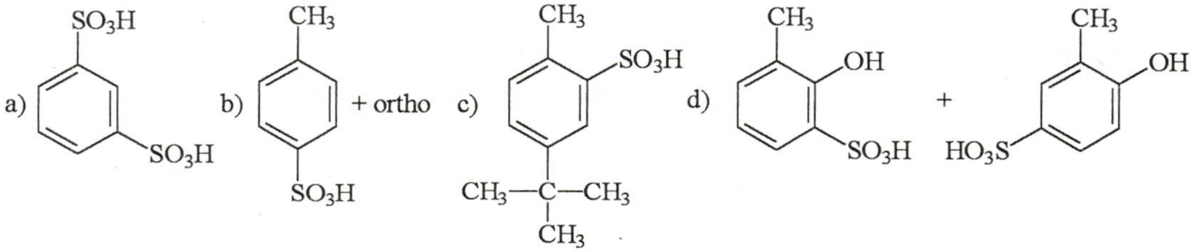

a) benzene ring with SO_3H and SO_3H (meta)

b) CH_3 ring with SO_3H (para) + ortho

c) CH_3 ring with SO_3H and $CH_3-C(CH_3)-CH_3$ (CH_3)

d) CH_3 ring with OH and SO_3H + CH_3 ring with OH and HO_3S

18.14

$CH_3CH_2CH_2-\ddot{C}l: + AlCl_3 \rightleftharpoons CH_3CH_2CH_2-\overset{\oplus}{\ddot{C}l}-\overset{\ominus}{AlCl_3} \longrightarrow CH_3\overset{\oplus}{C}HCH_2$ (H) $+ \overset{\ominus}{AlCl_4}$

$CH_3\overset{\oplus}{C}HCH_3$ + benzene $\longrightarrow$ arenium ion (H, $CHCH_3$ / CH_3, $\oplus$) $\xrightarrow{B:}$ CH_3CHCH_3 (benzene ring)

$CH_3\overset{\oplus}{C}HCH_2$ (H) + benzene $\longrightarrow$ arenium ion ($CH_3CH_2CH_2$, H, $\oplus$) $\xrightarrow{B:}$ $CH_2CH_2CH_3$ (benzene ring)

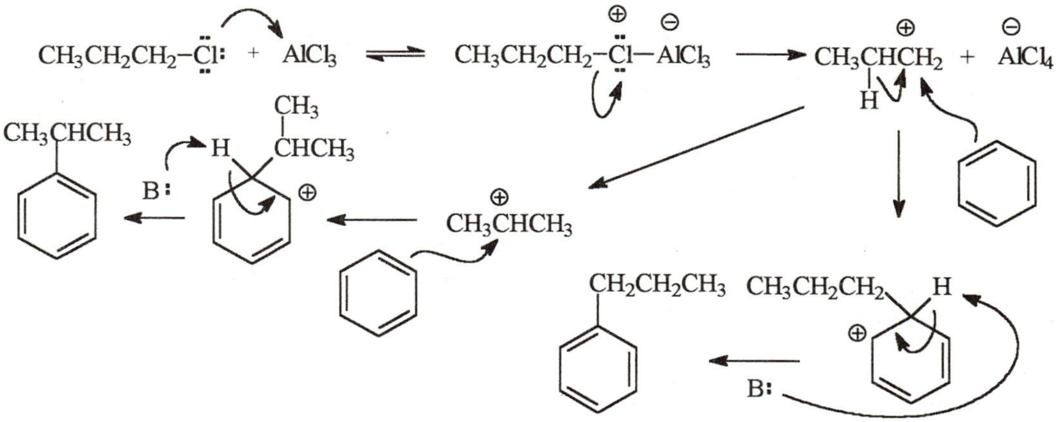

18.15

a) 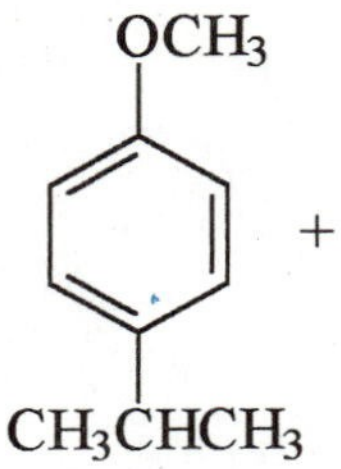 OCH_3 / $CH_3\overset{|}{C}HCH_3$ (para-substituted benzene) + ortho

b) 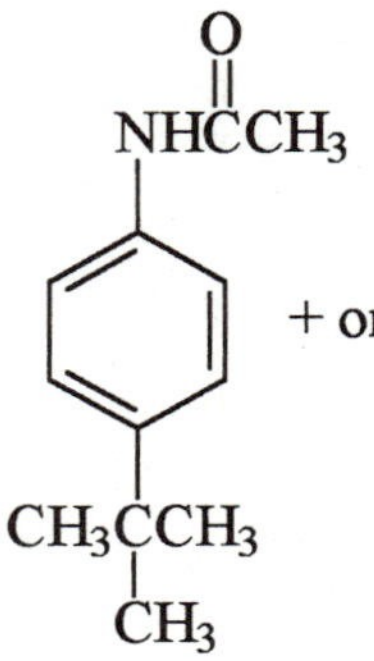 $NH\overset{O}{\overset{\|}{C}}CH_3$ / $CH_3\underset{CH_3}{\underset{|}{C}}CH_3$ (para-substituted benzene) + ortho

c) 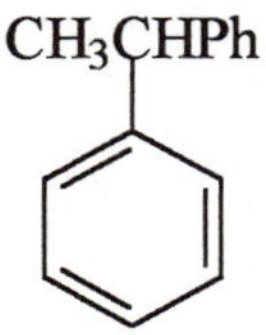 CH_3CHPh (on benzene)

d) 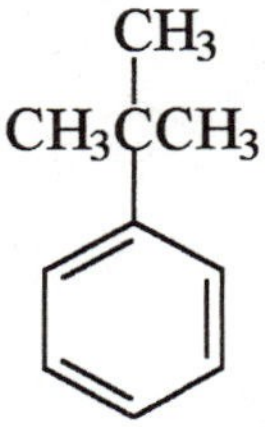 $CH_3\overset{CH_3}{\overset{|}{C}}CH_3$ (on benzene)

e) 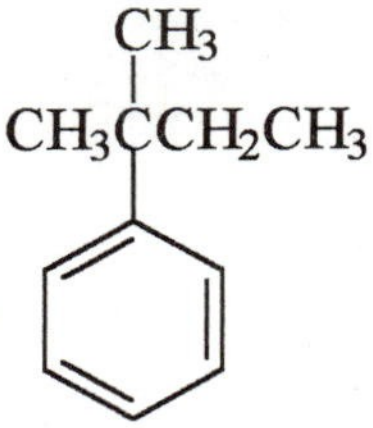 $CH_3\overset{CH_3}{\overset{|}{C}}CH_2CH_3$ (on benzene)

f) 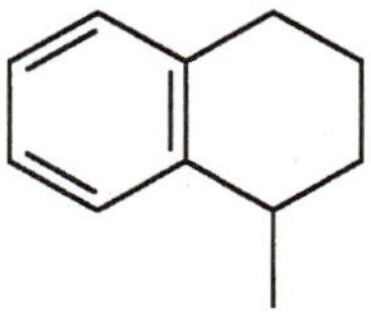 (1-methyltetralin)

g) no reaction

18.16

a) 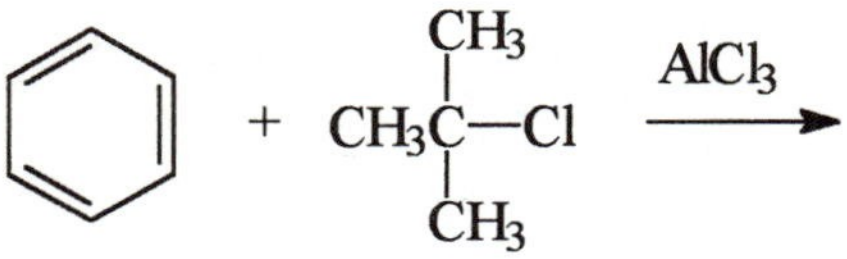 benzene + $CH_3\overset{CH_3}{\underset{CH_3}{C}}—Cl$ $\xrightarrow{AlCl_3}$

b) 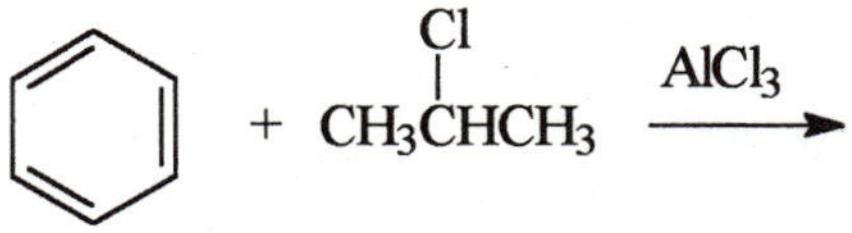 benzene + $CH_3\overset{Cl}{\overset{|}{C}}HCH_3$ $\xrightarrow{AlCl_3}$

18.18

$CH_3(CH_2)_7\underset{|}{C}H(CH_2)_5CH_3$ (on benzene)

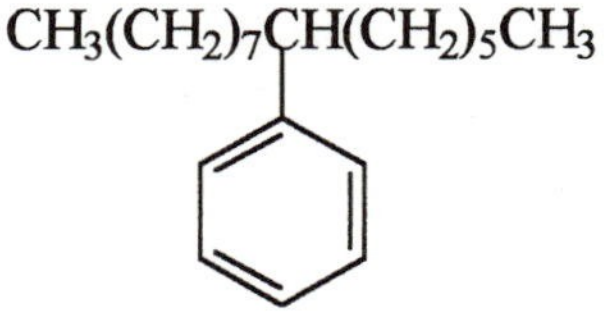

18.19 Both groups are weakly activating, but the position ortho to the methyl group is less sterically hindered than the position ortho to the isopropyl group. The Friedel-Crafts acylation reaction is very sensitive to steric effects.

18.20

a) 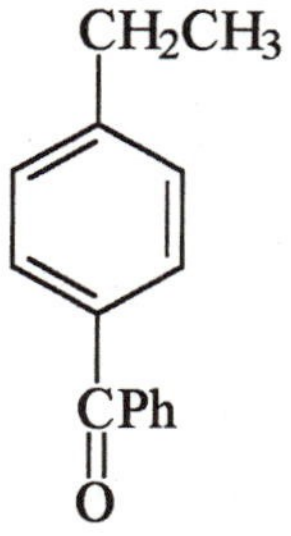 CH_2CH_3 / $\overset{O}{\underset{}{\|}}$CPh

b) 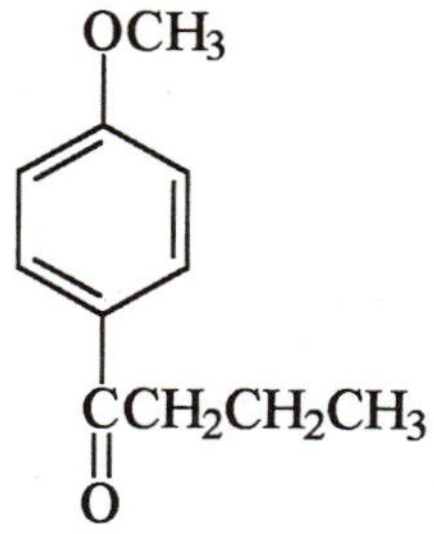 OCH_3 / $CCH_2CH_2CH_3$ (C=O)

c) no reaction

d) 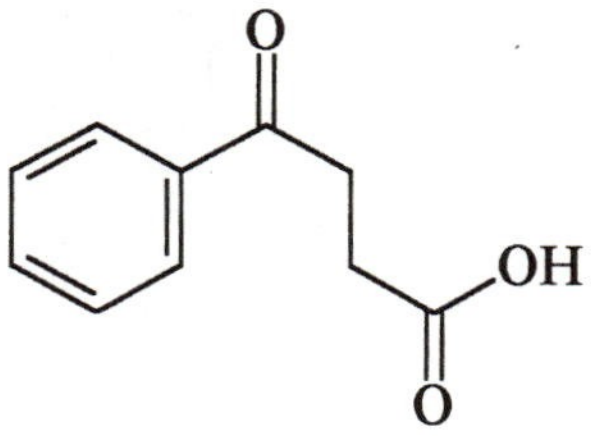 O / OH / O

e) 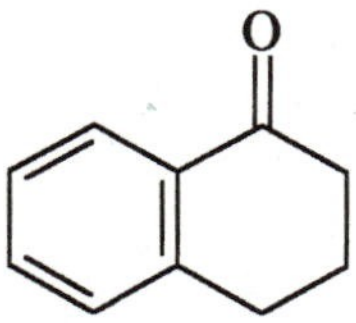 O

f) 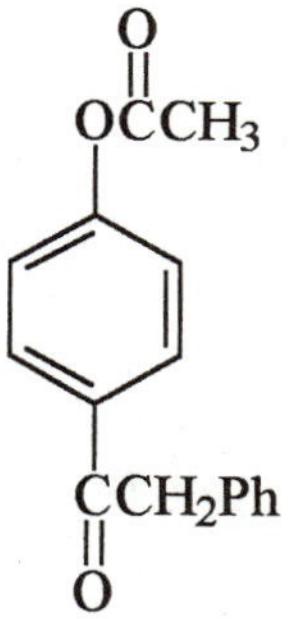 $O\overset{O}{\overset{\|}{C}}CH_3$ / CCH_2Ph (C=O)

18.21

a) 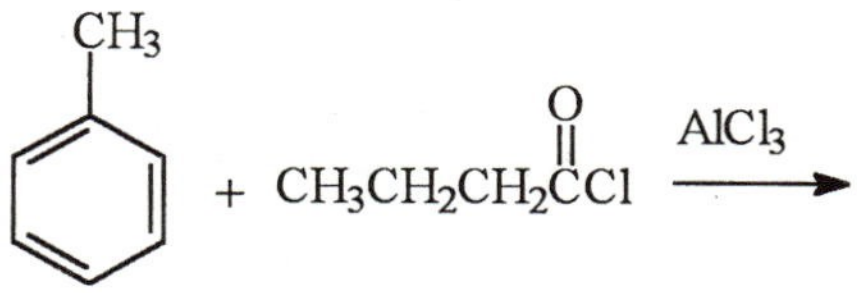 CH_3 + $CH_3CH_2CH_2\overset{O}{\overset{\|}{C}}Cl$ $\xrightarrow{AlCl_3}$

b) OCH_3 + $CH_3\overset{O}{\overset{\|}{C}}Cl$ $\xrightarrow{AlCl_3}$

c) + O O O $\xrightarrow{AlCl_3}$ O HO O $\xrightarrow[2)\ AlCl_3]{1)\ SOCl_2}$

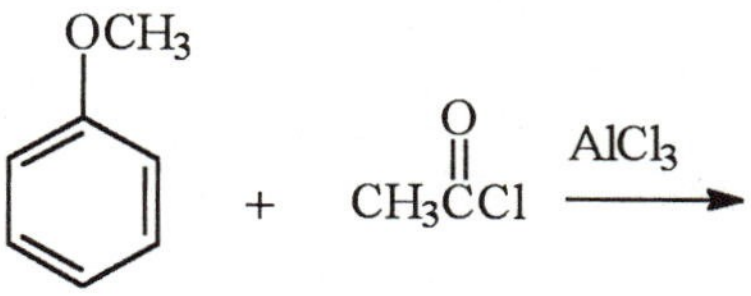

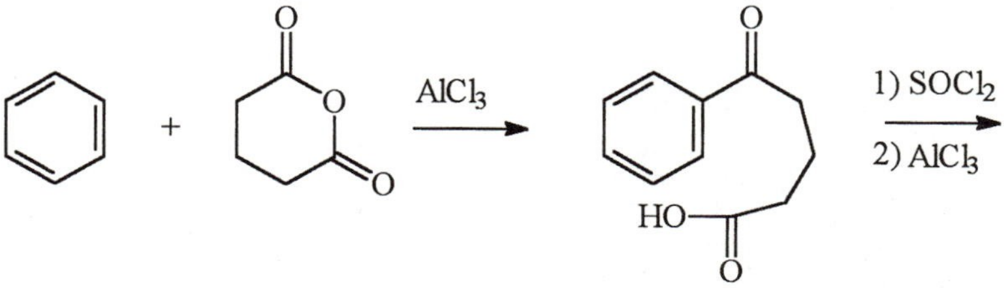

18.22

Br H ⊕ ⟷ Br H ⊕ ⟷ Br H ⊕ ⟷ Br H ⊕

Br H ⊕ ⟷ ⊕ Br H ⟷ Br H ⊕

⊕ Br H ⟷ ⊕ Br H ⟷ ⊕ Br H

Br H ⊕ ⟷ Br H ⊕ ⟷ ⊕ Br H

18.23

a) Cl

b) CH_3, $\overset{O}{\underset{}{C}}CH_3$

c) HO, O, O

18.24

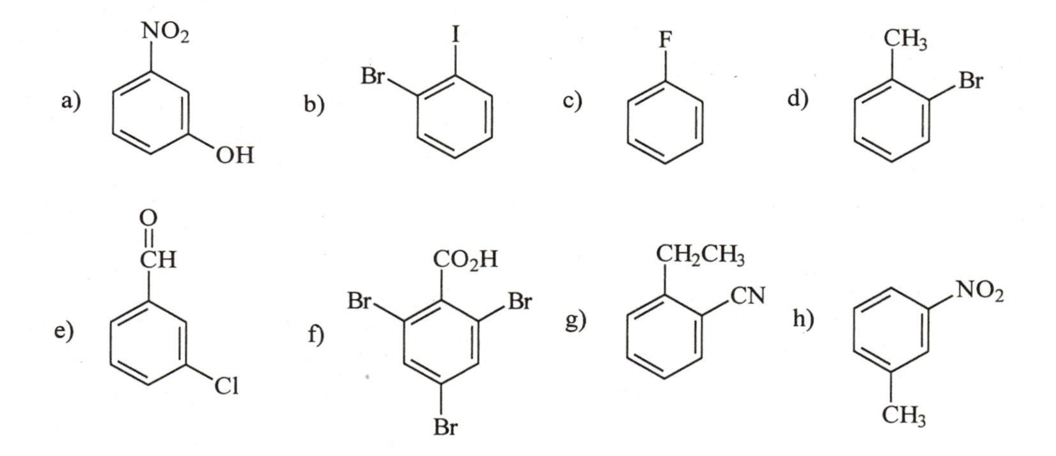

18.25 A nitro group in the para position is better at stabilizing the negative charge than is a nitro group in the meta position, but a nitro group in the meta position does help somewhat by its inductive effect.

Cl, NO_2 (meta) < Cl, NO_2 (para) < Cl, NO_2, NO_2

slowest   fastest

18.26

a) NO_2 ... $\oplus$ N–H $F^{\ominus}$

b) NO_2, OH, $\overset{O}{\overset{\|}{C}}CH_3$

18.27

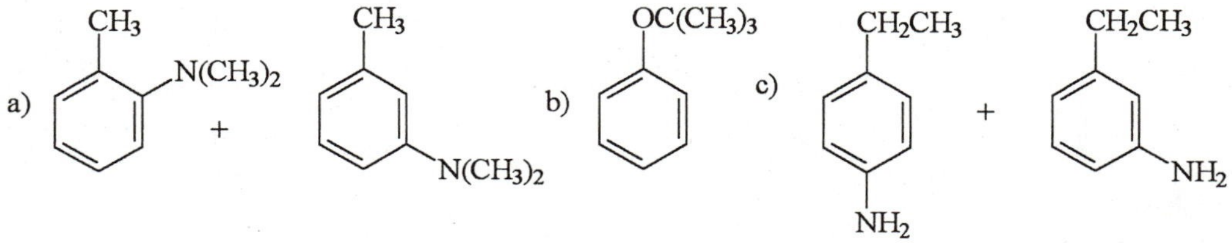

a)
CH3
N(CH3)2
+
CH3
N(CH3)2
b)
OC(CH3)3
c)
CH2CH3
NH2
+
CH2CH3
NH2

18.28

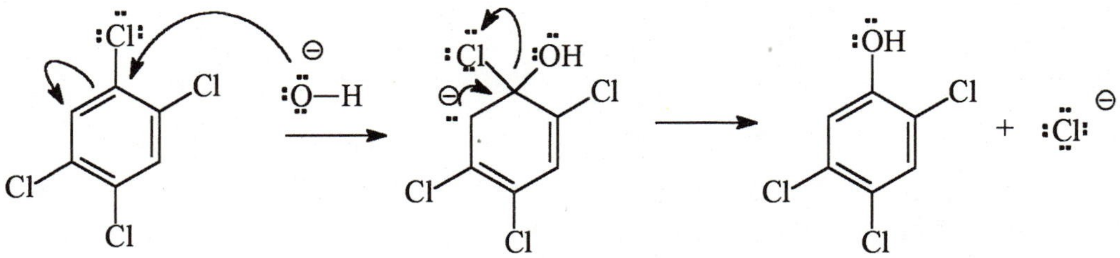

If a benzyne intermediate were involved, two products would be expected.

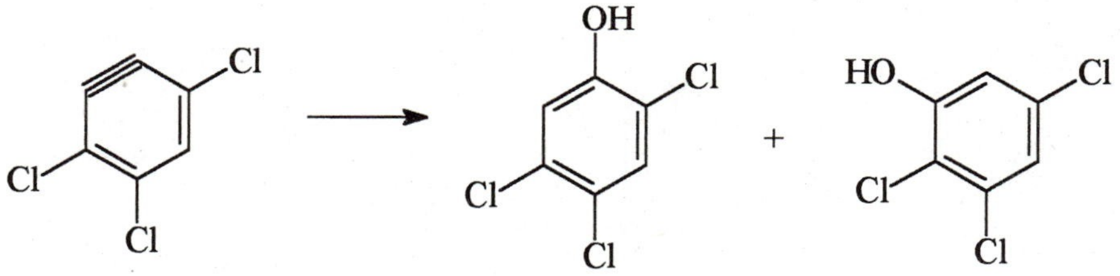

18.29

a) 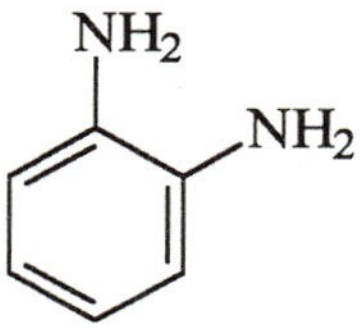 $C_6H_4(NH_2)_2$ (ortho: NH_2, NH_2)

b) $C_6H_5CH_2CH_3$

c) CO_2H, NO_2 (meta)

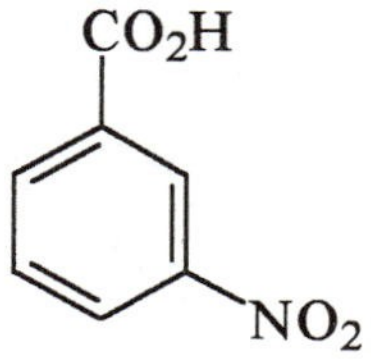

d) 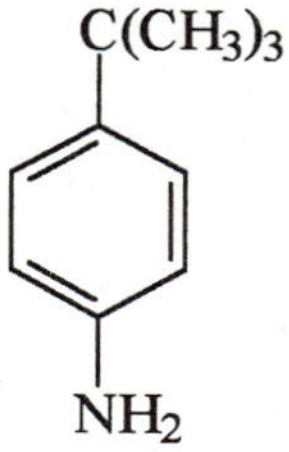 $C(CH_3)_3$, NH_2 (para)

e) 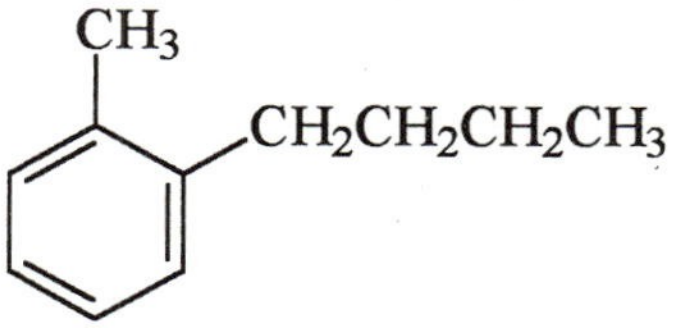 CH_3, $CH_2CH_2CH_2CH_3$ (ortho)

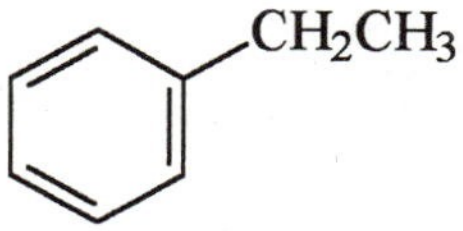

f) 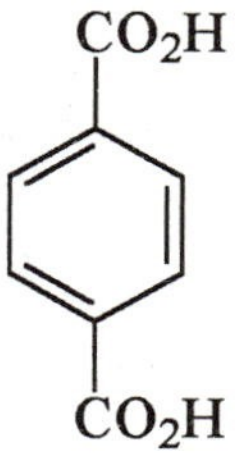 CO_2H, CO_2H (para)

g) 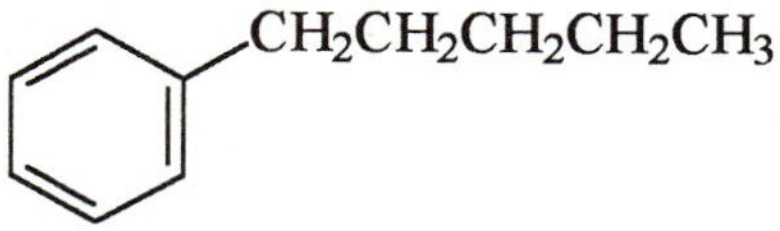 $C_6H_5CH_2CH_2CH_2CH_2CH_3$

18.30

a) 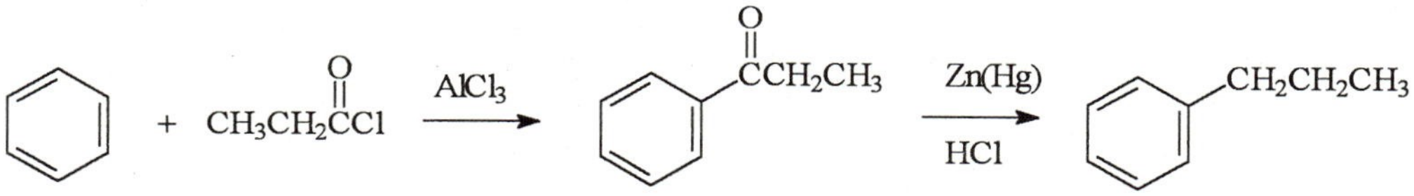 C_6H_6 + $CH_3CH_2\overset{O}{\overset{\|}{C}}Cl$ $\xrightarrow{AlCl_3}$ $C_6H_5\overset{O}{\overset{\|}{C}}CH_2CH_3$ $\xrightarrow[HCl]{Zn(Hg)}$ $C_6H_5CH_2CH_2CH_3$

b) 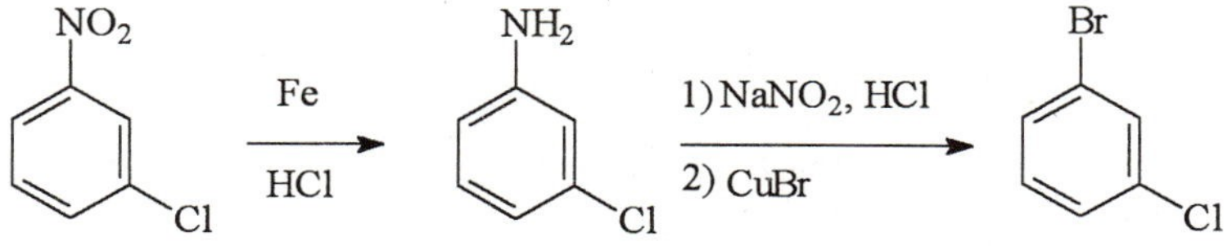 NO_2, Cl (meta) $\xrightarrow[HCl]{Fe}$ NH_2, Cl (meta) $\xrightarrow[2)\ CuBr]{1)\ NaNO_2,\ HCl}$ Br, Cl (meta)

18.31

a) benzene $\xrightarrow[AlCl_3]{CH_3\overset{O}{\overset{\|}{C}}Cl}$ $C_6H_5\overset{O}{\overset{\|}{C}}CH_3$ $\xrightarrow[AlCl_3]{Cl_2}$ m-$ClC_6H_4\overset{O}{\overset{\|}{C}}CH_3$

b) benzene $\xrightarrow[FeBr_3]{Br_2}$ C_6H_5Br $\xrightarrow{H_2SO_4}$ p-$BrC_6H_4SO_3H$

c) benzene $\xrightarrow{H_2SO_4}$ $C_6H_5SO_3H$ $\xrightarrow[FeBr_3]{Br_2}$ m-$BrC_6H_4SO_3H$

d) benzene $\xrightarrow[H_2SO_4]{HNO_3}$ $C_6H_5NO_2$ $\xrightarrow[AlBr_3]{Br_2}$ m-$BrC_6H_4NO_2$ $\xrightarrow[Pd]{H_2}$ m-$BrC_6H_4NH_2$

e) benzene $\xrightarrow[AlCl_3]{CH_3CH_2CH_2CH_2COCl}$ $C_6H_5COCH_2CH_2CH_2CH_3$ $\xrightarrow[HCl]{Zn(Hg)}$ $C_6H_5CH_2CH_2CH_2CH_2CH_3$

f) benzene $\xrightarrow[AlCl_3]{CH_3CH_2\overset{O}{\overset{\|}{C}}Cl}$ $C_6H_5\overset{O}{\overset{\|}{C}}CH_2CH_3$ $\xrightarrow[H_2SO_4]{HNO_3}$ m-$O_2NC_6H_4\overset{O}{\overset{\|}{C}}CH_2CH_3$

h) Benzene $\xrightarrow[H_2SO_4]{HNO_3}$ nitrobenzene (NO_2) $\xrightarrow[AlCl_3]{Cl_2}$ m-chloronitrobenzene (NO_2, Cl) $\xrightarrow[HCl]{Fe}$ m-chloroaniline (NH_2, Cl) $\xrightarrow[\text{2) CuCl}]{\text{1) } NaNO_2\text{, HCl}}$ m-dichlorobenzene (Cl, Cl)

i) Benzene $\xrightarrow[H_2SO_4]{HNO_3}$ nitrobenzene (NO_2) $\xrightarrow[HCl]{Fe}$ aniline (NH_2) $\xrightarrow{(CH_3\overset{O}{\overset{\|}{C}})_2O}$ acetanilide ($NH\overset{O}{\overset{\|}{C}}CH_3$) $\xrightarrow{HNO_3}$ p-nitroacetanilide ($NH\overset{O}{\overset{\|}{C}}CH_3$, NO_2) $\xrightarrow[H_2O]{KOH}$ p-nitroaniline (NH_2, NO_2) $\xrightarrow[\text{2) CuCN}]{\text{1) } NaNO_2\text{, HCl}}$ p-nitrobenzonitrile (CN, NO_2)

18.32

b) Benzene $\xrightarrow[AlCl_3]{Cl_2}$ chlorobenzene (Cl) $\xrightarrow{H_2SO_4}$ p-chlorobenzenesulfonic acid (Cl, SO_3H) $\xrightarrow[H_2SO_4,\ \Delta]{HNO_3}$ (O_2N, Cl, NO_2, SO_3H) $\xrightarrow[H_2SO_4,\ \Delta]{H_2O}$ (O_2N, Cl, NO_2)

c) Benzene $\xrightarrow[AlCl_3]{\text{succinic anhydride}}$ (HO, O; O) $\xrightarrow[Pd]{H_2}$ (HO, O) $\xrightarrow[\text{2) } AlCl_3]{\text{1) } SOCl_2}$ (O)

d) Benzene $\xrightarrow[AlCl_3]{CH_3I}$ toluene (CH_3) $\xrightarrow[H_2SO_4,\ \Delta]{HNO_3}$ (CH_3, O_2N, NO_2, NO_2) $\xrightarrow[NaOH]{KMnO_4}$ (CO_2H, O_2N, NO_2, NO_2)

18.33

a) 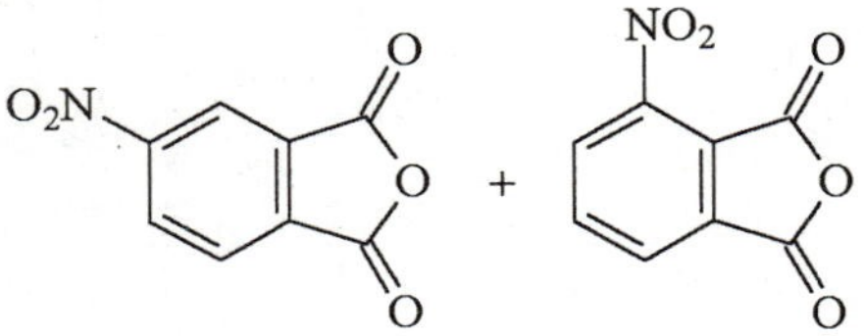 O_2N O O O + NO_2 O O O b) 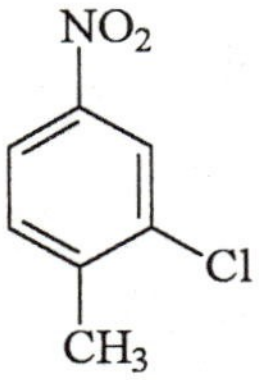 NO_2 Cl CH_3 c) 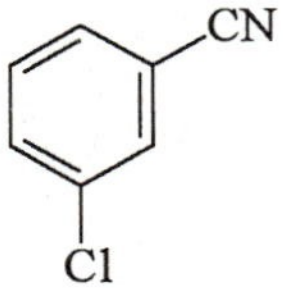 CN Cl

d) 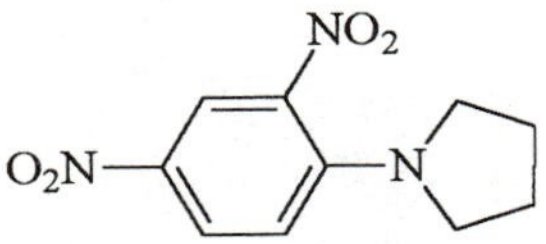 NO_2 O_2N— —N e) 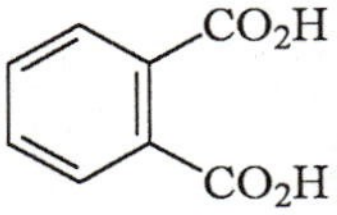 CO_2H CO_2H f) 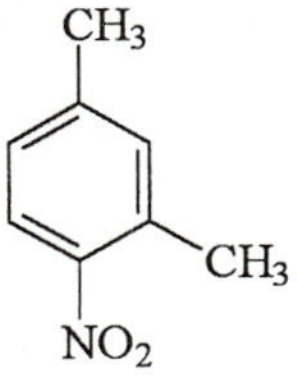 CH_3 CH_3 NO_2

g) 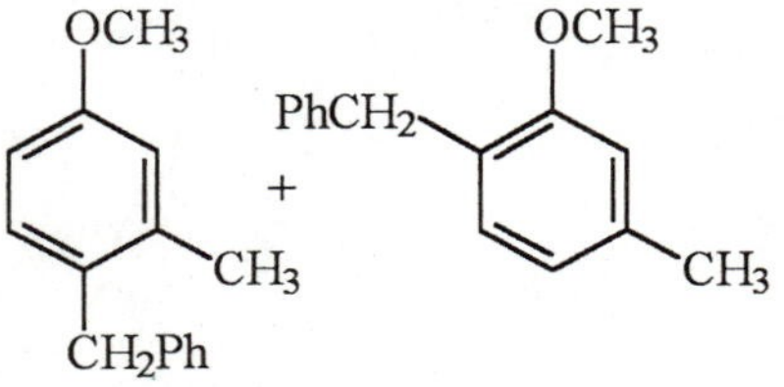 OCH_3 CH_3 CH_2Ph + OCH_3 $PhCH_2$ CH_3 h) 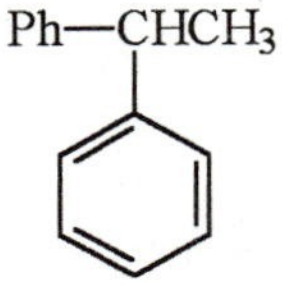 Ph—$CHCH_3$ i) 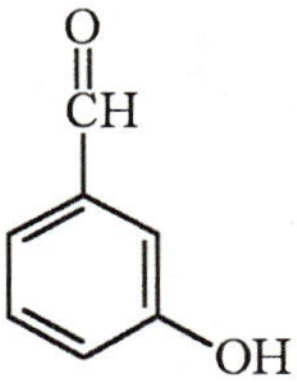 O ‖ CH OH

j)  O ‖ $NHCCH_3$ Br CH_3 k) 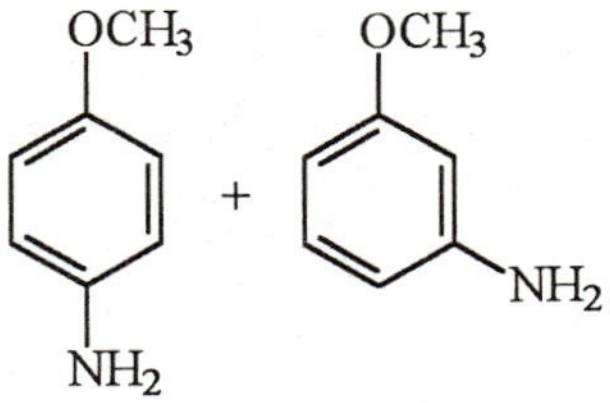 OCH_3 NH_2 + OCH_3 NH_2 l) 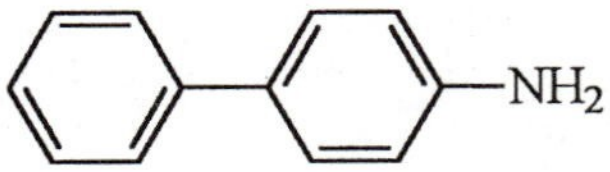 —NH_2

m) CH_3 / CH_3CHCH_3 + CH_3 CH_3 / $CHCH_3$

n) Br / SO_3H + Br / SO_3H

o)

p) O

q)

r) OH

18.34

a) NO_2 / NO_2

b) NO_2 / Cl

c) NH_2

d) no reaction

e) NO_2 / SO_3H

f) Br

18.35

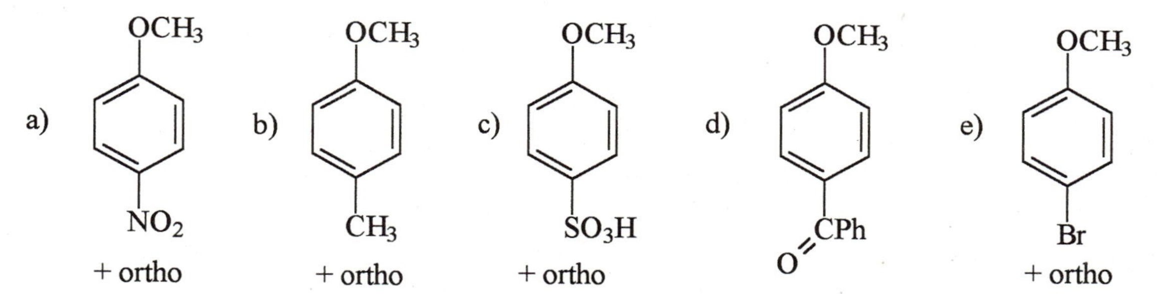

18.36

a) 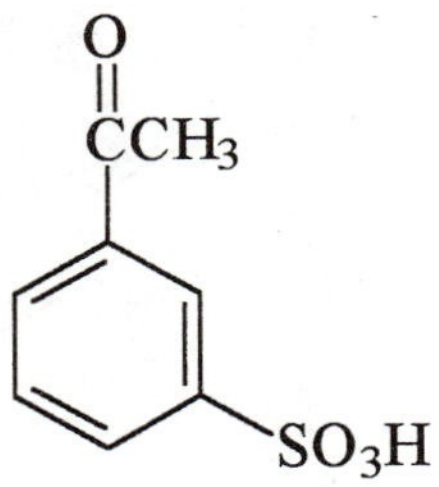 O ‖ CCH_3 SO_3H

b) 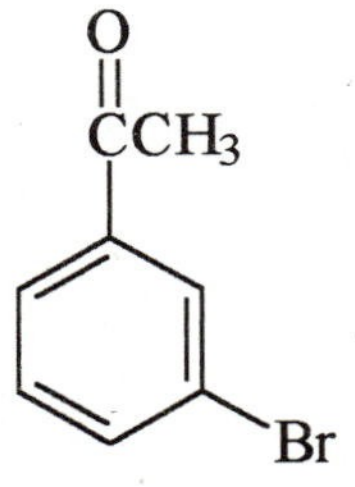 O ‖ CCH_3 Br

c) 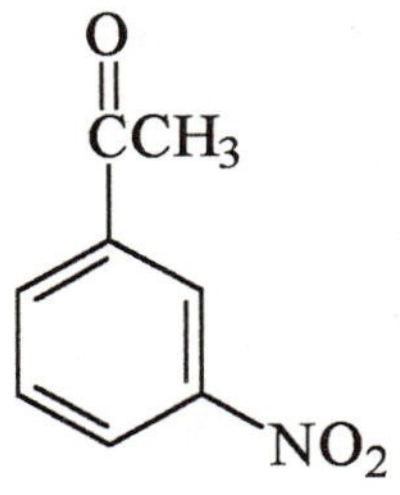 O ‖ CCH_3 NO_2

d) , e) , and f) 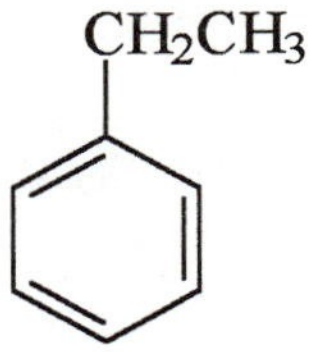 CH_2CH_3

18.37

a) 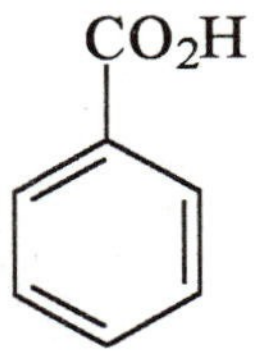 CO_2H

b) 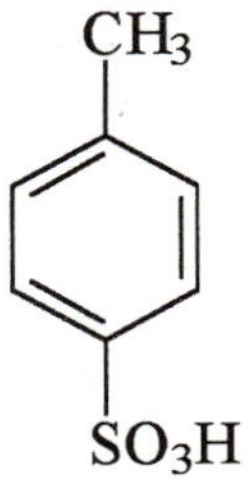 CH_3 SO_3H + ortho

c) 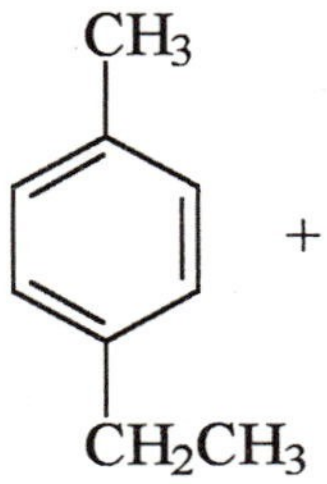 CH_3 CH_2CH_3 + ortho

d) 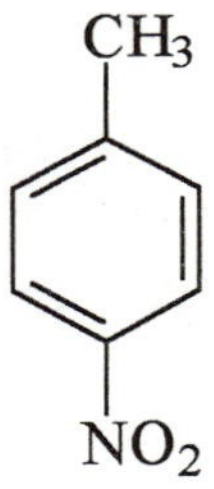 CH_3 NO_2 + ortho

e) 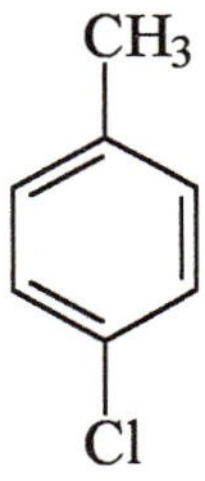 CH_3 Cl + ortho

f) 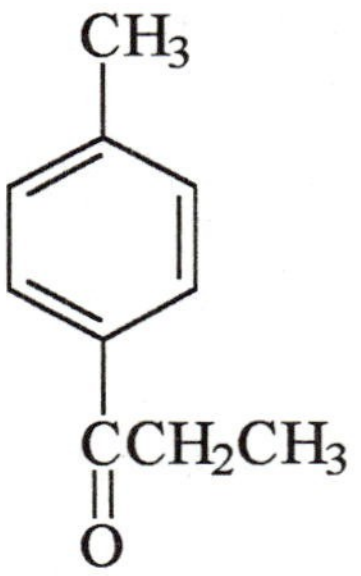 CH_3 CCH_2CH_3 ‖ O

18.38

a) 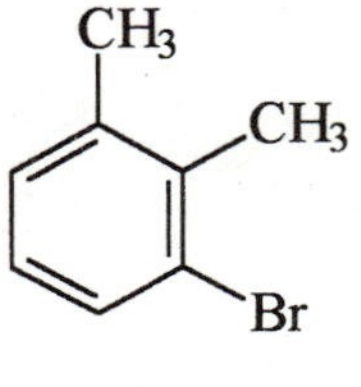 CH_3 CH_3 Br + CH_3 CH_3 Br

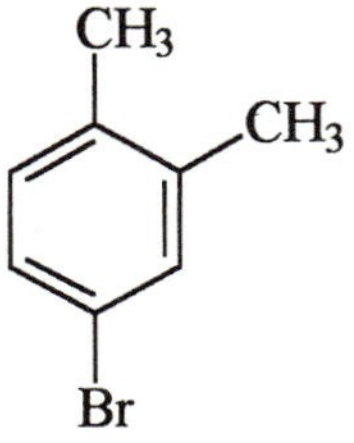

b) 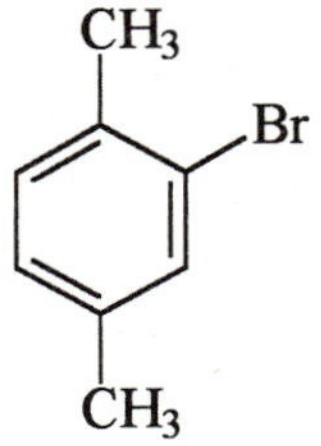 CH_3 Br CH_3

c) 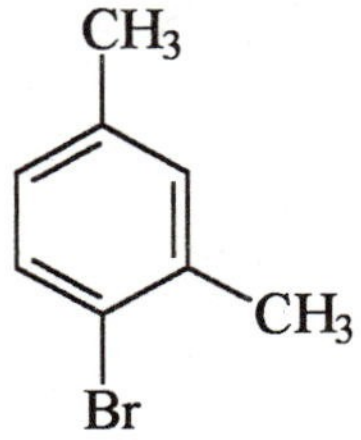 CH_3 CH_3 Br

d) 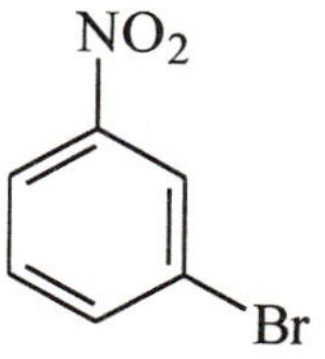 NO_2 … Br e) 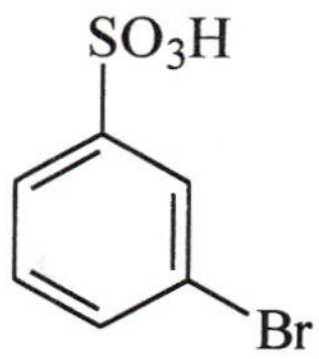 SO_3H … Br f) 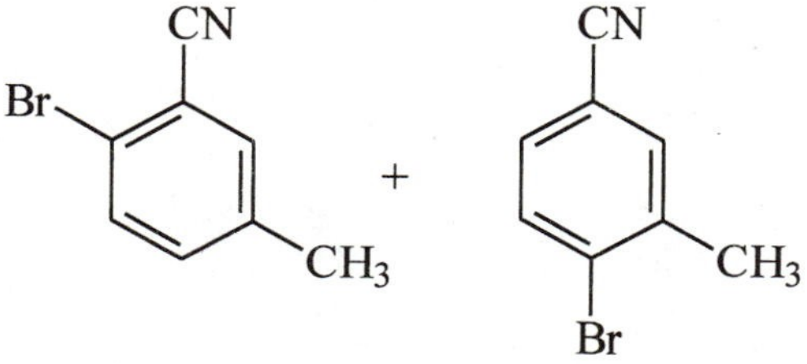 Br, CN, CH_3 + CN, CH_3, Br

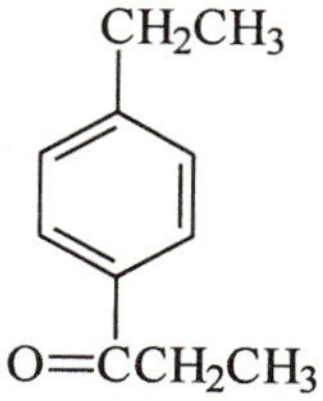

18.39

a) 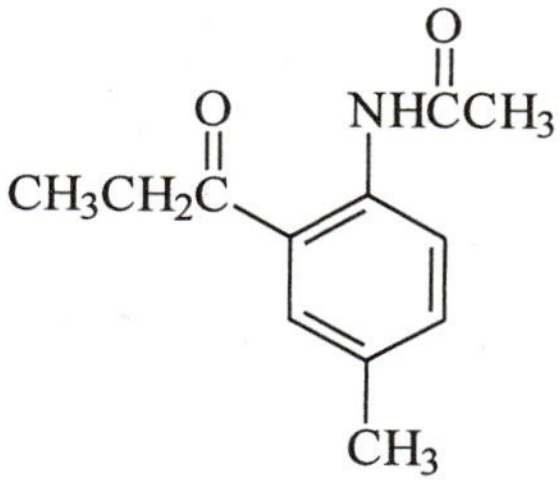 $CH_3CH_2\overset{O}{\overset{\|}{C}}$, $NH\overset{O}{\overset{\|}{C}}CH_3$, CH_3 b) CH_2CH_3, $O{=}CCH_2CH_3$ c) $\overset{O}{\overset{\|}{C}}CH_2CH_3$

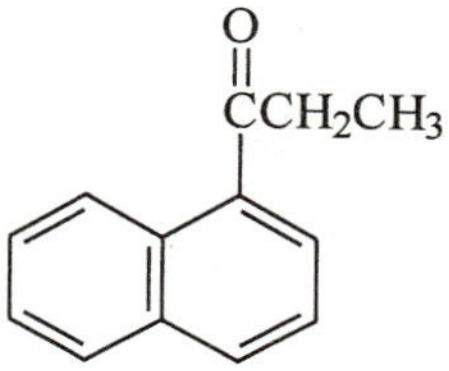

d) 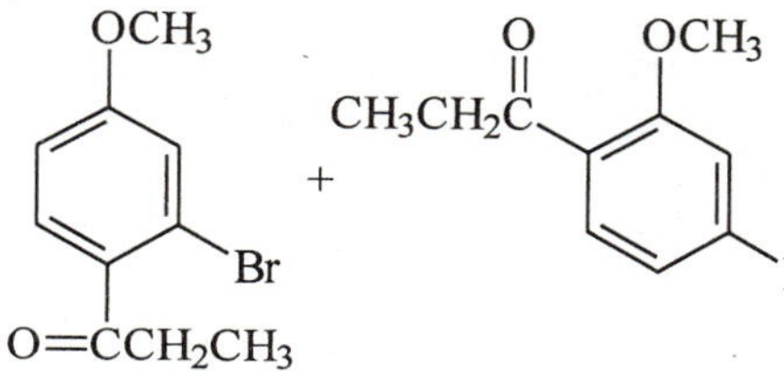 OCH_3, Br, $O{=}CCH_2CH_3$ + $CH_3CH_2\overset{O}{\overset{\|}{C}}$, OCH_3, Br e) 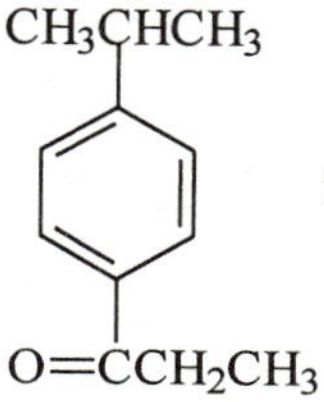 CH_3CHCH_3, $O{=}CCH_2CH_3$ f) 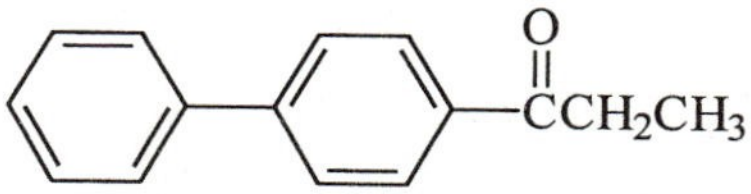 $\overset{O}{\overset{\|}{C}}CH_2CH_3$

18.40

a) H_2SO_4 b) 1) HNO_3, H_2SO_4 2) Fe, HCl

c) $CH_3\overset{Cl}{\overset{|}{C}}HCH_2CH_3$, $AlCl_3$ d) 1) $CH_3CH_2CH_2\overset{O}{\overset{\|}{C}}Cl$, $AlCl_3$ 2) Zn(Hg), HCl

e) 1) CH_3Cl, $AlCl_3$ 2) $KMnO_4$, NaOH, Δ 3) $H_3O^{\oplus}$

f) 1) H_2, Pt 2) $NaNO_2$, $H_3O^{\oplus}$ 3) CuCN g) Cl_2, $AlCl_3$ h) H_2SO_4, H_2O

18.41

a) 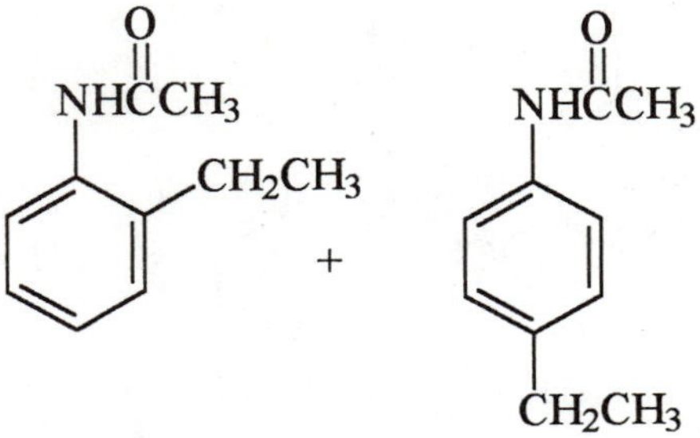 $NHCCH_3$ (O) ... CH_2CH_3 + $NHCCH_3$ (O) ... CH_2CH_3

b) 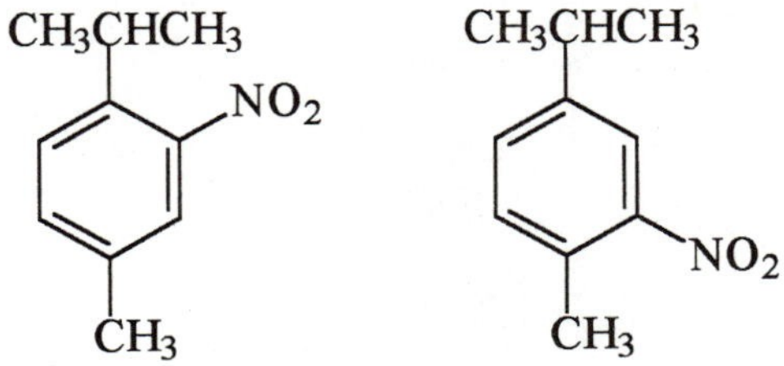 CH_3CHCH_3, NO_2, CH_3 CH_3CHCH_3, NO_2, CH_3

c) 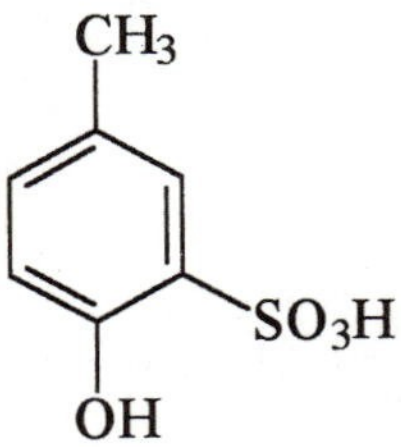 CH_3, SO_3H, OH

d) 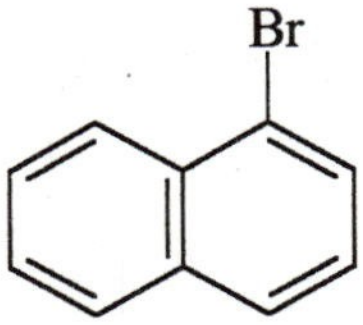 Br

e)

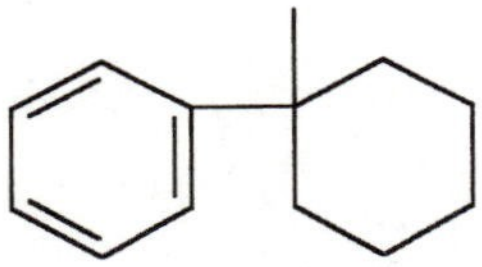

f) 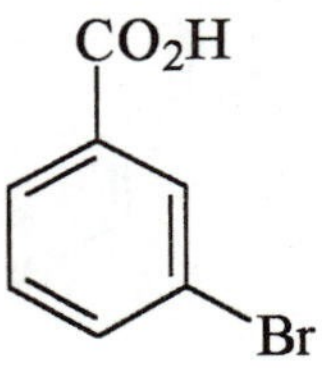 CO_2H, Br

g) 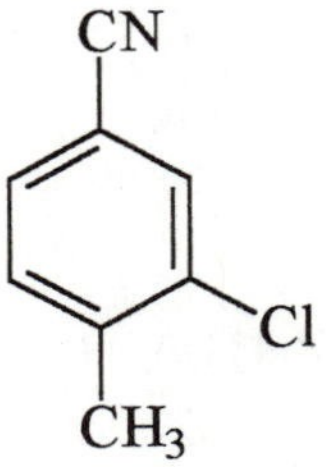 CN, Cl, CH_3

h) 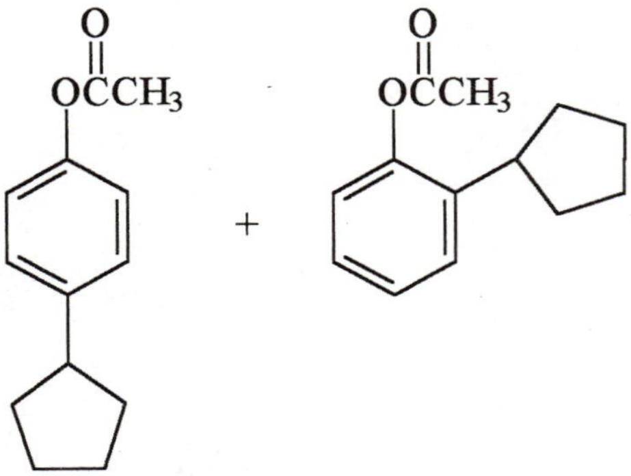 $OCCH_3$ (O) + $OCCH_3$ (O)

i) 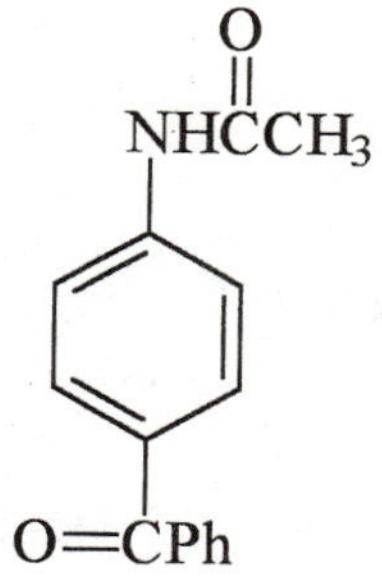 p-CH_3CONH–C_6H_4–$COPh$ (NHCCH$_3$ with C=O; O=CPh)

j) 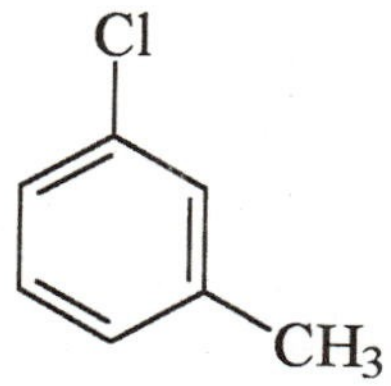 Cl, CH_3 (meta-chlorotoluene)

k) 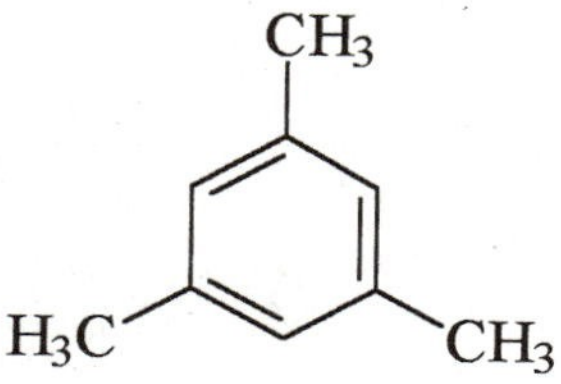 CH_3, H_3C, CH_3 (1,3,5-trimethylbenzene)

l) 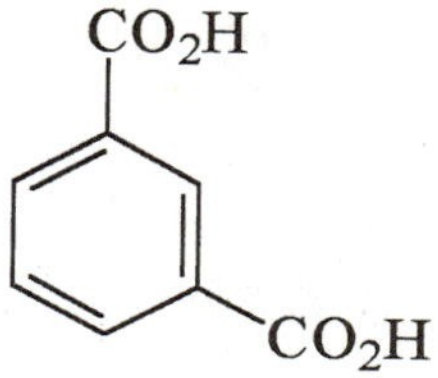 CO_2H, CO_2H (isophthalic acid)

18.42

a) 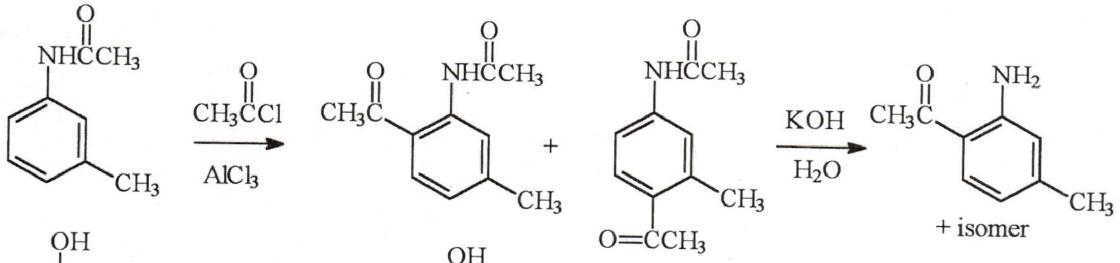 m-$CH_3C_6H_4NHCOCH_3$ (NHCCH$_3$, CH$_3$) $\xrightarrow[AlCl_3]{CH_3CCl}$ (CH$_3$C(=O), NHCCH$_3$, CH$_3$) + (NHCCH$_3$, CH$_3$, O=CCH$_3$) $\xrightarrow[H_2O]{KOH}$ (CH$_3$C(=O), NH$_2$, CH$_3$) + isomer

b) (OH, NH$_2$) $\xrightarrow[\text{2) CuCN}]{\text{1)}NaNO_2\text{, HCl}}$ (OH, CN)

c) (CCH$_2$CH$_3$ with C=O, SO$_3$H) $\xrightarrow[HCl]{Zn(Hg)}$ (CH$_2$CH$_2$CH$_3$, SO$_3$H)

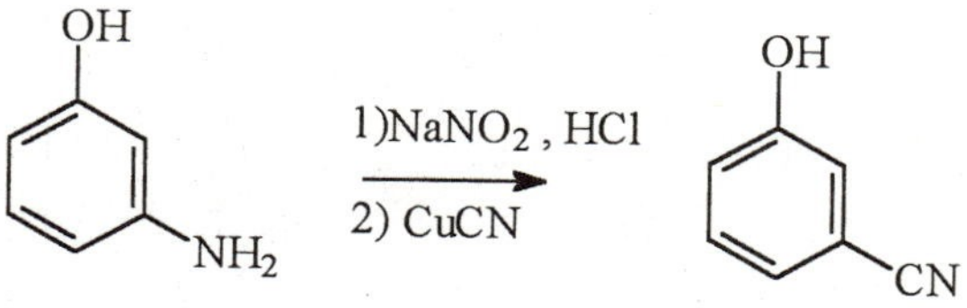

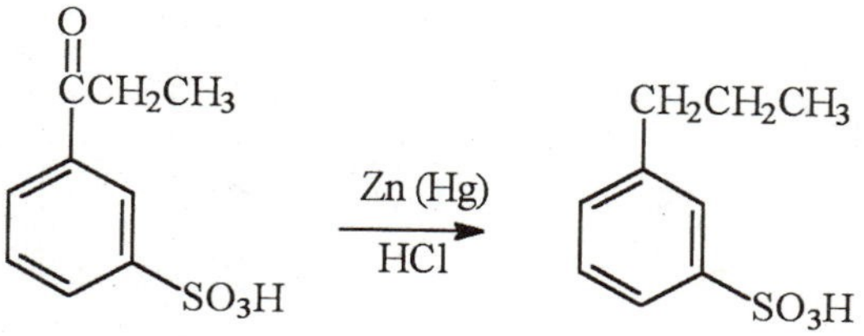

18.43

a) 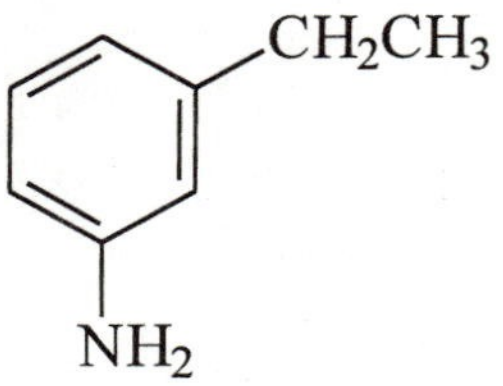 CH_2CH_3 / NH_2

b) 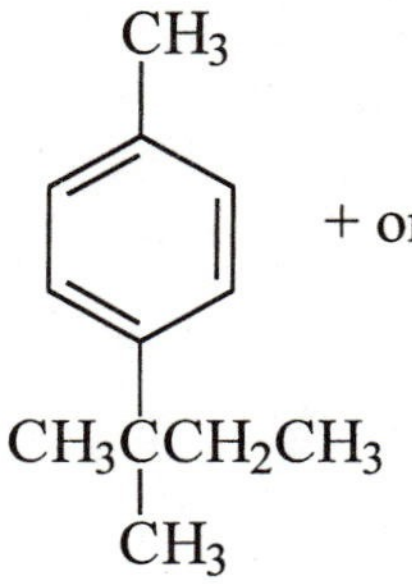 CH_3 / $CH_3\overset{|}{C}CH_2CH_3$ / CH_3 + ortho

c) 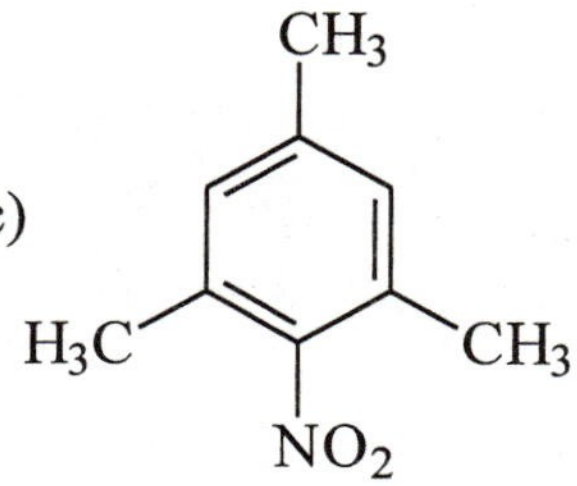 CH_3 / H_3C / CH_3 / NO_2

d) 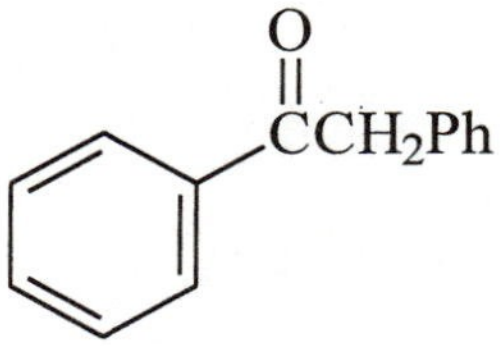 $\overset{O}{\overset{||}{C}}CH_2Ph$

e) 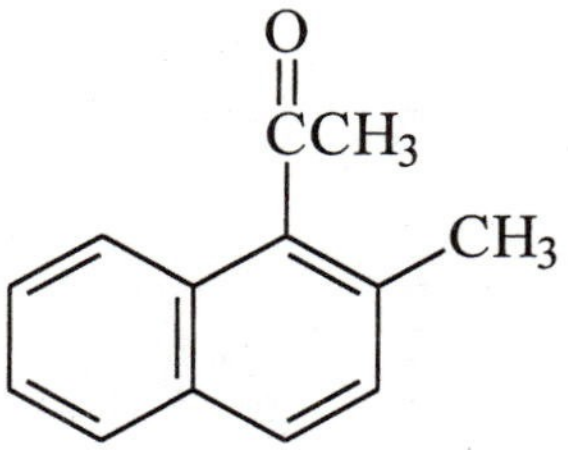 $\overset{O}{\overset{||}{C}}CH_3$ / CH_3

f) 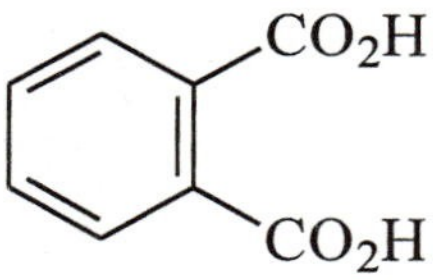 CO_2H / CO_2H

18.44

a) 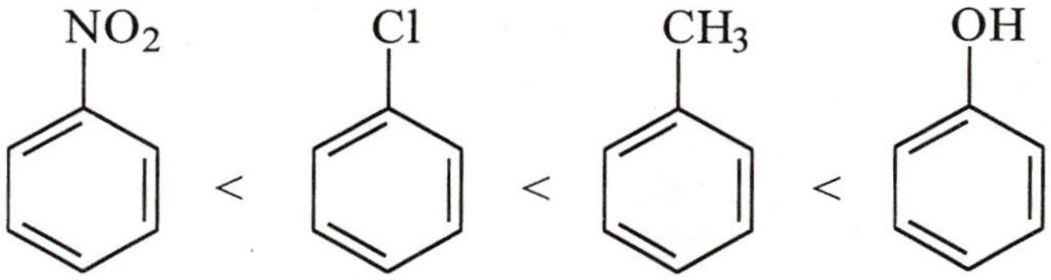 NO_2 < Cl < CH_3 < OH

slowest ... fastest

b) 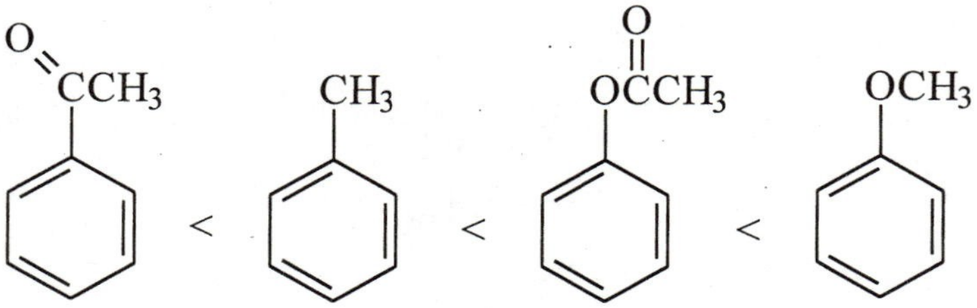 $\overset{O}{\overset{||}{C}}CH_3$ < CH_3 < $O\overset{O}{\overset{||}{C}}CH_3$ < OCH_3

slowest ... fastest

18.45

a) Benzene $\xrightarrow[AlCl_3]{CH_3Cl}$ toluene (CH_3) $\xrightarrow[H_2SO_4]{HNO_3}$ p-nitrotoluene (CH_3, NO_2) $\xrightarrow[2)\ H_3O^{\oplus}]{1)\ KMnO_4,\ NaOH,\ \Delta}$ p-nitrobenzoic acid (CO_2H, NO_2)

b) Benzene $\xrightarrow[AlCl_3]{CH_3CH_2COCl}$ propiophenone $\xrightarrow[FeBr_3]{Br_2}$ m-bromopropiophenone (Br) $\xrightarrow[Pd]{H_2}$ 1-bromo-3-propylbenzene (Br)

c) Benzene $\xrightarrow[H_2SO_4]{HNO_3}$ nitrobenzene (NO_2) $\xrightarrow[FeBr_3]{Br_2}$ m-bromonitrobenzene (NO_2, Br)

d) $\xrightarrow{\text{see (a)}}$ propiophenone $\xrightarrow[H_2SO_4]{HNO_3}$ m-nitropropiophenone (NO_2)

e) Benzene $\xrightarrow[AlCl_3]{(CH_3)_2CHCl}$ isopropylbenzene $\xrightarrow[\Delta]{H_2SO_4}$ p-isopropylbenzenesulfonic acid (SO_3H) $\xrightarrow[H_2SO_4]{HNO_3}$ 4-isopropyl-3-nitrobenzenesulfonic acid (NO_2, SO_3H) $\xrightarrow[H_2O,\ \Delta]{H_2SO_4}$ o-nitroisopropylbenzene (NO_2)

f) Benzene $\xrightarrow[AlCl_3]{CH_3CH_2Cl}$ ethylbenzene (CH_2CH_3) $\xrightarrow[\Delta]{H_2SO_4}$ p-ethylbenzenesulfonic acid (CH_2CH_3, SO_3H)

18.46

a) Toluene (CH_3 on a benzene ring) $\xrightarrow[H_2SO_4]{HNO_3}$ p-nitrotoluene (CH_3, NO_2) $\xrightarrow[HCl]{Fe}$ p-toluidine (CH_3, NH_2) $\xrightarrow{CH_3\overset{O}{\overset{\|}{C}}O\overset{O}{\overset{\|}{C}}CH_3}$ p-methylacetanilide (CH_3, $NH\overset{\|}{\underset{O}{C}}CH_3$)

$\xrightarrow{Cl_2}$ (CH_3, Cl, $NH\overset{\|}{\underset{O}{C}}CH_3$) $\xrightarrow[H_2O]{KOH}$ (CH_3, Cl, NH_2) $\xrightarrow[2)\ CuCN]{1)\ NaNO_2,\ H_3O^{\oplus}}$ (CH_3, Cl, CN)

b) Nitrobenzene (NO_2) $\xrightarrow[HCl]{Sn}$ aniline (NH_2) $\xrightarrow{CH_3\overset{O}{\overset{\|}{C}}O\overset{O}{\overset{\|}{C}}CH_3}$ acetanilide ($NH\overset{O}{\overset{\|}{C}}CH_3$)

$\xrightarrow[AlCl_3]{CH_3CH_2\overset{O}{\overset{\|}{C}}Cl}$ ($NH\overset{O}{\overset{\|}{C}}CH_3$, $O{=}CCH_2CH_3$) $\xrightarrow[H_2O]{KOH}$ (NH_2, $O{=}CCH_2CH_3$) $\xrightarrow[2)\ CuCN]{1)\ NaNO_2,\ H_3O^{\oplus}}$ (CN, $O{=}CCH_2CH_3$)

c)

NO_2 NO_2

HNO_3 / H_2SO_4 → ; Cl_2 / $AlCl_3$ → Cl

↓ Fe, HCl

OH NH_2

1) $NaNO_2$, $H_3O^{\oplus}$ ←
2) $H_3O^{\oplus}$, Δ

Cl Cl

d)

HNO_3 / H_2SO_4 → ; Fe, HCl →

NO_2 NH_2

↓ $CH_3\overset{O}{\overset{\|}{C}}O\overset{O}{\overset{\|}{C}}CH_3$

1) KOH, H_2O ←
2) $NaNO_2$, $H_3O^{\oplus}$
3) $NaBH_4$

Cl_2 ←

Cl Cl

$NH\underset{\overset{\|}{O}}{C}CH_3$ $NH\underset{\overset{\|}{O}}{C}CH_3$

e) NH_2 (aniline) —excess Br_2→ 2,4,6-tribromoaniline (NH_2, Br, Br, Br) —1) $NaNO_2$, $H_3O^{\oplus}$; 2) H_3PO_2→ 1,3,5-tribromobenzene (Br, Br, Br)

f) NO_2 (nitrobenzene) —Cl_2, $AlCl_3$→ NO_2, Cl (3-chloronitrobenzene) —Fe, HCl→ NH_2, Cl (3-chloroaniline)

—$CH_3\overset{O}{\overset{\|}{C}}O\overset{O}{\overset{\|}{C}}CH_3$→ $NH\overset{O}{\overset{\|}{C}}CH_3$, Cl —$Br_2$→ $NH\overset{O}{\overset{\|}{C}}CH_3$, Cl, Br

—1) KOH, H_2O; 2) $NaNO_2$, $H_3O^{\oplus}$; 3) KI→ I, Cl, Br

g)

CH_3Cl / $AlCl_3$ → CH_3 → H_2SO_4 → CH_3, SO_3H → HNO_3 / H_2SO_4 → CH_3, NO_2, SO_3H

H_2O / H_2SO_4 ↓

CH_3, NO_2 → Fe, HCl → CH_3, NH_2 → 1) $NaNO_2$, $H_3O^{\oplus}$ / 2) $H_3O^{\oplus}$, Δ → CH_3, OH

18.47

Br_3Fe $:\ddot{Br}-\ddot{Br}:$ → $\overset{\ominus}{Br_3Fe}-\overset{\oplus}{\ddot{Br}}-\ddot{Br}:$ → ⊕ Br, H

$\overset{\ominus}{Br_3Fe}-\ddot{Br}:$

↓

Br

$FeBr_3$ + HBr +

18.48

$H{-}\ddot{O}{-}\overset{\oplus}{N}(=\ddot{O}{:}){-}\ddot{O}{:}^{\ominus}$ $\rightleftharpoons$ $H{-}\overset{\oplus}{\ddot{O}}(H){-}\overset{\oplus}{N}(=\ddot{O}{:}){-}\ddot{O}{:}^{\ominus}$ $\rightleftharpoons$ ${:}\ddot{O}{=}\overset{\oplus}{N}{=}\ddot{O}{:}$ + H_2O

$H{-}\ddot{O}{-}S(=O)_2{-}OH$

CH_3

$HO{-}S(=O)_2{-}\ddot{O}{:}^{\ominus}$ H ${:}\ddot{O}{:}^{\ominus}{-}\overset{\oplus}{N}{=}\ddot{O}{:}$ $\oplus$ CH_3

NO_2 CH_3 + H_2SO_4 $\longleftarrow$

18.49

CH_3
$CH_3CHCH_2{-}\ddot{Cl}{:}$ $AlCl_3$ $\longrightarrow$ CH_3 $CH_3CHCH_2{-}\overset{\oplus}{\ddot{Cl}}{-}\overset{\ominus}{AlCl_3}$ $\longrightarrow$ CH_3 $CH_3{-}C(H){-}\overset{\oplus}{C}H_2$

$\downarrow$

CH_3 $CH_3{-}\overset{\oplus}{C}{-}CH_3$ (+ benzene) $\longleftarrow$ H_3C CH_3 H C CH_3 $\oplus$ $\longleftarrow$ $\overset{\ominus}{Cl_3Al}{-}\ddot{Cl}{:}$

$AlCl_3$ + HCl + $H_3C{-}C(CH_3){-}CH_3$ (phenyl)

18.50

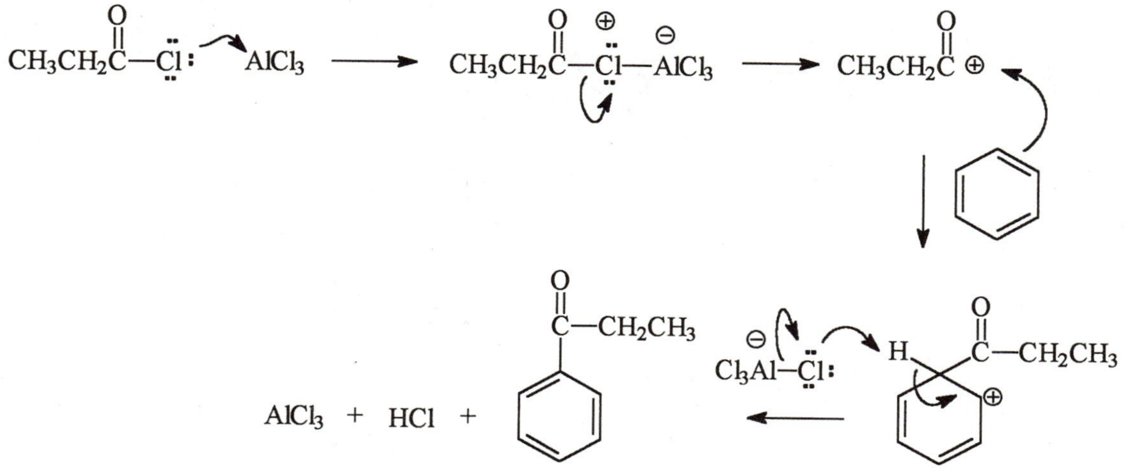

$$CH_3CH_2\overset{O}{\overset{\|}{C}}-\ddot{Cl}: + AlCl_3 \longrightarrow CH_3CH_2\overset{O}{\overset{\|}{C}}-\overset{\oplus}{Cl}-\overset{\ominus}{AlCl_3} \longrightarrow CH_3CH_2\overset{O}{\overset{\|}{C}}\oplus$$

$$CH_3CH_2\overset{O}{\overset{\|}{C}}\oplus + C_6H_6 \longrightarrow \text{arenium ion (H, } \overset{O}{\overset{\|}{C}}-CH_2CH_3\text{)}$$

$$\text{arenium ion} + Cl_3\overset{\ominus}{Al}-\ddot{Cl}: \longrightarrow AlCl_3 + HCl + C_6H_5\overset{O}{\overset{\|}{C}}-CH_2CH_3$$

18.51

$$\text{4-chloro-3-nitroacetophenone} + \ddot{N}H_2CH_2CH_3 \longrightarrow \text{carbanion intermediate (Cl, } \overset{\oplus}{N}H_2CH_2CH_3\text{, } NO_2\text{, } \overset{O}{\overset{\|}{C}}CH_3\text{)} \longrightarrow \overset{\ominus}{Cl}\ \ \oplus NH_2CH_2CH_3 \text{ (4-acetyl-2-nitrophenyl)}$$

$$\updownarrow$$

$$:\ddot{O}:^{\ominus}-\overset{}{C}CH_3 = \text{ring (Cl, } \oplus NH_2CH_2CH_3\text{, } NO_2\text{)}$$

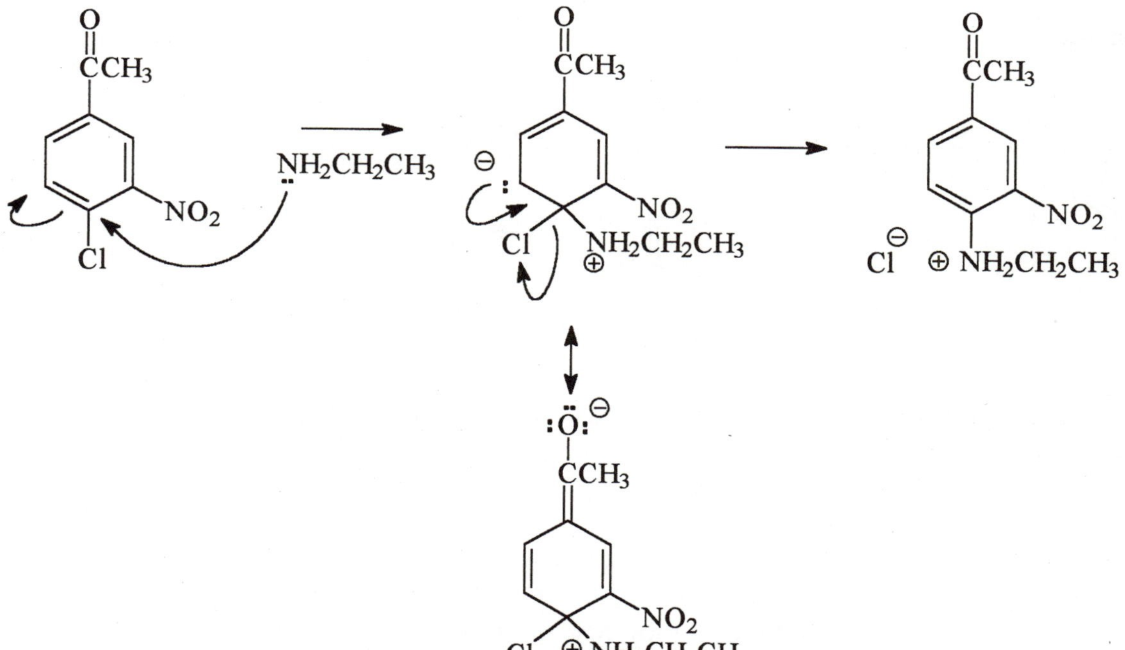

The acetyl group stabilizes the carbanion intermediate by resonance, as shown by the resonance structure above.

18.52

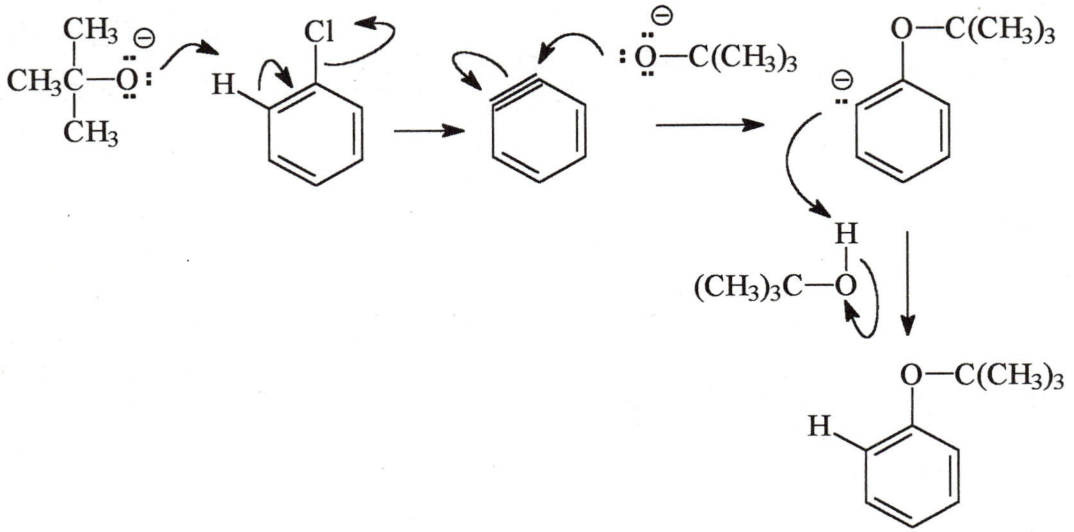

CH_3
$CH_3C—O:^{\ominus}$
CH_3
H
Cl
$:O—C(CH_3)_3$
$O—C(CH_3)_3$
H
$(CH_3)_3C—O$
$O—C(CH_3)_3$
H

18.53

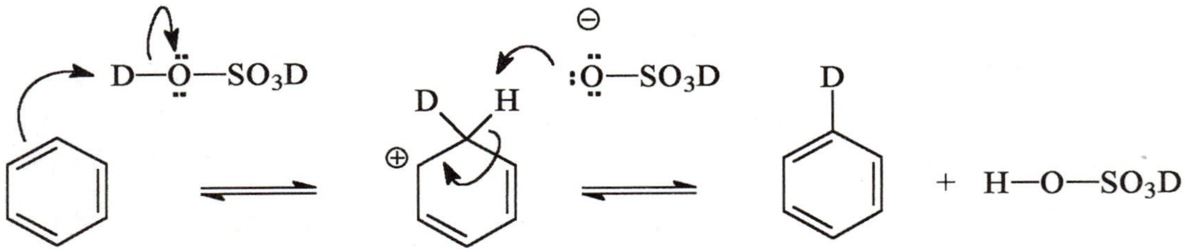

$D—O—SO_3D$
D
H
$:O—SO_3D$
D
$+ \ H—O—SO_3D$

18.54 The amount of ortho product decreases as the size (steric bulk) of the alkyl group increases, indicating that the bromination reaction is somewhat subject to steric effects.

18.55 Both of these groups have a positively charged atom and thus should be inductive electron withdrawing groups. The S has an unshared pair of electrons, but, because of its positive charge, it does not want to donate them by resonance. Therefore, both groups should be deactivating and direct substitution to the meta position.

18.56

$$\text{Benzene} \xrightarrow[\text{HCl, AlCl}_3]{H_2C{=}CH(CH_2)_{10}CH_3} C_6H_5\text{-}CH(CH_3)(CH_2)_{10}CH_3 \xrightarrow{H_2SO_4} p\text{-}HO_3S\text{-}C_6H_4\text{-}CH(CH_3)(CH_2)_{10}CH_3 \xrightarrow{NaOH}$$

$CH_3CH(CH_2)_{10}CH_3$ $CH_3CH(CH_2)_{10}CH_3$ SO_3H

18.57 The nitroso group withdraws electrons by its inductive effect and donates electrons by its resonance effect. As was the case with the halogens, the inductive effect is controlling the rate of reaction and deactivating the compound, but the ortho and para positions are less deactivated because of the donation of electrons by resonance.

18.58 The ring attached to the nitrogen is activated whereas the ring attached to the carbonyl group is deactivated.

$$C_6H_5\text{-}\overset{O}{\overset{\|}{C}}\text{-}NH\text{-}C_6H_4\text{-}Cl\ (ortho) \quad + \quad C_6H_5\text{-}\overset{O}{\overset{\|}{C}}\text{-}NH\text{-}C_6H_4\text{-}Cl\ (para)$$

19.59

a) $$\text{Benzene} \xrightarrow[\text{AlCl}_3]{CH_3Cl} C_6H_5CH_3 \xrightarrow[\text{2) } H_3O^{\oplus}]{\text{1) KMnO}_4\text{, NaOH, }\Delta} C_6H_5CO_2H \xrightarrow[H_2SO_4]{HNO_3} m\text{-}O_2N\text{-}C_6H_4\text{-}CO_2H$$

b)

O O O + OCH_3 $\xrightarrow{AlCl_3}$ COH (O) OCH_3 O

$\downarrow SOCl_2$

CCl (O) OCH_3 O $\xrightarrow{AlCl_3}$ O OCH_3 O

c)

$\xrightarrow[H_2SO_4]{HNO_3}$ NO_2 $\xrightarrow{Fe,\ HCl}$ NH_2 $\xrightarrow{CH_3\overset{O}{\overset{\|}{C}}O\overset{O}{\overset{\|}{C}}CH_3}$ $NH\overset{O}{\overset{\|}{C}}CH_3$

$\downarrow CH_3Cl,\ AlCl_3$

$NH\overset{O}{\overset{\|}{C}}CH_3$, CH_3 $\xrightarrow{Br_2}$ $NH\overset{O}{\overset{\|}{C}}CH_3$, Br, CH_3 $\xrightarrow[H_2O]{KOH}$ NH_2, Br, CH_3

d) Benzene $\xrightarrow[H_2SO_4]{HNO_3}$ nitrobenzene (NO_2) $\xrightarrow[FeBr_3]{Br_2}$ m-bromonitrobenzene (NO_2, Br) $\xrightarrow{Fe,\ HCl}$ m-bromoaniline (NH_2, Br)

$\xrightarrow{CH_3\overset{O}{\overset{\|}{C}}O\overset{O}{\overset{\|}{C}}CH_3}$ ($NH\overset{O}{\overset{\|}{C}}CH_3$, Br) $\xrightarrow[AlCl_3]{CH_3Cl}$ ($NH\overset{O}{\overset{\|}{C}}CH_3$, Br, CH_3) $\xrightarrow[H_2O]{KOH}$ (NH_2, Br, CH_3)

e) Aniline (NH_2) see (c) $\xrightarrow{3\ Br_2}$ (NH_2; Br, Br, Br) $\xrightarrow[2)\ CuCN]{1)\ NaNO_2,\ H_3O^{\oplus}}$ (CN; Br, Br, Br) $\xrightarrow[\Delta]{H_3O^{\oplus},\ H_2O}$ (CO_2H; Br, Br, Br)

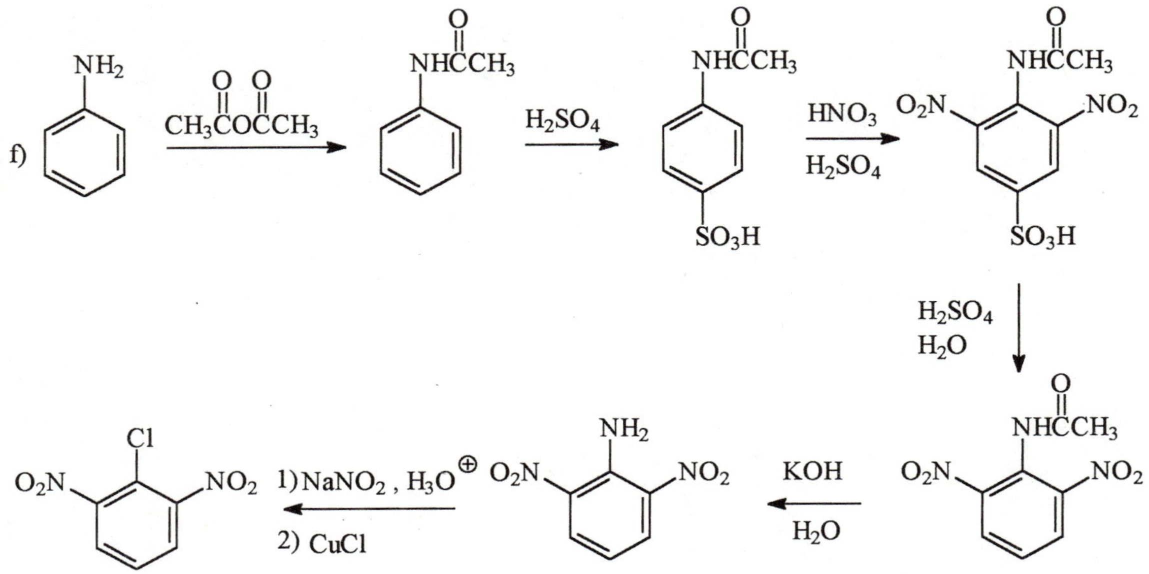

f)

NH_2

$CH_3\overset{O}{\overset{\|}{C}}O\overset{O}{\overset{\|}{C}}CH_3$

$NH\overset{O}{\overset{\|}{C}}CH_3$

H_2SO_4

$NH\overset{O}{\overset{\|}{C}}CH_3$

SO_3H

HNO_3

H_2SO_4

O_2N $NH\overset{O}{\overset{\|}{C}}CH_3$ NO_2

SO_3H

H_2SO_4

H_2O

O_2N $NH\overset{O}{\overset{\|}{C}}CH_3$ NO_2

KOH

H_2O

O_2N NH_2 NO_2

1) $NaNO_2$, $H_3O^{\oplus}$

2) CuCl

O_2N Cl NO_2

18.60

a)

O

Cl

$AlCl_3$

O

1) CH_3MgI

2) NH_4Cl, H_2O

O H

b)

O

Cl

$AlCl_3$

O

Cl_2

$AlCl_3$

O

Cl

$Ph_3P{=}CH_2$

Cl

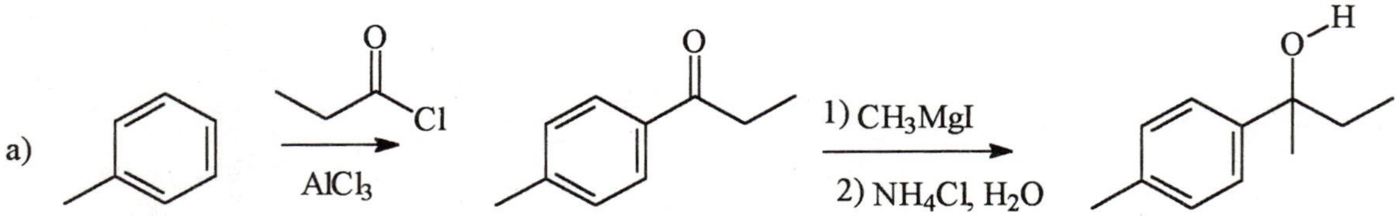

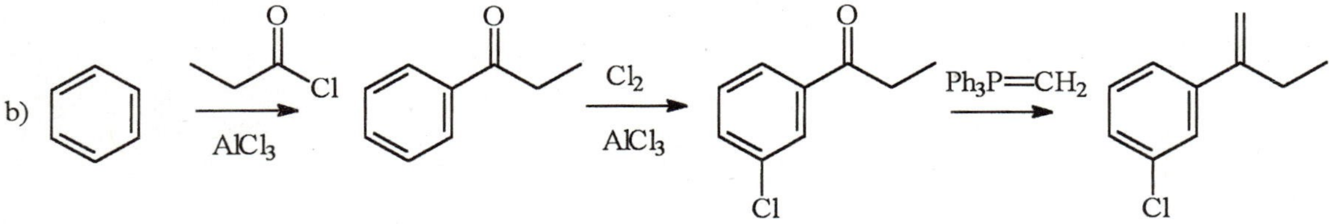

c)

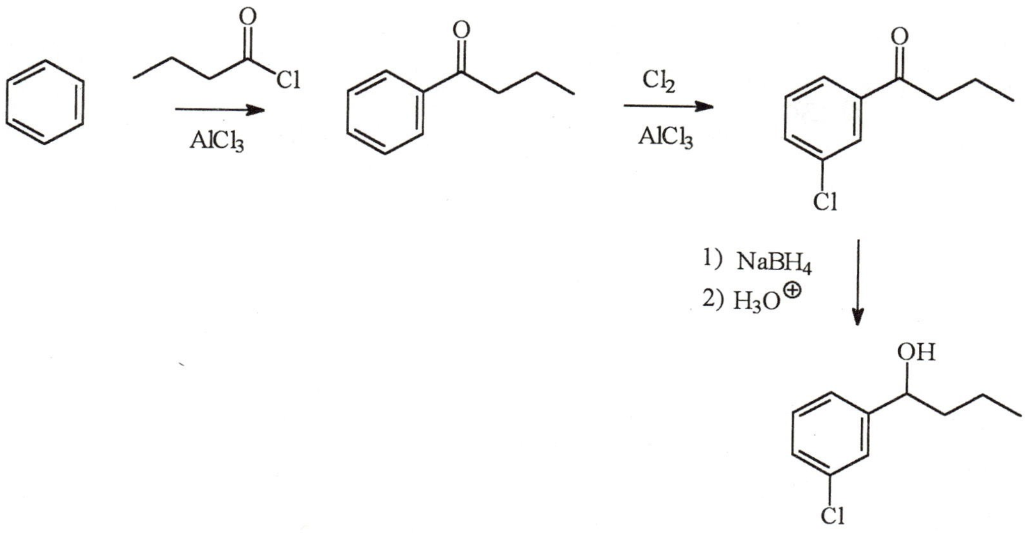

O

Cl

$\xrightarrow[AlCl_3]{}$

O

$\xrightarrow[AlCl_3]{Cl_2}$

O

Cl

1) $NaBH_4$
2) $H_3O^{\oplus}$

OH

Cl

d)

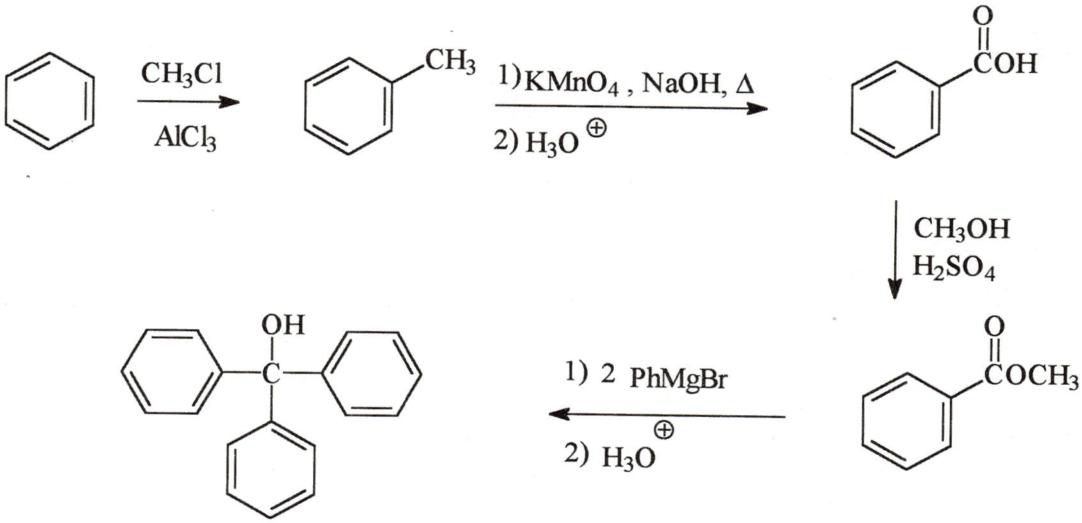

$\xrightarrow[AlCl_3]{CH_3Cl}$

CH_3

$\xrightarrow[2)\ H_3O^{\oplus}]{1)\ KMnO_4\ ,\ NaOH,\ \Delta}$

O

COH

CH_3OH
H_2SO_4

O

$COCH_3$

1) 2 PhMgBr
2) $H_3O^{\oplus}$

OH

C

e) Benzene $\xrightarrow[H_2SO_4]{HNO_3}$ NO_2 (nitrobenzene) $\xrightarrow[FeBr_3]{Br_2}$ NO_2, Br (m-bromonitrobenzene) $\xrightarrow{Fe, HCl}$ NH_2, Br (m-bromoaniline)

1) $NaNO_2$, $H_3O^{\oplus}$

2) $H_3O^{\oplus}$, Δ

$\downarrow$

OH, Br (m-bromophenol) $\xrightarrow[2)\ CH_3I]{1)\ KOH}$ OCH_3, Br (m-bromoanisole)

f) Benzene $\xrightarrow[H_2SO_4]{HNO_3}$ NO_2 (nitrobenzene) $\xrightarrow{Fe, HCl}$ NH_2 (aniline) $\xrightarrow{\text{excess } Br_2}$ NH_2, Br, Br, Br (2,4,6-tribromoaniline)

1) $NaNO_2$, $H_3O^{\oplus}$

2) $H_3O^{\oplus}$, Δ

$\downarrow$

OH, Br, Br, Br (2,4,6-tribromophenol)

g) benzene $\xrightarrow[AlCl_3]{\text{succinic anhydride}}$ 4-oxo-4-phenylbutanoic acid (HO, O) $\xrightarrow[2)\ H_3O^{\oplus},\ \Delta]{1)\ NaBH_4}$ (HO, O)

$\downarrow$ 1) $SOCl_2$ 2) $AlCl_3$

[(O)] $\xrightarrow{\text{spontaneous tautomerization}}$ (OH)

h) benzene $\xrightarrow[H_2SO_4]{HNO_3}$ (NO_2) $\xrightarrow{Fe,\ HCl}$ (NH_2) $\xrightarrow{CH_3\overset{O}{\overset{\|}{C}}O\overset{O}{\overset{\|}{C}}CH_3}$ ($NH\overset{O}{\overset{\|}{C}}CH_3$) $\xrightarrow[AlCl_3]{CH_3Cl}$ ($NH\overset{O}{\overset{\|}{C}}CH_3$, CH_3)

$\downarrow$ KOH, H_2O

(NH_2, CH_3) $\xrightarrow[2)\ H_3O^{\oplus},\ \Delta]{1)\ NaNO_2,\ H_3O^{\oplus}}$ (OH, CH_3) $\xrightarrow[2)\ H_3O^{\oplus}]{1)\ KMnO_4,\ NaOH,\ \Delta}$ (OH, CO_2H) $\xrightarrow{CH_3\overset{O}{\overset{\|}{C}}O\overset{O}{\overset{\|}{C}}CH_3}$ aspirin

i) Cl (chlorobenzene) $\xrightarrow[H_2SO_4]{HNO_3}$ 1-chloro-2,4-dinitrobenzene (Cl, NO_2, NO_2) $\xrightarrow{KOH}$ 2,4-dinitrophenol (OH, NO_2, NO_2)

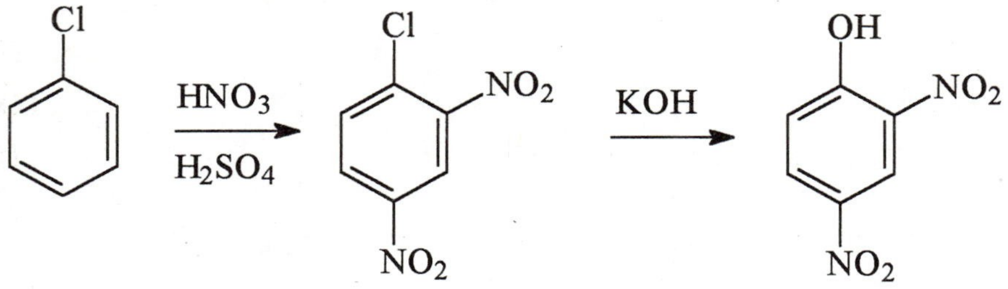

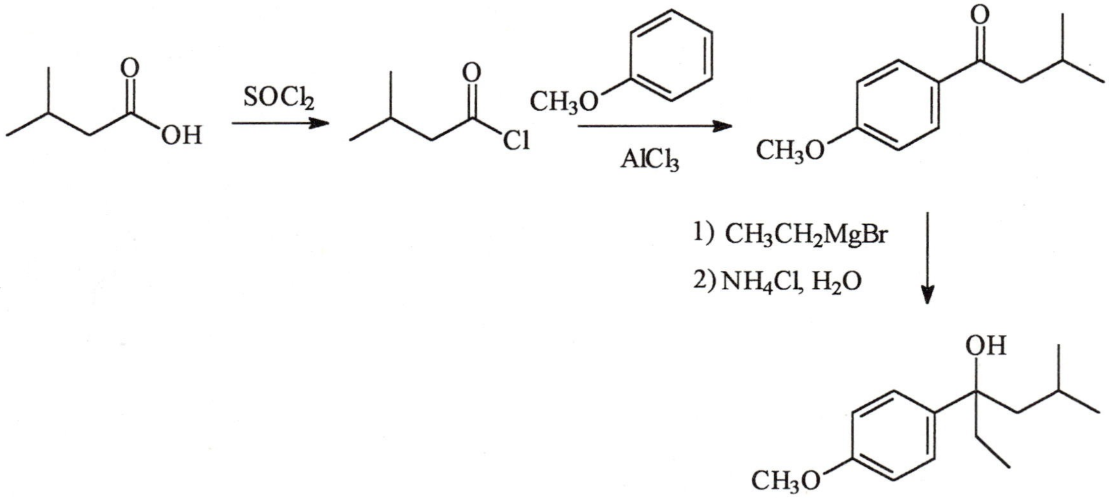

j) 3-methylbutanoic acid (O, OH) $\xrightarrow{SOCl_2}$ acid chloride (O, Cl) $\xrightarrow[AlCl_3]{CH_3O\text{-}C_6H_5}$ CH_3O-aryl ketone (O)

1) CH_3CH_2MgBr
2) NH_4Cl, H_2O

→ tertiary alcohol (OH, CH_3O)

18.61

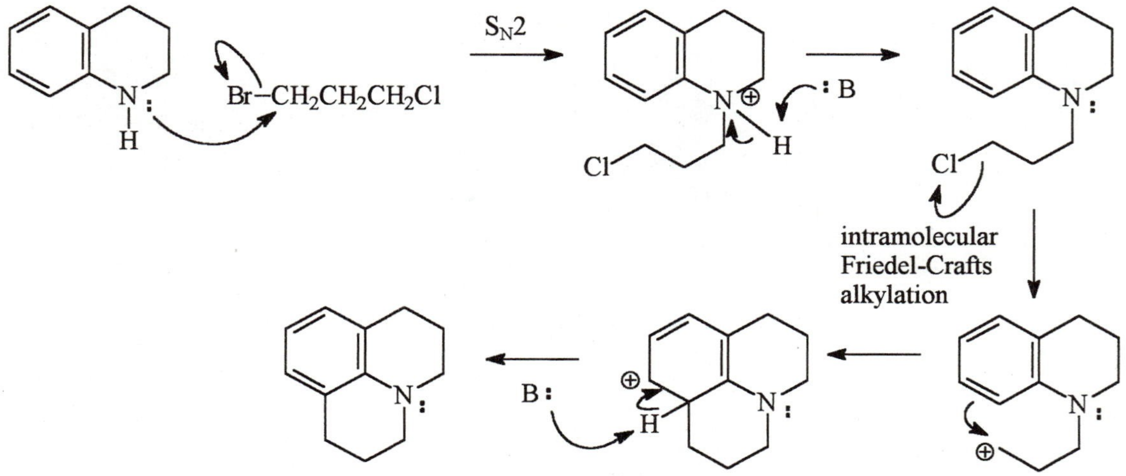

N–H (tetrahydroquinoline) + Br–$CH_2CH_2CH_2Cl$ $\xrightarrow{S_N2}$ $N^{\oplus}$–H, Cl, : B → N:, Cl

intramolecular
Friedel-Crafts
alkylation

→ N:, $\oplus$ → $\oplus$, H, N: → B: → N: (julolidine)

18.62

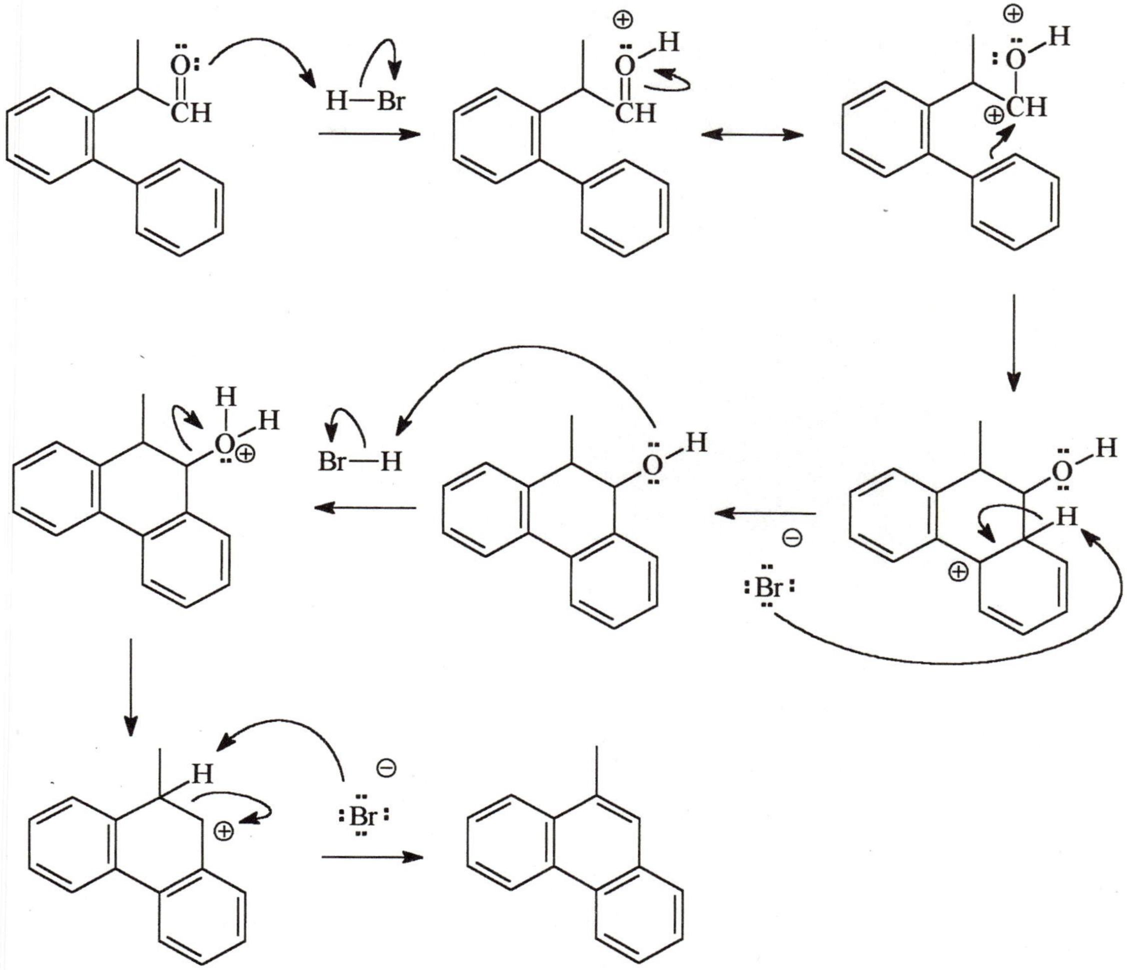

H—Br

Br—H

$:\ddot{Br}:^{\ominus}$

18.63 Nitrogen A is more nucleophilic than nitrogen B because the electron pair on nitrogen B is delocalized by resonance with the carbonyl group.

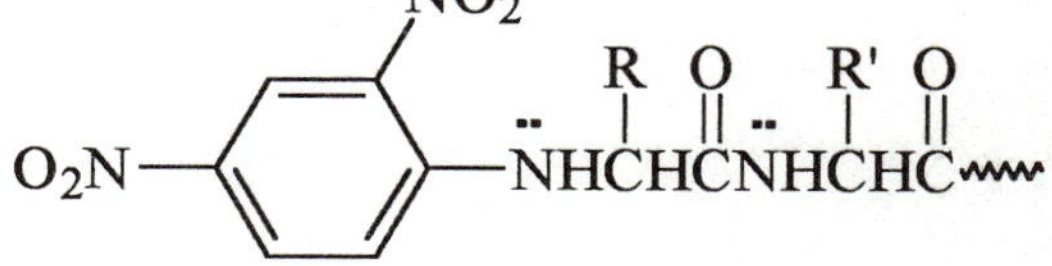

O_2N — NHCHCNHCHC (with NO_2, R, O, R', O substituents)

18.64 The nucleophile prefers to attack the benzyne intermediate at the meta position so the negative charge is located closer to the electron withdrawing oxygen.

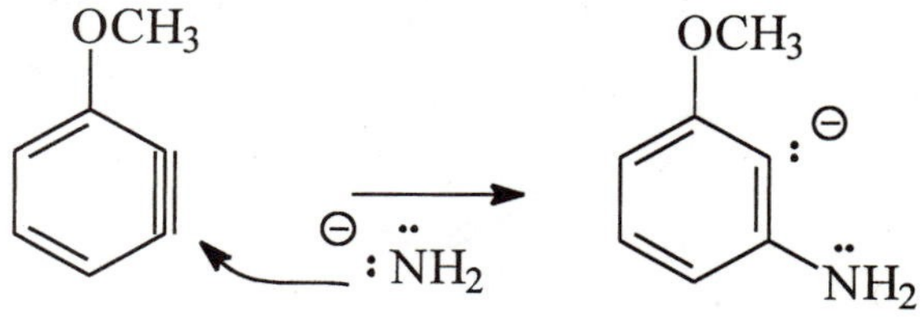

OCH3 OCH3 :NH2 NH2

18.65

H–C–H H–Cl H–C–H H–C–H OH OH H2C Cl H Cl Cl OH Cl Cl Cl :Cl:

H–OH OH H2C Cl Cl Cl Cl–H :OH OH H2C Cl Cl Cl OH H2C Cl Cl Cl OH Cl Cl Cl

OH H H H OH Cl C Cl Cl Cl Cl Cl :Cl: OH H H OH Cl C Cl Cl Cl Cl Cl Cl

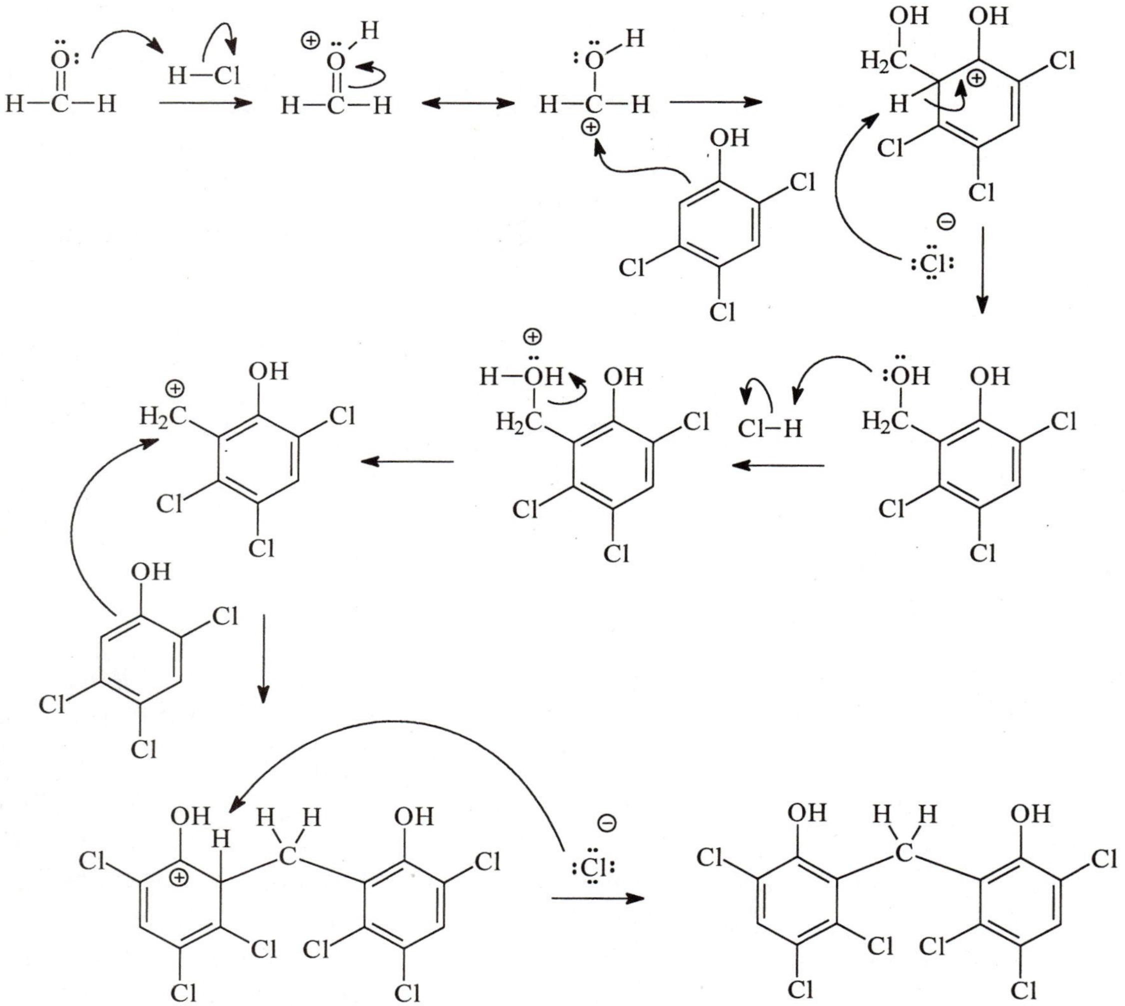

18.66

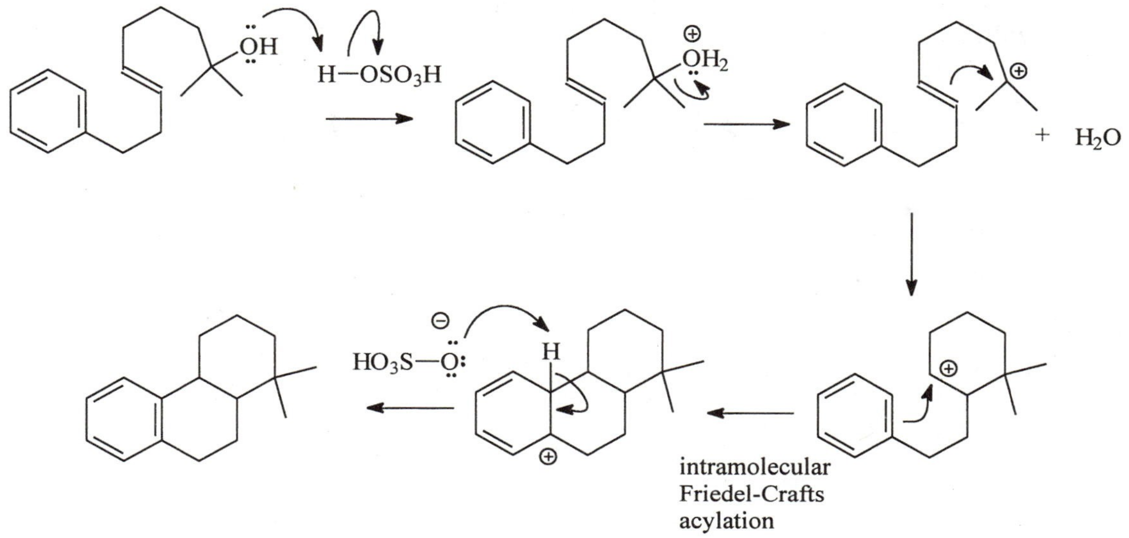

OH

H—OSO_3H

OH_2

+ H_2O

HO_3S—O:

H

intramolecular
Friedel-Crafts
acylation

18.67 This is a Friedel-Crafts alkylation reaction followed by an S_N1 reaction.

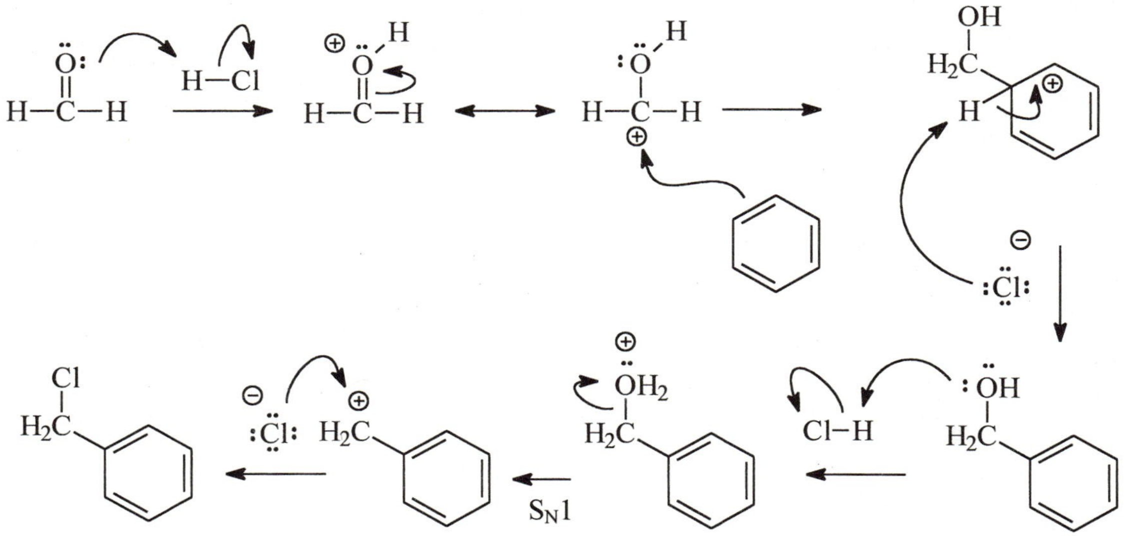

O:

H—C—H

H—Cl

O H

H—C—H

:O H

H—C—H

OH

H_2C

H

:Cl:

:OH

H_2C

Cl—H

OH_2

H_2C

S_N1

:Cl: H_2C

Cl

H_2C

18.68 Compound **B** has a DU of 5. The ratio of Hs in the NMR spectrum of **B** must be 4:4:2. Because there are four aromatic Hs, **B** must have a disubstituted aromatic ring. The quintuplet at 2 δ indicates that these two Hs are coupled to four Hs. These must be the four Hs at 2.8 δ, which appear as a triplet because they are split by two Hs. Therefore, the fragment $-CH_2CH_2CH_2-$ can be identified. Attaching this fragment to two positions of a benzene ring indicates that compound **B** is indan. The reaction that produces **B** from **A** is an intramolecular Friedel-Crafts alkylation reaction.

A

$AlCl_3$

$\ddot{C}l-AlCl_3$

$Cl_3Al-\ddot{C}l:$

H

B

indan

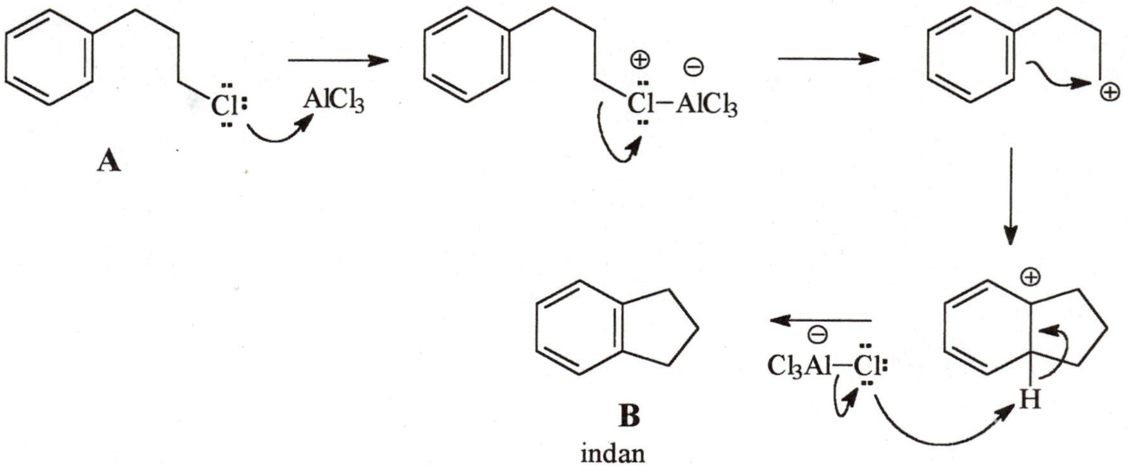

18.69 The product, which results from three Friedel-Crafts alkylation reactions, is triphenylmethane.

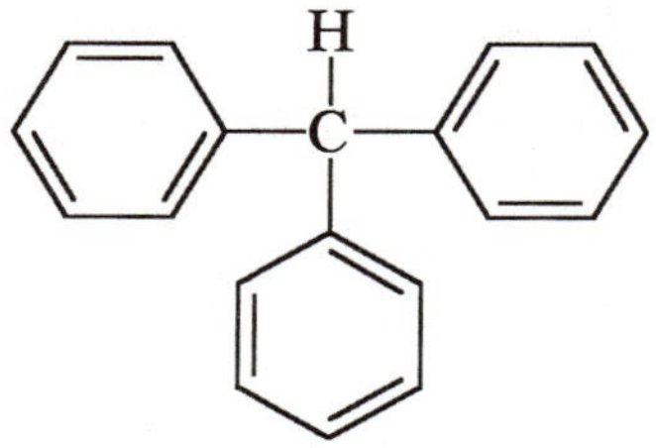

H

C

18.70

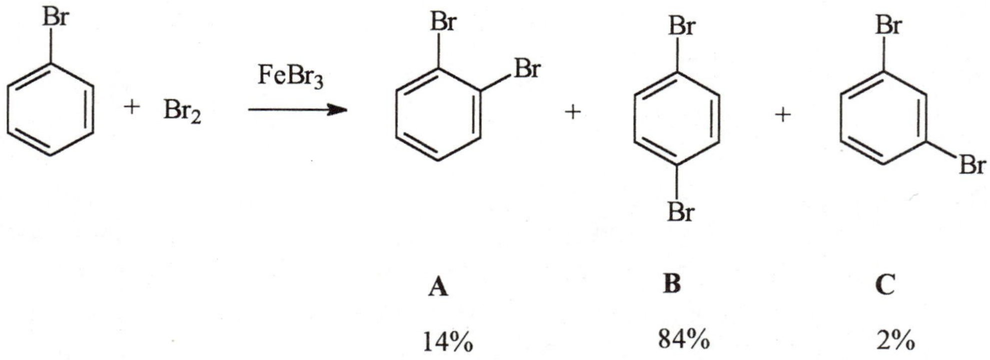

Br

+ Br_2

$FeBr_3$

Br Br

+

Br Br

+

Br Br

A 14%

B 84%

C 2%

After completing this chapter, you should be able to:

1. Show the products of any of the reactions discussed in this chapter.
   Problems 18.6, 18.8, 18.9, 18.11, 18.13, 18.15, 18.20, 18.23, 18.24, 18.26, 18.27, 18.29, 18.33, 18.34, 18.35, 18.36, 18.40, 18.41, 18.42.

2. Show the mechanisms for the reactions whose mechanisms were discussed.
   Problems 18.10, 18.12, 18.14, 18.46, 18.47, 18.48, 18.49, 18.50, 18.51, 18.52, 18.60, 18.61.

3. Predict the effect of a substituent on the rate and the regiochemistry of an electrophilic aromatic substitution reaction.
   Problems 18.1, 18.2, 18.3, 18.4, 18.5, 18.7, 18.19, 18.37, 18.38, 18.43, 18.53, 18.54, 18.55, 18.57.

4. Use these reactions to synthesize aromatic compounds.
   Problems 18.16, 18.17, 18.21, 18.30, 18.31, 18.32, 18.39, 18.44, 18.45, 18.58, 18.59.

Chapter 19
THE CHEMISTRY OF RADICALS

19.1

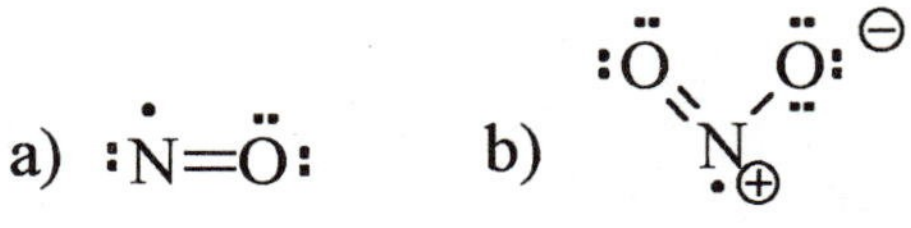

a) $:\dot{N}=\ddot{O}:$ b) $:\ddot{O}=N^{\oplus}(\cdot)-\ddot{O}:^{\ominus}$

19.2 The order of radical stability parallels carbocation stability.

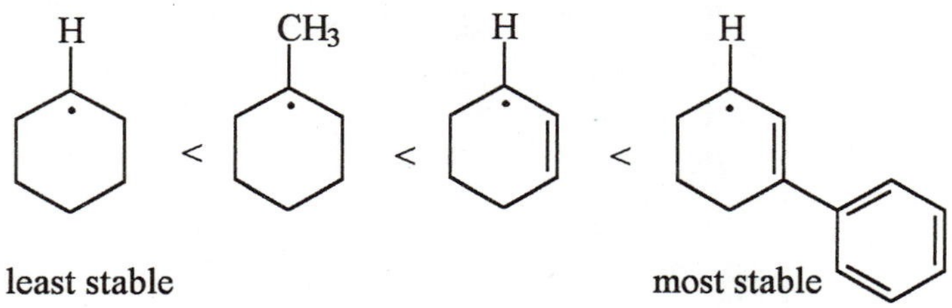

least stable < < < most stable

19.3

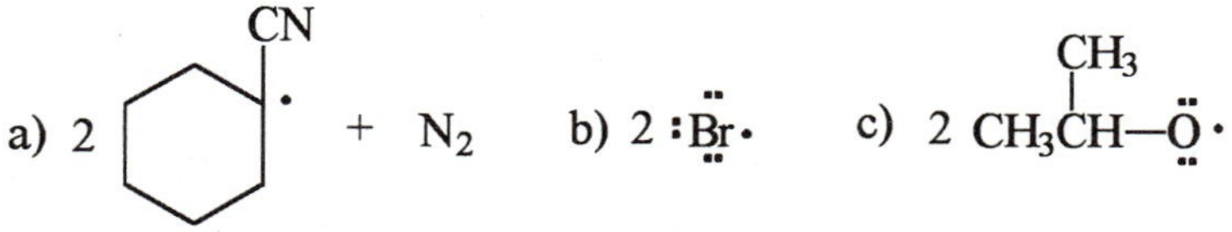

a) 2 (1-cyanocyclohexyl radical) + N_2 b) 2 $:\ddot{Br}\cdot$ c) 2 $CH_3CH(CH_3)-\ddot{O}\cdot$

19.4 a) Abstraction of a tertiary hydrogen occurs more readily than abstraction of a secondary hydrogen, so the second reaction is faster.

b) Formation of the resonance stabilized benzylic radical occurs more readily, so the first reaction is faster.

19.5

a) $CH_3\dot{C}HCH_2Br$ b) 2 $CH_3C(=O)\ddot{\underset{..}{O}}\cdot$ c) $CH_3(CH_2)_6CH_3$ + $CH_3CH_2CH_2CH_3$ + $CH_3CH_2CH{=}CH_2$

d) $(O{=})(^{-}O)\overset{\oplus}{N}{-}\overset{\oplus}{N}({=}O)(O^{-})$ e) 2 $CH_3C(CH_3)_2\cdot$ + N_2

19.6

a) $CH_3(CH_2)_{12}CH_3$ b) $CH_3CH_2OC(=O)(CH_2)_{16}C(=O)OCH_2CH_3$

19.7 One of the ways that radicals form stable products is by abstracting a hydrogen atom from the carbon atom next to another radical center. This reaction is called disproportionation and it competes with coupling.

19.8

$$:\ddot{Cl}{-}\ddot{Cl}: \xrightarrow{h\nu} 2\ :\ddot{Cl}\cdot$$

$$:\ddot{Cl}\cdot + H{-}CH_2Cl \longrightarrow :\ddot{Cl}{-}H + \cdot CH_2Cl$$

$$ClH_2C\cdot + :\ddot{Cl}{-}\ddot{Cl}: \longrightarrow CH_2Cl_2 + :\ddot{Cl}\cdot$$

+ various terminations

19.9

initiation $:\ddot{Cl}{-}CF_3$ + uv light $\longrightarrow$ $:\ddot{Cl}\cdot$ + $\cdot CF_3$

propagation $:\ddot{Cl}\cdot + O_3 \longrightarrow :\ddot{Cl}{-}\ddot{O}\cdot + O_2$

$:\ddot{Cl}{-}\ddot{O}\cdot + :\ddot{O}\cdot \longrightarrow O_2 + :\ddot{Cl}\cdot$

19.10 If the strength of a typical CH bond is taken as 98 kcal/mol (410 kJ/mol) (see Table 2.1), then the abstraction of the hydrogen atom by a fluorine atom is exothermic by 135 - 98 = 37 kcal/mol (155 kJ/mol) whereas a similar abstraction by an iodine atom is endothermic by 98 - 71 = 27 kcal/mol (113 kJ/mol).

19.11

a) 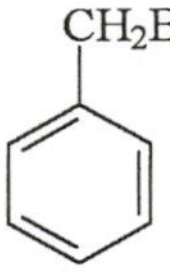 CH_2Br b) 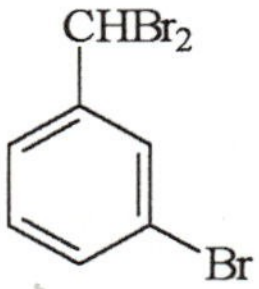 $CHBr_2$, Br c) 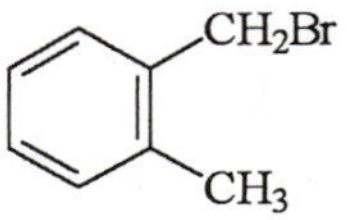 CH_2Br, CH_3 d) 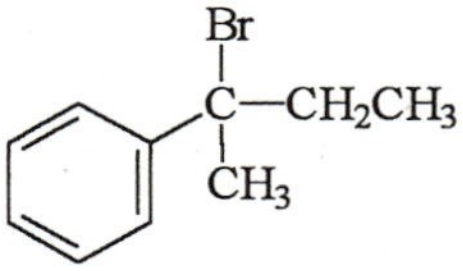 Br, C—CH_2CH_3, CH_3 racemic

e) 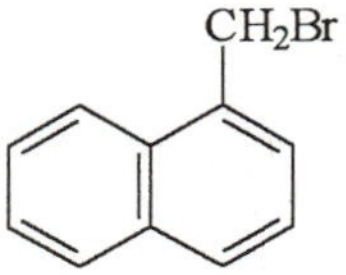 CH_2Br f) 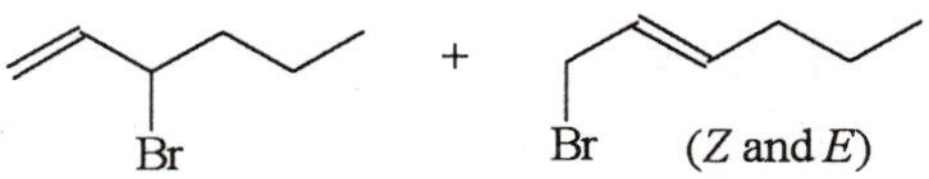 Br + Br (*Z* and *E*)

19.12 The major product is 1-phenyl-3-bromopropene because it is more stable due to conjugation. The other product, shown below, is not conjugated, so it is not formed.

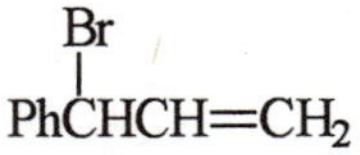

Br
PhCHCH=CH_2

19.13

a) 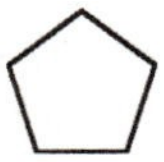 b) 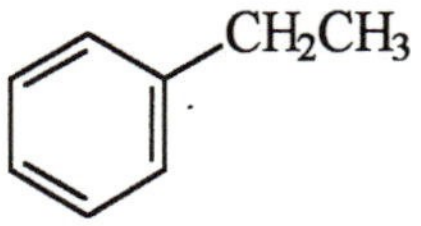 CH_2CH_3

19.14

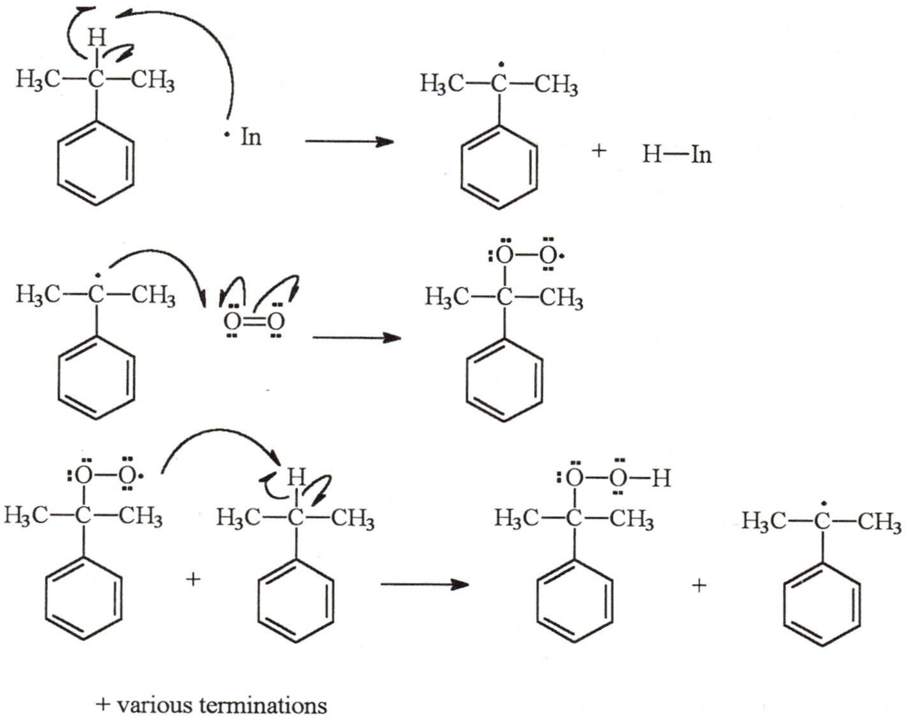

$H_3C-C(H)-CH_3$ (cumene) + $\cdot In$ ⟶ $H_3C-\dot{C}-CH_3$ + H—In

$H_3C-\dot{C}-CH_3$ + O=O ⟶ $H_3C-C(O-O\cdot)-CH_3$

$H_3C-C(O-O\cdot)-CH_3$ + $H_3C-C(H)-CH_3$ ⟶ $H_3C-C(O-O-H)-CH_3$ + $H_3C-\dot{C}-CH_3$

+ various terminations

This reaction is more selective than most autoxidations because the tertiary benzylic hydrogen is much more readily abstracted than any other hydrogen of cumene.

19.15 The radical produced from vitamin E is stabilized by resonance and its radical center is sterically hindered. Therefore it is not very reactive.

19.16

a) 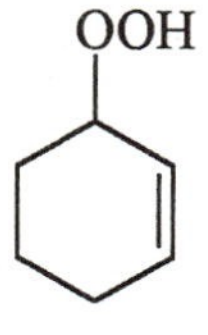 OOH

b) 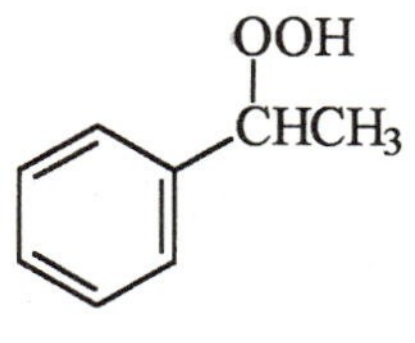 OOH
CHCH$_3$

19.17

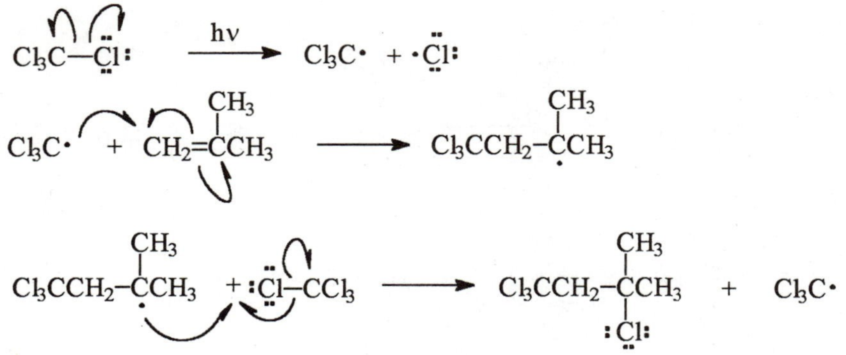

$$Cl_3C{-}Cl \xrightarrow{h\nu} Cl_3C\cdot + \cdot Cl$$

$$Cl_3C\cdot + CH_2{=}C(CH_3)CH_3 \longrightarrow Cl_3CCH_2{-}\dot{C}(CH_3)CH_3$$

$$Cl_3CCH_2{-}\dot{C}(CH_3)CH_3 + Cl{-}CCl_3 \longrightarrow Cl_3CCH_2{-}C(CH_3)(Cl)CH_3 + Cl_3C\cdot$$

+ various terminations

19.18

a) 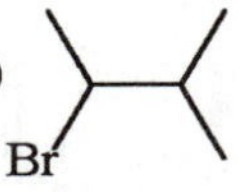 Br

b) 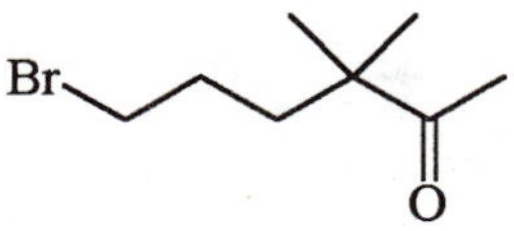 Br O

c) 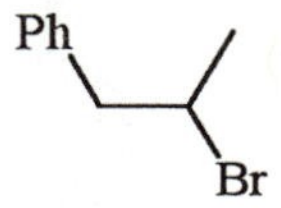 Ph Br

d) 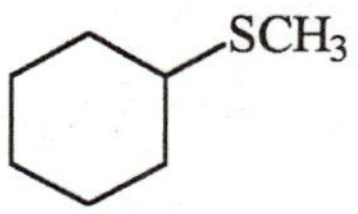 SCH_3

e) 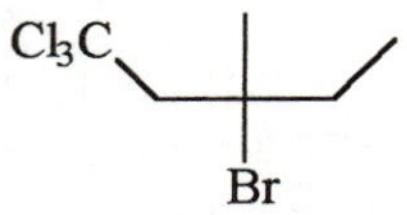 Cl_3C Br

f) 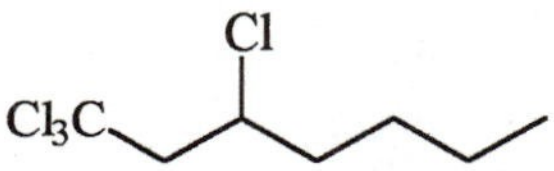 Cl_3C Cl

19.19

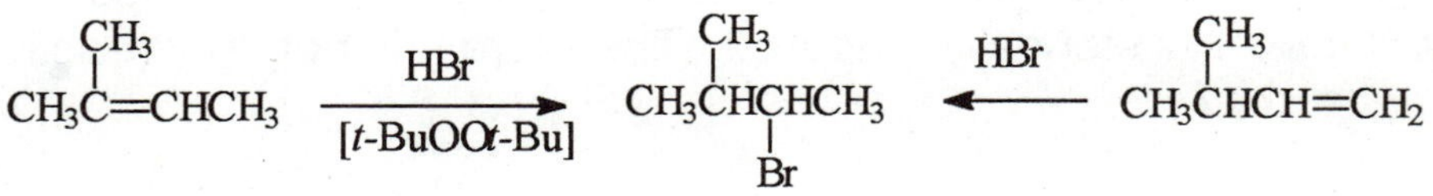

$$CH_3C(CH_3){=}CHCH_3 \xrightarrow[\text{[}t\text{-BuOO}t\text{-Bu]}]{HBr} CH_3CH(CH_3)CH(Br)CH_3 \xleftarrow{HBr} CH_3CH(CH_3)CH{=}CH_2$$

19.20

a) 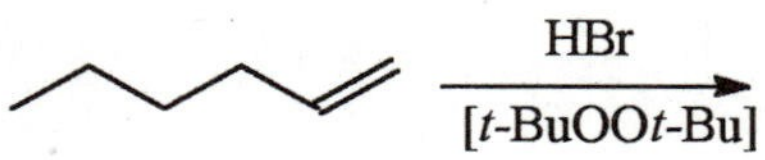 $\xrightarrow[\text{[}t\text{-BuOO}t\text{-Bu]}]{HBr}$

b) 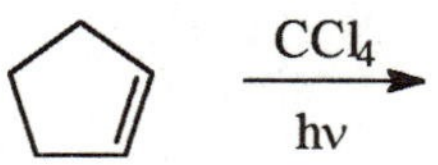 $\xrightarrow[h\nu]{CCl_4}$

c) 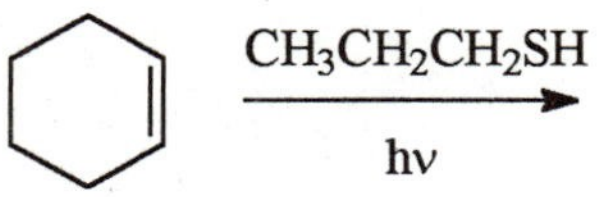 $\xrightarrow[h\nu]{CH_3CH_2CH_2SH}$

d) 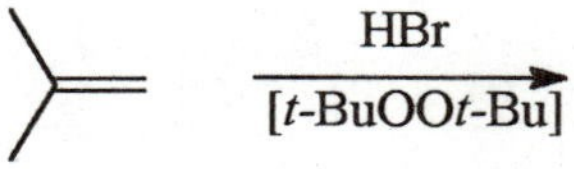 $\xrightarrow[\text{[}t\text{-BuOO}t\text{-Bu]}]{HBr}$

e) $PhCH{=}CH_2$ $\xrightarrow[\text{[}t\text{-BuOO}t\text{-Bu]}]{HBr}$

f) 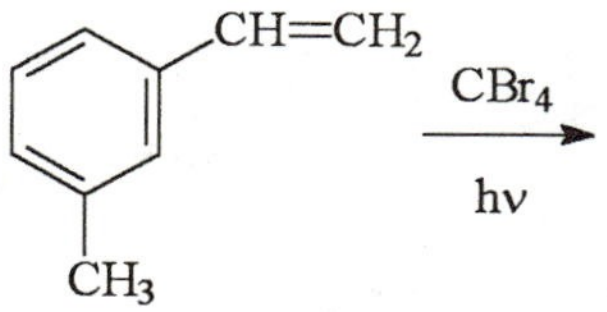 $CH{=}CH_2$, CH_3 $\xrightarrow[h\nu]{CBr_4}$

g) 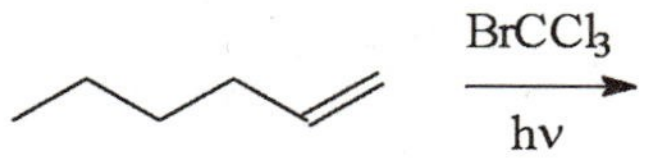 $\xrightarrow[h\nu]{BrCCl_3}$

19.21

a) 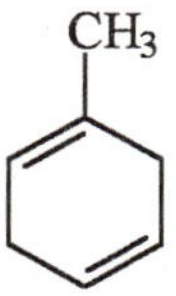 CH_3

b) 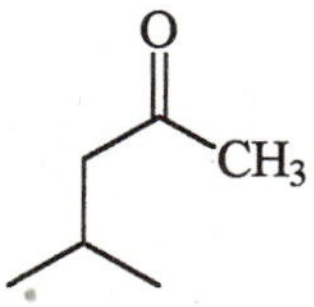 O, CH_3

c) 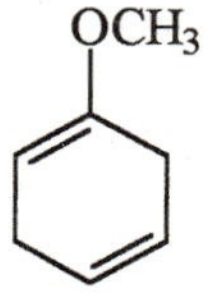 OCH_3

d) 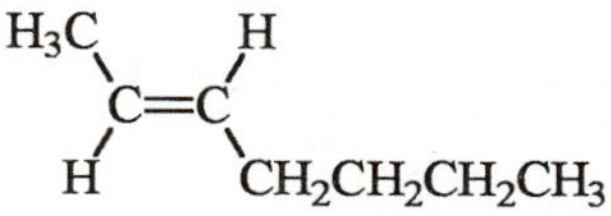 H_3C, H, C=C, H, $CH_2CH_2CH_2CH_3$

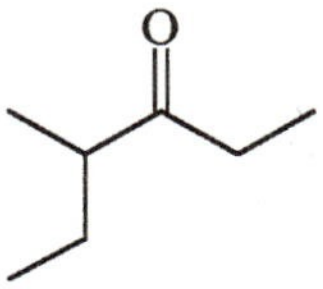

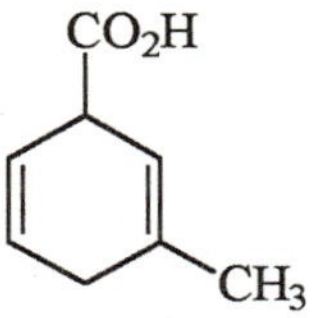

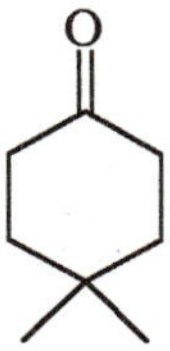

e) O

f) CO_2H, CH_3

g) O

h) O

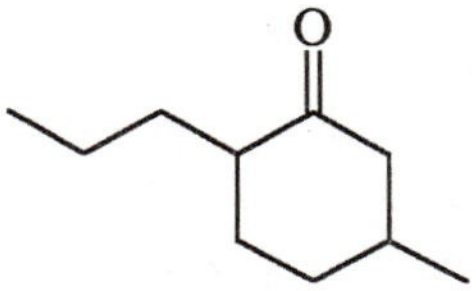

19.22 a) 1) Li, NH_3 (l); 2) $PhCH_2Br$
b) 1) Li, NH_3 (l); 2) $H_3O^{\oplus}$

19.23

a) 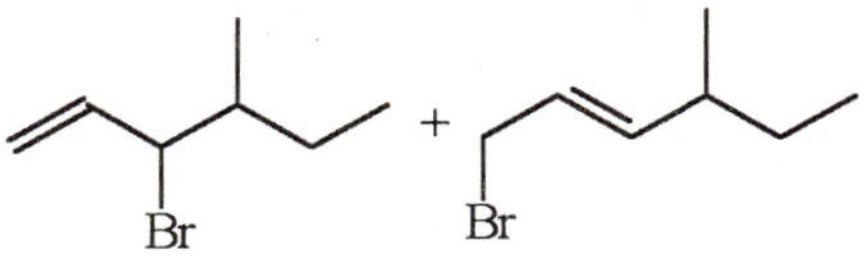 Br + Br

b) 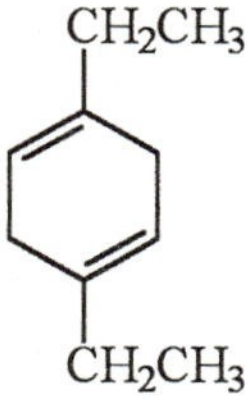 CH_2CH_3, CH_2CH_3

c) 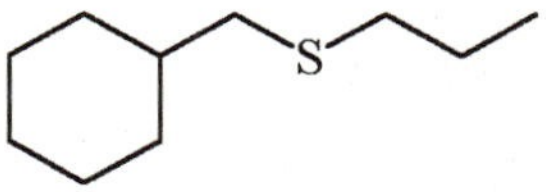 S

d) 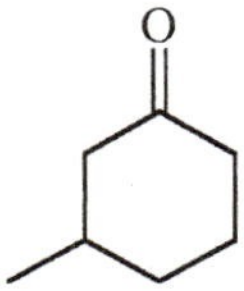 O

e) 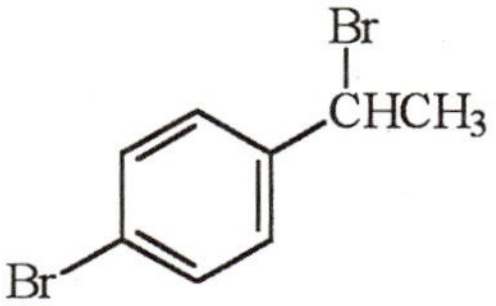 Br, $CHCH_3$, Br

f) 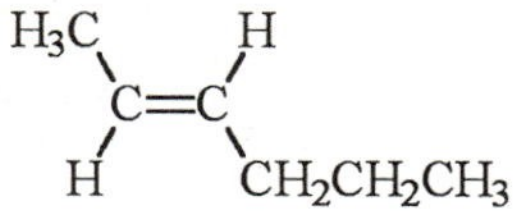 H_3C, H, C=C, H, $CH_2CH_2CH_3$

g) 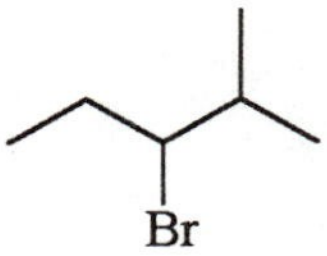 Br

h) 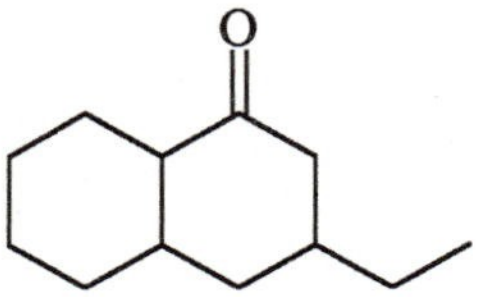 O

i)

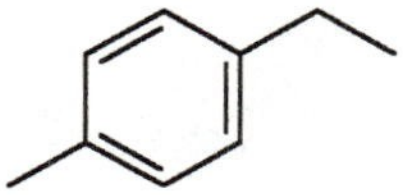

j) 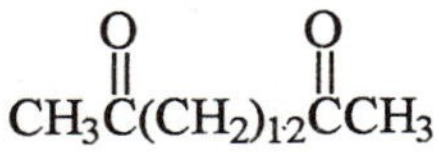 $CH_3\overset{O}{\overset{\|}{C}}(CH_2)_{12}\overset{O}{\overset{\|}{C}}CH_3$

k) 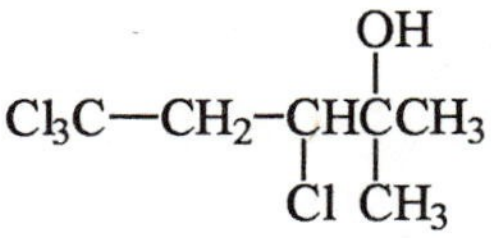 $Cl_3C-CH_2-\underset{Cl}{CH}\overset{OH}{\underset{CH_3}{C}}CH_3$

l) 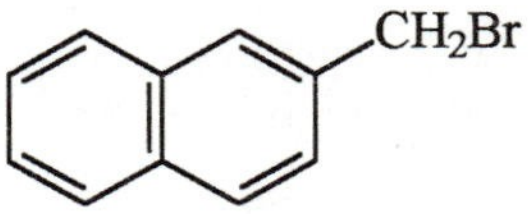 CH_2Br

m) 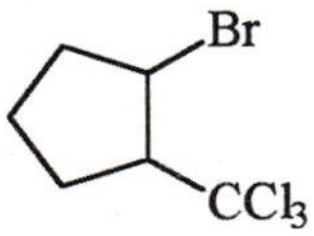 Br CCl_3

19.24

a) 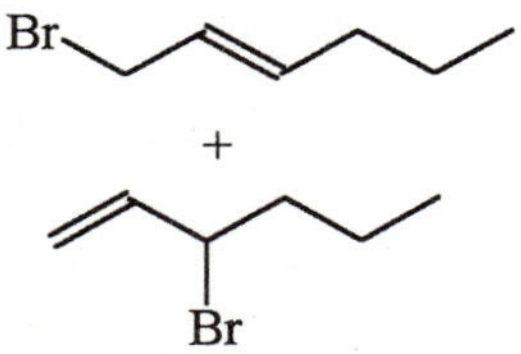 Br + Br

b)  Br

c) 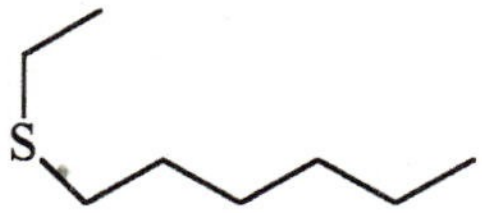 S

d) 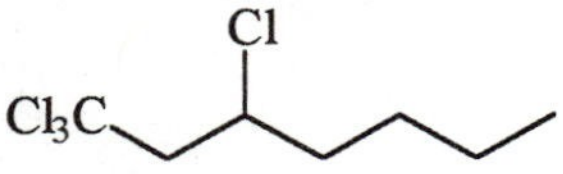 Cl_3C Cl

e) 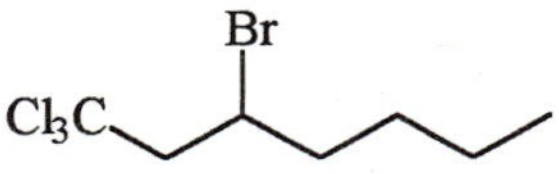 Cl_3C Br

19.25

a) $CH_3(CH_2)_{28}CH_3$

b) 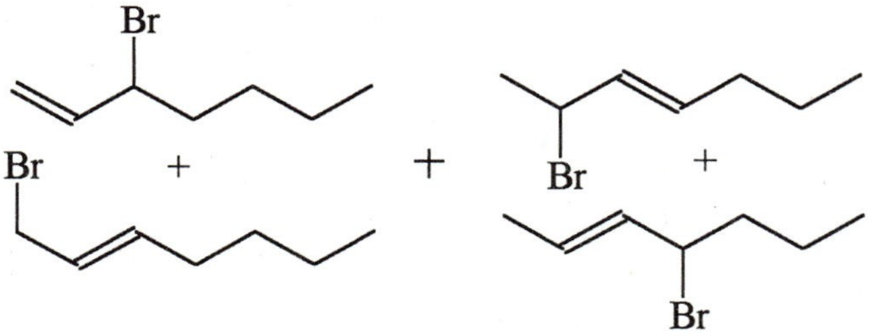 Br + Br + Br + Br

c) 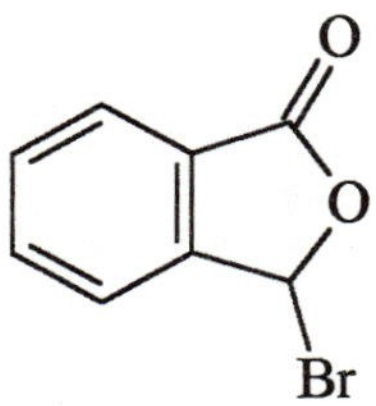 O O Br

d) 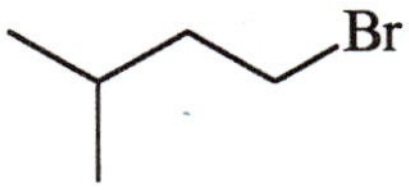 Br

e) 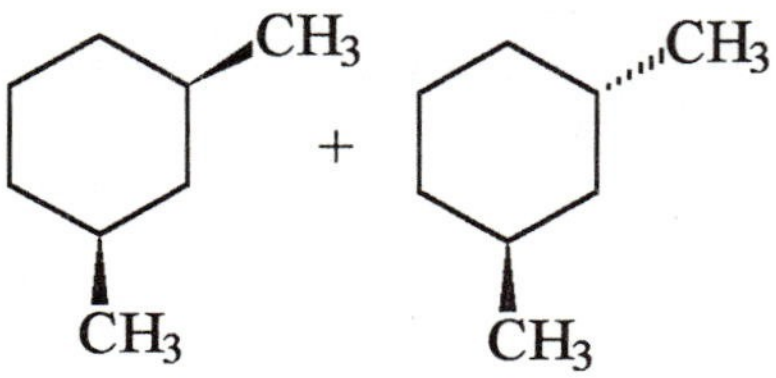 CH_3 CH_3 + CH_3 CH_3

f) 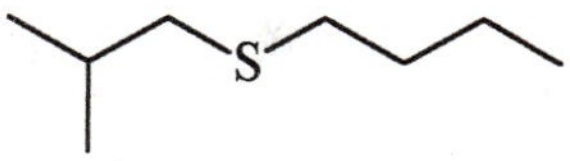 S

g) 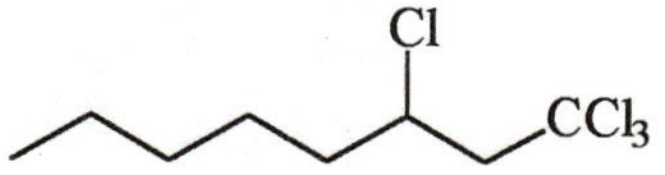 Cl CCl_3

h) 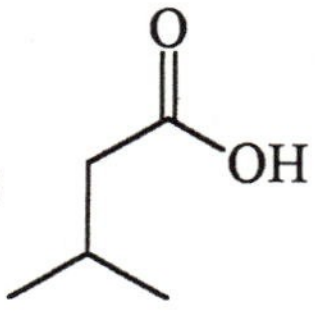 O OH

i) 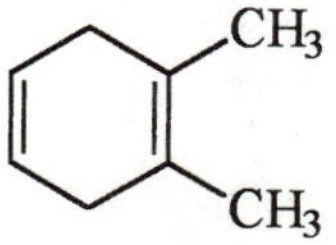 CH_3 CH_3

19.26

a) 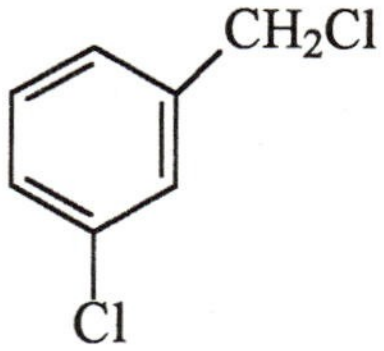 CH_2Cl Cl

b) 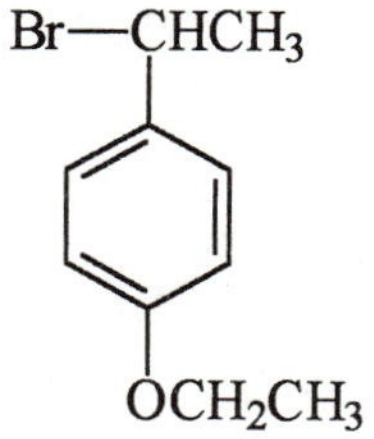 Br—$CHCH_3$ OCH_2CH_3

c) 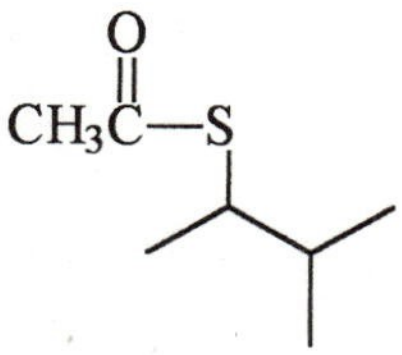 O CH_3C—S

d) 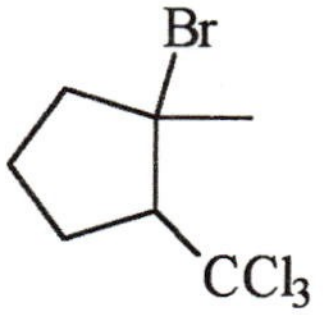 Br CCl_3

e)

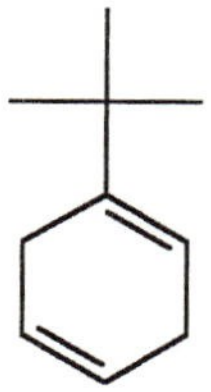

f) 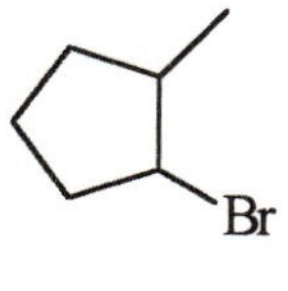 Br

g) 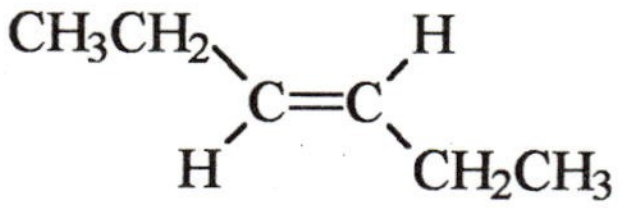 CH_3CH_2 H C=C H CH_2CH_3

h) O

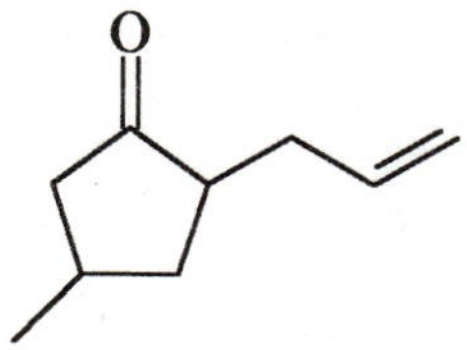

19.27

a) (toluene, CH_3) $\xrightarrow[\text{[AIBN]}]{\text{NBS}}$ (benzyl bromide, CH_2Br) $\xrightarrow[\text{[AIBN]}]{Bu_3SnD}$

b) (benzene) $\xrightarrow[AlCl_3]{CH_3(CH_2)_3CH_2\overset{O}{\overset{\|}{C}}Cl}$ (hexanophenone) $\xrightarrow[\text{2) Na, EtOH, } NH_3 \text{ (l)}]{\text{1) Zn (Hg), HCl}}$

c) $CH_3CH_2C{\equiv}CH$ $\xrightarrow[\text{2) } CH_3CH_2CH_2Br]{\text{1) } NaNH_2\text{, } NH_3}$ $CH_3CH_2C{\equiv}CCH_2CH_2CH_3$ $\xrightarrow{Na, NH_3 \text{ (l)}}$

d) (3-methoxytoluene, CH_3, OCH_3) $\xrightarrow[\text{[AIBN]}]{\text{NBS}}$ (3-methoxybenzyl bromide, CH_2Br, OCH_3) $\xrightarrow{\text{1) Mg, ether; 2) } CH_3\overset{O}{\overset{\|}{C}}H\text{; 3) } H_3O^{\oplus}}$

e) (1-pentene) $\xrightarrow[\text{[}t\text{-BuO}Ot\text{-Bu]}]{HBr}$ (1-bromopentane, Br) $\xrightarrow[EtOH]{NaOEt}$

f) (ethyl hexanoate, O, OEt) $\xrightarrow[\text{2) anode}]{\text{1) NaOH, } H_2O}$

g) $Ph\overset{O}{\overset{\|}{C}}H$ $\xrightarrow[NaOH]{\text{acetone}}$ (4-phenyl-3-buten-2-one, O, Ph) $\xrightarrow[\text{2) } CH_3CH_2Br]{\text{1) Li, } NH_3}$

19.28

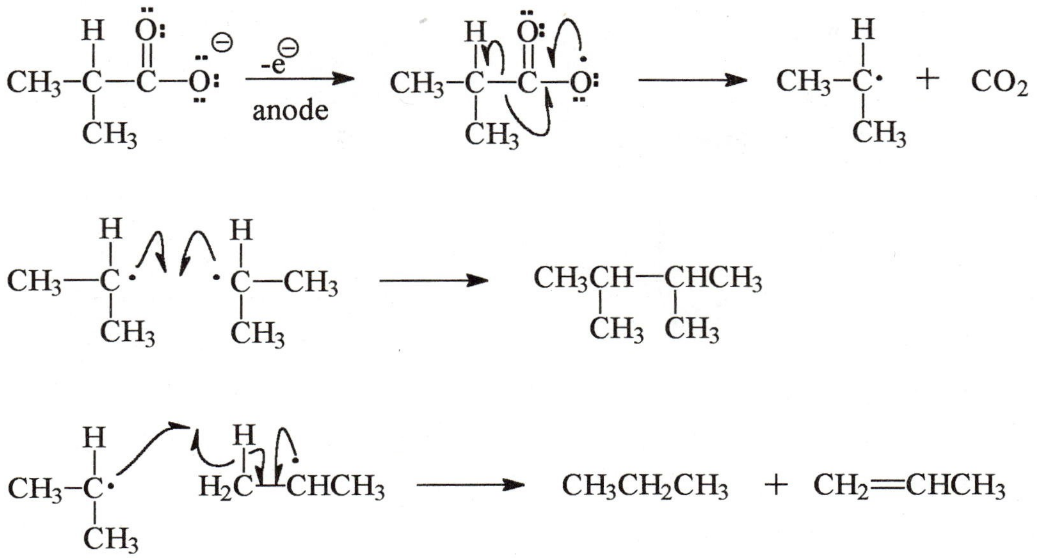

$$CH_3CH(CH_3)COO^- \xrightarrow[\text{anode}]{-e^-} CH_3CH(CH_3)COO\cdot \longrightarrow CH_3\dot{C}H(CH_3) + CO_2$$

$$CH_3\dot{C}H(CH_3) + \cdot CH(CH_3)CH_3 \longrightarrow CH_3CH(CH_3)-CH(CH_3)CH_3$$

$$CH_3\dot{C}H(CH_3) + H-CH_2\dot{C}HCH_3 \longrightarrow CH_3CH_2CH_3 + CH_2{=}CHCH_3$$

19.29

$$:\ddot{Br}-\ddot{Br}: \xrightarrow{h\nu} 2\ :\ddot{Br}\cdot$$

$$C_6H_5CH_2-H + \cdot\ddot{Br}: \longrightarrow C_6H_5\dot{C}H_2 + HBr$$

$$C_6H_5\dot{C}H_2 + :\ddot{Br}-\ddot{Br}: \longrightarrow C_6H_5CH_2Br + \cdot\ddot{Br}:$$

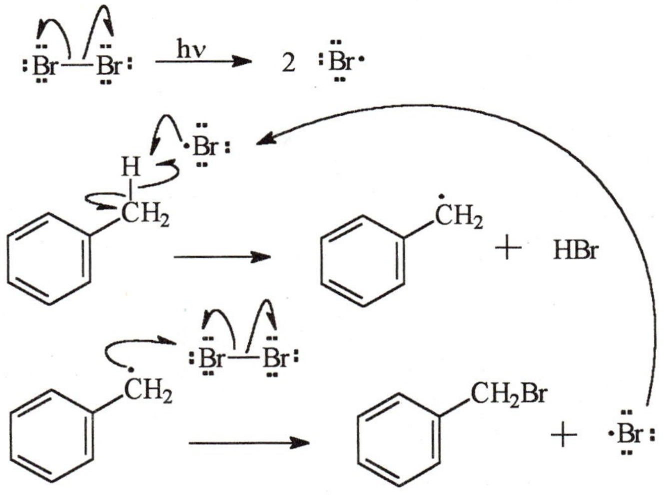

and various termination steps

19.30

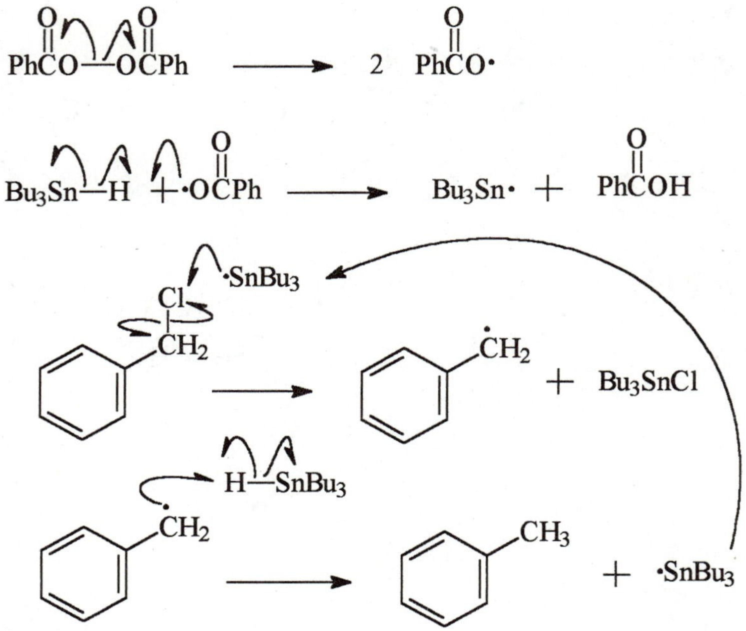

$$PhC(=O)O{-}OC(=O)Ph \longrightarrow 2\ PhC(=O)O\cdot$$

$$Bu_3Sn{-}H + \cdot OC(=O)Ph \longrightarrow Bu_3Sn\cdot + PhC(=O)OH$$

$$C_6H_5CH_2{-}Cl + \cdot SnBu_3 \longrightarrow C_6H_5\dot{C}H_2 + Bu_3SnCl$$

$$C_6H_5\dot{C}H_2 + H{-}SnBu_3 \longrightarrow C_6H_5CH_3 + \cdot SnBu_3$$

and various termination steps

19.31

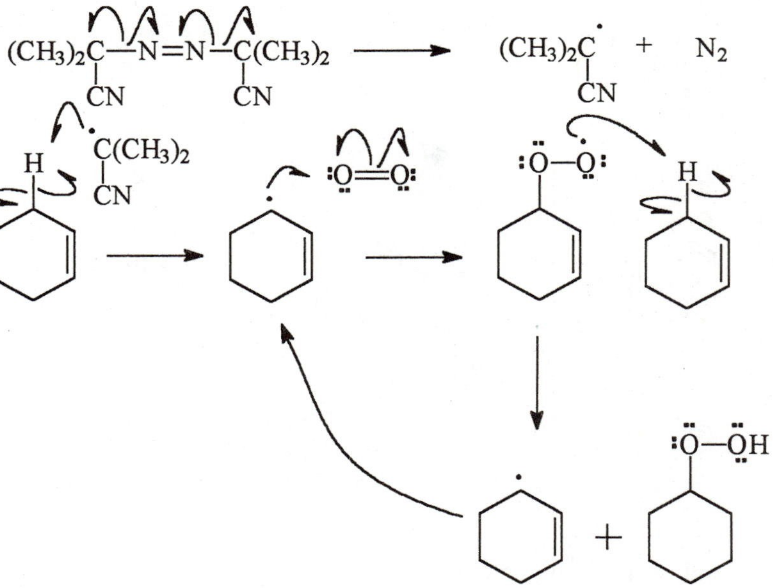

$$(CH_3)_2C(CN){-}N{=}N{-}C(CN)(CH_3)_2 \longrightarrow (CH_3)_2\dot{C}(CN) + N_2$$

$$\dot{C}(CH_3)_2CN$$

and various termination steps

19.32

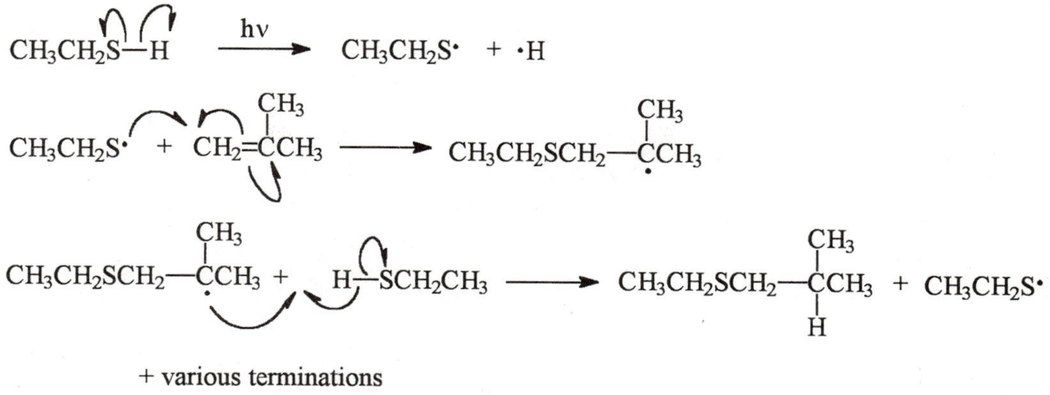

$$CH_3CH_2S{-}H \xrightarrow{h\nu} CH_3CH_2S\cdot + \cdot H$$

$$CH_3CH_2S\cdot + CH_2{=}C(CH_3)CH_3 \longrightarrow CH_3CH_2SCH_2{-}\dot{C}(CH_3)CH_3$$

$$CH_3CH_2SCH_2{-}\dot{C}(CH_3)CH_3 + H{-}SCH_2CH_3 \longrightarrow CH_3CH_2SCH_2{-}CH(CH_3)CH_3 + CH_3CH_2S\cdot$$

+ various terminations

19.33

CH3 · Li → CH3 ⊖ H—OEt → CH3 H H •Li → CH3 ⊖ ↔ CH3 ⊖ EtO—H → CH3

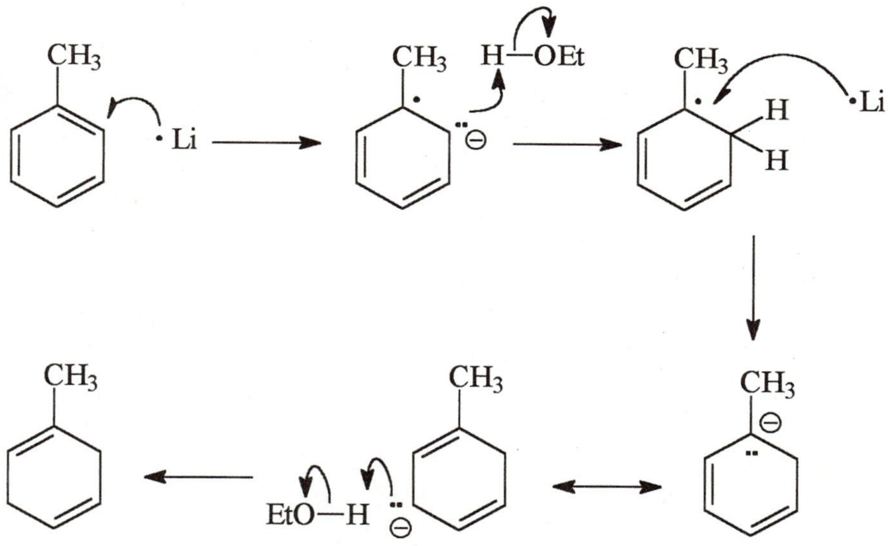

19.34

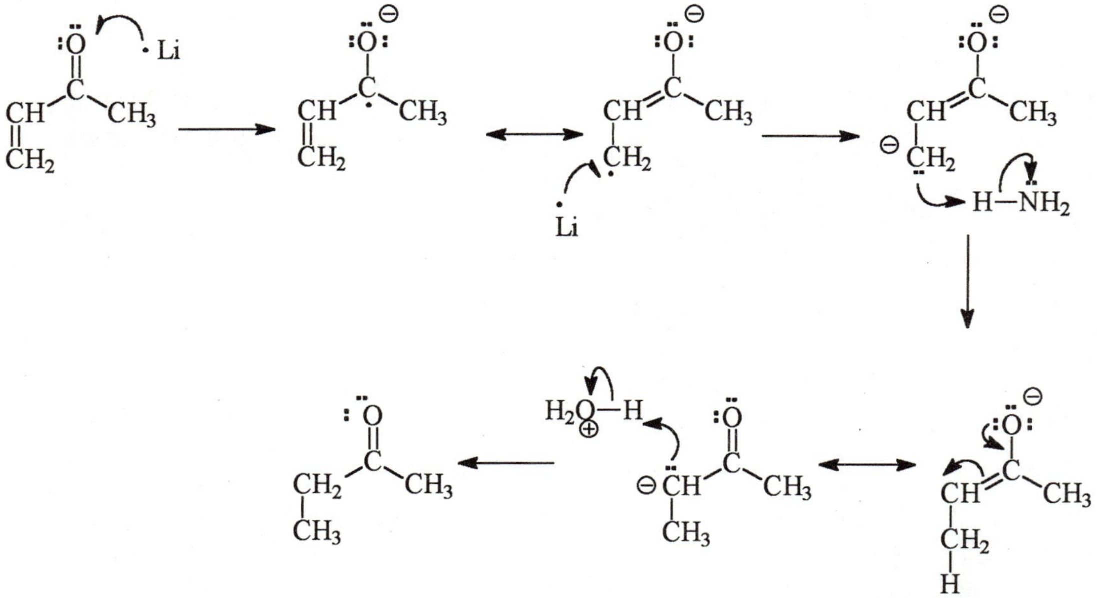

19.35

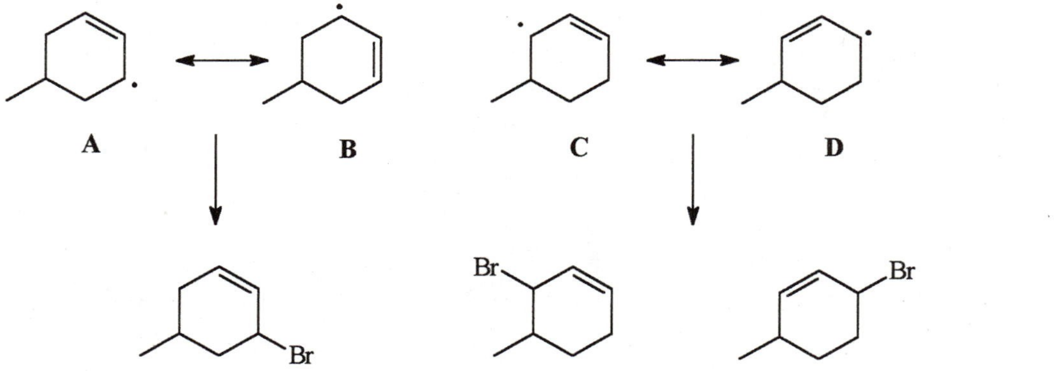

A B C D

There are two types of allylic hydrogens available to be abstracted in this compound. Each of the resulting allylic radicals (A and C) provide two radical sites that can abstract a bromine atom. The radical sites of resonance structures A and B are identical, so only one product is formed. However, the radical sites of the resonance structures C and D are different, so two products are formed.

19.36 The heat energy evolved = (95 kcal/mol - 92 kcal/mol) = 3 kcal/mol (12 kJ/mol)

19.37 Azobenzene is not a useful radical initiator because the product, the phenyl radical, is not stable. Therefore dissociation is quite slow.

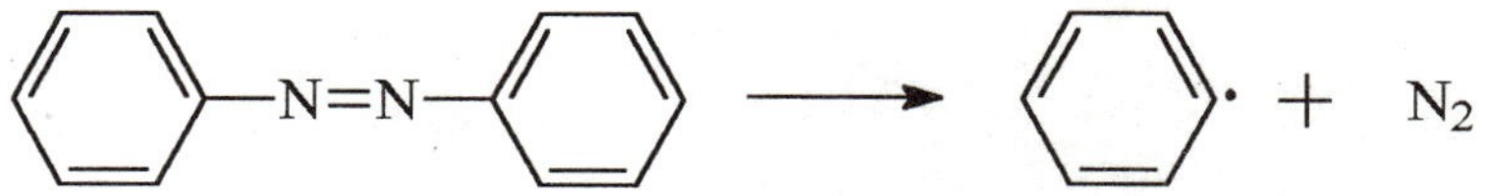

Ph–N=N–Ph ⟶ Ph· + N_2

19.38 The triphenylmethyl radicals react with O_2 to form peroxide radicals.

19.39 The π antibonding MO of a 1,4-cyclohexadiene is too high in energy for an electron from Li atom to be readily added to it because it is not conjugated.

19.40

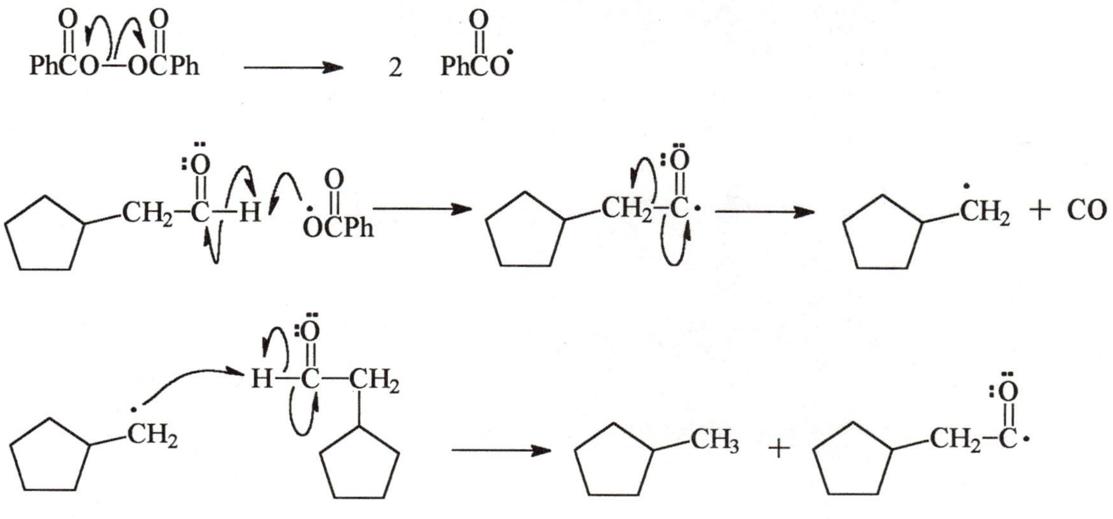

PhC(=O)O–OC(=O)Ph ⟶ 2 PhC(=O)O·

(cyclopentyl)–CH_2–C(=O)–H + ·OC(=O)Ph ⟶ (cyclopentyl)–CH_2–C(=O)· ⟶ (cyclopentyl)–·CH_2 + CO

(cyclopentyl)–·CH_2 + H–C(=O)–CH_2–(cyclopentyl) ⟶ (cyclopentyl)–CH_3 + (cyclopentyl)–CH_2–C(=O)·

19.41

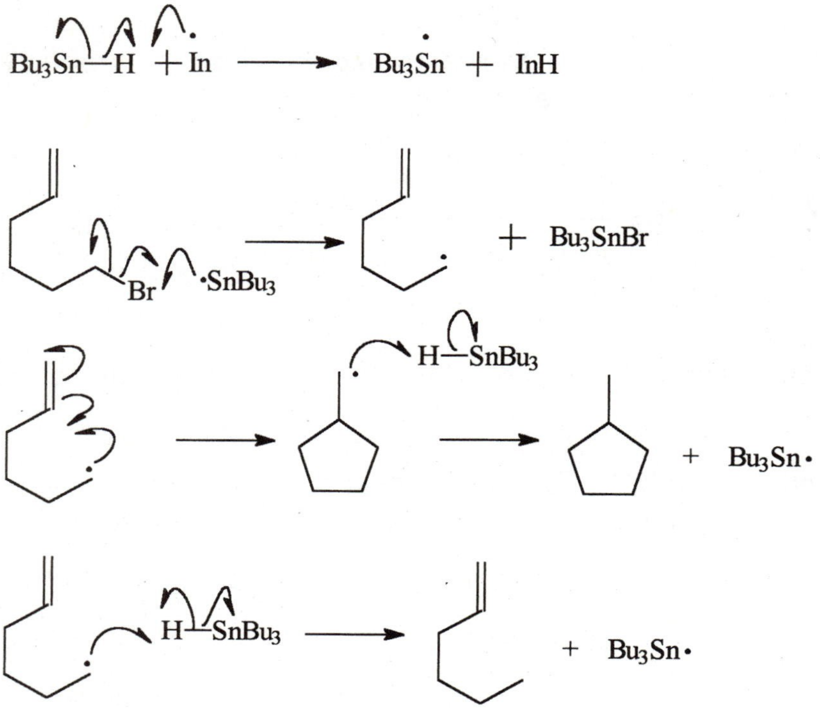

$$Bu_3Sn{-}H + In\cdot \longrightarrow Bu_3Sn\cdot + InH$$

Br $\cdot SnBu_3$ → + Bu_3SnBr

$H{-}SnBu_3$

→ → + $Bu_3Sn\cdot$

$H{-}SnBu_3$ → + $Bu_3Sn\cdot$

19.42 The hydroxy group loses a proton form its conjugated base in the first step.

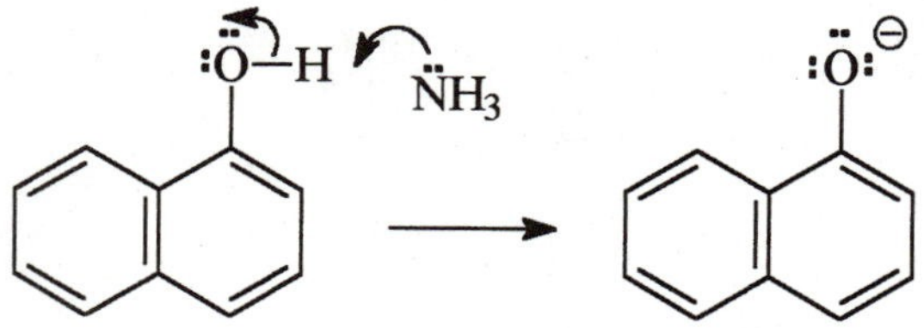

$:\ddot{O}{-}H$ $\ddot{N}H_3$ → $:\ddot{O}:^{\ominus}$

This makes the substituent electron donating, so the reduction occurs in the other ring.

19.43

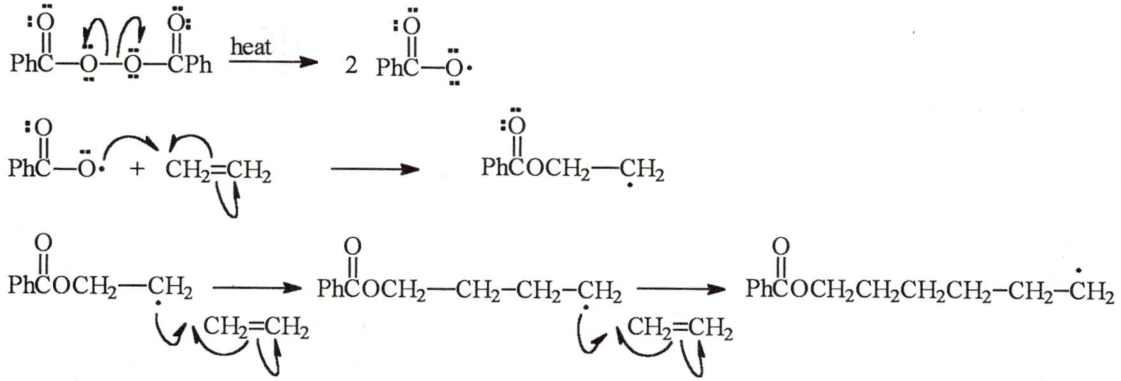

19.44

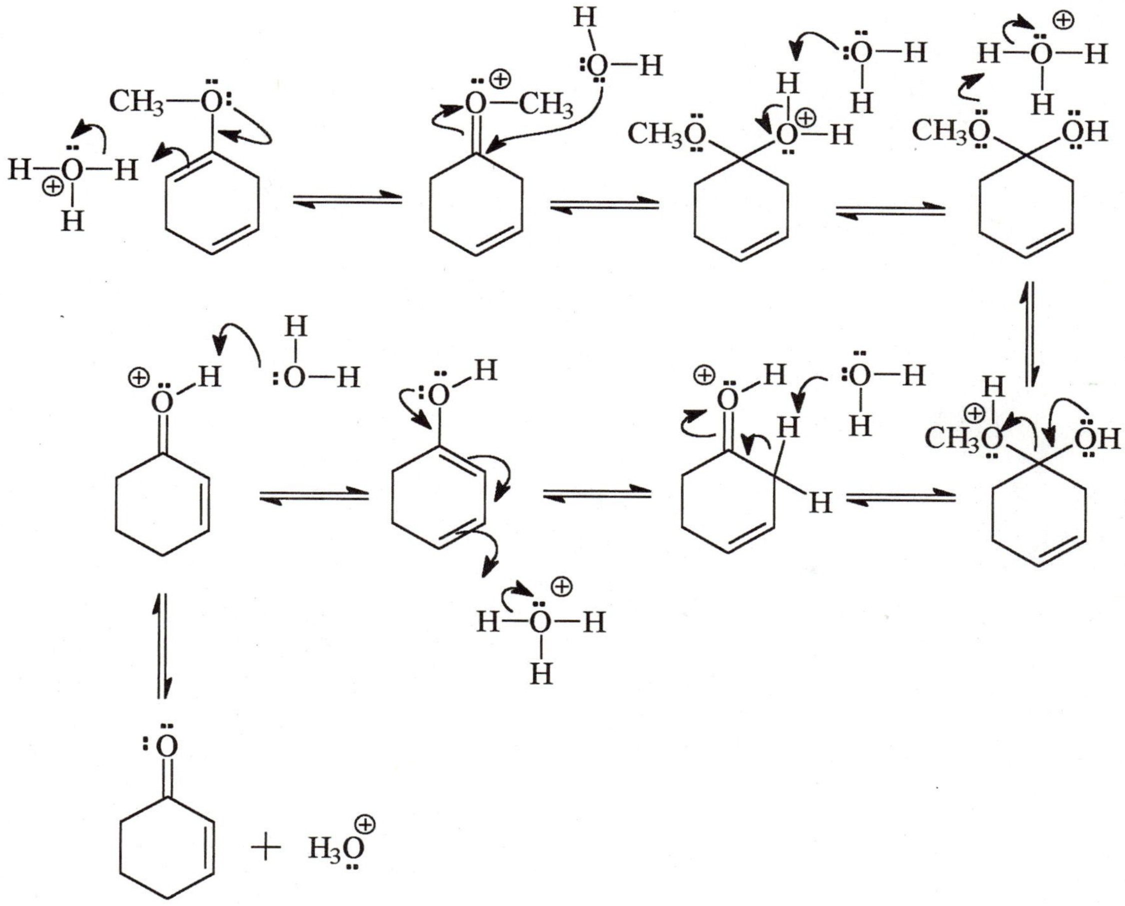

19.45

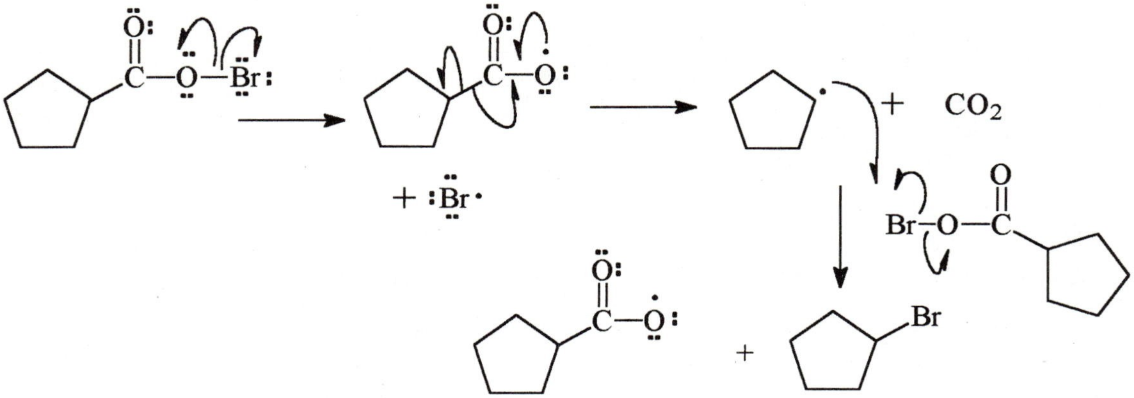

C—O—Br → C—O + Br· → cyclopentyl· + CO_2

Br—O—C(=O)—cyclopentyl → cyclopentyl—C(=O)—O· + cyclopentyl—Br

19.46

$C_6H_5CH_2$—C(=O)—$CH_2C_6H_5$ $\xrightarrow{h\nu}$ $C_6H_5\dot{C}H_2$ + ·C(=O)—$CH_2C_6H_5$

$C_6H_5CH_2$—C(=O)· → $C_6H_5\dot{C}H_2$ + CO

$C_6H_5\dot{C}H_2$ + ·$CH_2C_6H_5$ → $C_6H_5CH_2CH_2C_6H_5$

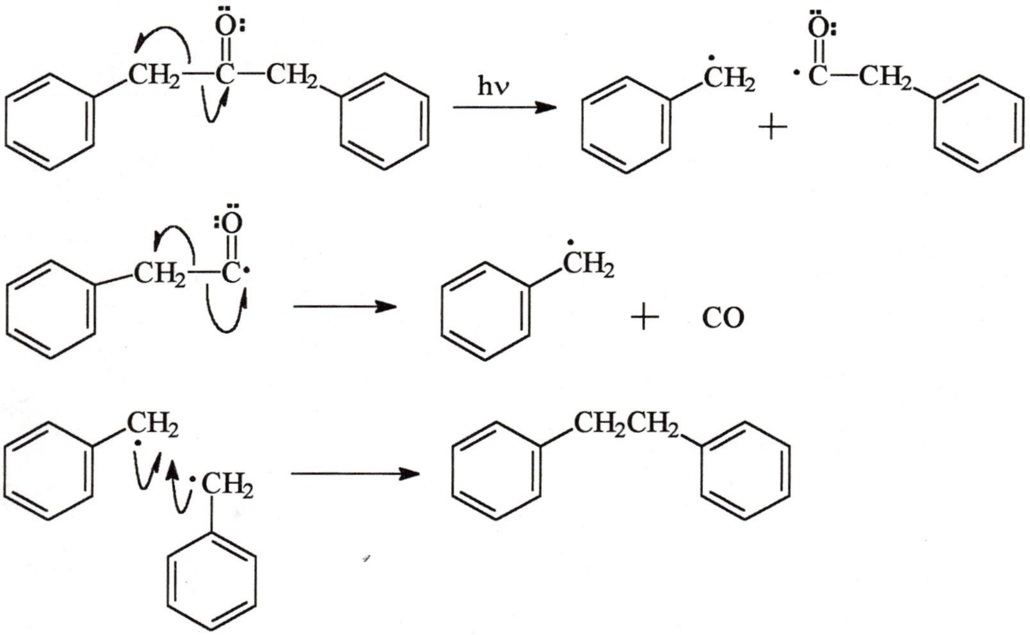

19.47

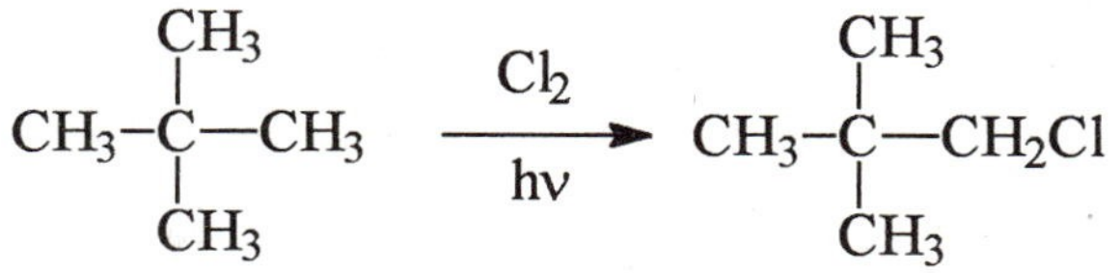

$$CH_3-\underset{CH_3}{\overset{CH_3}{C}}-CH_3 \xrightarrow[h\nu]{Cl_2} CH_3-\underset{CH_3}{\overset{CH_3}{C}}-CH_2Cl$$

19.48

:Ö
C—ÖCH₂CH₃ (ring with N=N) ⟶ C—OCH₂CH₃ (diradical) + N_2 ⟶ C—OCH₂CH₃ (cyclobutane)

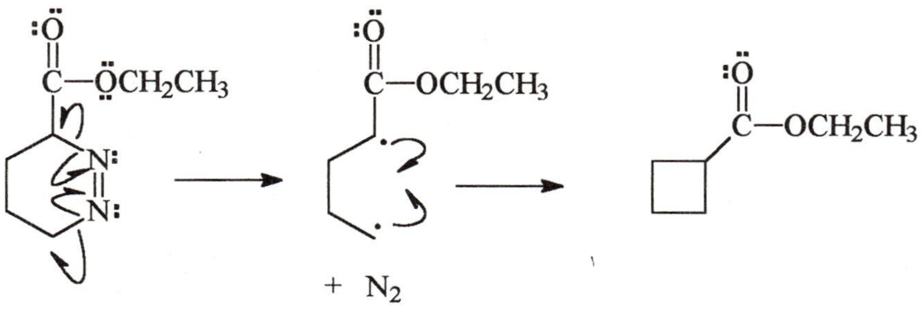

After completing this chapter, you should be able to:

1. Understand the effect of the structure of a radical on its stability and explain how this affects the rate and regiochemistry of radical reactions. Problems 19.1, 19.2, 19.4, 19.10, 19.21.

2. Show the products of the reactions discussed in this chapter. Problems 19.3, 19.5, 19.6, 19.11, 19.12, 19.13, 19.16, 19.18, 19.23, 19.24, 19.25, 19.26, 19.35.

3. Show the mechanisms of these reactions. Problems 19.8, 19.17, 19.28, 19.29, 19.30, 19.31, 19.32, 19.33, 19.34, 19.40, 19.41, 19.45, 19.46.

4. Use these reactions in synthesis. Problems 19.19, 19.20, 19.22, 19.27.

Chapter 20
PERICYCLIC REACTIONS

20.1 a) conrotation b) disrotation

20.2

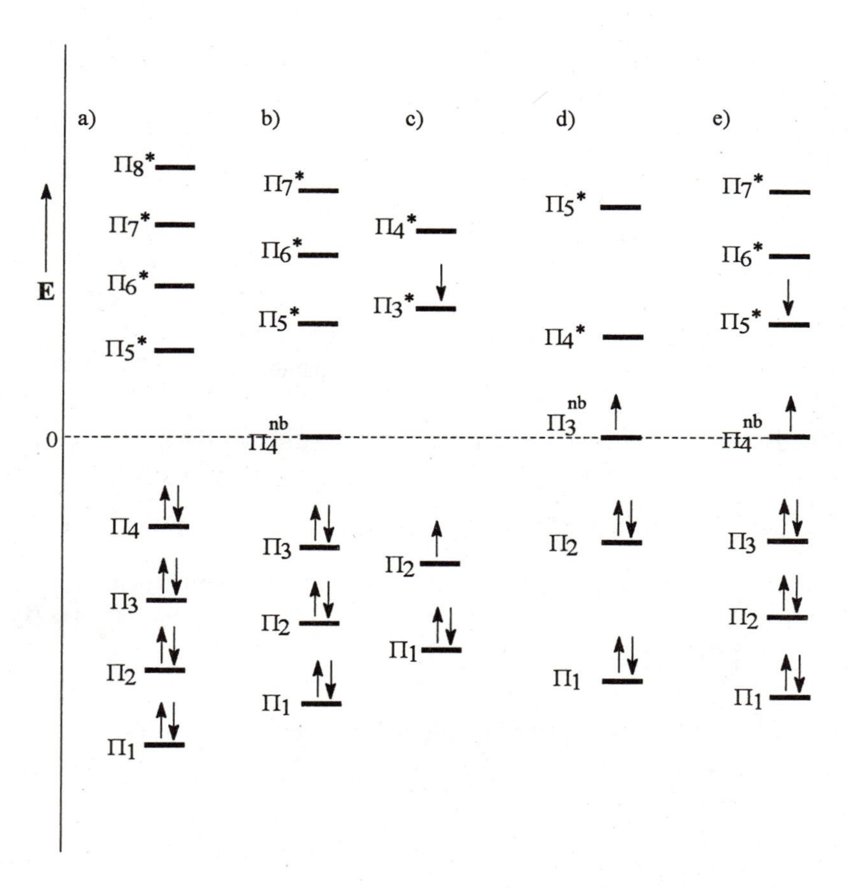

20.3

b) $\Pi 2$
1 node

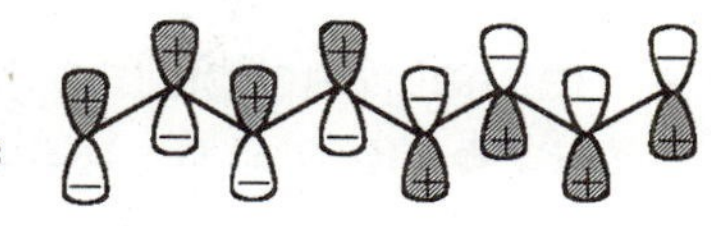

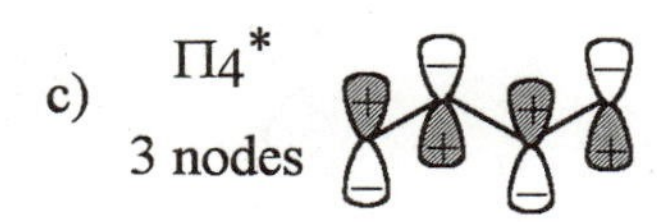

d) $\Pi 4^{nb}$
3 nodes

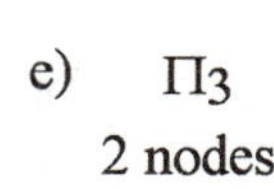

20.4 For the photochemical reaction, the HOMO of diene is π_3^*, so conrotation is forbidden.

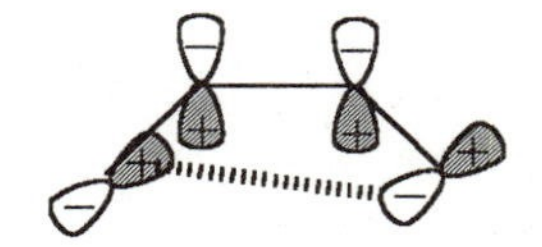

20.5 For the thermal reaction, the HOMO of triene is π_3, so conrotation is forbidden.

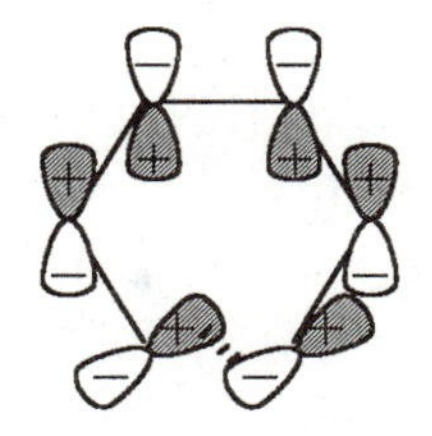

For the photochemical reaction, the HOMO of triene is π_4^*, so disrotation is forbidden.

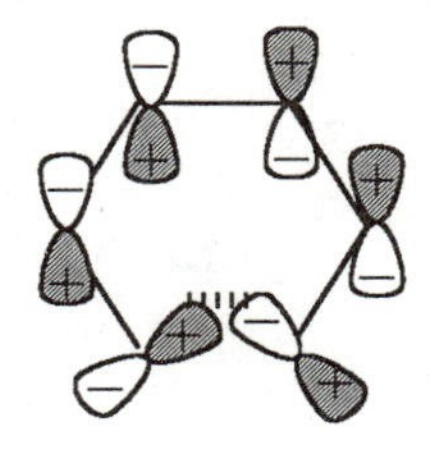

20.6 b) This cation has three p orbitals, so there are three MOs. The three MOs are occupied by two electrons, so for the thermal reaction the HOMO is π_1. Therefore disrotation is allowed.

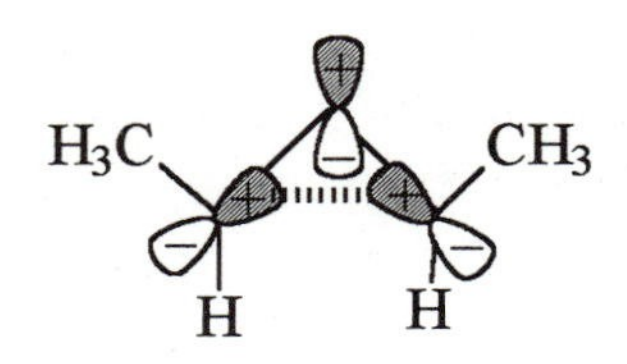

c) This cation has seven p orbitals occupied by six electrons. For the photochemical reaction, the HOMO is π_4^{nb}. The disrotatory closure here is forbidden.

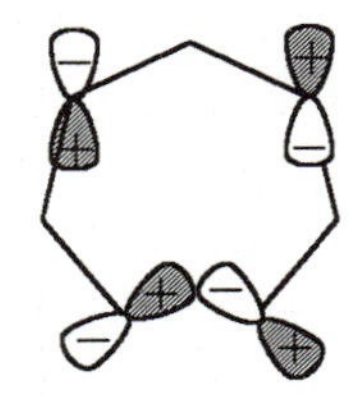

20.7 a) For an odd number of electron pairs (3), conrotation is photochemically allowed and thermally forbidden.

b) For an odd number of electron pairs (1), disrotation is thermally allowed.

c) For an odd number of electron pairs (3), disrotation is photochemically forbidden.

20.8 b) There are two electron pairs involved and the reaction occurs by disrotation. According to the chart, a disrotation involving an even number of electron pairs is photochemically allowed.

c) There are three electron pairs involved and the reaction occurs by conrotation. According to the chart this reaction is thermally forbidden.

d) This reaction involves two electron pairs and occurs by disrotation, so it is thermally forbidden.

e) This reaction involves four electron pairs and occurs by disrotation, so it is photochemically allowed

20.9

a) CH_3, CH_3 b) CH_3, CH_3 c) CH_3, CH_3 d) CH_3, CH_3

e) CH_2CH_3, CH_3 f) CH_3, CH_3 g) H, CH_3, H, CH_3 + H_3C, CH_3

20.10 For a cycloaddition reaction, if the overlaps of the orbitals of the HOMO of one component and LUMO of the other component are bonding then the reaction is allowed; but if these overlaps are antibonding the reaction is forbidden. In this orbital drawing of the photochemical [4 +2] cycloaddition, one overlap is bonding and the other is antibonding, so the reaction is forbidden.

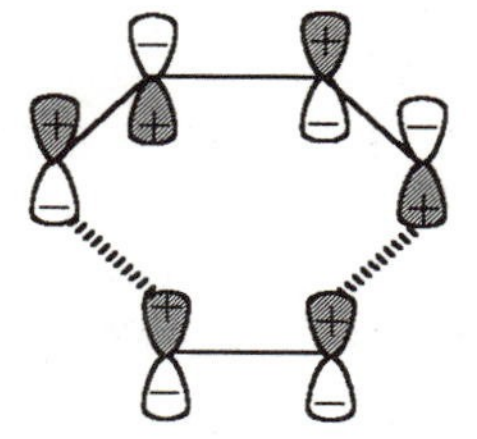

LUMO of the excited diene

HOMO of the alkene

20.11 In this orbital drawing of the thermal [4 +4] cycloaddition, one overlap is bonding and the other is antibonding so the reaction is forbidden.

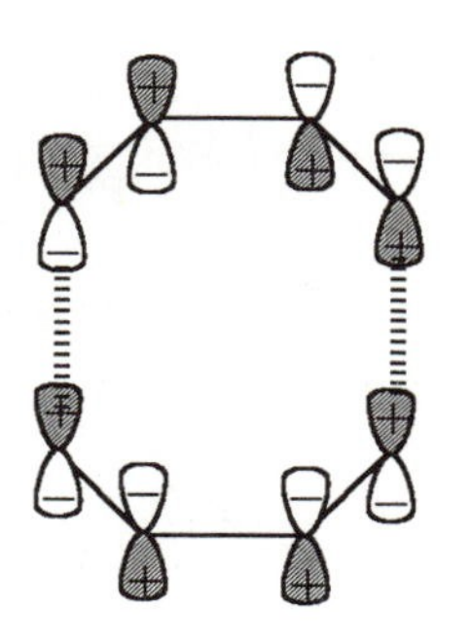

HOMO of one diene

LUMO of the other diene

20.12 a) This is a [4 + 4] cycloaddition involving four electron pairs, so it is photochemically allowed.

b) This is a [6 + 2] cycloaddition involving four electron pairs, so it is photochemically allowed but thermally forbidden.

20.13 The left diene is more reactive because its s-*cis* conformation is less hindered and is therefore present in higher concentration.

20.14

a)

b) HO_2C CO_2H

c) O_2CCH_3 CN CN O_2CCH_3

d)

e)

f)

g) 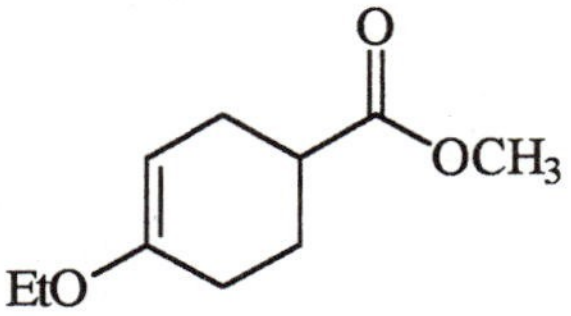 EtO, O, OCH_3

h) 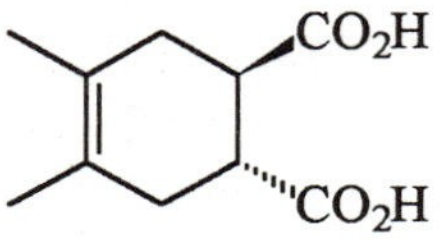 CO_2H, CO_2H

i) 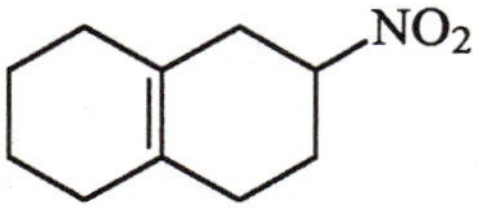 NO_2

j) 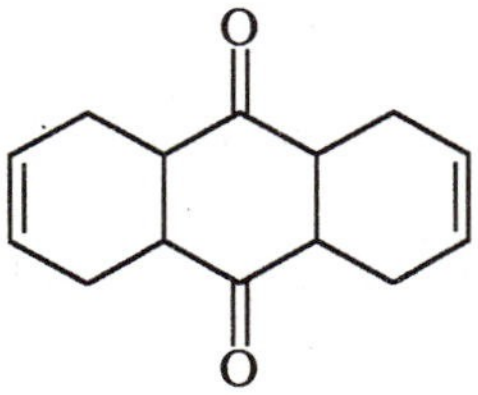 O, O

20.15 This is a [10 + 2] cycloaddition involving six electron pairs, so it is photochemically allowed.

20.16

a)

b) 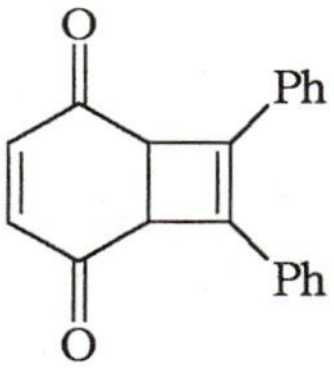 O, O, Ph, Ph

c) 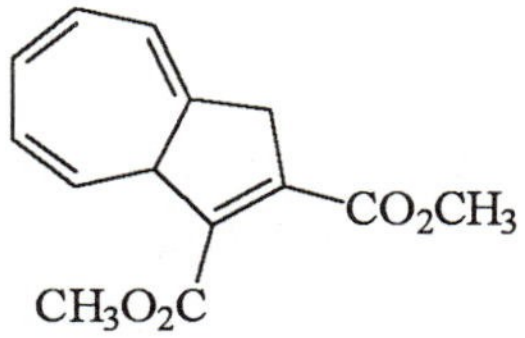 CO_2CH_3, CH_3O_2C

d) 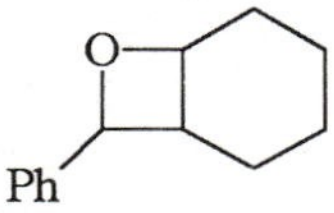 O, Ph

20.17 A sigmatropic rearrangement is an intramolecular reaction involving migration of a σ-bonded group over a conjugated π system. The i and j values in the notation of sigmatropic rearrangements refer to the number of bonds separating the migration origin and the migration terminus in each component.

a) [3,3] sigmatropic rearrangement b) [1,7] sigmatropic rearrangement
c) [3,5] sigmatropic rearrangement

20.18 The two fragments of a [1,5] sigmatropic rearrangement are a hydrogen atom and a pentadienyl radical. The HOMO of the pentadienyl radical for the photochemical reaction is π_4*. According to the orbital drawing one overlap is bonding and one is antibonding, so the reaction is photochemically forbidden.

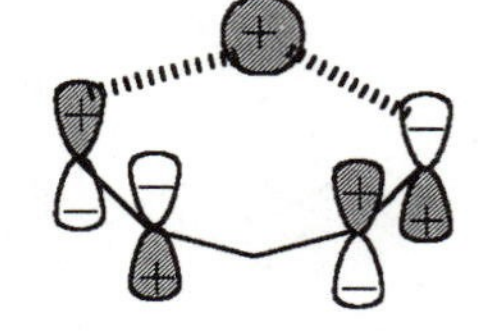

20.19 For the thermal reaction the HOMO of the hexatrienyl radical is π_4^{nb}. In the orbital drawing shown, one overlap is bonding and one is antibonding, so the reaction is forbidden.

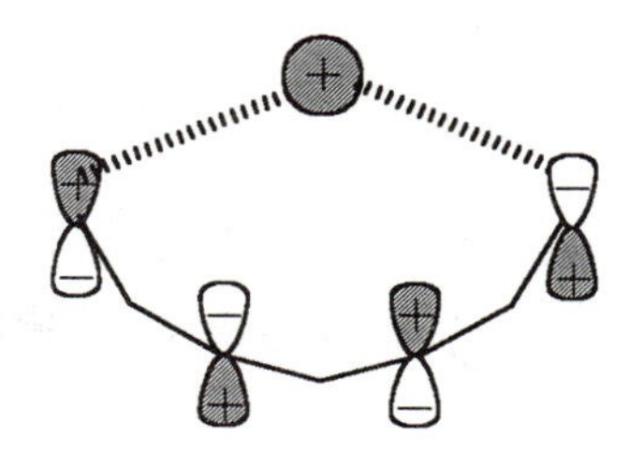

20.20 a) This a [3,3] sigmatropic rearrangement involving an odd number of electron pairs, so it is thermally allowed.

b) This a [1,7] sigmatropic rearrangement involving an even number of electron pairs, so it is photochemically allowed.

c) This a [3,5] sigmatropic rearrangement involving an even number of electron pairs, so it is photochemically allowed.

20.21

a) 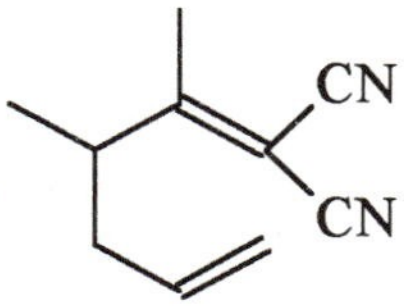 CN, CN

b)

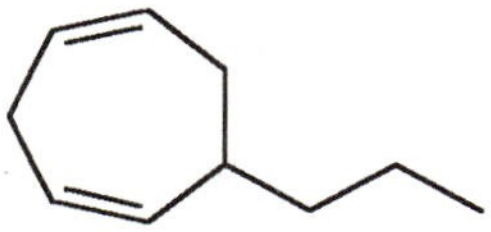

c) 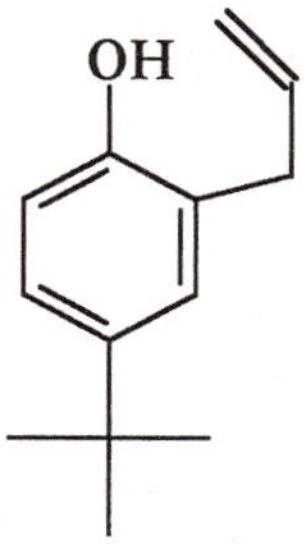 OH

d) 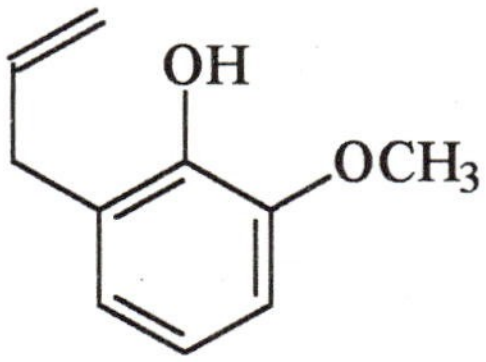 OH, OCH_3

e) 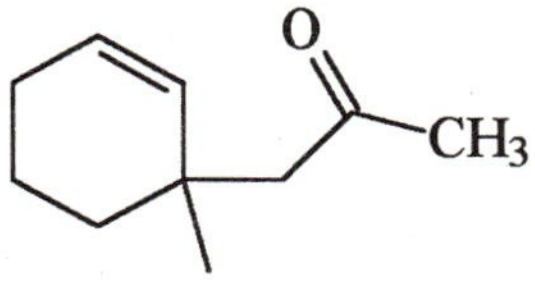 O, CH_3

20.22 This reaction is a [3,3]-sigmatropic rearrangement involving an odd number of electron pairs, so it is thermally allowed. The reactant is favored at equilibrium because it has less ring strain.

20.23

a) 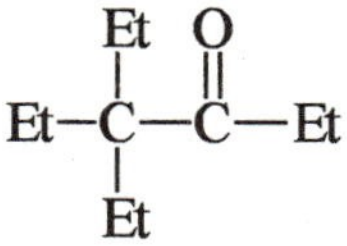 $Et{-}C(Et)_2{-}C({=}O){-}Et$ b) PhNHC(=O)Ph c) O, O

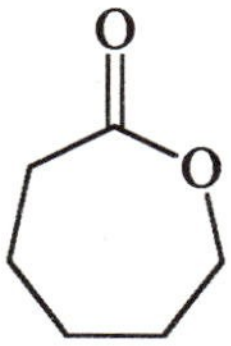

20.24

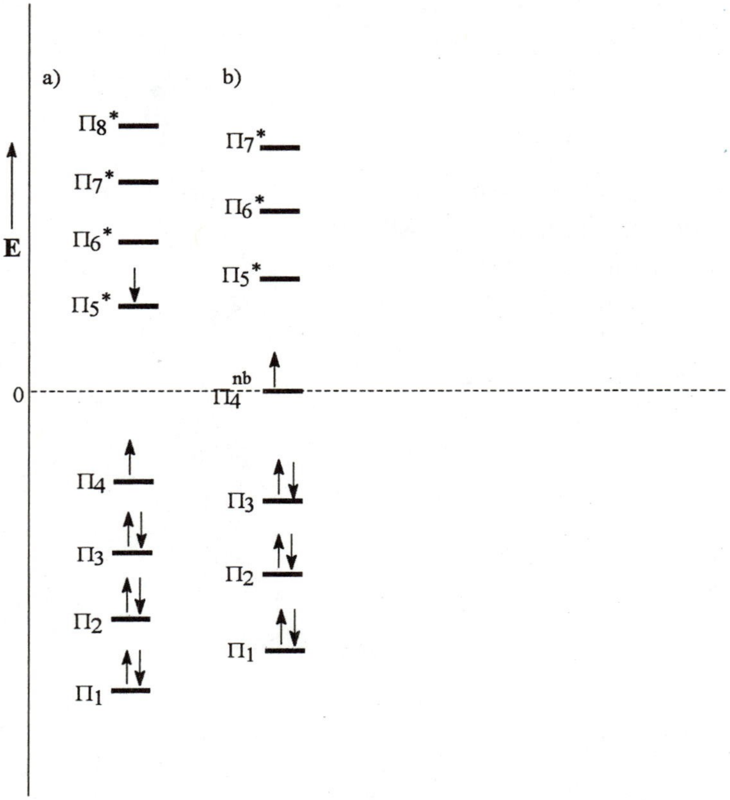

a)

Π_8^*

Π_7^*

Π_6^*

Π_5^* ↓

Π_4 ↑

Π_3 ↑↓

Π_2 ↑↓

Π_1 ↑↓

b)

Π_7^*

Π_6^*

Π_5^*

Π_4^{nb} ↑

Π_3 ↑↓

Π_2 ↑↓

Π_1 ↑↓

E

0

20.25

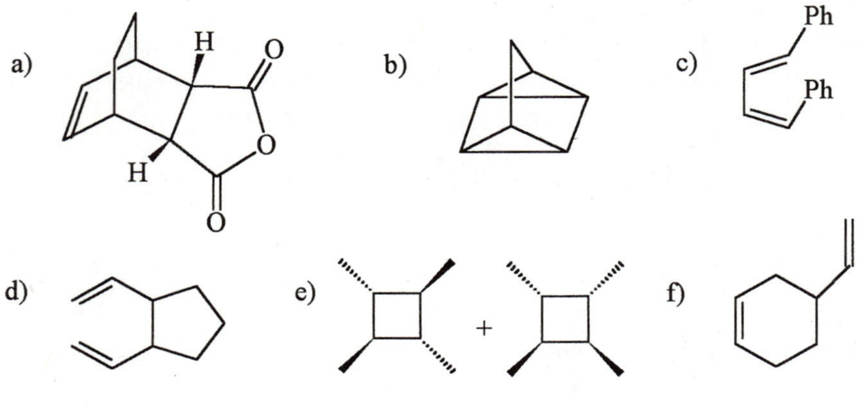

a) H H O O O

b)

c) Ph Ph

d)

e) +

f)

g) 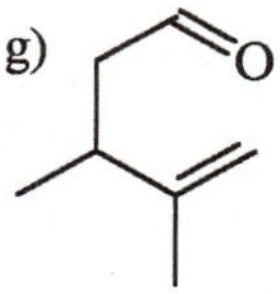 O

h) 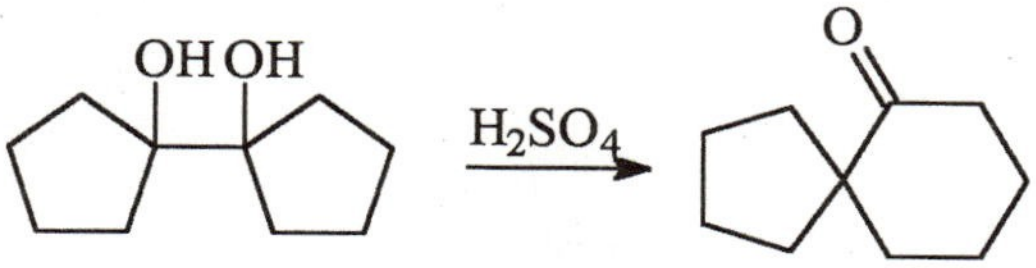 OH OH $\xrightarrow{H_2SO_4}$ O

i) $CH_3CH_2CH_2\overset{O}{\overset{\|}{C}}OCH_2CH_2CH_3$

20.26

a) 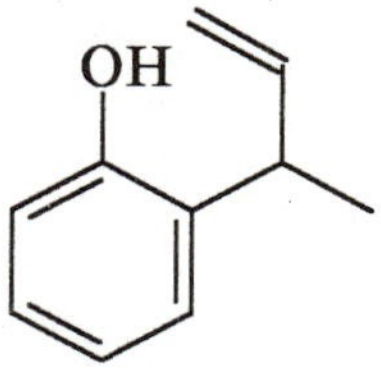 OH

b) 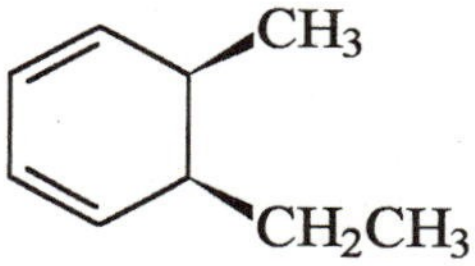 CH_3 CH_2CH_3

c) 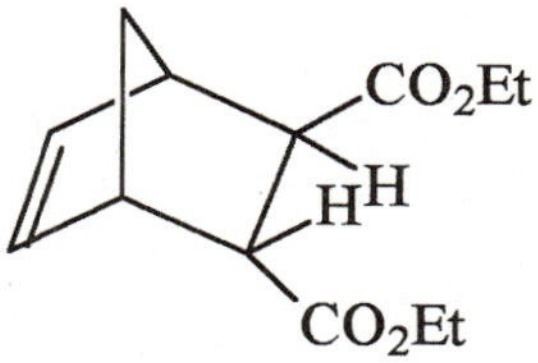 CO_2Et H H CO_2Et

d) 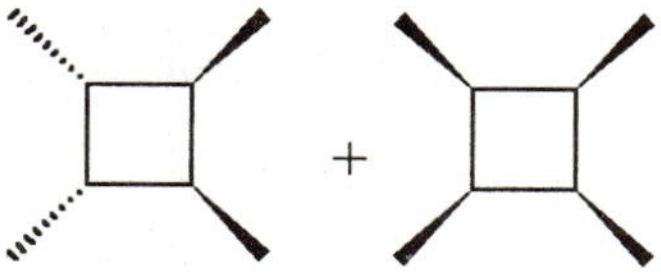 +

e) 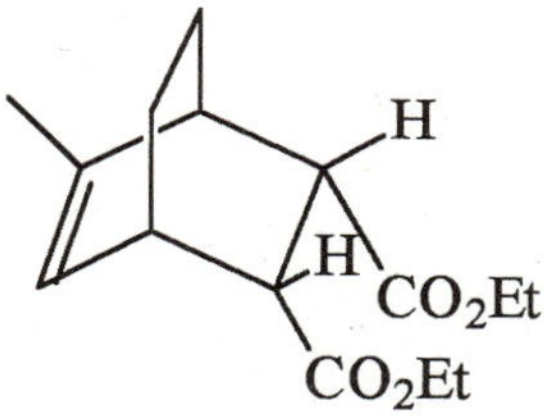 H H CO_2Et CO_2Et

f) 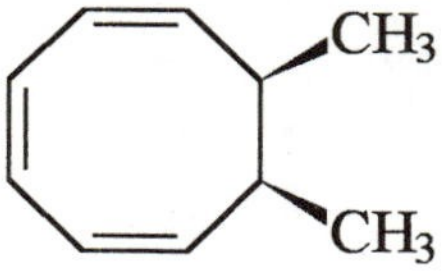 CH_3 CH_3

g)

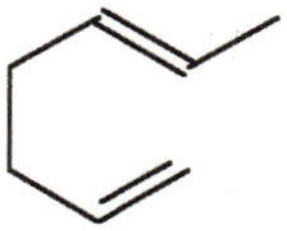

h) 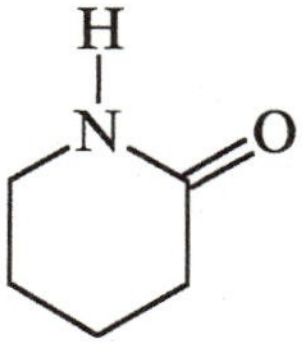 H N O

i) 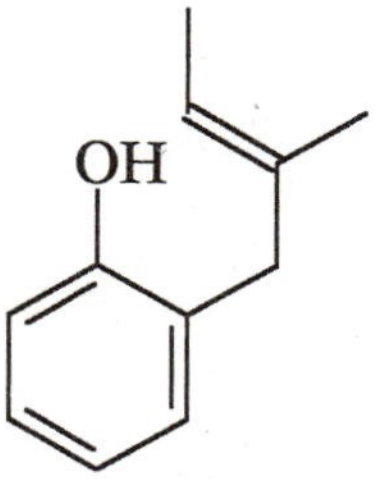 OH

20.27

a) 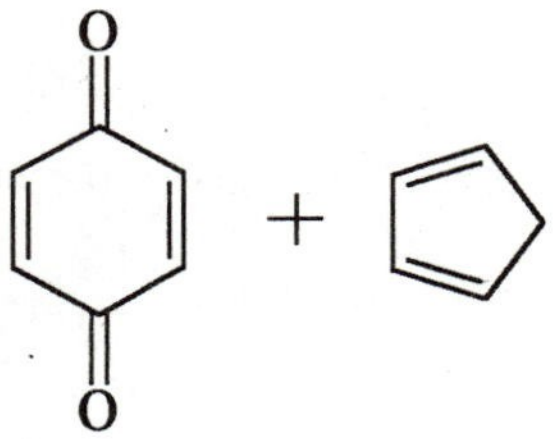 O, O +

b) 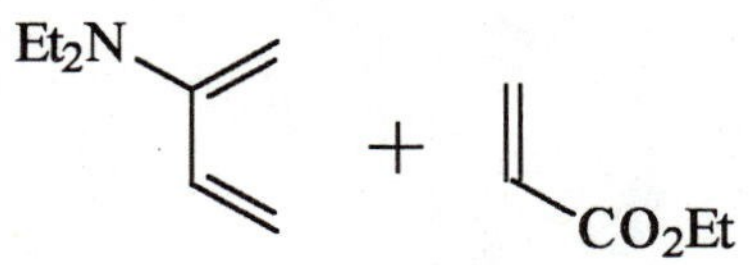 Et_2N + CO_2Et

c) 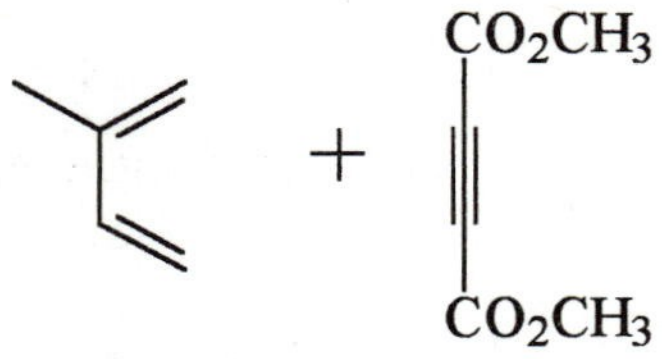 + CO_2CH_3 CO_2CH_3

d) 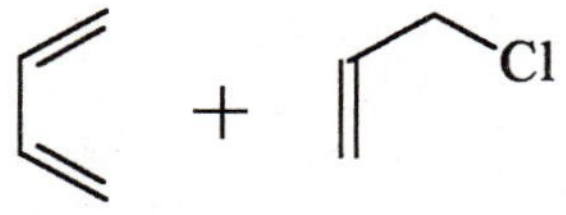 + Cl

e) 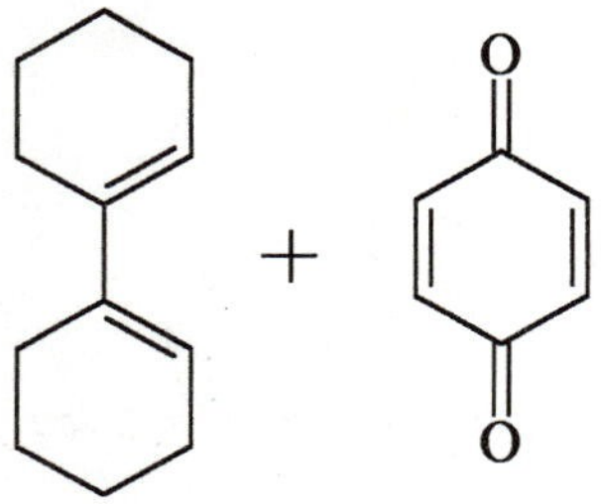 + O O

f) 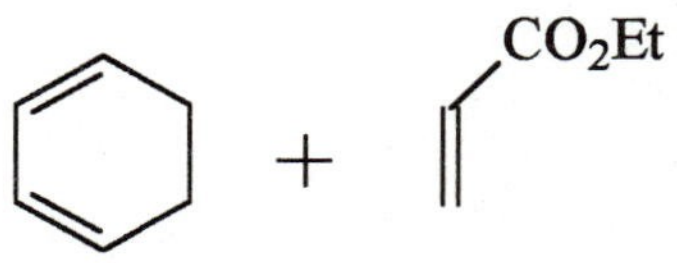 + CO_2Et

g) 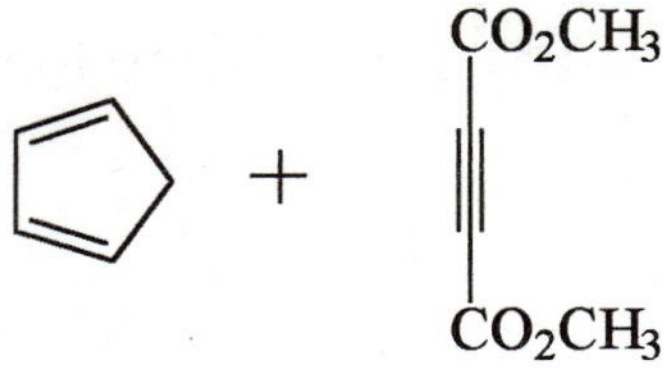 + CO_2CH_3 CO_2CH_3

20.28 a) For the thermal [6+2] cycloaddition, one overlap is bonding and the other is antibonding, so the reaction is forbidden.

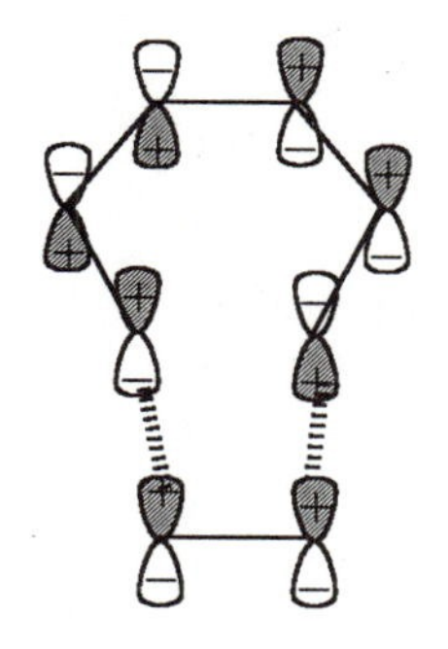

LUMO of the ground state triene

HOMO of the alkene

b) For the photochemical [6+2] cycloaddition, both overlaps are bonding, so this reaction is allowed.

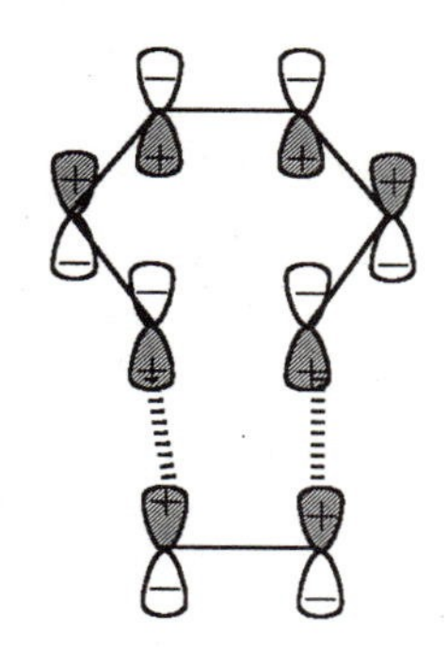

LUMO of the excited state triene

HOMO of the alkene

c) For the thermal reaction, the HOMO = π_3 ; conrotation is forbidden.

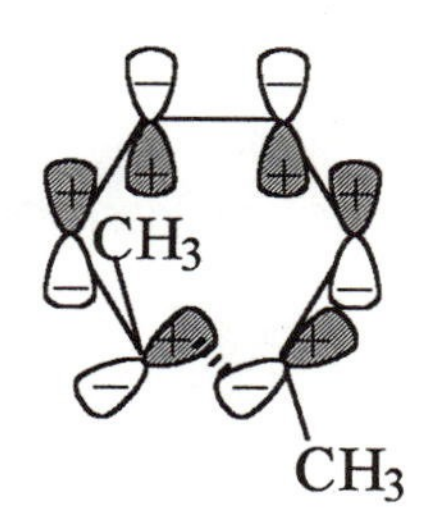

HOMO of the ground state triene

d) For the photochemical [4+4] cycloaddition both overlaps are bonding, so the reaction is allowed.

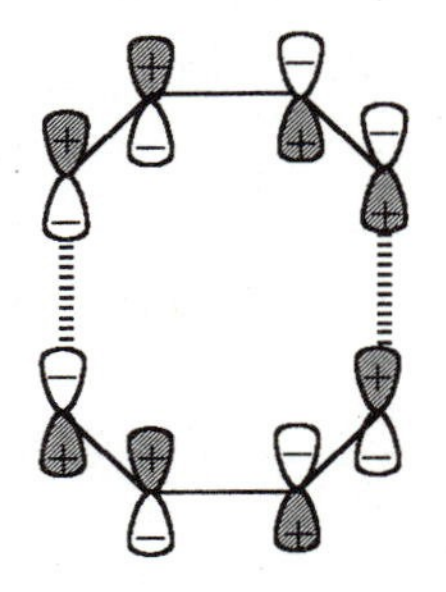

HOMO of the ground state diene

LUMO of the excited diene

e) For the thermal reaction the HOMO = π_4 ; conrotation is allowed.

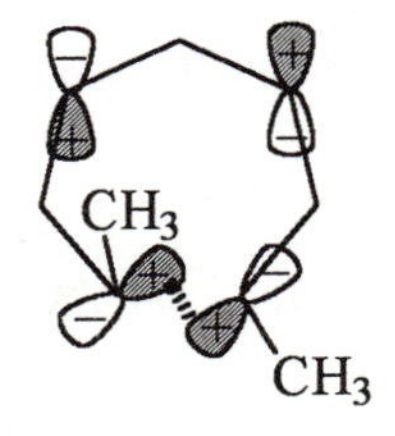

HOMO of the ground state trienyl anion

f) For the thermal [3,5] sigmatropic rearrangement one overlap is bonding and the other is antibonding, so the reaction is forbidden.

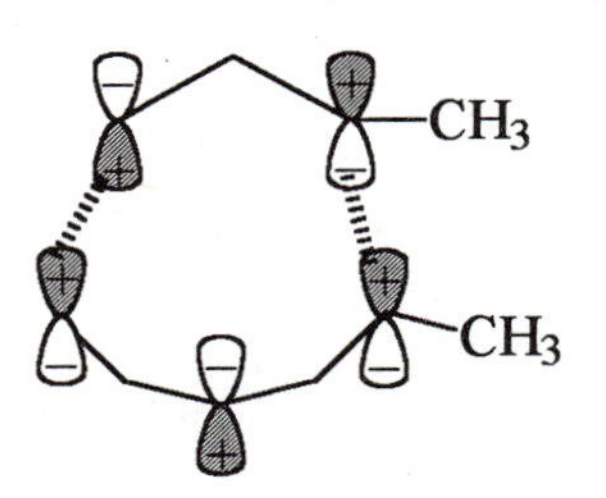

HOMO of the allyl radical

HOMO of the pentadienyl radical

g) For the thermal [1,5] sigmatropic rearrangement, both overlaps are bonding therefore the reaction is allowed.

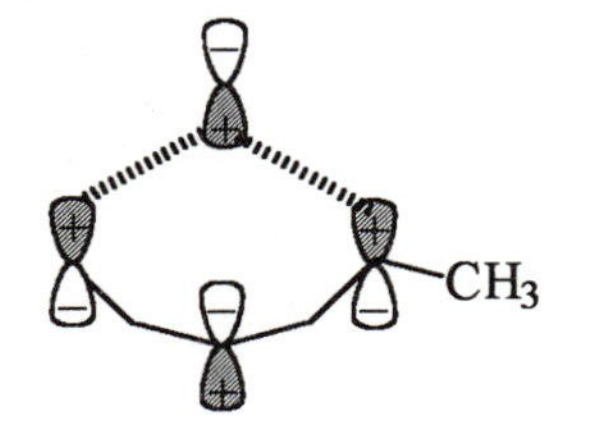

HOMO of the methyl radical

HOMO of the pentadienyl radical

20.29 a) This is a [6+2] cycloaddition involving an even number of electron pairs, so the reaction is photochemically allowed.

b) This is a [1,5] sigmatropic rearrangement involving an odd number of electron pairs, so the reaction is thermally allowed.

c) This is a [1,7] sigmatropic rearrangement involving an even number of electron pairs, so the reaction is photochemically allowed.

d) This is a [6+2] cycloaddition involving an even number of electron pairs, so the reaction is photochemically allowed.

e) This is a [2+2] cycloaddition involving an even number of electron pairs, so the reaction is photochemically allowed.

f) This is a [6+4] cycloaddition involving an odd number of electron pairs, so the reaction is thermally allowed.

g) This is a [3,7] sigmatropic rearrangement involving an odd number of electron pairs, so the reaction is thermally allowed.

h) This is a [6+4] cycloaddition involving an odd number of electron pairs, so the reaction is thermally allowed.

20.30

a) hν; H_2 / Pt

b) CN, NC; Δ; CN, H, H, CN; H_2 / Pt

c) CO_2Et; Δ; CO_2Et; 1) 2 EtMgBr 2) NH_4Cl, H_2O

d) CO_2CH_3, CO_2CH_3; Δ

e)  + $h\nu$

f) 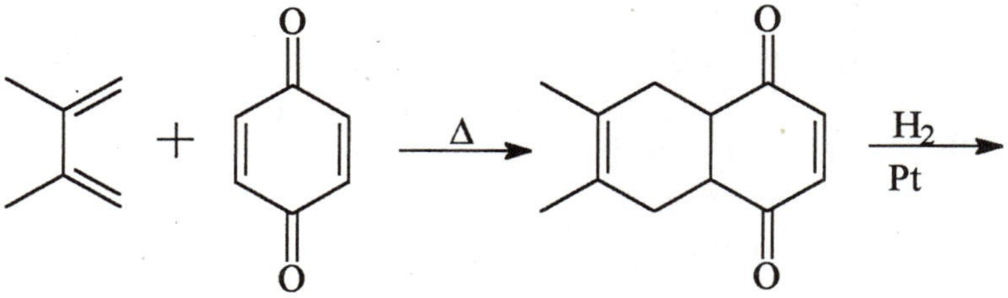 O O + Δ O O H_2 Pt

g) 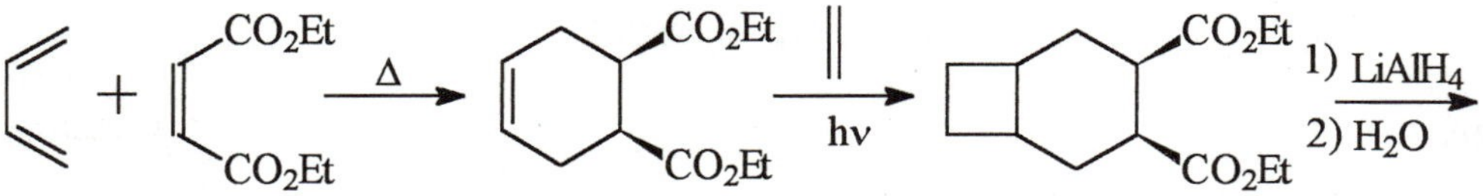 + CO_2Et CO_2Et Δ CO_2Et CO_2Et $h\nu$ CO_2Et CO_2Et 1) $LiAlH_4$ 2) H_2O

20.31

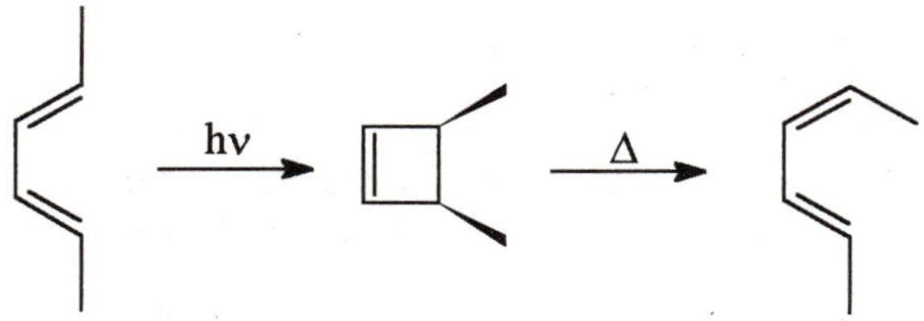

$h\nu$ Δ

20.32 a) The left compound is more reactive because the double bonds of the diene are held fixed in a s-*cis* conformation whereas the right compound can interconvert to a s-*trans* conformation.

b) The right compound is more reactive because the equilibrium favors the s-*cis* conformation due to the large steric strain present in the s-*trans* conformation.

c) The left compound is more reactive because the double bonds are fixed in an s-*cis* conformation. The right compound cannot react as a diene in a Diels-Alder reaction because the double bonds are held s-*trans*.

20.33

$$\text{2-methoxyphenol} \xrightarrow{NaOH} \text{2-methoxyphenoxide} \xrightarrow{CH_2=CHCH_2Cl} \text{allyl 2-methoxyphenyl ether} \xrightarrow{\Delta} \text{2-allyl-6-methoxyphenol}$$

20.34 The thermal ring opening of *trans*-3,4-dimethylcyclobutene occurs with a conrotatory motion. Conrotation can be either clockwise or counterclockwise. Clockwise opening produces the (2*E*,4*E*)-isomer, while the counterclockwise opening produces the (2*Z*,4*Z*)-isomer. The formation of the (2*Z*,4*Z*)-isomer is disfavored because of the steric interaction between the two methyl groups in the transition state.

20.35

$$HC(OH)H{-}C(OH)Ph_2 \xrightarrow{H-OH_2^+} HC(OH)H{-}C(OH_2^+)Ph_2 \xrightarrow{-H_2O} HC(OH)H{-}C^+Ph_2 \xrightarrow{\text{hydride shift}} H{-}C^+(OH){-}CHPh_2 \xrightarrow{:OH_2} H{-}C(=O){-}CHPh_2$$

The OH that is lost produces a more stable carbocation intermediate than the carbocation that would result if the other OH is lost.

20.36

H—$\ddot{O}H_2^{\oplus}$

$\ddot{N}$—$\ddot{O}H$

CH_3—C—CH_3 → $\ddot{N}$—$\ddot{O}H_2^{\oplus}$, CH_3—C—CH_3 → CH_3—$\ddot{N}$=$\overset{\oplus}{C}$—CH_3 ⟷ $\overset{\oplus}{N}$—CH_3, CH_3—C

:$\ddot{O}H_2$

↓

$\ddot{N}$—CH_3

CH_3—C—$\ddot{O}$—H, H (⊕), $H_2\ddot{O}$:

← $\ddot{N}$—CH_3, CH_3—C—$\ddot{O}H$ ⇌ H—$\ddot{N}$—CH_3, CH_3—C=$\ddot{O}$

20.37

$\ddot{O}$:

CH_3—C—CH_3

$CF_3C\ddot{O}$—$\ddot{O}H$

$\ddot{O}$:

→ → :$\ddot{O}$—H

CH_3—C—CH_3

$CF_3C\ddot{O}$—$\ddot{O}$:

$\ddot{O}$:

→ $\ddot{O}$:

CF_3C—$\ddot{O}$: ⊖

H

:O ⊕

CH_3—C—$\ddot{O}CH_3$

↓

$\ddot{O}$:

$CF_3C\ddot{O}H$ + $CH_3C\ddot{O}CH_3$

20.38

[4+2]
Δ

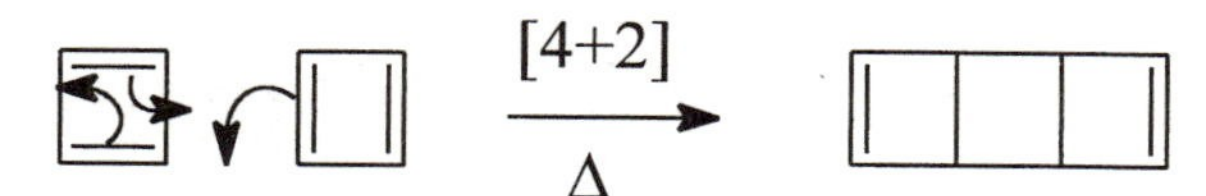

The dimer of cyclobutadiene is formed by a thermally allowed [4+2] cycloaddition reaction. One molecule of cyclobutadiene reacts as the diene and the other as the dienophile in a Diels-Alder reaction.

20.39

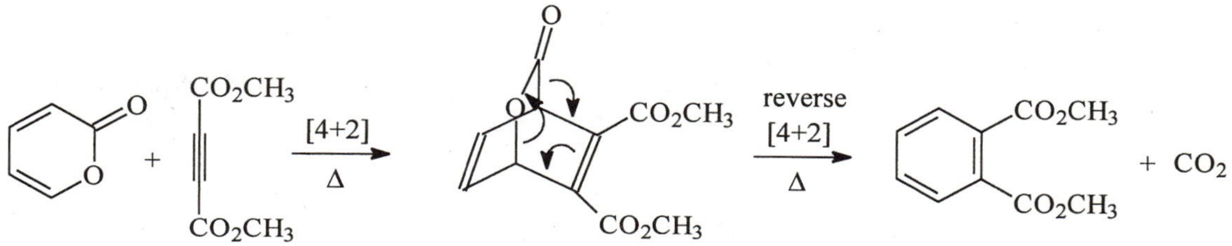

O
O
+
CO_2CH_3
CO_2CH_3
[4+2]
Δ
O
O
CO_2CH_3
CO_2CH_3
reverse
[4+2]
Δ
CO_2CH_3
CO_2CH_3
+ CO_2

The first step of this reaction is a thermally allowed [4+2] cycloaddition (Diels-Alder reaction) to produce an intermediate bicyclic compound as shown above. This intermediate rapidly undergoes a thermally allowed reverse Diels-Alder reaction to produce the stable aromatic diester and carbon dioxide.

20.40 This reaction is related to the pinacol rearrangement.

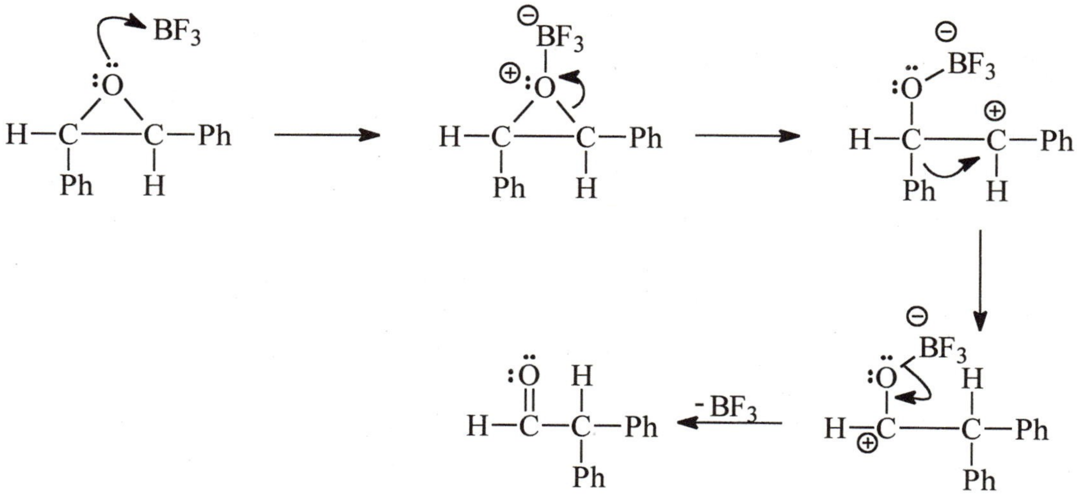

BF_3
H—C—C—Ph
Ph
H
- BF_3

20.41

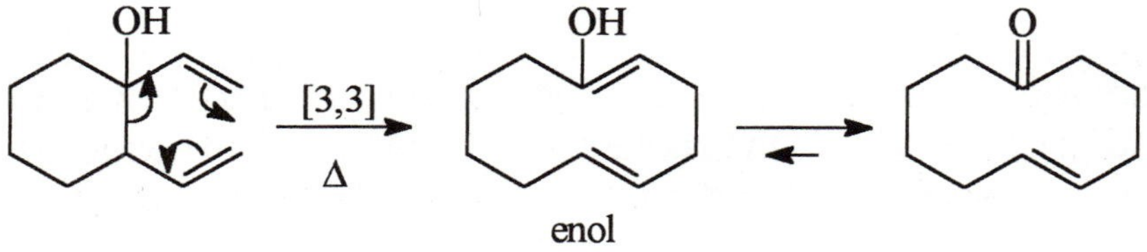

OH ... [3,3] Δ ... OH ... O

enol

20.42 This is a thermally allowed [1,5] sigmatropic rearrangement reaction involving the migration of a hydrogen atom. Since there is free rotation at the migration origin, there can be two possible transitions states producing two products as shown below.

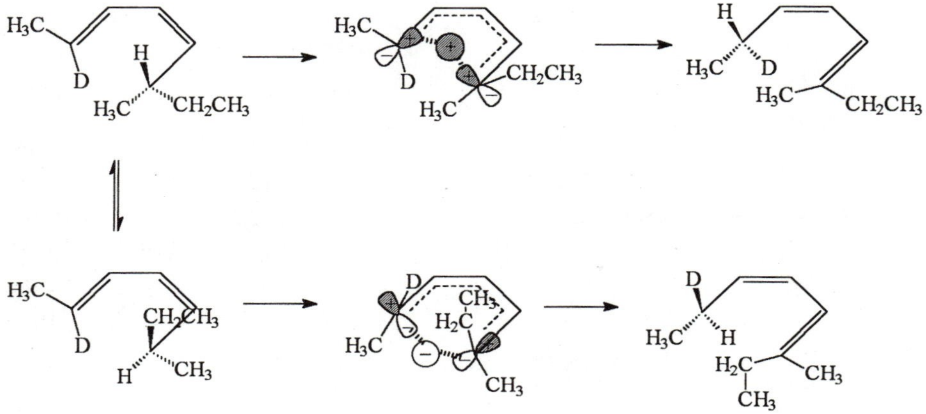

20.43

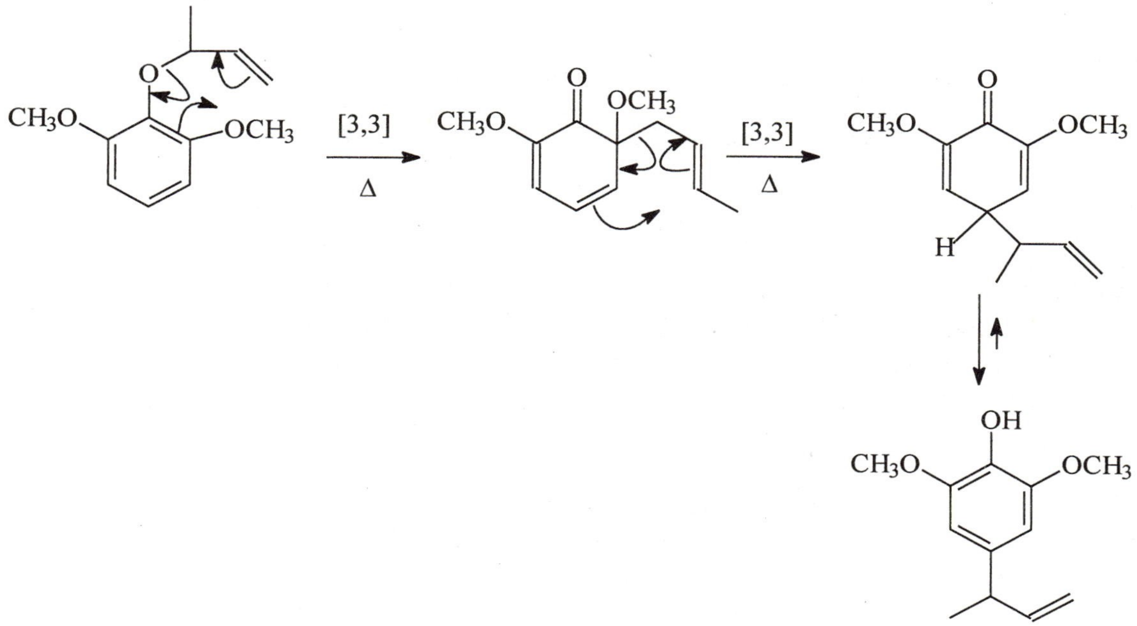

The product is formed by a Claisen rearrangement followed by a Cope rearrangement as shown above. These reactions are [3,3] sigmatropic rearrangements involving an odd number of electron pairs, so the reactions are thermally allowed.

20.44

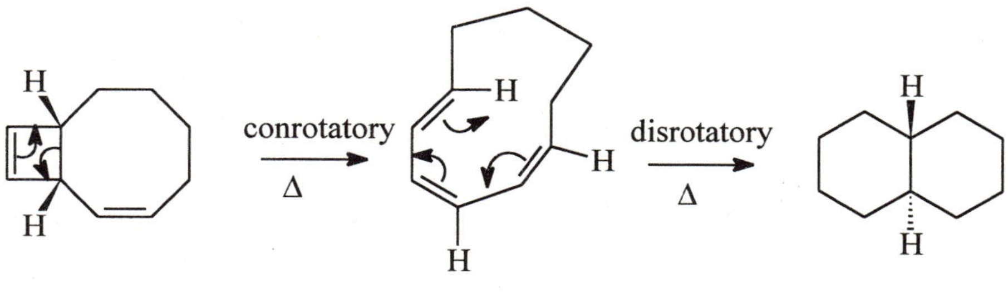

20.45 The reaction is very exothermic because it produces very stable aromatic compounds. But it is a reverse [2+2] cycloaddition reaction and, as such, it is thermally forbidden, so the reactant is stable at room temperature.

20.46

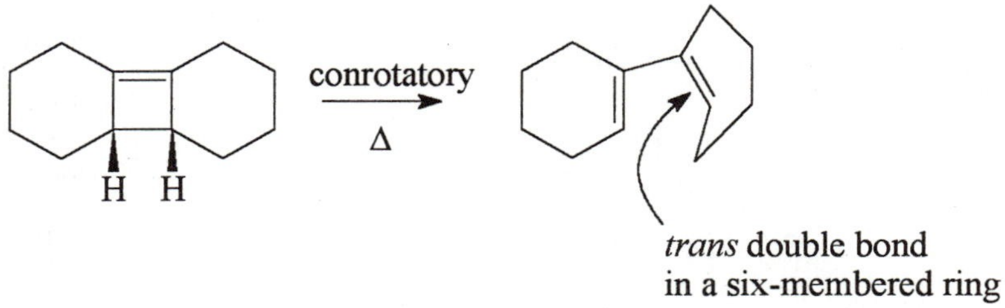

conrotatory
Δ

H H

trans double bond
in a six-membered ring

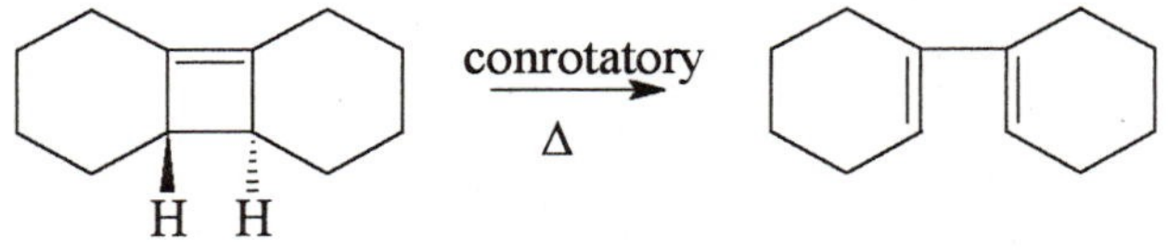

conrotatory
Δ

H H

The allowed thermal ring opening of a cyclobutene occurs with a conrotatory motion. The ring opening of the *cis* isomer would produce a highly strained six-membered ring with a *trans* double bond. Therefore ring opening by a pericyclic process is disfavored for the *cis* isomer. The formation of the observed product from the *cis* isomer must be due to some other high energy reaction pathway, perhaps a nonconcerted reaction.

20.47 Cyclopropyl carbocations are not very stable and react rapidly to form more stable allyl carbocation. This is a thermally allowed reaction because it is an electrocyclic process involving odd number of electron pairs (1). The thermal ring opening of the *cis*-dimethylcyclopropyl carbocation occurs with a disrotatory motion.

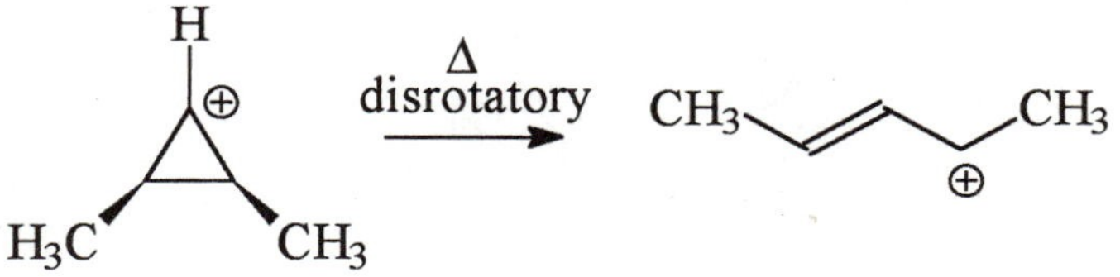

H
⊕
H_3C CH_3
Δ
disrotatory
CH_3 CH_3
⊕

20.48 The product is a Diels-Alder adduct. Anthracene reacts like a diene and benzyne as a dienophile to form this symmetric compound.

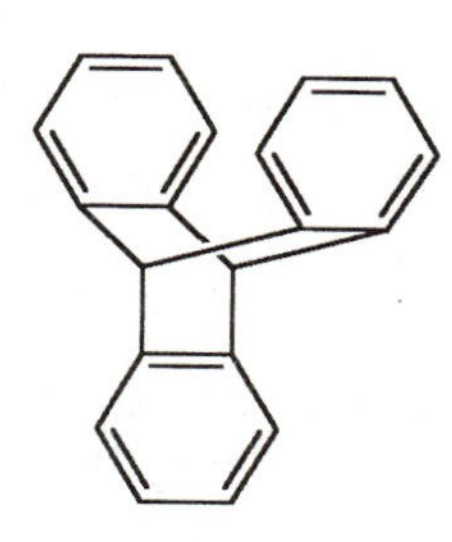

20.49

$$[\text{cyclohexene}]^{+\bullet} \xrightarrow[\text{[4+2]}]{\text{reverse}} [\text{butadiene}]^{+\bullet} + CH_2{=}CH_2$$

m/z 82   m/z 54

20.50

$$CH_3-\overset{\overset{\displaystyle O}{\|}}{C}-O\underset{\underset{\displaystyle CH_3}{|}}{C}HCH_3$$

After completing this chapter, you should be able to:

1. Show the energies and nodal properties of the pi MOs of a small conjugated system, whether it is composed of an even or odd number of orbitals.
   Problems 20.2, 20.3, 20.24.

2. Classify reactions as electrocyclic reactions, [x + y] cycloadditions, or [i,j] sigmatropic rearrangements and determine whether they are allowed or forbidden..
   Problems 20.7, 20.8, 20.12, 20.15, 20.17, 20.20, 20.22, 20.29, 20.38.

3. Use the pi MOs to explain whether an electrocyclic reaction should occur by a conrotatory or a disrotatory motion.
   Problems 20.4, 20.5, 20.6, 20.28.

4. Use the pi MOs to explain whether a cycloaddition is thermally or photochemically allowed.
   Problems 20.10, 20.11, 20.28.

5. Use the pi MOs to explain whether a sigmatropic rearrangement is thermally or photochemically allowed.
   Problems 20.18, 20.19, 20.28.

6. Show the products, including stereochemistry, of any of these reactions.
   Problems 20.9, 20.14, 20.16, 20.21, 20.23, 20.25, 20.16.

7. Show the mechanisms of the pinacol, Beckmann, and Baeyer-Villiger rearrangements.
   Problems 20.35, 20.36, 20.37, 20.40.

8. Use these reactions in synthesis.
   Problems 20.27, 20.30, 20.31, 20.33.

Chapter 21
THE SYNTHESIS OF ORGANIC COMPOUNDS

21.1

a)  O—CH_2OCH_3

b) 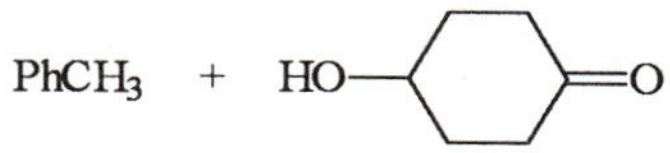 $PhCH_3$ + HO—⟨ring⟩=O

c) 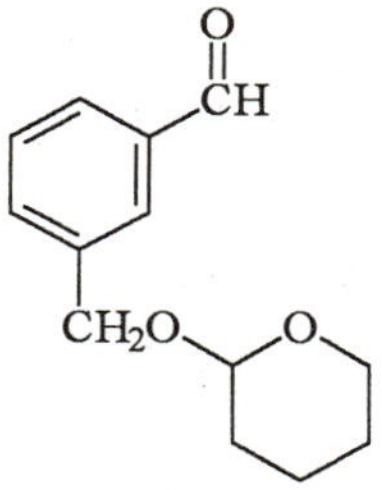 O ‖ CH; CH_2O O

d) 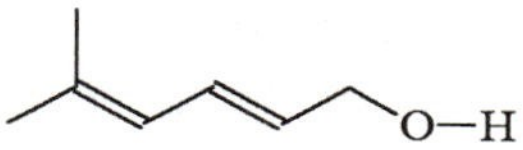 O—H

e) 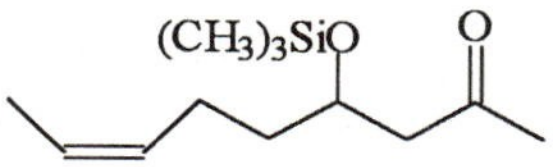 $(CH_3)_3SiO$ O

f) 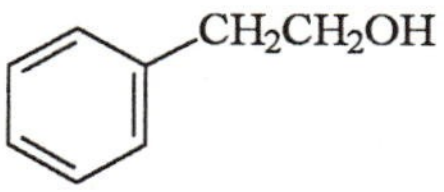 CH_2CH_2OH

g) 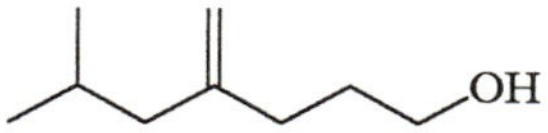 OH

21.2

Ph; ⊕ $H—\ddot{O}H_2$; $\ddot{O}—CH_2—\ddot{O}—CH_3$ → Ph; H; ⊕ $\ddot{O}—CH_2—\ddot{O}—CH_3$ → Ph; $\ddot{O}—H$

\+

H; $H—\ddot{O}$:; ⊕ $CH_2=\ddot{O}—CH_3$

↓

$H—\ddot{O}—CH_2—\ddot{O}—CH_3$ (⊕, H); $H—\ddot{O}$: H

← $H—\ddot{O}—CH_2—\ddot{O}—CH_3$; ⊕ $H—\ddot{O}H_2$

← $H—\ddot{O}—CH_2—\ddot{O}—CH_3$ (H, ⊕)

→ $H\ddot{O}CH_3$ + ⊕ $H—\ddot{O}=CH_2$; $H—\ddot{O}$: H → $H_2C=\ddot{O}$

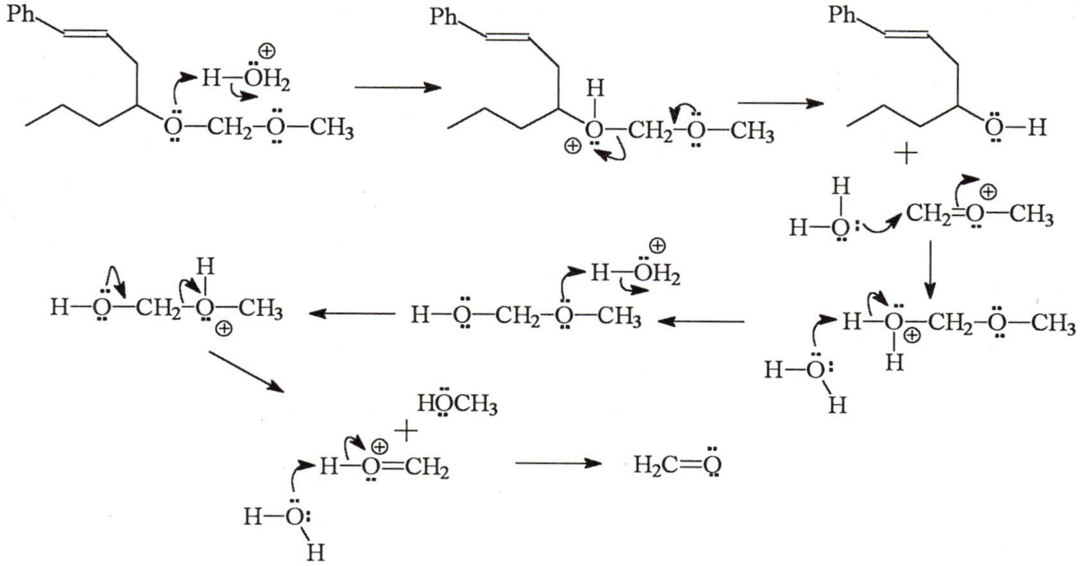

21.3

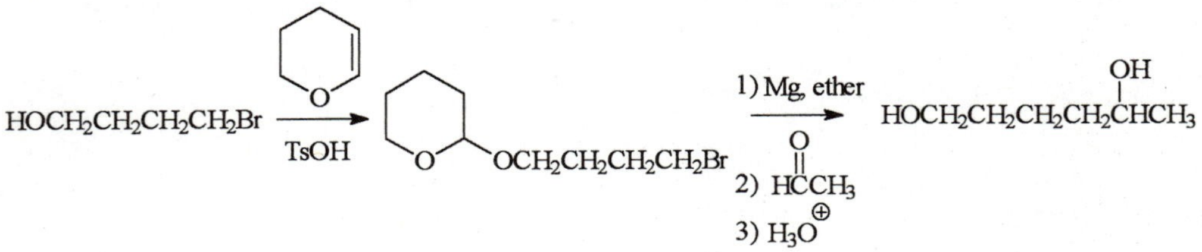

$HOCH_2CH_2CH_2CH_2Br \xrightarrow{TsOH}$ $OCH_2CH_2CH_2CH_2Br$ $\xrightarrow{\text{1) Mg, ether; 2) } \overset{O}{\overset{\|}{HCCH_3}}\text{; 3) } H_3O^{\oplus}}$ $HOCH_2CH_2CH_2CH_2\overset{OH}{\overset{|}{C}}HCH_3$

21.4

a)

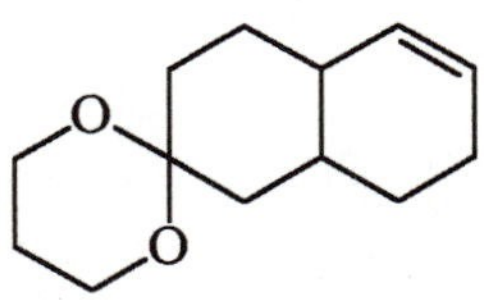

b) 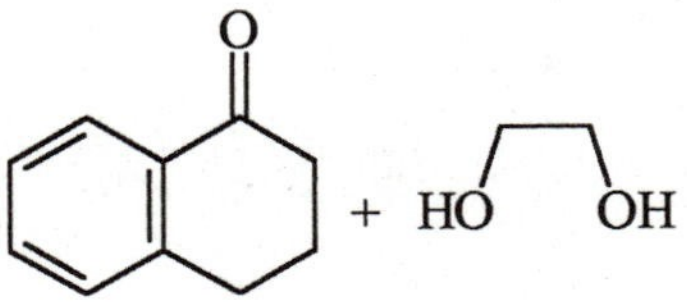 + HO OH

21.5

a) 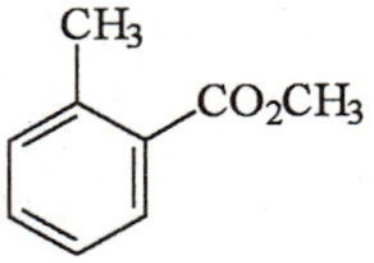 CH_3, CO_2CH_3

b) 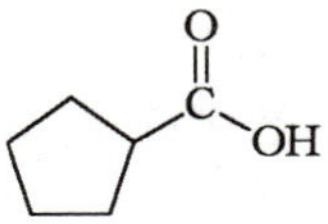 $\overset{O}{\overset{\|}{C}}$ OH

c) $(CH_3)_3CO\overset{O}{\overset{\|}{C}}CH_2CH_2CH_2\overset{O}{\overset{\|}{C}}OH \xrightarrow[H_2O]{HCl} HO\overset{O}{\overset{\|}{C}}CH_2CH_2CH_2\overset{O}{\overset{\|}{C}}OH$

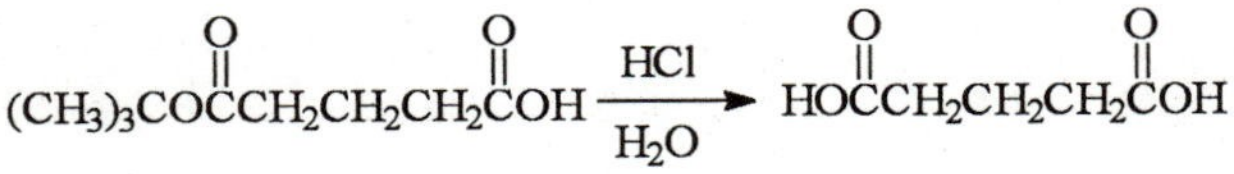

d) 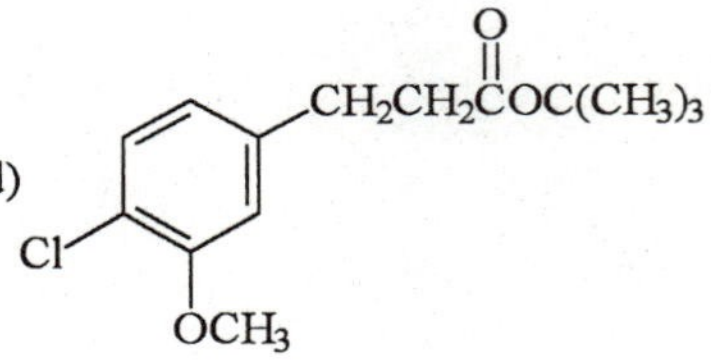 $CH_2CH_2\overset{O}{\overset{\|}{C}}OC(CH_3)_3$, Cl, OCH_3

21.6

$CH_3\overset{CH_3}{\overset{|}{C}}=CH_2$ + $H-\ddot{O}-SO_3H$ $\longrightarrow$ $CH_3\overset{CH_3}{\overset{|}{\underset{\oplus}{C}}}-\underset{H}{\underset{|}{CH_2}}$

$H-\ddot{O}-\overset{O}{\overset{\|}{C}}CH_2CH_2CH_2CH_3$ $\longrightarrow$ $CH_3\overset{CH_3}{\overset{|}{C}}-CH_3$, $H-\overset{\oplus}{O}-\overset{}{C}CH_2CH_2CH_2CH_3$ (=O)

$^{\ominus}:\ddot{O}-SO_3H$

$\downarrow$

$CH_3\overset{CH_3}{\overset{|}{\underset{CH_3}{\underset{|}{C}}}}-\ddot{O}-\overset{O}{\overset{\|}{C}}CH_2CH_2CH_2CH_3$

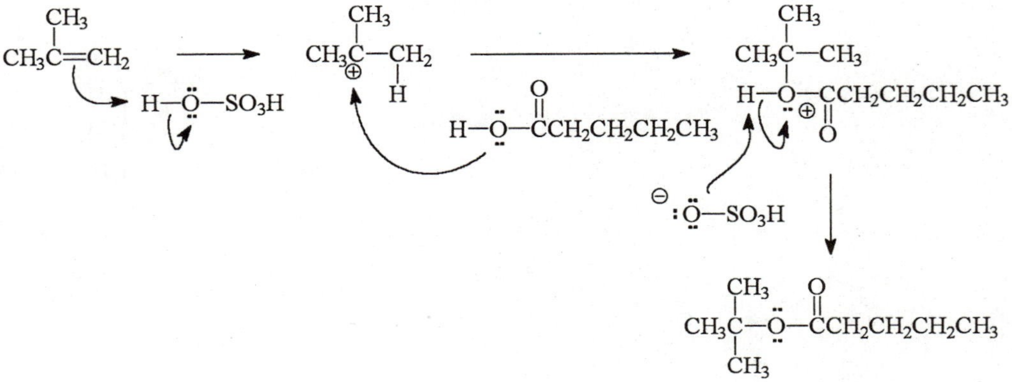

21.7

a) 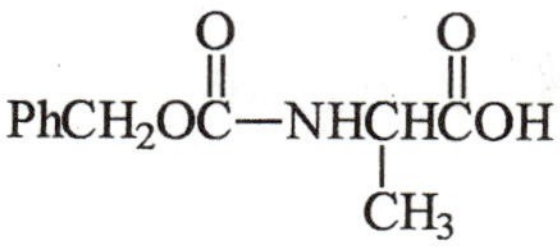 $PhCH_2O\overset{O}{\overset{\|}{C}}{-}NH\underset{CH_3}{\underset{|}{C}}H\overset{O}{\overset{\|}{C}}OH$ b) 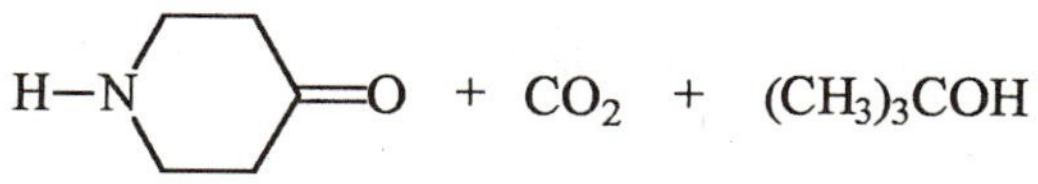 H−N(piperidine ring)=O + CO_2 + $(CH_3)_3COH$

c) $PhCH_3$ + $NH_2CH_2CH_2Ph$ + CO_2 d)  $(CH_3)_3CO\overset{O}{\overset{\|}{C}}NHCH_2CO_2H$

21.8

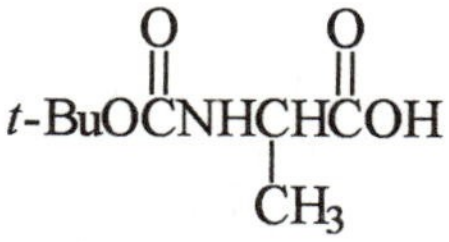

$t\text{-}BuO\overset{O}{\overset{\|}{C}}NH\underset{CH_3}{\underset{|}{C}}H\overset{O}{\overset{\|}{C}}OH$
A

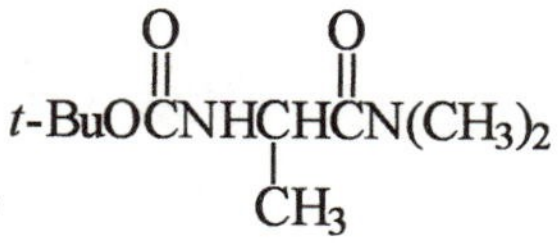

$t\text{-}BuO\overset{O}{\overset{\|}{C}}NH\underset{CH_3}{\underset{|}{C}}H\overset{O}{\overset{\|}{C}}N(CH_3)_2$
B

The amide cannot be prepared directly because the unprotected amino group will react with the acyl chloride group.

21.9 The ester bond of the carbamate group is hydrolyzed more rapidly than either amide bond. The resulting carbamic acid then eliminates carbon dioxide to produce the amine. Conditions drastic enough to hydrolyze an amide bond are never employed in this process.

21.10

b) 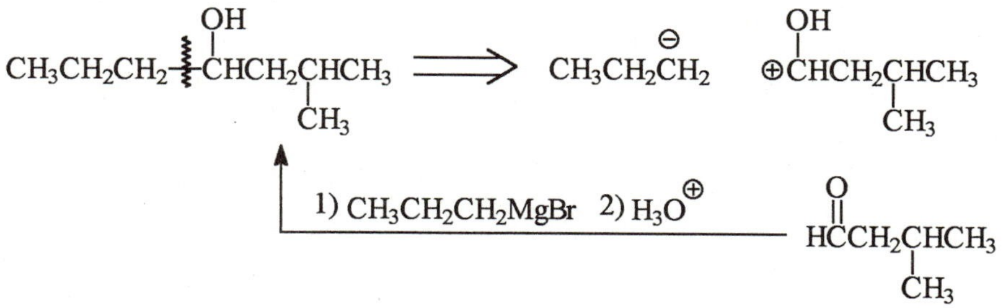 $CH_3CH_2CH_2{-}\!\!\wr\!\!-\overset{OH}{\overset{|}{C}}HCH_2\underset{CH_3}{\underset{|}{C}}HCH_3 \Longrightarrow CH_3CH_2\overset{\ominus}{C}H_2 \quad \oplus\overset{OH}{\overset{|}{C}}HCH_2\underset{CH_3}{\underset{|}{C}}HCH_3$

$H\overset{O}{\overset{\|}{C}}CH_2\underset{CH_3}{\underset{|}{C}}HCH_3$ —— 1) $CH_3CH_2CH_2MgBr$ 2) $H_3O^{\oplus}$ ——→ (alcohol above)

c) $PhC{\equiv}C$-$CH_2CH_2CH_2CH_3$ $\Longrightarrow$ $Ph\overset{\ominus}{C{\equiv}C}$ $\overset{\oplus}{C}H_2CH_2CH_2CH_3$

$PhC{\equiv}C{-}H$ $\xrightarrow{1)\ NaNH_2\quad 2)\ CH_3CH_2CH_2CH_2Br}$ $PhC{\equiv}C$-$CH_2CH_2CH_2CH_3$

d) $PhCH$-$CH\overset{O}{\overset{\|}{C}}CH_3$ $\Longrightarrow$ $\overset{\oplus}{Ph}CH$ $\overset{\ominus}{C}H\overset{O}{\overset{\|}{C}}CH_3$

$Ph\overset{O}{\overset{\|}{C}}H + CH_3\overset{O}{\overset{\|}{C}}CH_3$ $\xrightarrow[H_2O\ \Delta]{NaOH}$ $PhCH$-$CH\overset{O}{\overset{\|}{C}}CH_3$

21.11

a) $H{-}C{\equiv}C{-}H$ $\xrightarrow[2)\ CH_3CH_2Br]{1)\ NaNH_2}$ $CH_3CH_2{-}C{\equiv}C{-}H$ $\xrightarrow[2)\ CH_3CH_2Br]{1)\ NaNH_2}$ $CH_3CH_2{-}C{\equiv}C{-}CH_2CH_3$

$\xrightarrow[\text{Lindlar catalyst}]{H_2}$ cis-$CH_3CH_2(H)C{=}C(H)CH_2CH_3$ (H and H on the same side; CH_3CH_2 and CH_2CH_3 on the same side)

b) $Ph\overset{O}{\overset{\|}{C}}Cl + (CH_3)_2CuLi \longrightarrow Ph\overset{O}{\overset{\|}{C}}CH_3$ $\xrightarrow[NaOH,\ H_2O,\ \Delta]{Ph\overset{O}{\overset{\|}{C}}H}$ $PhCH{=}CH\overset{O}{\overset{\|}{C}}Ph$

c) $PhCH_2Br \xrightarrow[DMSO]{KCN} PhCH_2CN \xrightarrow[\Delta]{H_3O^{\oplus}} PhCH_2\overset{O}{\overset{||}{C}}OH \xrightarrow[H_2SO_4,\ \Delta]{CH_3OH} PhCH_2\overset{O}{\overset{||}{C}}OCH_3$

$\downarrow$ 1) 2 CH_3MgI 2) NH_4Cl, H_2O

$PhCH_2C(OH)(CH_3)CH_3$

d) $2\ CH_3CH_2\overset{O}{\overset{||}{C}}OEt \xrightarrow[2)\ H_3O^{\oplus}]{1)\ NaOEt} CH_3CH_2\overset{O}{\overset{||}{C}}-CH(CH_3)-\overset{O}{\overset{||}{C}}OCH_2CH_3 \xrightarrow[2)\ CH_3CH_2Br]{1)\ NaOEt} CH_3CH_2\overset{O}{\overset{||}{C}}-C(CH_3)(CH_2CH_3)-\overset{O}{\overset{||}{C}}OCH_2CH_3$

e) benzonitrile (Ph–CN) $\xrightarrow[2)\ H_3O^{\oplus}]{1)\ CH_3CH_2MgBr}$ PhC(=O)CH₂CH₃ $\xrightarrow{Ph_3P{=}CH_2}$ PhC(=CH₂)CH₂CH₃

or

benzene $\xrightarrow{CH_3CH_2\overset{O}{\overset{||}{C}}Cl,\ AlCl_3}$ PhC(=O)CH₂CH₃

f) butadiene + $CH_3CH{=}CHCO_2CH_2CH_3$ $\xrightarrow{\Delta}$ cyclohexene with $CO_2CH_2CH_3$ (wedge) and CH_3 (hash) $\xrightarrow[2)\ H_3O^{\oplus}]{1)\ LiAlH_4}$ cyclohexene with CH_2OH (wedge) and CH_3 (hash)

g) $CH_3\overset{O}{\overset{||}{C}}CH_2\overset{O}{\overset{||}{C}}OEt \xrightarrow[2)\ CH_3CH_2Br]{1)\ NaOEt} CH_3\overset{O}{\overset{||}{C}}CH(CH_2CH_3)\overset{O}{\overset{||}{C}}OEt \xrightarrow[2)\ CH_3I]{1)\ NaOEt} CH_3\overset{O}{\overset{||}{C}}-C(CH_3)(CH_2CH_3)-\overset{O}{\overset{||}{C}}OEt$

$\downarrow$ 1) NaOH, H_2O 2) $H_3O^{\oplus}$, Δ

$CH_3CH_2CH(CH_3)\overset{O}{\overset{||}{C}}CH_3$

h) Benzene $\xrightarrow[AlCl_3]{CH_3CH_2CH_2COCl}$ butyrophenone $\xrightarrow[Pd]{H_2}$ butylbenzene $\xrightarrow[AlCl_3]{CH_3CCl(=O)}$ 4-butylacetophenone

21.12

a) $CH_3OH \xrightarrow{HI} CH_3I$

$CH_3CH_2OH \xrightarrow{Na} CH_3CH_2ONa$

$CH_3I + CH_3CH_2ONa \longrightarrow CH_3OCH_2CH_3$

b) $CH_3CH_2OH \xrightarrow[H_2SO_4]{Na_2Cr_2O_7} CH_3CO_2H \xrightarrow[H_2SO_4]{CH_3OH} CH_3CO_2CH_3$

c) $CH_3CH_2OH \xrightarrow{PCC} CH_3\overset{O}{\overset{\|}{C}}H \xrightarrow[\Delta]{NaOH} CH_3CH{=}CH\overset{O}{\overset{\|}{C}}H \xrightarrow[Pt]{H_2} CH_3CH_2CH_2\overset{O}{\overset{\|}{C}}H$

d) butanal $\xrightarrow[2)\ H_3O^{\oplus}]{1)\ LiAlH_4}$ 1-butanol $\xrightarrow{SOCl_2}$ 1-chlorobutane $\xrightarrow{1)\ Mg,\ ether;\ 2)\ \text{butanal};\ 3)\ H_3O^{\oplus}}$ 5-octanol (OH) $\xrightarrow[H_2SO_4]{Na_2Cr_2O_7}$ 4-octanone (O)

e) $CH_3CH_2OH \xrightarrow{HBr} CH_3CH_2Br \xrightarrow{Ph_3P} Ph_3\overset{\oplus}{P}{-}CH_2CH_3\ Br^{\ominus} \xrightarrow{1)\ BuLi;\ 2)\ \text{butanal}}$ 3-heptene

f) butanal $\xrightarrow[\text{2) } H_3O^{\oplus}]{\text{1) } Ag_2O\text{, NaOH}}$ butanoic acid (OH) $\xrightarrow[H_2SO_4]{CH_3CH_2OH}$ ethyl butanoate

g) 2 $CH_3\overset{O}{\overset{\|}{C}}OCH_3$ $\xrightarrow[\text{2) } H_3O^{\oplus}]{\text{1) } NaOCH_3\text{, } CH_3OH}$ $CH_3\overset{O}{\overset{\|}{C}}CH_2\overset{O}{\overset{\|}{C}}OCH_3$ $\xrightarrow[\text{2) 1-chlorobutane (Cl)}]{\text{1) } NaOCH_3}$ $CH_3\overset{O}{\overset{\|}{C}}\underset{CH_2CH_2CH_2CH_3}{CH}\overset{O}{\overset{\|}{C}}OCH_3$ $\xrightarrow[\text{2) } H_3O^{\oplus}]{\text{1) NaOH, } H_2O}$ 2-heptanone

h) benzene $\xrightarrow[AlCl_3]{CH_3I}$ toluene (CH_3) $\xrightarrow[\text{2) } H_3O^{\oplus}]{\text{1) } KMnO_4\text{, NaOH}}$ benzoic acid (CO_2H) $\xrightarrow[\text{2) } LiAlH(OtBu)_3]{\text{1) } SOCl_2}$ benzaldehyde (O=CH)

i) butanal (H) $\xrightarrow[\text{2) } H_3O^{\oplus}]{\text{1) } Ag_2O\text{, NaOH}}$ butanoic acid (OH) $\xrightarrow[\text{2) benzene, } AlCl_3]{\text{1) } SOCl_2}$ 1-phenyl-1-butanone

j) 1-phenyl-1-butanone $\xrightarrow[H_2SO_4]{HNO_3}$ 1-(3-nitrophenyl)-1-butanone (NO_2)

k) toluene $\xrightarrow[AlCl_3]{\text{butanoyl chloride (Cl)}}$ 1-(4-methylphenyl)-1-butanone (O)

l) 1-phenyl-1-butanone $\xrightarrow[Pd]{H_2}$ butylbenzene $\xrightarrow[AlCl_3]{Cl_2}$ 1-butyl-4-chlorobenzene (Cl)

m) $\xrightarrow[\text{HCl}]{\text{Sn}}$

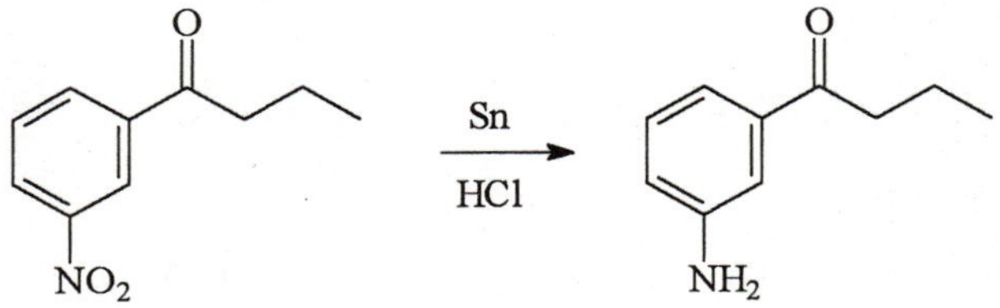

NO_2 NH_2

n) $\xrightarrow[\text{KOH, }\Delta]{NH_2NH_2}$ $\xrightarrow[\text{HCl}]{\text{Sn}}$

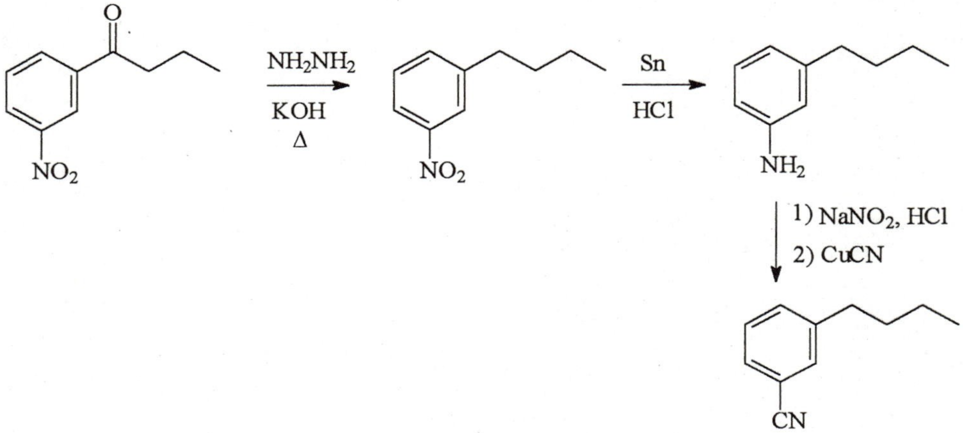

NO_2 NO_2 NH_2

1) $NaNO_2$, HCl
2) CuCN

CN

o) $CH_3I + Ph_3P \longrightarrow Ph_3\overset{\oplus}{P}{-}CH_3 \; I^{\ominus}$ $\xrightarrow[\text{2)}]{\text{1) BuLi}}$

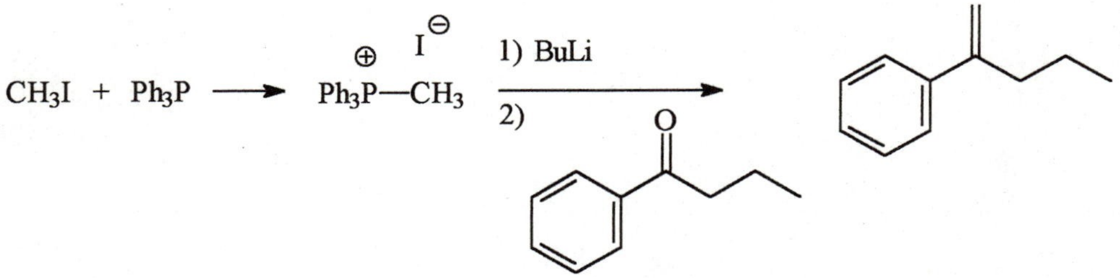

O

p) see (g)

q) O O OCH_3 $\xrightarrow[\text{2)}]{\text{1) } NaOCH_3}$ Cl → O O OCH_3

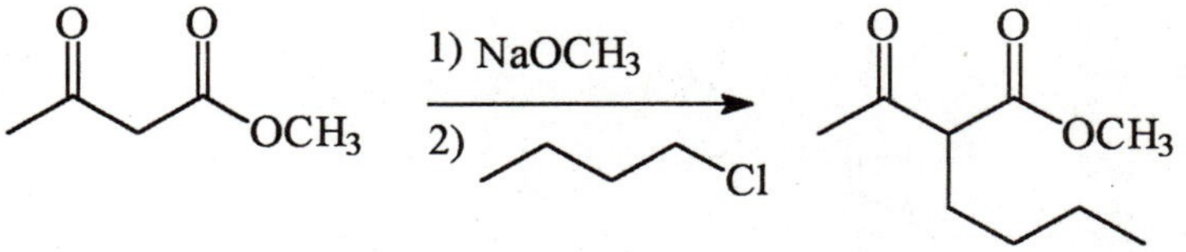

r) 2 O H $\xrightarrow[H_2O,\ \Delta]{NaOH}$ O H $\xrightarrow[Pt]{H_2}$ O H

s) $\xrightarrow[AlCl_3]{CH_3\overset{O}{\overset{\|}{C}}Cl}$ O $\xrightarrow[NaOH,\ \Delta]{Ph\overset{O}{\overset{\|}{C}}H}$ O Ph $\xrightarrow{PhCO_3H}$ O O Ph

t) $\xrightarrow[H_2SO_4]{HNO_3}$ NO_2 $\xrightarrow[HCl]{Sn}$ NH_2 $\xrightarrow[2)\ H_3O^{\oplus},\ \Delta]{1)\ NaNO_2,\ HCl}$ OH

H_2, Pt
high pressure

OH $\xrightarrow{PCC}$ O $\xrightarrow[2)\ NH_4Cl,\ H_2O]{1)\ CH_3CH_2MgBr}$ HO

21.13

a) 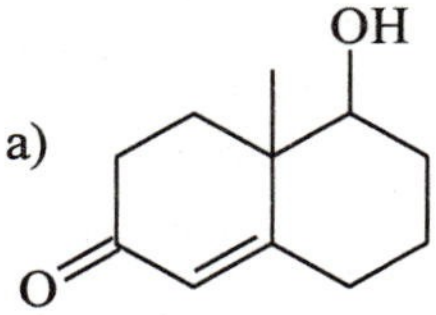

b) 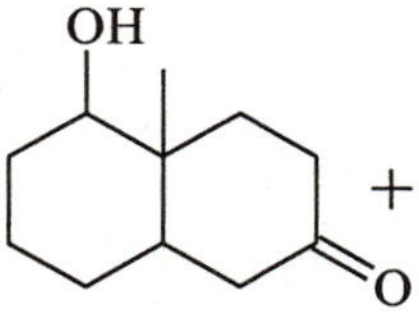 + $PhCH_3$

c) 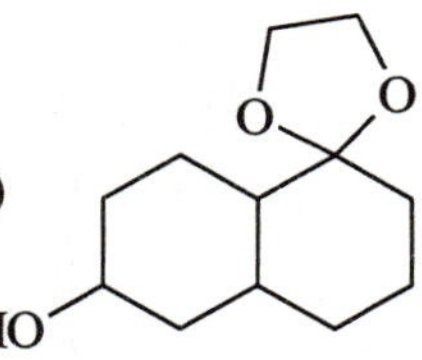

d) 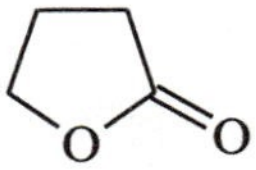

e) 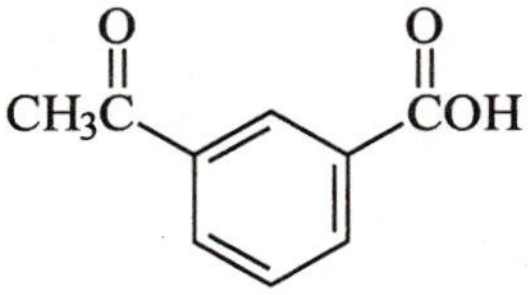 + $(CH_3)_3COH$

f) 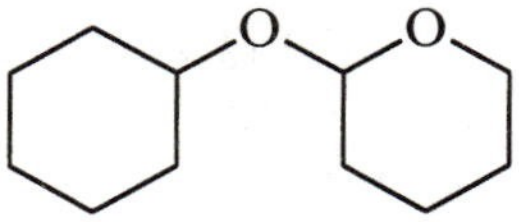

g) 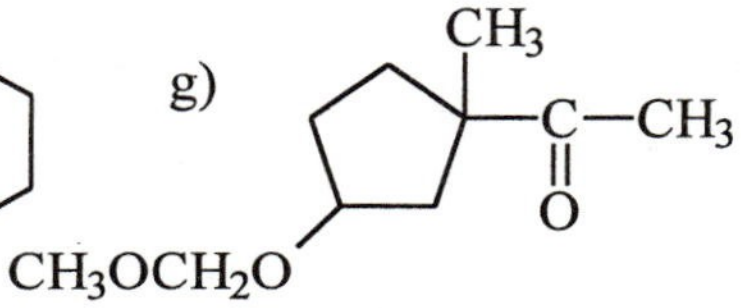

h) 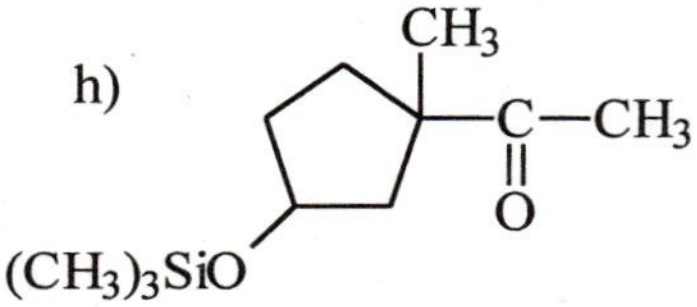

i) 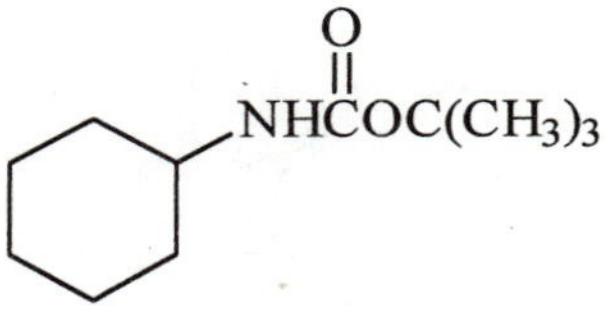

j) 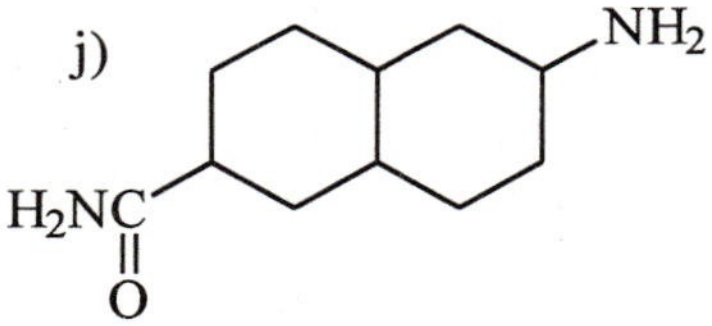

k) 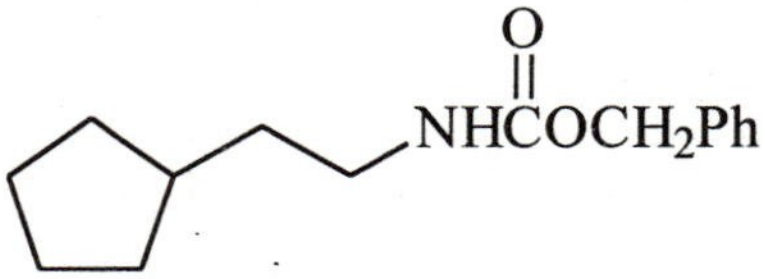

l) 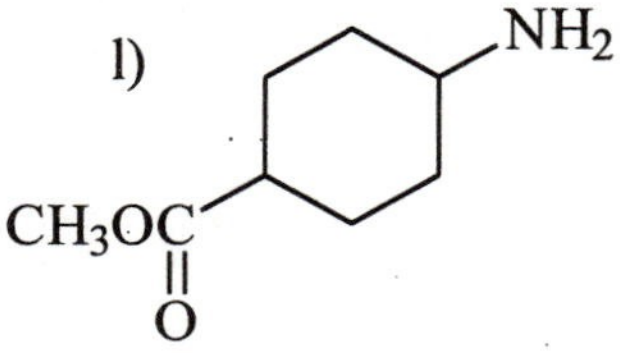 + $PhCH_3$

21.14

a)

$$CH_3CH_2CH_2CH_2 \vdots C(=O) \vdots CH(CH_2CH_3)CH_2CH_3 \Longrightarrow$$

$CH_3CH_2CH_2CH_2$ (⊕ or ⊖) + $C(=O)-CH(CH_2CH_3)CH_2CH_3$ (⊖ or ⊕)

$CH_3CH_2CH_2CH_2-C(=O)$ (⊕ or ⊖) + $CH(CH_2CH_3)CH_2CH_3$ (⊖ or ⊕)

$CH_3CH_2CH(CH_2CH_3)CH(=O)$ —1) HS–$CH_2CH_2CH_2$–SH, BF_3→ 2-[$CH(CH_2CH_3)CH_2CH_3$]-1,3-dithiane (C–H) —1) BuLi; 2) $CH_3CH_2CH_2CH_2-Cl$→ 2-[$CH(CH_2CH_3)CH_2CH_3$]-2-($CH_2CH_2CH_2CH_3$)-1,3-dithiane —$HgCl_2$, H_2O→

1,3-dithiane (CH_2, H H) —1) BuLi; 2) $CH_3CH_2CHCl(CH_2CH_3)$→ 2-[$CH(CH_2CH_3)CH_2CH_3$]-1,3-dithiane

$CH_3CH_2CH(CH_2CH_3)CH(=O)$ —1) $CH_3CH_2CH_2CH_2MgBr$; 2) $H_3O^{\oplus}$→ $CH_3CH_2CH_2CH_2CH(OH)CH(CH_2CH_3)CH_2CH_3$ —PCC→

b) $CH_3CH_2 \mid C{\equiv}C \mid CH_2CH_2 \mid \overset{O}{\overset{\|}{C}} \mid CH_3 \Longrightarrow$

$\overset{\oplus}{CH_3CH_2}$ $\quad \overset{\ominus}{C}{\equiv}CCH_2CH_2\overset{O}{\overset{\|}{C}}CH_3$

$CH_3CH_2C{\equiv}\overset{\ominus}{C}$ $\quad \overset{\oplus}{CH_2}CH_2\overset{O}{\overset{\|}{C}}CH_3$

$CH_3CH_2C{\equiv}CCH_2\overset{\oplus \text{ or } \ominus}{CH_2}$ $\quad \underset{\ominus \text{ or } \oplus}{\overset{O}{\overset{\|}{C}}CH_3}$

$CH_3CH_2C{\equiv}CCH_2CH_2\underset{\oplus \text{ or } \ominus}{\overset{O}{\overset{\|}{C}}}$ $\quad \overset{\ominus \text{ or } \oplus}{CH_3}$

$CH_3\overset{O}{\overset{\|}{C}}CH_2CH_2C{\equiv}CH \xrightarrow[\text{TsOH}]{HOCH_2CH_2OH}$ $CH_3C(OCH_2CH_2O)CH_2CH_2C{\equiv}CH \xrightarrow[\text{2) } CH_3CH_2Cl \text{ 3) } H_3O^{\oplus}]{\text{1) } NaNH_2}$

$CH_3CH_2C{\equiv}CCH_2CH_2Cl \xrightarrow[\text{2) } H_3O^{\oplus}]{\text{1) 2-methyl-1,3-dithiane anion } (H_3C\text{–}\overset{\ominus}{C}S_2)}$

$CH_3\overset{O}{\overset{\|}{C}}H \xrightarrow[\text{2) } H_3O^{\oplus}]{\text{1) } CH_3CH_2C{\equiv}CCH_2CH_2MgBr}$ $CH_3CH_2C{\equiv}CCH_2CH_2\overset{OH}{\overset{|}{C}}HCH_3 \xrightarrow{PCC}$

c) Cyclohexyl–CH(OH)–CH_2CH_3 $\Longrightarrow$ cyclohexyl ⊖ or ⊕ + OH / $CHCH_2CH_3$ ⊖ or ⊕

cyclohexyl–CH(OH) ⊕ or ⊖ + CH_2CH_3 ⊖ or ⊕

$CH_3CH_2\overset{O}{\overset{\|}{C}}H$ —1) cyclohexyl-MgBr; 2) $H_3O^{\oplus}$→

Cyclohexyl–$\overset{O}{\overset{\|}{C}}H$ —1) CH_3CH_2MgBr; 2) $H_3O^{\oplus}$→

Cyclohexyl–$\overset{O}{\overset{\|}{C}}H$ —1) HS–$CH_2CH_2CH_2$–SH, BF_3; 2) BuLi→ cyclohexyl–C(H)(1,3-dithiane, S / S) —1) BuLi; 2) CH_3CH_2Br; 3) $HgCl_2$, H_2O→ cyclohexyl–$\overset{O}{\overset{\|}{C}}CH_2CH_3$ —$NaBH_4$→

d)

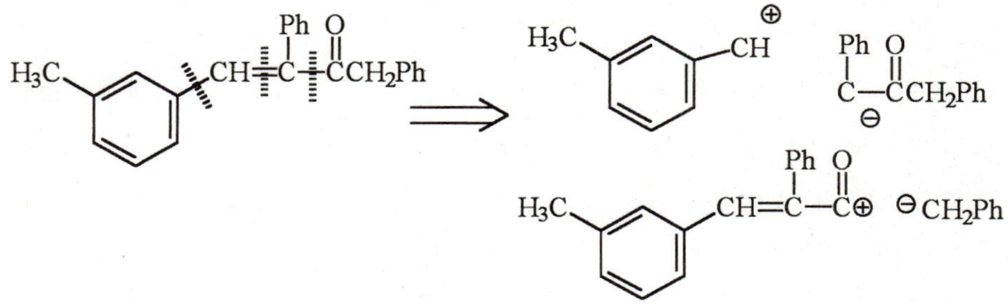

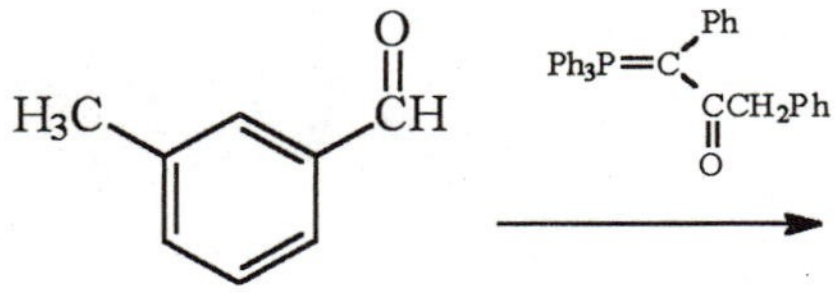

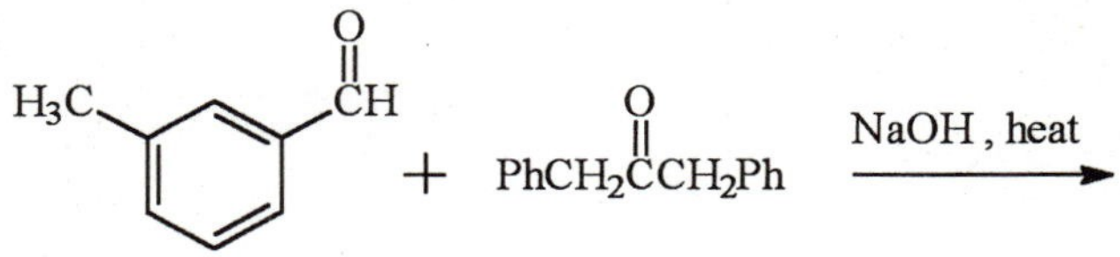

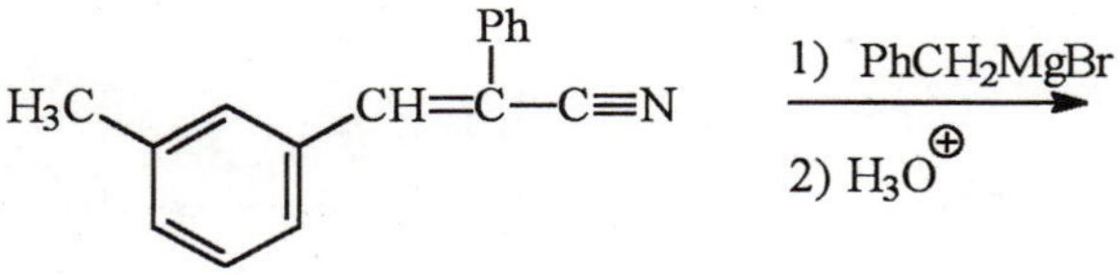

H_3C — CH═C(Ph)—C(═O)—CCH_2Ph ⟹ H_3C — $\overset{\oplus}{C}H$ Ph O $\overset{\ominus}{C}$—$\overset{\|}{C}CH_2Ph$

H_3C — CH═C(Ph)—$\overset{O}{\overset{\|}{C}}\oplus$ $\ominus CH_2Ph$

H_3C — $\overset{O}{\overset{\|}{C}}H$ $\xrightarrow{Ph_3P{=}C(Ph)\overset{O}{\overset{\|}{C}}CH_2Ph}$

H_3C — $\overset{O}{\overset{\|}{C}}H$ + $PhCH_2\overset{O}{\overset{\|}{C}}CH_2Ph$ $\xrightarrow{\text{NaOH , heat}}$

H_3C — CH═C(Ph)—C≡N $\xrightarrow[\text{2) } H_3O^{\oplus}]{\text{1) } PhCH_2MgBr}$

21.15

a)

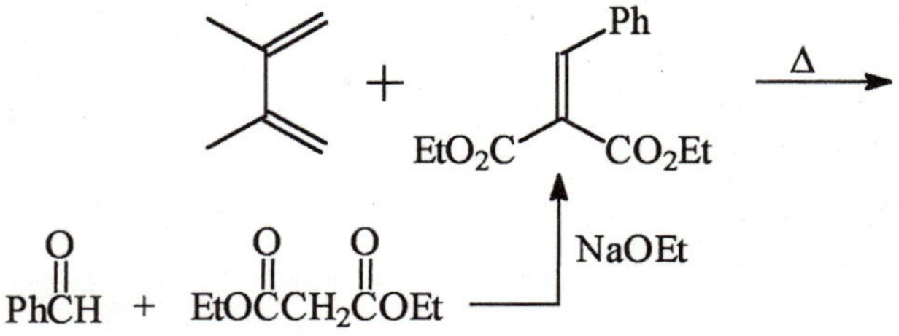

(2,3-dimethyl-1,3-butadiene) + EtO_2C—C(═CH—Ph)—CO_2Et $\xrightarrow{\Delta}$

$Ph\overset{O}{\overset{\|}{C}}H$ + $EtO\overset{O}{\overset{\|}{C}}CH_2\overset{O}{\overset{\|}{C}}OEt$ $\xrightarrow{\text{NaOEt}}$ (to the alkene above)

b)

H_2N—C_6H_5 $\xrightarrow{CH_3\overset{O}{\overset{\|}{C}}Cl}$ CH_3CONH—C_6H_5 $\xrightarrow[AlCl_3]{CH_3CH_2CH_2\overset{O}{\overset{\|}{C}}Cl}$ CH_3CONH—C_6H_4—$\overset{O}{\overset{\|}{C}}CH_2CH_2CH_3$

$\downarrow NaBH_4$

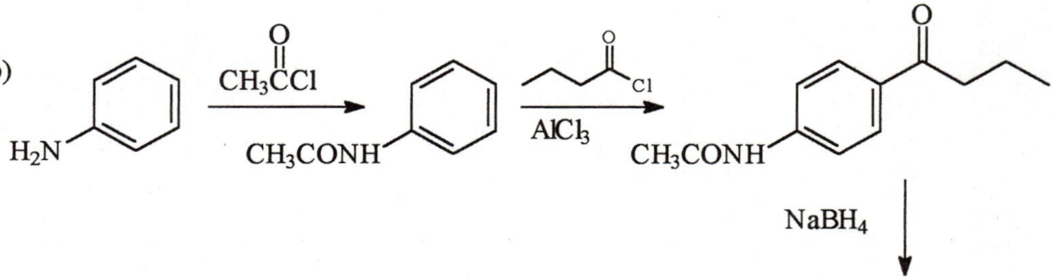

c)

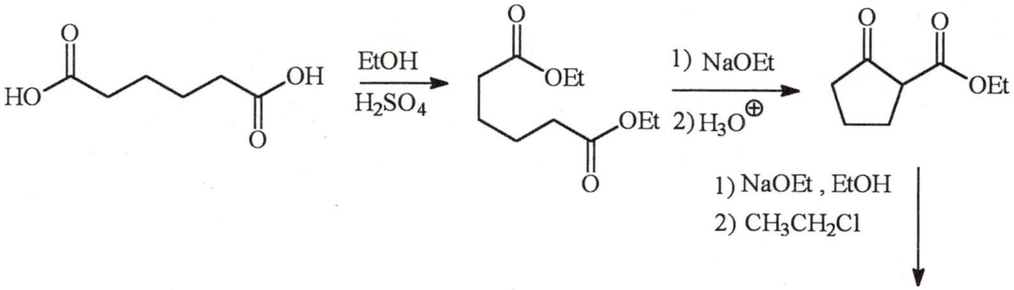

HO, OH

EtOH

H_2SO_4

OEt

OEt

1) NaOEt

2) $H_3O^{\oplus}$

OEt

1) NaOEt , EtOH

2) CH_3CH_2Cl

d)

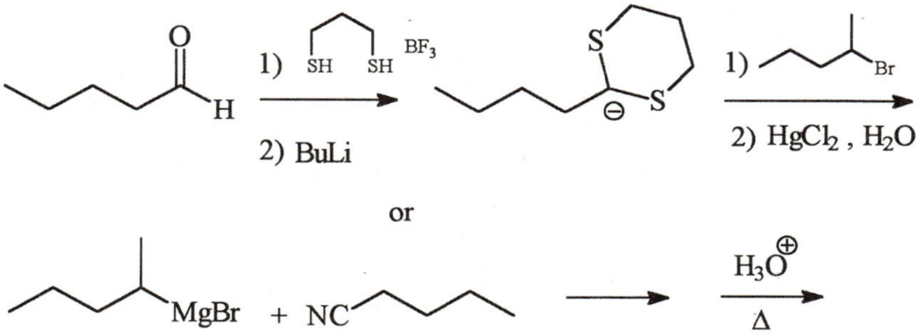

H

1) SH SH BF_3

2) BuLi

S

S

1) Br

2) $HgCl_2$, H_2O

or

MgBr + NC

$H_3O^{\oplus}$

Δ

e) 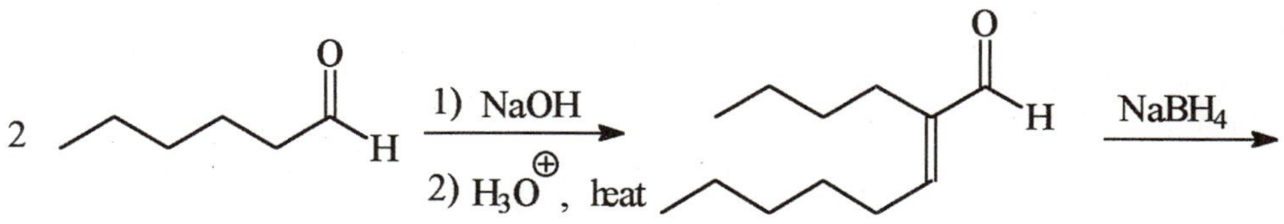 2

H

1) NaOH

2) $H_3O^{\oplus}$, heat

H

$NaBH_4$

f)

O

H

NaOH

H_2O

heat

H

H_2

Pd

H

1) CH_3MgBr

2) $H_3O^{\oplus}$

3) PCC

$Ph_3P=CH_2$

CH_2

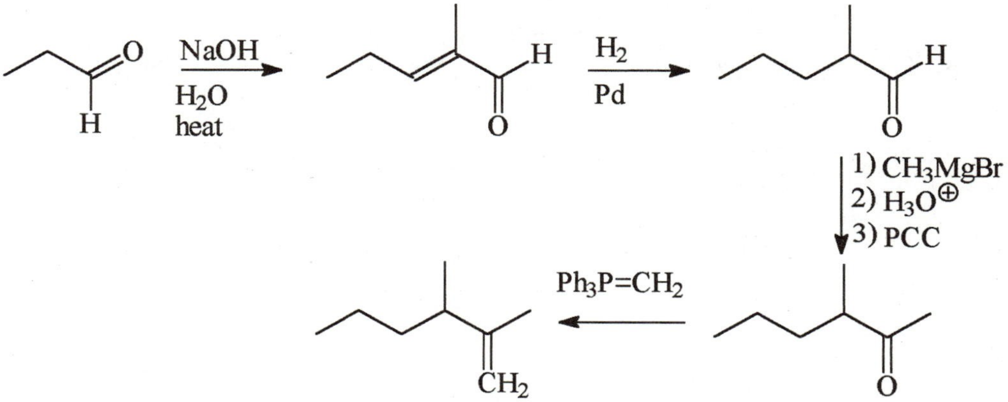

g)

1) NaOEt, EtOH
2) (1-bromopropane)

1) NaOEt, EtOH
2) (bromoethane)

OEt

1) NaOH, H_2O
2) $H_3O^{\oplus}$, heat

1) CH_3MgBr
2) NH_4Cl, H_2O

h)

Br

HO OH
TsOH

O O

Br

Ph_3P

$Br^{\ominus}$
$^{\oplus}PPh_3$

1) BuLi
2) (acetone)
3) $H_3O^{\oplus}$, heat

i)

1) HNO_3
H_2SO_4
2) H_2 / Pt

NH_2

$NaNO_2$
$H_3O^{\oplus}$ heat

OH

1) NaOH
2) CH_3Cl

OCH_3

Cl (isobutyryl chloride)
$AlCl_3$

O

CH_3O

1) $LiAlH_4$
2) $H_3O^{\oplus}$
3) $SOCl_2$

21.16

a) 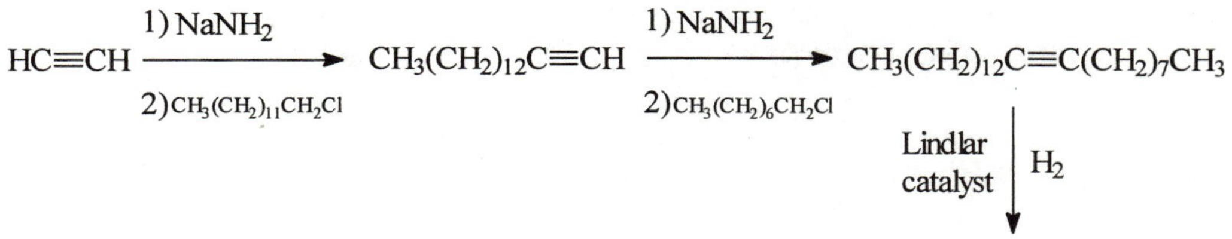 $HC\equiv CH$ —(1) $NaNH_2$; 2) $CH_3(CH_2)_{11}CH_2Cl$)→ $CH_3(CH_2)_{12}C\equiv CH$ —(1) $NaNH_2$; 2) $CH_3(CH_2)_6CH_2Cl$)→ $CH_3(CH_2)_{12}C\equiv C(CH_2)_7CH_3$

↓ Lindlar catalyst, H_2

b) 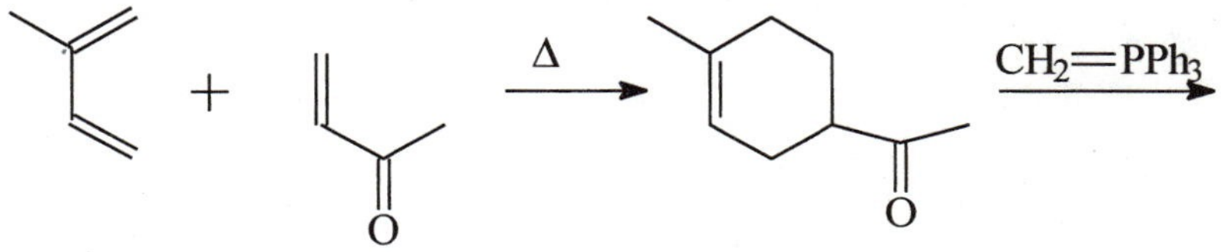 + Δ → $CH_2{=}PPh_3$ →

O O

c) O

1) NaOEt, O O

2) $H_3O^{\oplus}$

O

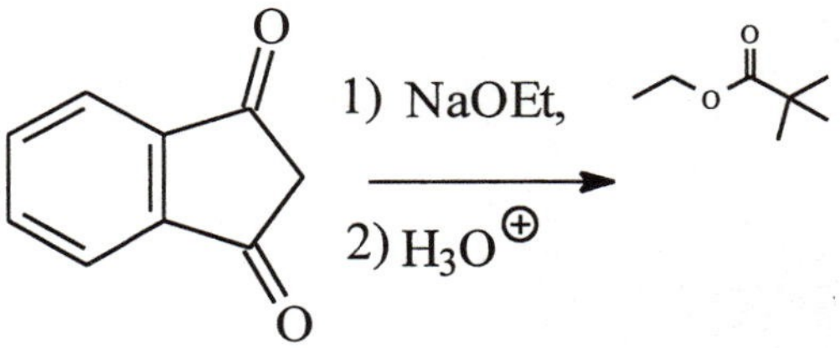

21.17

a) 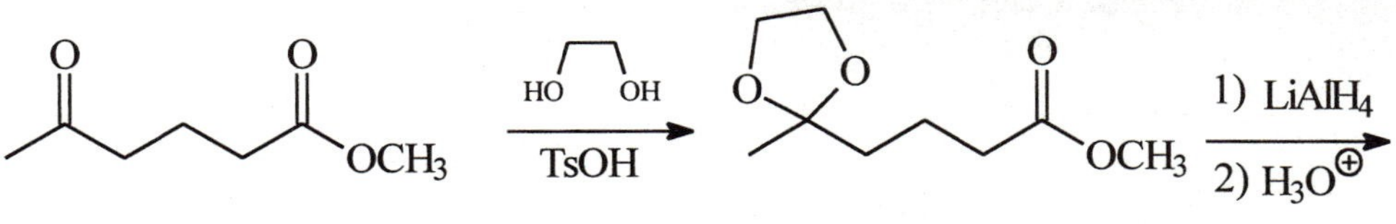 O O OCH_3 HO OH TsOH O O O OCH_3 1) $LiAlH_4$ 2) $H_3O^{\oplus}$

b) 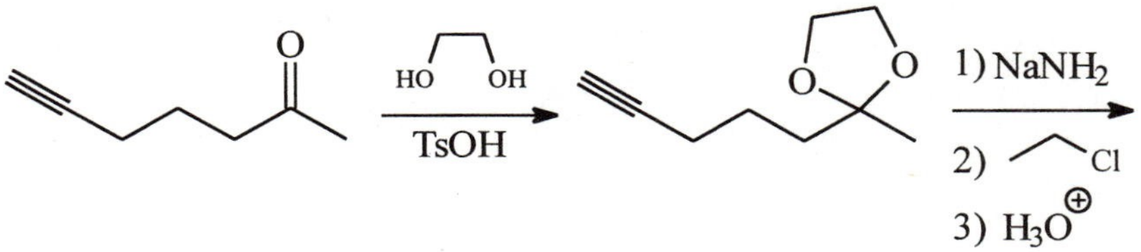 O HO OH TsOH O O 1) $NaNH_2$ 2) Cl 3) $H_3O^{\oplus}$

c) Br ... (O) ... 1) HO⁀OH, TsOH; 2) Mg, ether → BrMg ... (1,3-dioxolane) ... → 1) C_6H_5CHO (O, C, H); 2) $H_3O^{\oplus}$ →

d) (O, O, OCH_3) 1) HO⁀OH, TsOH; 2) $LiAlH_4$; 3) $H_3O^{\oplus}$ → (O, OH) $CH_3\overset{O}{\overset{\|}{C}}Cl$ →

e) $HOCH_2C{\equiv}CH$ (dihydropyran, O), TsOH → (O) $OCH_2C{\equiv}CH$ 1) CH_3CH_2MgBr; 2) CO_2; 3) $H_3O^{\oplus}$ →

f) CO_2H, NH_2 → t-BuO$\overset{O}{\overset{\|}{C}}O\overset{O}{\overset{\|}{C}}Ot$-Bu → CO_2H, NH$\overset{O}{\overset{\|}{C}}Ot$-Bu → $SOCl_2$ → $\overset{O}{\overset{\|}{C}}Cl$, NH$\overset{O}{\overset{\|}{C}}Ot$-Bu → 1) $CH_3CH_2NH_2$; 2) $H_3O^{\oplus}$ →

g) CO_2H, O → CH_3OH, H_2SO_4 → CO_2CH_3, O → $CH_2{=}PPh_3$ → CO_2CH_3, CH_2 → 1) KOH, H_2O; 2) $H_3O^{\oplus}$ →

21.18

a) $HC{\equiv}CH \xrightarrow[\text{2) } CH_3CH_2CH_2Cl]{\text{1) } NaNH_2} CH_3(CH_2)_2C{\equiv}CH \xrightarrow[\text{2) } CH_2O]{\text{1) EtMgBr}} CH_3(CH_2)_2C{\equiv}CCH_2OH$

$\xrightarrow{CH_3\overset{O}{\overset{\|}{C}}Cl} CH_3(CH_2)_2C{\equiv}CCH_2O\overset{O}{\overset{\|}{C}}CH_3 \xrightarrow[H_2]{\text{Lindlar catalyst}}$

b) $HC{\equiv}C(CH_2)_2C{\equiv}CH \xrightarrow[\text{2) } CH_3(CH_2)_2CH_2Cl]{\text{1) } NaNH_2} CH_3(CH_2)_3C{\equiv}C(CH_2)_2C{\equiv}CH$

$\xrightarrow{\text{1) } NaNH_2\text{; 2) } ClCH_2(CH_2)_5OCH_2OCH_3\text{; 3) Lindlar catalyst, } H_2} CH_3(CH_2)_3CH{=}CH(CH_2)_2CH{=}CH(CH_2)_6OCH_2OCH_3$

$\xrightarrow{\text{1) } HCl,\ H_2O\text{; 2) } CH_3\overset{O}{\overset{\|}{C}}Cl}$

c) 1-hexyne (drawn as skeletal structure) $\xrightarrow[\text{2) Cl(CH}_2)_6\text{O-THP}]{\text{1) } NaNH_2}$ $\xrightarrow[\text{3) } H_2\text{, Lindlar catalyst}]{\text{1) TsOH; 2) } CH_3COCl}$

Cl(CH₂)₆OH (Cl~~~~~~OH) + dihydropyran, TsOH → Cl~~~~~~O-THP

d)

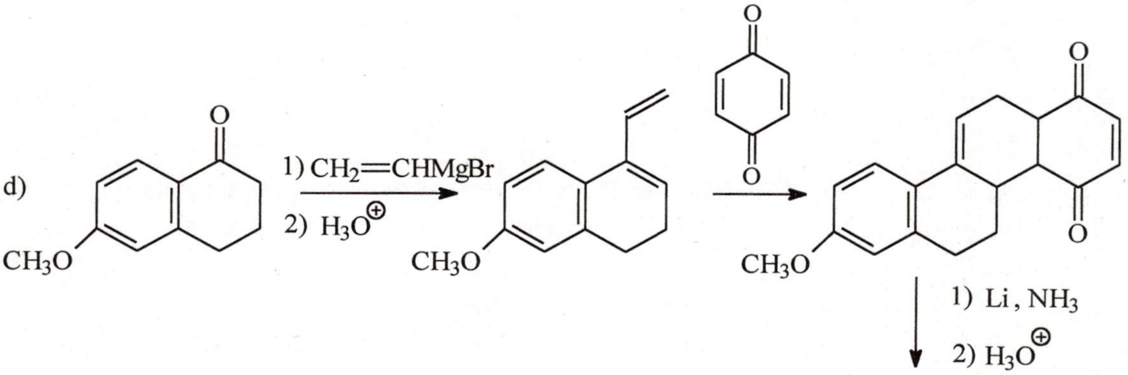

O

CH_3O

1) $CH_2{=}CHMgBr$
2) $H_3O^{\oplus}$

CH_3O

O

O

O

O

CH_3O

1) Li, NH_3
2) $H_3O^{\oplus}$

21.19

a)

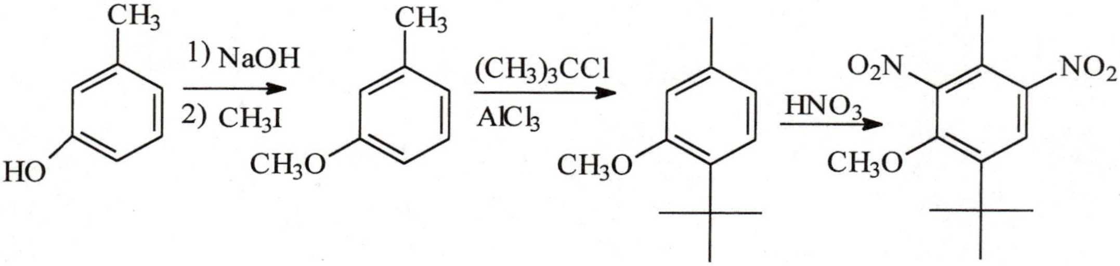

CH_3

HO

1) NaOH
2) CH_3I

CH_3

CH_3O

$(CH_3)_3CCl$
$AlCl_3$

CH_3O

HNO_3

O_2N

NO_2

CH_3O

b)

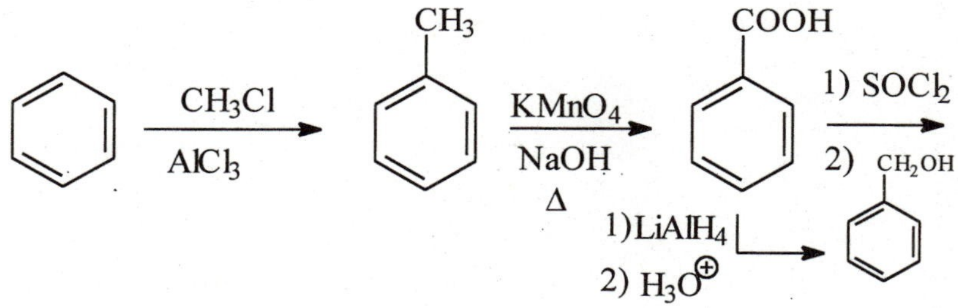

CH_3Cl
$AlCl_3$

CH_3

$KMnO_4$
NaOH
Δ

COOH

1) $SOCl_2$
2) CH_2OH

1) $LiAlH_4$
2) $H_3O^{\oplus}$

c)

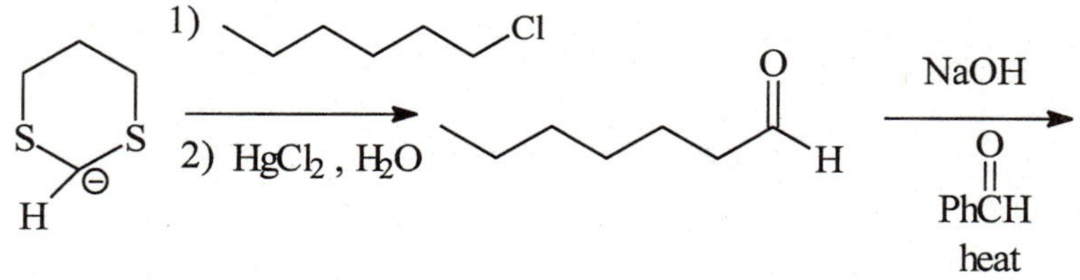

S

S

H

⊖

1) Cl
2) $HgCl_2$, H_2O

O

H

NaOH

O
‖
PhCH
heat

d) Br, H_3C — 1) Mg, ether 2) CO_2 3) $H_3O^{\oplus}$ → COOH, H_3C — 1) $SOCl_2$ 2) $(CH_3CH_2)_2NH$ →

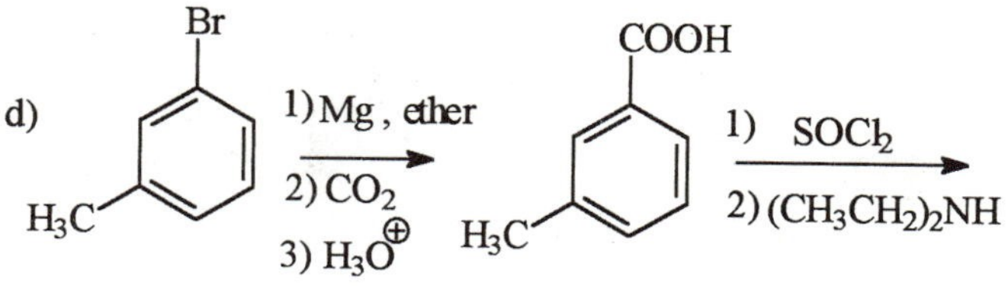

21.20

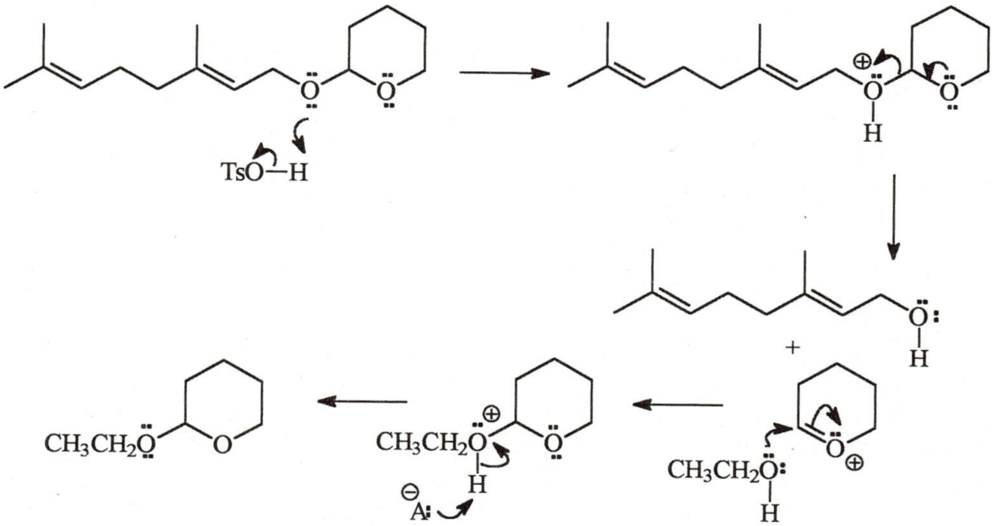

TsO—H

$CH_3CH_2\ddot{O}$

CH_3CH_2O

H

A:

+

$CH_3CH_2\ddot{O}$:

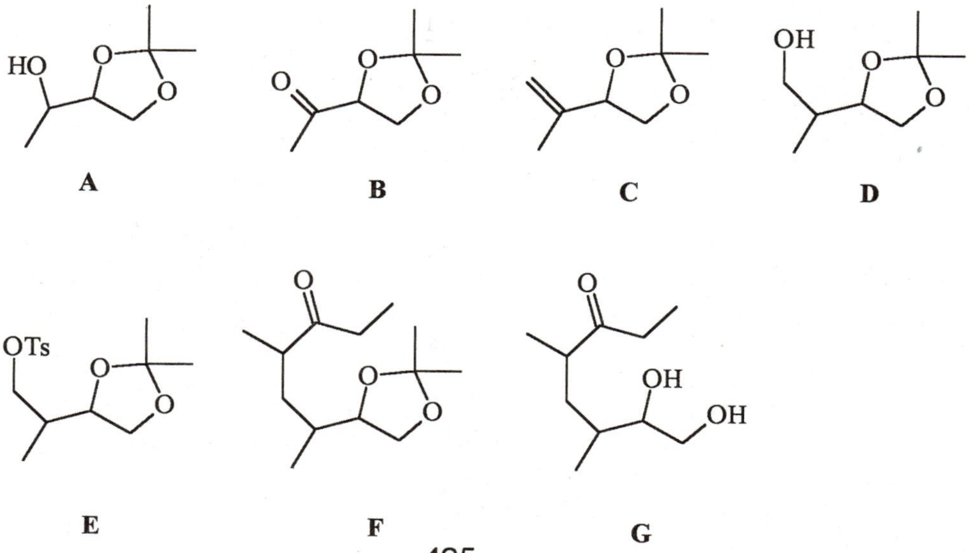

H

21.21

HO

A

B

C

OH

D

OTs

E

F

OH

OH

G

21.22

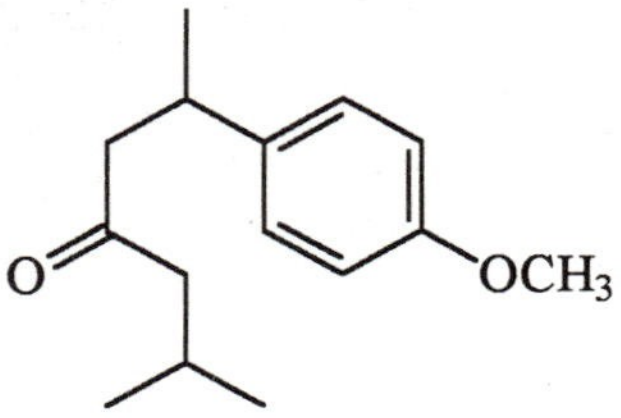

A 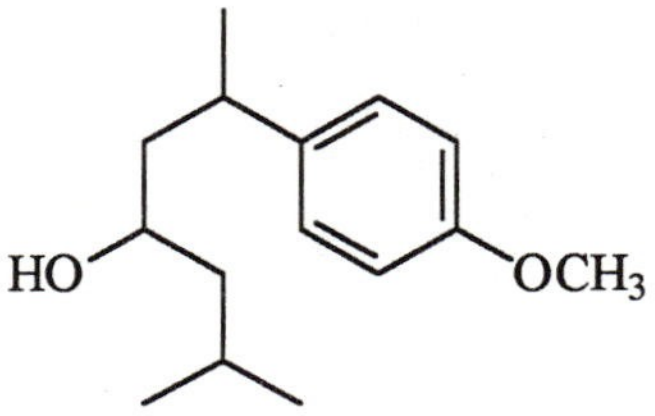 B 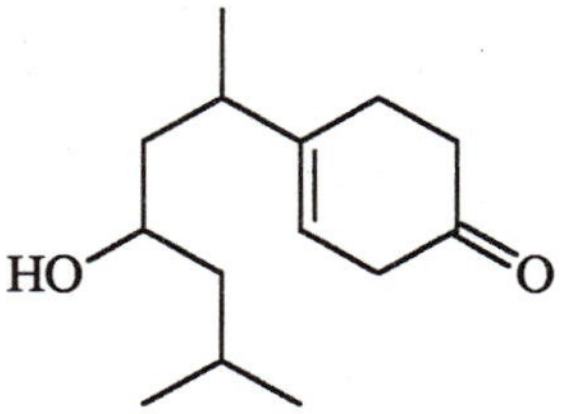 C

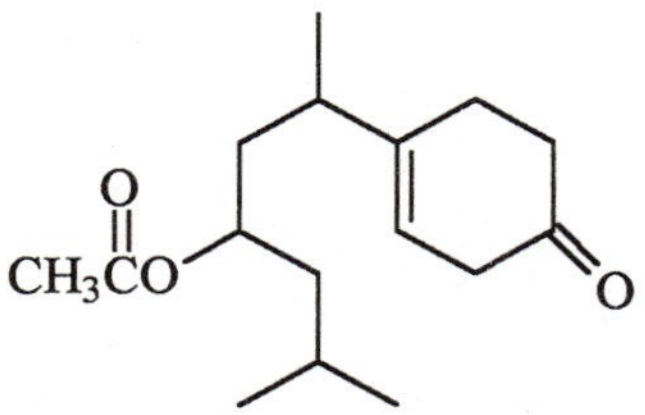

D 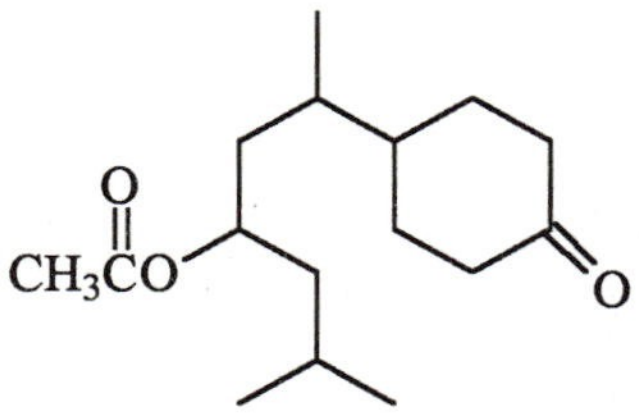 E 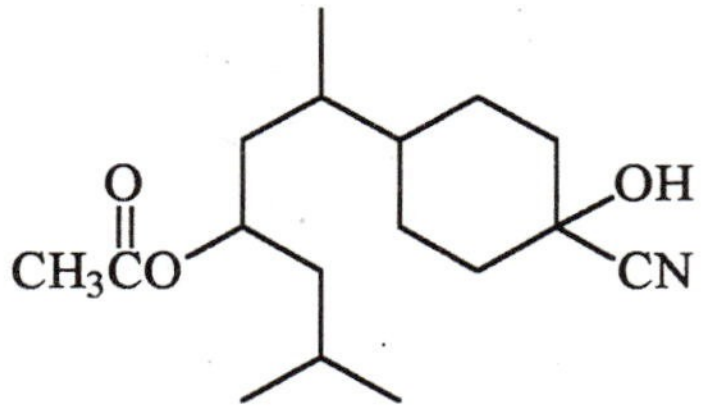 F

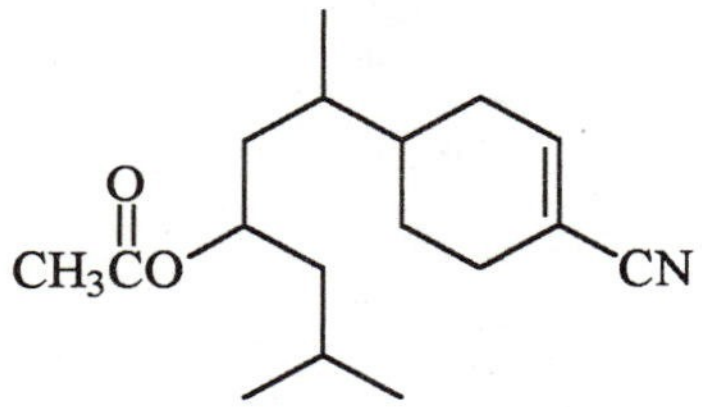

G 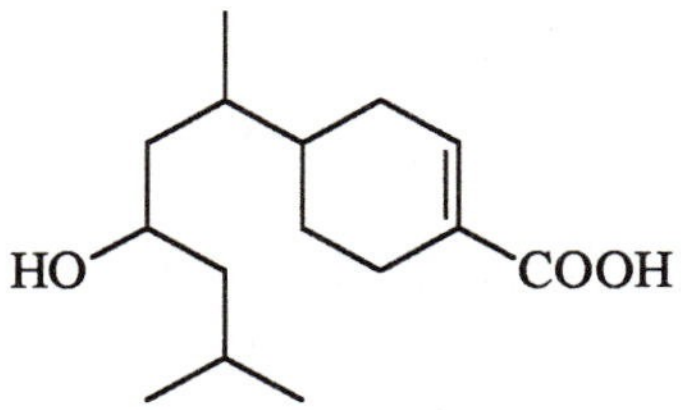 H 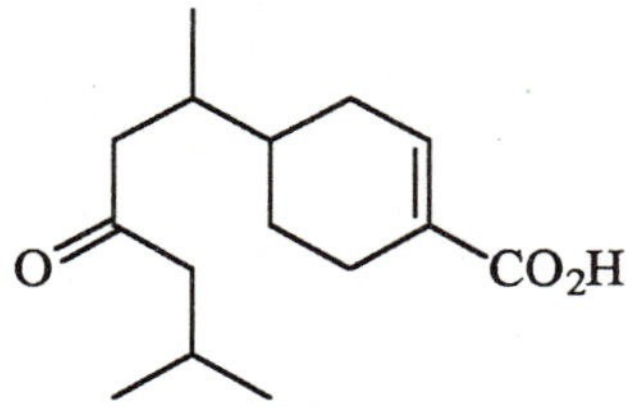 I

After completing this chapter, you should be able to:

1. Show the products of the new reactions introduced in this chapter.
   Problems 21.1, 21.4, 21.5, 21.7, 21.13.

2. Show the mechanisms of these reactions.
   Problems 21.2, 21.6, 21.20.

3. Recognize when a protective group is needed and how to use it in the synthesis of an organic compound.
   Problems 21.3, 21.8, 21.9, 21.17.

4. Use retrosynthetic analysis to design syntheses of more complex organic compounds.
   Problems 21.10, 21.11, 21.12, 21.14, 21.15, 21.16, 21.18, 21.19.

Chapter 22
INDUSTRIAL ORGANIC CHEMISTRY

22.1

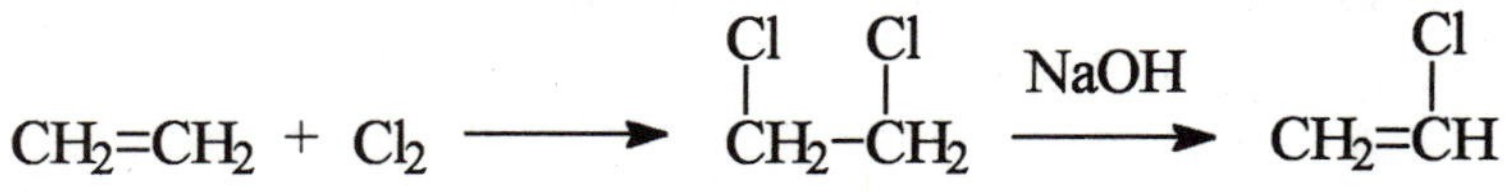

$$CH_2{=}CH_2 + Cl_2 \longrightarrow \underset{}{CH_2(Cl){-}CH_2(Cl)} \xrightarrow{NaOH} CH_2{=}CH(Cl)$$

22.2 1) $KMnO_4$, NaOH 2) $H_3O^{\oplus}$

22.3

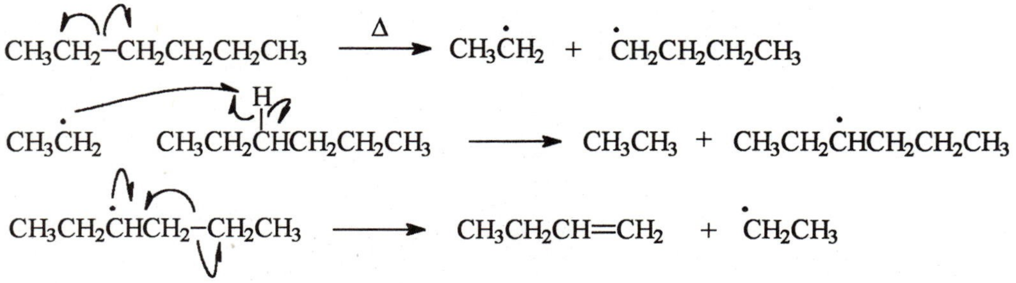

$$CH_3CH_2{-}CH_2CH_2CH_2CH_3 \xrightarrow{\Delta} CH_3\dot{C}H_2 + \dot{C}H_2CH_2CH_2CH_3$$

$$CH_3\dot{C}H_2 \quad CH_3CH_2CH(H)CH_2CH_2CH_3 \longrightarrow CH_3CH_3 + CH_3CH_2\dot{C}HCH_2CH_2CH_3$$

$$CH_3CH_2\dot{C}HCH_2{-}CH_2CH_3 \longrightarrow CH_3CH_2CH{=}CH_2 + \dot{C}H_2CH_3$$

22.4

$$CH_3CH_2{-}CH_2CH_2CH_2CH_2CH_3 \xrightarrow{\Delta} CH_3\dot{C}H_2 + \dot{C}H_2CH_2CH_2CH_2CH_3$$

$$CH_3\dot{C}H_2 \quad CH_3CH_2CH(H)CH_2CH_2CH_2CH_3 \longrightarrow CH_3CH_3 + CH_3CH_2\dot{C}HCH_2CH_2CH_2CH_3$$

$$CH_3{-}CH_2{-}\dot{C}HCH_2CH_2CH_2CH_3 \longrightarrow \dot{C}H_3 + CH_2{=}CHCH_2CH_2CH_2CH_3$$

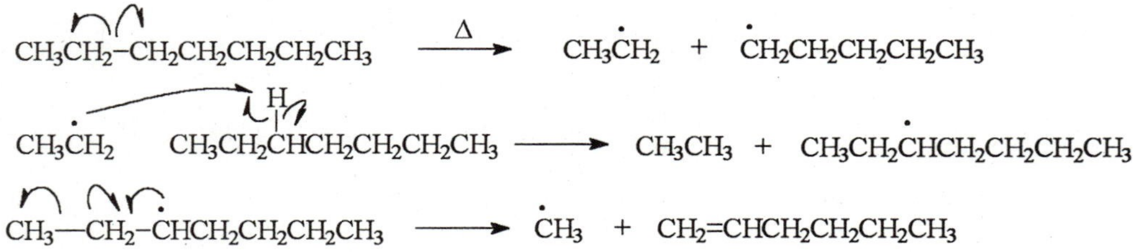

This process is not very favorable because it requires the formation of a less stable methyl radical in the β-scission process.

22.5

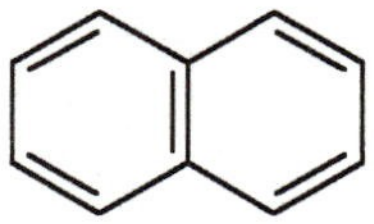

22.6 Reaction of an alkene with a peracid is the most common method for preparing epoxides in the laboratory. The use of an unstable and relatively expensive reagent, such as a peracid, is avoided by industry whenever possible.

22.7

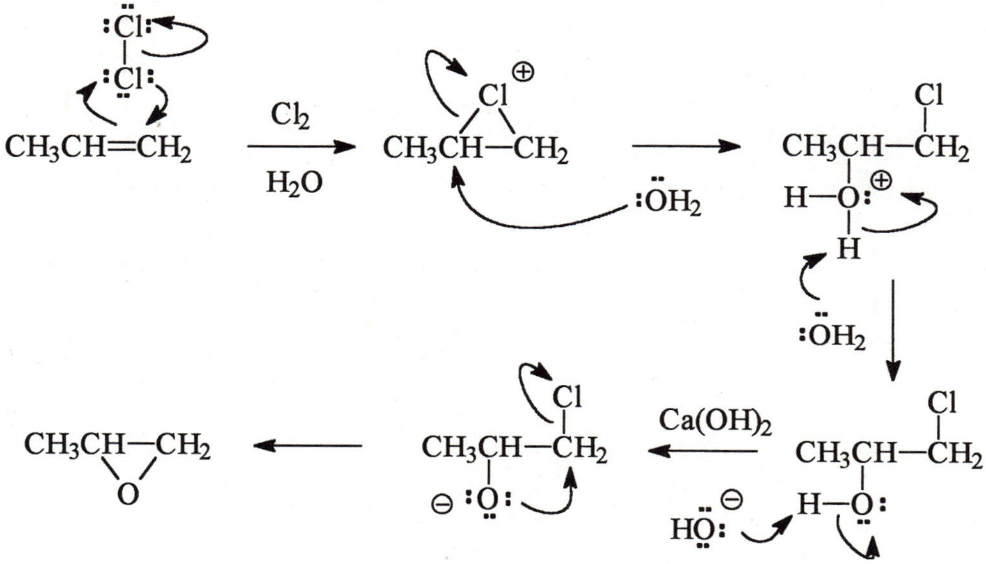

$$CH_3CH{=}CH_2 \xrightarrow[H_2O]{Cl_2} CH_3\overset{\overset{Cl^{\oplus}}{\wedge}}{CH{-}CH_2} \xrightarrow{:\ddot{O}H_2} CH_3CH(\overset{\oplus}{O}H_2){-}CH_2Cl \xrightarrow{:\ddot{O}H_2} CH_3CH(OH){-}CH_2Cl$$

$$CH_3CH(OH){-}CH_2Cl \xrightarrow[H\ddot{O}:^{\ominus}]{Ca(OH)_2} CH_3CH(O^{\ominus}){-}CH_2Cl \longrightarrow CH_3\underset{O}{CH{-}CH_2}$$

22.8

$$H_2\overset{\overset{:\ddot{O}}{\wedge}}{C{-}CH_2} \xrightarrow{:C{\equiv}N:^{\ominus}} H_2C(:\ddot{O}:^{\ominus}){-}CH_2CN \xrightarrow{H{-}C{\equiv}N:} H_2C(:\ddot{O}H){-}CH_2CN + :C{\equiv}N:^{\ominus}$$

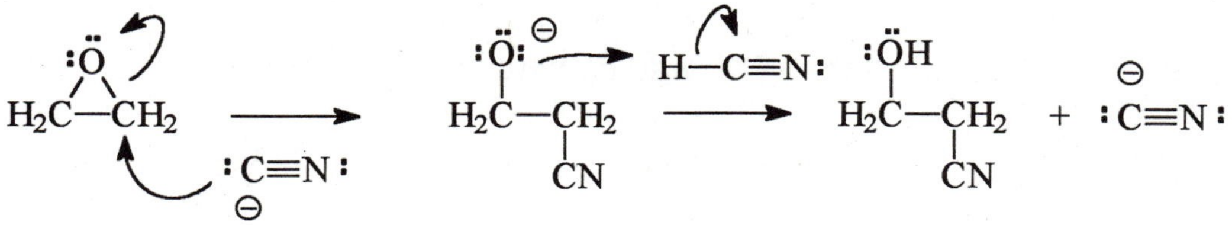

22.9

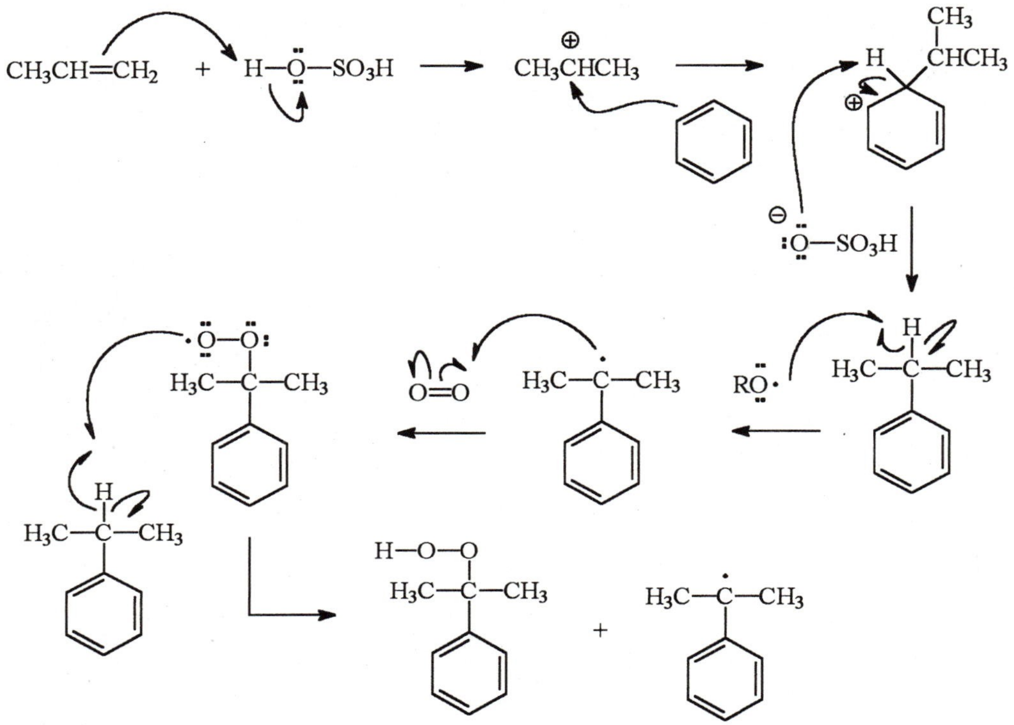

$$CH_3CH{=}CH_2 + H{-}O{-}SO_3H \longrightarrow CH_3\overset{\oplus}{C}HCH_3$$

See p. 923 in the textbook for the rest of this answer.

22.10 Use an aldol condensation followed by catalytic hydrogenation.

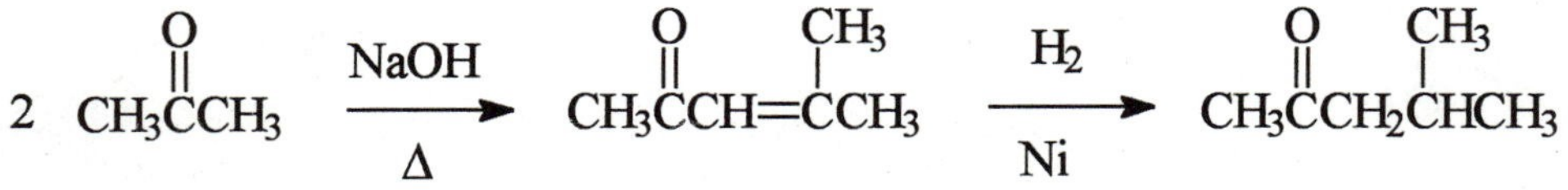

$$2\ CH_3\overset{O}{\overset{\|}{C}}CH_3 \xrightarrow[\Delta]{NaOH} CH_3\overset{O}{\overset{\|}{C}}CH{=}\overset{CH_3}{\overset{|}{C}}CH_3 \xrightarrow[Ni]{H_2} CH_3\overset{O}{\overset{\|}{C}}CH_2\overset{CH_3}{\overset{|}{C}}HCH_3$$

22.11 An aromatic ring is formed in the product.

22.12 Cyclohexanol and cyclohexanone have a larger molecular mass and are more polar than cyclohexane, so their boiling points are much higher that that of cyclohexane. Therefore, distillation can be used for the separation.

22.13

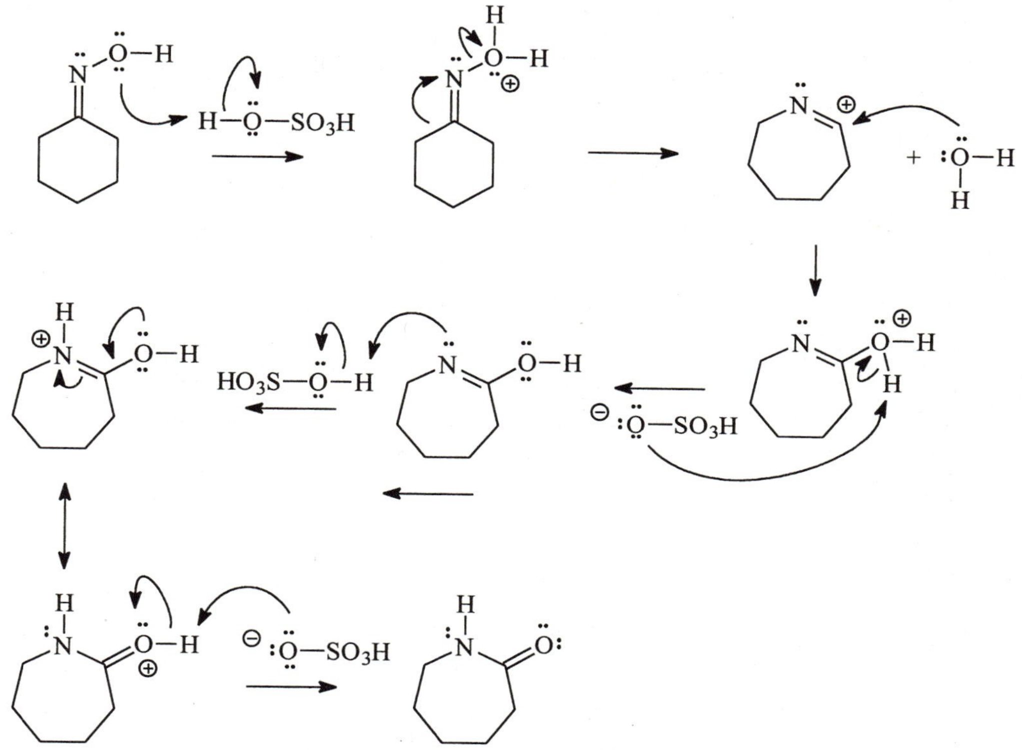

22.14

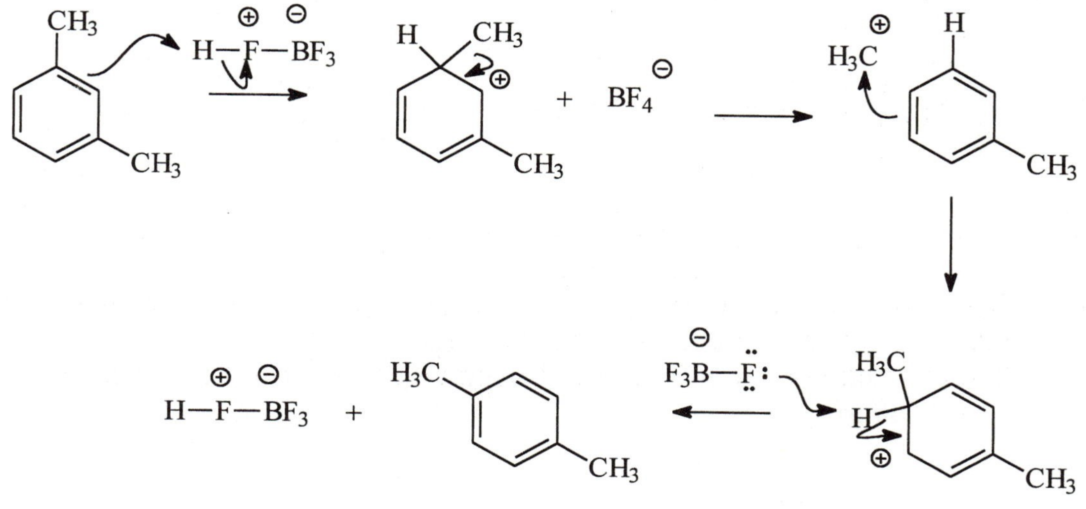

22.15

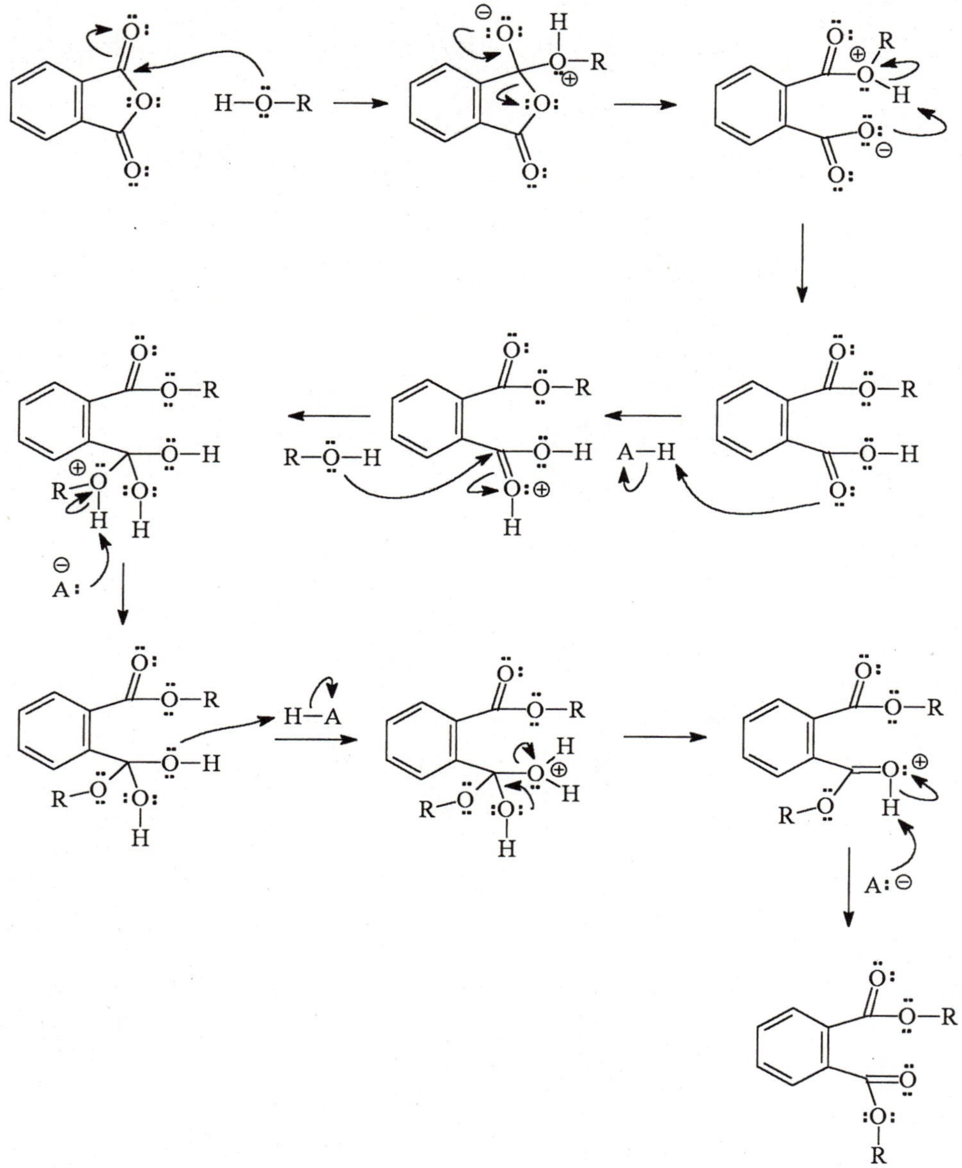

The first esterification involves the reaction of an alcohol with an anhydride, a reaction that is very favorable. The second esterification involves the reaction of an alcohol with a carboxylic acid (Fischer esterification), a less favorable reaction that does require an acid catalyst.

22.16

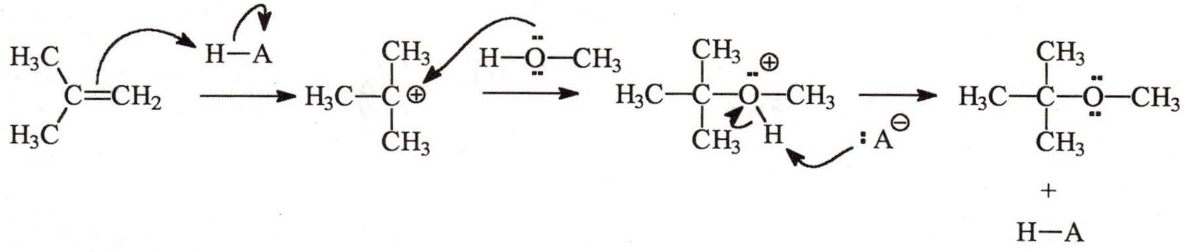

$$(H_3C)_2C{=}CH_2 + H{-}A \longrightarrow H_3C{-}\overset{CH_3}{\underset{CH_3}{C^{\oplus}}} \xrightarrow{H{-}\ddot{O}{-}CH_3} H_3C{-}\overset{CH_3}{\underset{CH_3}{C}}{-}\overset{\oplus}{O}(H){-}CH_3 \xrightarrow{:A^{\ominus}} H_3C{-}\overset{CH_3}{\underset{CH_3}{C}}{-}\ddot{O}{-}CH_3 + H{-}A$$

22.17 This is an E2 elimination reaction.

22.18

a)  $CH_2ClCHClCH_2CH_3$

b) 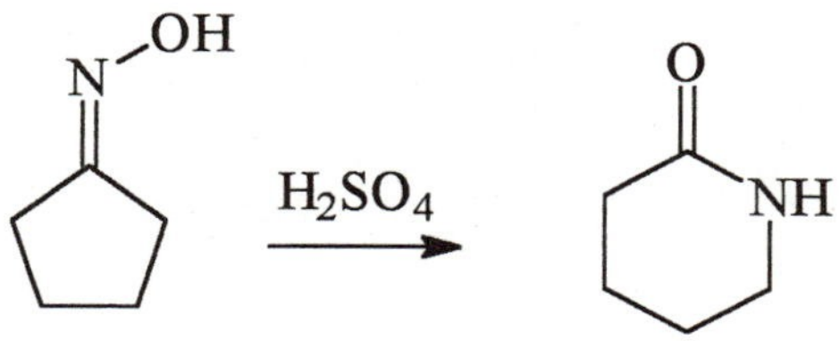 cyclopentanone oxime $\xrightarrow{H_2SO_4}$ 2-piperidinone (cyclic amide with O and NH)

c) naphthalene + 2 H_2

d) $CH_3CH(OH)CH_2C{\equiv}N$

e) anthracene

f) $H_3C{-}C(CH_3)_2{-}OCH_2CH_3$

g) $CH_3CHClCHClCH_3 \xrightarrow{Ca(OH)_2} CH_2{=}CH{-}CH{=}CH_2$

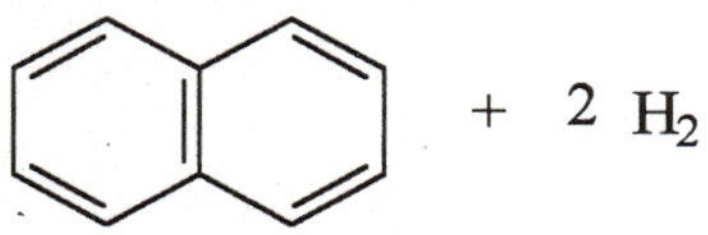

h) 3-phenylpentane + 2-phenylpentane

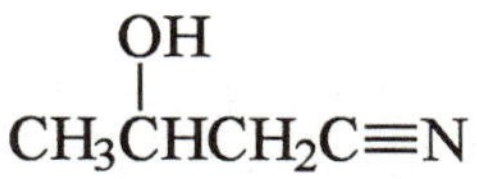

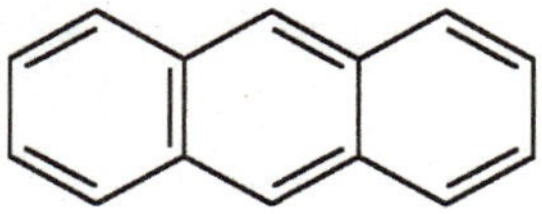

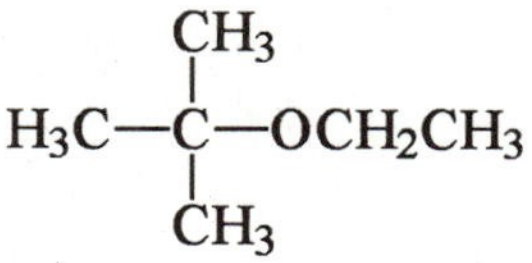

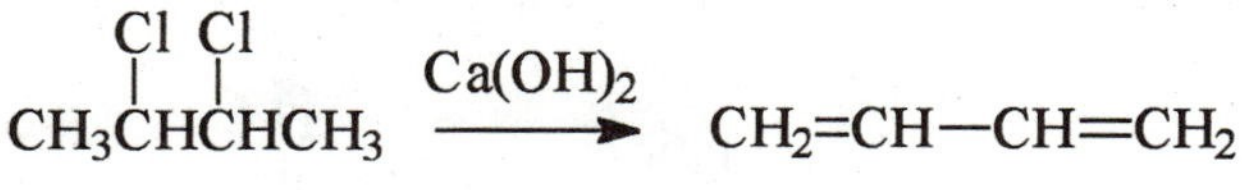

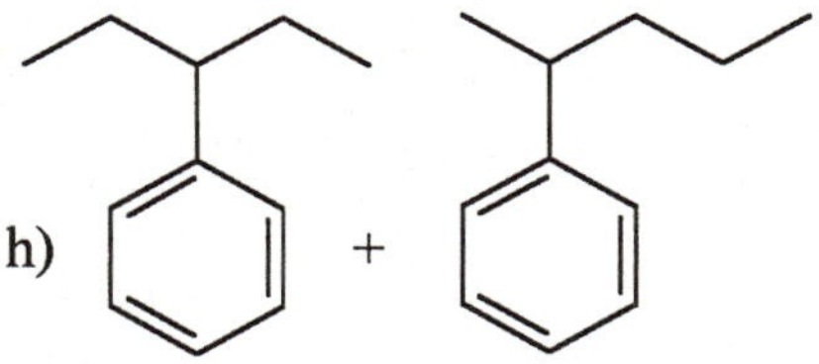

22.19

$$\text{(CH}_3)_2\text{C=CHCH}_3 \xrightarrow{Cl_2} \text{(CH}_3)_2\text{CCl-CHClCH}_3 \xrightarrow{Ca(OH)_2} \text{CH}_2\text{=C(CH}_3)\text{CH=CH}_2$$

22.20 Butanal can be prepared by oxidation of ethylene to ethanal, an aldol condensation of ethanal to produce 2-butenal, and reduction of the double bond by catalytic hydrogenation.

$$H_2C{=}CH_2 + 1/2\ O_2 \xrightarrow[CuCl_2]{PdCl_2} CH_3\overset{O}{\overset{\|}{C}}H \xrightarrow[\Delta]{NaOH} CH_3CH{=}CH\overset{O}{\overset{\|}{C}}H \xrightarrow[Ni]{H_2} CH_3CH_2CH_2\overset{O}{\overset{\|}{C}}H$$

22.21

$$\text{benzene} \xrightarrow[AlCl_3]{(CH_3)_2CHCOCl} C_6H_5COCH(CH_3)_2 \xrightarrow[Pd]{H_2} C_6H_5CH_2CH(CH_3)_2 \xrightarrow[AlCl_3]{CH_3COCl} p\text{-}CH_3COC_6H_4CH_2CH(CH_3)_2$$

22.22

$$H_2C{=}CH_2 + 1/2\ O_2 \xrightarrow{Ag} H_2C\overset{O}{-}CH_2 \xrightarrow[H_2SO_4]{CH_3OH} H_2C(OH){-}CH_2(OCH_3) \xrightarrow[H_2SO_4]{CH_3OH} H_2C(OCH_3){-}CH_2(OCH_3)$$

22.23 Methyl ethyl ketone is prepared by hydration of the butene mixture followed by oxidation. The mixture of alkenes is a satisfactory starting material because the all give the same product, 2-butanol.

$$CH_3CH_2CH{=}CH_2 \xrightarrow[H_2SO_4]{H_2O} CH_3CH_2\overset{OH}{\overset{|}{C}}HCH_3 \xrightarrow{\text{oxidation}} CH_3CH_2\overset{O}{\overset{\|}{C}}CH_3$$

22.24

$$H_2C{=}CH_2 \xrightarrow{Cl_2} H_2\overset{Cl}{C}{-}\overset{Cl}{C}H_2 \xrightarrow{NaOH} H_2C{=}\overset{Cl}{C}H \xrightarrow{Cl_2} H_2\overset{Cl}{C}{-}\overset{Cl}{\underset{Cl}{C}}H \xrightarrow{NaOH} H_2C{=}\overset{Cl}{\underset{Cl}{C}}$$

22.25

$$CH_2{=}CHCH_2Cl \xrightarrow[H_2O]{Cl_2} ClCH_2CH(OH)CH_2Cl \xrightarrow{Ca(OH)_2} \overset{O}{\triangle}{-}CH_2Cl$$

22.26

$$C_6H_5CH_2CH_3 \xrightarrow[h\nu]{Cl_2} C_6H_5CH(Cl)CH_3 \xrightarrow{Ca(OH)_2} C_6H_5CH{=}CH_2$$

This route is unattractive to industry because it uses chlorine although there is no chlorine in the final product.

22.27

$$Cl_2 \xrightarrow{300\,^{\circ}C} 2\ :\ddot{C}l\cdot$$

$$:\ddot{C}l\cdot + CH_2{=}CH{-}CH{=}CH_2 \longrightarrow \overset{Cl}{CH_2}{-}\dot{C}H{-}CH{=}CH_2 \longleftrightarrow \overset{Cl}{CH_2}{-}CH{=}CH{-}\dot{C}H_2$$

$$\xrightarrow{Cl{-}Cl} :\ddot{C}l\cdot + \overset{Cl}{CH_2}{-}CH{=}CH{-}\overset{Cl}{CH_2} + \overset{Cl}{CH_2}{-}\overset{Cl}{CH}{-}CH{=}CH_2$$

22.28

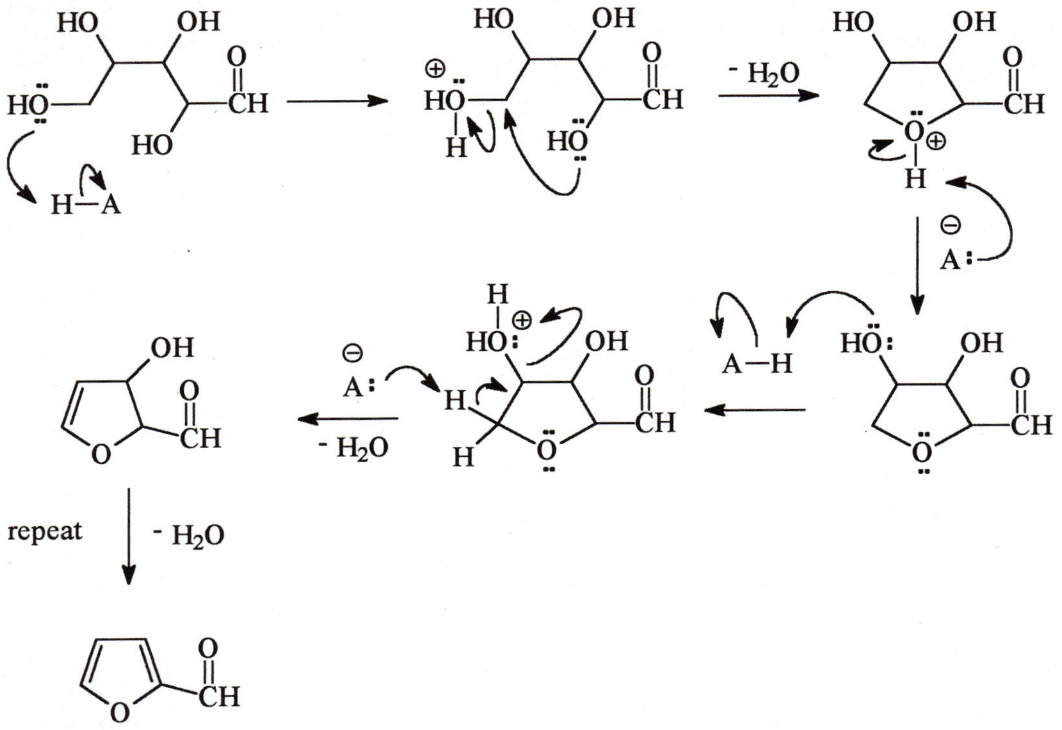

HO OH O
HÖ CH
HO
H—A

→

⊕ HO OH O
HÖ CH
H HÖ

$- H_2O$ →

HO OH O
CH
Ö ⊕
H
⊖
A:

↓

HÖ: OH O
CH
Ö

A—H

←

H
⊕
HÖ: OH O
H CH
H Ö

⊖
A:

←
$- H_2O$

OH O
CH
O

repeat ↓ $- H_2O$

O
CH
O

22.29

O NaBH₄ → OH HCl → Cl

↓ ⊖CN

CN

← H_2O / H_2SO_4

CO_2H

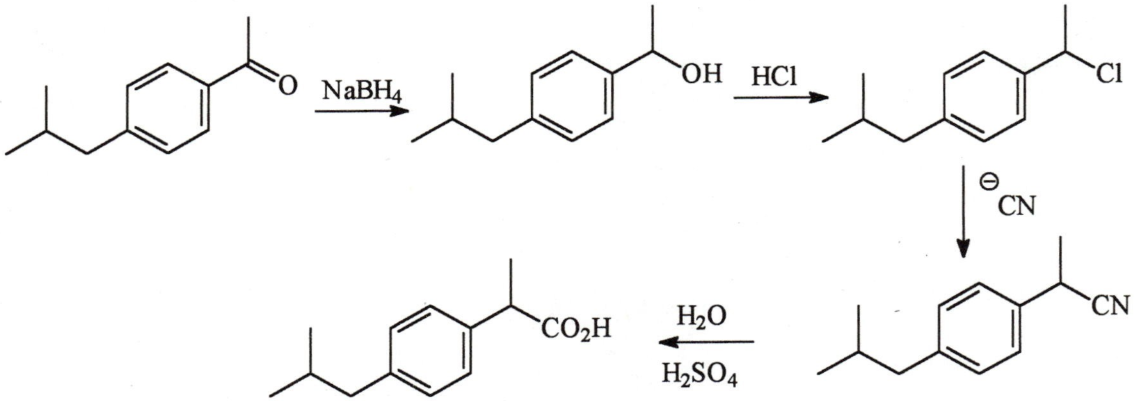

22.30 This would probably not be a good method because the chlorine atom is not very selective and should abstract both the hydrogen on C-1 of butane as well as the hydrogen on C-2. Therefore the oxime of butanal should be formed along with the oxime of 2-butanone.

22.31

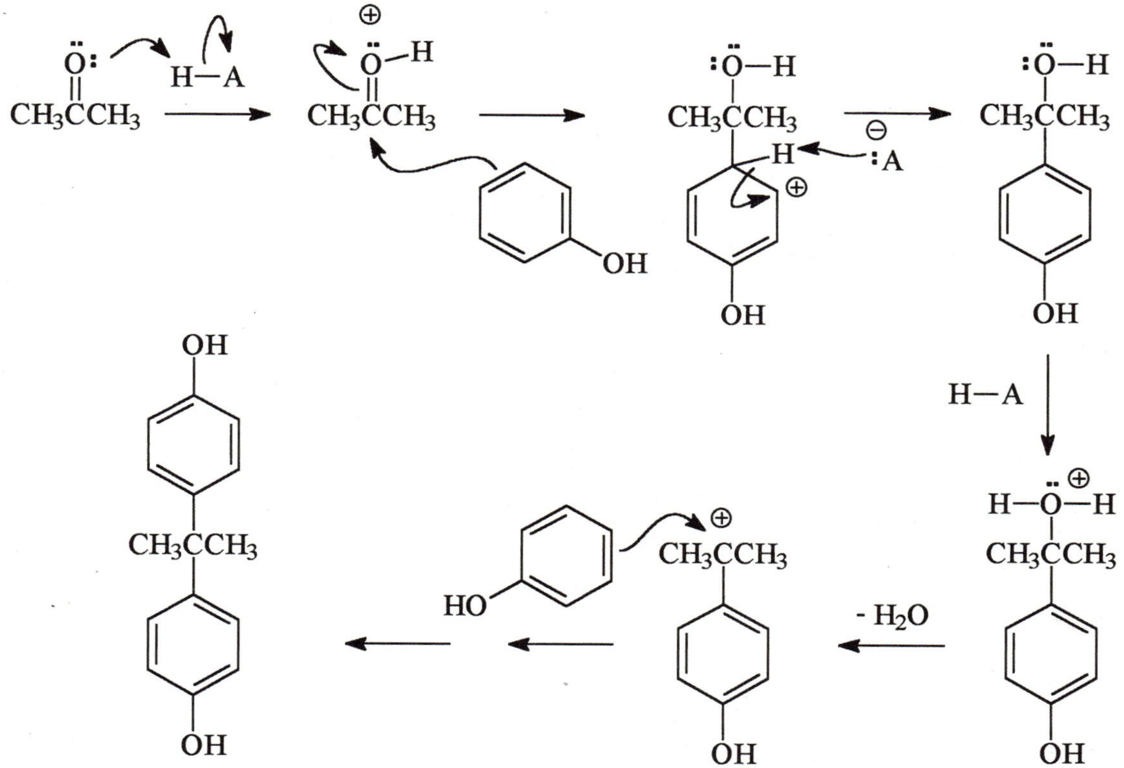

22.32

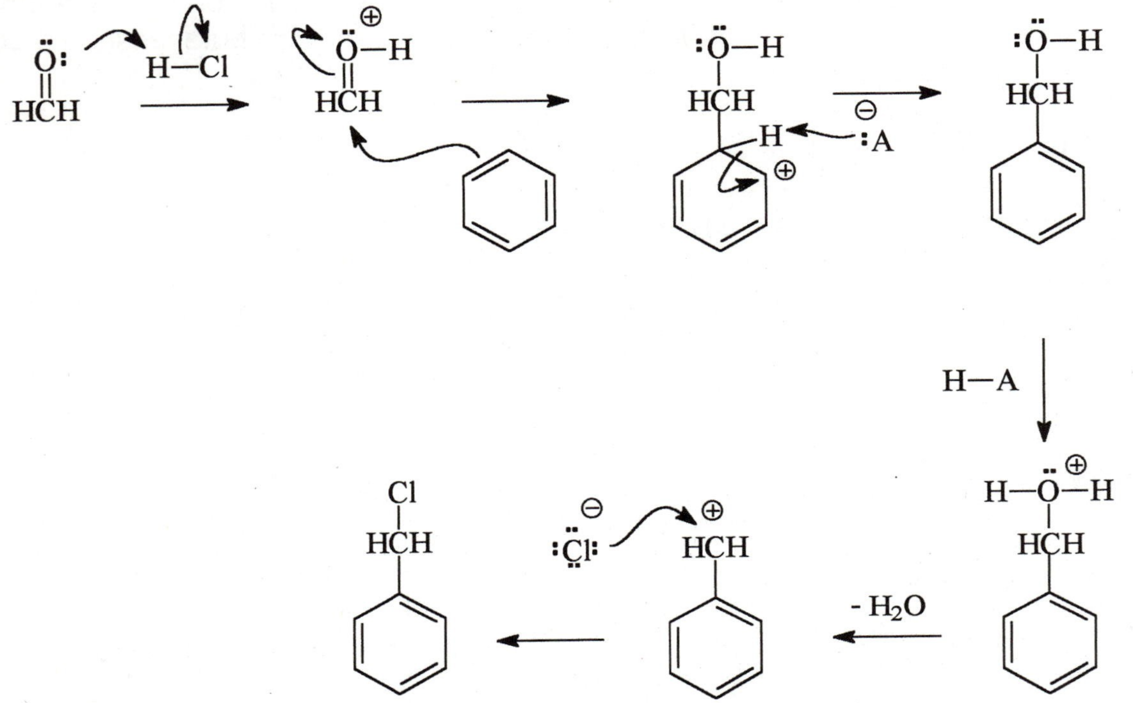

22.33

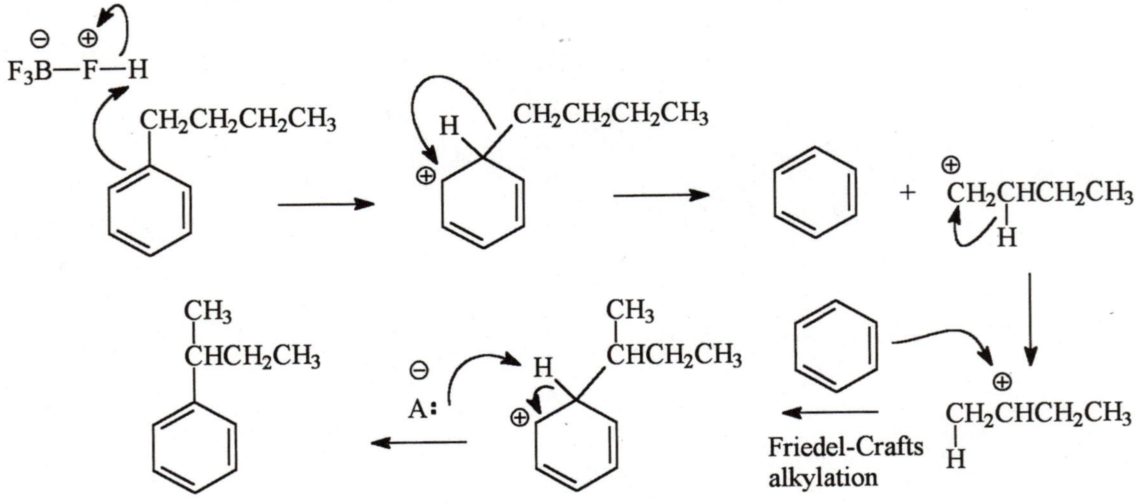

22.34

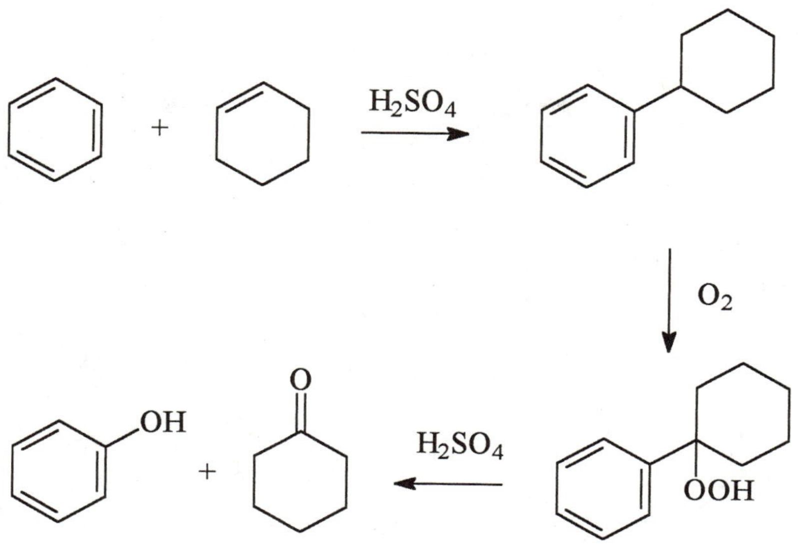

$$+ \xrightarrow{H_2SO_4}$$

$$\downarrow O_2$$

OH $+$ O $\xleftarrow{H_2SO_4}$ OOH

The cyclohexanone can be converted back to cyclohexene by reduction followed by elimination of water.

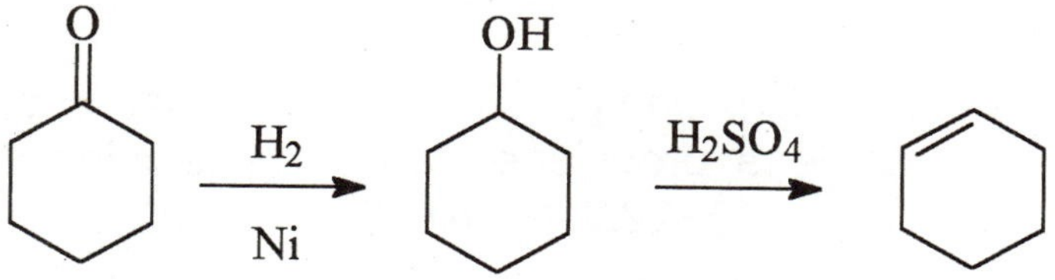

O $\xrightarrow[Ni]{H_2}$ OH $\xrightarrow{H_2SO_4}$

22.35

$$2\ CH_3(CH_2)_6CH{=}CH_2 \xrightarrow{\text{catalyst}} CH_3(CH_2)_6CH{=}CH(CH_2)_6CH_3 + H_2C{=}CH_2$$

22.36

2 metathesis catalyst

22.37

CH_3 CH_3 oxidation O O O

After completing this chapter, you should be able to:

1. Identify the major sources for organic chemicals.

2 Know the major reactions and products that are used to produce important chemicals from ethylene, propylene, benzene, the xylenes, and butylene. Problems 22.1, 22.2, 22.5, 22.7, 22.10, 22.11, 22.18, 22.19, 22.20, 22.21, 22.24, 22.29.

3. Understand some of the differences between reactions conducted in the research laboratory and those conducted industrially in terms of philosophy, goals, methodology, processing, and outcome. Problems 22.6, 22.12, 22.22, 22.23, 22.25, 22.26, 22.30, 22.34.

4. Write mechanisms for these reactions where appropriate. Problems 22.3, 22.4, 22.8, 22.9, 22.13, 22.14, 22.15, 22.16, 22.17, 22.27, 22.28, 22.31, 22.32, 22.33.

Chapter 23
SYNTHETIC POLYMERS

23.1

$$PhC(=O)\ddot{O}-\ddot{O}C(=O)Ph \longrightarrow 2\ PhC(=O)\ddot{O}\cdot$$

$$PhC(=O)\ddot{O}\cdot + CH_2{=}CH(CH_3) \longrightarrow PhC(=O)\ddot{O}-CH_2-\dot{C}H(CH_3)$$

$$PhC(=O)\ddot{O}-CH_2-\dot{C}H(CH_3) + CH_2{=}CH(CH_3) \longrightarrow PhC(=O)\ddot{O}-CH_2-CH(CH_3)-CH_2-\dot{C}H(CH_3) \xrightarrow{\text{etc.}} \left(-CH_2-CH(CH_3)-\right)_n$$

Addition of the radical always occurs at C-1 of propene so that the more stable secondary radical is formed.

23.2

a) $\left(-CH_2-CH(Cl)-\right)_n$ b) $\left(-CH_2-CH(CN)-\right)_n$ c) $\left(-CH_2-CCl_2-\right)_n$

23.3

a) $CH_2{=}CH-OC(=O)CH_3$ b) $CH_2{=}CH(CH_3)$

23.4

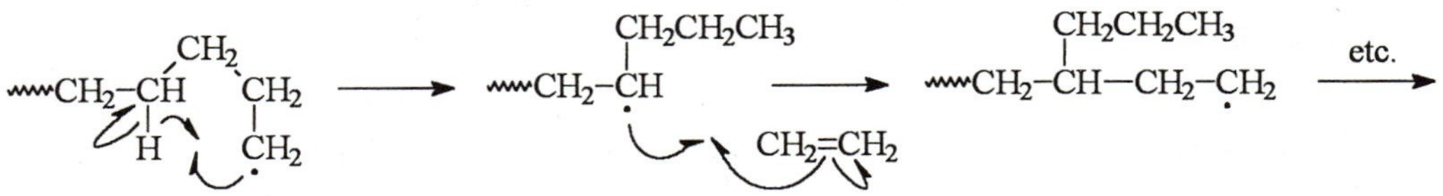

$$\sim CH_2-CH(-H)-CH_2-CH_2-\dot{C}H_2 \longrightarrow \sim CH_2-\dot{C}H(CH_2CH_2CH_3) \xrightarrow{CH_2=CH_2} \sim CH_2-CH(CH_2CH_2CH_3)-CH_2-\dot{C}H_2 \xrightarrow{\text{etc.}}$$

23.5 Both (a) and (b) can form atactic polymers whereas (c) cannot.

23.6 The primary carbocation that is required as an intermediate in the cationic polymerization of ethylene is too unstable.

23.7

$$\sim CH_2-\overset{C\equiv N}{\underset{\ominus}{\ddot{C}H}}$$

intermediate in the anionic polymerization of acrylonitrile

$$\sim CH_2-\overset{CH_3}{\underset{CH_3}{C}}:\ominus$$

intermediate in the anionic polymerization of isobutylene

The anionic intermediate formed in the polymerization of acrylonitrile is stabilized by resonance and is readily formed whereas the anionic intermediate formed in the polymerization of isobutylene is quite unstable and is difficult to form.

23.8 THF is much less reactive towards nucleophiles than is ethylene oxide because no ring strain is relieved when THF reacts.

23.9 a) The alkene on the right produces a more crystalline polymer because it has no chiral centers and is more stereoregular.

b) The alkene on the right produces a more crystalline polymer because its side chain is more polar.

c) Coordination polymerization produces a less branched and more stereoregular polymer which is more crystalline.

d) The alkene on the right produces a more crystalline polymer because it has no chiral centers and is more stereoregular.

23.10 Teflon has no chiral centers, so it has no stereochemical complications. Nor does it have any hydrogens that can be abstracted from the interior of the chain, so it is linear.

23.11 Poly(methyl methacrylate) cannot be prepared by cationic polymerization because the carbonyl group destabilizes the intermediate carbocation. It can be produced by anionic polymerization because the carbonyl group stabilizes the carbanion intermediate by resonance.

23.12 Radicals prefer to add to C-1 of isoprene because the resulting radical is the most stable of the four possibilities. It is stabilized by resonance and the odd electron is on a tertiary carbon in one resonance structure and a primary carbon in the other. In contrast, addition at C-2 or C-3 produces less stable radicals because they have no resonance stabilization. Addition at C-4 produces a resonance stabilized radical, but the odd electron is on a secondary carbon and a primary carbon in the two resonance structures.

23.13

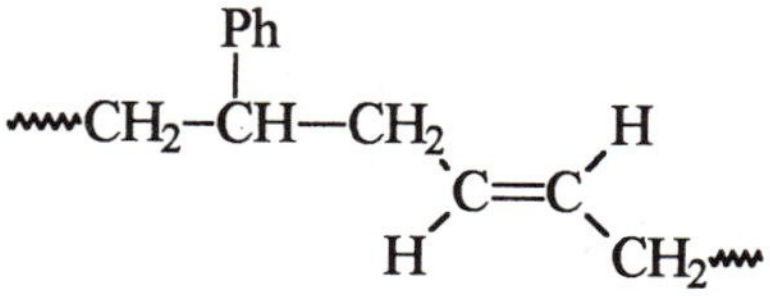

$$\sim CH_2-\underset{}{\overset{Ph}{\overset{|}{C}}}H-CH_2-C(H)=C(H)-CH_2\sim$$

Trans double bonds predominate because they are more stable than *cis* double bonds.

3.14

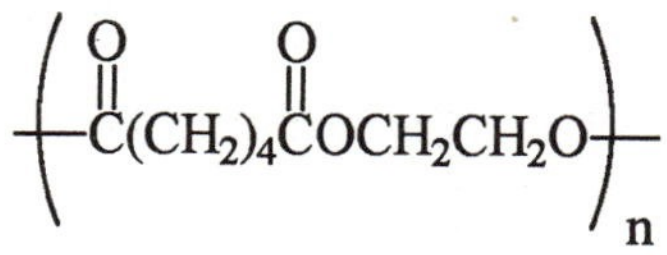

$$\left(-\overset{O}{\overset{||}{C}}(CH_2)_4\overset{O}{\overset{||}{C}}OCH_2CH_2O- \right)_n$$

23.15

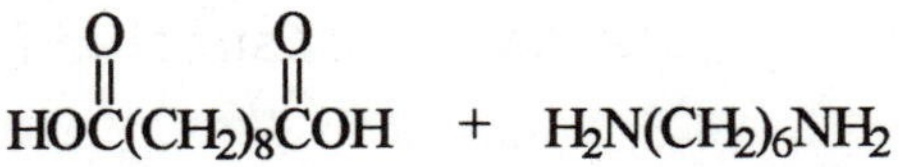

$$HO\overset{O}{\overset{\|}{C}}(CH_2)_8\overset{O}{\overset{\|}{C}}OH \quad + \quad H_2N(CH_2)_6NH_2$$

23.16 If the polyester is to be formed from more diol units, then each polymer chain must terminate with a diol unit at each end. To accommodate even a small excess of diol units, there must be many chains, so the chains must be relatively short.

23.17 Direct preparation of poly(vinyl alcohol) would require an enol as the monomer. However, recall from Section 10.7 that most enols cannot be isolated because they spontaneously isomerize to the carbonyl tautomer.

23.18

a) 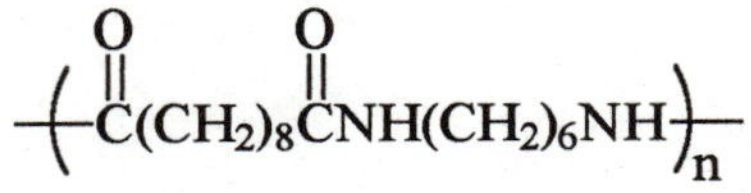 $\left(\overset{O}{\overset{\|}{C}}(CH_2)_8\overset{O}{\overset{\|}{C}}NH(CH_2)_6NH\right)_n$

heat

b) 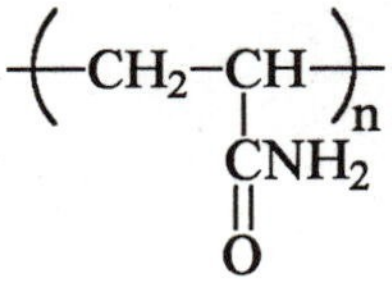 $\left(CH_2-\underset{\underset{\underset{O}{\|}}{CNH_2}}{\underset{|}{CH}}\right)_n$ radical or anionic polymerization

c) 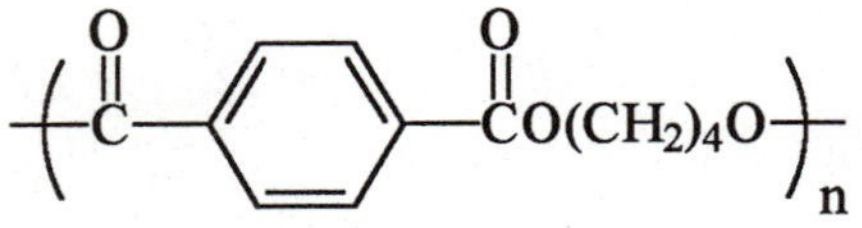 $\left(\overset{O}{\overset{\|}{C}}-C_6H_4-\overset{O}{\overset{\|}{C}}O(CH_2)_4O\right)_n$

heat

d) 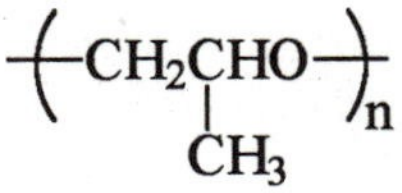 $\left(CH_2\underset{\underset{CH_3}{|}}{C}HO\right)_n$ anionic polymerization

e) 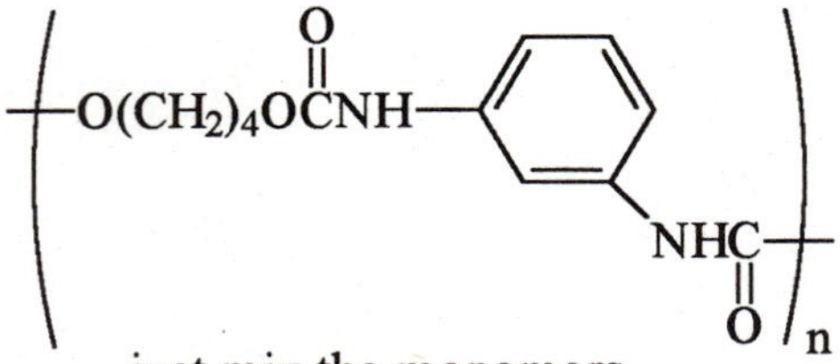 $\left(O(CH_2)_4O\overset{O}{\overset{\|}{C}}NH-C_6H_4-NH\underset{\underset{O}{\|}}{C}\right)_n$

just mix the monomers

f) 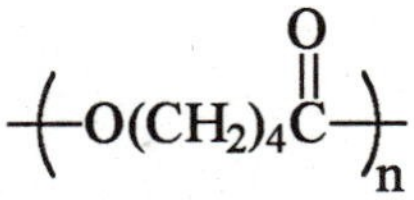 $\left(O(CH_2)_4\overset{O}{\overset{\|}{C}}\right)_n$ add catalytic amount of $OH^{\ominus}$ or H_2O and heat

23.19

a) 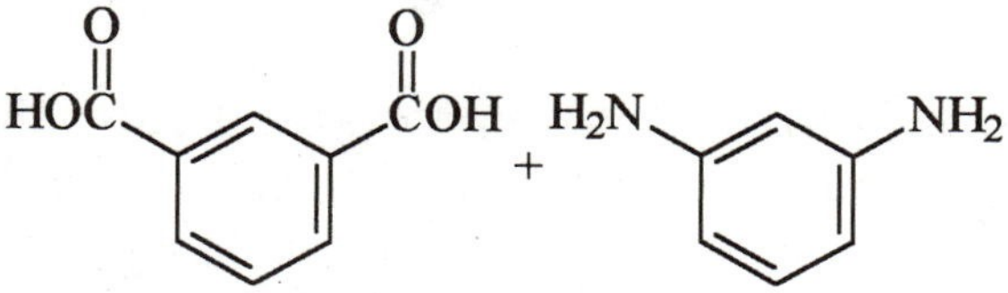 HOC(=O)– and –C(=O)OH on a benzene ring (meta) + H_2N– and –NH_2 on a benzene ring (meta)

b) $H_2N(CH_2)_{11}\overset{O}{\overset{\|}{C}}OH$

c) 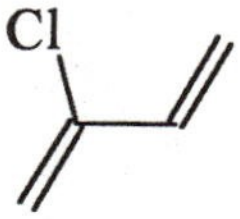 Cl (2-chloro-1,3-butadiene, line structure)

d) 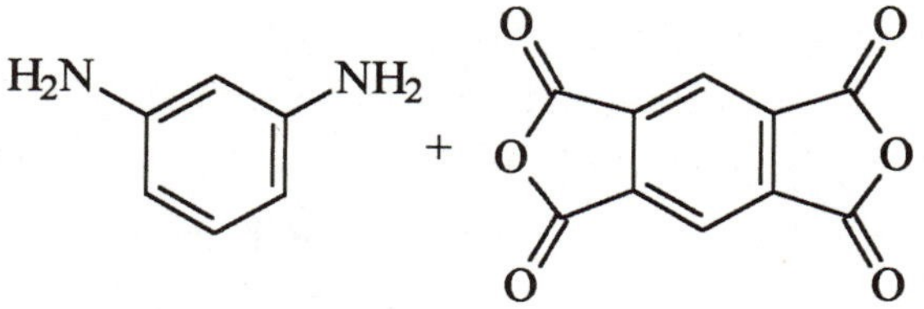 H_2N– and –NH_2 on a benzene ring (meta) + O, O, O, O, O, O (pyromellitic dianhydride, line structure)

e) $CH_2{=}CHCH_2CH(CH_3)CH_3$

f) H_2N–C_6H_4–$\overset{O}{\overset{\|}{C}}OH$ (para)

g) CH_3–$C{=}CH_2$ on a benzene ring

h) $HOCH_2{-}C(CH_3)_2{-}COOH$ or O, O (lactone ring with two methyl groups, line structure)

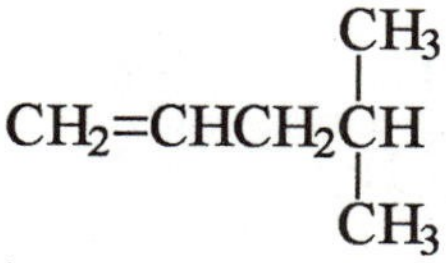

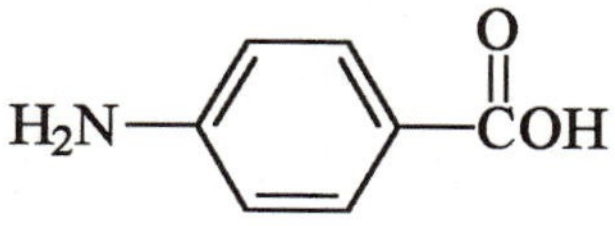

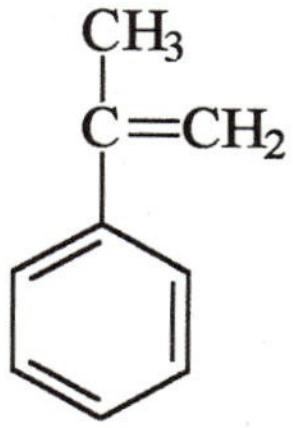

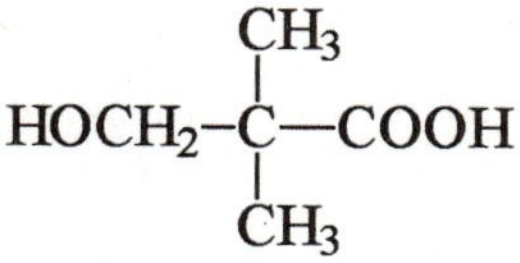

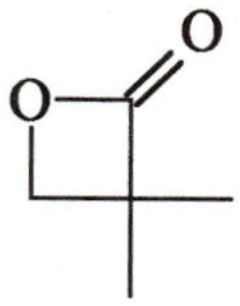

i) 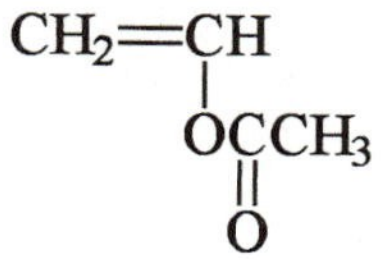 $CH_2{=}CH{-}O\overset{\|}{\underset{O}{C}}CH_3$

j) 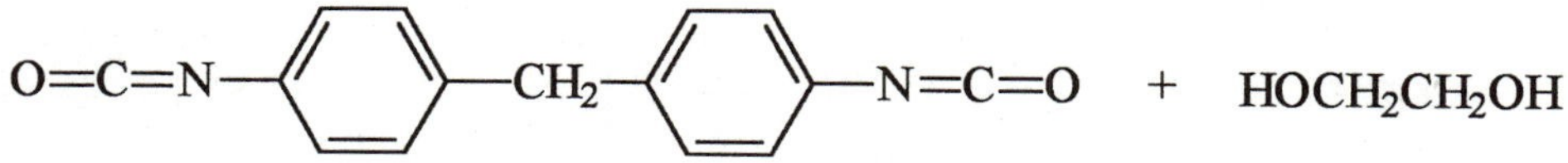 $O{=}C{=}N$–C_6H_4–CH_2–C_6H_4–$N{=}C{=}O$ + $HOCH_2CH_2OH$

23.20

$$\mathrm{Ph\overset{O}{\overset{\|}{C}}O{-}O\overset{O}{\overset{\|}{C}}Ph \longrightarrow 2\ Ph\overset{O}{\overset{\|}{C}}O^{\bullet}}$$

$$\mathrm{Ph\overset{O}{\overset{\|}{C}}O^{\bullet} \quad CH_2{=}\underset{Cl}{\underset{|}{C}}H \longrightarrow Ph\overset{O}{\overset{\|}{C}}O{-}CH_2{-}\underset{Cl}{\underset{|}{\dot{C}}}H \quad CH_2{=}\underset{Cl}{\underset{|}{C}}H}$$

$$\longleftarrow \mathrm{Ph\overset{O}{\overset{\|}{C}}OCH_2\underset{Cl}{\underset{|}{C}}H{-}CH_2{-}\underset{Cl}{\underset{|}{\dot{C}}}H \quad CH_2{=}\underset{Cl}{\underset{|}{C}}H}$$

23.21

$$\mathrm{CH_2{=}\overset{Ph}{\overset{|}{C}}H \quad F_3B{-}\overset{H}{\overset{|}{\ddot{O}}}{}^{\oplus}{-}H \longrightarrow CH_3{-}\overset{Ph}{\overset{|}{C}}{}^{\oplus}H \quad CH_2{=}\overset{Ph}{\overset{|}{C}}H \longrightarrow CH_3\overset{Ph}{\overset{|}{C}}H{-}CH_2{-}\overset{Ph}{\overset{|}{C}}{}^{\oplus}H \quad CH_2{=}\overset{Ph}{\overset{|}{C}}H}$$

$$\downarrow$$

23.22

$$\mathrm{CH_2{=}\overset{O}{\overset{\|}{\overset{COCH_3}{\overset{|}{C}}}}H \quad :\ddot{N}H_2^{\ominus} \longrightarrow H_2N{-}CH_2\overset{COCH_3}{\overset{|}{\ddot{C}}}H^{\ominus} \quad CH_2{=}\overset{COCH_3}{\overset{|}{C}}H \longrightarrow H_2NCH_2\overset{COCH_3}{\overset{|}{C}}HCH_2{-}\overset{COCH_3}{\overset{|}{\ddot{C}}}H^{\ominus} \quad CH_2{=}\overset{COCH_3}{\overset{|}{C}}H}$$

$$\downarrow$$

23.23

a) $CH_2{=}CH_2 \xrightarrow{Cl_2} ClCH_2CH_2Cl \xrightarrow{Ca(OH)_2} CH_2{=}CHCl \xrightarrow[\text{initiator}]{\text{radical}}$

b) benzene $\xrightarrow[AlCl_3]{CH_3Cl}$ p-$CH_3C_6H_4CH_3$ $\xrightarrow{KMnO_4}$ p-$HOOCC_6H_4COOH$ $\xrightarrow[H_2SO_4]{CH_3OH}$ p-$CH_3OC(=O)C_6H_4C(=O)OCH_3$

$CH_2{=}CH_2 \xrightarrow[2)\,NaOH]{1)\,Cl_2,\,H_2O}$ $CH_2{-}CH_2$ (epoxide, O) $\xrightarrow{H_3O^{\oplus}} HOCH_2CH_2OH \xrightarrow[\Delta]{}$

c) $CH_2{-}CH_2$ (epoxide, O) $\xrightarrow[CN^{\ominus}]{HCN} HOCH_2CH_2CN \xrightarrow{H_3O^{\oplus}} CH_2{=}CHCN \xrightarrow[\text{initiator}]{\text{radical}}$

23.24

R· + ~~~$CH_2{-}CH(C_6H_5)$~~~ $\longrightarrow$ ~~~$CH_2{-}\dot{C}(C_6H_5)$~~~ + $CH_2{=}CH(C_6H_5)$ $\longrightarrow$ ~~$\dot{C}H(C_6H_5){-}CH_2{-}C(C_6H_5)$~~ (branch: ~CHCH~) $\xrightarrow{\text{etc.}}$

23.25

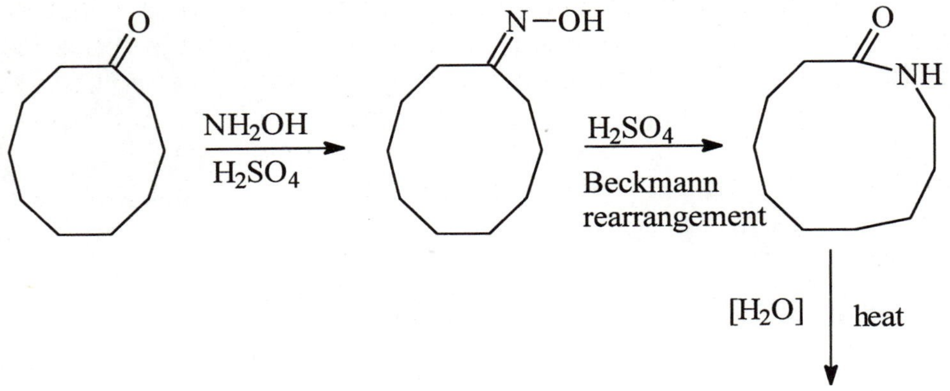

O; N—OH; O; NH

NH_2OH / H_2SO_4

H_2SO_4 / Beckmann rearrangement

$[H_2O]$ heat

23.26

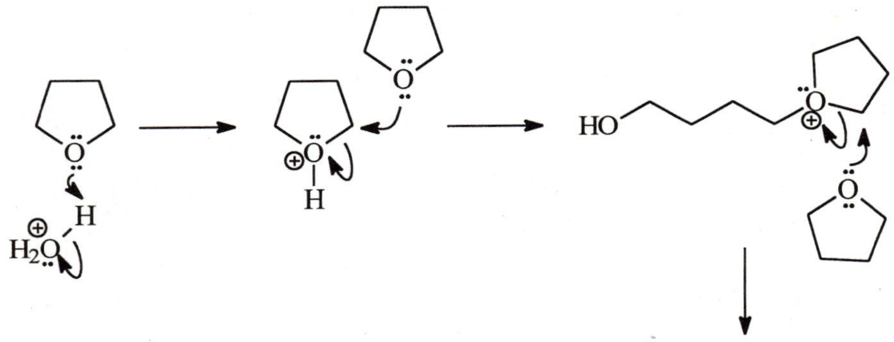

HO; H; H_2O

23.27

$$\text{COCH}_2\text{CHCH}_2\text{OC}$$

n

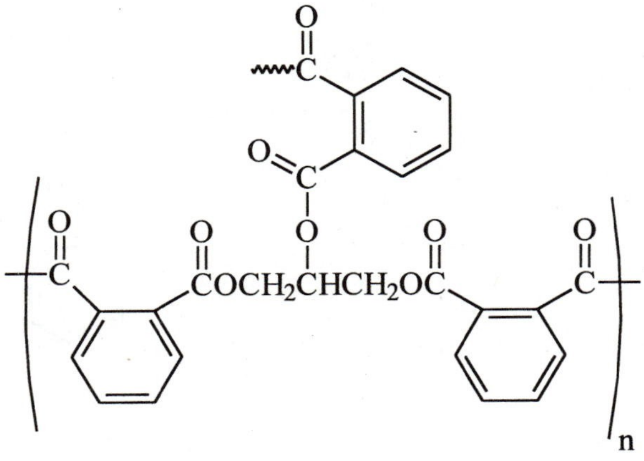

This polymer would be thermosetting because it forms a three dimensional network.

23.28 Radical polymerization is initiated by trace amounts of a radical initiator. Styrene can readily undergo radical chain polymerization in the presence of a radical impurity. BHT is a radical scavenger. It reacts with radicals to produce sterically hindered, resonance stabilized radicals that are not reactive enough to continue the chain. Therefore, addition of BHT prevents the polymerization of styrene due to radical impurities by terminating the chains.

23.29

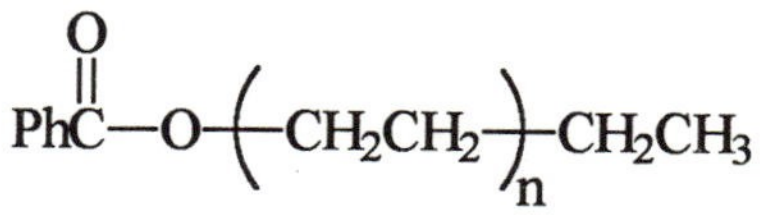

23.30

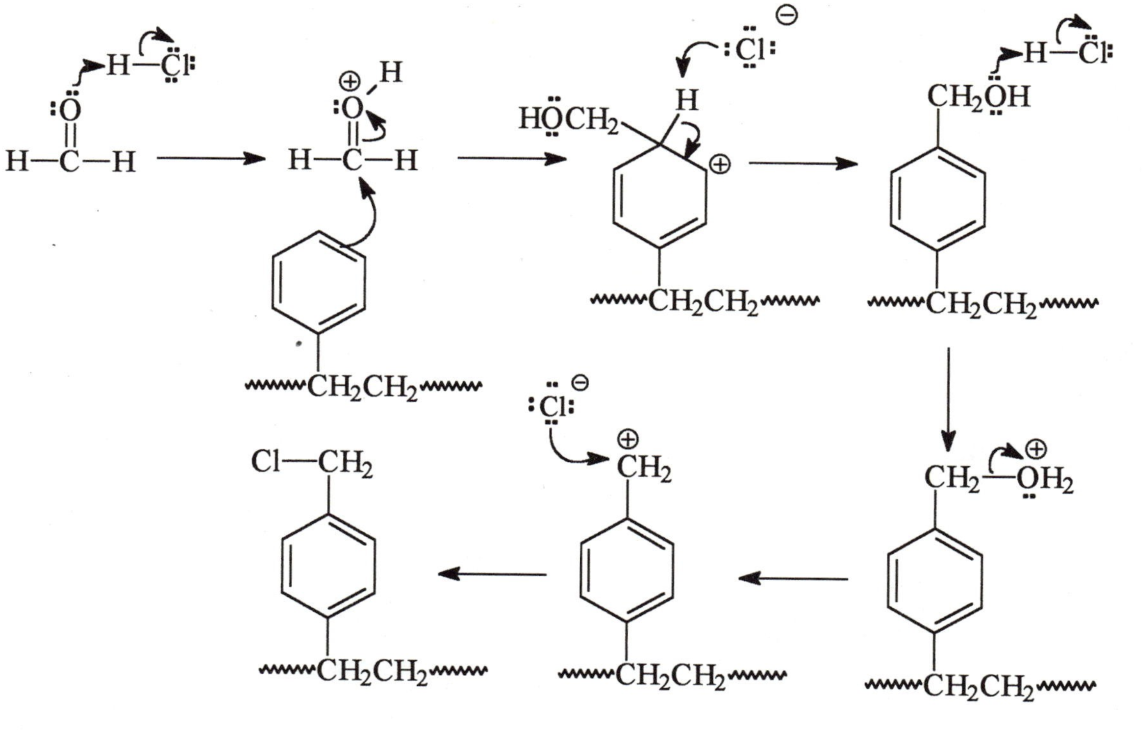

23.31

Benzene $\xrightarrow[\text{AlCl}_3,\ \text{HCl}]{CH_2{=}CH_2}$ $C_6H_5CH_2CH_3$ $\xrightarrow[\text{Cr}_2O_3,\ 600^\circ C]{Fe_2O_3}$ $C_6H_5CH{=}CH_2$ $\xrightarrow{\text{radical polymerization}}$ ~~~CH_2CH_2~~~ (with phenyl substituent)

$\xrightarrow[HCl]{HCHO}$ ~~~CH_2CH_2~~~ (with $C_6H_4CH_2OH$) $\xrightarrow{HCl}$ ~~~CH_2CH_2~~~ (with $C_6H_4CH_2Cl$) $\xrightarrow{(CH_3)_3N}$ ~~~CH_2CH_2~~~ (with $C_6H_4CH_2\overset{\oplus}{N}(CH_3)_3\ Cl^{\ominus}$)

23.32

~~~$C(=O)$–C_6H_4–$C(=\ddot{O}{:})$–OCH_2CH_2O~~~ + $CH_3\ddot{O}{:}^{\ominus}$ $\longrightarrow$ ~~~$C(=O)$–C_6H_4–$C(\ddot{O}{:}^{\ominus})(OCH_3)$–$OCH_2CH_2O$~~~

$\downarrow$

~~~$C(=O)$–C_6H_4–$C(=\ddot{O}{:})$–OCH_3 + ${:}\ddot{O}CH_2CH_2$~~~ $^{\ominus}$

$CH_3\ddot{O}$–H + ${:}\ddot{O}CH_2CH_2$~~~$^{\ominus}$ $\longrightarrow$ $HOCH_2CH_2$~~~

23.33 As T increases, TΔS gets larger. This causes the more disordered state (the more random state) to be more favored, and the rubber band shrinks.

23.34 The polymer of pure vinylidiene chloride is more crystalline because it has no chiral centers. Crystalline polymers are strong and stiff and are not suitable for applications such as thin film wraps. Copolymerizing with vinyl chloride makes the polymer more amorphous, and more suitable for the intended application.

23.35

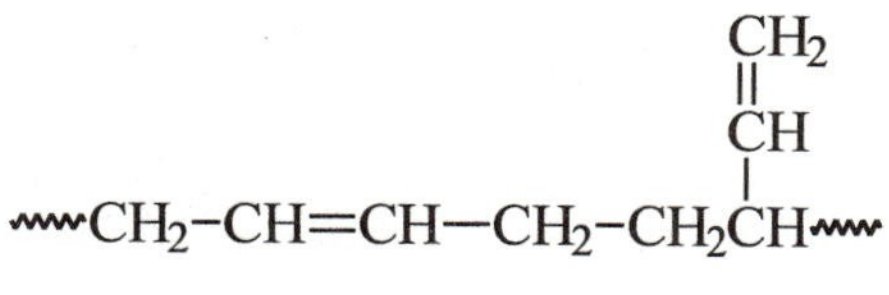

$$\sim\sim CH_2{-}CH{=}CH{-}CH_2{-}CH_2\underset{\displaystyle |}{C}H\sim\sim \quad (\text{CH bears } {-}CH{=}CH_2)$$

This polymer can be either *cis* or *trans*. In addition, its monomers can react by both 1,2- and 1,4- addition.

23.36 Coordination polymerization of ethylene with small amounts of a long chain alkene will produce a less regular polymer because of the random placement of the three-carbon branches.

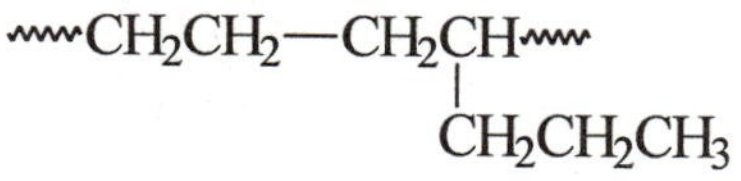

$$\sim\sim CH_2CH_2{-}CH_2\underset{CH_2CH_2CH_3}{\underset{|}{C}H}\sim\sim$$

After completing this chapter, you should be able to:

1. Show the repeat unit for any addition or condensation polymer. Problems 23.2, 23.3, 23.14, 23.15, 23.18, 23.19.

2. Write the mechanisms for the formation of addition polymers that are prepared by radical, anionic, or cationic initiation. Problems 23.1, 23.6, 23.7, 23.8, 23.10, 23.12, 23.20, 23.21, 23.22, 23.26, 23.28, 23.29, 23.30, 23.32.

3. Discuss the structure and stereochemistry of polymers in terms of both regular and irregular features. Problems 23.4, 23.5, 23.11, 23.13, 23.24, 23.35.

4. Discuss how the physical properties of a polymer are related to its structure.
   Problems 23.9, 23.27, 23.34, 23.36.

5. Discuss the chemical properties of a polymer.
   Problem 23.31.

Chapter 24
CARBOHYDRATES

24.1

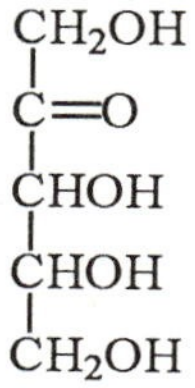

$$
\begin{array}{l}
CH_2OH \\
| \\
C{=}O \\
| \\
CHOH \\
| \\
CHOH \\
| \\
CH_2OH
\end{array}
$$

24.2 An aldopentose has three stereocenters , so there are 2^3 = eight stereoisomeric aldopentoses. A 2-ketopentose has two stereocenters, so there are four stereoisomeric 2-ketopentoses.

24.3

a)

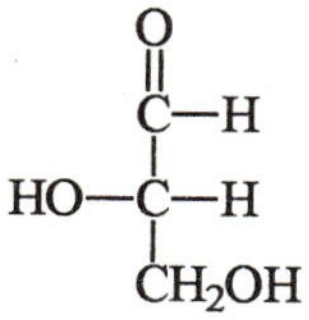

$$
\begin{array}{c}
O \\
\| \\
C{-}H \\
| \\
HO{-}C{-}H \\
| \\
CH_2OH
\end{array}
$$

b)

$$
\begin{array}{c}
O \\
\| \\
C{-}H \\
| \\
H{-}C{-}OH \\
| \\
H{-}C{-}OH \\
| \\
HO{-}C{-}H \\
| \\
HO{-}C{-}H \\
| \\
CH_2OH
\end{array}
$$

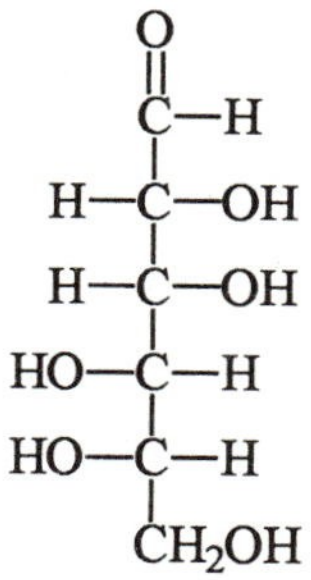

24.4 a) L-erythrose b) L-gulose c) D-altrose

24.5

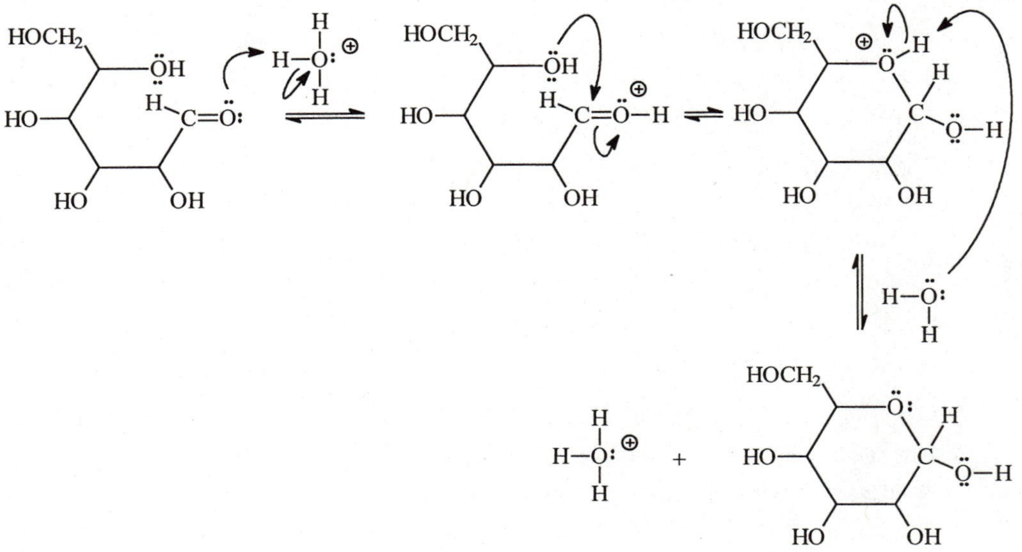

24.6

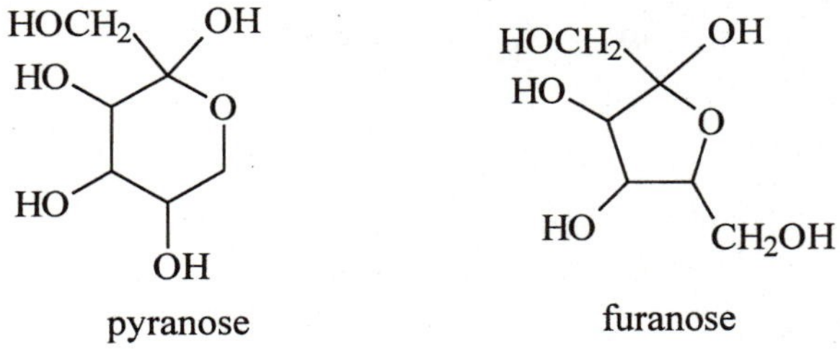

pyranose

furanose

D-Fructose has more of its uncyclized form present at equilibrium than does D-glucose because its ketone carbonyl group is less reactive than the aldehyde carbonyl of glucose.

24.7 The distribution of the different forms in L-glucose is identical to that of D-glucose.

24.8

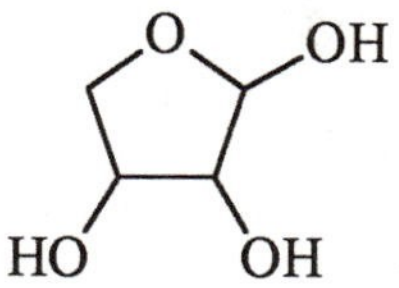

24.9

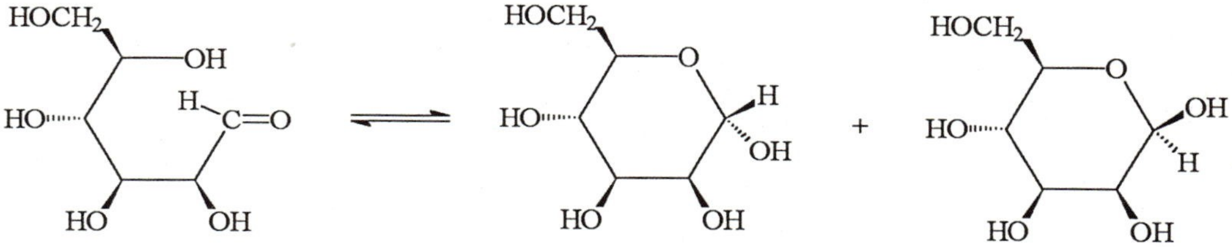

α-D-mannopyranose

β-D-mannopyranose

24.10

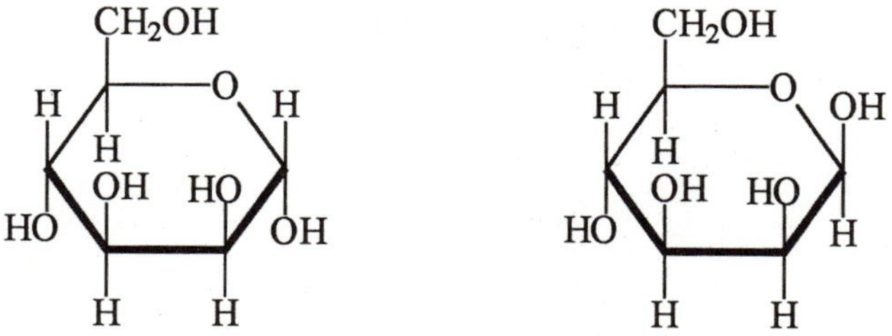

α-D-mannopyranose

β-D-mannopyranose

24.11

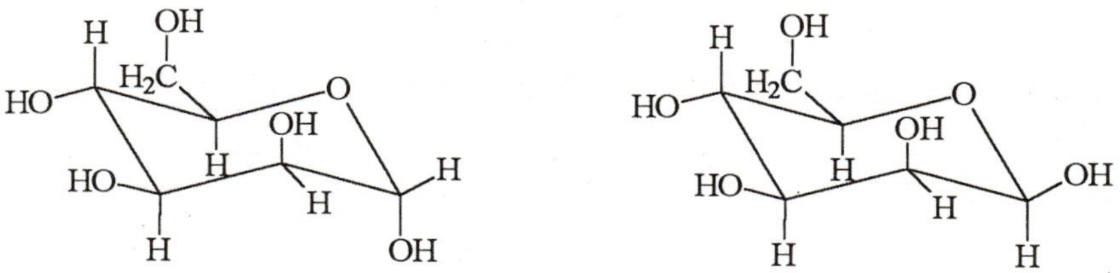

α-D-mannopyranose

β-D-mannopyranose

24.13 67.4% α, 32.6% β.

24.14 Galactaric acid has a plane of symmetry, so it is a is a meso compound.

24.15 On oxidation, D-ribose and D-xylose will give diacids that are meso compounds.

24.16 D and L designations are given to optically active compounds. Xylitol is not given a D designation because it is a meso compound.

24.17 D-allose and D-galactose.

24.18 Under these reaction conditions, α-D-glucopyranose undergoes isomerization to the β-isomer.

24.19

24.20

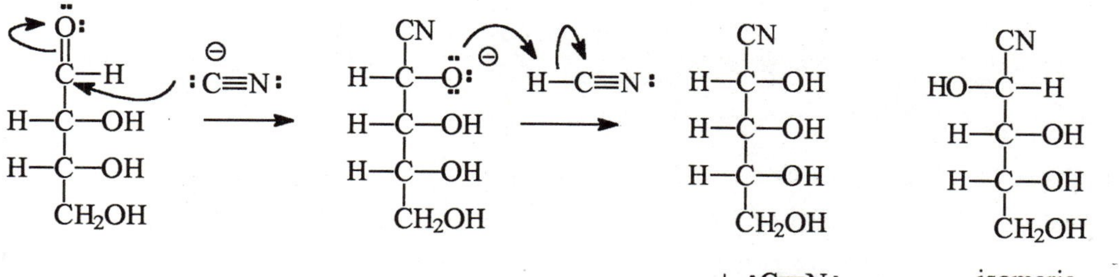

:Ö: ‖ C—H, H—C—OH, H—C—OH, CH_2OH + $:\overset{\ominus}{C}\equiv N:$ ⟶

CN, H—C—Ö:$^{\ominus}$, H—C—OH, H—C—OH, CH_2OH + H—C≡N: ⟶

CN, H—C—OH, H—C—OH, H—C—OH, CH_2OH + $:\underset{\ominus}{C}\equiv N:$

CN, HO—C—H, H—C—OH, H—C—OH, CH_2OH

isomeric product

24.21

O ‖ CH, H—C—OH, H—C—OH, H—C—OH, CH_2OH

D-ribose

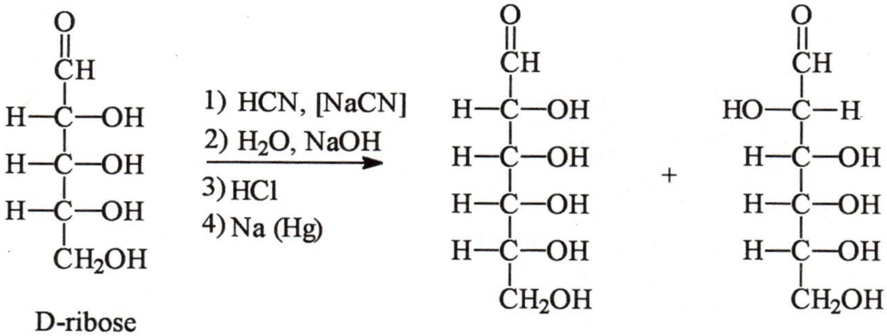

1) HCN, [NaCN]
2) H_2O, NaOH
3) HCl
4) Na (Hg)

⟶

O ‖ CH, H—C—OH, H—C—OH, H—C—OH, H—C—OH, CH_2OH

D-allose

+

O ‖ CH, HO—C—H, H—C—OH, H—C—OH, H—C—OH, CH_2OH

D-altrose

24.22

a) 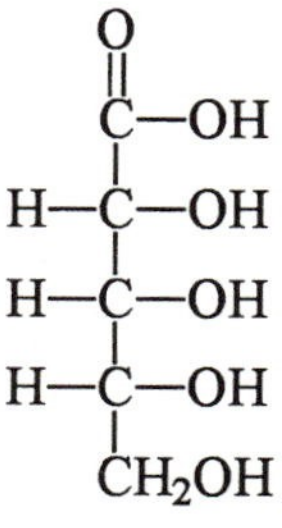 O ‖ C—OH, H—C—OH, H—C—OH, H—C—OH, CH_2OH

b) O ‖ CH, H—C—OH, H—C—OH, HO—C—H, H—C—OH, CH_2OH + O ‖ CH, HO—C—H, H—C—OH, HO—C—H, H—C—OH, CH_2OH

c) CH_2OH, H—C—OH, HO—C—H, H—C—OH, H—C—OH, CH_2OH

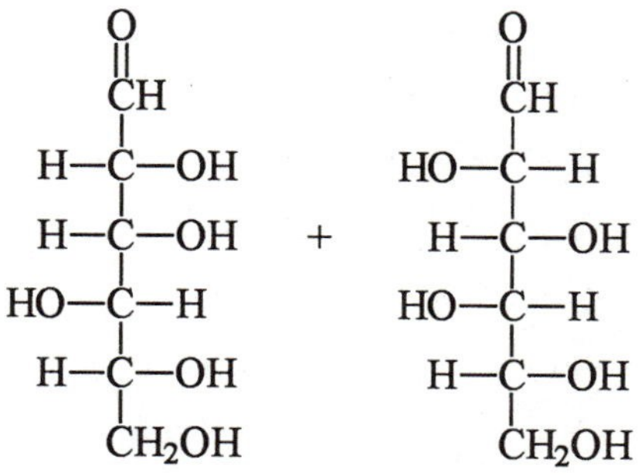

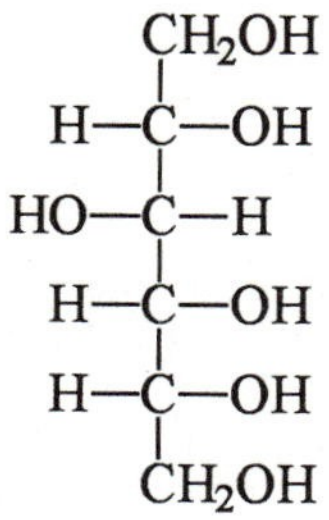

d) [Haworth structures: methyl β-glycopyranoside + methyl α-glycopyranoside; ring substituents CH_2OH, O, H, OH, HO, OCH_3]

CH_2OH / H / H / OH / HO / HO / OCH_3 / H / H / H + CH_2OH / H / H / OH / HO / HO / OCH_3 / H / H / H

e)

$$\begin{array}{c} COOH \\ H-C-OH \\ HO-C-H \\ H-C-OH \\ H-C-OH \\ COOH \end{array}$$

f) [Haworth structures: CH_2OH / H / H / OH / HO / H / OH / H / OH / H + CH_2OH / OH / H / H / OH / HO / H / H / H / OH]

24.24

$$\begin{array}{c} CHO \\ H-C-OH \\ H-C-OH \\ CH_2OH \\ \textbf{A} \end{array} \longrightarrow \begin{array}{c} CHO \\ HO-C-H \\ H-C-OH \\ H-C-OH \\ CH_2OH \\ \textbf{C} \end{array} + \begin{array}{c} CHO \\ H-C-OH \\ H-C-OH \\ H-C-OH \\ CH_2OH \\ \textbf{D} \end{array}$$

$$\begin{array}{c} CO_2H \\ H-C-OH \\ H-C-OH \\ CO_2H \\ \textbf{B} \end{array} \qquad \begin{array}{c} CO_2H \\ HO-C-H \\ H-C-OH \\ H-C-OH \\ CO_2H \\ \textbf{E} \end{array} \qquad \begin{array}{c} CO_2H \\ H-C-OH \\ H-C-OH \\ H-C-OH \\ CO_2H \\ \textbf{F} \end{array}$$

(A → B, C → E, D → F)

24.25 Maltose and cellobiose both have a hemiacetal group that is in equilibrium with an aldehyde group in aqueous solution. It is the aldehyde group that gives a positive test for a reducing sugar. Sucrose does not have a hemiacetal group (it has two acetal groups) so there is no aldehyde group present at equilibrium in a solution of sucrose.

24.26

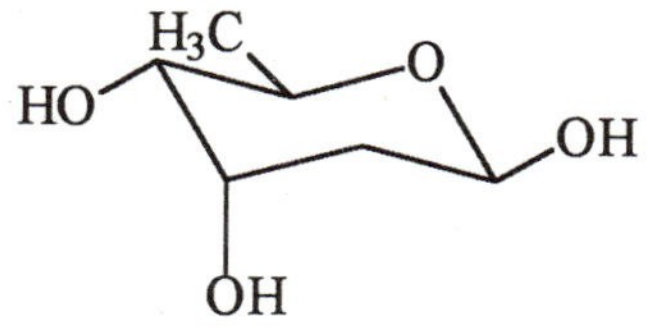

There are no hydroxy groups present on C-2 and C-6.

24.27

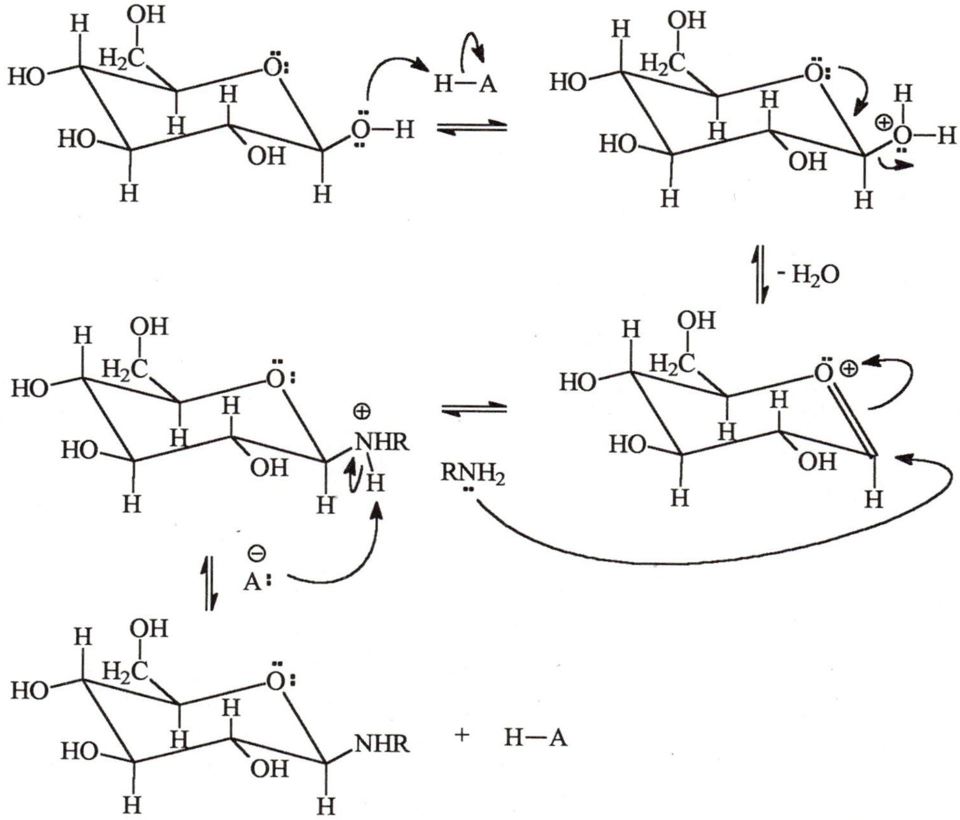

24.28

a)

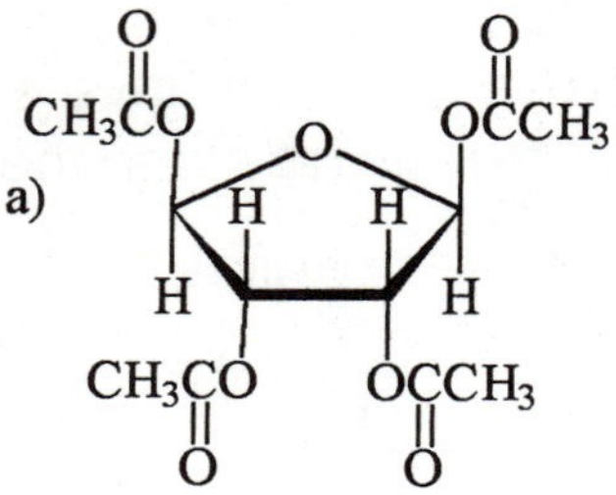

b)

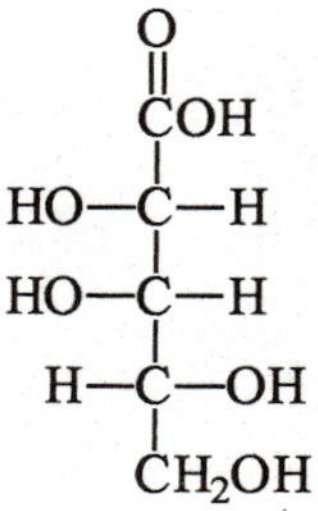

c)

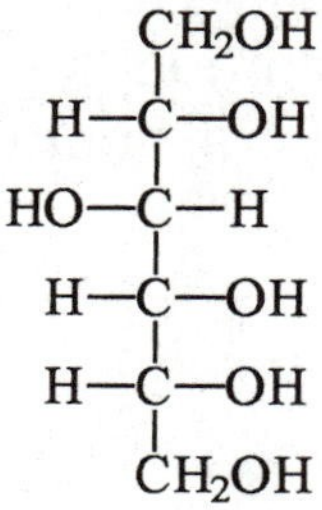

d) 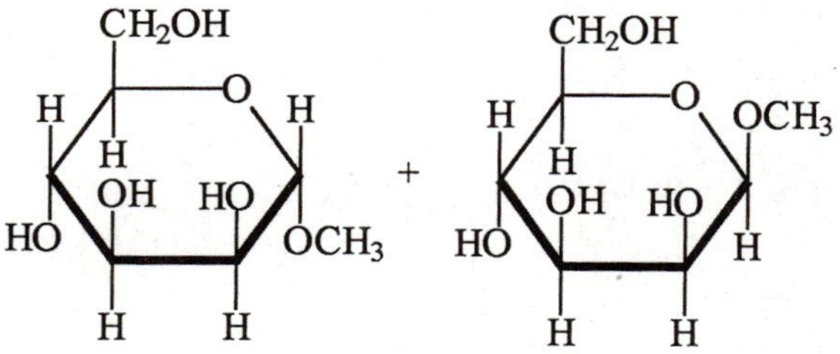 +

e) 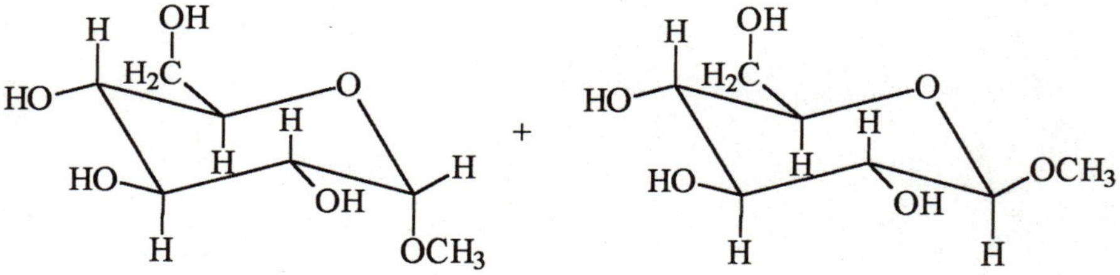 +

f) 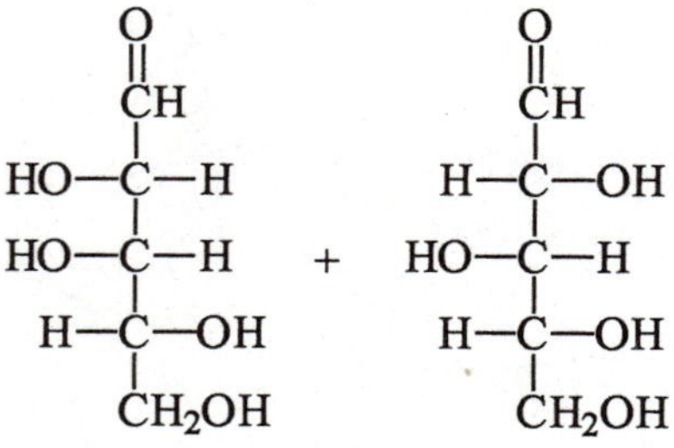 +

g) 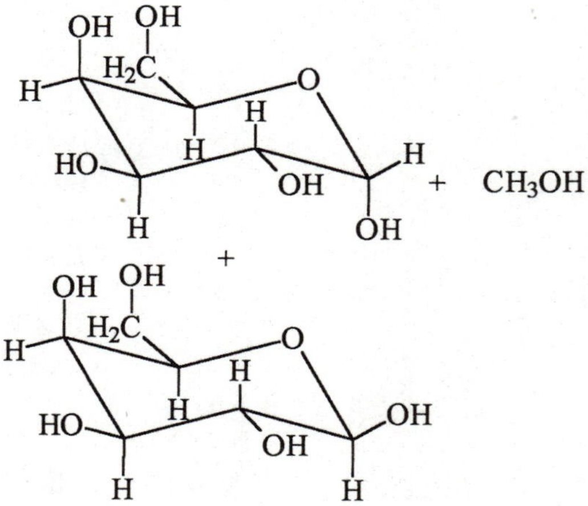 + CH_3OH

+

24.29

CH₂OH H—A O :ÖH H H OH HO HO H H H ⇌ CH₂OH O ⊕ :ÖH₂ H H OH HO HO H H H ⇌ CH₂OH Ö H H OH HO HO ⊕ H H H

↕

CH₂OH O OCH₂CH₃ H H OH HO HO H H H ⇌ CH₂OH :A⊖ H Ö :OCH₂CH₃ ⊕ H H OH HO HO H H H ⇌ HÖCH₂CH₃ CH₂OH Ö⊕ H H OH HO HO H H H

24.30

HOCH₂ ÖH HO C=Ö: H H—O: ⊕ H HO HO OH ⇌ HOCH₂ ÖH HO C=Ö—H ⊕ HO HO OH ⇌ HOCH₂ ⊕ O H :OH C Ö—H :Ö—H H HO HO OH

⇌

HOCH₂ Ö: C—Ö H ⊕ HO HO OH ⇌ HOCH₂ Ö: :Ö—H C Ö—H ⊕ H HO HO OH ⇌ HOCH₂ Ö: :Ö—H C Ö—H HO HO OH H H—O: ⊕ H

↕

HOCH₂ Ö: C=O ⊕ H :OH₂ HO HO OH ⇌ HOCH₂ Ö: C=O HO HO OH

24.31 Methyl-α-D-glucoside is an acetal and acetals are very stable to nucleophiles and bases. However, acetals react with acids to form aldehydes.

24.32 Lactose is a disaccharide consisting of D-galactose and D-glucose joined by a 1,4-β-glycosidic bond. The anomeric carbon on the D-glucose unit of lactose has a hemiacetal group, therefore it exhibits mutarotation in basic solution. Sucrose does not have a hemiacetal group, so it does not exhibit mutarotation in basic solution.

24.33 Let x = the decimal fraction of α-D-galactopyranose that is present at equilibrium.

$$+80.2 = x(+150.7) + (1-x)(+52.8)$$

$$x = 0.279$$

Therefore 27.9% of the α-isomer and 72.1% of the β-isomer are present at equilibrium.

24.34

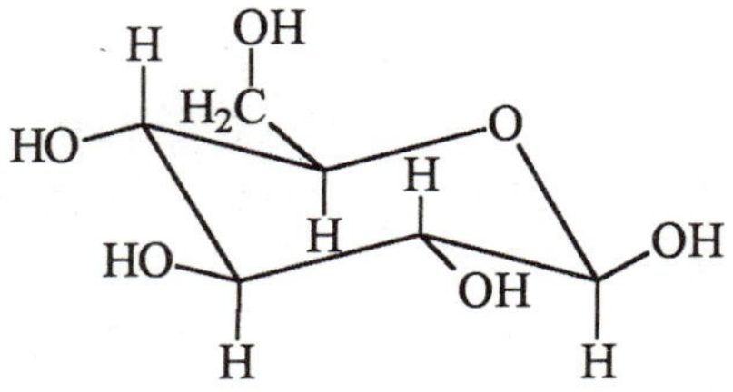

β-D-glucopyranose

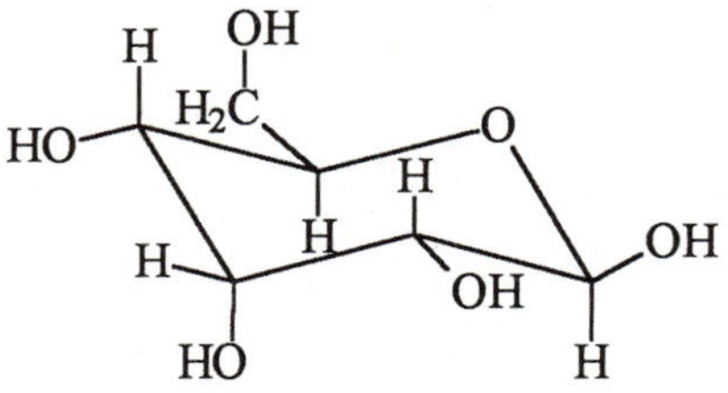

β-D-allopyranose

All substituents are equatorial in the stable chair conformation of β-D-glucopyranose while one hydroxy group is axial in the stable chair conformation of β-D-allopyranose. Therefore β-D-allopyranose is less stable than β-D-glucopyranose.

24.35 The reduction product of β-D-glucose is an enantiomer of the reduction product of β-D-gulose.

24.36

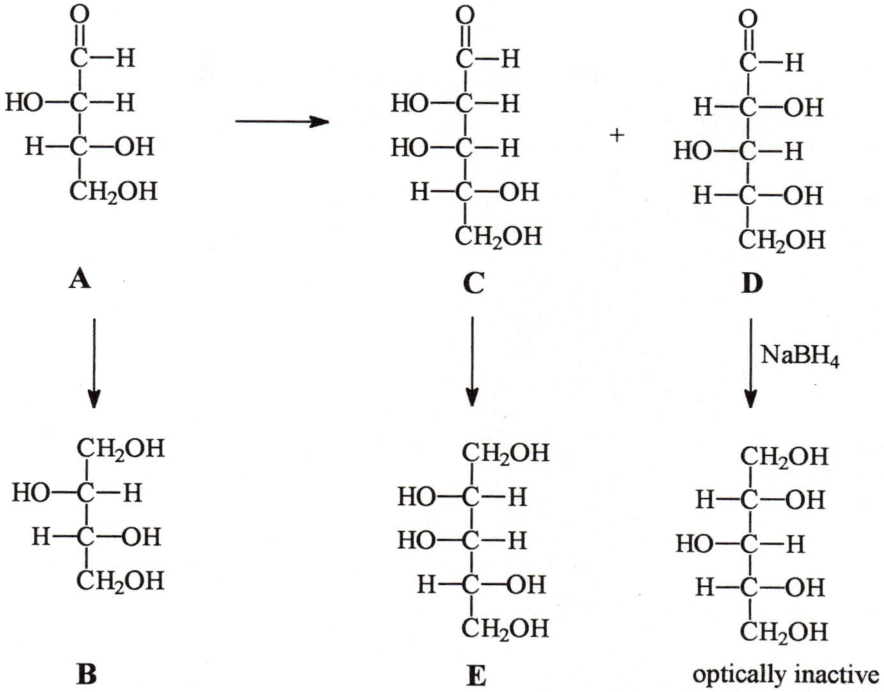

A → C + D

A → B

C → E

D → ($NaBH_4$) → optically inactive

The reduction product of D has a plane of symmetry, so it will not rotate plane polarized light.

24.37

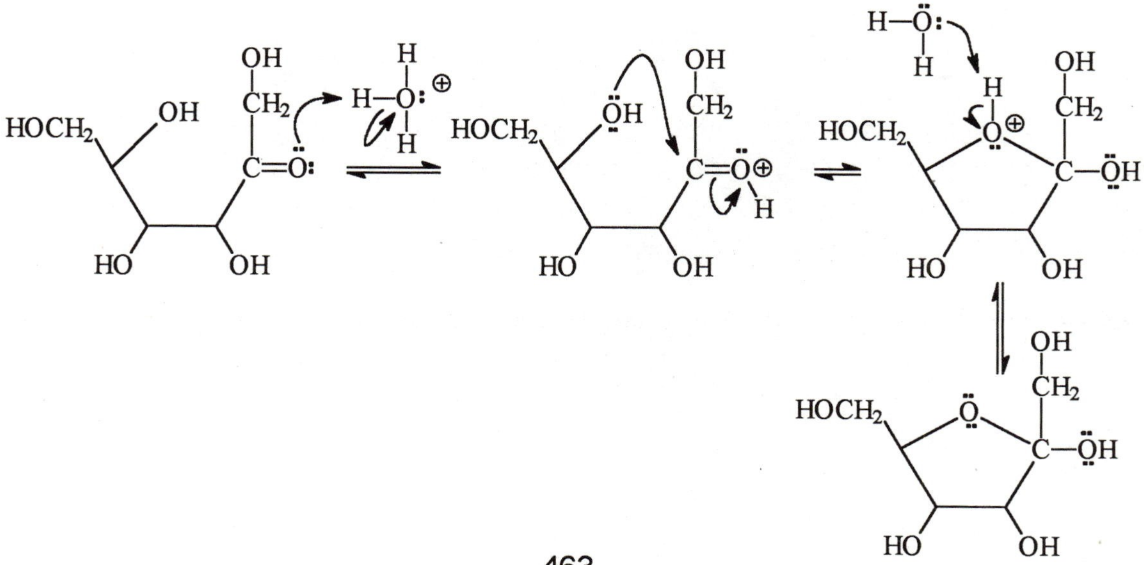

24.38

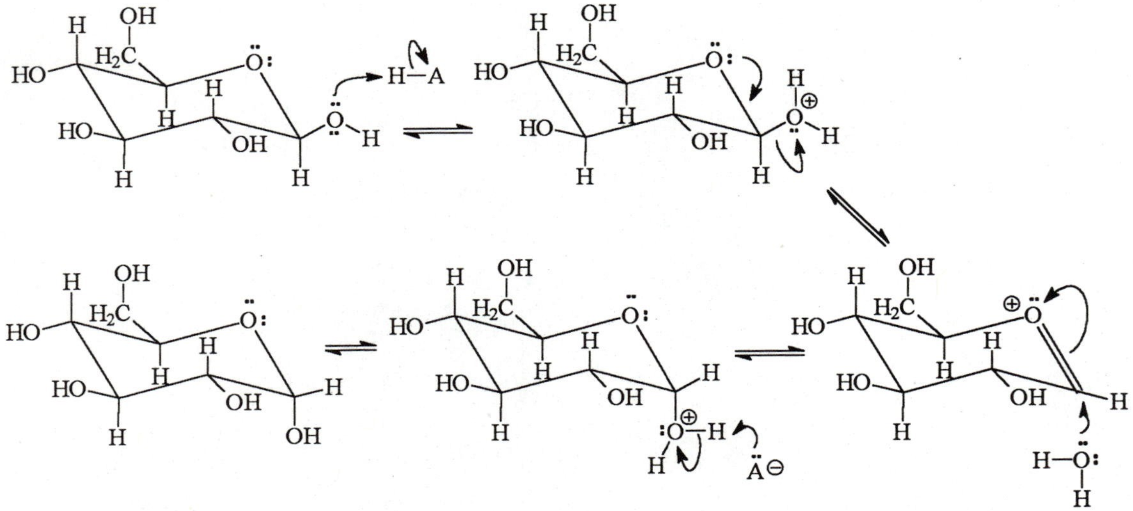

24.39

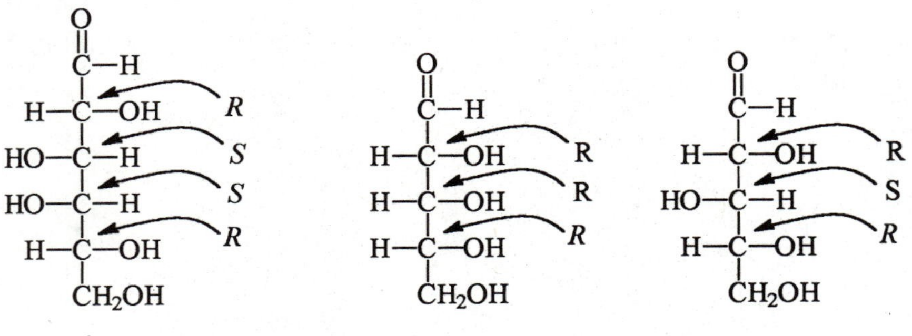

D-galactose

D-ribose

D-xylose

24.40

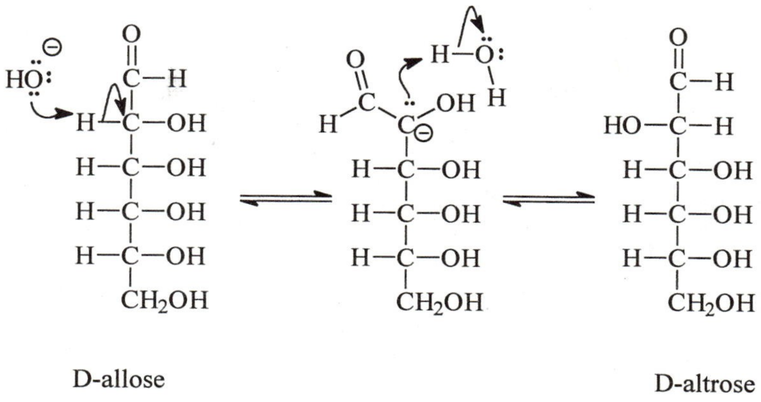

D-allose ⇌ [enolate intermediate] ⇌ D-altrose

D-allose

D-altrose

24.41 Three successive Kiliani-Fisher syntheses starting from D-glyceraldehyde produce the D-isomers of all eight aldohexoses. These are all D-isomers because the configuration at C-2 of glyceraldehyde is maintained throughout all of the reactions and this carbon ends up as C-5 of all of the aldohexoses. Comparison of the glucose thus synthesized with natural glucose shows them to be the same enantiomer, so natural glucose is D.

24.42 D-galactose can be converted to D-talose by treatment with a base. The reaction proceeds through a planar enolate anion.

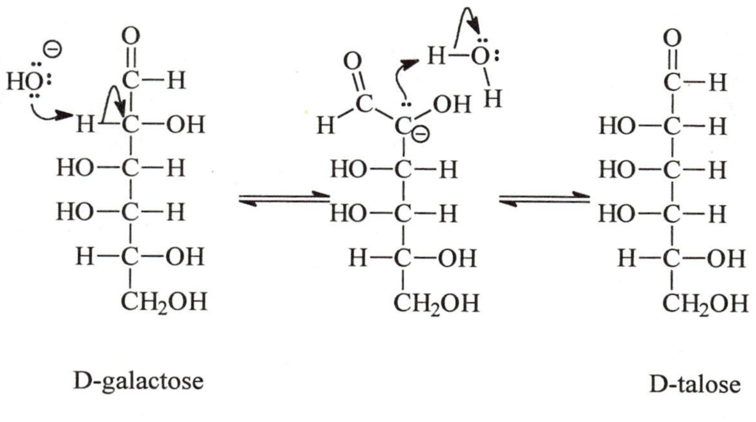

D-galactose

D-talose

24.43

```
      O
      ‖
      C—H
      |
  HO—C—H
      |
  HO—C—H
      |
   H—C—OH
      |
      CH2OH

      X
```

```
      O
      ‖
      C—H
      |
  HO—C—H
      |
   H—C—OH
      |
      CH2OH

      Y
```

24.44

```
      O                       O                     O
      ‖                       ‖                     ‖
      C—H                     C—H                   C—H
      |                       |                     |
   H—C—OH     ——→        HO—C—H          +      H—C—OH
      |                       |                     |
   H—C—OH                  H—C—OH               H—C—OH
      |                       |                     |
   H—C—OH                  H—C—OH               H—C—OH
      |                       |                     |
      CH2OH                H—C—OH               H—C—OH
                              |                     |
                              CH2OH                 CH2OH

      A                       C                     D
      |                       |                     |
      ↓                       ↓                     ↓

      COOH                    COOH                  COOH
      |                       |                     |
   H—C—OH                 HO—C—H                H—C—OH
      |                       |                     |
   H—C—OH                  H—C—OH               H—C—OH
      |                       |                     |
   H—C—OH                  H—C—OH               H—C—OH
      |                       |                     |
      COOH                 H—C—OH               H—C—OH
                              |                     |
                              COOH                  COOH

      B                       E                     F
```

24.45

HOCH$_2$ ... H—A ... HO ... HO ... OH ... O—H ⇌ ... ⇌ ... HOCH$_2$ ... OH ... HO ... H—C=O$^{\oplus}$—H ... HO ... OH

H—O:—H ... OH ... HOCH$_2$ ... OH ... HO ... OH ⇌ ... OH ... HOCH$_2$ ... O$^{\oplus}$—H ... OH ... HO ... OH

24.46

$^{\ominus}$B: ... H—C(H)—OH, C=O, $CH_2OPO_3^{2\ominus}$ ⇌ :C$^{\ominus}$(H)—OH, C=O:, $CH_2OPO_3^{2\ominus}$ ⟷ C(H)—OH, C—O:$^{\ominus}$, $CH_2OPO_3^{2\ominus}$ ... H—B

H—C(H)=O, H—C—OH, $CH_2OPO_3^{2\ominus}$ ⇌ B—H ... C(H)=O, :C$^{\ominus}$—OH, $CH_2OPO_3^{2\ominus}$ ⟷ C(H)—O:$^{\ominus}$, C—OH, $CH_2OPO_3^{2\ominus}$ ⇌ C(H)—O—H ... B$^{\ominus}$, C—OH, $CH_2OPO_3^{2\ominus}$

24.47

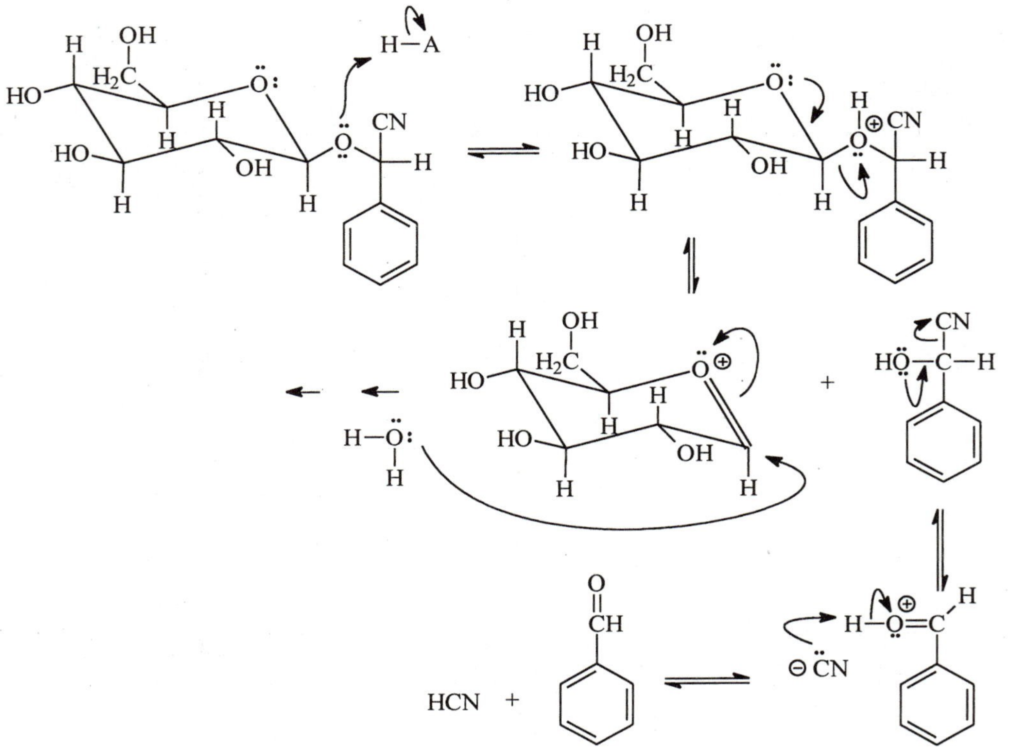

24.48 The two α-D-glucopyranose units in trehalose are connected by a 1,1'-glycosidic linkage. Trehalose is not a reducing sugar because there is no hemiacetal group in this structure.

24.49 This is an aldol condensation.

$$HO^{-} + \begin{array}{c} CH_2OH \\ | \\ C{=}O \\ | \\ H{-}C{-}OH \\ | \\ H \end{array} \rightleftharpoons \begin{array}{c} CH_2OH \\ | \\ C{=}O \\ | \\ {:}\overset{\ominus}{C}{-}OH \\ | \\ H \end{array} + \begin{array}{c} H \\ | \\ C{=}O \\ | \\ H{-}C{-}OH \\ | \\ CH_2OH \end{array} \rightleftharpoons \begin{array}{c} CH_2OH \\ | \\ C{=}O \\ | \\ HO{-}C{-}H \\ | \\ H{-}C{-}O^{\ominus} \\ | \\ H{-}C{-}OH \\ | \\ CH_2OH \end{array} + H_2O$$

$$\rightleftharpoons \begin{array}{c} CH_2OH \\ | \\ C{=}O \\ | \\ HO{-}C{-}H \\ | \\ H{-}C{-}OH \\ | \\ H{-}C{-}OH \\ | \\ CH_2OH \end{array}$$

24.50

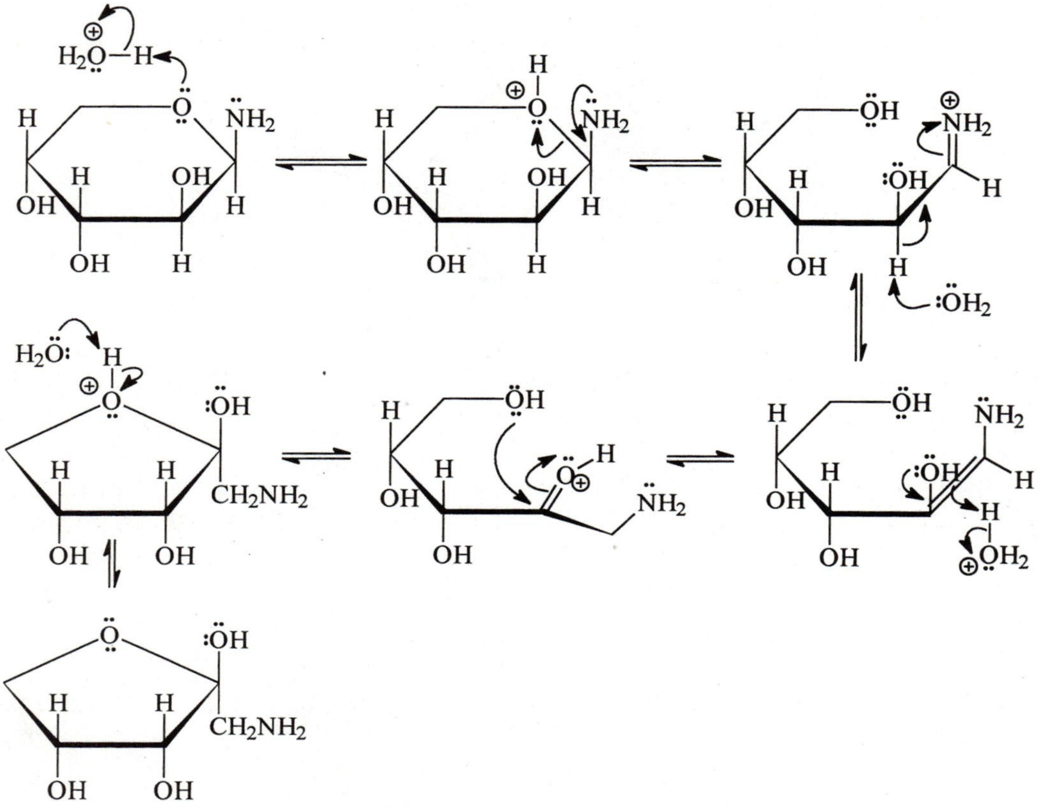

$H_2O^{\oplus}$–H

$\rightleftharpoons$ $\rightleftharpoons$ $\rightleftharpoons$ $\rightleftharpoons$ $\rightleftharpoons$

:OH_2 H_2O: CH_2NH_2 NH_2

24.51

D-talose:

	O=C–H
HO–	C –H
HO–	C –H
HO–	C –H
H–	C –OH
	CH_2OH

$\xrightarrow{NaBH_4}$

	CH_2OH
HO–	C –H
HO–	C –H
HO–	C –H
H–	C –OH
	CH_2OH

$\xleftarrow{NaBH_4}$

D-altrose:

	O=C–H
HO–	C –H
H–	C –OH
H–	C –OH
H–	C –OH
	CH_2OH

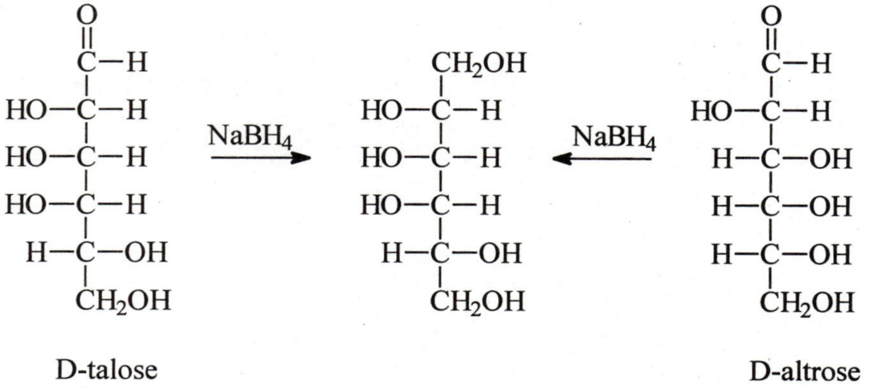

24.52 The hydrogen on the carbon 2 is equatorial in the stable chair conformations of the stereoisomers of D-mannopyranose. Therefore, the anomer configuration of D-mannopyranose cannot be determined by the NMR method described in the Elaboration.

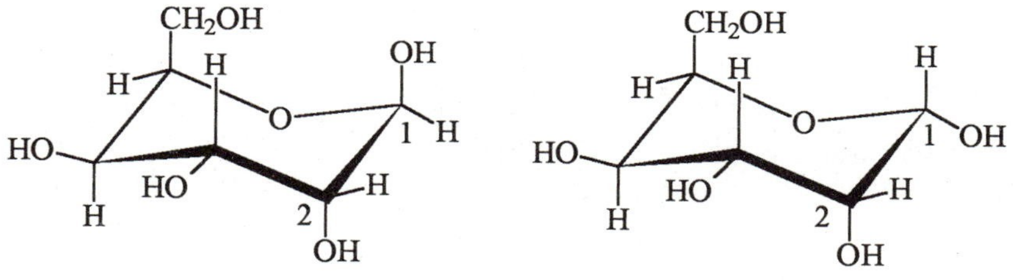

α-D-mannopyranose β-D-mannopyranose

24.53 The presence of a carbonyl group is readily apparent in an IR spectrum. Carbonyl groups have an absorption band in the general region of 1700 cm^{-1}. The open chain form of D-glucose has a aldehyde group and the cyclic structure does not. The absence of any absorption in the carbonyl region of the IR spectrum of D-glucose indicates that the open chain form is not present and supports the cyclic structure.

After completing this chapter, you should be able to:

1. Show the general structures for carbohydrates including the variations that occur.
   Problem 24.1.

2. Discuss the stereochemistry of carbohydrates, including the use of D or L to designate absolute stereochemistry.
   Problems 24.2, 24.3, 24.4, 24.39.

3. Understand the cyclization of monosaccharides to form pyranose and furanose rings.
   Problems 24.6, 24.7, 24.8, 24.9, 24.10, 24.11, 24.31, 24.34.

4. Show the common reactions of monosaccharides that were presented in this chapter: oxidation with nitric acid; oxidation with bromine; reduction with sodium borohydride; esterification; glycoside formation; and the Kiliani-Fischer synthesis.
   Problems 24.14, 24.15, 24.16, 24.17, 24.18, 24.21, 24.22, 24.28, 24.35, 24.41, 24.51.

5. Understand Fischer's structure proof for glucose and apply this type of reasoning to other stereochemical problems.
   Problems 24.23, 24.24, 24.36, 24.43, 24.44.

6. Understand the general structural features of disaccharides and polysaccharides.
   Problems 24.25, 24.26, 24.32, 24.48.

7. Show the mechanisms of these reactions where appropriate.
   Problems 24.5, 24.19, 24.20, 24.27, 24.29, 24.30, 24.37, 24.38, 24.40, 24.42, 24.45, 24.47, 24.49, 24.50.

Chapter 25
AMINO ACIDS, PEPTIDES, AND PROTEINS

25.1 isoleucine and threonine

25.2 The positively charged nitrogen acts as an inductive electron withdrawing group, and therefore makes the carboxylic acid group a stronger acid.

25.3 The ester is the stronger acid because the inductive electron-withdrawing effect of the ester group increases the acid strength of the ammonium group. The carboxylate anion is an electron-donating group and decreases the acid strength of the ammonium group in the amino acid.

25.4 The inductive electron withdrawing effect of the positively charged nitrogen, which increases acid strength, is stronger at the main carboxylic acid group because the distance separating the groups is smaller. Inductive effects decrease rapidly with distance.

25.5 The carboxylic acid group in the side chain of glutamic acid is farther from the electron-withdrawing positive nitrogen than is the case with aspartic acid.

25.6

a) $\overset{\oplus}{N}H_2$ (ring), $CO_2^{\ominus}$

c) $^{\ominus}S{-}CH_2\underset{}{C}H(\overset{\oplus}{N}H_3){-}CO_2^{\ominus}$

d) $\overset{\oplus}{H_2N}{=}C(NH_2)NHCH_2CH_2CH_2CH(\overset{\oplus}{N}H_3)CO_2^{\ominus}$

25.7

a) $H_3\overset{\oplus}{N}CH_2CH_2CH_2CH_2CH(\overset{\oplus}{N}H_3)CO_2H$ $\quad$ $H_3\overset{\oplus}{N}CH_2CH_2CH_2CH_2CH(\overset{\oplus}{N}H_3)CO_2^{\ominus}$

$H_3\overset{\oplus}{N}CH_2CH_2CH_2CH_2CH(NH_2)CO_2^{\ominus}$ $\quad$ $H_2NCH_2CH_2CH_2CH_2CH(NH_2)CO_2^{\ominus}$

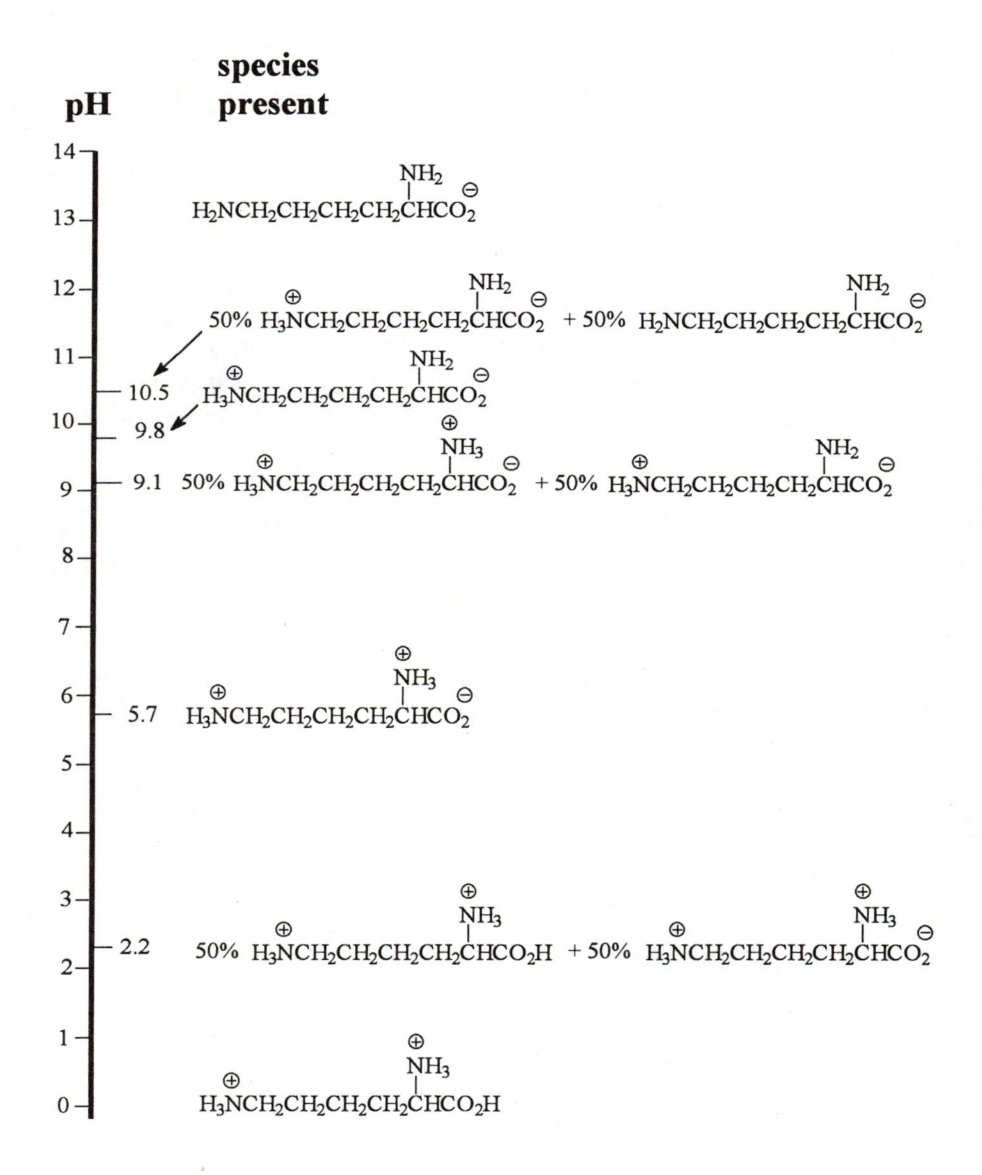

25.8

a) 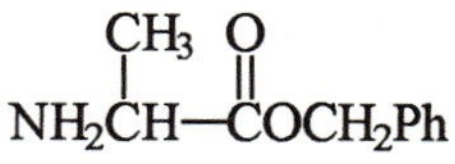 $NH_2CH(CH_3)-C(=O)OCH_2Ph$

b) $PhC(=O)-NHCH(CH_2CH(CH_3)CH_3)-C(=O)OH$

c) $CH_3C(=O)-NHCH(CH_3)-C(=O)OH$

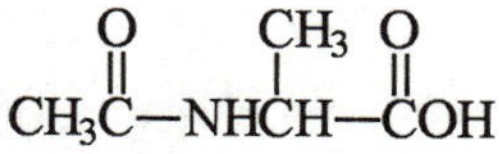

25.9

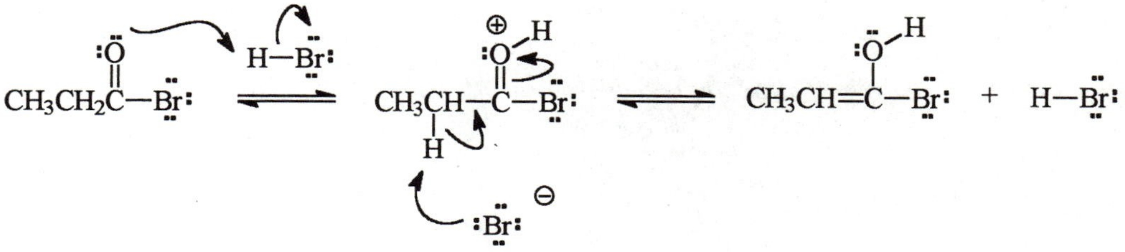

$$CH_3CH_2C(=\ddot{O}:)-\ddot{B}r: + H-\ddot{B}r: \rightleftharpoons CH_3CH(H)-C(=\overset{\oplus}{O}H)-\ddot{B}r: + :\ddot{B}r:^{\ominus} \rightleftharpoons CH_3CH=C(\ddot{O}H)-\ddot{B}r: + H-\ddot{B}r:$$

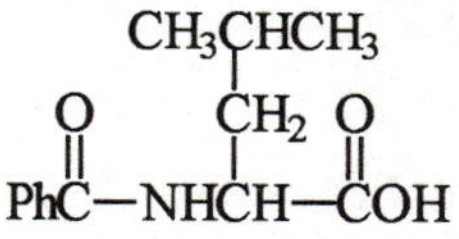

25.10

a) 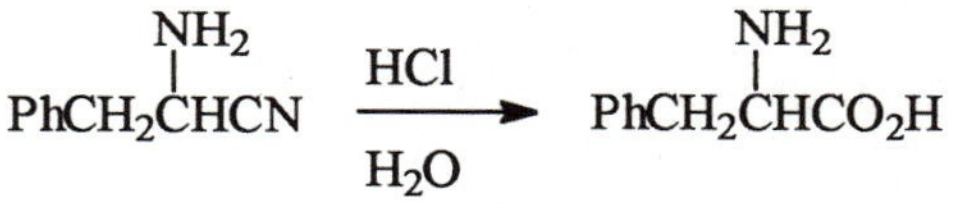 $$PhCH_2CH(NH_2)CN \xrightarrow[H_2O]{HCl} PhCH_2CH(NH_2)CO_2H$$

b) 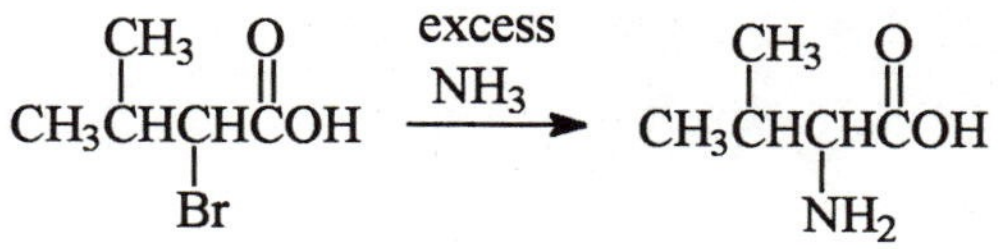 $$CH_3CH(CH_3)CH(Br)C(=O)OH \xrightarrow{\text{excess } NH_3} CH_3CH(CH_3)CH(NH_2)C(=O)OH$$

c) $$\text{PhthN}-C(CO_2Et)_2-CH_2CH(CH_3)CH_3 \xrightarrow[H_2O,\ \Delta]{HCl} CH_3CH(CH_3)CH_2CH(NH_2)CO_2H$$

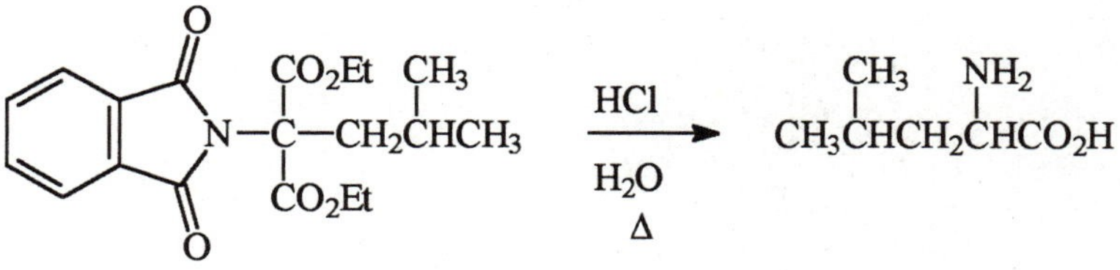

25.11

a) 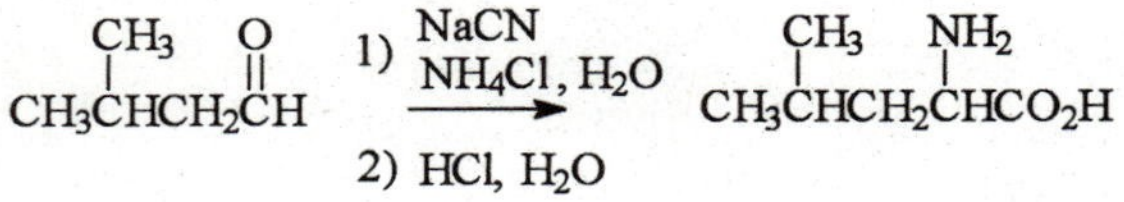 $CH_3CH(CH_3)CH_2CHO$ $\xrightarrow[\text{2) HCl, } H_2O]{\text{1) NaCN, } NH_4Cl\text{, } H_2O}$ $CH_3CH(CH_3)CH_2CH(NH_2)CO_2H$

b) 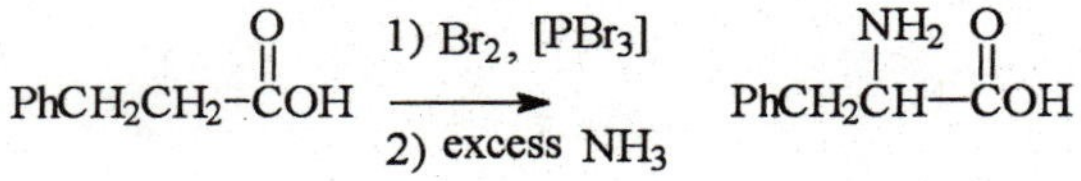 $PhCH_2CH_2-C(=O)OH$ $\xrightarrow[\text{2) excess } NH_3]{\text{1) } Br_2\text{, [}PBr_3\text{]}}$ $PhCH_2CH(NH_2)-C(=O)OH$

c) $Br-CH(CO_2Et)_2$ + phthalimide anion → N-phthalimido$-CH(CO_2Et)_2$

1) NaOEt 2) 3-(bromomethyl)indole (CH_2Br) 3) HCl, H_2O, Δ

→ 3-indolyl$-CH_2CH(NH_2)CO_2H$ (N–H indole)

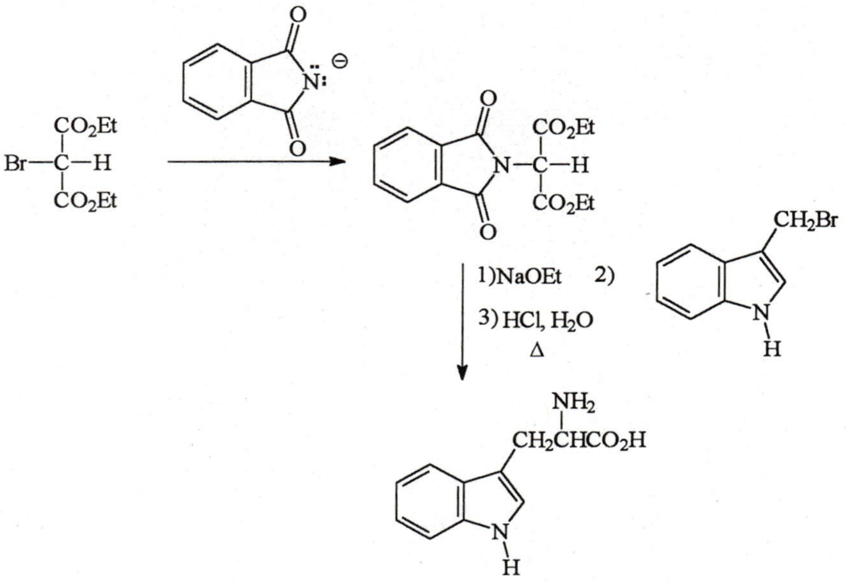

25.13

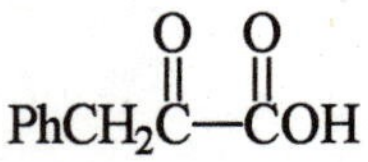

$PhCH_2C(=O)-C(=O)OH$

25.14

$NH_2CH(CH_2Ph)-C(=O)-NHCH(CH(CH_3)CH_3)-C(=O)-NHCH(CH_2CO_2H)-C(=O)OH$

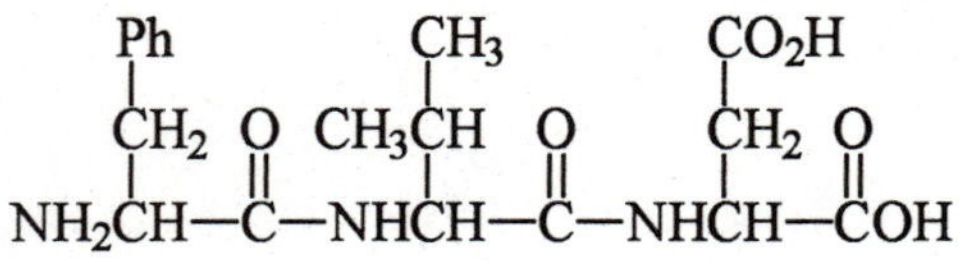

25.15 The amino acids are leucine, cysteine, tyrosine, and glutamic acid. Leu-Cys-Tyr-Glu.

25.16 The amide groups in the side chains of asparagine and glutamine residues in a peptide are cleaved under the same conditions that hydrolyze the amide bonds of the peptide. As a result, aspartic acid and glutamic acid are isolated instead.

25.17

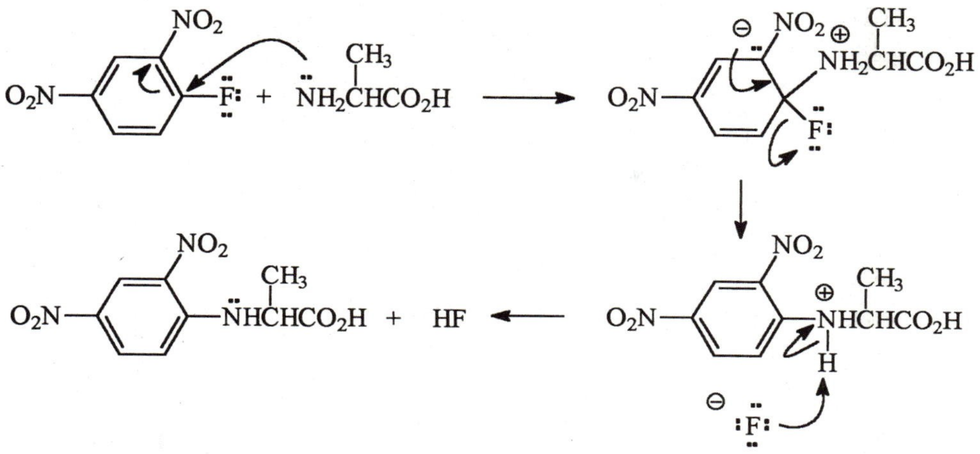

NO_2 CH_3
O_2N—(benzene ring)—$\ddot{F}$: + $\ddot{N}H_2\overset{|}{C}HCO_2H$ ⟶ O_2N—(ring)$^{\ominus}$ NO_2 $^{\oplus}$ CH_3 $NH_2\overset{|}{C}HCO_2H$ $\ddot{F}$:

↓

NO_2 CH_3 NO_2 $\oplus$ CH_3
O_2N—(ring)—$\ddot{N}H\overset{|}{C}HCO_2H$ + HF ⟵ O_2N—(ring)—$NH\overset{|}{C}HCO_2H$ with H
$\ominus$:$\ddot{F}$:

This is a nucleophilic aromatic substitution proceeding via the addition-elimination mechanism (see Section 18.12). The nitro groups are necessary to stabilize the intermediate with the negative charge in the benzene ring.

25.18

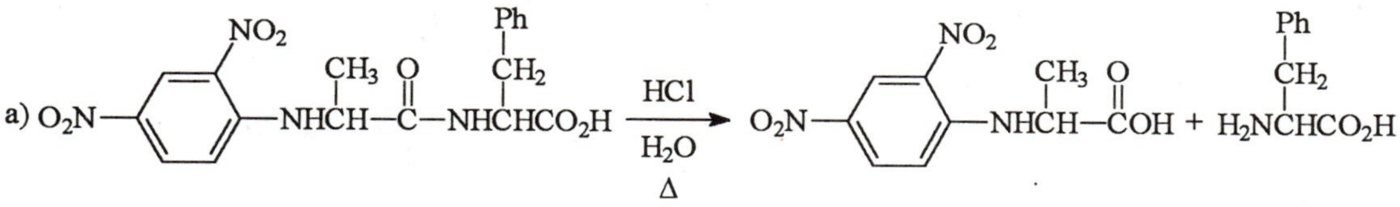

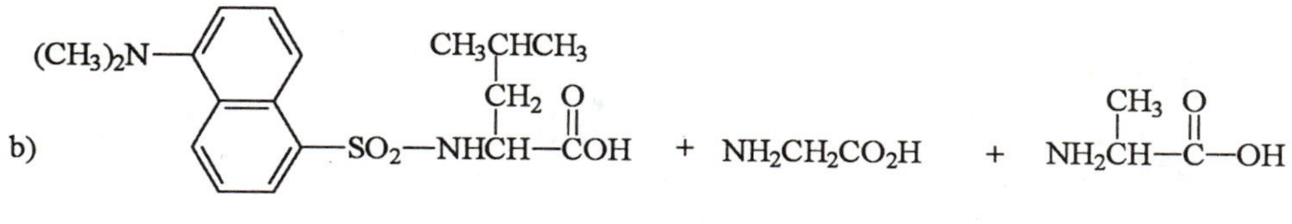

a) O_2N—(ring with NO_2)—$NH\overset{CH_3}{\overset{|}{C}}H$—$\overset{O}{\overset{||}{C}}$—$NH\overset{Ph-CH_2}{\overset{|}{C}}HCO_2H$ $\xrightarrow[H_2O,\ \Delta]{HCl}$ O_2N—(ring with NO_2)—$NH\overset{CH_3}{\overset{|}{C}}H$—$\overset{O}{\overset{||}{C}}OH$ + $H_2N\overset{Ph-CH_2}{\overset{|}{C}}HCO_2H$

b) $(CH_3)_2N$—(naphthalene)—SO_2—$NH\overset{CH_3CHCH_3-CH_2}{\overset{|}{C}}H$—$\overset{O}{\overset{||}{C}}OH$ + $NH_2CH_2CO_2H$ + $NH_2\overset{CH_3}{\overset{|}{C}}H$—$\overset{O}{\overset{||}{C}}$—OH

c)

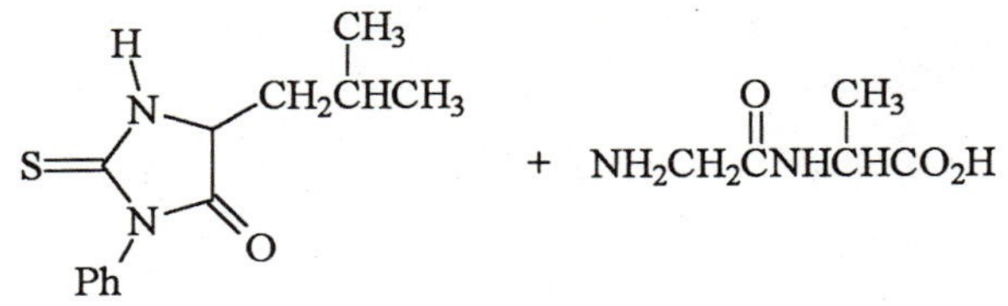

H
CH3
N
CH2CHCH3
S=
N
O
Ph
+ NH2CH2CNHCHCO2H (O above C; CH3 above CH)

$NH_2CH_2\overset{O}{\overset{\|}{C}}NH\overset{CH_3}{\overset{|}{C}}HCO_2H$

d)

NO2
NO2
NH
NO2
(CH2)4 O
O2N– –NHCH—COH + NH2CHCO2H (CH3 above CH)

$O_2N\text{–}C_6H_3(NO_2)\text{–}NH\overset{(CH_2)_4NH\text{–}C_6H_3(NO_2)_2}{\overset{|}{C}}H\text{–}\overset{O}{\overset{\|}{C}}OH + NH_2\overset{CH_3}{\overset{|}{C}}HCO_2H$

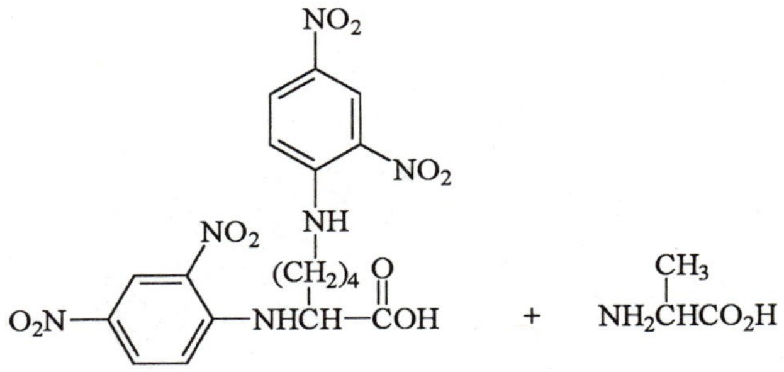

25.19 Phe-Phe-Ala or Phe-Ala-Phe

25.21

a)

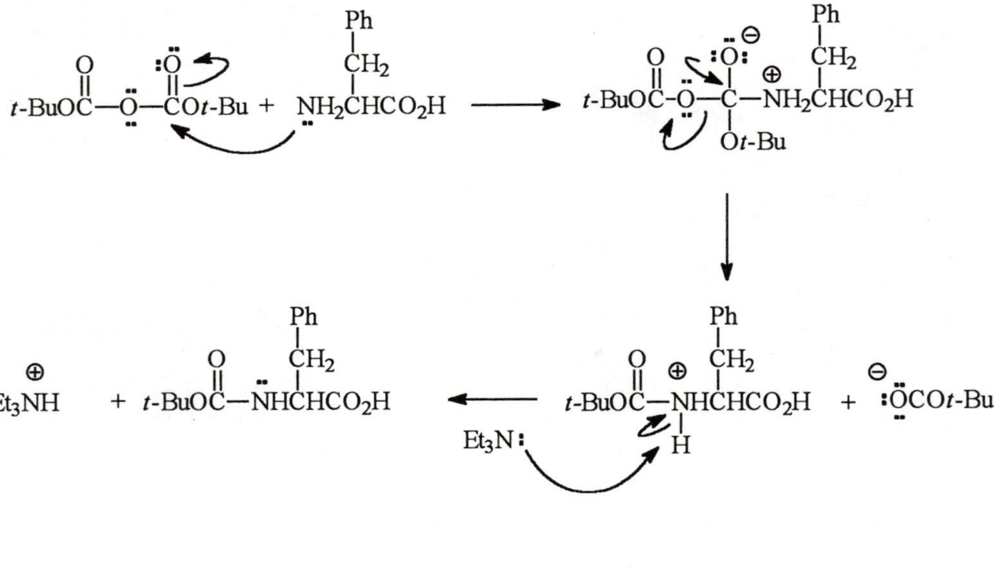

Ph
CH2
O :Ö
t-BuOC—Ö—COt-Bu + NH2CHCO2H ⟶

Ph
⊖
O :Ö: CH2
⊕
t-BuOC-Ö—C—NH2CHCO2H
Ot-Bu

$t\text{-BuO}\overset{O}{\overset{\|}{C}}\text{–}\ddot{O}\text{–}\overset{:\ddot{O}}{\overset{\|}{C}}\text{O}t\text{-Bu} + \ddot{N}H_2\overset{CH_2Ph}{\overset{|}{C}}HCO_2H \longrightarrow t\text{-BuO}\overset{O}{\overset{\|}{C}}\text{-}\ddot{O}\text{–}\overset{:\ddot{O}:^{\ominus}}{\overset{|}{C}}(\text{O}t\text{-Bu})\text{–}\overset{\oplus}{N}H_2\overset{CH_2Ph}{\overset{|}{C}}HCO_2H$

↓

Ph
O ⊕ CH2 ⊖
t-BuOC—NHCHCO2H + :ÖCOt-Bu
H
Et3N:

Ph
⊕ O CH2
Et3NH + t-BuOC—NHCHCO2H ⟵

$Et_3\overset{\oplus}{N}H + t\text{-BuO}\overset{O}{\overset{\|}{C}}\text{–}\ddot{N}H\overset{CH_2Ph}{\overset{|}{C}}HCO_2H \longleftarrow t\text{-BuO}\overset{O}{\overset{\|}{C}}\text{–}\overset{\oplus}{N}H(H)\overset{CH_2Ph}{\overset{|}{C}}HCO_2H + :\overset{\ominus}{\ddot{O}}COt\text{-Bu}$ (with $Et_3N:$)

b)

$$t\text{-Bu}-\ddot{O}-C(=O)-NHCH(CH_2Ph)CO_2H + H-Cl \longrightarrow (CH_3)_3C-\overset{\oplus}{\ddot{O}}(H)-C(=O)-NHCH(CH_2Ph)CO_2H$$

$$\longrightarrow (CH_3)_2C^{\oplus}-CH_3 + H-\ddot{O}-C(=O)-NHCH(CH_2Ph)CO_2H$$

$$H_2\ddot{O}: + H_3C-C^{\oplus}(CH_3)-CH_2-H \longrightarrow H_3\overset{\oplus}{O}: + H_3C-C(CH_3)=CH_2$$

$$H-O-C(=O)-NHCH(CH_2Ph)CO_2H + H-Cl \longrightarrow H-O-C(=O)-\overset{\oplus}{N}H_2CH(CH_2Ph)CO_2H$$

$$H_2\ddot{O}: + H-O-C(=O)-\overset{\oplus}{N}H_2CH(CH_2Ph)CO_2H \longrightarrow H_3\overset{\oplus}{O}: + CO_2 + NH_2CH(CH_2Ph)CO_2H$$

c)

$$PhCH_2O-C(=\ddot{O}:)-Cl + \ddot{N}H_2CH_2CO_2H \longrightarrow PhCH_2O-C(:\ddot{O}:^{\ominus})(:\ddot{Cl}:)-\overset{\oplus}{N}H_2CH_2CO_2H$$

$$\longrightarrow PhCH_2O-C(=\ddot{O}:)-\overset{\oplus}{N}(H)HCH_2CO_2H + {}^{\ominus}:\ddot{Cl}:$$

$$\longrightarrow HCl + PhCH_2O-C(=\ddot{O}:)-NHCH_2CO_2H$$

25.22 a)

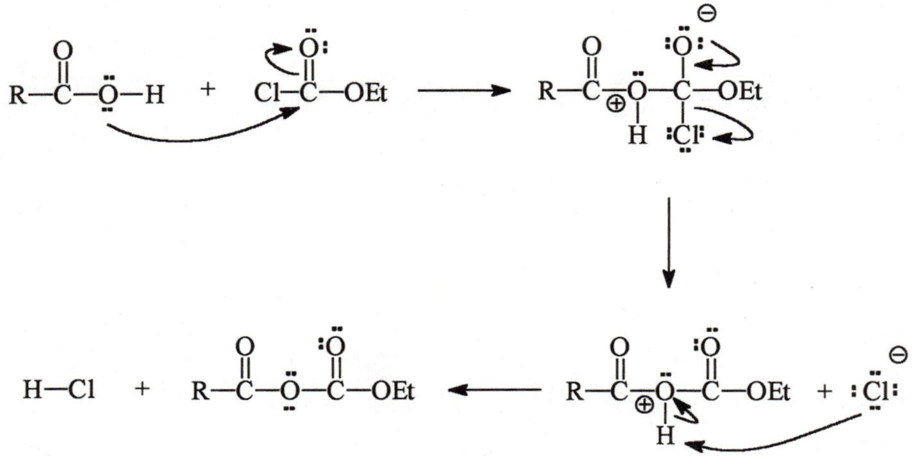

R—C(=O)—Ö—H + Cl—C(=Ö:)—OEt ⟶ R—C(=O)—$\oplus$O(H)—C(:Ö:$^{\ominus}$)(Cl:)—OEt

⟶ R—C(=O)—$\oplus$O(H)—C(=:Ö)—OEt + :Cl:$^{\ominus}$ ⟶ H—Cl + R—C(=O)—Ö—C(=:Ö)—OEt

The second product is not formed because chloride ion is a better leaving group than is ethoxide ion.

b)

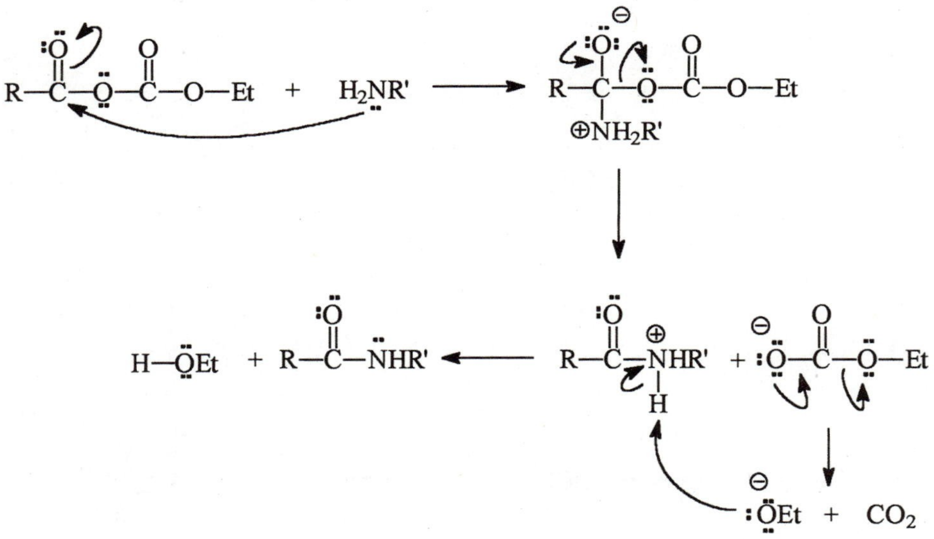

R—C(=:Ö)—Ö—C(=O)—O—Et + H_2NR' ⟶ R—C(:Ö:$^{\ominus}$)($\oplus NH_2R'$)—Ö—C(=O)—O—Et

⟶ R—C(=:Ö)=$\oplus$NHR'(H) + $^{\ominus}$:Ö—C(=O)—Ö—Et ⟶ :ÖEt$^{\ominus}$ + CO_2

⟶ H—ÖEt + R—C(=:Ö)—NHR'

A carbonyl group substituted with an oxygen is less reactive than a carbonyl group substituted with a carbon due to resonance stabilization.

$$R{-}\overset{\overset{\textstyle :\ddot{O}}{\|}}{C}{-}\ddot{\underset{\cdot\cdot}{O}}{-}R' \longleftrightarrow R{-}\overset{\overset{\textstyle \ominus\,:\ddot{O}:}{|}}{C}{=}\overset{\oplus}{\underset{\cdot\cdot}{O}}{-}R'$$

The left carbonyl group has one such resonance interaction to stabilize it whereas the right carbonyl group has two.

25.23

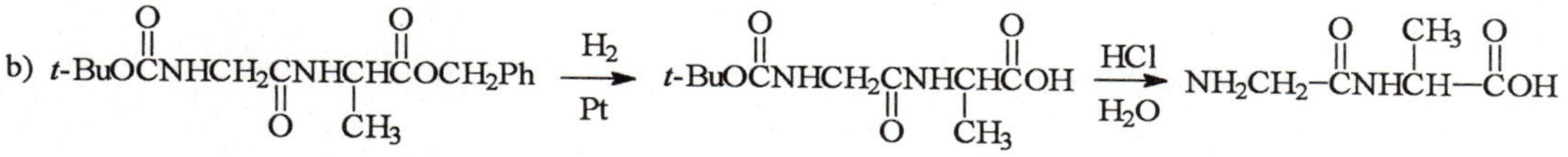

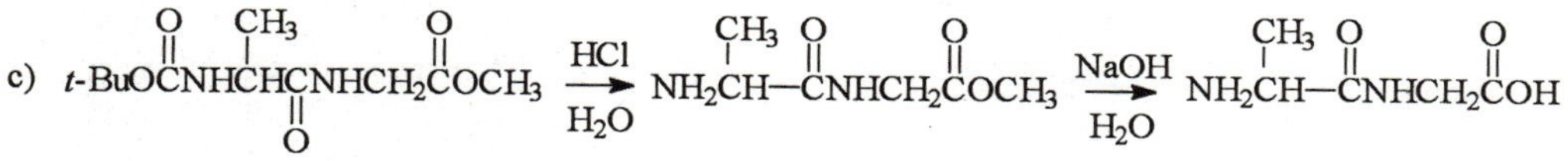

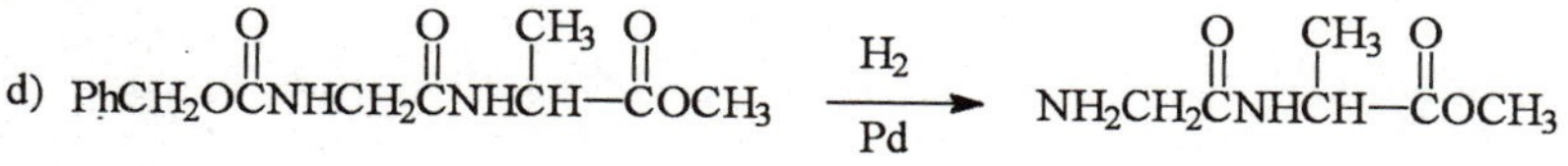

a) $t\text{-BuOC(=O)NHCH(CH}_3\text{)CO}_2\text{H}$

b) $t\text{-BuOC(=O)NHCH}_2\text{C(=O)NHCH(CH}_3\text{)C(=O)OCH}_2\text{Ph} \xrightarrow[\text{Pt}]{\text{H}_2} t\text{-BuOC(=O)NHCH}_2\text{C(=O)NHCH(CH}_3\text{)C(=O)OH} \xrightarrow[\text{H}_2\text{O}]{\text{HCl}} \text{NH}_2\text{CH}_2\text{–C(=O)NHCH(CH}_3\text{)–C(=O)OH}$

c) $t\text{-BuOC(=O)NHCH(CH}_3\text{)C(=O)NHCH}_2\text{C(=O)OCH}_3 \xrightarrow[\text{H}_2\text{O}]{\text{HCl}} \text{NH}_2\text{CH(CH}_3\text{)–C(=O)NHCH}_2\text{C(=O)OCH}_3 \xrightarrow[\text{H}_2\text{O}]{\text{NaOH}} \text{NH}_2\text{CH(CH}_3\text{)–C(=O)NHCH}_2\text{C(=O)OH}$

d) $\text{PhCH}_2\text{OC(=O)NHCH}_2\text{C(=O)NHCH(CH}_3\text{)–C(=O)OCH}_3 \xrightarrow[\text{Pd}]{\text{H}_2} \text{NH}_2\text{CH}_2\text{C(=O)NHCH(CH}_3\text{)–C(=O)OCH}_3$

25.24

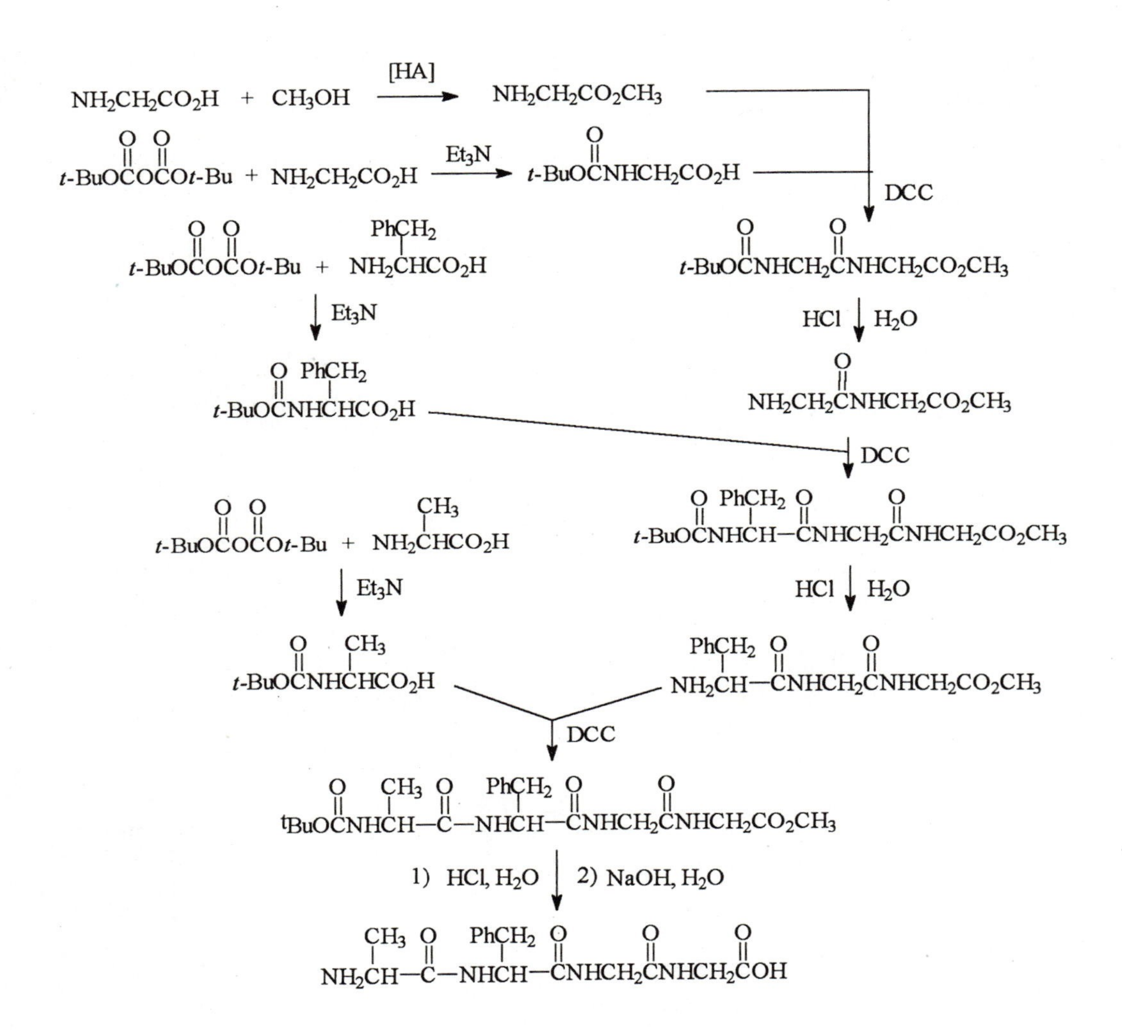

[HA]
NH2CH2CO2H + CH3OH
NH2CH2CO2CH3
t-BuOCOCOt-Bu + NH2CH2CO2H
Et3N
t-BuOCNHCH2CO2H
DCC
t-BuOCOCOt-Bu + NH2CHCO2H
PhCH2
t-BuOCNHCH2CNHCH2CO2CH3
Et3N
HCl H2O
t-BuOCNHCHCO2H
NH2CH2CNHCH2CO2CH3
DCC
t-BuOCOCOt-Bu + NH2CHCO2H
CH3
t-BuOCNHCH—CNHCH2CNHCH2CO2CH3
Et3N
HCl H2O
t-BuOCNHCHCO2H
NH2CH—CNHCH2CNHCH2CO2CH3
DCC
tBuOCNHCH—C—NHCH—CNHCH2CNHCH2CO2CH3
1) HCl, H2O
2) NaOH, H2O
NH2CH—C—NHCH—CNHCH2CNHCH2COH

25.25

$$BOC-NHCH_2-\overset{O}{\overset{\|}{C}}-OH + Cl-CH_2-C_6H_4-\text{polymer}$$

$$\downarrow Et_3N$$

$$BOC-NHCH_2-\overset{O}{\overset{\|}{C}}-O-CH_2-C_6H_4-\text{polymer}$$

$$\downarrow \text{1) } CF_3CO_2H \quad \text{2) } BOC-NH-CH_2-\overset{O}{\overset{\|}{C}}-OH,\ DCC$$

$$BOC-NHCH_2-\overset{O}{\overset{\|}{C}}-NHCH_2-\overset{O}{\overset{\|}{C}}-O-CH_2-C_6H_4-\text{polymer}$$

$$\downarrow \text{1) } CF_3CO_2H \quad \text{2) } BOC-NH-\overset{PhCH_2}{\overset{|}{C}H}-\overset{O}{\overset{\|}{C}}-OH,\ DCC$$

$$BOC-NH\overset{PhCH_2}{\overset{|}{C}H}-\overset{O}{\overset{\|}{C}}-NHCH_2-\overset{O}{\overset{\|}{C}}-NHCH_2-\overset{O}{\overset{\|}{C}}-O-CH_2-C_6H_4-\text{polymer}$$

$$\downarrow \text{1) } CF_3CO_2H \quad \text{2) } BOC-NH-\overset{CH_3}{\overset{|}{C}H}-\overset{O}{\overset{\|}{C}}-OH,\ DCC$$

$$BOC-NH\overset{CH_3}{\overset{|}{C}H}-\overset{O}{\overset{\|}{C}}-NH\overset{PhCH_2}{\overset{|}{C}H}-\overset{O}{\overset{\|}{C}}-NHCH_2-\overset{O}{\overset{\|}{C}}-NHCH_2-\overset{O}{\overset{\|}{C}}-O-CH_2-C_6H_4-\text{polymer}$$

$$\downarrow HF$$

$$NH_2\overset{CH_3}{\overset{|}{C}H}-\overset{O}{\overset{\|}{C}}-NH\overset{PhCH_2}{\overset{|}{C}H}-\overset{O}{\overset{\|}{C}}-NHCH_2-\overset{O}{\overset{\|}{C}}-NHCH_2-\overset{O}{\overset{\|}{C}}-OH$$

25.26 The nitrogen of the side chain group of tryptophan is not very basic because the lone pair of electrons are in a p orbital that is part of the conjugated cycle of the aromatic ring.

25.27 a)

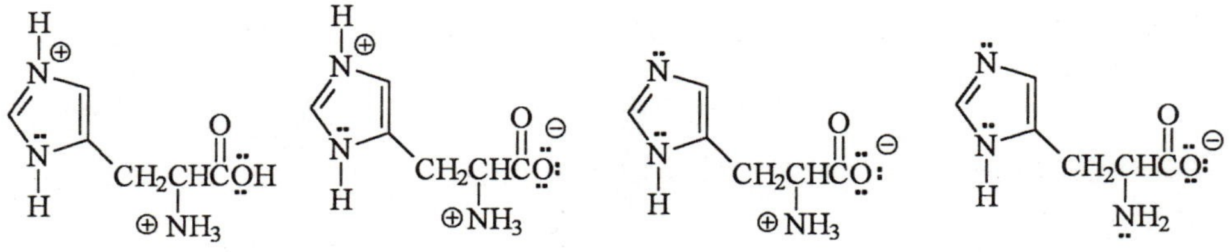

H N ⊕ N H CH_2CHCOH O ⊕ NH_3

dicationic form

H N ⊕ N H $CH_2CHCO^{\ominus}$ O ⊕ NH_3

cationic form

N N H $CH_2CHCO^{\ominus}$ O ⊕ NH_3

dipolar ion

N N H $CH_2CHCO^{\ominus}$ O NH_2

anionic form

b)

pH

species present

14, 13, 12, 11, 10, 9, 8, 7, 6, 5, 4, 3, 2, 1, 0

pH 14–9.3: N N H $CH_2CHCO^{\ominus}$ O NH_2

9.3: 50% N N H $CH_2CHCO^{\ominus}$ O NH_2 + 50% N N H $CH_2CHCO^{\ominus}$ O ⊕ NH_3

7.6: N N H $CH_2CHCO^{\ominus}$ O ⊕ NH_3

6: 50% H N ⊕ N H $CH_2CHCO^{\ominus}$ O ⊕ NH_3 + 50% N N H $CH_2CHCO^{\ominus}$ O ⊕ NH_3

3.9: H N ⊕ N H $CH_2CHCO^{\ominus}$ O ⊕ NH_3

1.8: 50% H N ⊕ N H CH_2CHCOH O ⊕ NH_3 + 50% H N ⊕ N H $CH_2CHCO^{\ominus}$ O ⊕ NH_3

pH 1.8–0: H N ⊕ N H CH_2CHCOH O ⊕ NH_3

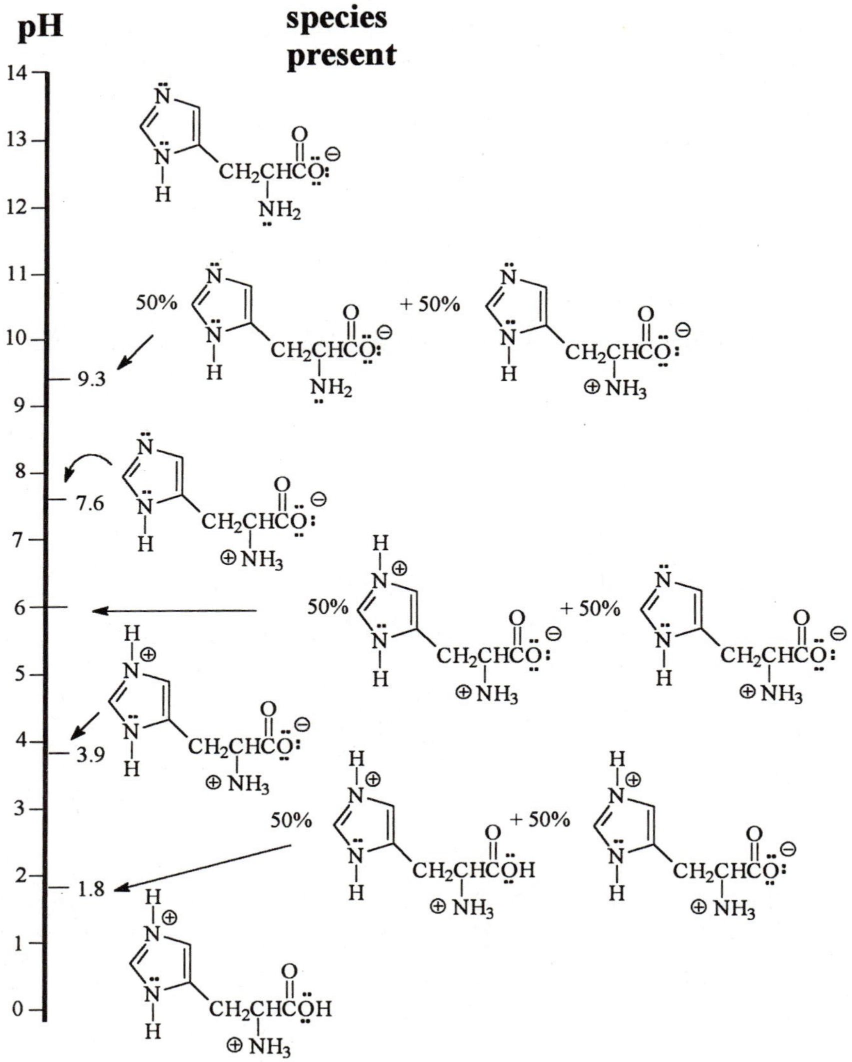

25.28 The nitrogens in the side chains of glutamine and asparagine are not very basic because they are bonded directly to a carbonyl group, so the electron pairs are delocalized and are not readily available for protonation.

25.29 The protonated side chain of arginine is resonance stabilized, so it is a weak acid.

25.30 Nitrogen 3 of the side chain ring of histidine is protonated in the monocationic form because the electron pair on this nitrogen is in a sp^2 orbital. The electron pair on nitrogen 1 is in a p orbital that is part of the aromatic cycle, so it is less available for reaction.

3 N, 2, 1 N–H; CH_2CHCOH (C=O), NH_2

25.31

$CH_3CH{=}NH$ + $^{\ominus}{:}C{\equiv}N{:}$ ⟶ $CH_3CH(C{\equiv}N)\text{–}NH^{\ominus}$ (+ H–O–H) ⟶ $CH_3CH(C{\equiv}N)\text{–}NH_2$

25.32

$CH_3CHBrC(=O)Br$ + $CH_3CH_2C(=O)OH$ ⇌ $CH_3CHBrC(O^{\ominus})(Br)\text{–}O^{\oplus}(H)\text{–}C(=O)CH_2CH_3$ ⇌ $CH_3CHBrC(=O)\text{–}O^{\oplus}(H)\text{–}C(=O)CH_2CH_3$ + $:Br:^{\ominus}$

⇅

$CH_3CHBrC(=O)OH$ + $:Br\text{–}C(=O)CH_2CH_3$ ⇌ $CH_3CHBrC(=O)\text{–}O^{\oplus}(H)\text{–}C(O^{\ominus})(Br)CH_2CH_3$

25.33

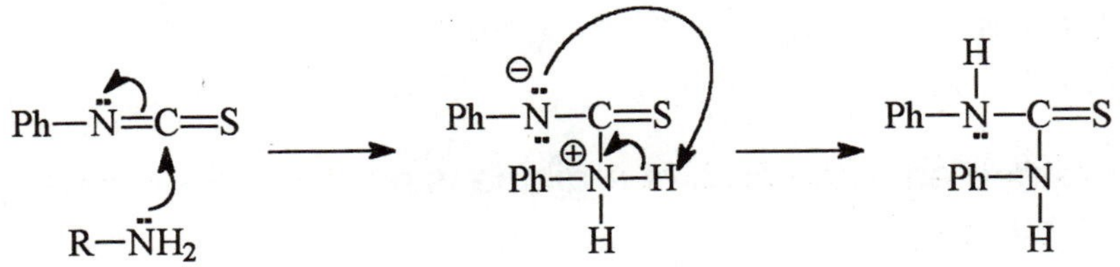

$Ph{-}\ddot{N}{=}C{=}S$ + $R{-}\ddot{N}H_2$ $\longrightarrow$ $Ph{-}\overset{\ominus}{\ddot{N}}{-}C{=}S$ / $Ph{-}\overset{\oplus}{N}H{-}H$ $\longrightarrow$ $Ph{-}N(H){-}C({=}S){-}N(H){-}Ph$

25.34

a) 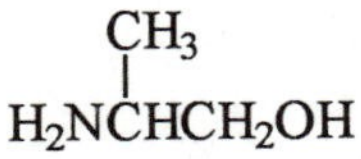 $H_2NCH(CH_3)CH_2OH$

b) 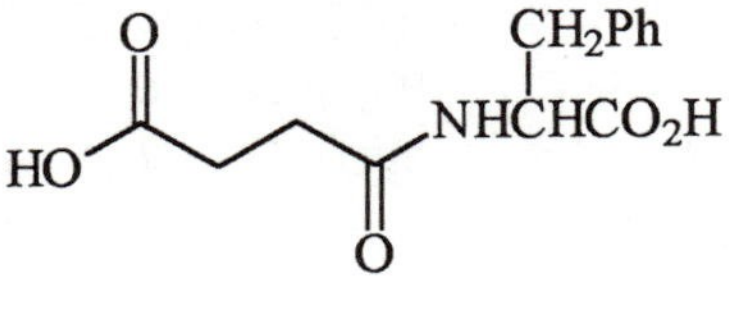 $HO_2CCH_2CH_2C(=O)NHCH(CH_2Ph)CO_2H$

c) 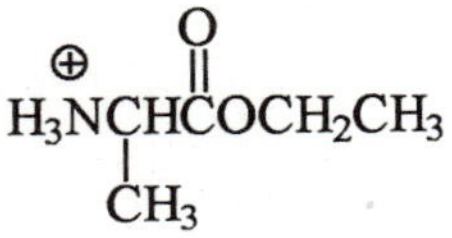 $H_3\overset{\oplus}{N}CH(CH_3)COCH_2CH_3$ (C=O)

d) 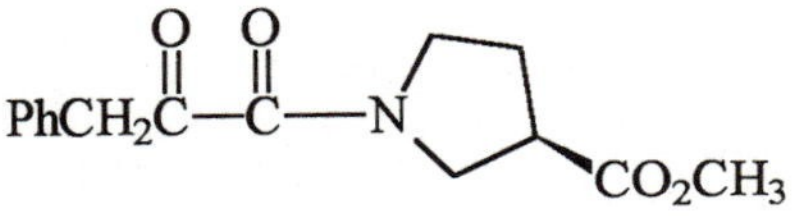 $PhCH_2C(=O){-}C(=O){-}N$ (pyrrolidine ring) with CO_2CH_3

25.35

a) 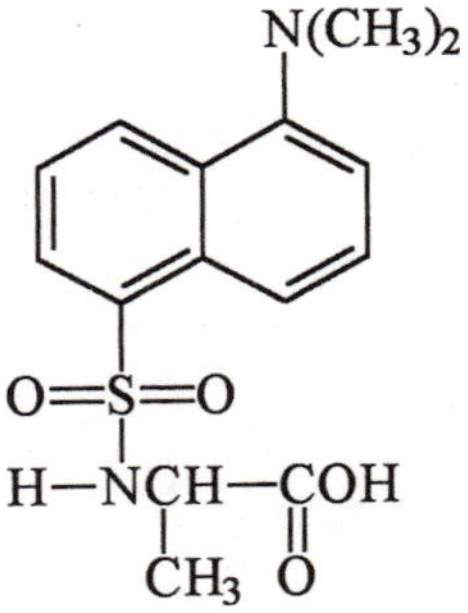 5-$N(CH_3)_2$-naphthalene-1-SO_2: $O{=}S{=}O$, $H{-}NCH(CH_3){-}COH$ (C=O)

b) O_2N–(2-NO_2-phenyl)–$NHCH(CH_2Ph)CO_2H$ + $H_3\overset{\oplus}{N}CH(CH_3)COH$ (C=O) + $H_3\overset{\oplus}{N}CH_2COH$ (C=O)

c) 5-$N(CH_3)_2$-naphthalene-1-SO_2: $O{=}S{=}O$, $H{-}NCH{-}COH$ (C=O), $H_3C{-}CH{-}CH_3$ + $H_3\overset{\oplus}{N}CH_2COH$ (C=O)

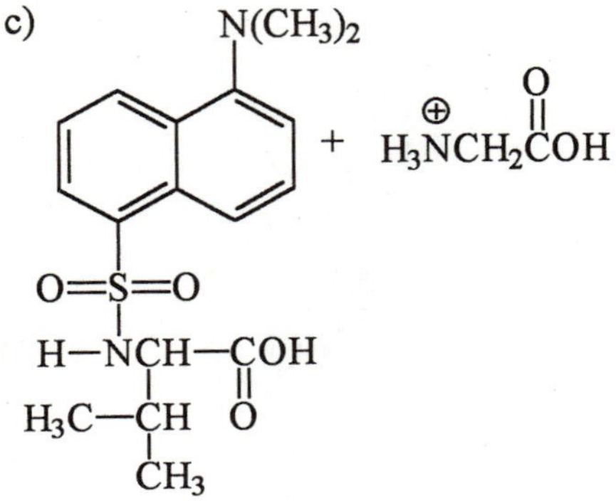

d) 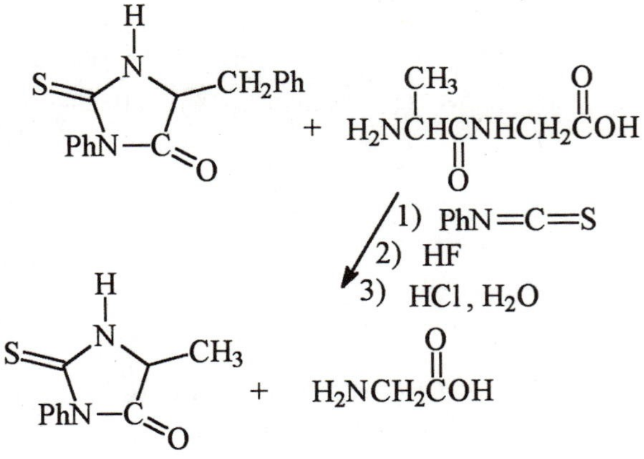 Thiohydantoin: $S{=}$, N–H, CH_2Ph, $PhN{-}C{=}O$ + $H_2NCH(CH_3)C(=O)NHCH_2COH$ (C=O)

1) $PhN{=}C{=}S$
2) HF
3) HCl, H_2O

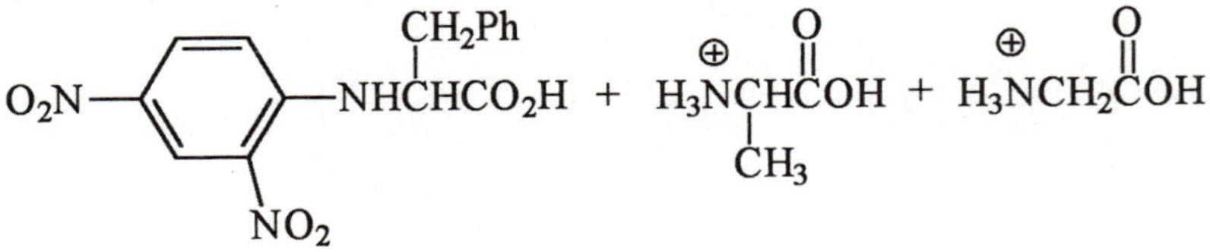

Thiohydantoin: $S{=}$, N–H, CH_3, $PhN{-}C{=}O$ + H_2NCH_2COH (C=O)

25.36

a)

$$t\text{-BuOC(=O)NHCH(CH}_2\text{Ph)C(=O)NHCH(CH}_3\text{)C(=O)OCH}_3 \xrightarrow[\text{H}_2\text{O}]{\text{HCl}} \text{H}_2\text{NCH(CH}_2\text{Ph)C(=O)NHCH(CH}_3\text{)C(=O)OCH}_3$$

$$\downarrow \text{NaOH, H}_2\text{O}$$

$$\text{H}_2\text{NCH(CH}_2\text{Ph)C(=O)NHCH(CH}_3\text{)C(=O)OH}$$

b)

$$\text{PhCH}_2\text{OC(=O)NHCH}_2\text{C(=O)OH} \xrightarrow{\text{ClC(=O)OEt}} \text{PhCH}_2\text{OC(=O)NHCH}_2\text{C(=O)OC(=O)OEt}$$

$$\downarrow \text{H}_2\text{NCH}_2\text{C(=O)OCH}_3$$

$$\text{H}_2\text{NCH}_2\text{C(=O)NHCH}_2\text{C(=O)OCH}_3 \xleftarrow[\text{Pt}]{\text{H}_2} \text{PhCH}_2\text{OC(=O)NHCH}_2\text{C(=O)NHCH}_2\text{C(=O)OCH}_3$$

$$\downarrow \text{NaOH, H}_2\text{O}$$

$$\text{H}_2\text{NCH}_2\text{C(=O)NHCH}_2\text{C(=O)OH}$$

25.37

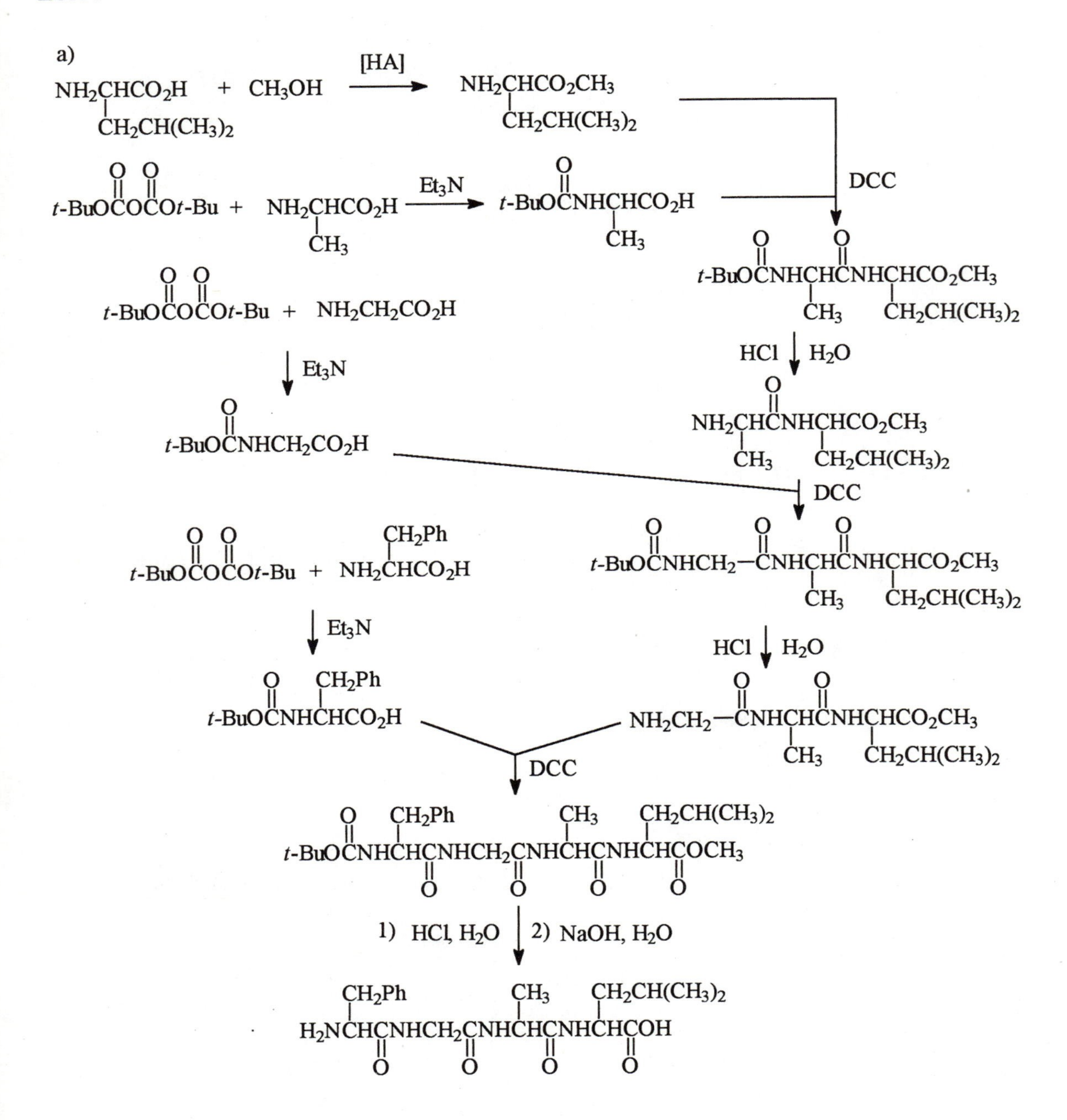

a)
[HA]
NH_2CHCO_2H
$CH_2CH(CH_3)_2$
+ CH_3OH
$NH_2CHCO_2CH_3$
$CH_2CH(CH_3)_2$
Et_3N
t-BuOCOCOt-Bu + NH_2CHCO_2H
CH_3
t-BuOCNHCHCO$_2$H
DCC
t-BuOCNHCHCNHCHCO$_2CH_3$
t-BuOCOCOt-Bu + $NH_2CH_2CO_2H$
HCl
H_2O
t-BuOCNHCH$_2CO_2$H
NH_2CHCNHCHCO$_2CH_3$
CH_2Ph
t-BuOCOCOt-Bu + NH_2CHCO_2H
t-BuOCNHCH$_2$—CNHCHCNHCHCO$_2CH_3$
t-BuOCNHCHCO$_2$H
NH_2CH_2—CNHCHCNHCHCO$_2CH_3$
t-BuOCNHCHCNHCH$_2$CNHCHCNHCHCOCH$_3$
1) HCl, H_2O
2) NaOH, H_2O
H_2NCHCNHCH$_2$CNHCHCNHCHCOH

b)

$$\text{BOC—HNCH(CH}_2\text{CH(CH}_3)_2)\text{C(=O)OH}$$

$$\downarrow \quad \text{ClCH}_2\text{—C}_6\text{H}_4\text{—(polymer)}, \ \text{Et}_3\text{N}$$

$$\text{BOC—HNCH(CH}_2\text{CH(CH}_3)_2)\text{C(=O)O—CH}_2\text{—C}_6\text{H}_4\text{—(polymer)}$$

$$\downarrow \quad \text{CF}_3\text{COOH , CH}_2\text{Cl}_2$$

$$\text{H}_2\text{NCH(CH}_2\text{CH(CH}_3)_2)\text{C(=O)O—CH}_2\text{—C}_6\text{H}_4\text{—(polymer)}$$

$$\downarrow \quad \text{1) BOC—HNCH(CH}_3)\text{C(=O)OH, DCC} \quad \text{2) CF}_3\text{COOH , CH}_2\text{Cl}_2$$

$$\text{H}_2\text{NCH(CH}_3)\text{C(=O)NHCH(CH}_2\text{CH(CH}_3)_2)\text{C(=O)O—CH}_2\text{—C}_6\text{H}_4\text{—(polymer)}$$

$$\downarrow \quad \text{1) BOC—HNCH}_2\text{C(=O)OH, DCC} \quad \text{2) CF}_3\text{COOH , CH}_2\text{Cl}_2$$

$$\text{H}_2\text{NCH}_2\text{C(=O)NHCH(CH}_3)\text{C(=O)NHCH(CH}_2\text{CH(CH}_3)_2)\text{C(=O)O—CH}_2\text{—C}_6\text{H}_4\text{—(polymer)}$$

$$\downarrow \quad \text{1) BOC—HNCH(CH}_2\text{Ph)C(=O)OH, DCC} \quad \text{2) HF}$$

$$\text{H}_2\text{NCH(CH}_2\text{Ph)C(=O)NHCH}_2\text{C(=O)NHCH(CH}_3)\text{C(=O)NHCH(CH}_2\text{CH(CH}_3)_2)\text{C(=O)—OH}$$

25.39 Component 1 = Gly-Val
Component 2 = Phe-Ala-Leu
Component 3 = Leu-Gly
The structure of pentapeptide is Phe-Ala-Leu-Gly-Val.

25.39 The amino acid sequence of bradykinin is
Arg-Pro-Pro-Gly-Phe-Ser-Pro-Phe-Arg.

After completing this chapter, you should be able to:

1. Show the general structure, including stereochemistry, for an amino acid.
   Problem 25.1.

2. Recognize nonpolar, polar, acidic, and basic side chains on amino acids.

3. Understand the acid-base reactions of amino acids.
   Problems 25.2, 25.3, 25.4, 25.5, 25.6, 25.7, 25.26, 25.27, 25.28, 25.29, 25.30.

4. Understand the general chemical reactions of amino acids.
   Problems 25.8, 25.34.

5. Show syntheses of amino acids by the Strecker method, α-substitution by NH_3, or the Gabriel/malonate method.
   Problems 25.10, 25.11.

6. Understand how amino acids combine to form a peptide.
   Problems 25.14, 25.15.

7. Understand how peptides are sequenced, using hydrolysis, Sanger's reagent, the Edman degradation, and enzymatic hydrolysis.
   Problems 25.16, 25.18, 25.19, 25.20, 25.35, 25.38, 25.39.

8. Understand the use of protective groups and coupling reagents in peptide synthesis.
   Problems 25.23, 25.36.

9. Show a reaction scheme for the synthesis of a peptide in solution.
   Problems 25.24, 25.37.

10. Show a reaction scheme for the synthesis of a peptide using the solid-phase method.
    Problems 25.25, 25.37.

11. Understand some of the general principles of enzyme catalysis.

12. Show the mechanisms of these reactions where appropriate.
    Problems 25.9, 25.12, 25.17, 25.21, 25.22, 25.31, 25.32, 25.33.

Chapter 26
NUCLEOTIDES AND NUCLEIC ACIDS

26.1

a)

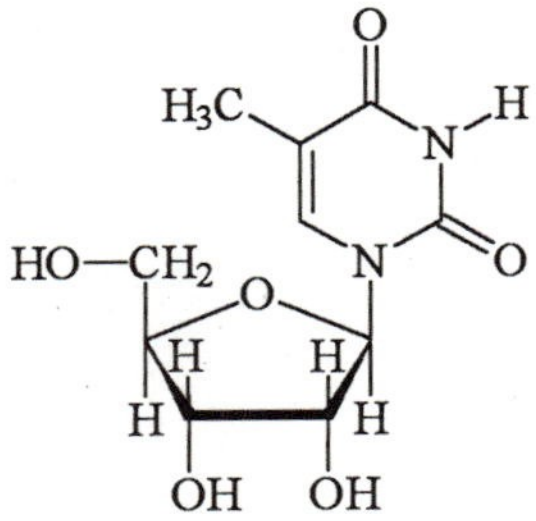

b)

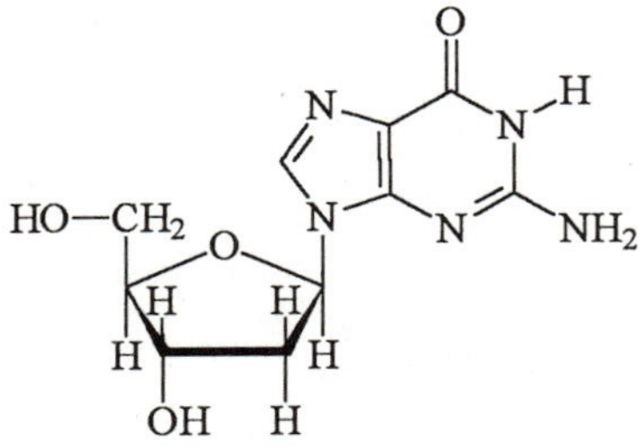

c)

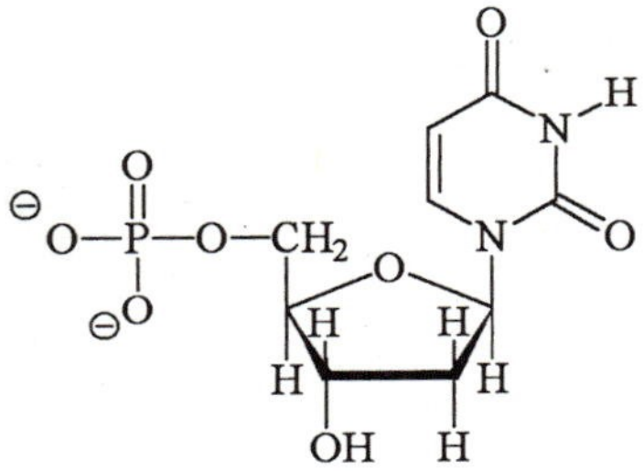

d)

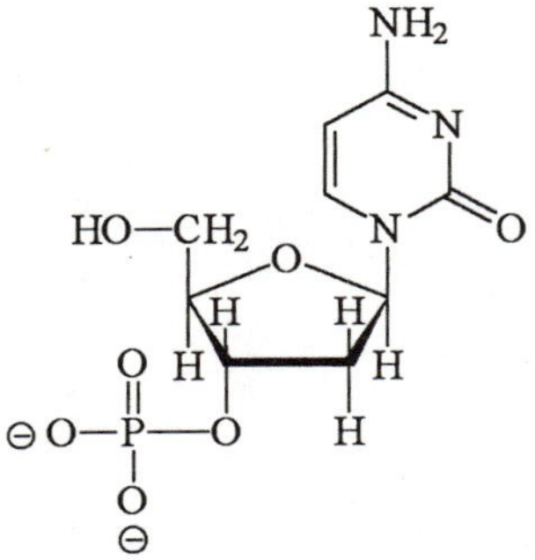

26.2 a) deoxyguanosine b) uridine 5'-monophosphate
c) deoxycytidine 5'-monophosphate

26.3 Pyrimidine has six electrons in its cyclic pi system and therefore fits Huckel's rule. The unshared electrons on the nitrogens are not part of the pi system. The six-membered ring of purine is aromatic for the same reason as pyrimidine. The five-membered ring is also aromatic. Like imidazole (see Chapter 17), it has a total of six electrons in its cyclic pi system. The electrons on N-9 are part of the pi system, whereas the electrons on N-7 are not.

26.4

a)

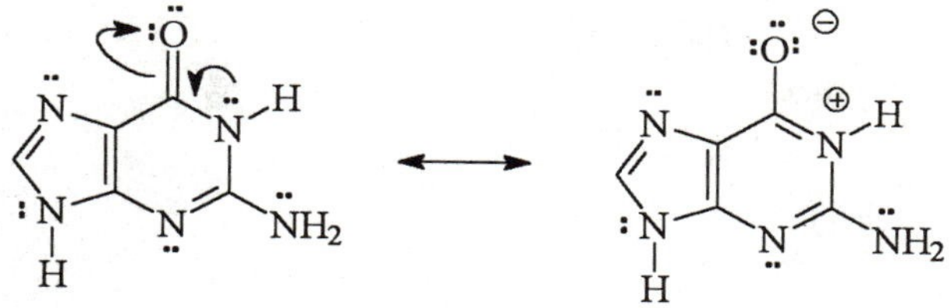

b)

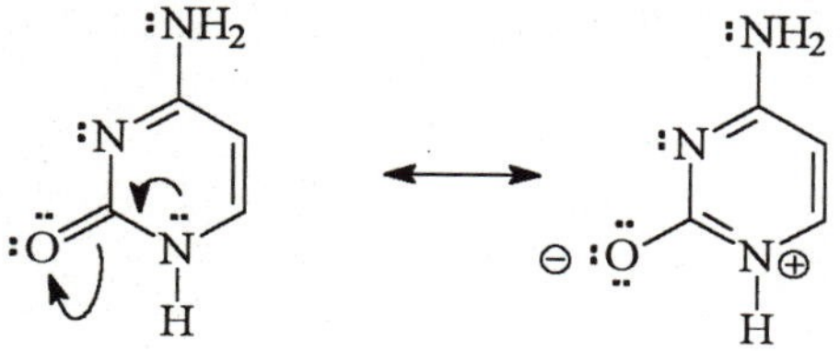

c)

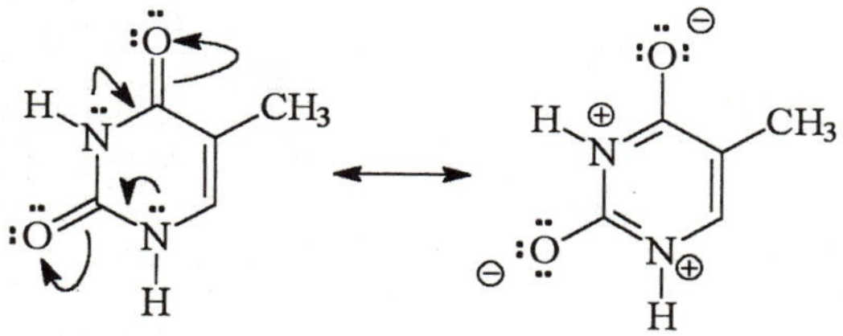

26.5 5'-end TAACGCTCG 3'-end

26.6

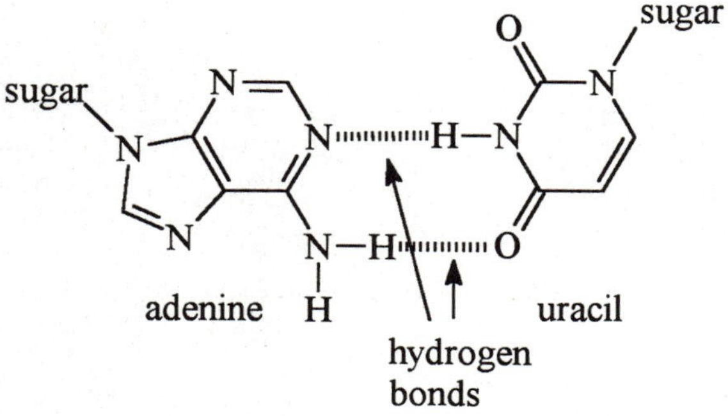

26.7

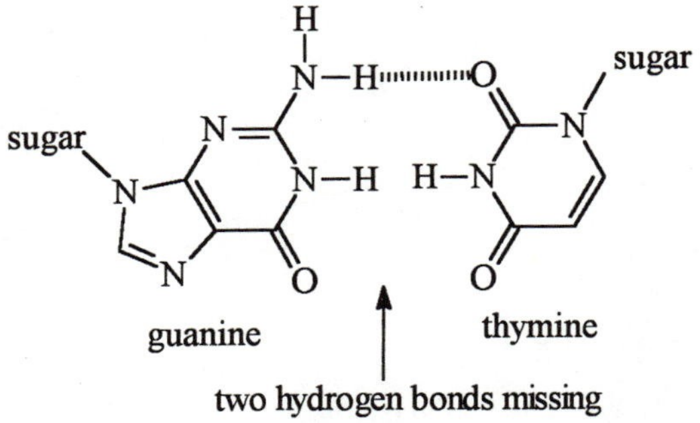

sugar
guanine
sugar
thymine
two hydrogen bonds missing

26.8

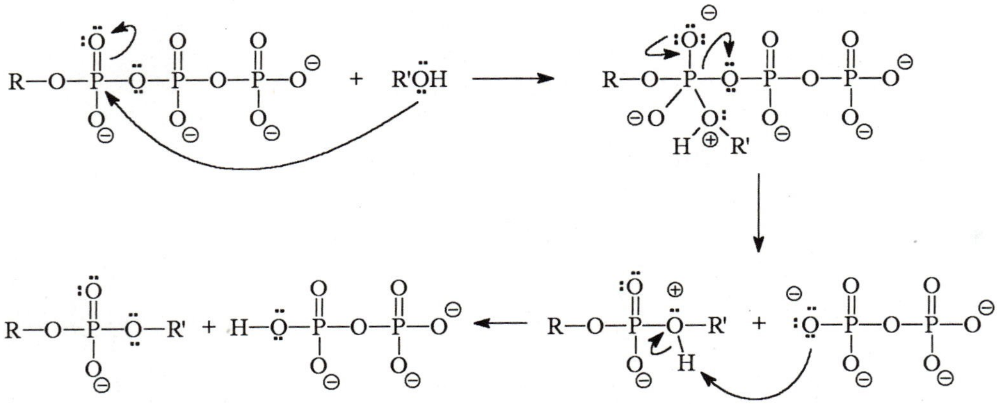

26.9

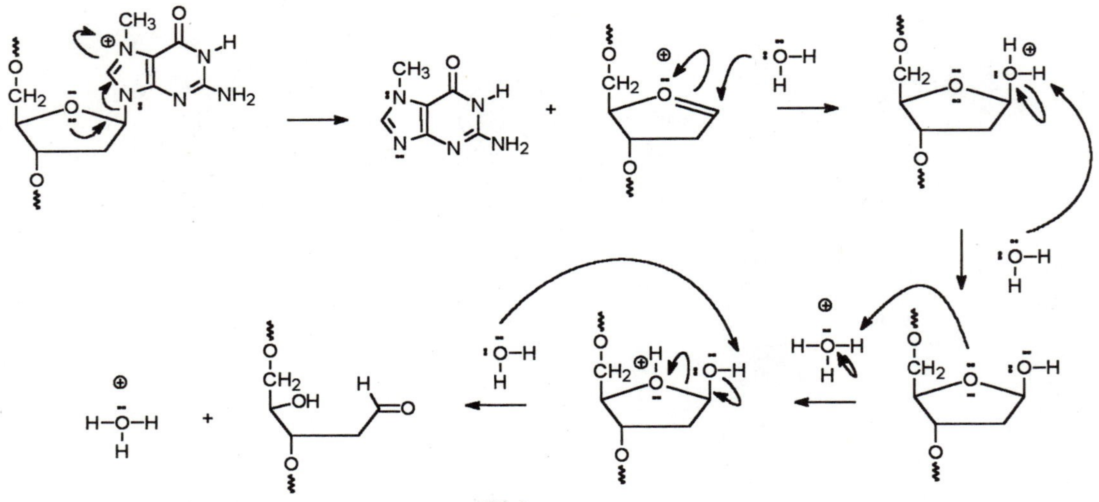

26.10

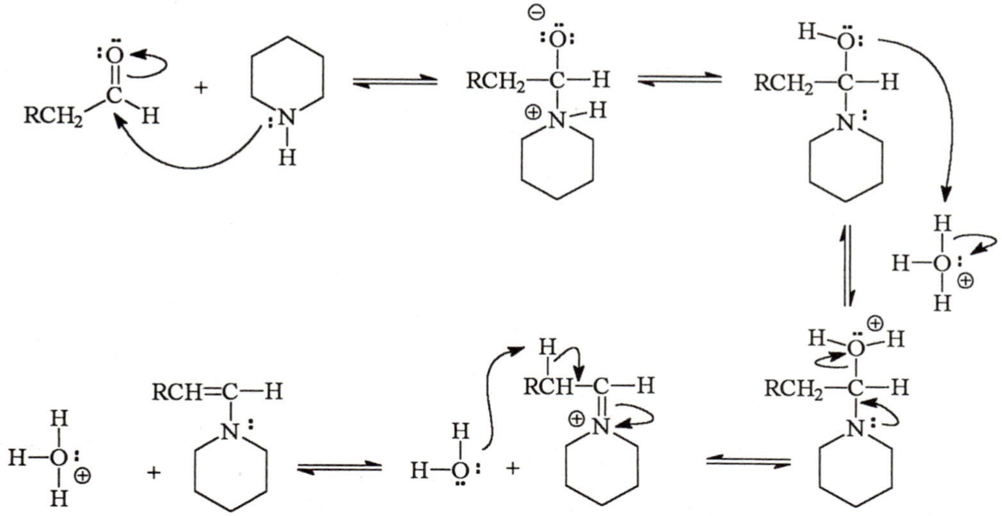

RCH$_2$ C=O H + piperidine ⇌ RCH$_2$—C(O$^-$)—H with $^+$N—H ⇌ RCH$_2$—C(OH)—H with N: ; H$_3$O$^+$

H$_3$O$^+$ + RCH=C—H (N:) ⇌ H—O: (H) + RCH—C—H (=N$^+$) ⇌ RCH$_2$—C(O$^+$H$_2$)—H

26.11

sugar + :NH$_2$NH$_2$ → CH$_3$, NH$_2$ → sugar

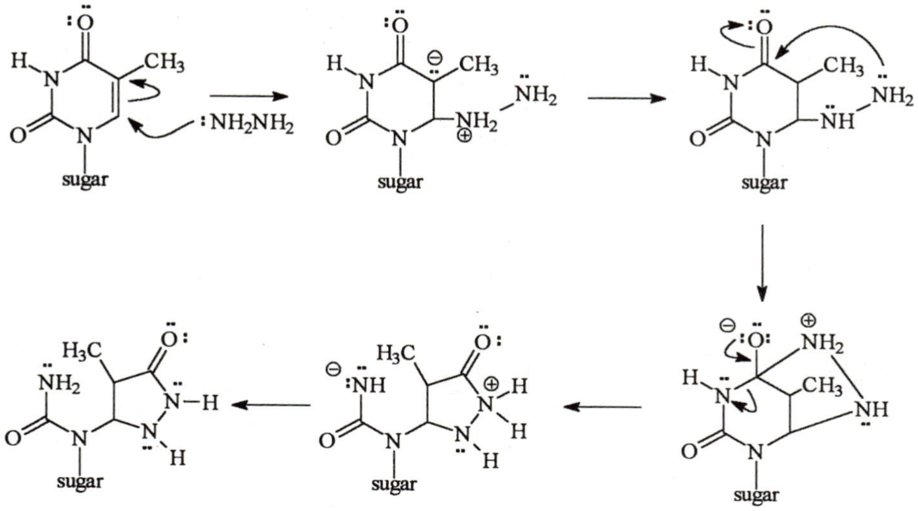

26.12

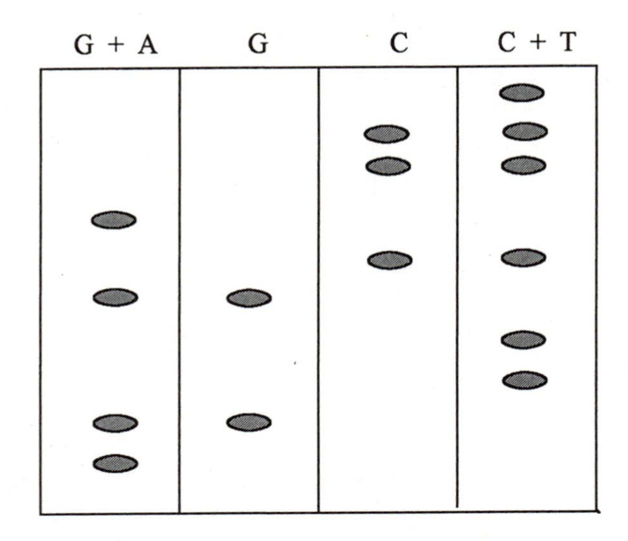

26.13

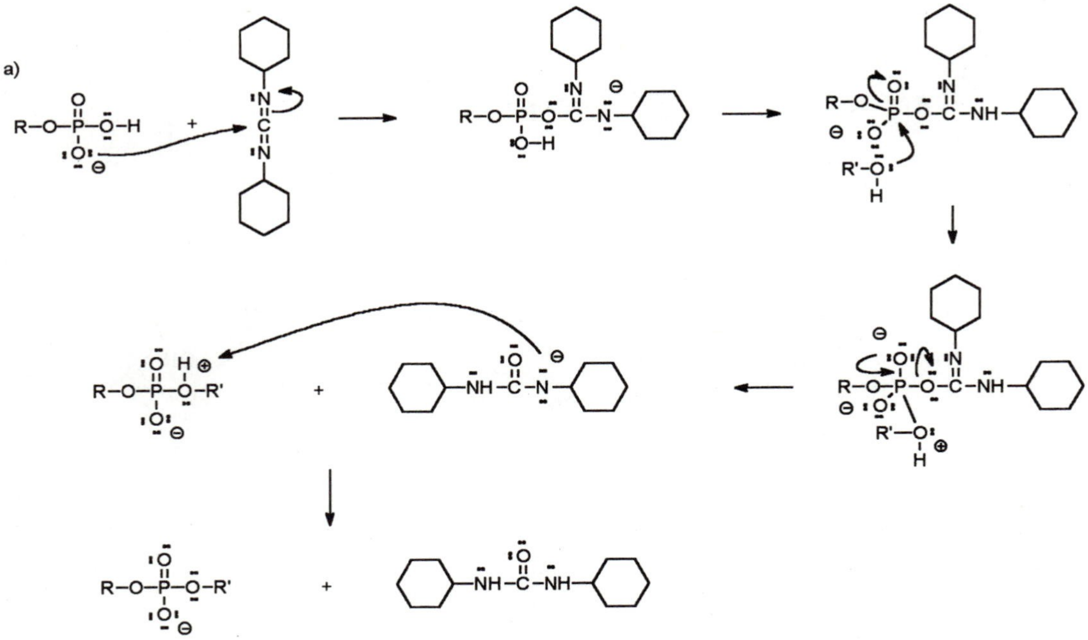

a)

b)

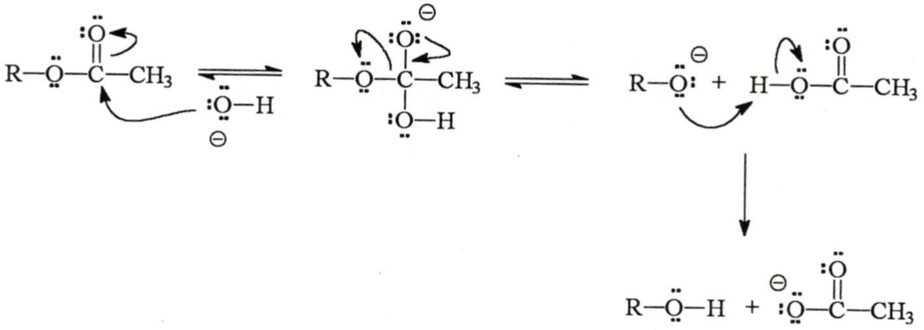

$$R{-}O{-}C(=O){-}CH_3 + {}^{\ominus}O{-}H \rightleftharpoons R{-}O{-}C(O^{\ominus})(O{-}H){-}CH_3 \rightleftharpoons R{-}O^{\ominus} + H{-}O{-}C(=O){-}CH_3$$

$$\downarrow$$

$$R{-}O{-}H + {}^{\ominus}O{-}C(=O){-}CH_3$$

c)

$$RNH{-}C(=O){-}R' + NH_3 \rightleftharpoons RNH{-}C(O^{\ominus})(NH_3^{\oplus}){-}R' \rightleftharpoons RNH_2^{\oplus}{-}C(O^{\ominus})(NH_2){-}R'$$

$$\rightleftharpoons$$

$$RNH_2 + NH_2{-}C(=O){-}R'$$

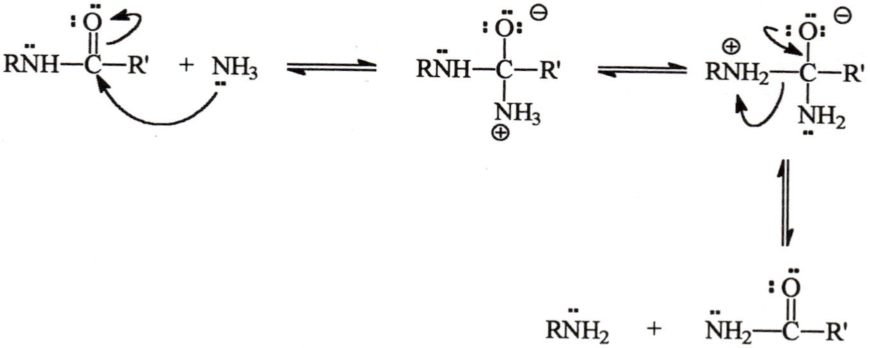

d)

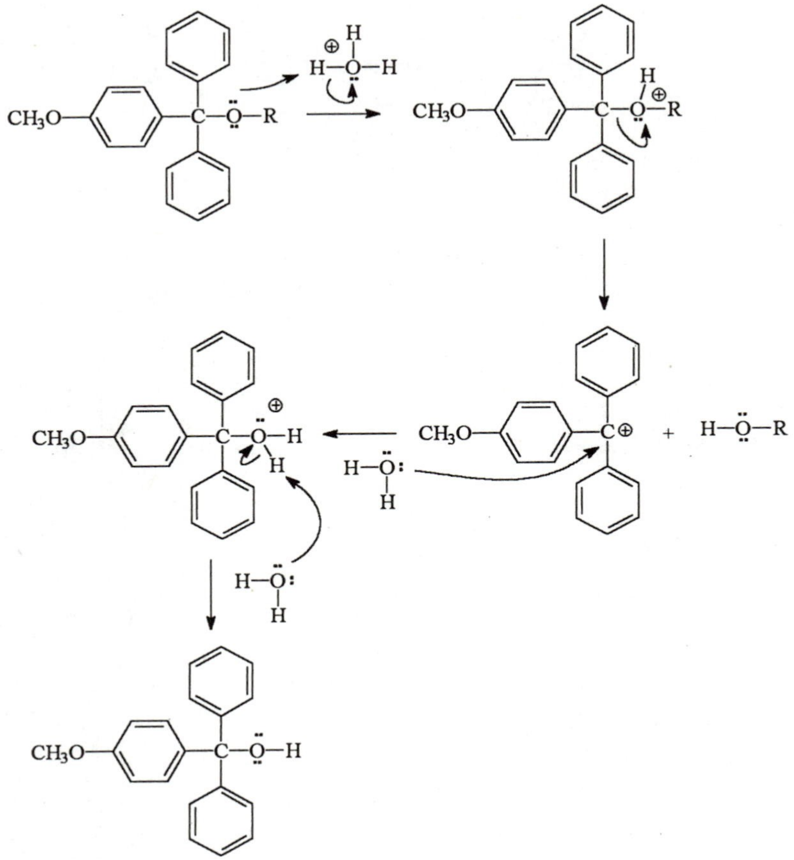

CH₃O-C₆H₄-C(C₆H₅)₂-Ö-R → (H₃O⁺) CH₃O-C₆H₄-C(C₆H₅)₂-O⁺(H)-R → CH₃O-C₆H₄-C⁺(C₆H₅)₂ + H-Ö-R

CH₃O-C₆H₄-C⁺(C₆H₅)₂ + H₂O → CH₃O-C₆H₄-C(C₆H₅)₂-O⁺H₂ → (H₂O) CH₃O-C₆H₄-C(C₆H₅)₂-Ö-H

26.14 The methoxy group provides additional resonance stabilization of the carbocation intermediate.

26.15

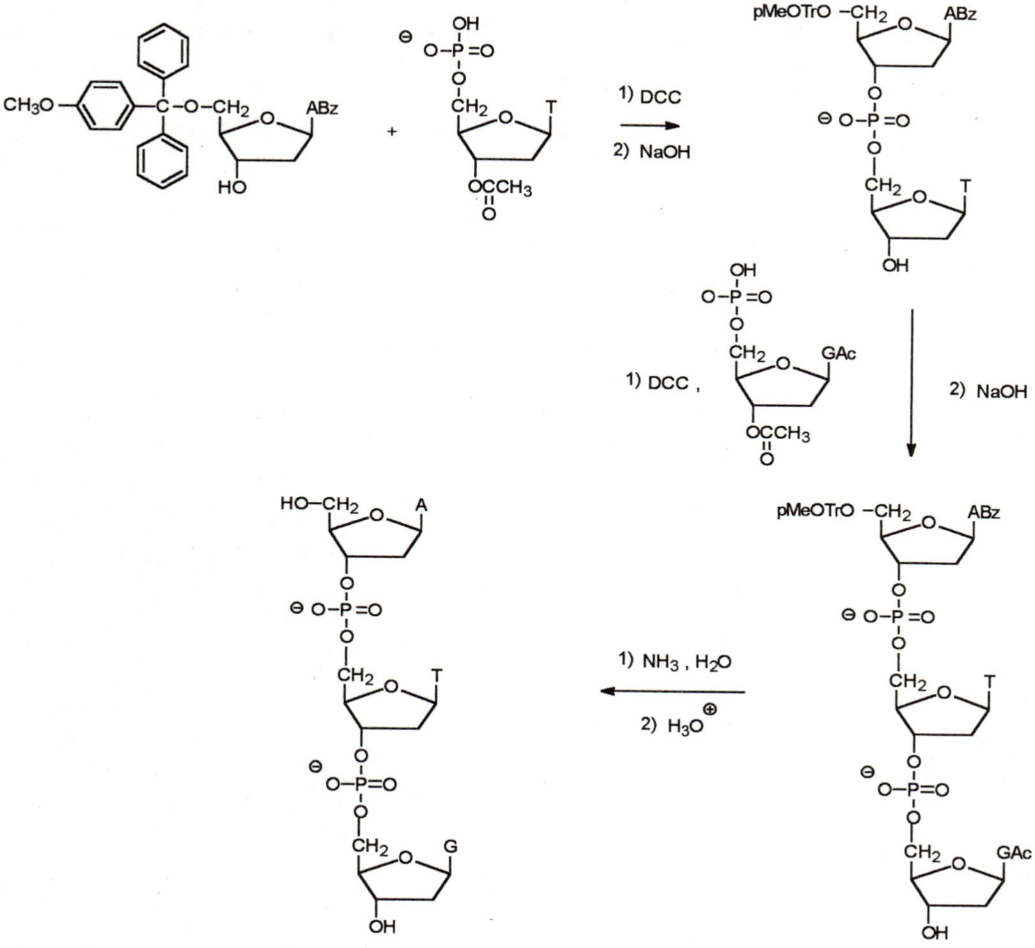

1) DCC

2) NaOH

1) DCC ,

2) NaOH

1) NH_3 , H_2O

2) $H_3O^{\oplus}$

26.16

S_N2

$$PhS^{\ominus} + CH_3{-}O{-}P(=O)(O\text{-sugar})_2 \xrightarrow{S_N2} PhS{-}CH_3 + {}^{\ominus}O{-}P(=O)(O\text{-sugar})_2$$

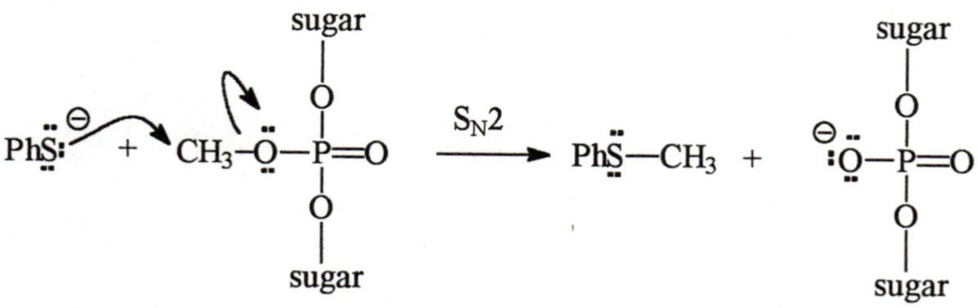

The S_N2 reaction is much faster at methyl than at the more hindered carbons of the other ester groups.

26.17 The DMTr protective group is cleaved more readily because it produces a more stable carbocation in the cleavage process.

26.18

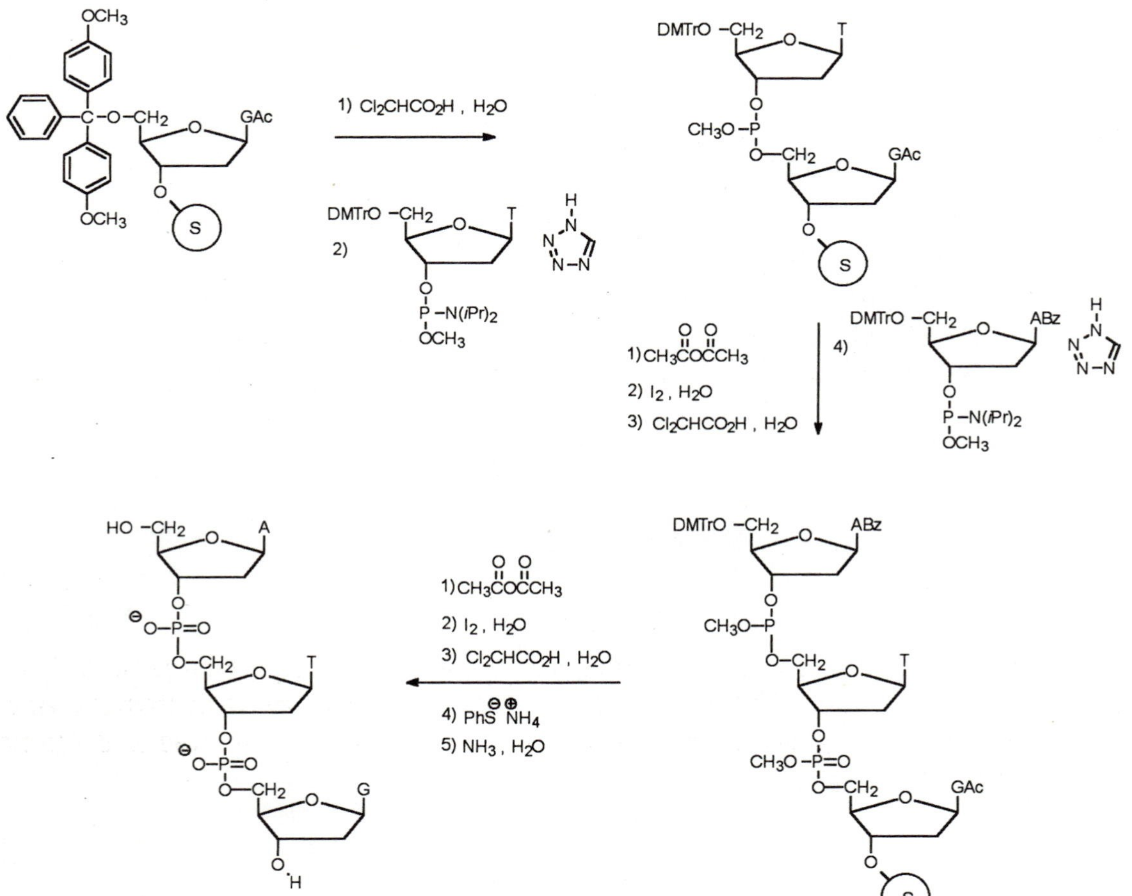

1) Cl_2CHCO_2H, H_2O

2) DMTrO–CH2 (T nucleoside 3'-O–P–N(iPr)2, OCH3), tetrazole

1) CH_3COCCH_3 (acetic anhydride)

2) I_2, H_2O

3) Cl_2CHCO_2H, H_2O

4) DMTrO–CH2 (ABz nucleoside 3'-O–P–N(iPr)2, OCH3), tetrazole

1) CH_3COCCH_3 (acetic anhydride)

2) I_2, H_2O

3) Cl_2CHCO_2H, H_2O

4) $PhS^{\ominus}$ $NH_4^{\oplus}$

5) NH_3, H_2O

26.19

a) 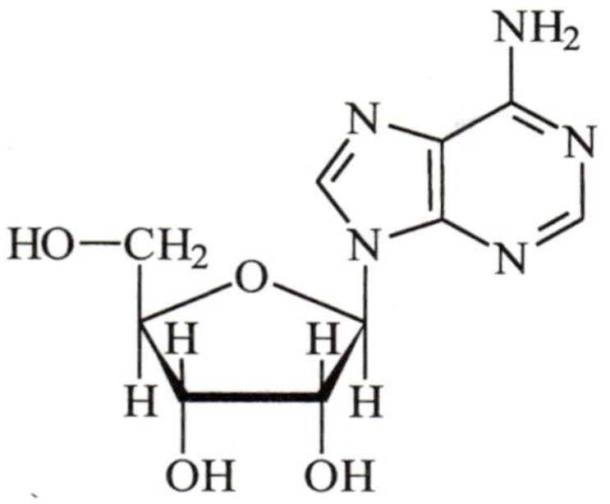 HO—CH$_2$ (adenine nucleoside)

NH$_2$
N N
N N
O
H H H H
OH OH

b) 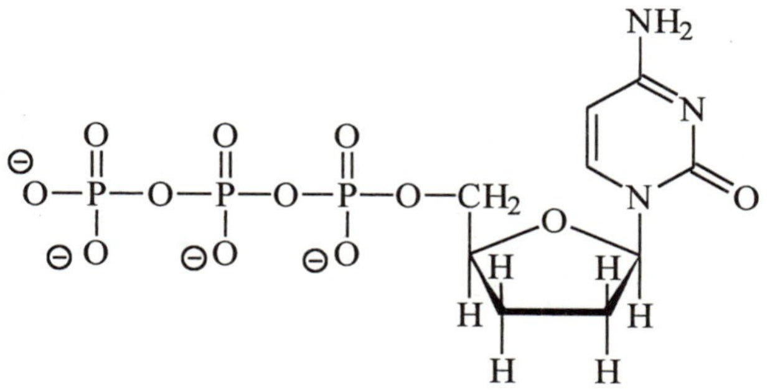 $^{\ominus}$O—P(=O)(O$^{\ominus}$)—O—P(=O)(O$^{\ominus}$)—O—P(=O)(O$^{\ominus}$)—O—CH$_2$ (cytosine nucleoside)

NH$_2$
N
N O
O
H H H H
H H

c) 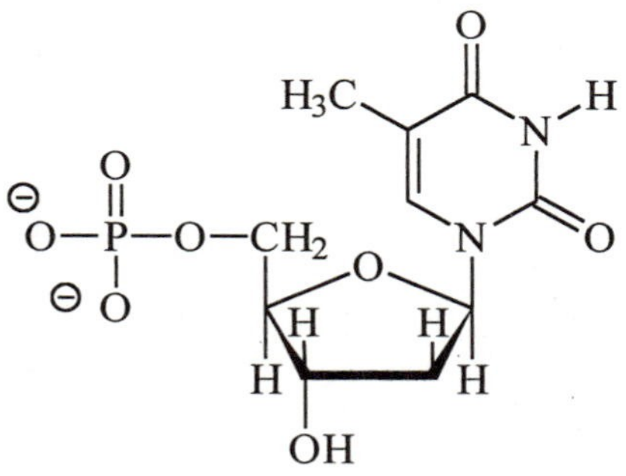 $^{\ominus}$O—P(=O)(O$^{\ominus}$)—O—CH$_2$ (thymine nucleoside)

O
H$_3$C N—H
N O
O
H H H H
OH

26.20 a) deoxyadenosine 5'-monophosphate.
b) uridine.

26.21 The lone electron pair on the NH_2 attached to the ring of cytosine is relatively basic. The lone electron pairs on the nitrogens of thymine and uracil are conjugated with carbonyl groups. They are amides and are not very basic.

26.22

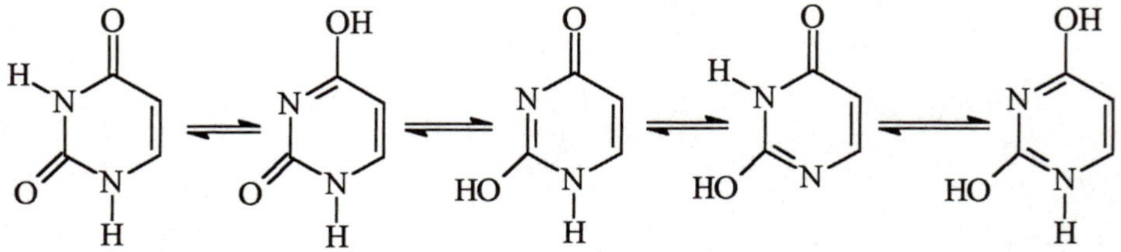

26.23 The complementary DNA is 3'-end-CGAATACG-5'-end and the complimentary RNA is 3'-end-CGAAUACG-5'-end.

26.24

a)

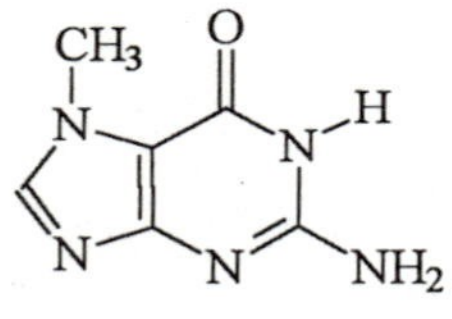

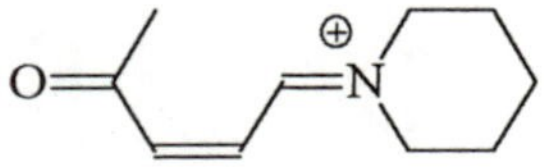

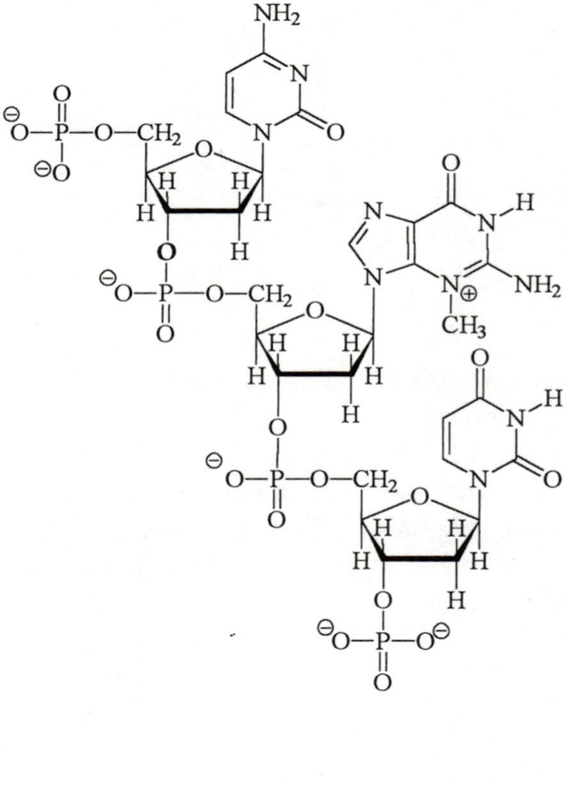

b)

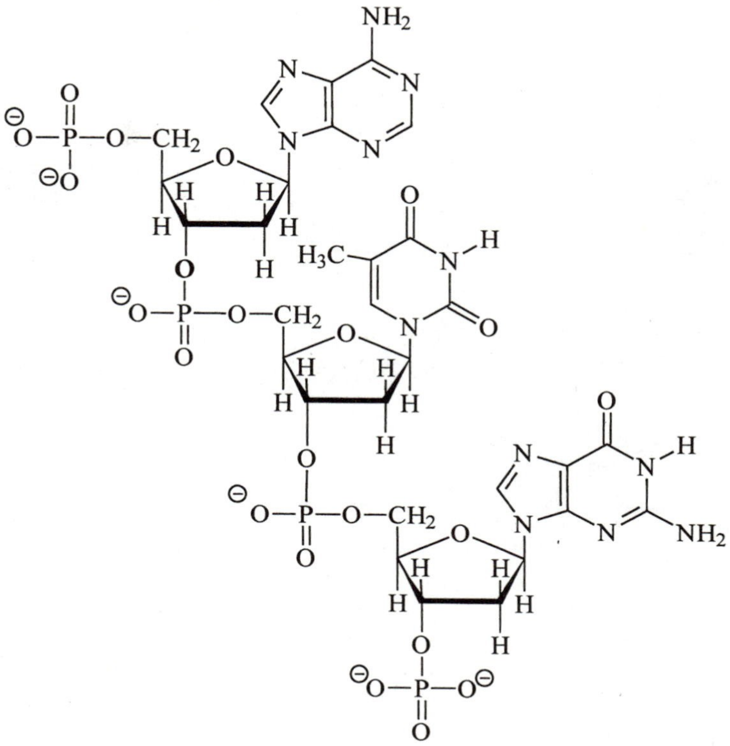

c)

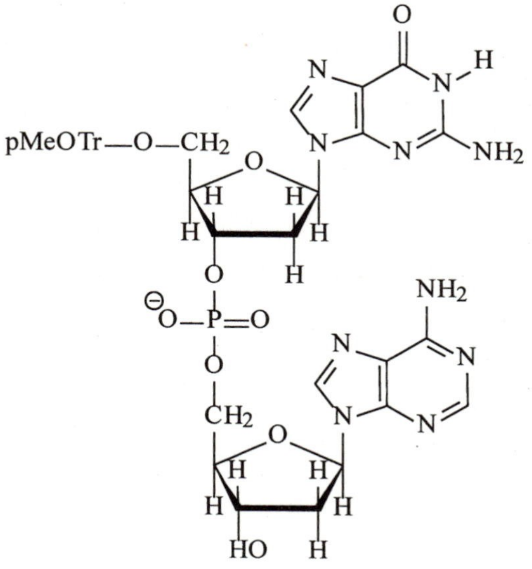

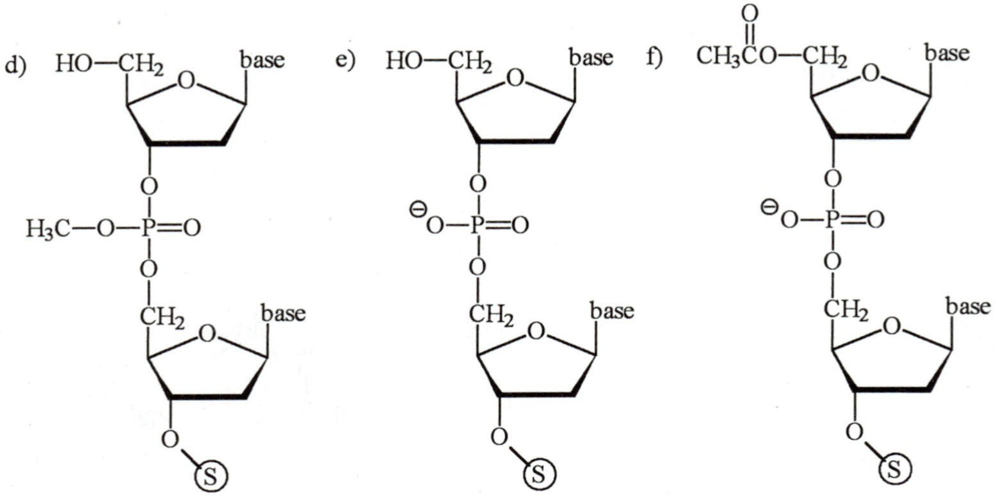

26.25 This should have little effect on the formation of base pairs because the methylated nitrogen atom still has a hydrogen that can hydrogen bond to thymine.

26.26

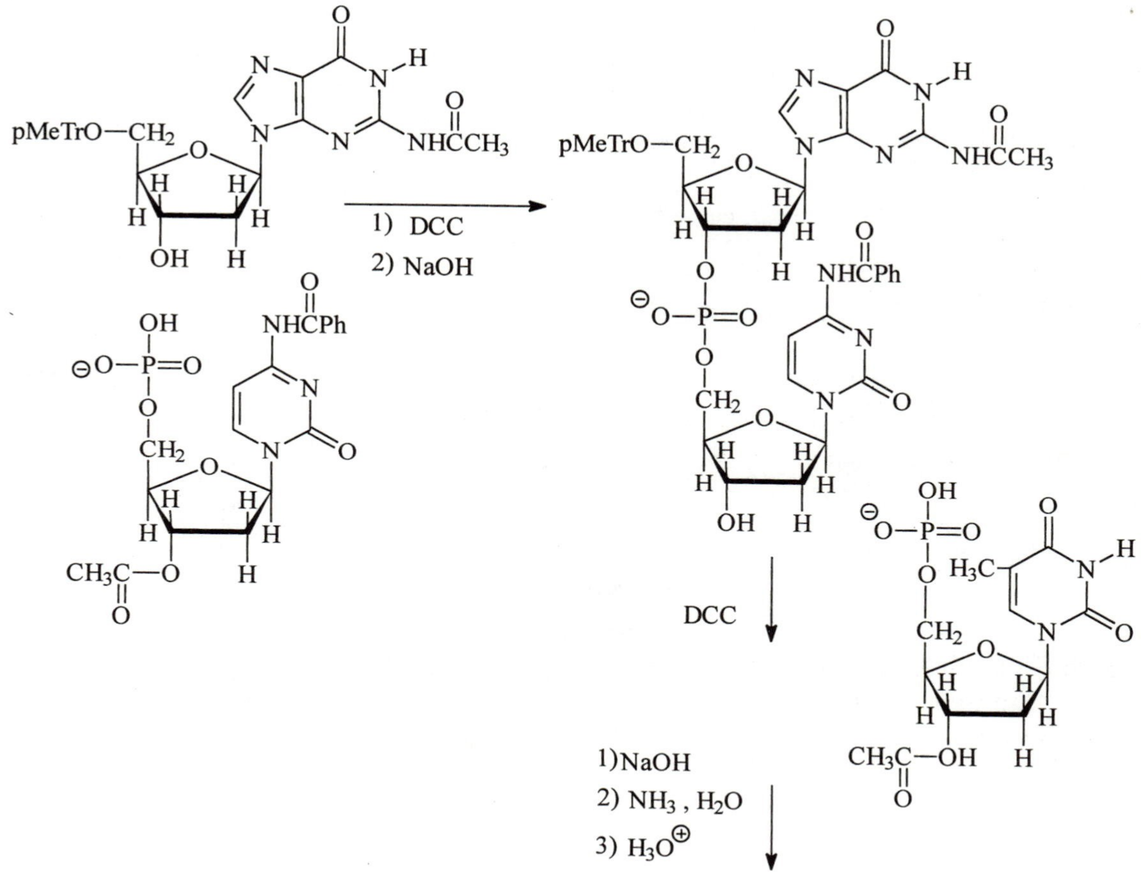

pMeTrO—CH_2 ... $NHCCH_3$ ... $\xrightarrow{\text{1) DCC} \quad \text{2) NaOH}}$ pMeTrO—CH_2 ... $NHCCH_3$

OH ... NHCPh ... $\ominus$O—P=O ... O ... CH_2 ... CH_3C—O

DCC

OH ... $\ominus$O—P=O ... O ... H_3C ... CH_2 ... CH_3C—OH

1) NaOH
2) NH_3 , H_2O
3) $H_3O^{\oplus}$

5'-end GCT 3'-end

26.27

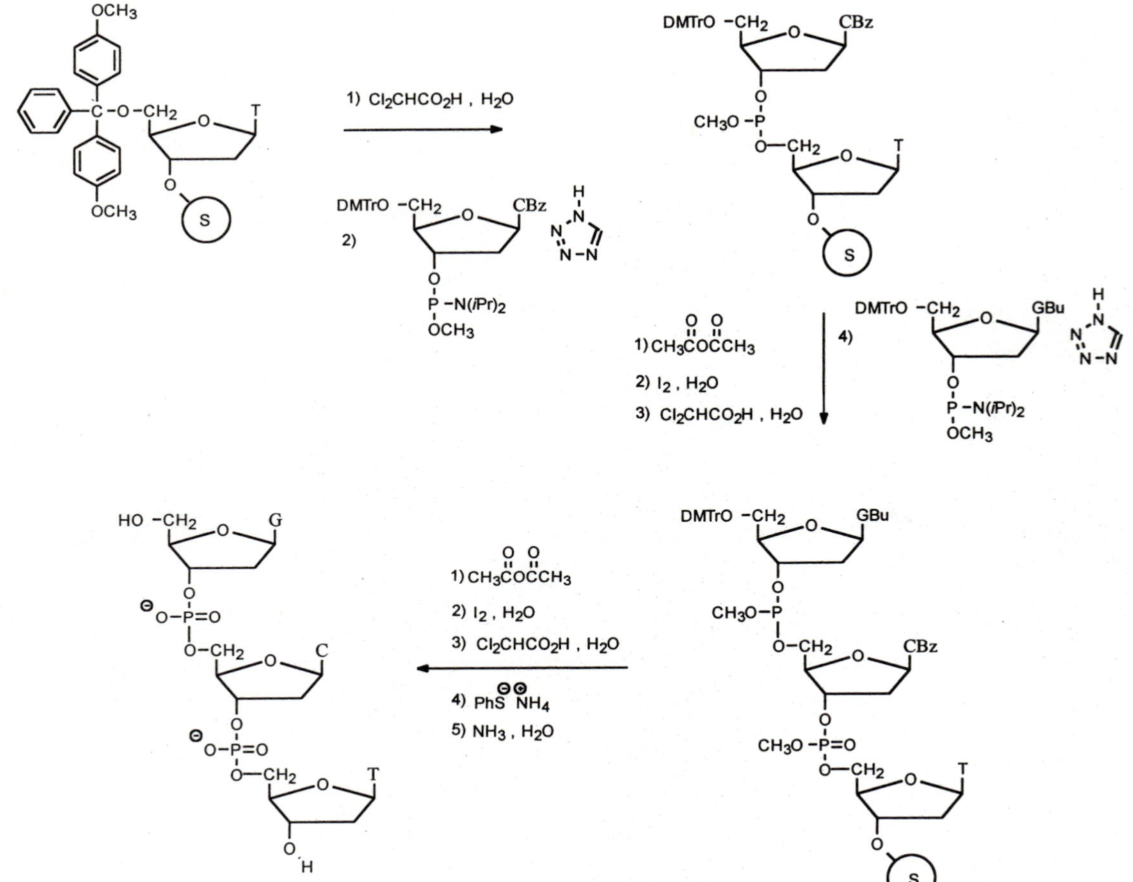

1) Cl_2CHCO_2H, H_2O

2) DMTrO–CH_2 ... CBz, P–N(iPr)$_2$, OCH_3; tetrazole

1) CH_3COCCH_3 (acetic anhydride)

2) I_2, H_2O

3) Cl_2CHCO_2H, H_2O

4) DMTrO–CH_2 ... GBu, P–N(iPr)$_2$, OCH_3; tetrazole

1) CH_3COCCH_3 (acetic anhydride)

2) I_2, H_2O

3) Cl_2CHCO_2H, H_2O

4) $PhS^{\ominus}\ NH_4^{\oplus}$

5) NH_3, H_2O

26.28

$R{-}O{-}P(=O)(OR){-}O{-}R + HO^{\ominus} \longrightarrow$ pentacoordinate intermediate $\longrightarrow R{-}O{-}P(=O)(OH){-}O{-}R + RO^{\ominus}$

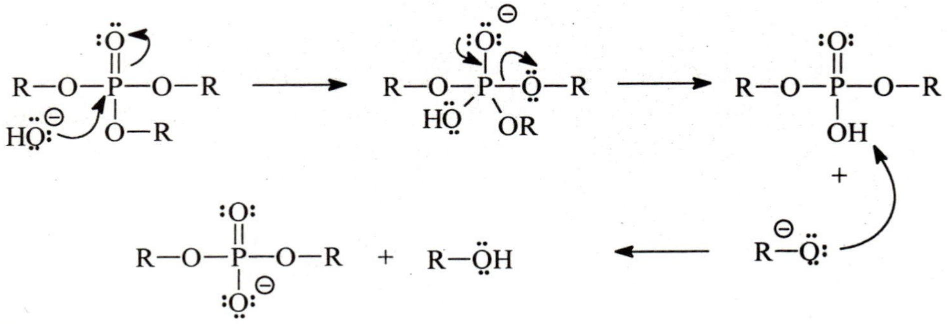

$R{-}O^{\ominus} + R{-}O{-}P(=O)(OH){-}O{-}R \longrightarrow R{-}O{-}P(=O)(O^{\ominus}){-}O{-}R + R{-}OH$

26.29

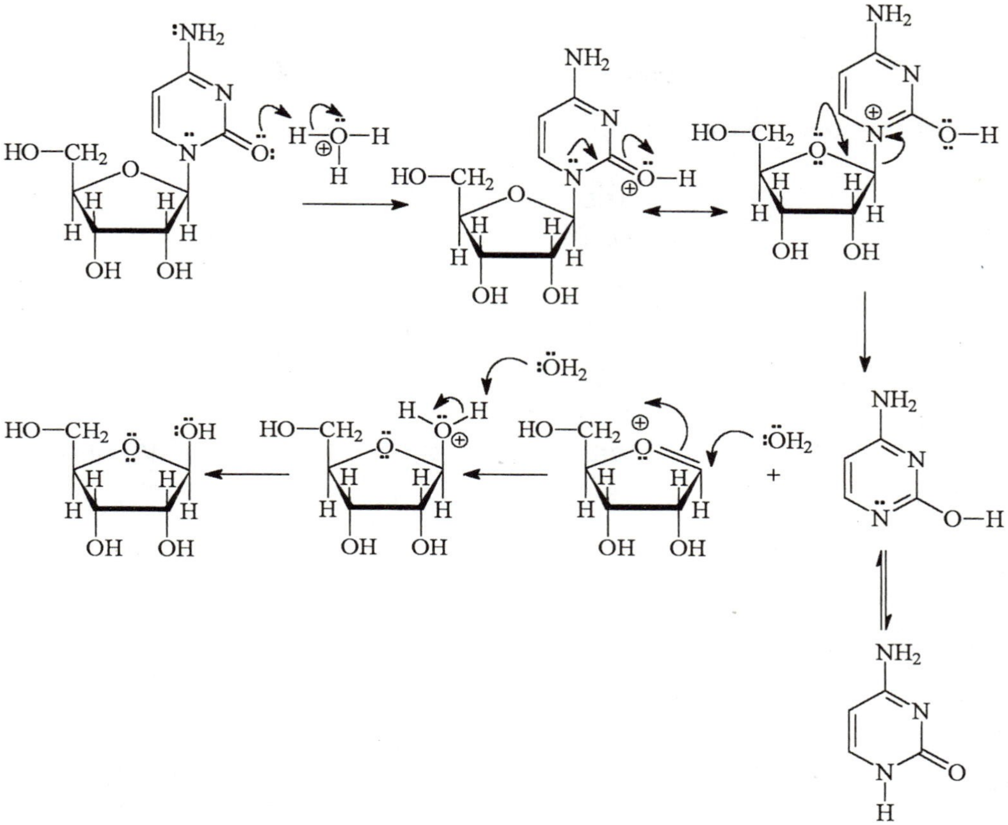

26.30 The reaction with NaOH and the phosphodiester is an S_N2 reaction at carbon and is favored at the primary carbon site. The 5'-end of this cyclic phosphodiester is a primary carbon, so it is cleaved preferentially over the 3'-end, which is secondary.

26.31 The O-methylated guanine can form only two hydrogen bonds, similar to adenine. Therefore, when DNA with O-methylated guanine replicates the complimentary base pair of adenine (thymine) is incorporated in the new DNA strand.

26.32 Hypoxanthine can form two hydrogen bonds but orientation is reversed from adenine. Therefore, the affinity of hypoxanthine for thymine will be much less and another base can easily be incorporated in the place of thymine.

After completing this chapter, you should be able to:

1. Show the general structures of nucleosides, nucleotides, DNA, and RNA. Problems 26.1, 26.2, 26.3, 26.4, 26.19, 26.20, 26.21, 26.22.

2. Show the hydrogen bonding that occurs between adenine and thymine or uracil and between guanine and cytosine. Problems 26.6, 26.7, 26.25, 26.31, 26.32.

3. Understand the general features of replication, transcription, and translation. Problems 26.5, 26.23.

4. Understand the chemical cleavage method and the chain-terminator method for determining the sequence of DNA. Problems 26.12, 26.24.

5. Show a reaction scheme for the synthesis of a polynucleotide. Problems 26.15, 26.18, 26.26, 26.27.

6. Show the mechanisms for these reactions where appropriate. Problems 26.8, 26.9, 26.10, 26.11, 26.13, 26.14, 26.16, 26.17, 26.28, 26.29, 26.30.

Chapter 27
OTHER NATURAL PRODUCTS

27.1

a) 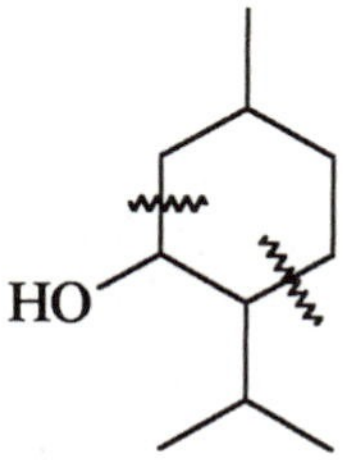 HO

monoterpene

b)

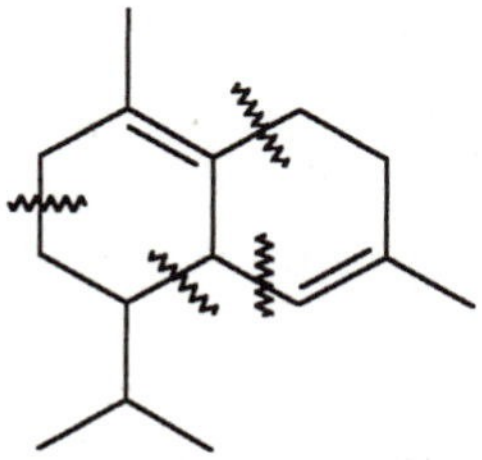

sesquiterpene

c)

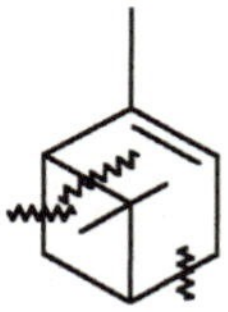

monoterpene

d) 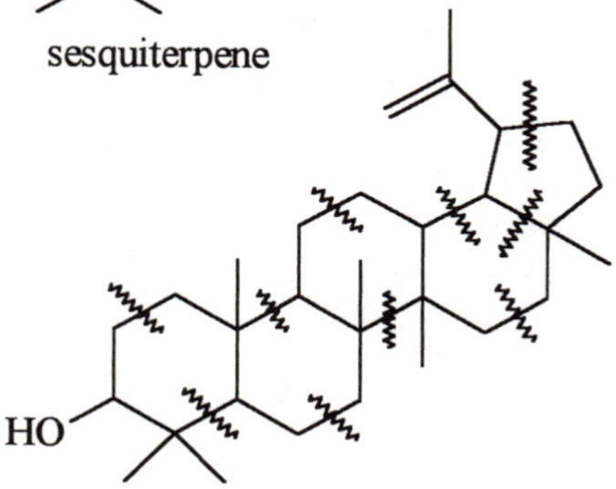 HO

triterpene

27.2

$$CH_2CSCoA \rightleftharpoons \ominus CH_2CSCoA \rightleftharpoons CH_3C(O^-)(SCoA)-CH_2CSCoA \rightleftharpoons CH_3C(=O)-CH_2C(=O)SCoA + {}^{\ominus}SCoA$$

H :B

CH_3CSCoA

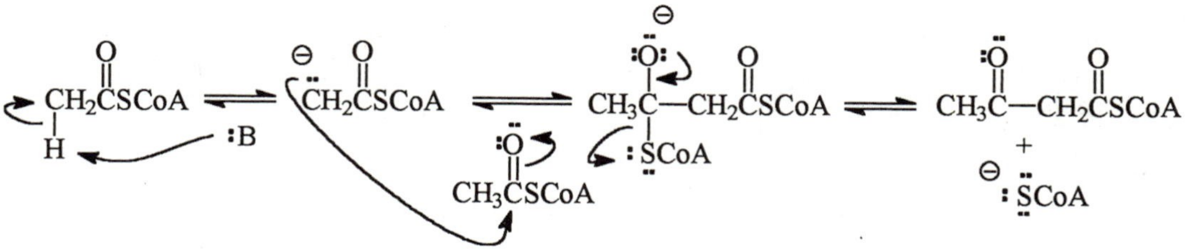

27.3

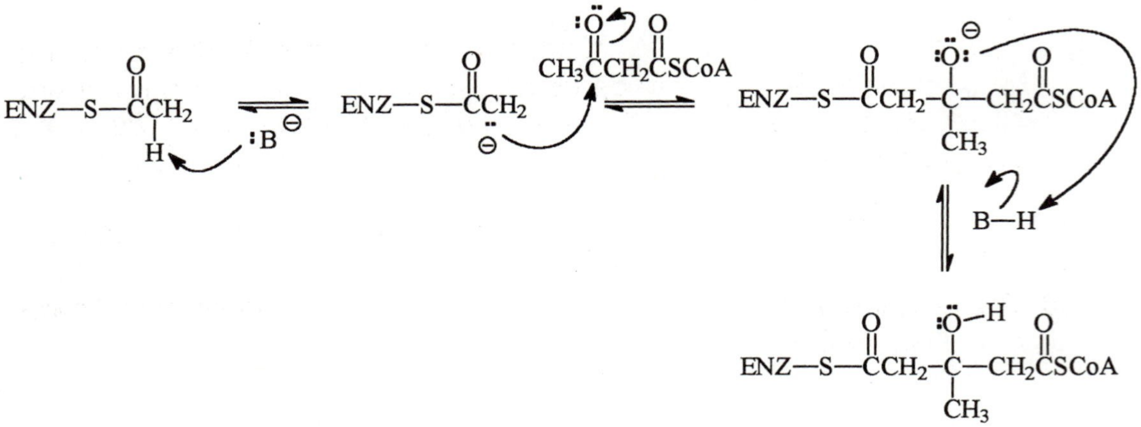

ENZ—S—$\overset{O}{\overset{\|}{C}}CH_2$—H + :$B^{\ominus}$ ⇌ ENZ—S—$\overset{O}{\overset{\|}{C}}\ddot{C}H_2^{\ominus}$ + $CH_3\overset{:\ddot{O}}{\overset{\|}{C}}CH_2\overset{O}{\overset{\|}{C}}SCoA$ ⇌ ENZ—S—$\overset{O}{\overset{\|}{C}}CH_2$—$\overset{:\ddot{O}:^{\ominus}}{C}$($CH_3$)—$CH_2\overset{O}{\overset{\|}{C}}SCoA$

B—H

ENZ—S—$\overset{O}{\overset{\|}{C}}CH_2$—$\overset{:\ddot{O}—H}{C}$($CH_3$)—$CH_2\overset{O}{\overset{\|}{C}}SCoA$

27.4 This anion is a good leaving group because it is a weak base.

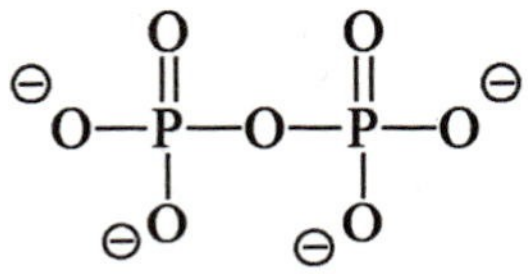

$^{\ominus}O—\overset{O}{\overset{\|}{P}}(O^{\ominus})—O—\overset{O}{\overset{\|}{P}}(O^{\ominus})—O^{\ominus}$

27.5

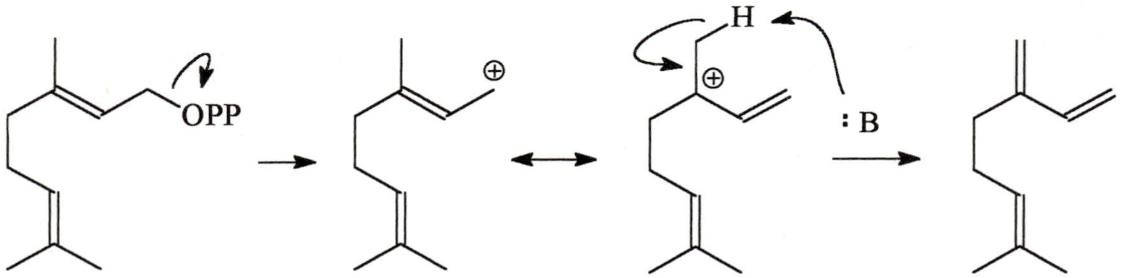

OPP → $\oplus$ ↔ $\oplus$ H :B →

27.6

OPP → $\oplus$ → H $\oplus$ → $\oplus$:$\ddot{O}$—H H → $\oplus$ $\ddot{O}$—H H :B H → $\ddot{O}$—H H

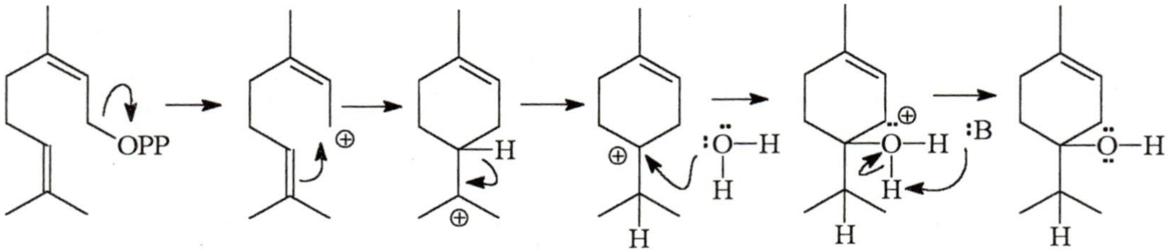

27.7 The cyclization occurs so as to form the more stable tertiary carbocation.

27.8 Path A is disfavored because it produces a secondary carbocation. It is favored because the 5-membered rings that are formed have lower ring strain. Path B is favored because it produces a tertiary carbocation. It is disfavored due to the strain in the 4-membered ring of the product.

27.9 The cyclization occurs so as to form the more stable tertiary carbocation.

27.10 The reaction proceeds as shown in Figure 27.4 to produce the following carbocation, which then cyclizes as shown.

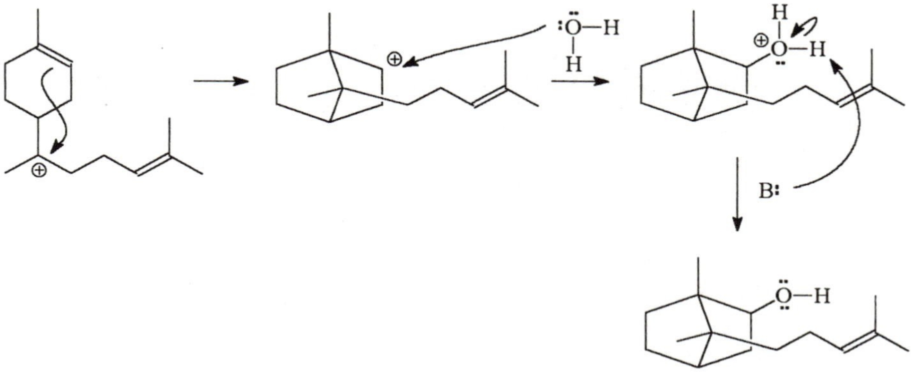

27.11 Both cyclizations occur so as to form the more stable tertiary carbocations.

27.12

H
C
H
H—A
H_3C
H_3C
$H-CH_2$
CH_2
A:
H_3C

27.13

OPP
H
A—H
H
H
OPP
OPP
OPP
CH_2
H
A:
OPP
CH_2

27.14

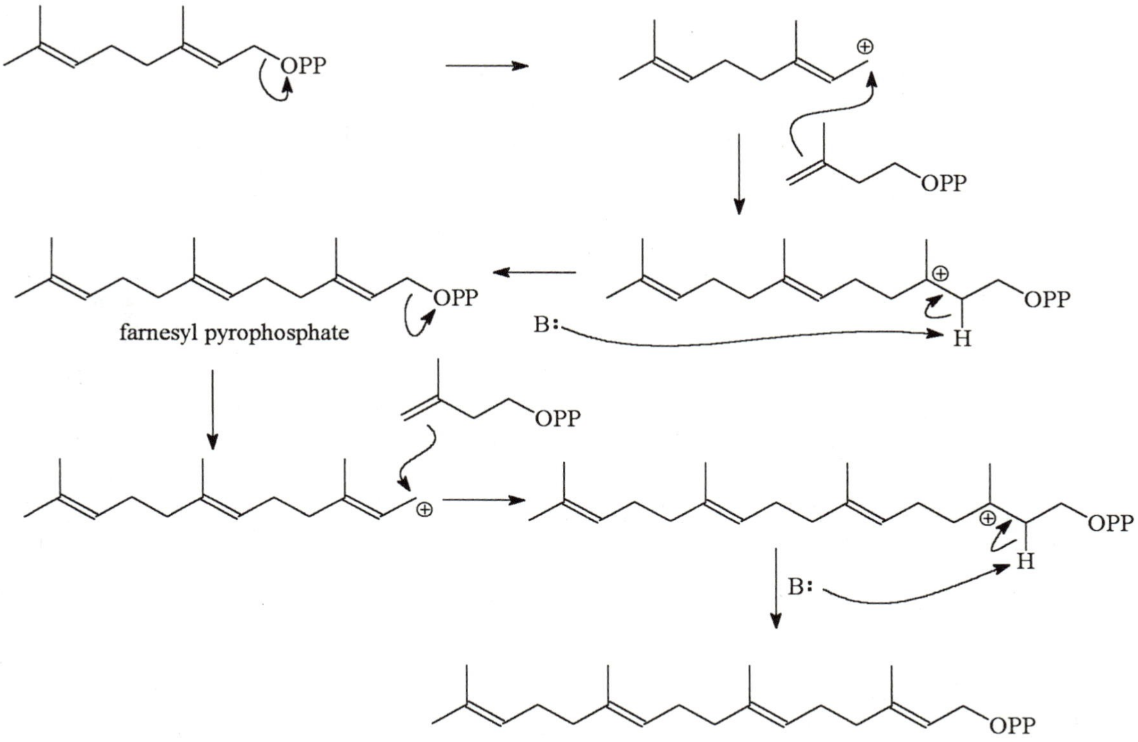

OPP

farnesyl pyrophosphate

B:

H

OPP

27.15 Different protons are removed in the elimination step in order to produce α- and β-carotene. More β-carotene is formed because the new double bond is more stable when it is conjugated.

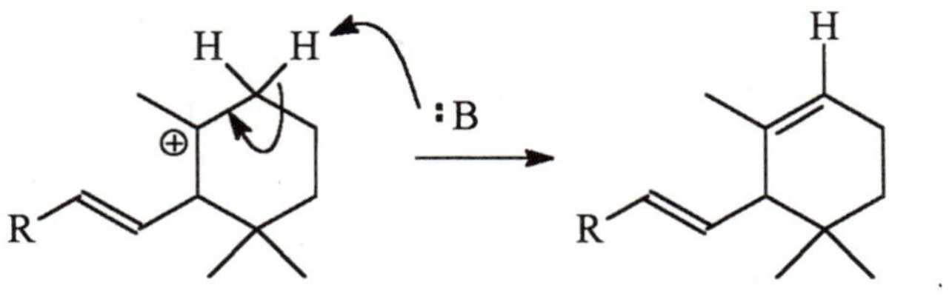

H H :B R R H

α-carotene

27.16 The epoxide opens so as to form the more stable tertiary carbocation.

27.17

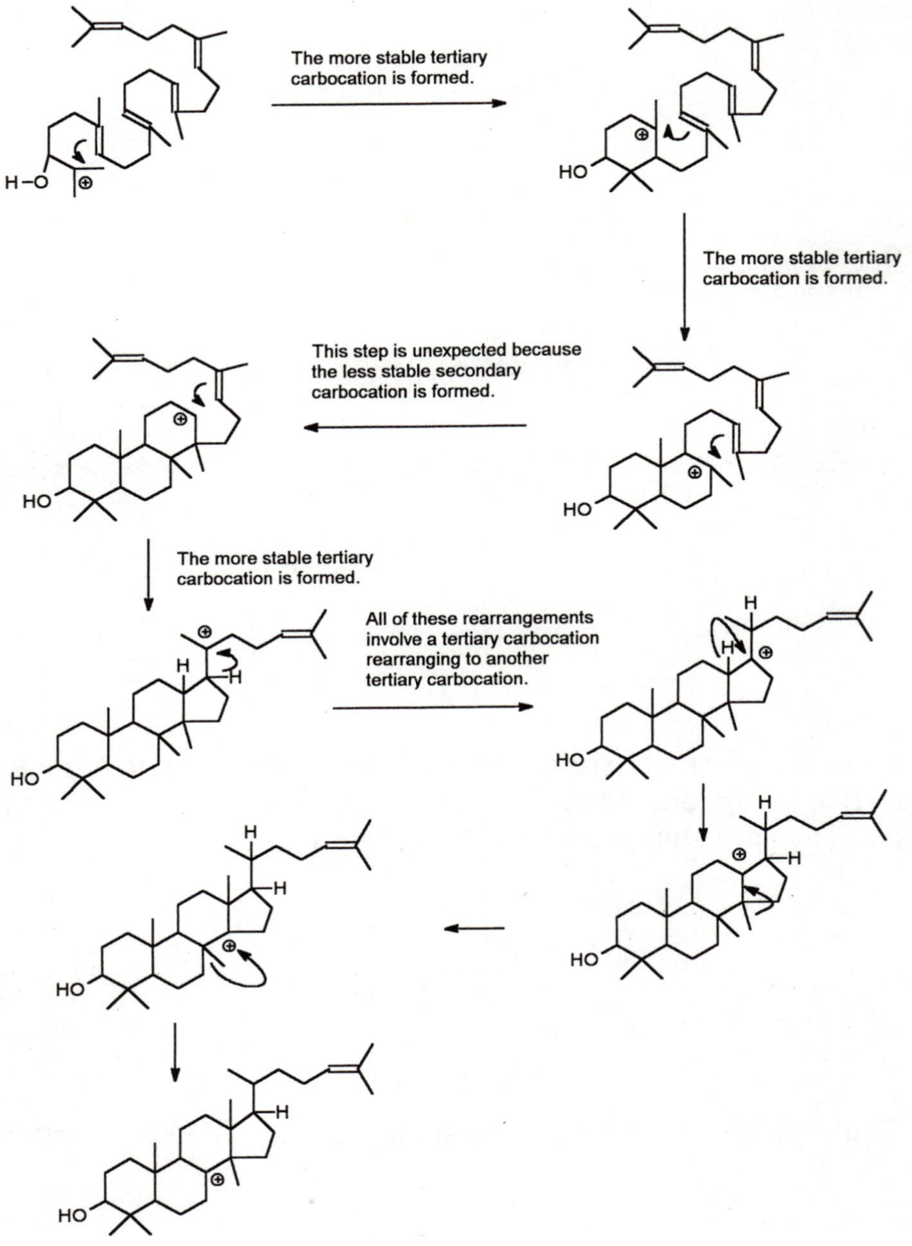

The more stable tertiary carbocation is formed.

The more stable tertiary carbocation is formed.

This step is unexpected because the less stable secondary carbocation is formed.

The more stable tertiary carbocation is formed.

All of these rearrangements involve a tertiary carbocation rearranging to another tertiary carbocation.

27.18

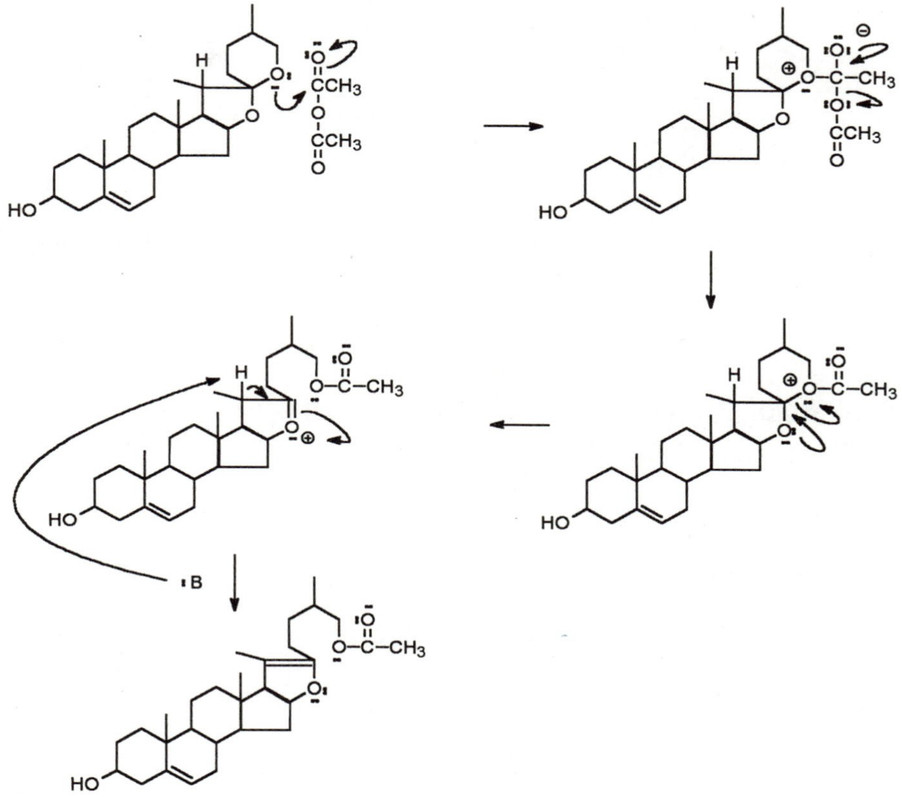

27.19

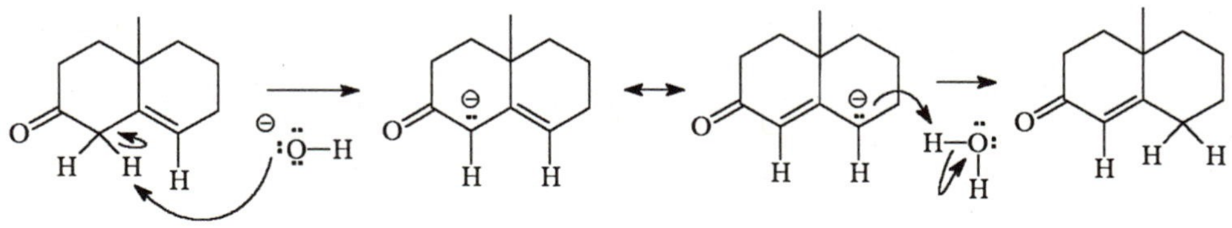

27.20

$$CH_3-\overset{O}{\overset{\|}{C}}-SR + \overset{\ominus}{C}H_2-\overset{O}{\overset{\|}{C}}-SR \longrightarrow CH_3-\overset{O^{\ominus}}{\overset{|}{\underset{SR}{\underset{|}{C}}}}-CH_2-\overset{O}{\overset{\|}{C}}-SR \longrightarrow CH_3-\overset{O}{\overset{\|}{C}}-CH_2-\overset{O}{\overset{\|}{C}}-SR + :\ddot{S}R^{\ominus}$$

27.21

$$CH_3\overset{OH}{\overset{|}{C}H}-\underset{H}{\underset{|}{C}H}-\overset{O}{\overset{\|}{C}}SR \xrightarrow{\ ^{\ominus}:\ddot{O}-H\ } CH_3\overset{OH}{\overset{|}{C}H}-\overset{\ominus}{C}H-\overset{O}{\overset{\|}{C}}SR \longrightarrow CH_3CH{=}CH-\overset{O}{\overset{\|}{C}}SR + {}^{\ominus}:\ddot{O}-H$$

27.22 a) alkaloid b) terpene c) steroid d) prostaglandin
e) fat f) terpene

27.23

a) 3 H–CO–CH₂–CO–H + H–CO–CH₂CH₂CH₂CH₂CH₃ + H–CO–CH₂CH₂CH₂–CO–OH

b) H_3C, CH_3, H, H, H, O

c) 2 $H-\overset{O}{\overset{\|}{C}}-H$ + $CH_3-\overset{O}{\overset{\|}{C}}-CH_3$ + H–CO–CH₂CH₂–CO–CO–H

d) 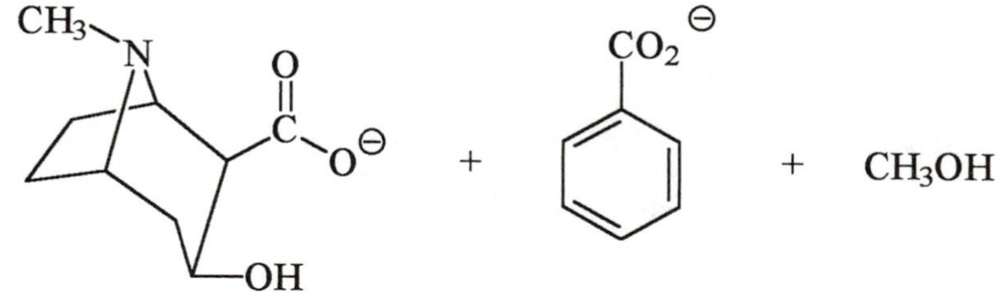 [tropine-type bicyclic structure: CH_3–N bridge, $-\overset{O}{\overset{||}{C}}-O^{\ominus}$, –OH] + [benzene ring–$CO_2^{\ominus}$] + CH_3OH

e) 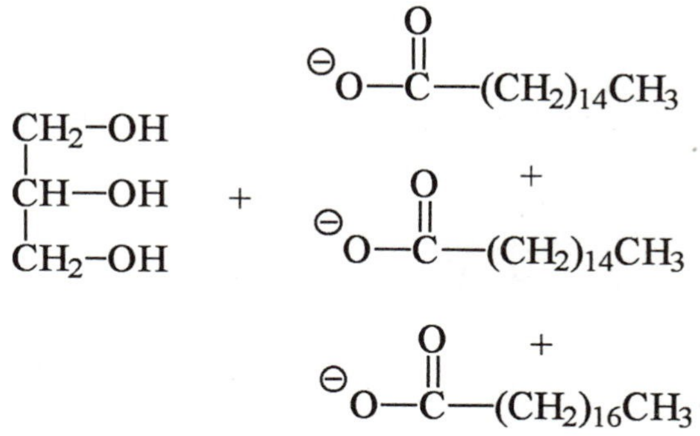 $$\begin{array}{l} CH_2-OH \\ | \\ CH-OH \\ | \\ CH_2-OH \end{array} + \begin{array}{c} {}^{\ominus}O-\overset{O}{\overset{||}{C}}-(CH_2)_{14}CH_3 \\ + \\ {}^{\ominus}O-\overset{O}{\overset{||}{C}}-(CH_2)_{14}CH_3 \\ + \\ {}^{\ominus}O-\overset{O}{\overset{||}{C}}-(CH_2)_{16}CH_3 \end{array}$$

27.24 Estradiol contains a phenol group which is relatively acidic. When treated with NaOH, estradiol forms a salt which is water soluble. Therefore the separation can be accomplished by dissolving the mixture in an organic solvent, such as ether or dichloromethane, and extracting with aqueous sodium hydroxide solution. Acidification of the aqueous solution causes the estradiol to precipitate, and the progesterone can be isolated from the organic phase by evaporation of the solvent.

27.25 a)

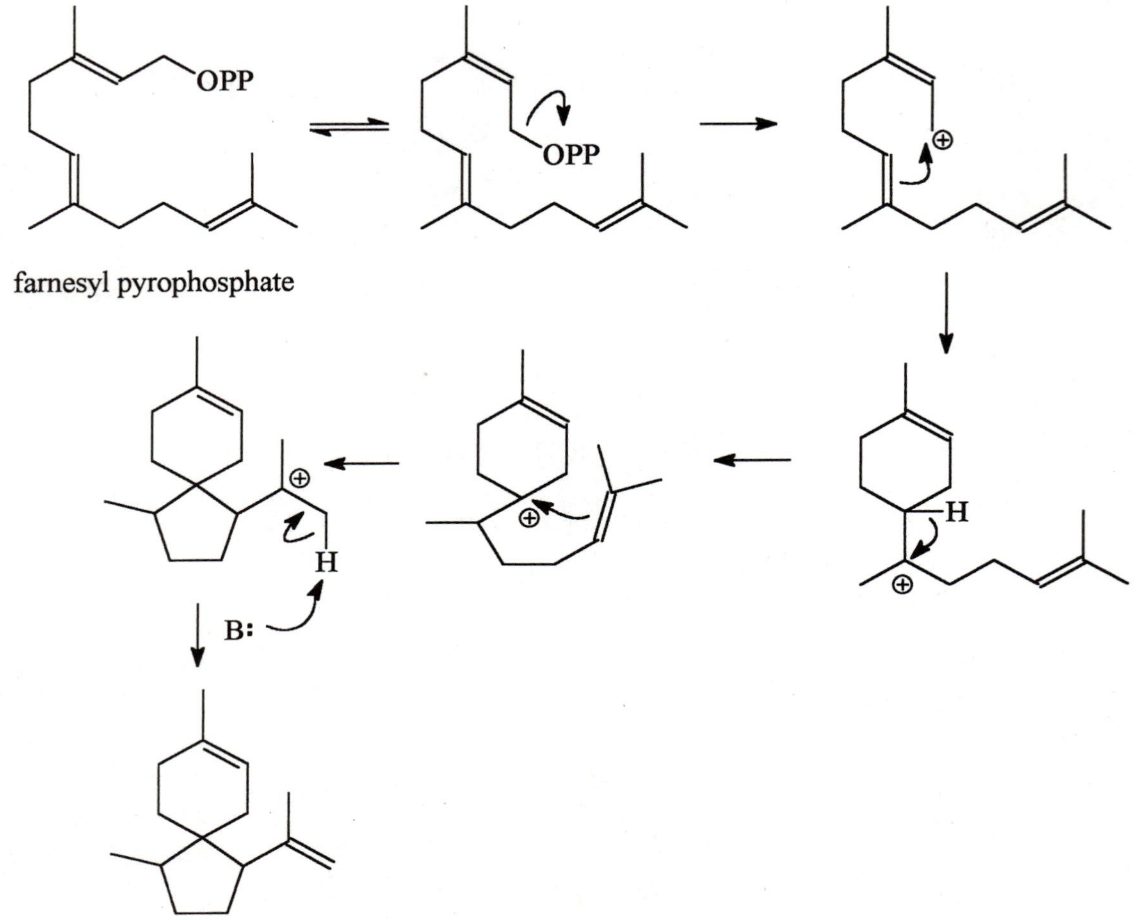

OPP

OPP

farnesyl pyrophosphate

H

H

B:

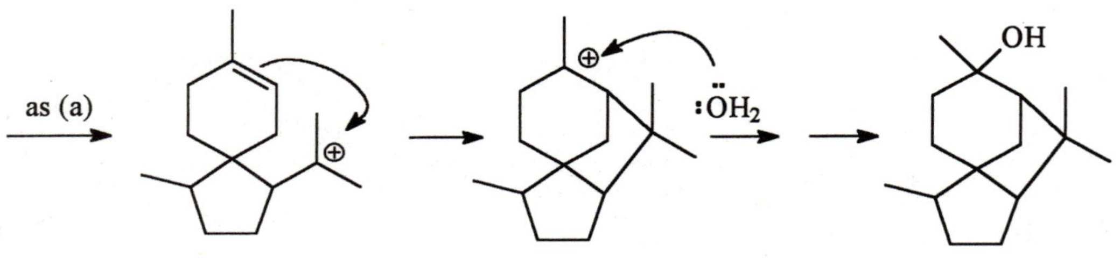

b) as (a)

:$\ddot{O}H_2$

OH

27.26

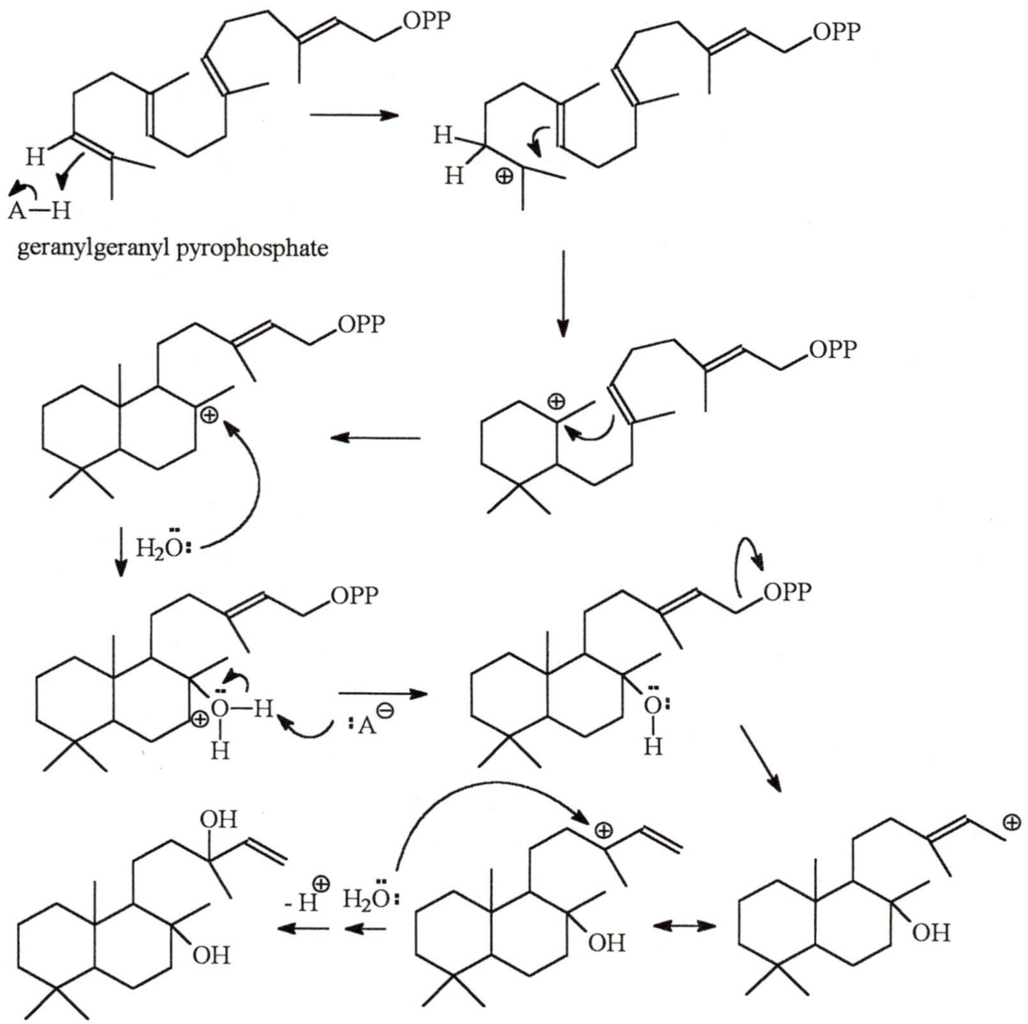

geranylgeranyl pyrophosphate

OPP

H

A—H

$H_2\ddot{O}$:

:A⊖

- H⊕

27.27

OPP

H

:B

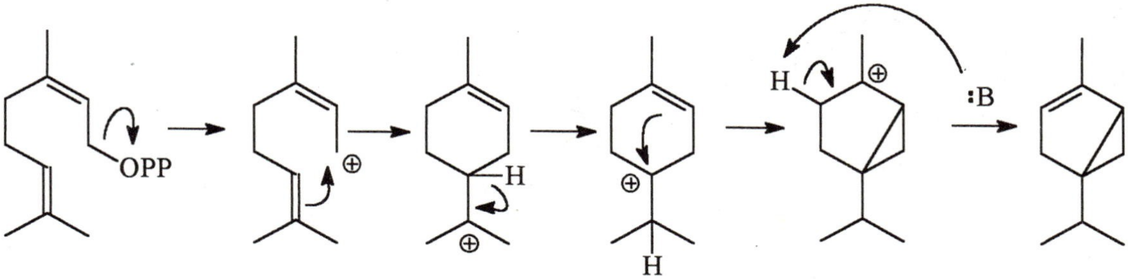

27.28

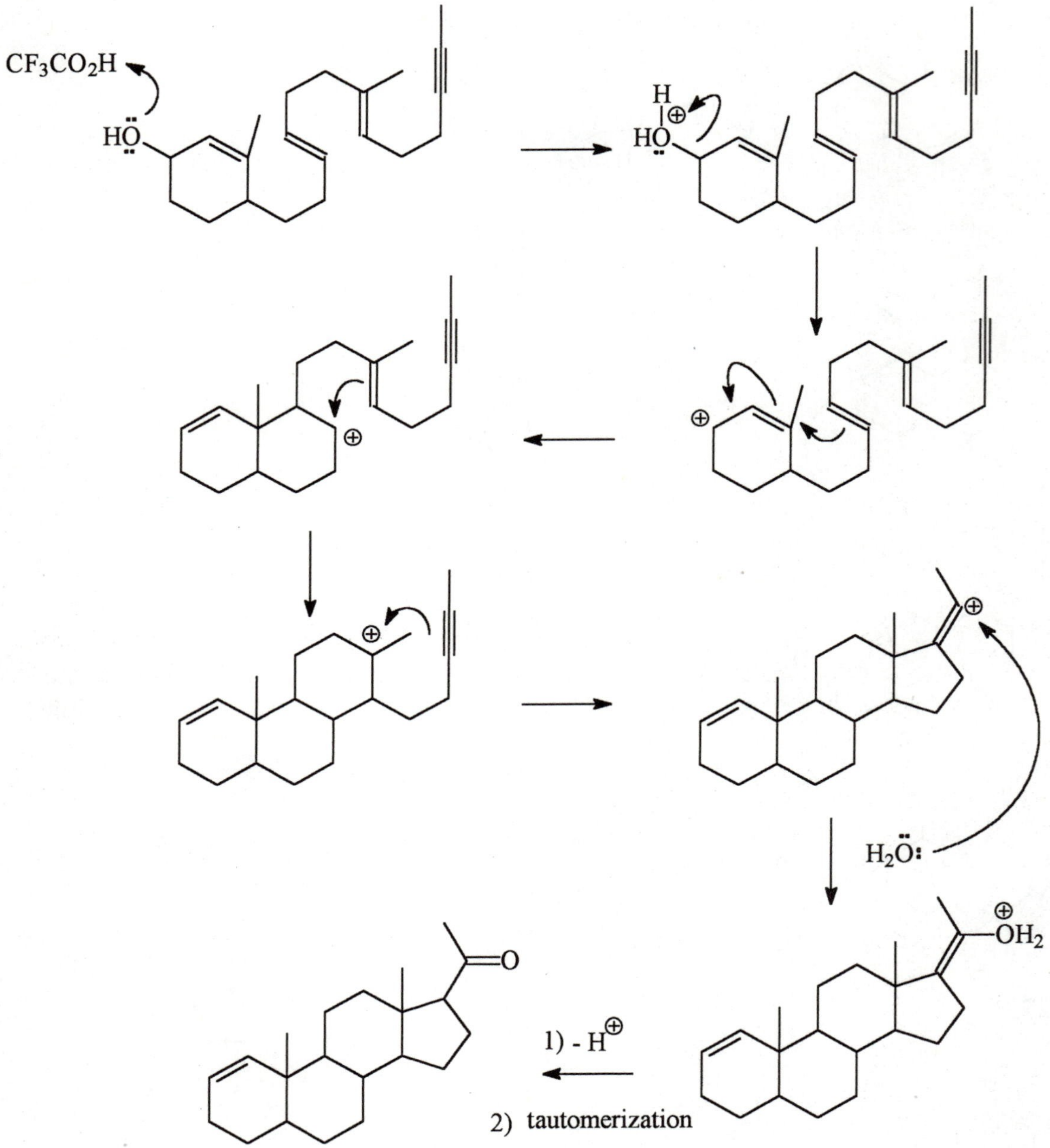

CF_3CO_2H

HÖ

H
HÖ⊕

⊕

⊕

⊕

⊕

$H_2\ddot{O}$:

⊕
OH_2

O

1) - $H^{\oplus}$

2) tautomerization

27.29 This is an aldol condensation.

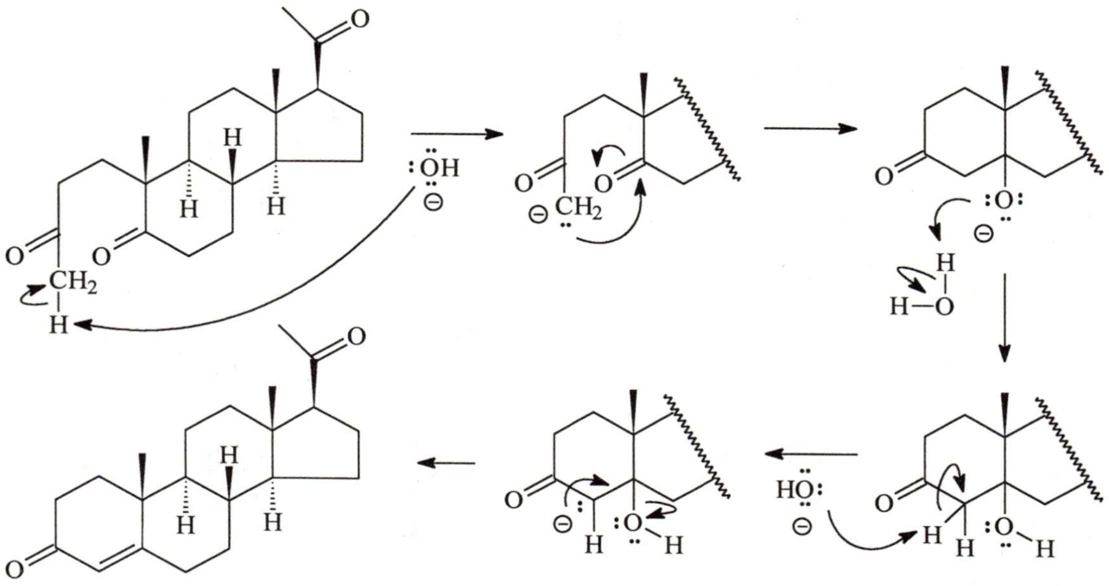

27.30 The first cyclization occurs so as to produce a six-membered ring, rather than a less stable five-membered ring. The next two cyclizations occur so as to produce the more stable carbocations (the more highly substituted carbocation).

27.31

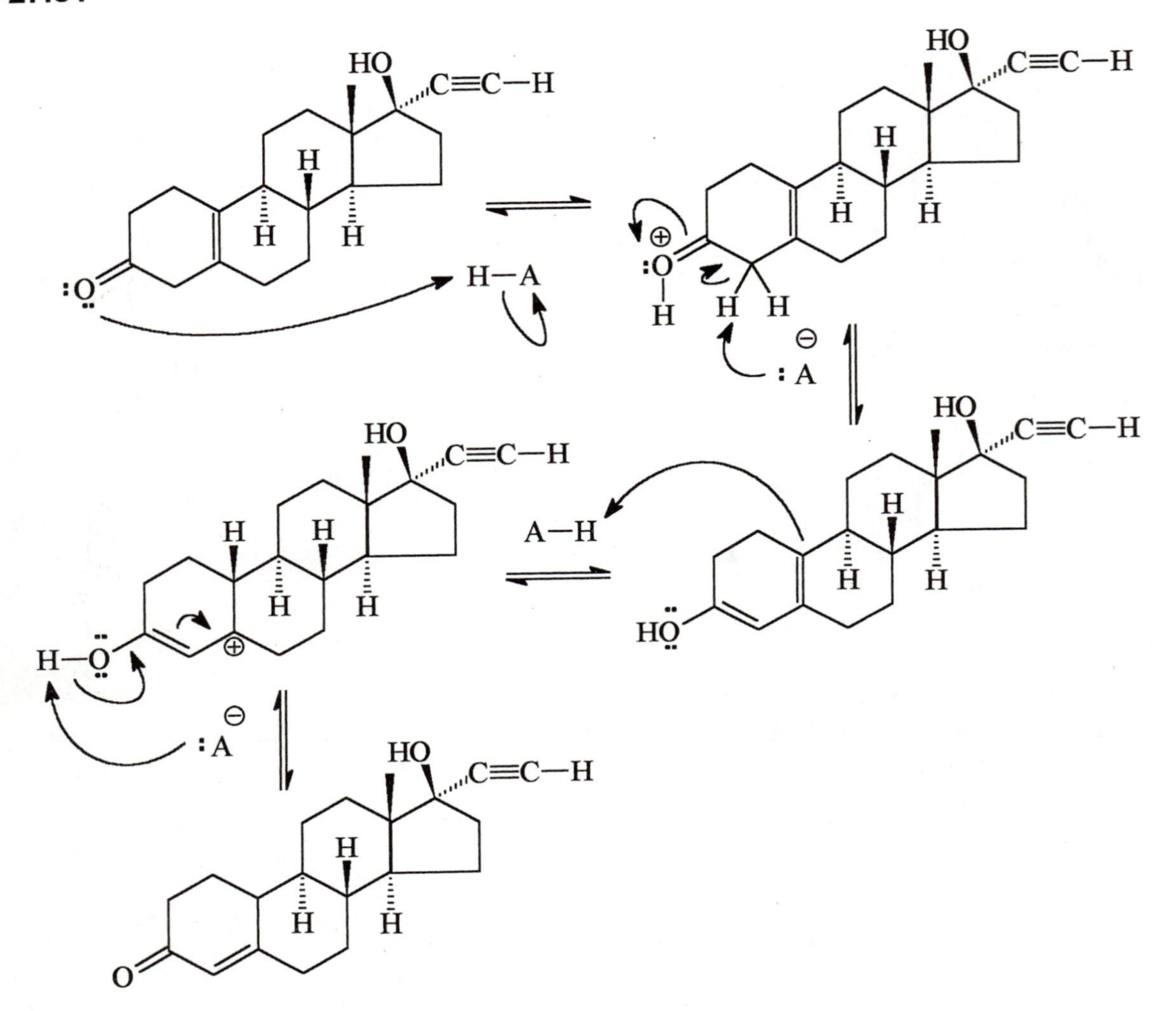

HO
C≡C—H
H
H
H
:O
H—A
HO
C≡C—H
H
H
H
:O
H
H
H
:A
HO
C≡C—H
H
H
H
HO
A—H
HO
C≡C—H
H
H
H
H
H
H—O
:A
HO
C≡C—H
H
H
H
O

27.32

H—Ö—H (⊕) ... ÖH → ... $\ddot{O}H_2$ (⊕) → ... ⊕ → ... ⊕

$H_2\ddot{O}$:

$H_2\ddot{O}$: ... ⊕OH—H → ...OH → H—O—H (⊕) ... OH → $H_2\ddot{O}$: ⊕ ... OH → - H⊕ → OH ... OH

27.33

H—Ö—H (⊕) ... O → ⊕ ... O

↓

O ... ← $H_2\ddot{O}$: ⊕ ... O ... H ... H

27.34 This is a reverse Diels-Alder reaction (a reverse [4 + 2] cycloaddition).

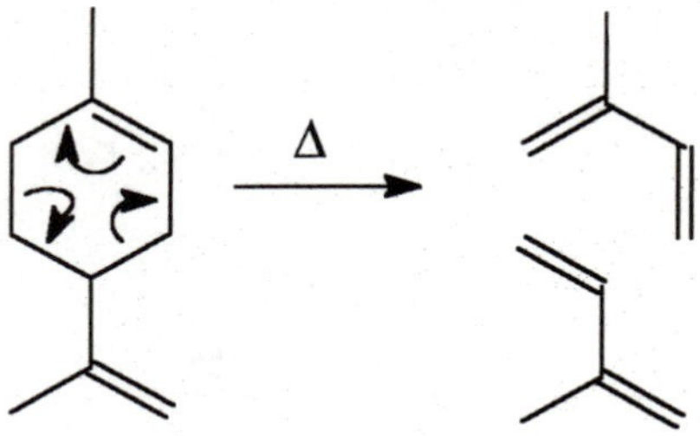

Δ

27.35

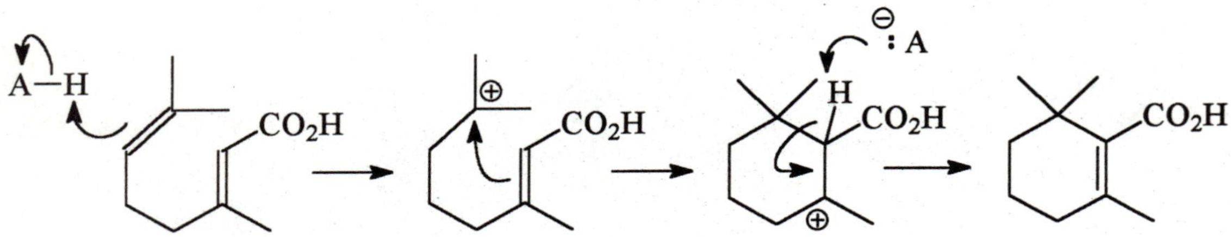

A—H

CO_2H

B and C are unreactive because they would form less stable carbocations (primary or secondary rather than tertiary) in the first step of the mechanism.

27.36

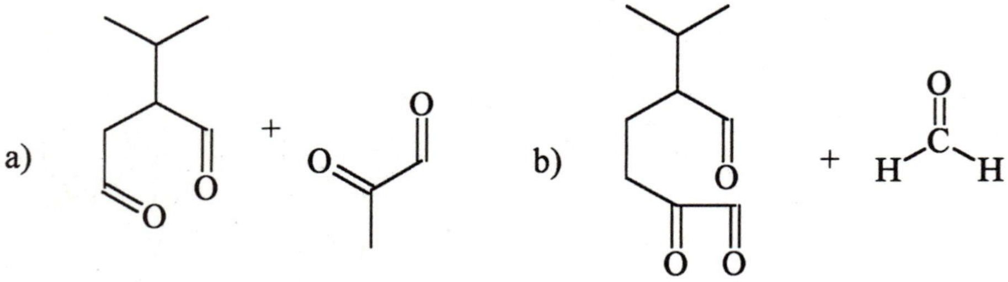

a) +

b) +

27.37

OH

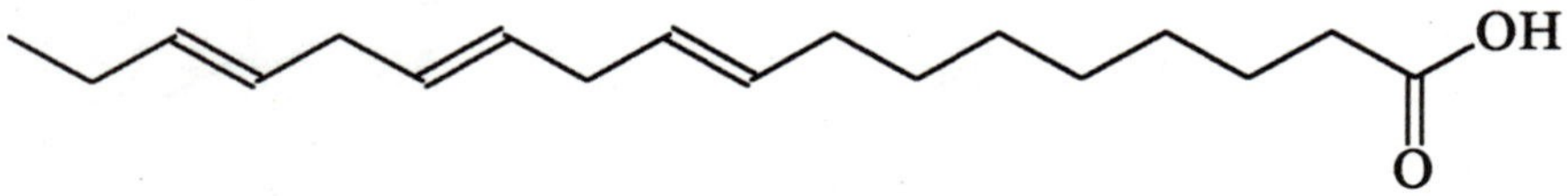

27.38

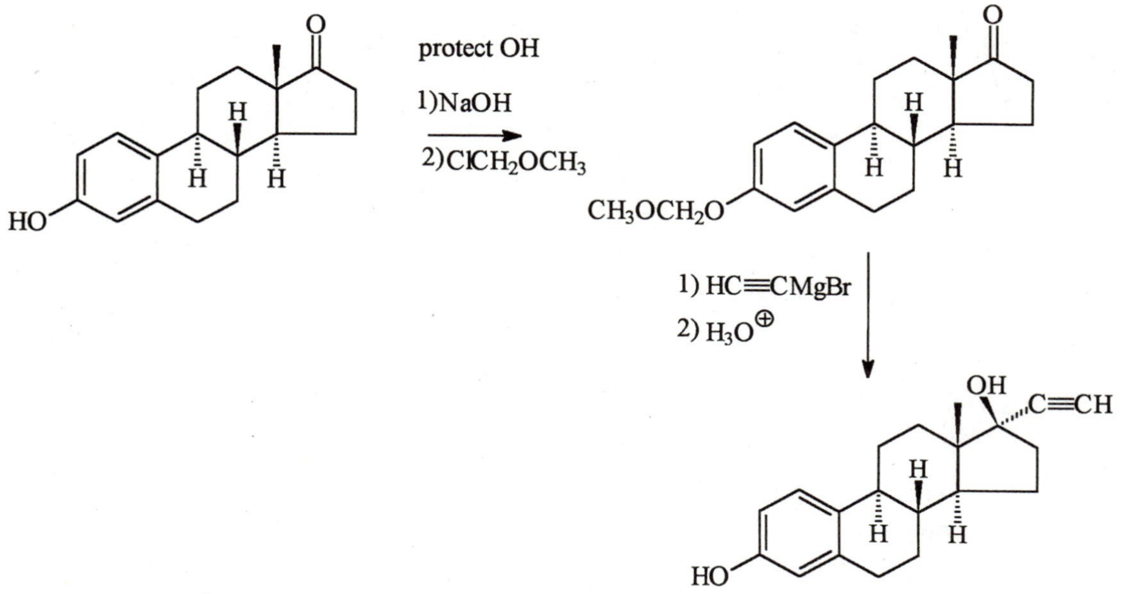

protect OH

1) NaOH

2) $ClCH_2OCH_3$

CH_3OCH_2O

1) $HC\equiv CMgBr$

2) $H_3O^{\oplus}$

OH

$C\equiv CH$

HO

27.39

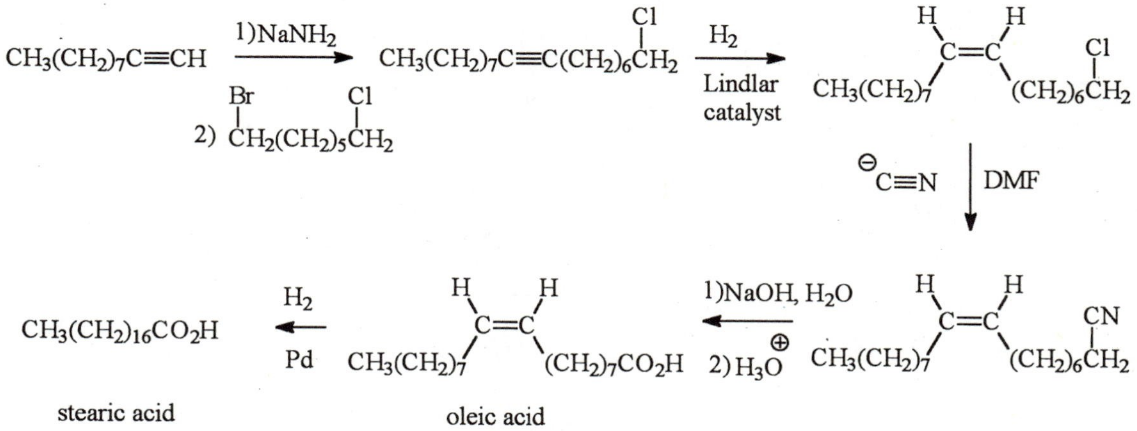

$CH_3(CH_2)_7C\equiv CH$ → 1) $NaNH_2$; 2) $BrCH_2(CH_2)_5CH_2Cl$ → $CH_3(CH_2)_7C\equiv C(CH_2)_6CH_2Cl$

→ H_2, Lindlar catalyst → cis $CH_3(CH_2)_7CH=CH(CH_2)_6CH_2Cl$

→ $^{\ominus}C\equiv N$, DMF → cis $CH_3(CH_2)_7CH=CH(CH_2)_6CH_2CN$

→ 1) NaOH, H_2O; 2) $H_3O^{\oplus}$ → cis $CH_3(CH_2)_7CH=CH(CH_2)_7CO_2H$

oleic acid

→ H_2, Pd → $CH_3(CH_2)_{16}CO_2H$

stearic acid

27.40

1) LDA
2) CH_3I

CH_3

1) LDA
2) Br

CH_3

$Ph_3P{=}CH_2$

CH_3

27.41

a)

step 4	step 5	step 7	step 8	step 9
1) $LiAlH_4$ 2) $H_3O^{\oplus}$	PCC	$H_3O^{\oplus}$	NaOH, H_2O Δ (aldol condensation)	1) CH_3MgI 2) NH_4Cl, H_2O

b)

BrMg

A

O O

PPh$_3$

O O

B

c)

EtO O

a [3,3] sigmatropic rearrangement

step 3

EtOC
O

d)

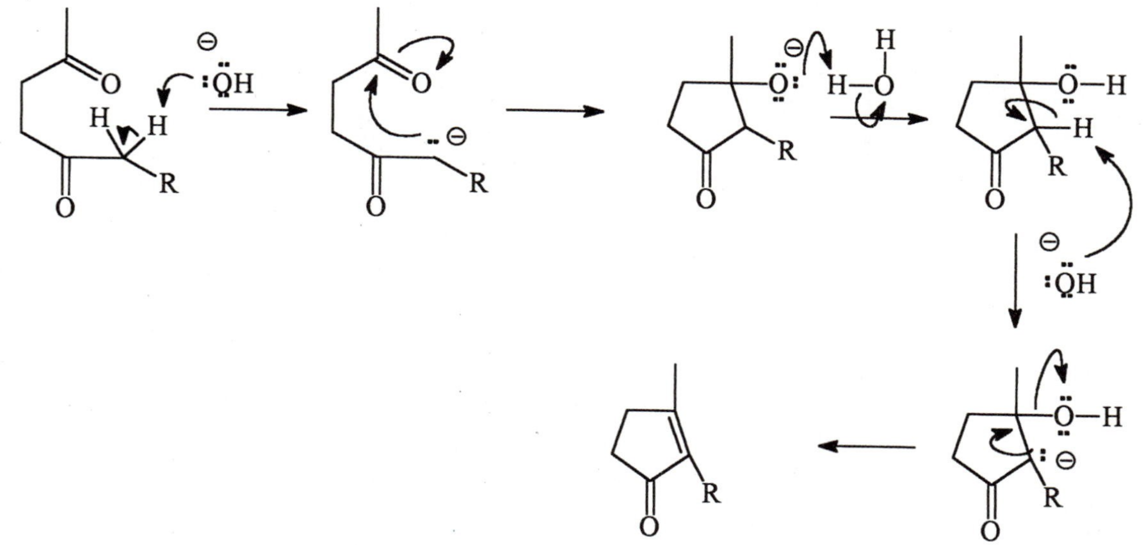

27.42 a) This is a Diels-Alder reaction.

A

b) The *trans*-isomer is more stable because the substituents can be equatorial on both rings.

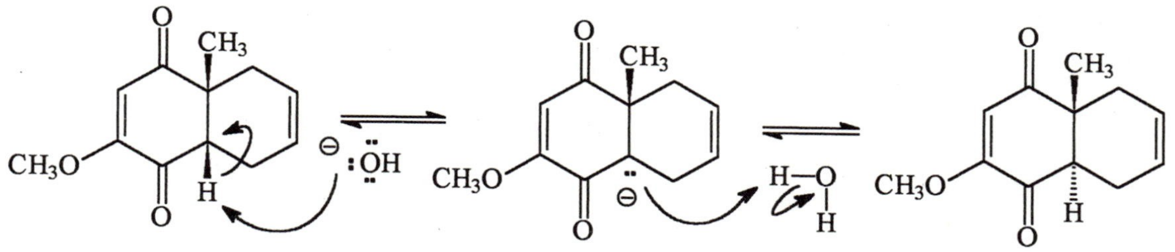

c) Lithium aluminum hydride can be used to reduce the carbonyl groups.

d)

:ÖH
CH_3
H—A
CH_3O
OH
H

⊕
:OH_2
CH_3
CH_3O
OH
H

⊕
CH_3
$CH_3\ddot{O}$
OH
H

$H_2\ddot{O}$:
⊕
CH_3
$CH_3\ddot{O}$
OH
H

A:⊖
H
⊕
H—Ö
CH_3
$CH_3\ddot{O}$
OH
H

CH_3
H—Ö
$CH_3\ddot{O}$
OH
H
H—A

CH_3
H—Ö
⊕
$CH_3\ddot{O}$
H
OH
H

CH_3
H—⊕O
OH
H
: A⊖

CH_3
O
OH
H

27.43

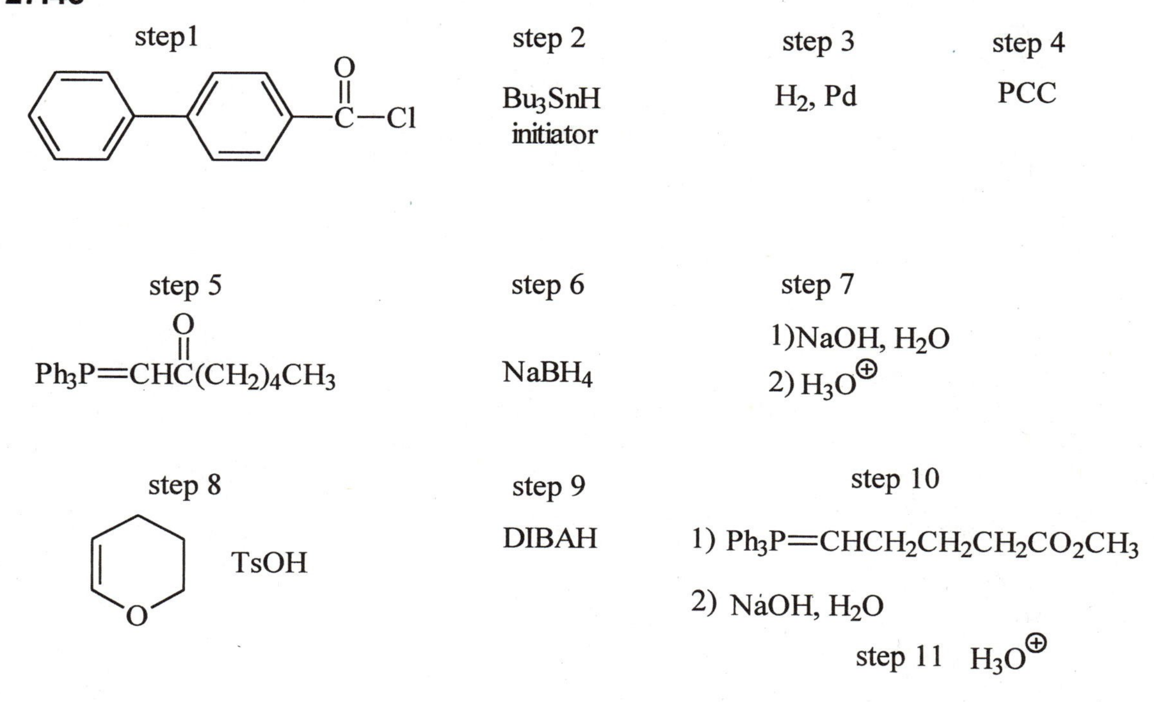

27.44

After completing this chapter, you should be able to:

1. Recognize the general structural features associated with terpenes and how they can be viewed as being formed from isoprene units. Problem 27.1.

2. Understand the hypothetical mechanisms by which terpenes are formed. Problems 27.2, 27.3, 27.4, 27.5, 27.6, 27.7, 27.8, 27.9, 27.10, 27.11, 27.12, 27.13, 27.14, 27.15, 27.25, 27.26, 27.27, 27.32, 27.33, 27.34, 27.35, 27.36, 27.37.

3. Do the same for steroids, prostaglandins and fats.
   Problems 27.16, 27.17, 27.18, 27.19, 27.20, 27.21, 27.28, 27.29, 27.30, 27.31.

4. Recognize the general structural features of the natural products discussed in this chapter.
   Problem 27.22.